STRUCTURAL SAFETY & RELIABILITY

Volume III

Proceedings of ICOSSAR '89, the 5th International
Conference on Structural Safety and Reliability,
San Francisco, August 7-11, 1989.

Edited by
A. H-S. Ang
University of California, Irvine, CA, USA

M. Shinozuka
Princeton University, Princeton, NJ, USA

and

G.I. Schuëller
University of Innsbruck, Innsbruck, Austria

Published by the
American Society of Civil Engineers
345 East 47th Street
New York, New York 10017-2398

ABSTRACT

This proceedings of the 5th International Conference on Structural Safety and Reliability (ICOSSAR '89) held on August 7-11, 1989 in San Francisco, California contains almost 340 papers representing the works of authors from 30 countries. These papers cover a wide range of development in structural safety and reliability that concern all types of structures from space structures to land-based facilities and ocean/offshore systems. Some of the topics included are wind engineering, seismic design, building performance, redundancy, fatigue, load modeling, optimal design, and risk analysis. These volumes should serve as a valuable reference on recent developments in structural safety/reliability and probabilistic mechanics.

Library of Congress Cataloging-in-Publication Data

International Conference on Structural Safety and Reliability (5th: 1989: San Francisco, Calif.)
    Structural safety and reliability: proceedings of ICOSSAR '89, the 5th International Conference on Structural Safety and Reliability, San Francisco, August 7-11, 1989/edited by A. H-S. Ang, M. Shinozuka, and G.I. Schuëller.
      p.   cm.
    Includes bibliographical references.
    ISBN 0-87262-743-8
    1. Structural stability—Congresses.   2. Reliability (Engineering)—Congresses. 3. Structural design—Congresses. I. Ang. Alfredo Hua-Sing, 1930- II. Shinozuka, Masanobu. III. Schuëller, Gerhart I. IV. American Society of Civil Engineers. V. Title.
TA656.5.I57 1989   89-18539
624.1'7—dc20   CIP

PREFACE

The 5th International Conference on Structural Safety and Reliability (ICOSSAR '89) that was held in San Francisco on 7–11 August 1989 shows that these conferences have grown to become the premier gatherings of international experts and students in the field of structural reliability and probabilistic mechanics. Almost 340 papers were presented representing the works of authors from 30 countries. At this conference, the Freudenthal Lecture was inaugurated as the lead keynote lecture of the Conference, honoring the lecturer for his/her distinguished contributions to structural safety and reliability.

The 1989 Conference covered a wide range of developments in structural safety and reliability—from new research to practical applications of reliability principles. The papers examined all aspects of the modern view of structural safety and reliability (i.e. from the standpoint of risk acceptance) covering all types of structures and large technological systems, ranging from space structures to land-based facilities and ocean/offshore systems.

The papers presented at the Conference are contained in these volumes, and represent a permanent record of the proceedings of the ICOSSAR '89, consisting of keynote lectures, technical papers, and ongoing research papers. These volumes should serve as a valuable reference on recent developments in structural safety/reliability and probabilistic mechanics. It is our hope that the material in these volumes will help to further advance the state of science and practice of structural reliability in the safety assurance and design of structures.

The Editors

A. H-S. Ang
M. Shinozuka
G. I. Schuëller

CONFERENCE ORGANIZATION

Organizing Institutions

International Association for Structural Safety and Reliability (IASSAR)
American Society of Civil Engineers (ASCE)

Conference Co-Chairmen

A. H-S. Ang, University of California, Irvine, USA
M. Shinozuka, Princeton University, USA

Conference Scientific Committee

G. I. Schuëller, University of Innsbruck, Austria (Chairman)
J. T-P. Yao, Texas A & M University, USA (Vice-Chairman)

M. Amin, Sargent & Lundy Engineers, USA
A. H-S. Ang, University of California, Irvine, USA (Ex-Officio)
D. Blockley, University of Bristol, England
F. Casciati, University of Pavia, Italy
H. Ishikawa, Kagawa University, Japan
M. Ito, University of Tokyo, Japan
W. D. Iwan, California Institute of Technology, USA
H. Kameda, Kyoto University, Japan
F. Kozin, Polytechnic Institute of New York, USA
Y. K. Lin, Florida Atlantic University, USA
F. Moses, Case-Western Reserve University, USA
A. Nishimura, Kobe University, Japan
W. Schiehlen, University of Stuttgart, FR-Germany
H. Shibata, University of Tokyo, Japan
M. Shinozuka, Princeton University, USA (Ex-Officio)
N. Shìraishi, Kyoto University, Japan
P. D. Spanos, Rice University, USA
J. Spencer, US Coast Guard, USA
E. H. Vanmarcke, Princeton University, USA
Y. Yamada, Kyoto University, Japan
J-N. Yang, George Washington University, USA

Conference Advisory Committee

K. S. Pister, University of California, Berkeley, USA (Chairman)

M. S. Agbabian, University of Southern California, USA
G. Albright, National Science Foundation, USA
M. P. Gaus, National Science Foundation, USA
A. H. Hadjian, Bechtel Power Corporation, USA
S-C. Liu, National Science Foundation, USA
T. L. Moser, Lyndon B. Johnson Space Center, USA
L. R. Shaffer, US Army, CERL, USA
H. C. Shah, Stanford University, USA
W. Sirignano, University of California, Irvine, USA
S. Stiansen, American Bureau of Shipping, USA
H. Yang, Purdue University, USA

Conference Organizing Committee

A. Der Kiureghian, University of California, Berkeley, USA (Chairman)

A. Agogino, University of California Berkeley, USA
F. Filippou, University of California, Berkeley, USA

A. S. Kiremidjian, Stanford University, USA
L. E. Malik, URS, USA
A. Mansour, University of California, Berkeley, USA
J. L. Sackman, University of California, Berkeley, USA
J. Savy, Lawrence Livermore National Lab., USA
N. Sitar, University of California, Berkeley, USA
W. H. Tang, University of Illinois, Urbana, USA
A. K. Vaish, PMB, USA

Conference Secretariat

ICOSSAR '89 Secretariat
c/o ASCE, Attn: Ms. Elizabeth Yee
345 East 47th Street
New York, NY 10017-2398, USA

SUPPORTING INSTITUTIONS/ORGANIZATIONS

National Science Foundation, Washington, DC, USA
National Center for Earthquake Engineering Research, Buffalo, NY, USA
U.S. Army Construction Engineering Research Laboratory, Champaign, IL, USA
U.S. Coast Guard, Washington, DC, USA
University of California, Irvine, CA, USA
Princeton University, Princeton, NJ, USA
University of Innsbruck, Innsbruck, Austria
University of California, Berkeley, CA, USA
Stanford University, Stanford, CA, USA
University of Illinois, Urbana, IL, USA
University of Southern California, Los Angeles, CA, USA
Florida Atlantic University, Boca Raton, FL, USA
Rice University, Houston, TX, USA
Bechtel Power Corporation, Los Angeles, CA, USA
Flour Daniel, Inc., Irvine, CA, USA
Sargent & Lundy Engineers, Chicago, IL, USA
Woodward-Clyde Consultants, Plymouth Meeting, PA, USA
Hanshin Expressway Public Corporation, Osaka, Japan
Soartech, Inc., Takamatsu, Japan
Kajima Corporation, Tokyo, Japan
Shimizu Corporation, Tokyo, Japan
Ohbayashi Corporation, Tokyo, Japan
Taisei Corporation, Tokyo, Japan
Takenaka Corporation, Tokyo, Japan
Kawasaki Heavy Industries, Ltd, Kobe, Japan
Mitsubishi Heavy Industries, Ltd, Nagasaki, Japan
The Toyko Electric Power Company, Inc., Tokyo, Japan
Tokyo Electric Power Services Company, Ltd., Tokyo, Japan
Shikoku Research Institute Inc., Takamatsu, Japan
Nippon Telegraph and Telephone Company, Tokyo, Japan
Tokyo Gas Company, Tokyo, Japan
Osaka Gas Co., Ltd., Osaka, Japan

Officers of IASSAR and Organizers of ICOSSAR '89

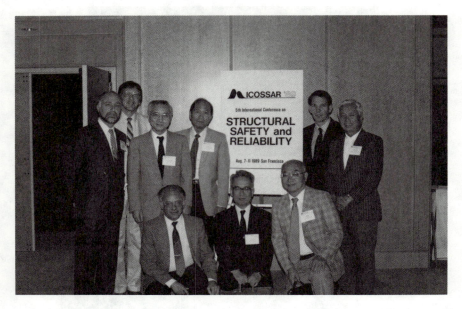

Standing L. to R.: A. der Kiureghian, Chm. Conf. Organizing Comm.; G. I. Schuëller, Chm Exec Bd. & Sc. Comm.; M. Shinozuka, Exec. Vice Pres. & Conf. Co-chm.; A. H-S. Ang, Pres. & Conf. Co-chm.; R. E. Melchers, Exec. Bd. member; S. L. Lee, Exec. Bd. member;
Kneeling: L. Esteva, Exec. Bd. member; H. Shibata, Exec. Bd. member; J. T. P. Yao, Vice Chm. Sc. Comm.

OPENING SESSION

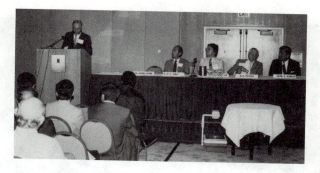

Presided by: Prof. A. Der Kiureghian
Seated L. to R.:   A. H-S. Ang; G. I. Schuëller; K. S. Pisfer, Dean of Eng., UC. Berkeley; The Honorable A. Agnos, Mayor of San Francisco.

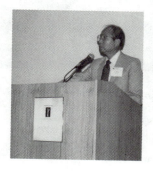

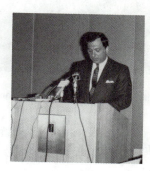

Prof. A. H-S. Ang formally opening the ICOSSAR '89

The Honorable A. Agnos welcoming the participants to San Francisco.

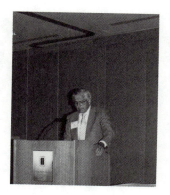

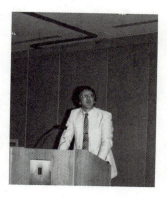

Dean K. S. Pisfer—Opening remarks as Chm. of Advisory Comm.

Prof. G. I. Schuëller—Opening remarks as Chm. of Scientific Comm.

CONTENTS

OFFSHORE AND MARINE STRUCTURES

GEOTECHNICAL ENGINEERING

GEOTECHNICAL ENGINEERING (Ongoing Research)

EARTHQUAKE ENGINEERING

DAMAGE

VOLUME 2

SYSTEM RELIABILITY

COMPUTATIONAL METHODS

RANDOM VIBRATION

FATIGUE AND FRACTURE RELIABILITY

FATIGUE AND FRACTURE RELIABILITY (Ongoing Research)

VOLUME 3

PROBABILISTIC ANALYSIS

FUZZY SET

FUZZY SET (Ongoing Research)

BRIDGES AND INDUSTRIAL FACILITIES

BRIDGES AND INDUSTRIAL FACILITIES (Ongoing Research)

AEROSPACE

AEROSPACE (Ongoing Research)

FREUDENTHAL LECTURE

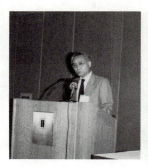 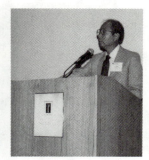

Prof. M. Shinozuka of Princeton University being introduced by A. H-S. Ang as the inaugural Freudenthal lecturer.

KEYNOTE LECTURES

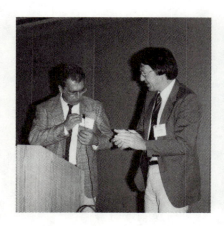 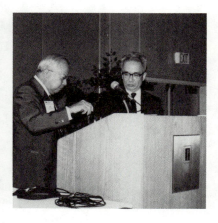

Dr. P. Kafka of GRS, FR Germany being introduced by G. I. Schueller as a keynote lecturer

Dr. J. Kondo of the Japan Science Council, Japan being introduced by H. Shibata as a keynote lecturer

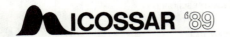
OPEN WORLD PROBLEMS IN STRUCTURAL RELIABILITY

David Ian Blockley
Department of Civil Engineering, University of Bristol, BS8 1TR, UK

Abstract

A closed world model represents total knowledge
about everything in a particular system. An open world
model represents partial knowledge where some things are
known to be true, some are known to be false, some are
unknown and some are inconsistent. In the paper problems
are classified into four types involving decision making
under certainty, risk, risk with vagueness, and partial (open)
knowledge. It is argued that many structural failures are
due to unintended and unforeseen consequences. In order to
develop a structural reliability theory to deal with these
types of failure, the problems of open world modelling need
to be addressed. The theories of support logic and interval
probability theory are introduced and their potential as a
basis for a new theory of structural reliability is discussed.

Keywords:Reliability, open-world, interval probability, risk, structures.

1. INTRODUCTION

We will begin by distinguishing two kinds of theoretical models.
Following Reiter, (Sowa 1974) a closed world model represents total
knowledge about everything in a particular system and an open world model
represents partial knowledge where some things are known to be true, some
are known to be false and others are simply unknown. Thus in a closed world
every concept is either true or false and no undefined or inconsistent states
are possible. In a closed world the information is complete in that all and
only the relationships that can possibly hold among concepts are those
implied by the given information. In an open world model there are four
possible states of a concept, true, false, unknown and inconsistent with
degrees of uncertainty between these extremes. Most logicians like to forbid
inconsistencies but in practical problems the finding and settling of
inconsistency is an important element of the problem solving process.

The relationship of these models with structural reliability can be
clarified by firstly considering a classification of problem types.

2. TYPES OF PROBLEM

A problem will be characterised as a doubtful or difficult question to which there may be a number of possible answers. Each possible answer is a conjectural solution which has to be considered and evaluated in the decision making process. Four types of problem are (following Collingridge 1980):-

Type 1   Where all of the consequences of adopting a conjectural solution are known for certain.

Type 2   Where all of the consequences of adopting a conjectural solution have been precisely identified but only the probabilities of occurrence are known.

Type 3   Where all of the consequences of adopting a conjectural solution have been approximately identified so that only the possibilities of ill defined or fuzzy consequences are known.

Type 4   Where only some of the consequences (precise or fuzzy) of adopting a conjectural solution have been identified.

The simplest decision problem is one where there is a single decision maker with a single objective, all of the options are known and an objective function can be defined whose values offer a measure of the extent to which the objective is reached by a particular conjectural solution. The best solution is then clearly the one which maximises the extent to which the objective is reached. This is the well known starting point for decision theory, it is the Type 1 problem of decision making under certainty ie determinism.

The Type 2 problem is the extension of the Type 1 problem to cases where an objective probability distribution over the set of possible consequences of a particular conjectural solution is available. This is decision making under conditions of risk and the decision rule is that the decision maker chooses the solution which maximises the expected value of the objective function. If, in a more complex case the probability density function cannot be measured objectively then it may be obtained subjectively by a betting strategy and the problem is solved with the same procedure. These developments enable this Bayesian approach to cope with problems where the factual and measurable information is more and more sparse. Multi-attribute decision theory attempts to deal with decision problems involving a number of objectives and game theory has been used to tackle problems where members of a group are rivals. Nevertheless the fundamental assumption in Type 2 problems is that all of the consequences of a conjectural solution have been precisely identified.

Type 3 problems are an extension of Type 2 problems with an explicit consideration of vagueness or lack of information about the precise definition of a given consequence of a conjectural solution. The explicit treatment of such vagueness is the objective of Zadeh's fuzzy set theory.

The theoretical developments are impressive but all involve the restrictive closed world assumption that all of the consequences (fuzzy or precise) of a conjectural solution are known. Type 4 problems are those of real world problem solving. Since only some of the consequences are identified it is necessary to make an open world assumption since the entire sample space is not known. This has been termed decision making under ignorance (Collingridge 1980); it might preferably be defined as open world decision making. There is always the possibility of unforeseen and unwanted consequences occurring after the adoption of a particular conjectural solution.

3. STRUCTURAL RELIABILITY

It has been shown (eg Pidgeon et al 1986, 1988) that many structural failures are due to unintended and unforeseen consequences of the particular structural solutions adopted. For example the collapse of a factory roof in the UK under snow loading has been described (Pidgeon et al 1986). Although superficially this failure seemed to be a simple overload due to drifting snow it was shown that the failure occurred from the unintended consequences of progress in our understanding of the structural behaviour of cold formed steel purlins. As a result of this progress certain types of structure have become very much more sensitive to assumptions in the modelling of snow load and in particular to the drifting of the snow. The adoption of a structural solution with a step in the roof and with cold formed steel purlins designed according to modern methods led to snow overload and roof collapse.

Thus in a certain general sense this particular failure is an example of human error of a rather subtle kind. However in another more restricted sense it is a technical failure, an overloading due to snow. Of course this demonstrates that in a fundamental way all error is human error (Blockley 1980) and at the deepest level of analysis it is artificial to separate the influence of human error from the traditional concerns of reliability theory.

Thus in a comprehensive treatment of reliability it is often impossible to separate the role of human error from technical error, the two are intimately bound up in a socio-technical framework. The failure of the roof under snow loading illustrates clearly the open world nature of real structural design. The actual underlying causes of the failure were not considered in the design and analysis process, although of course the possibility of snow overload was explicitly considered under the terms of the code of practice. The calculation of a probability of failure by traditional means would have been at best inadequate and possibly misleading.

The requirement in open world structural reliability problems is to be able to manipulate evidential support for a conclusion and evidential support against a conclusion quite separately ie to decouple them so that no assumptions about the unknown and unintended consequences of a conjectural solution to a problem are built in to the mathematics.

4. COMPLETENESS

A mathematical formalism is said to be complete if every formula which (in accordance with its intended interpretation) is provable within the formalism embodies a true proposition. Conversely the formalism is complete if every true proposition is embodied in a provable formula (Korner 1968). Unfortunately Godel has shown that this deepest level of mathematical completeness is unattainable in any formal system rich enough to contain arithmetic. However in structural reliability theory this deep aspect of incompleteness need not be of concern in practical application.

What is of concern however is the need to allow in our theoretical framework for the open world incompleteness of Type 4 problems. The central relationship is that between the measure of probability (support, belief, confidence etc.) p(A) for the truth or dependability of a proposition (event) A and the measure of probability for the truth or dependability of the negation of the proposition p($\bar{A}$). In standard probability theory if p(A) is known, the theory forces the value of p($\bar{A}$) to be 1-p(A), this can only be justified if the sample space from which A is taken is known or <u>it is sufficiently accurate for all practical purposes</u> to assume that it is accurately known (Type 3 problem). In any real engineering problem (Type 4) decisions have to be made under the incompleteness of partial ignorance since 'How can we know what we do not know?'.

5. TOWARDS AN OPEN WORLD RELIABILITY THEORY

Two recent theories of support logic (Blockley, Baldwin 1987) and interval probability (Cui, Blockley 1989) have been suggested. There are differences in the mathematical manipulations of these theories, but in essence they are so very close that both can be discussed for the purposes of this paper in terms of support logic.

In this theory two measures are associated with a concept, proposition or event. The necessary support N(A), is a measure of the truth or dependability of the concept and the possibility, P(A), is a quite separate measure of the truth or dependability of the negation of the concept such that N($\bar{A}$) = 1-P(A). In general N(A) + N($\bar{A}$) $\leq$ 1. In this way the measure of support for a concept is decoupled (for Type 4 problems) from the measure of support against the concept; the two measures are assessed quite separately. Three extreme states are possible therefore ie support pair values of [0, 0], [1, 1], [0, 1]. The first represents absolutely false since there is no evidence for and maximum evidence against. The second represents absolutely true since there is maximum evidence for and no evidence against. The third represents 'don't know' since there is no evidence for and no evidence against. The theory is therefore able to model three of the four states set out in the Introduction. The fourth requirement is the ability to model inconsistency. This can be done with support logic pairs since if two solutions are found for the support for a proposition A then any inconsistencies between those solutions can be found. For example if we find A with support [0.3, 0.6] and quite independently (implying the use of the multiplication model (Blockley, Baldwin 1987, Cui, Blockley 1989) we find A with support [0.4, 0.7] then

from Figure 1 the support for A is [0.58, 88] and the conflict (A) = [N(A∩Ā), P(A∩Ā)] = [0.25, 0.58].

	0.3 = N(A)	0.4 = N(Ā)	0.3 = N(Au)
0.4 = N(A)	N(A∩A) = 0.12	N(A∩Ā) = 0.16	N(A∩Au) = 0.12
0.3 = N(Ā)	N(Ā∩A) = 0.09	N(Ā∩Ā) = 0.12	N(Ā∩Au) = 0.09
0.3 = N(Au)	N(Au∩ A) = 0.09	N(Au∩Ā) = 0.12	N(Au∩Au) = 0.09

Figure 1.

Thus an interval probability theory such as support logic is a theory which can be used as a basis for an open world probability theory, because it can model the four possible states of true, false, don't know and inconsistent as set out in the Introduction.

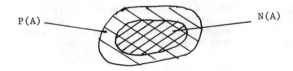

Figure 2.

Figure 2 is a Venn diagram for an open world set A with supports N(A), P(A). Notice that the underlying assumption is that the sets A and Ā are crisp but that the boundary between A and Ā is not known precisely. In this sense the spirit of a fuzzy set is captured. It would be possible to extend the definition of a probability interval p(A) = [N(A), P(A)] to a fuzzy probability through the inclusion of a membership function over the interval so that p(A) = $\chi_{p(A)}$ (q) where q ϵ [N(A), P(A)] and χ is the fuzzy membership function, however this will not be pursued further for two reasons. Firstly it introduces a level of complication to the algebra which is unwarranted and is the reason why the earlier work on fuzzy logic was abandoned. Secondly and more deeply, the fuzziness inherent in any problem can be represented more effectively by the use of the concept of hierarchical modelling (Blockley, Henderson 1988). Any discussion of uncertainty modelling is based upon choices made in the modelling of the phenomena under consideration and in particular the choice of sample space. If we think of all concepts as holons in a hierarchically structured knowledge base then by

looking upwards towards the infinite vague unity of the universe any concept (holon) is a part, and looking downwards, to the precise infinitesimals of the universe any concept (holon) is a whole. The sample space represents a choice of holons which are convenient for solving a problem. In problems where the sample space consists for all practical purposes of mutually exclusive non-interactive concepts than classical probability is sufficient. However whilst it is difficult to choose a sample space with these characteristics, then estimates of their inter-dependencies have to be obtained. In a sense fuzzy set theory is an attempt to change the level of definition (in fact to artificially induce vagueness) in order to enable better problem solving. Thus although the mathematical distinction between the membership function of fuzzy set theory and the probability function is clear and the conceptual distinction is valid, it is possible to model fuzziness without the use of fuzzy set theory. The algebra of interval probability or support logic could be applied to a fuzzy membership function or to a probability measure. The latter will involve one mapping from the power set to the interval [0, 1] and the former will involve two mappings (the membership function from the sample space as well as the probability function from the power set) to [0, 1]. The problem with this approach is that it can lead to an infinite regress since the membership function can itself be defined in terms of other concepts which also have membership functions.

For the purposes of practical computation the interval probability of support logic as a measure on concepts (holons) arranged in a hierarchically structured knowledge base, is a simple open world model of concepts which are more or less precisely defined. At high levels in a knowledge base concepts will be vague and therefore will tend to attract high levels of evidential support but at the expense of information content. At low levels the concepts will be precise and therefore of high information content but attracting consequently lower levels of support.

6. CONCLUSION

An open world model represents partial knowledge where some things are true, some are false, some are unknown and some are inconsistent. Classical reliability theory based on traditional probability theory allows only closed world modelling. It has been shown that many structural failures occur because of the unwanted and unintended consequences of the structural design solutions adopted. In order to deal with such problems a mathematics of open world modelling is required. The theories of interval probability and support logic have been introduced and are shown to have at least some of the characteristics required for the basis of a new theory of open world structural reliability.

APPENDIX

1.   Blockley D.I., The Nature of Structural Design and Safety, Ellis Horwood, Chichester, 1980.

2.   Blockley D.I., Baldwin J.F., "Uncertain Inference in Knowledge Based Systems", Proc. Am. Soc. Civ. Engrs, Eng Mechs Div, April 1987.

3.   Blockley D.I., Henderson J.R. "Knowledge Base for Risk and Cost Benefit Analysis of Limestone Mines in the West Midlands", Proc. Instn. Civ. Engrs., Part 1, 84, 539-564 June, 1988.

4.   Collingridge, D. "The Social Control of Technology", Open University Press, Milton Keynes, 1980.

5.   Cui W.C., Blockley D.I. "Interval Probability Theory for Evidential Support", to be published in Int. Journal Intelligent Systems.

6.   Korner S., "The Philosophy of Mathematics", Dover Pubs, New York, 1968.

7.   Sowa J.F. "Conceptual Structures", Addison-Wesley, New York, 1974.

8.   Pidgeon N.F., Blockley D.I., Turner, B.A., "Design Practice and Snow loading-Lessons from a Roof Collapse", The Structural Engineer, Vol. 64A, No.3 March 1986.

9.   Pidgeon N.F., Blockley D.I., Turner B.A., "Site Investigations - Lessons from a Late Discovery of Hazardous Waste", The Structural Engineer, Vol. 66, No. 19, Oct. 1988.

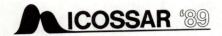

5th International Conference on
Structural Safety and Reliability

TWO PRINCIPLES FOR DATA BASED PROBABILISTIC SYSTEM ANALYSIS

Vicente Solana* and Niels C. Lind**

*   Instituto de Matemáticas, National Research Council of Spain
     Serrano 123, Madrid, Spain. E-28006.
**   Institute for Risk Research, University of Waterloo
     Waterloo, Ontario, Canada N2L 3G1

Abstract

A major problem in probabilistic system analysis is to find a self-consistent method of inference about probabilities based on a random sample of system state variables. Such a method is formalized as a two-phase process. Invariance requirements are reviewed and expressed as a strong invariance principle. A principle of data monotonicity is restated on the basis of propositional probability formalism. Distributions may be assigned by minimization of cross-entropy using fractile constraints. It is shown that this method satisfies the requirements of both principles.

Keywords

Inference method, invariance principles, data monotonicity principle, minimum cross-entropy, fractile constraints, data encoding, probability assignement, entropy methods.

1.Introduction

Many statistical methods are available to make inferences from observations, but not all are self-consistent. If a method admits more than one way to derive a result from a set of data, they should lead to the same result. This paper considers the estimation (i.e. the assignment) of a probability distribution to a random variable on the basis of a random sample of observations of the variable. It is shown that there are two stringent requirements that must be imposed on the method of inference. First, the resulting distribution should be invariant under any non-singular transformation of the variables. Second, the distribution function should be a monotonically non-increasing function of all data elements. Finally, it is shown that there exists one method of inference that does not violate these requirements, namely the minimization of cross-entropy subject to fractile constraint encoded data.

"At the beginning of every problem in probability theory there arises the need to assign some initial probability distribution" (Jaynes [9]). In conventional probability system analysis (PSA) methods the form of the probability distribution is postulated. But a postulated distribution is neither the " true distribution" nor an approximation to an "unknown-but-existing" distribution of probabilities of the system states. This is a common major criticism of conventional PSA methods.

This important first step has largely been taken for granted during the recent major expansion of PSA methods. Yet the output, i.e. the calculated probabilities about any function of states of a system, can be only as good as the input probabilities.

Probabilities should be assigned exclusively on the basis of information that is available and is objective, namely the data or observations of the system. Usually, the data refer to experimental values and measurements of the quantities related to the state variables, whether obtained separately or associated.

There are generally many ways in statistics to express, or *encode*, the observations. In each way, a particular set of constraints on the distributions arises from the data, representing functional dependencies among probabilities. There are usually many distributions that satisfy a given set of constraints derived from observations.

The fundamental problem is to construct a method to select a unique probability distribution that satisfies a set of constraints that encodes data of observations and is preferred over all other distributions in some sense. This problem raises two concerns:

(1) Given the observations, find the information that objectively encodes this data as constraints.

(2) Given the constraints, find a best probability distribution among all possible distributions satisfying these constraints.

Objectivity in the encoding process requires that there be a one-to-one correspondence between a set of observations and the set of constraints. Finite sequences of moments calculated from data do not satisfy this requirement.

The problem considered here calls for an approach to inductive inference in the manner (2), that has been proposed by Jeffreys [3], Jaynes [9], Shore and Johnson [7], Tikochinsky, Tishby and Levine[13] and Skilling [10]. The aspect of data encoding, (1), has not been considered previously, and it poses a quite different problem of inference.

In probabilistic engineering mechanics the initial distribution is defined over the space of "basic" random variables. Accordingly, the first task in modelling a problem probabilistically is to choose what random variables to consider as "basic". A problem in mechanics can generally be formulated in different ways that are physically equivalent. For example, wind velocity or wind pressure may with equal validity be chosen as a basic random variable in a problem of structural response. This poses an awkward problem of choice. On the other hand, if the choice makes no difference, one variable may be just as basic as another. The invariance principle states that the inference of probabilities of all analyzed events should be invariant under non-singular transformation of the basic random variables. In effect, this invariance principle states that there are no random variables that are "basic", and it poses a strict and fundamental requirement on any PSA method.

Lind and Chen [1] proposed a principle of reliability consistency, stating that the value of any distribution function of any scalar basic random variable, for any value of the argument, should be monotonically non-increasing when considered as a function of the elements of

a random sample of the variables used in its estimation. Some common methods of estimation violate this requirement. On the other hand, cross-entropy estimation with fractile constraints produces posterior distributions that satisfy these principles (Solana and Lind [2]).

2.The Strong Invariance Principle

Concepts of invariance have proved to be useful tools in the theory of probability inference as in many other branches of science. Invariance has been used to characterize certain features of statistical methods that remain unchanged under a class of transformations (usually coordinate system changes).

Major contributions employing invariance principles for inference of probabilities are due to Jeffreys, who developed the invariance theory of parametric estimation in [3] and [4], and to Shore and Johnson, who derived the minimum cross-entropy inference method in an axiomatic way in [7]. The latter authors stated four axioms, expressing uniqueness, invariance of the choice of coodinate system, system independence and subset independence, which are all based on one fundamental invariance principle: "if a problem can be solved in more than one way, the results should be consistent". Other authors, such as Kullback [8] and Jaynes [9], obtained properties of functionals that are invariant under variable transformations. Skilling [10] developed a parallel axiomatic based on invariance of distributions for physical quantities other than probabilities.

In this section we first state a strong invariance principle and describe an inference method based on data using operators for encoding data of observations and for choosing probability density function. Next, the strong invariance principle is applied to inference based on data using non-singular transformations of the random variables.

Different formulations of invariance in probability inference methods can be reduced in essence to one principle of invariance similar to the fundamental principle of Shore and Johnson [7]. It is here called the *strong invariance principle*:

"In any probability assignment, alternative ways of using the same data, should lead to the same result."

This principle requires self-consistency of any inference method. That is, it must not be possible to obtain contradictory results from the same procedure for any sample of data [3]. In the most immediate application of this principle, if an inference method based on data is invariant under a certain set of transformations, then the result should also be invariant under the same set of transformations. It is most important when applying this principle to assure that the alternative ways of application of the same data convey exactly the same information.

Let $x_j, j = 1, 2, ..., r$, be a vector sequence of data of observations of the vector random variable X, $x_j \in R^r$, and let $(x_i)_j$ be the observations (data) of the scalar random variable X_i, $i = 1, 2, ..., n$, corresponding to a component of x_j, $(x_i)_j \in R$.

A general method of inference about probabilities based on data should consist of two distinct steps. In the first step the method creates a set of constraints from the data. Let H be an operator that encodes the data $\{x_1, ..., x_j, ..., x_r\}$ into constraints; then this step may be represented symbolically by

$$I_X(x_1, ..., x_j, ..., x_r) = H[\{x_1, ..., x_j, ..., x_r\}] \tag{1}$$

where $I_X(x_1,\ldots,x_j,\ldots,x_r)$ denotes the encoded information about the constraints. For example, if a Weibull distribution is fitted by the conventional method of moments, then the information $I(.)$ about the data is extracted in the form of sample mean and variance; from this information the distribution parameters may be estimated.

In the second step a probability distribution of the random variable X is selected from among all possible distributions $q_X(x)$ subject to the constraints in $I_X(x_1,\ldots,x_j,\ldots,x_r)$ encoding the data. Let E be the inference operator that selects the probability distribution. Then, this step of the method is represented by

$$q_X(x|x_1,\ldots,x_j,\ldots,x_r) = E[I_X(x_1,\ldots,x_j,\ldots,x_r),p_X(x)] \qquad (2)$$

where $p_X(x)$ is a *reference* probability distribution, which is sometimes called the prior distribution (Kullback, [8])or a measure function (Jaynes, [9]). Some methods do not employ a reference distribution.

Now consider an (external) operator G applied to this method, for example a transformation of variables, or a filter. There are then two alternative ways of application explained below. The invariance principle stipulates that the same result should be obtained by using the same information in both ways of application, and the method is then said to be invariant with respect to G.

We apply the strong invariance principle to the case of a non-singular transformation of variables, G. Let $Y = f(X) : R^r \rightarrow R^r$, be a vector state function Y that corresponds to the transformation

$$\{y_1,\ldots,y_j,\ldots,y_r\} = G[x_1,\ldots,x_j,\ldots,x_r]. \qquad (3)$$

In the phase of encoding of an inference method, the invariance principle requires that the operators H and G are commutative

$$G[H[x_1,\ldots,x_j,\ldots,x_r]] = H[G[x_1,\ldots,x_j,\ldots,x_r]]. \qquad (4)$$

Correspondingly, the principle states in terms of constraint-based information that

$$I_Y(y_1,\ldots,y_j,\ldots,y_r) = G[I(x_1,\ldots,x_j,\ldots,x_r)], \qquad (5)$$

where $I_Y(.)$ refers to the new constraints that arise from the transformed vector data points using the same operator H.

For the probability distribution selection process of an inference method, the invariance principle requires that the operator G be distributive on the inference operator E. The principle states, in terms of the constraint information based on data,

$$E[G[I_X(x_1,\ldots,x_j,\ldots,x_r)];G[p_X(x)]] = G[E[I_X(x_1,\ldots,x_j,\ldots,x_r)];p_X(x)], \qquad (6)$$

or, in terms of the selected probability distributions,

$$q_Y[y|y_1,\ldots,y_j,\ldots,y_r] = G[q_X[x|x_1,\ldots,x_j,\ldots,x_r]], \qquad (7)$$

where $q_Y[y|.]$ has been chosen such that

$$q_Y[y|y_1,\ldots,y_j,\ldots,y_r] = E[I_Y\{y_1,\ldots,y_j,\ldots,y_r\};p_Y(y)]. \qquad (8)$$

Note that this application of the strong invariance principle differs from those of Shore and Johnson [7], Kullback [8] and Skilling [10]. Here invariance is not only applied to the process

of selection of a probability distribution, but also to the process of encoding of data that provide constraints.

3. The Principle of Data Monotonicity

Data monotonicity is a necessary attribute of any distribution assigned on the basis on data of observations. The principle of data monotonicity was formulated by Lind and Chen [1] as a condition of reliability consistency for structural reliability analysis. Solana and Lind [2] showed that data monotonicity is a property of the probability distributions obtained by applying the minimum cross-entropy principle with fractile constraints.

In this section the principle of data monotonicity is first restated on the basis of the algebra of propositions for inference about probabilities. Next, its application is formulated for univariate and multivariate distributions. This principle becomes an essential consistency requirement for data-based inference about the probabilities of the input and output of any system.

According to Cox [5] probability is a measure related to the assignment of a proposition as an inference from a given proposition. Probabilities so defined can be assigned on a set $\{(g/h)\}$ of pairs of propositions, in which every element, denoted by (g/h), is a proposition g inferred from the given proposition h. The measure that constitutes these probabilities is a real-valued function, usually denoted $P(g/h)$ on the set $\{(g/h)\}$.

Here we follow Jeffreys' axiomatic formulation [3] of probabilities as an inductive inference, as well as the axiomatic derivation by Cox [5, 6] using Boolean algebra. In particular, each of these formulations imply the properties of classical probabilities.

For the present purpose, this section deals only with Jeffreys' fifth axiom [3] which, in the notation of the present paper, states that

"The set $\{g/h\}$ of propositions g on the given data h, when ordered in terms of more or equal probable than, can be put in one-to-one correspondence with a set of real numbers in non-decreasing order."

The same axiom will be called Jeffreys' monotonicity axiom, carefully to be distinguished from the principle of data monotonicity stated in this paper.

Let the different propositions g be of the class $x \leq x_0$ for any x and $x_0 \in R$, and the fixed proposition h be the given data, namely a sequence of observations $x_j, j = 1, 2, ..., n, x_j \in R$. Then Jeffreys' monotonicity axiom specializes to the well-known form

$$P[x \leq x_0 | x_1, x_2, ..., x_n] \leq P[x \leq x_0 + m_0 | x_1, x_2, ..., x_n] \ , \text{ for any } m_0 > 0 \ . \tag{9}$$

Probability axioms have been stated by all authors with respect to the sets $\{(g_s/h)\}$ of elements constituted by *different* propositions g_s on the *same* given proposition h about data. Cox's definition of probabilities allows us to consider sets of propositions in an alternative and apparently new way. This corresponds to the sets $\{(g/h_s)\}$, in which the elements are a *fixed* proposition g on *different* given data propositions h_s. Then a new principle can be introduced in a way parallel to Jeffreys' monotonicity axiom:

Data monotonicity principle. The set $\{(g/h)\}$ of a proposition g on given different data propositions h_s, when ordered as increasing data values in terms of more or equal probable than, can be put in one-to-one correspondence with a set of real numbers, the assigned probabilities, in non-increasing order.

Next, let g again be of the class $x \leq x_0$, $x \in R$ (or the class of vectors $\mathbf{x} \leq \mathbf{x}_0, \mathbf{x} \in R^n$) and let the propositions h_s be the given data of observations as a scalar sequence of $x_j, j = 1, 2, ..., r$, $x_j \in R$ (or a vector sequence $\mathbf{x}_j, j = 1, 2, ..., r$, such that $\mathbf{x}_j = \{(x_1)_j, ..., (x_i)_j, ..., (x_n)_j\}, \mathbf{x}_j \in R^n, (x_i)_j \in R)$.

In the case of a sequence of r observations of a random variable X, the data monotonicity principle specializes to

$$P[x \leq x_0 | x_1 + m_1, \ldots, x_j + m_j, \ldots, x_r + m_r] \leq P[x \leq x_0 | x_1, \ldots, x_j, \ldots, x_r], \; m_j > 0. \quad (10)$$

Moreover, in the case of a sequence of r observations of a vector random variable $\mathbf{X}$, the principle specializes to

$$P[\mathbf{x} \leq \mathbf{x}_0 | \mathbf{x}_1 + \mathbf{m}_1, \ldots, \mathbf{x}_j + \mathbf{m}_j, \ldots, \mathbf{x}_r + \mathbf{m}_r] \leq P[\mathbf{x} \leq \mathbf{x}_0 | \mathbf{x}_1, \ldots, \mathbf{x}_j, \ldots, \mathbf{x}_r], \quad (11)$$

where $\mathbf{m}_j = \{(m_i)_j\}$ and $(m_i)_j > 0$, $i = 1, 2, \ldots, n$, $j = 1, 2, ..., r$.

Since the random variable X (or $\mathbf{X}$) is defined in a continuous domain R (or R^n), and since it is differentiable in this domain, the principle of data monotonicity may be expressed alternatively by the non-negativeness of the partial derivatives of the probability distribution functions. For a scalar random variable

$$\frac{\partial F_{\mathbf{X}}(x | x_1, \ldots, x_j, \ldots, x_r)}{\partial x_j} \leq 0, \quad (j = 1, 2, ..., r), \quad (12)$$

and for a vector random variable

$$\frac{\partial F_{\mathbf{X}}(\mathbf{x} | \mathbf{x}_1, \ldots, \mathbf{x}_j, \ldots, \mathbf{x}_r)}{\partial (x_i)_j} \leq 0, \quad (i = 1, 2, ..., n \; j = 1, 2, ..., r). \quad (13)$$

This is the form in which the principle was initially stated [1, 2]. This formulation is suitable for checking a particular distribution.

4. Cross-Entropy Method Using Fractile Constraints

There exists a data-based method of inference about probabilities that satisfies both requirements the invariance principle and the data monotonicity principle. This method was developed by Lind and Solana in [11] and [12].

The method is presented here for only the case of a scalar random variable. The two phases of the inference method are in accordance with section 2, data encoding and probability selection.

(1) Data encoding

The basis of the data encoding process is the following well-known property of random sample. Given a random sample of data of size r, $(x)_j$, $j = 1, 2, ...r$, a new observation of the random variable X has equal probability of falling in each of the $r + 1$ intervals in which the elements of the sample divides the domain of X.

This may be expressed in the form of the following *SampleRule:* "*The elements of a random sample of size r of a random variable are the $j/(r+1)$ fractiles, $(j = 1, 2, ...r)$ of the distribution of the random variable*" [11].

Let $Q_X(x | x_1, x_2, ...x_r)$ be the distribution of X inferred from the sample of data, having density function $q_X(x |.)$. Let the random variable X have the finite or infinite domain $I = [x_0, x_{r+1}]$ partitioned into the $r + 1$ subintervals $I_0 = [x_0, x_1), I_1 = [x_1, x_2), ..., I_r = [x_r, x_{r+1}]$. The sample rule prescribes the fractiles at $x = x_j$ points as the following fractile-pair set of constraints,

$$(x; Q_X(x|x_1, x_2, \ldots, x_r))_j = (x_j, Q_j), \ j = 1, 2, \ldots, r \tag{14}$$

where $Q_j = j/(r+1)$

(2) Probability selection

Given is a reference distribution $P_X(x)$ having density $p_X(x)$ that is positive everywhere in I. The value of $P_X(x)$ at $x = x_j$ is denoted by P_j. The probability selection process is to find a posterior distribution $Q_X(x|x_1, \ldots, x_j, \ldots, x_r)$ that minimizes the cross-entropy (Kullback's functional)

$$D(q, p) = \int_I q_X(x|.)[\log q_X(x|.) - \log p_X(x)] \, dx \tag{15}$$

and satisfies the fractile constraints (14).

The general solution, determined by using discontinuity functions and Lagrange's multiplier method (Lind and Solana [11]), may be expressed in terms of the interval multipliers

$$\mu_j = [(P_{j+1} - P_j)(r+1)]^{-1} \ j = 0, \ldots, r. \tag{16}$$

The solution is

$$q_X(x|x_1, x_2, \ldots, x_r) = \mu_j \, p_X(x), \ x \in I_j, \ j = 1, 2, \ldots, r, \tag{17}$$
$$Q_X(x|x_1, x_2, \ldots, x_r) = [j/(r+1)] + \mu_j[P(x) - Pj], \ x \in I_j, \ j = 1, 2, \ldots, r; \tag{18}$$

the cross-entropy functional (15) takes the minimum value

$$D_{\min} = \log(r+1) - (r+1)^{-1} \sum_0^r \log(P_{j+1} - P_j). \tag{19}$$

The posterior density function $q_X(x|x_1, x_2, \ldots, x_r)$ has piecewise the form of the reference density function $p_X(x)$ scaled over each interval by the constant factor μ_j. The minimum value of the cross-entropy function is functionally independent of the reference density $p_X(x)$ and of the reference distribution $P_X(x)$ except for the values assumed at the points x_i.

The cross-entropy inference method using fractile constraints thus satisfies the requirements of the principle of invariance together with the data monotonicity principle. Apparently, there is no other method available that generally satisfy the requirements imposed by the invariance principle and the data monotonicity principle.

5. Conclusions

Methods of inference assigning probability distributions of random variables from observed realizations are considered in this paper. If such a method is to be self-consistent, it must be invariant under all admissible variable transformations (the strong invariance principle). Moreover the assigned probabilities should satisfy the Jeffreys' monotonicity axiom. As functions of observations the distributions functions, in particular, are then shown to be subject to the requirement that every partial derivative with respect to an observation should be negative or zero (the data monotonicity principle). These two principles impose stringent requirements of admissibility on methods of estimation. Many such methods do not satisfy these requirements. However, it is shown that cross-entropy estimation based on fractiles derived by the sample rule does indeed satisfy both principles in the one-dimensional case.

References

[1] LIND, N.C., and CHEN, X.: "Consistent Distribution Parameter Estimation for Reliability Analysis", Structural Safety, Vol. 4, 141-149, 1987.

[2] SOLANA, V., and LIND, N.C.: "A Monotonic Property of Distributions Based on Entropy with Fractile Constraints", 8th International Maximum Entropy Conference, St. John's College, Cambridge, England, August 1988. John Skilling (ed.), Kluwer Academic Publications, Dordrecht, Netherland, 1989 (in press).

[3] JEFFREYS, H.: Theory of Probability, 2nd ed., Oxford University Press, London, 1948.

[4] JEFFREYS, H.: "An Invariant form of the Prior Probability in Estimation Problems", Proceedings of the Royal Society, A, vol. 186, 1946.

[5] COX, R.T.: The Algebra of Probable Inference, Oxford University Press, London, 1961.

[6] COX, R.T.: "Probability, Frequency and Reasonable Expectation", American Journal of Physics, Vol.14,1,1-13, 1946.

[7] SHORE, J.E., and JOHNSON, R.W.: "Axiomatic Derivation of Principle of Maximum Entropy and the Principle of Minimum Cross-Entropy", IEEE Transactions on Information Theory, IT-26, 1, 26-37, 1980.

[8] KULLBACK, S.: Information Theory and Statistics, Wiley, New York, 1969.

[9] JAYNES, E.T.: Papers on Probability, Statistical Physics and Statistics, Rosencrantz, R.D. (ed.), Reidel Publishing Co., Dordrecht, Netherlands, 1983.

[10] SKILLING, J.: "The Axioms of Maximum Entropy", G.J. Erickson and C.R. Smith (eds). Maximum-Entropy and Bayesian Methods in Science and Engineering, Vol. 1, 173-187, Kluwer Academic Publications, 1988.

[11] LIND, N.C., and SOLANA, V." Estimation of Random Variables with Fractile Constraints", IRR paper No. 11, Institute for Risk Research, University of Waterloo, Waterloo, Ontario, Canada, 1988.

[12] LIND, N.C., and SOLANA, V.: "Cross-Entropy Estimation of Distributions Based on Scarce Data", Society for Risk Analysis Annual Conference, Washington, D.C., 1988.

[13] TIKOCHINSKY, Y., TISHBY, N.Z., LEVINE, R.D.: "Consistent Inference Probabilities for Reproducible Experiments", Physics Review Letters, V.51, 1357-1360, 1984.

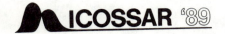
RELIABILITY ANALYSIS
FOR NONLINEAR LIMIT STATES

AVINASH M. NAFDAY

EQE ENGINEERING, INC.
3150 BRISTOL, SUITE 350
COSTA MESA, CALIFORNIA 92626

ABSTRACT

An analytical procedure, incorporating individual
probability distributions of random variables, is proposed
for generating moments of any arbitrary order for the
nonlinear limit state functions. The results are applicable
for estimating reliability measures in second or higher
moments based schemes. Exact probability of failure can
also be calculated by contour integration in a complex
plane. The formulation is based on the use of operational
transforms and does not require any linearization,
equivalent normal transformations or iterative numerical
algorithms.

KEYWORDS

Analytical Formulation, Moment Generating Functions,
Nonlinear Limit States, Operational Transforms, Reliability
Evaluation

1. INTRODUCTION

Second moment approach is now in routine use for the solution of structural
reliability analysis problems. Basic principles of second moment concepts
are well known and extensively covered in the literature [Ditlevsen,
1979; Dolinski, 1983; Hasofer and Lind, 1974; Rackwitz and Fiessler, 1978;

Schuller and Stix, 1987; Shinozuka, 1983]. The problem is generally formulated in terms of a limit state function $g(\overline{X})$ of random variables $\overline{X} = (X_1, X_2, \ldots, X_n)$, describing loads, strengths and geometry; such that $g(\overline{X})$ represents a failure surface dividing the space of variables into two sets denoting safety and failure. In general, $g(\overline{X}) > 0$ means that the structure is safe and $g(\overline{X}) < 0$ implies that the structure has failed. The reliability index β is then defined as the ratio of the mean μ_g of the limit state function $g(\overline{X})$ to its standard deviation σ_g.

Frequently, information about the higher moments or distribution of random variables is available and this information can also be incorporated in the computation of reliability index [Rackwitz and Fiesseler, 1978; Winterstein and Bjerager, 1987]. In such an extended format, each random variable may have any specified normal or non-normal probability distribution. Non-normal distributions are usually transformed to their "equivalent" normal form by suitable transformations and any of the several variations of advanced second moment methods are used to find the reliability index [Schuller and Stix, 1987]. Except for the case of linear limit state function with normal random variables, the form of the function $g(\overline{X})$ is nonlinear. Therefore, most of the current procedures resort to linearization of the limit state function by Taylor's series expansion or some other approximation and solve for the reliability index by an iterative numerical algorithm.

The aim of this paper is to present an analytical method based on the use of operational transforms, that leads to the moment generating function for the nonlinear limit state function of random variables incorporating their individual probability distributions. Thus, the method can either be used to compute an 'exact' second moment reliability index or serve as an input to any of the several methods based on higher moments [Grigoriu and Lind, 1980; Winterstein and Bjerager, 1987]. The proposed procedure does not involve any linearization, equivalence transformations of non-normal distributions or iterative numerical algorithms. Moreover, it can lead to an exact probability of failure by inverse transformation of the moment generating function. In case analytical inversion is difficult, several approximate numerical schemes are possible to estimate the probability of failure.

2. LIMITATIONS OF CURRENT APPROACHES

There are several limitations to the current procedures for extended second moment reliability analysis, particularly for nonlinear limit states. The accuracy of the current second moment methods depends on the form of the nonlinear functions and linearization can become increasingly inaccurate for highly nonlinear limit states. Approximate procedures involving second order derivatives of the performance function have been suggested for such cases [Fiesseler et. al., 1979; Der Kiureghian et. al., 1987]. However, this requires that the function has continuous derivatives of the first and higher orders.

In case distributions of the random variables are known, these are usually transformed into an equivalent normal form. There are many alternative proposals for such transformations [Paloheimo, 1973; Rackwitz and Fiessler, 1978]. For example, Rackwitz and Fiesseler [1978] estimate the parameters of the equivalent normal distribution by matching the cumulative distribution functions and probability density functions of the actual variables with the approximate normal distribution at the failure checking point. Such transformations usually distort the original expressions and for large

coefficient of variations, non-negative variables often have negative "equivalent" normal mean. Moreover, the transformations of non-normal variables to approximate normal variables is not unique and depends on the ordering of basic variables. All such transformations become increasingly inaccurate with skewness of original distributions and increasing dimensions of the problems [Ayubb and Haldar, 1984; Schuller and Stix, 1987].

Algorithms for advanced second moment analysis using transformed variables become cumbersome with increasing number of basic variables; involving linearization, computation of derivatives, convergence problems in iterations and may lead to a local optima in nonlinear optimization. These problems can be avoided by the proposed analytical formulation.

3. ANALYTICAL FORMULATION

The reliability for any limit state function $g(\overline{X})$ is the probability content of the safe domain or the value of the distribution of $g(\overline{X})$ at the zero value. Using traditional procedures, the evaluation of this distribution or even the moments of the scalar control variable $g(\overline{X})$ is not feasible due to requirement of numerical integrations in spaces of large dimensions [Grigoriu and Lind, 1980]. Grigoriu [1982/83] evaluated these from approximation of $g(\overline{X})$ by a polynomial. The method proposed here is, however, based on developing a moment generating function of $g(\overline{X})$ for finding the exact moments of any arbitrary order for use in moments-based procedures. The inversion of the moment generating function determines the distribution function of $g(\overline{X})$. For ease of understanding, the procedure will be illustrated with the assumption of independence of random variables. This is not a restrictive assumption since any vector with a set of dependent variables can be transformed into a vector with independent components [Hohenbichler and Rackwitz, 1981].

Moment generating function for the specified performance function is determined by use of operator transform algebra [Giffin, 1976]. Two fundamental operators L/F and M are defined, denoting operations of sum and product of random variables, respectively. The properties of these operators allows one to express any limit state function in terms of convolution of operators. For example, a limit function $f = X_1 + X_2$ using L/F operator, becomes $L/F(f) = L/F(X_1) \, L/F(X_2)$. Similarly, $g = X_1 X_2$ can be transformed to $M(g) = M(X_1) \, M(X_2)$. A more complicated limit state function of the form, say, $h = X_1 X_2 + X_3 X_4$ can be expressed as

$$L/F(h) = L/F[\{ M(X_1) \, M(X_2) \}^{-1}] \, L/F[\{ M(X_3) \, M(X_4) \}^{-1}] \qquad (1)$$

where the values for $L/F(..)$ and $M(..)$, depending on the probability distributions, are found from standard tables [Bateman, 1954]. Rules of operations for more complex limit state functions can similarly be specified [Giffin, 1976] and are not reproduced here because of space restrictions.

It will immediately be recognized that L/F refers to Laplace or Fourier operator because of their convolution property for sums of random variables. M denotes Mellin transform operator. Although not as well known as the other two, it plays an analogous important role for the products and quotients of random variables.

There is one problem with such operations,- e.g., everytime there is a combination of two operators, inversion of the resultant operator is required. As an illustration, in the above example for limit state function 'h',

inversion of M operator is necessary before L/F operator can be used. This can become quite tedious for complicated expressions. However, recently developed mathematical theorems relating these transform operators and their inverses allows one to bypass the inversion operations at each stage. It can be shown that the L/F(h) turns out to be a moment generating function of 'h' and mean, variance or even higher moments can easily be obtained. Thus, procedures that incorporate more than two moments can be used.

Even exact probability of failure can be computed by inversion of L/F(h) to recover the distribution of 'h'. This, however, involves contour integration in a complex plane and use of the theorem of residues. Alternatively, a distribution is fitted to the moments generated from L/F(h) to evaluate the reliability.

Transform methods provide an elegant means of analyzing nonlinear functions of random variables. The possible use of computational procedures, such as Fast Fourier Transforms, for this purpose are currently under investigation. The use of operational transforms will be illustrated here with a simple example involving only M-operators. Further details of this procedure, including comparision with Monte Carlo simulation, for larger and more complex problems will be published separately.

4. EXAMPLE

In offshore structure design, the determination of wave forces on cylindrical piles is of paramount importance. The expression for maximum inertia force I in shallow water can be expressed as $I = C\ G\ H\ T^{-1}$. C, the inertia coefficient is lognormally distributed. H, the wave height, has a Rayleigh distribution and the wave period T usually has a two parameter Weibull distribution. G is a geometrical factor treated here as a constant. The probability density functions and Mellin transforms for each variable are given in Table 1.

Using the convolution property [Giffin, 1976] of product transform operator M on the expression $I = C\ G\ H\ T^{-1}$ and the values of individual transforms from Table 1,

$$M_I(s) = G^{s-1}\ M_C(s)\ M_H(s)\ M_T(2-s) \qquad (2)$$

$$= (\frac{G\theta}{T_s})^{s-1} \exp\ [(s-1)\ \lambda + \frac{1}{2}\ (s-1)^2\ \zeta^2]\ \Gamma(\frac{s+1}{2})\ \Gamma(\frac{5-s}{4}) \qquad (3)$$

Based upon this equation and Table 2, moments of any order can be generated. For example, the mean and variance of the inertia force are

$$\mu_I = M_I(2) = (\frac{G\theta}{T_s})\ C\ \Gamma(\frac{3}{2})\ \Gamma(\frac{3}{4}) = 0.77\ \frac{G\ H_s\ C}{T_s} \qquad (4)$$

where H_s = Significant wave height

$$\sigma_I^2 = M_I(3) - M_I^2(2) = (\frac{G\theta}{T_s})^2\ C^2\ [\Gamma(2)\ \Gamma(\frac{1}{2})\ \exp\ \zeta^2 - \Gamma^2\ (\frac{3}{2})\ \Gamma^2(\frac{3}{4})] \qquad (5)$$

TABLE 1

M-TRANSFORMS FOR THE RANDOM VARIABLES

Variable	Density Function	M - Transform
c	$f(c) = \dfrac{1}{\zeta c \sqrt{2\pi}} \exp\left[-\dfrac{1}{2}\left(\dfrac{\ln c - \lambda}{\zeta}\right)^2\right]$ λ and ζ are the parameters	$M_c(s) = \exp\left[(s-1)\,\lambda + 1/2\,(s-1)^2\,\zeta^2\right]$
H	$f(H) = \gamma\left(\dfrac{1}{\theta}\right)^\gamma H^{\gamma-1}\exp\left[-\left(\dfrac{H}{\theta}\right)^\gamma\right]$ γ is a parameter, $\theta = Hs/\sqrt{2}$ where H_s = Significant wave height	$M_H(s) = \left(\dfrac{1}{\theta}\right)^{1-s}\cdot\Gamma\left(\dfrac{s-1}{\gamma}+1\right)$
T	$f(T) = 4\,\dfrac{T^3}{T_s^4}\exp\left[-\left(\dfrac{T}{T_s}\right)^4\right]$ T_s = a data based parameter	$M_T(2\text{-}s) = \left(\dfrac{1}{T_s}\right)^{s-1}\Gamma\left(\dfrac{5-s}{4}\right)$

TABLE 2

RELATIONSHIP BETWEEN COMMON STATISTICAL PROPERTIES AND M-TRANSFORMS

Statistics	Symbol	M-Transform Expression
Mean	μ	$M(2)$
Variance	σ^2	$M(3) - M^2(2)$
Skewness	μ_3	$M(4) - 3\,M(3)\,M(2) + 2\,M^2(2)$
Fourth Central Moment	μ_4	$M(5) - 4M(4)\,M(2) + 6M(3)\,M^2(2)$ $- 3\,M^4(2)$

Similarly, higher order moments can be obtained. These moments are useful in the estimation of reliability measures.

5.   CONCLUSIONS

The proposed analytical procedure for reliability analysis is applicable to nonlinear limit state functions and the random variables may have non-normal distributions. Different random variables can have different distributions and there is no need to transform them to equivalent normal form since the distributions are used directly. Linearization and approximation are not used at any stage in the procedure and thus the associated problems of invariance, existence and continuity of derivatives, convergence of iterations in numerical algorithms etc. do not arise. The methodology is general and can be used in second or higher moments context and can even be extended for finding the exact value of the probability of failure.

6.   REFERENCES

Ayubb, B. M. and Haldar, A.,"Practical Structural Reliability Techniques," Journal of Structural Engineering, American Society of Civil Engineers, Vol. 110, No. 8, August, 1984.

Bateman, H., "Tables of Integral Transforms", McGraw Hill, New York, 1954.

Der Kiureghian, A., Lin, H. Z. and Hwang, S. J.,"Second Order Reliability Approximations," Journal of Engineering Mechanics, American Society of Civil Engineers, Vol. 113, No. 8, 1987, pp. 1208-1225.

Ditlevsen, O., "Generalized Second Moment Reliability Index," Journal of Structural Mechanics, Vol. 7, No. 4, 1979, pp. 435-451.

Dolinski, K., "First Order Second Moment Approximation in Reliability of Structural Systems: Critical Review and Alternative Approach," Structural Safety, Vol. 1, pp. 211-241, 1983.

Fiesseler, B., Neumann, H. J., and R. Rackwitz, "Quadratic Limit States in Structural Reliability," Journal of Engineering Mechanics Division, American Society of Civil Engineers, Vol. 105, No. EM4, pp. 661-676, 1979.

Giffin, W.C., "Transform Techniques for Probability Modeling", Academic Press, New York, 1976.

Grigoriu, M. and Lind, N. C., "Optimal estimation of Convolution Integrals," Journal of Engineering Mechanics Division, American Society of Civil Engineers, Vol. 106, No. EM6, pp. 1349-1364, 1980.

Grigoriu, M., "Methods for Approximate Reliability Analysis," Structural Safety, Vol. 1, pp 155-165, 1982/83.

Hasofer, A. M. and Lind, N. C., "Exact and Invariant Second Moment Code Format," Journal of Engineering Mechanics Division, American Society of Civil Engineers, Vol. 100, No. EM1, pp.111-121, 1974.

Hohenbichler, M. and Rackwitz, R., "Non-normal Dependent Vectors in Structural Safety, Journal of Engineering Mechanics, ASCE, Vol. 107, No. EM6, 1981, pp. 1227-1241.

Paloheimo, E., "Eine Bemessungsmethode die sich aut variierende Frankilen grundet," Arbeitstagung des Deutschen Beton vereins Sicherheit von Betonbauten, Berlin, 1973.

Rackwitz, R. and Fiessler, B., " Structural Reliability Under Combined Random Load Sequences," Computers and Structures, Vol. 9, pp 489-494, 1978.

Schuller, G. I. and Stix, R., "A Critical Appraisal of Methods to Determine Failure Probabilities," Structural Safety, Vol. 4, 1987, pp. 293 - 309.

Shinozuka, M., "Basic Analysis of Structural Safety," Journal of Structural Engineering, American Society of Civil Engineers, Vol. 109, No. 3, pp. 721-740, 1983.

Winterstein, S., and Bjerager, P.,"The Use of Higher Moments in Reliability Estimation," Proceedings of the Fifth ICASP in Soil and Structural Engineering, Vancouver, Canada, 1987.

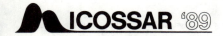
ON THE APPLICATION OF CONDITIONAL
INTEGRATION IN STRUCTURAL RELIABILITY ANALYSIS

W. OUYPORNPRASERT, C.G. BUCHER, G.I. SCHUËLLER

Institute of Engineering Mechanics,
University of Innsbruck,
Innsbruck, Austria

ABSTRACT

An accurate, and efficient computational procedure is proposed for structural reliability analysis. The suggested advanced Monte-Carlo simulation utilizes the concept of conditional integration which takes into account the shape of limit-state functions in the generation of samples. An efficient computational scheme to determine approximate analytical limit-state functions, from which the samples will be generated, is also suggested for those cases where no analytical limit-state function is available. Numerical examples show the applicability of the proposed procedure to analyze structural systems in terms of failure probability.

KEYWORDS

Structural Reliability, Monte Carlo Simulation, Conditional Integration, Structural Systems

1. INTRODUCTION

It is well known that since material properties of structures as well as loadings to which they are exposed to during their life time reveal statistical characteristics, structrural safety may be more realistically expressed in quantitative terms such as the reliability which is defined as the probability that the structural load effects will not exceed the structural resistance. This complementary term i.e. the failure probability may be expressed for time-invariant structural reliability problems as:

$$p_f = \int\limits_{D_f} f_{\underline{X}}(\underline{x}) \, d\underline{x} \qquad (1)$$

where $f_{\underline{X}}(\underline{x})$ is the joint probability density function of n random variables $\underline{X}$ and D_f represents the failure domain which can be determined from the limit-state function, $Z=g(\underline{X})$, defined such that $Z\leq0$ denotes the failure state and $Z>0$ the safe state, respectively. In general, the integral of eq. (1) is quite involved. This is due to the complexity of the joint probability density function as well as the shape of the failure domain [1]. By applying the concept of conditional integration [2-3], which takes into account the shape of limit-state functions in the generation of samples, with remarkable high efficiency most accurate results can already be obtained. The significant numerical advantage of simulation methods applying the concept of conditional integration over existing accurate methods e.g. [4-6] has been shown in [2-3]. However, in many cases e.g. where the limit-state functions cannot be expressed explicitly i.e. in analytical form, or where the a single sample requires a complete finite element analysis, approximate analytical expressions for limit-state functions, which will be called below as response surfaces , are required.

The objective of this paper is to show the efficient use of conditional integration in structural reliability analysis utilizing response functions. First the concept of conditional integration will be summarized. Then, an efficient computational scheme to determine response surfaces [7] is be presented. In the following, two frame structures are analyzed by applying the proposed procedure. Finally, some conclusions will be drawn from the numerical results.

2. CONCEPT OF CONDITIONAL INTEGRATION

First a failure function, W, such that $W=-Z\geq0$ is defined. Then the probability density function of W, $f_W(z)$, is the truncated probability density function of Z, $f_Z(z)$. Therefore, $f_W(z)$ is also the optimum importance sampling density function of Z, i.e. $f_W(z)=f_Z(z)/p_f$ [8]. Since p_f is not known in advance, it is difficult to determine $f_W(z)$. Fortunately, for failure calculations, samples may be generated from an importance sampling density function of W, $h_W(z)$, which is similar to the shape of $f_W(z)$. Since $h_W(w)$ is a probability density function, ep. (1) may be rewritten as:

$$P_f = \int_{D_f} f_{\underline{X}}(\underline{x}) \, d\underline{x} \int_0^\infty h_W(w) \, dw \qquad (2)$$

The integration of the integral on the left hand side conditional on $W=w$ leads to:

$$P_f = \int_0^\infty [\int_{S_W} \frac{f_{\underline{X}}(\underline{x})}{h_W(w)} \, d\underline{x}] \, h_W(w) \, dw \qquad (3)$$

where S_W is the conditional surface defined by $W=w$ and $h_W(w)$ of the inner integral. It may be interpreted as the probability of occurrence of surface S_W. Generally speaking, the inner integral is , in fact, the surface integral of the failure probability conditional on the given surface S_W. Note that the values of n-1 variables, $\underline{X}^R=\{X_1,X_2,...,X_{k-1},X_{k+1},...,X_n\}$, can be generated freely; whereas the values of X_k are to be analyzed such that the $\underline{x}$ lie on the surface S_W. Furthermore, since $h_W(w)$ is not the optimum importance sampling density function, the numerical efficiency can be improved by applying a regular importance sampling function for $\underline{X}_R$, $h_Y(\underline{x}_R)$. Thus, eq. (3) becomes:

$$P_f = \int_0^\infty [\int_{S_W} \frac{f_{\underline{X}}(\underline{x})}{h_{\underline{Y}}(\underline{x}^R)h_W(w)} h_{\underline{Y}}(\underline{x}^R) d\underline{x}^R dx_k] h_W(w) dw \qquad (4)$$

The advantage of conditional integration as described above may be listed as follows: First, for $h_W(w)$ as well as $h_{\underline{Y}}(\underline{x}^R)$ respectively any probability density function can be chosen from which samples can easily be generated. Second, since W is the failure function, samples obtained from the simulation procedure automatically lie within the failure domain. Third, $h_{\underline{Y}}(\underline{x}^R)$ can be selected such that samples concentrate in the region which contributes most to the failure probability. Fourth, a conditional importance sampling density function can be adjusted freely so that the ratio $f_{\underline{X}}(\underline{x})/(h_{\underline{Y}}(\underline{x}^R)h_W(w))$ is nearly constant for most $\underline{x}$. These facts result in substantial advantages with respect to numerical efficiency.

3. EFFICIENT SCHEME TO DETERMINE RESPONSE FUNCTIONS

3.1. Basic response functions

Generally, an actual limit-state function, $g(\underline{X})$, of n random variables $\underline{X}$ can be expanded in terms of a Taylor series as:

$$g(\underline{X}) = a_0 + \sum_{i=1}^n b_i X_i + \sum_{i=1}^n \sum_{j=1}^n c_{ij} X_i X_j + \sum_{i=1}^n \sum_{j=1}^n \sum_{k=1}^n d_{ijk} X_i X_j X_k + \ldots \qquad (5)$$

where a_0, b_i, c_{ij} and d_{ijk} are coefficients related to the values of partial derivatives. In most cases for structural reliability problems, $g(\underline{X})$ can be represented sufficiently well by a complete second-order polynomial, $\bar{g}(\underline{X})$, as:

$$\bar{g}(\underline{X}) = (a_0 + \sum_{i=1}^n b_i X_i + \sum_{i=1}^n c_{ii} X_i^2) + (\sum_{i=1}^n \sum_{j=i+1}^n 2c_{ij} X_i X_j) \qquad (6)$$

where the first parenthesis, $\bar{g}_1(\underline{X})$, and the second parenthesis, $\bar{g}_2(\underline{X})$, represent $g(\underline{X})$ along and between axes, respectively. It can be seen that the numbers of unknown coefficients for $\bar{g}_1(\underline{X})$ and $\bar{g}_2(\underline{X})$ are $2n+1$ and $n(n-1)/2$, respectively. Since the number of random variables involved in structural systems can be very large - e.g. n maybe as high as 100 - the determination of $\bar{g}(\underline{X})$ through regression analysis can be prohibitively expensive. It has been shown in [7] that $\bar{g}(\underline{X})$ can be determined most efficiently by simple interpolation procedures without solving any large systems of linear equations. This fact makes the proposed procedure, as presented below, most attractive. Since for complex structural systems, the limit-state functions may be available only in form of characteristic functions, the interpolation procedures should utilize the points on the failure surfaces and the iterative procedures to find those points should utilize the characteristic function, instead of the values of the limit-state functions. The response functions should be defined in the *original* space so that the quality the response functions can be judged by qualified engineers.

3.2. Suggested computational scheme

For convenience, the interpolation procedure will now be explained relative to mean values, $\underline{x}_M$. The suggested computational scheme can be summarized as follows:

Step 1 : $\bar{g}_1(\underline{X})$ is obtained by using points along axes X_i. Since the number of points required for each axis is three, the points x_i are chosen to be , while keeping the values of the remaining variables at mean values, $x_{Mi}+f_{Li}\sigma_{Xi}$, x_{Mi} and $x_{Mi}+f_{Ri}\sigma_{Xi}$ where f_{Li} and f_{Ri} are real numbers determined from iterative procedure satisfying the condition $g(\underline{X})=0$ and σ_{Xi} is the standard deviation of X_i. L and R refer to the left and to the right hand side of mean values, respectively. Using $2n+1$ interpolation points, the coefficients for $\bar{g}_1(\underline{X})$ can be determined separately for each axis.

Step 2 : The interpolation for mixed terms $\bar{g}_2(\underline{X})$ is performed. For a particular pair of X_i and X_j - keeping the values of the remaining variables at mean values- the values of x_i and x_j are considered from the points on the failure surface in the more important parts for X_i and X_j. The respective coefficient c_{ij} can be calculated independently from the other c_{ij}'s. The response function $\bar{g}(\underline{X})$ is obtained by simply adding $\bar{g}_2(\underline{X})$ to $\bar{g}_1(\underline{X})$.

For details on the numerical procedure to determine coefficients of a standard complete second-order polynomial and an efficient iterative procedure to determine the points on the failure surfaces, it is referred to [7].

4. APPLICATION TO STRUCTURAL SYSTEMS RELIABILITY

4.1. General remarks

The concept of conditional integration is now used to determine the failure probabilities of two frame structures utilizing response functions. The probability integration can be carried out by simulation procedures utilizing conditional importance density density functions [3] which is implemented within a general purpose computer code, i.e. the ISPUD code [4]. The conditional importance sampling density functions can be determined from the simulation method with a small number of simulations. For the numerical examples, $h_W(w)$ is modeled by an exponential distribution. Actually, finite element analysis may be required to obtain response functions. In these cases, where actual limit-state functions are not available in explicit form, the computational effort for simulation procedures utilizing response functions can be much lower than that utilizing actual limit-state functions.The results, as given below, intend to show the quality of response functions in terms of induced errors only. Accurate results can be obtained with 1,000 simulations (CPU-time ≈ 1minute). Other accurate methods [4-6] require over 8,000 simulations to achieve the same degree of accuracy [6].

4.2. Numerical Examples

Example 1

A steel portal frame as sketched in Fig. 1 is analyzed. The plastic moment capacity X_1-X_5 and loads X_6-X_7 are assumed to be uncorrelated and lognormally distributed. Their statistical properties, which are taken from [9], are listed in Table 1.

An estimate of p_f utilizing a response function is $3.918 \cdot 10^{-3}$. Comparing with the exact result using actual limit-state function ($p_f = 4.046 \cdot 10^{-3}$), the error is 3.14 %. Note that CPU-time required to obtain response function is 2.0 sec.

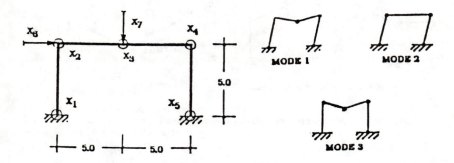

Fig. 1 A steel portal frame with possible collapse mechanisms

Table 1. Statistical properties of random variables for example 1[9]

Variables	Type of Distr.	Units	Mean	Stand. Dev.
X_1-X_5	Lognormal	kNm	134.9	6.745
X_6	Lognormal	kN	50.0	15.0
X_7	Lognormal	kN	40.0	12.0

Example 2

A gable frame as sketched in Fig. 2 can be modeled with 8 random variables. The correlation matrix is given in Table 2. Their statistical properties are listed in Table 3.

Utilizing response function, the estimate of p_f is $6.615 \cdot 10^{-2}$. The error is 0.08% when compared with the exact value for $p_f = 6.610 \cdot 10^{-2}$ i.e. by using actual limit-state functions. CPU-time to obtain the response function is 2.36 seconds.

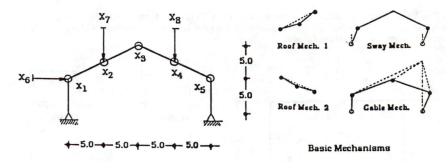

Fig. 2. A gable frame structure and its basic collapse mechanisms

Table 2. Correlation matrix of random variables for example 2

$\rho_{X_i X_j}$	X_1	X_2	X_3	X_4	X_5	X_6	X_7	X_8
X_1	1.0							
X_2	0.5	1.0						
X_3	0.5	0.5	1.0					
X_4	0.2	0.2	0.5	1.0		Sym.		
X_5	0.2	0.2	0.5	0.5	1.0			
X_6	0	0	0	0	0	1.0		
X_7	0	0	0	0	0	0.1	1.0	
X_8	0	0	0	0	0	0.1	0.3	1.0

Table 3. Statistical properties of random variables for example 2

Variables	Type of Distr	Units	Mean	Stand.Dev.
X_1-X_5	Lognormal	kNm	600.0	60.0
X_6	Gumbel	kN	150.0	45.0
X_7-X_8	Gumbel	kN	100.0	30.0

5. CONCLUSIONS

From numerical results it can be concluded that the concept of conditional integration can be applied most efficiently in conjunction with response functions for structural reliability problems. Since the solutions utilizing response functions are very close to the results utilizing actual limit-state functions, the response functions obtained from the proposed procedure can represent the actual limit-state functions very well, at least in the region which contributes most to the failure probabilities. Finally it schould be mentioned that the accuracy of results utilizing response functions is not sensitive to the type of distributions of random variables.

6. ACKNOWLEDGEMENT

This research is partially supported by the Austrian Research Council (FWF) under Contract No. P7286 which is gratefully acknowledged by the authors.

7. REFERENCES

[1]   Schuëller,G. I., Stix, R.,A critical appraisal of methods to determine failure probabilities, *Structural Safety* 4/4, 293-309 (1987).
[2]   W. Ouypornprasert, Effiziente Integrationsmethoden zur genauen Zuverlässigkeitsanalyse von Tragwerken, Report No. 21-88, Institut für Mechanik, Universität Innsbruck, 1988.
[3]   W. Ouypornprasert, Efficient computational methods for structural reliability analysis based on conditional importance sampling functions, *ZAMM* 69, No.4/5, T560 (1989). (to appear)
[4]   U. Bourgund, C.G. Bucher, Importance sampling procedure using design points (ISPUD)- a user's manual, Inst.Eng.Mech., Univ.Innsbruck, Report No.8-86 (1986).
[5]   C.G. Bucher, Adaptive sampling- an iterative fast Monte Carlo procedure, *Structural Safety* 5/3, 119-126 (1988).
[6]   W. Ouypornprasert, Adaptive numerical integration for reliability analysis, in Report No.12, Institute of Engineering Mechanics, University of Innsbruck, 17-48 (1987).
[7]   W. Ouypornprasert, C.G. Bucher, An efficient scheme to determine response functions for reliability analyses, Inst.Eng.Mech., Univ.Innsbruck, Int.Work.Rep.No. 30, 1988.
[8]   R.Y. Rubinstein, *Simulation and Monte Carlo Method*, John Wiley&Sons, New York (1981).
[9]   M.J. Grimmelt, A method to calculate reliability of structural systems under combined loadings Diss.,(in German), Tech.Univ.Munich, FR Germany (1983).

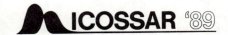

RELIABILITY OF PLASTIC SLABS

P. Thoft-Christensen
University of Aalborg
Sohngaardsholmsvej 57, DK-9000 Aalborg, Denmark

Abstract

In the paper it is shown how upper and lower bounds for the reliability of plastic slabs can be determined. For the fundamental case it is shown that optimal bounds of a deterministic and a stochastic analysis are obtained on the basis of the same failure mechanisms and the same stress fields.

Keywords

Structural reliability, plastic slabs, upper and lower bounds.

1. Introduction

Upper and lower bounds of the reliability of plastic slabs can be determined on the basis of the upper and lower bound theorems of general plasticity. Upper bounds are in this paper obtained on the basis of a geometrically possible failure mechanism and lower bounds by considering statically admissible stress fields corresponding to stresses within or on the yield surface. In the paper it is shown that upper and lower bounds of the reliability index β and the reliability R of plastic slabs can easily be found. It is also shown that the best bounds of the stochastic analysis and of the corresponding deterministic analysis are obtained for the same failure mechanisms and the same stress fields in the so-called fundamental case.

2. Upper Bounds

Upper bounds of the reliability of a plastic slab can easily be obtained by using the so-called yield line theory, see Thoft-Christensen & Pirzada [1]. In the yield line theory a collapse pattern consisting of straight yield lines is assumed

and the external work W_e performed by the loads and the internal work W_i dissipated in the yield lines are calculated. For simplicity, let the loading and the bending moment capacities be modelled by normally distributed random variables $P_1, \ldots, P_n$ and $M_{p_1}, \ldots, M_{p_m}$ and let all remaining parameters such as dimensions be deterministic quantities. A convenient safety margin M_u then is

$$M_u(\overline{x}) = W_i(\overline{x}) - W_e(\overline{x}) = \sum_{i=1}^{m} f_i(\overline{x})M_{p_i} - \sum_{i=1}^{n} g_i(\overline{x})P_i \tag{1}$$

where f_i and g_i are deterministic functions of the variable $\overline{x} = (x_1, \ldots, x_k)$ defining a set of yield patterns. The safety margin M_u is linear in the normally distributed random variables, so upper bound $\beta_u(\overline{x})$ of the reliability index $\beta(\overline{x})$ and an upper bound $R_u(x)$ of the reliability $R(\overline{x})$ can easily be calculated for any admissible set of variables $\overline{x}$

$$\beta_u(\overline{x}) = E[M_u(\overline{x})]/(D[M_u(\overline{x})])^{1/2} \tag{2}$$

$$R_u(\overline{x}) = 1 - \Phi(-\beta(\overline{x})) \tag{3}$$

where $E[\cdot], D[\cdot]$ and Φ are the expected value, the standard deviation and the one-dimensional distribution function, respectively.

The most significant failure mode within a set of failure modes (yield patterns) given by $\overline{x}$ is then obtained by minimizing $\beta_u(\overline{x})$ with regard to $\overline{x}$.

In the corresponding deterministic analysis an upper bound $p_u(\overline{x})$ of the load-carrying capacity $p(\overline{x})$ (only one load) is obtained for any set of parameters from the work equation

$$W_i(\overline{x}) = W_e(\overline{x}) \tag{4}$$

and the optimal yield pattern corresponds to the set of $\overline{x}$ parameters resulting in a minimum value of $p_u(\overline{x})$.

It is shown by Thoft-Christensen & Pirzada [1] that the deterministic optimal yield pattern and the corresponding stochastic most significant failure mode (yield pattern) are generally different. However, in the so-called *fundamental case* the two yield patterns are equal. In the fundamental case $M_{p_1}, \ldots, M_{p_m}$ are fully correlated random variables and the loading is modelled by a single random variable P. The proof of this statement is straightforward and is similar to the corresponding proof for lower bounds given in section 4.

3. Lower Bounds

It is generally believed that the optimal upper bounds β_u and R_u determined by the method in section 2 are so close to the exact values that they can be used for practical design of slabs. From a theoretical point of view it is more satisfactory to use lower bounds. The problem is, however, that good lower bounds β_l and R_l for β and R are generally more difficult to obtain.

The equilibrium equation in Cartesian coordinates for a slab can be written

$$\frac{\partial^2 m_x(x,y)}{\partial x^2} - 2\frac{\partial^2 m_{xy}(x,y)}{\partial x \partial y} + \frac{\partial^2 m_y(x,y)}{\partial y^2} = -p(x,y) \qquad (5)$$

where m_x and m_y are bending moments, m_{xy} is the torsional moment and p is the load intensity perpendicular to the slab. For the sake of simplicity it is assumed that the load $p(x,y)$ is a constant p.

To illustrate how lower bounds of the load-carrying capacity p (the deterministic case) or for β and R (the stochastic case) can be estimated a rectangular slab is considered. For such a slab it has been suggested, see e.g. Nielsen [2], to construct safe moment fields of the form

$$\left.\begin{array}{l} m_x = \alpha_1 + \alpha_2 x + \alpha_3 x^2 \\ m_y = \alpha_4 + \alpha_5 y + \alpha_6 y^2 \\ m_{xy} = \alpha_7 + \alpha_8 x + \alpha_9 y + \alpha_{10} xy \end{array}\right\} \qquad (6)$$

where $\overline{\alpha} = (\alpha_1, \alpha_2, \ldots, \alpha_{10})$ are parameters which are chosen so that the boundary conditions are satisfied in such a way that the moment field is safe, i.e. within or on the yield surface at any point of the slab.

In a deterministic analysis a lower bound P_l of the load-carrying capacity p can be determined by inserting (6) into (5)

$$p_l = -2(\alpha_3 - \alpha_{10} + \alpha_6) \qquad (7)$$

where the parameters α_3, α_6 and α_{10} will depend on the geometry of the slab and the bending moment capacities. As an example the following expression

$$p_l = 8\frac{1}{ab}\left(1 + \frac{a}{b} + \frac{b}{a}\right)m_p \qquad (8)$$

is obtained for a simply supported isotropic slab with side lengths a and b and bending moment capacity m_p, see Nielsen [2].

In general the equation (7) for the lower bound p_l will have the form

$$p_l = \sum_{i=1}^{m} f_i(\overline{\alpha})m_{p_i} \qquad (9)$$

where m_{p_i}, $i = 1, \ldots, m$ are bending moment capacities and where $f_i, i = 1, \ldots, m$ are deterministic functions.

The optimal lower bound p_l of a given set of moment fields as (6) is obtained by choosing the parameters $\overline{\alpha}$ so that the maximum of p is determined under the constraints from the boundary conditions and the yield condition.

A lower bound of the reliability of a plastic slab with the load P and the bending moment capacities $M_{p_i}, i = 1, \ldots, m$ modelled as normally distributed random variables can be estimated on the basis of the safety margin

$$M_l(\overline{\alpha}) = \sum_{i=1}^{m} f_i(\overline{\alpha}) M_{p_i} - P \qquad (10)$$

Note that the safety margin M_l in (10) has the same form as M_u in (1). It is therefore a trivial matter to estimate the lower bounds $\beta_l(\overline{\alpha})$ and $R_l(\overline{\alpha})$ as functions of $\overline{\alpha}$ for the reliability index β and the reliability of a plastic slab. The optimal lower bound within a set of moment fields is then obtained by maximizing e.g. $\beta_l(\overline{\alpha})$ with regard to $\overline{\alpha}$ under the boundary conditions and the yield condition as constraints. An interesting question is then whether the optimal lower bounds p_l and β_l are obtained for the same $\overline{\alpha}$.

4. Lower Bounds for the Fundamental Case

Let the bending moment capacities of the plastic slab be modelled by fully correlated random variables $M_{p_1}, \ldots, M_{p_m}$ and the load by a single random variable P. Furthermore, assume that a set of moment fields is defined by a set of parameters $\overline{\alpha}$ given by (6). The corresponding reliability problem is called *the fundamental case*. It will now be shown that the corresponding deterministic analysis (when all variables are deterministic) has the same optimal moment fields (or $\overline{\alpha}$ values) as the stochastic case. The proof is almost identical to the proof for the upper bounds given in Thoft-Christensen & Pirzada [1]. The safety margin M can be written

$$M_l = f(\overline{\alpha})M_p - P = k(M_p, P, \overline{\alpha}) \qquad (11)$$

where f is a function only depending on $\overline{\alpha}$ and $M_p = M_{p_1}$. The sensitivity of β_l with regard to each of the parameters α_i is then determined by, see Madsen,

Krenk & Lind [3]

$$\frac{\partial \beta_l(\overline{\alpha}_0)}{\partial \alpha_i} = \frac{\partial k(m_p^*, p^*, \overline{\alpha}_0)}{\partial \alpha_i} \frac{1}{|\nabla k(m_p^*, p^*, \overline{\alpha}_0|} \tag{12}$$

where m_p^* and p^* signify the design point and $\overline{\alpha}_0$ are the values of $\overline{\alpha}$ for which β_l is calculated. The maximum value of β_l is obtained for

$$\frac{\partial \beta_l(\alpha_0)}{\partial \alpha_i} = 0 \quad \Longleftrightarrow \quad \frac{\partial f(\overline{\alpha}_0)}{\partial \alpha_i} = 0 \tag{13}$$

In the deterministic analysis the lower bound p_l for the load-carrying capacity p for the corresponding problem is given by

$$p_l(\overline{\alpha}) = f(\overline{\alpha}) m_p \tag{14}$$

it is seen that the optimal lower bound obtained by

$$\frac{\partial p_l(\overline{\alpha})}{\partial \alpha_i} = 0 \tag{15}$$

results in the same equation (13) as for the stochastic analysis. Therefore, the optimal moment fields are equal.

5. Example

Consider a square slab supported at two adjacent edges (see figure 1). An upper bound for the reliability can be obtained by a simple yield pattern given by a single yield line defined by the variable x_0 shown in figure 1. The external and internal work corresponding to a unit downward displacement of point A are

$$W_e(x) = \frac{1}{2} a(3a - x_0) p \tag{16}$$

$$W_i(x) = (\frac{a}{x_0} + \frac{x_0}{a}) m_p \tag{17}$$

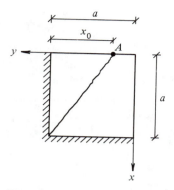

Figure 1. Square slab supported at two adjacent edges. Uniformly distributed load $P \sim N(4, 0.4)$ and isotropic bending moment capacity $M_p \sim (1, 0.1)$.

Setting $W_e(x_0) = W_i(x_0)$ the deterministic load-carrying capacity p as a function of x_0 can be determined. The minimum value of p is obtained for $x_0 = 0.72a$. The corresponding optimal yield pattern gives the optimal upper bound for the reliability index for this set of yield patterns. The upper-bound safety margin is

$$M_u = W_i(0.72a) - W_e(0.72a) = 5.55M_p - P \tag{18}$$

and the corresponding reliability index $\beta_u = 2.26$.

A lower bound can be determined on the basis of the moment field given by (see Nielsen [3])

$$m_x = 2\alpha\left(\frac{x}{a} - \frac{x^2}{a^2}\right) \tag{19}$$

$$m_y = 2\alpha\left(\frac{y}{a} - \frac{y^2}{a^2}\right) \tag{20}$$

$$m_{xy} = \alpha\left(\frac{x}{a} - \frac{xy}{a^2} + \frac{y}{a}\right) \tag{21}$$

where α is a parameter. By inserting (19) - (21) into the equilibrium equation (5) the load-bearing capacity p_l is obtained as a function of α

$$p_l(\alpha) = 6\alpha \frac{1}{a^2} \tag{22}$$

The boundary conditions are fulfilled by (19) - (21). It can be shown (see Nielsen [3]) that the maximum numerical value of the principal moments is equal to $\frac{4}{3}\alpha$. The optimal lower bound p_l corresponding to the moment field (19) - (21) is then obtained from (22) by inserting $\alpha = \frac{3}{4}m_p$. The result is $p_l = 9m_p/2a^2$. The optimal lower bound for the reliability index can then be calculated from the corresponding safety margin ($a = 1$)

$$M_l = 4.50M_p - P \tag{25}$$

The corresponding reliability index is $\beta_l = 0.83$. Note the big gap between the optimal lower bound β_l and the optimal upper bound β_u.

6. Conclusions

In the paper it is shown that upper and lowr bounds can easily be determined on the basis of the upper and lower bound theorems of general plasticity. For the fundamental case where the bending moment capacities are modelled by fully correlated random variables and the load by a single random variable it is shown that optimal (bent) bounds of the stochastic analysis and the corresponding deterministic analysis are obtained for the same yield patterns and moment fields.

7. References

[1]   THOFT-CHRISTENSEN, P. & G. B. PIRZADA: "Upper-Bound Estimate of the Reliability of Plastic Slabs". In Probabilistic Methods in Civil Engineering (editor P. D. Spanos), ASCE, N. Y., 1988, pp. 98 - 103.

[2]   NIELSEN, M. P.: "Limit Analysis and Concrete Plasticity". Prentice-Hall, Inc., Englewood Cliffs, 1984.

[3]   NIELSEN, M. P.: "Limit Analys of Reinforced Concrete Slabs". Acta Polytechn. Scand., Civil Eng. Build. Constr. Ser. No 26, 1964.

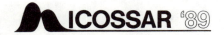
Utility of Reliability Index for Structural Design

Takashi Chou

Department of Civil Engineering, Shinshu University,
Nagano, Japan

ABSTRACT

Relations between design results, such as configurations, and the reliability indices, in accordance with the Advanced First Order Second Moment
(AFOSM) method, are numerically investigated with
the help of some fundamental problems and practical
examples. In general, the AFSOM method contains
some intrinsic errors. When the index value is
large, the error in the failure probability
increases significantly, even if the error in the
index is small. It is explained that the design
result is not affected to the same degree as the
error in the failure probability by the error in
the index, but it is usually affected to the same
or somewhat larger degree as the error of the index.

KEYWORDS

Computation; design; failure; probability;
reliability; safety; structural engineering.

1. INTRODUCTION

The reliability index , β , in accordance with the Advanced
First Order Second Moment (AFSOM) method (see Appendix Ⅱ)
(Rackwitz et al. 1978; Parkinson 1978; Ang et al. 1984;
Ramachandran 1984; Chou 1982 and 1987) is considered to be a
very powerful tool for the structural reliability problems
which can have any nonlinear or non-closed algebraic
performance function with nonnormal basic random variables.
This is because the calculation of the failure probability, p_f,
from $\Phi(-\beta)$ usually provides a reasonable estimate, i.e.,

$$p_f \simeq \Phi(-\beta) \qquad (1)$$

without much computational effort, where $\Phi(.)$ denotes the

standard normal cumulative probability function.

In general, the AFOSM method contains some intrinsic errors. According to the writer's experience, the error in β to $-\Phi^{-1}(p_f)$ is usually estimated to be less than 5% for many practical engineering problems, where $\Phi^{-1}(.)$ denotes the inverse function of $\Phi(.)$. As the value of β increases, however, the error of $\Phi(-\beta)$ to p_f increases significantly, even if the error in β is small. In order to observe these circumstances, the following probability effect ratio, γ, is defined as shown below:

$$\gamma = \Phi(-\alpha\beta)/\Phi(-\beta) \tag{2}$$

where α is a coefficient that takes values of 0.95-1.05 corresponding to the extent of the error in β. The results of γ are shown in Fig.1, from which it can be seen that γ amounts, for example, to 3.5 when β =5 and α = 0.95. This means that only a 5% error in β produces a 250% error in the failure probability. Therefore, it may be safely said that the reliability index, β, is a reasonably accurate estimation of $-\Phi^{-1}(p_f)$, i.e.,

$$\beta \simeq - \Phi^{-1}(p_f) \tag{3}$$

in comparison with Eq.1.

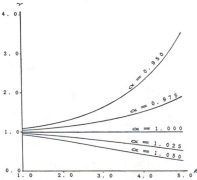

Fig. 1. Probability Effect Ratio, γ.

A question may be raised how the utility of the reliability index is due to improper correspondence to the failure probability in case of a large β value. As a matter of fact, it is undesirable to use this index with relation to such problems with which the probability itself is important and its variance brings about directly some serious consequences. The reliability index is usually utilized for structural design problems. In this paper, the relations between design results, such as configurations, and the reliability indices will be numerically demonstrated with the help of some fundamental problems and practical examples. It will be indicated that it is preferable to use the reliability index rather than the failure probability, and the reliability index is available for the structural design problem, even if the value of β is large and its compatibility to the failure probability is not so good.

2. INFLUENCE OF ERROR IN RELIABILITY INDEX ON DESIGN RESULTS

Case 1.

Let us consider a performance function

$$g(\underline{x})=x_1 -x_2 \tag{4}$$

where $\underline{x}$ is a vector of the random variables having a strength, x_1, and a load term, x_2. $g(\underline{x})\leq 0$ identifies failure, and $g(\underline{x})> 0$

corresponds to a safe state. When x_1 and x_2 are statistically independent and they are normally distributed variables, the central safety factor, ν, required by the reliability index, β_D, can be expressed as:

$$\nu(\beta_D) = \left[1 + \sqrt{1-(1-\beta_D{}^2 V_1{}^2)(1-\beta_D{}^2 V_2{}^2)} \right] \Big/ (1-\beta_D{}^2 V_1{}^2) \quad (5)$$

where V_1 and V_2 are coefficients of variation (COV) of x_1 and x_2, respectively. In order to observe the influence of the error of β_D on ν, the following central safety factor ratio, δ, representing the effect of the error in the reliability index on the design result, is defined:

$$\delta = \nu(\alpha\beta_D) / \nu(\beta_D) \quad (6)$$

Calculating results of δ with relation to a combination of COV are shown in Table 1, compared with γ. From many results

Table 1. Examples of δ and γ for Case 1 with $V_1 = 0.25$ and $V_2 = 0.3$

	β_D	1.0		2.0		3.0		3.8	
		δ	γ	δ	γ	δ	γ	δ	γ
	0.950	0.98	1.08	0.95	1.26	0.88	1.62	0.53	2.12
α	0.975	0.99	1.04	0.98	1.12	0.93	1.28	0.69	1.46
	1.025	1.01	0.96	1.03	0.89	1.08	0.78	1.88	0.68
	1.050	1.02	0.93	1.05	0.79	1.16	0.60	1.84	0.46

as in Table 1 it is revealed that the values of δ are not affected to the extent of the values of γ by α. When the value of β is close to the reciprocal of the COV of the strength term, the design result is considerably affected by the error of the reliability index. Under such a condition, however, the reliability can not be increased beyond some fixed level by strengthening the structure, and therefore it can hardly be said to be realistic. This means that the probability distribution of the strength term, x_1, which is usually expressed as a nonlinear function of the basic random variables, may not be considered as the normal distribution, even if they are normally distributed variables.

Case 2.
When x_1 and x_2 in Eq.4 are lognormally distributed variables, the central safety factor, ν, can be expressed as:

$$\nu(\beta_D) = \exp\left[\beta_D \sqrt{\ln\{(1+V_1{}^2)(1+V_2{}^2)\}} - \ln\sqrt{(1+V_2{}^2)/(1+V_1{}^2)} \right] \quad (7)$$

It is noticed that Eq.7, which is led with the use of the reliability index, holds approximately. This will also be noticed in the succeeding cases. Table 2 shows results from Eq.6 using Eq.7 as the central safety factor for a example, compared with γ from Eq.2. It can be observed from many examples as in Table 2 using Eqs.6-7 that the design results are affected to the same or somewhat larger extent as the errors in reliability indices and none of the unrealistic situations such as in Case 1 can be

Table 2.Examples of δ and γ for Case 2 with $V_1=0.25$ and $V_2=0.3$

β_D	1.0		2.0		3.0		4.0		5.0	
	δ	γ	δ	γ	δ	γ	δ	γ	δ	γ
0.950	0.98	1.08	0.96	1.26	0.94	1.62	0.93	2.28	0.91	3.55
α 0.975	0.99	1.04	0.98	1.12	0.97	1.28	0.96	1.52	0.95	1.90
1.025	1.01	0.96	1.02	0.89	1.03	0.78	1.04	0.65	1.05	0.52
1.050	1.02	0.93	1.04	0.79	1.06	0.60	1.08	0.42	1.10	0.27

seen.

Case 3-9.
   With regard to the performance function in Eq.4, let us consider various types of combinations of the distribution shown in Table 3. Notations denote in the table that N=Normal, L=Lognormal, E=Type I asympototic, W=Weibull with a lower limit, x_1, as:

$$x_1 = x_m - 2\sigma \tag{8}$$

where x_m is a mean value and σ is a standard deviation, B=Beta with a lower limit the same as Eq.8 and an upper limit, x_u, as:

$$x_u = x_m + 3\sigma \tag{9}$$

and NB, for example, denotes that x_1=Normal and x_2=Beta. Table 3

Table 3. ν and δ for Cases 1-9

Case No.		1	2	3	4	5	6	7	8	9
Distributions		NN	LL	NL	LN	NE	LE	WE	NB	BB
0.95		3.92	2.94	3.84	2.73	3.84	3.00	3.05	4.00	2.80
α 1.00	ν	4.47	3.12	4.38	2.87	4.34	3.19	3.23	4.56	2.89
1.05		5.21	3.30	5.06	3.01	5.06	3.39	3.42	5.31	2.98
0.95		0.88	0.94	0.88	0.95	0.88	0.94	0.92	0.88	0.97
α 1.00	δ	1.00	1.00	1.00	1.00	1.00	1.00	1.00	1.00	1.00
1.05		1.16	1.06	1.16	1.05	1.17	1.06	1.08	1.16	1.03

shows the central safety factors, which are obtained numerically by an algorithm shown in Appendix III, and the central safety factor ratios for Cases 3-9 with Cases 1 and 2, where β_D=3.0, α =0.95, 1.00, 1.05, V_1=0.25, and V_2=0.3. To examine the compatibility of the reliability indices for the central safety factors in Table 3 with β_D, the Monte Carlo simulations were executed and the results are shown in Table 4. These tables suggest that in case x_1=Normal, even x_2=Nonnormal, the influence of the error in the reliability index is similar to that in Case 1, whereas in case x_1=Lognormal, even x_2=Nonlognormal, it is similar to that in Case 2, and the realized reliabilities,β, are nearly equal to the design target,β_D.

Case 10-11.
   The performance function of the plastic flexural problem of a steel beam section may be given as:

Table 4. Monte Carlo Simulation Results for Cases 1-9 with $\beta_D = 3$

Case No.	1	2	3	4	5	6	7	8	9
Distributions	NN	LL	NL	LN	NE	LE	WE	NB	BB
β	3.00	2.98	2.94	3.04	2.93	2.98	3.06	3.01	3.16

$$g(\underline{x}) = x_1 x_2 x_3 - (x_4 + x_5) x_6 \qquad (10)$$

where x_1 =plastic section modulus of the beam, x_2 =yield strength of steel, x_3 =correction variable for resistance, x_4 =bending moment due to dead loads, x_5 =bending moment due to live loads and x_6 =correction variable for load effect. Determine mean value of x_1, for $\beta_D = 4.0$ and $\alpha = 0.95 - 1.05$, supposing the follwing mean vector, $\underline{m}$, in which elements are given by dimensionless values, COV vector, $\underline{V}$, and distribution vector, $\underline{D}$. It is assumed that all random variables are statistically independent. The results are listed in Table 5. It can be deduced from the table that the influence, i.e., δ, of the error in the

$$\underline{m} = (x_{m1} \quad 2.0 \quad 1.0 \quad 1.0 \quad 1.0 \quad 1.0) \qquad (11)$$
$$\underline{V} = (0.1 \quad 0.15 \quad 0.1 \quad 0.05 \quad 0.35 \quad 0.1) \qquad (12)$$
$$\underline{D} = (N \ N \ N \ N \ N \ \overline{N}) \text{ for Case 10, } (L \ L \ L \ L \ L \ L) \text{ for Case 11} \qquad (13)$$

Table 5. x_{m1}, α, γ and δ for Cases 10 and 11

	β_D	3.8	3.9	4.0	4.1	4.2
	α	0.950	0.975	1.000	1.025	1.050
	γ	2.28	1.52	1.0	0.65	0.42
Case 10	x_{m1}	3.17	3.28	3.40	3.53	3.66
	δ	0.93	0.96	1.0	1.04	1.08
Case 11	x_{m1}	3.15	3.26	3.37	3.48	3.60
	δ	0.93	0.97	1.0	1.03	1.07

reliability index, i.e., α, on the design result, i.e., x_{m1}, is somewhat larger than α. When the strength term, i.e., $x_1 x_2 x_3$, and the load effect term, i.e., $(x_4 + x_5) x_6$, are normally distributed variables, 0.80 (for $\alpha = 0.95$) and 1.34 (for $\alpha = 1.05$) as the values of δ are produced from Eqs.4-5. The corresponding values in Case 10 are 0.93 and 1.08. As mentioned in Case 1, it is not advisable to assume that the strength term is a normally distributed variable.

Case 12-13.
The performance function of a singly reinforced concrete rectangular section subjected to bending moment can be expressed as:

$$g(\underline{x}) = [x_1 x_2 \{x_3 - (x_1 x_2)/(1.7 x_4 x_5)\}] x_6 - (x_7 + x_8) x_9 \qquad (14)$$

where x_1 =area of steel reinforcement, x_2 =yield strength of steel, x_3 =effective depth, x_4 =width of section, x_5 =compressive strength of concrete, x_6 =correction variable for strength term, x_7 =bending moment due to dead loads, x_8 =bending moment due to

live loads and x_9=correction variable for load effect term.
Assume that the variables are uncorrelated, and means, COVs, and
distributions are as follows:

$$\underline{m}=(x_{m_1}\ 323.4\text{MPa}\ 84.92\text{cm}\ 100\text{cm}\ 28.22\text{MPa}\ 1.0\ 59.4\text{MN-cm}\ 25.1\text{MN-cm}\ 1.0) \quad (15)$$

$$\underline{V}=(0.03\ 0.04\ 0.05\ 0.05\ 0.2\ 0.1\ 0.05\ 0.35\ 0.1) \quad (16)$$

$$\underline{D}=(\text{N N N N N N N N N}) \text{ for Case 12,}$$
$$(\text{L L L L L L L L L}) \text{ for Case 13} \quad (17)$$

Fig. 2 shows the design results for the area of steel
reinforcement required by β_D=2.5-5.5. If a changing range of β_D
is limited to be the extent of , say, 1, the design results are
almost proportional to β_D, while the failure probabilities
associated with the extent of β_D are logarithmically affected.

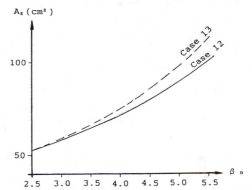

Fig. 2. Design Results of Cases 12 and 13.

3. CONCLUSIONS

Some fundamental and practical examples of influence of the
error in the reliability index on the design result are presented,
from which the following conclusions can be drawn:
1. The design result is not affected to the same degree as
the error in the failure probability by the error in the
reliability index, but it is usually affected to the same or
somewhat larger degree as the error in the index, probably less
than 5%. This level of accuracy in the index seems to be
practically sufficient, with the lack of some statistical data
taken into consideration.
2. When the probability distribution of the strength term
is assumed as the normal variate and the value of the reliability
index is close to the reciprocal of COV of the term, the design
result is considerably affected by the error in the index. Such
a condition, however, is not realistic, because the reliability
is not increased beyond a certain fixed level by strengthening
the structure. It is not advisable to assume that the strength
term is a normally distributed variable.
3. It is preferable to use the reliability index rather than
the failure probability, because the design result is not directly
affected by the probability.

4. The reliability index is available for structural design problems, even if the value of the index is large and its compatibility to the failure probability is not so good.

Appeendix I - References

Ang,A.H-S. and Tang,W.H.(1984). Probability Concepts in Engineering Planning and Design Vol.Ⅱ, John Wiley & Sons Inc., pp.350-383.

Chou, T.(1982). "Some Considerations On Safety Index," Proc. of JSCE, Vol.324, pp.41-50, (in Japanese).

Chou, T.(1987). "Reliability Index For Structural Reliability Analysis," Proc. of JCOSSAR, Vol.1, pp.335-340, (in Japanese).

Parkinson,D.B.(1978). "Solution for Second Moment Reliability Index," Proc. of ASCE, EM5-104, pp.1267-1275.

Rackwitz, R., and Fiessler, B.,(1978) "Structural Reliability Under Combined Random Load Sequences," Journal of Computers and Structures, Vol.9, pp.489-494.

Ramachandran,K.(1984)."Discussion to Basic Analysis of Structural Safety," Proc. of ASCE, SE10-110, pp.2554-2556.

APPENDIX Ⅱ - AFSOM METHOD

A flow chart of the procedure in accordance with AFSOM method is shown in Fig. A1. Symbols used in the figure are as shown below:

$\underline{F}$: a vector of the cumulative distribution function having elements F_i,

$\underline{f}$: a vector of the density function having elements f_i,

$\underline{\rho}$: a correlation matrix having elements ρ_{ij},

$\underline{x}^0$: a vector having the elements as initial values of n basic random variables ,

ϕ : the density function of the standard normal variate.

It should be emphasized that F_i, f_i and $g(\underline{x})$ are not necessarily expressed as a closed form and can be just evaluated numerically for the fixed random values. In the sequential procedure, vibration may occur. This can be avoided by decreasing a step taken.

APPENDIX Ⅲ - DESIGN BY NEWTON-RAPHSON METHOD USING NUMERICAL DIFFERENTIATING

Fig. A2 shows a flow chart to determine a mean value of a design variable, x_{mD}, by the Newton-Raphson method with which the numerical differentiating is used. Meanwhile, $x_{mD.0}$ is a initial value of x_{mD} and $\triangle x_{mD.k}$ is a infinitesimal increasing quantity of $x_{mD.k}$, say, $x_{mD.k}/100$. Furthermore, $\beta(\underline{x})$ is the reliability index for $\underline{x}$, and ε is a small positive value, say, 0.0001. In accordance with the writer's experience, convergence of this sequential procedure is obtained by less than 5 or 6 iterations.

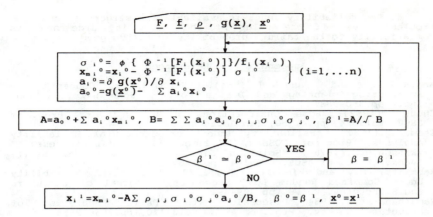

$$\underline{F}, \ \underline{f}, \ \underline{\rho}, \ g(\underline{x}), \ \underline{x}^0$$

$$\sigma_i^0 = \phi \{ \Phi^{-1}[F_i(x_i^0)] \} / f_i(x_i^0)$$
$$x_{mi}^0 = x_i^0 - \Phi^{-1}[F_i(x_i^0)] \ \sigma_i^0 \quad\} \ (i=1,\ldots n)$$
$$a_i^0 = \partial \ g(\underline{x}^0)/\partial \ x_i$$
$$a_0^0 = g(\underline{x}^0) - \ \Sigma \ a_i^0 x_i^0$$

$$A = a_0^0 + \Sigma \ a_i^0 x_{mi}^0, \ B = \Sigma \ \Sigma \ a_i^0 a_j^0 \rho_{ij} \sigma_i^0 \sigma_j^0, \ \beta^1 = A/\sqrt{B}$$

$$\beta^1 \simeq \beta^0$$ YES $$\beta = \beta^1$$

NO

$$x_i^1 = x_{mi}^0 - A\Sigma \ \rho_{ij} \sigma_i^0 \sigma_j^0 a_j^0/B, \quad \beta^0 = \beta^1, \ \underline{x}^0 = \underline{x}^1$$

Fig. A1. Flow Chart of AFOSM Method

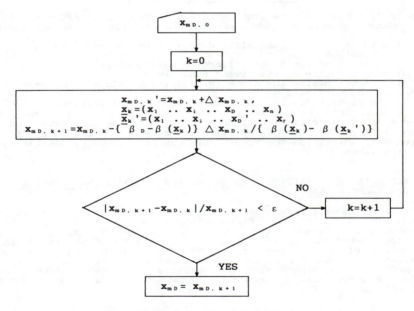

$$x_{mD.0}$$

$$k=0$$

$$x_{mD.k}' = x_{mD.k} + \triangle x_{mD.k},$$
$$\underline{x}_k = (x_1 \ .. \ x_i \ .. \ x_D \ .. \ x_n)$$
$$\underline{x}_k' = (x_1 \ .. \ x_i \ .. \ x_D' \ .. \ x_r)$$
$$x_{mD.k+1} = x_{mD.k} - \{ \beta_D - \beta(\underline{x}_k) \} \triangle x_{mD.k}/\{ \beta(\underline{x}_k) - \beta(\underline{x}_k') \}$$

$$|x_{mD.k+1} - x_{mD.k}|/x_{mD.k+1} < \varepsilon$$ NO $$k=k+1$$

YES

$$x_{mD} = x_{mD.k+1}$$

Fig. A2 Flow Chart of Design

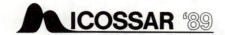

5th International Conference on
Structural Safety and Reliability

SAFETY LEVELS OF REINFORCED CONCRETE BOX CULVERTS
FOR INSTALLING POWER SUPPLY CABLES

Noboru YASUDA* and Tomoaki TSUTSUMI**

* Assistant Researcher, Civil Engineering & Archi-
tecture Department, Engineering Research Center,
The Tokyo Electric Power Co., Inc.
** Research Staff Senior Engineer, Civil Engineering
& Architecture Department, Engineering Research
Center, The Tokyo Electric Power Co., Inc.

ABSTRACT

The authors have examined the applicability of
reliability-based design to the existing reinforced
concrete box culverts for installing power supply
cables. The lowest safety index of the members of
existing box culverts for each failure mode by
bending moment and shear force is designated as the
possible level of target safety index(β_T) of the
mode. Thickness and reinforcement of each member to
meet the corresponding failure modes are computed.
The results of the redesign show that concrete
volume and reinforcement can be reduced 13% and
15%, respectively.

KEYWORD

Reliability-based Design, Safety Level, Target
Safety Index, Reinforced Concrete Structure.

1. INTRODUCTION

Various methods for applying the reliability-based design to
structures have recently been proposed and their usefulness is
also being recognized. However, there are only few reports on how
to establish the target safety index(β_T) of structures and how to
determine the dimensions of their members to meet the index.
Therefore, this report examines the safety index of box culverts
with power supply cables installed inside which The Tokyo Electric
Power Co., Inc. possess, by investigating the actual variation of
random variables relevant to soil parameters, material strength,
etc. Based on the results, we propose to establish an optimal

design   method based on the target β_T , and then examine the
efficiency of the method through the redesign of existing box
culverts.

2. OBJECTIVE STRUCTURES AND RANDOM VARIABLES

Objective structures, in this study, are the existing box
culverts of Tokyo Electric Power Co., Inc. All of the box
culverts are constructed by the cut-and-cover method. They are
made of reinforced concrete with a rectangular cross section that
has a relatively shallow cover of fill, as shown in Figs. 1(a)
through (c).

The box culverts, as a part of the electric power supply
facilities, are equipped with high-voltage(66 to 125 kv) power
cables. They also have been designed to prevent any possible
damage or deterioration of the cables, to provide an needed space
for routine periodic inspection work, and to discharge the heat
generated by the live cables.

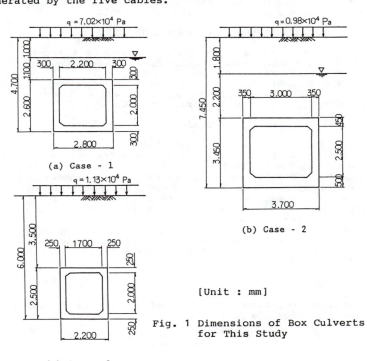

(a) Case - 1

(b) Case - 2

(c) Case - 3

[Unit : mm]

Fig. 1 Dimensions of Box Culverts
       for This Study

The random variables as shown below are selected to for use in
the box culverts design:

(1) variables related to the soil properties of the surrounding ground (including the ground water levels).
(2) variables related to the material properties of reinforced concrete.
(3) variables related to the surcharge loads on the ground surface.
(4) variables related to the correction factor[3] for estimating shear capacity of reinforced concrete
(5) variables related to the error of structural dimensions due to construction.

Among these, the variables are selected through sensitivity analysis in terms of a wider range of variation and a relatively significant effect on the results of the design.

Here, the sensitivity of random variables are estimated by the relative magnitude of deviation(Δ) between safety index(β) and β' as

$$\Delta = \left| \frac{\beta - \beta'}{\beta} \right| \times 100 (\%)$$

where β defines the safety index when randomness of all variable are taken into account and β' defines the safety index when the randomness of variables other than each objective variable are considered.

Five random variables which gave relative deviation larger than 10% for each mode of bending or shear failure(see Table 1) are selected as objective valiables; the coefficient of earth pressure at rest(Ko), unit weight of soil(γ_t), yield strength of rein-forcing bar(σ_{sy}), clear cover distance for reinforcement (cover; d') and correction factor in the equation for shear capacity(α_Q). The other variables which gave a small Δ are considered as deterministic values.

Table 1 Result of Sensitivity Analysis

Sensitivity / Mode	over 10%	1 - 10 %	under 1%
Bending failure	Ko, τt, σsy, d'	q	τc, f_c
Shear failure	Ko, α_Q	τ_t, τc	f_c, σsy, d', q

Notes
Ko: Coefficient of Earth Pressure at Rest $\quad$ σsy: Yield Strength of Reinforcement
τt: Unit Weight of Soil $\qquad\qquad\qquad\quad$ d': Clear Cover
q: Live Load $\qquad\qquad\qquad\qquad\qquad\quad$ f_c: Compressive Strength of Concrete
τc: Unit Weight of Reinforced Concrete $\quad$ α_Q: Correction Factor in Equation for Shear
$\qquad\qquad\qquad\qquad\qquad\qquad\qquad\qquad$ Capacity

The statistical characteristics of the random variables are obtained directly from available field data. In cases where no field data is available, they are selected from the published data

obtained by literature survey [1],[2],[3].

Table 2 Statistical Characteristics of Random Variables

Random Variables	Case No.	A.S.D.M[*]	Statistical Values		Estimation Method
			Mean	Standard Deviation	
Ko: Coefficient of Earth Pressure at Rest	1	0.50	0.40	0.22	In-situ test results
	2	0.50	0.50	0.28	
	3	0.50	0.40	0.22	
τ_t: Unit Weight of Soil	1	1.80	1.80	0.09	In-situ test results
	2	1.95	1.95	0.10	
$(\times 10^3$ kg/m^3)	3	1.80	1.80	0.09	
σsy: Yield Strength of Reinforcement	1	2,940	3,765	146	Literature Survey [1] (JSCE. 1983)
	2	2,940	3,667	219	
$(\times 10^5$ Pa)	3	2,940	3,663	209	
d': Concrete Cover for Reinforcement	1	7.0	7.0	0.65	Literature Survey [2] (Takeyama. 1981)
	2	6.0/10.0	6.0/10.0	0.65	
(cm)	3	5.0	5.0	0.65	
α_Q: Correction Factor in The Equation for Shear Capacity	1				Literature Survey [3] (Okamura. 1980)
	2	-	1.11	0.12	
	3				

[*]: Using Values in Allowable Stress Design Method

3. RELIABILITY ANALYSIS

As the modes of failure, both bending failure and shear failure occurring at a cross section of reinforced concrete members for static loading cases are selected in this study. The limit state for bending failure is defined as a state in which the bending moment of a member subjected to external forces exceeds the ultimate bending moment i.e., the bending moment that will produce a state in which the maximum strain at the extreme fibers of the concrete reaches the value of 0.0035. The performance function(Z) is expressed as,

$$Z = Mu - M$$

where   Mu : Ultimate bending moment [4]

$$Mu = 0.85 \cdot fc \cdot b \left\{ \frac{1}{2} - \frac{1}{12} \cdot \left(\frac{0.002}{\varepsilon cu} \right)^2 \right\} \cdot X^2$$
$$+ 0.85 \cdot fc \cdot b \left(1 - \frac{1}{3} \cdot \frac{0.002}{\varepsilon cu} \right) \cdot X \cdot \left(\frac{t}{2} - X \right)$$
$$+ \varepsilon cu \cdot \frac{X-d'}{X} \cdot Es \cdot As1 \left(\frac{t}{2} - d' \right) + \varepsilon cu \cdot \frac{X-d}{X} \cdot Es \cdot As2 \left(\frac{t}{2} - d \right)$$

x : Distance from neutral axis to compressive extreme fiber (cm).
b,t : Width of member, and thickness of member(cm).
d',d : Concrete cover for reinforcement

of compressive side and
effective height(cm).

A_{s1}, A_{s2} : Axial reinforcement volume(cm^2).

f_c : Compressive strength of
concrete(Pa).

Es : Young's modulus of reinforce-
ment(Pa).

ε cu : Ultimate compressive strain of
concrete(= 0.0035).

N : Axial force(N).

M : Working bending moment.

On the other hand, the limit state for shear failure is
defined as a state in which the shear force of a member subjected
to external forces exceeds the ultimate shear capacity. The
performance function is expressed as,

$$Z = Qu - Q$$

where   Qu : Ultimate shear capacity [4]

$$Qu = \alpha_0 \left\{ 0.94 \ f_c^{\frac{1}{3}} \ (1 + \beta d + \beta p + \beta n) \cdot b \cdot d \right.$$

$$+ Asw \cdot \sigma sy \ \frac{d \ (Sin \ \theta + Cos \ \theta)}{1.15 \ s}$$

α_0 : Correction factor in the equation
for shear capacity[3].

βd : Influence coefficient of scale
effect.

βp : Influence coefficient of axial
steel ratio.

βn : Influence coefficient of axial
compressive force.

Asw : A pair of shear reinforcing bar
volume(cm^2).

s : Pitch of shear reinforcing
bar(cm)

θ : Angle between shear reinforcing
bar and axis member.

Q : Working shear force.

In this study, safety indices are computed using an advanced
first order second moment method in which the performance function
is expanded in a Taylor series about a failure point.

4. ESTABLISHMENT OF TARGET SAFETY INDEX

In this study, a target safety index is established based
upon the results of reliability analysis applied to the existing

structures. The procedure is described below.

(1) Analysis of Reliability of Existing Structures.

The safety indices of the three box culverts are computed by using the random variables(eight kinds) of each box culvert. In order to find the minimum safety index of each member(top slab, side wall, and bottom slab) of the box culverts, three to six check points are chosen for each member(see Fig. 2). From the analysis mentioned above, the safety indices obtained for the respective modes of bending failure and shear failure for each member in the box culverts are shown in Table 3.

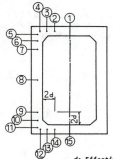

Failure Mode	Member	Conputation Point
Bending	Top Slab Side Wall Bottom Slab	①③④ ⑤⑥⑧⑩⑪ ⑫⑬⑮
Shear	Top Slab Side Wall Bottom Slab	②③④ ⑤⑥⑦⑨⑩⑪ ⑫⑬⑭

d: Effective Height

Fig. 2 Check Points of Safety Index

Table 3 Safety Index

	Member	Bending Failure Mode	Shear Failure Mode
Case-1	Top Slab Side Wall Bottom Slab	11.3 10.9 10.4	9.1 5.3 9.6
Case-2	Top Slab Side Wall Bottom Slab	10.6 12.9 11.5	4.6 4.0 4.2
Case-3	Top Slab Side Wall Bottom Slab	9.6 16.7 6.2	5.9 6.3 5.4

(2) Establishment of Target Safety Index for Each Failure Mode

Many box culverts have been designed and constructed by the company in accordance with the allowable stress method and they have had no experience of functional failure at present. Hence, it is assumed that the level of safety maintained in these existing structures is adequate. According to this, it may be justified to consider the lowest value among the safety indices of these existing box culverts as the permissible lower bound of the

safety indices within the framework of current design practice based on the allowable stress method.

Therefore, the minimum value among these safety indices for each mode of failure(bending failure and shear failure) is designated as the corresponding target safety index. The resultant indices are as:

β_T = 6.2 as the target safety index against bending failure

β_T = 4.0 as the target safety index against shear failure

These values are rather higher than those mentioned in the literature[5]. However, since we do not consider such loads as seismic loads, thermal loads from cables, and so on that are usually considered in actual design, these minimum values may be a possible candidate in this preliminary investigation of the efficiency of the redesign concept. In order to establish more reliable target safety indices for actual design, further case studies will be required.

5. REDESIGN WITH USING TARGET SAFETY INDICES

The cross sections of case 2 are redesigned by using the target safety indices obtained in section 4; thickness and reinforcement in each member are designed to meet the corresponding target safety index for the failure modes. The results of redesign show that the concrete volume and reinforcement can be reduced 13% and 15%, respectively. In this study, the reinforcement was determined to meet the practice in the placement at the site. Therefore, each member was slightly over designed to have some allowance for the target safety index(see Fig. 3 and Table 4).

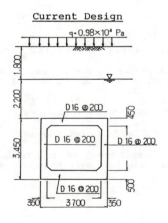
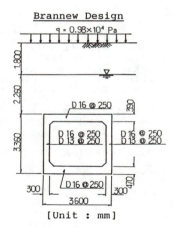

[Unit : mm]

Fig. 3 Dimensions of Current Design and Brannew Design

Table 4 Comparison between Current and Brannew Design

	Current Design				Brannew Design			
	Thickness (cm)	Reinforce- ment (cm²/m)	Safety Index		Thickness (cm)	Reinforce- ment (cm²/m)	Safety Index	
			Bending Failure	Shear Failure			Bending Failure	Shear Failure
Top Slab	45	6.57	10.6	4.6	39	6.62	7.0	4.3
Side Wall	35	8.20	12.9	4.0	30	6.62	6.5	4.8
Bottom Slab	50	5.35	11.5	4.2	47	7.60	7.3	4.1
Quantity of Material	Concrete Volume 5.38 m³/m	Reinforce Volume 327 kg/m	———		Concrete Volume 4.68 m³/m	Reinforce Volume 278 kg/m	———	
Target Satety Index	———						6.2	4.0

6. CONCLUSION

The following conclusion is obtained from the results of this study on safety levels of the existing box culverts in accordance with a reliability-based design concept:

(1) The box culverts, designed by allowable stress design method and constructed with this in mind, have a certain range of safety factor depending on the structures or their members when the variation of each random variable is taken into account.

(2) By selecting the minimum safety indices obtained for exist- ing box culverts as the respective target safety index for each failure mode, it is possible to reduce the excessive volume of concrete and reinforcing bars even within the limit of current allowable stress design.

7. REFERENCES

[1] JSCE: "Guideline of Limit-States Design Method for Concrete Structures(Draft)",1983.
[2] Takeyama Y. and Suzuki M.: " A Study on Real Condition of Dispersion of Structural Variables and Strength Estimate", Proc. of JSCE, Vol. 4, No.36, pp.7-8, 1981.
[3] Okamura H. et al : "Proposed Design Equation for Shear Strength of Reinforced Concrete Beams without Web Reinforce- ment", Proc. of JSCE, Vol.300, 1980.
[4] JSCE: "Standard Specification for Unreinforced and Reinforced Concrete", 1986.
[5] U.S. DEPARTMENT OF COMMERCE : "Development of a Probability Based Load Criterion for American National Standard A58", 1980.

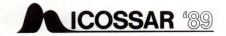
DYNAMIC RELIABILITY ANALYSIS OF 3D FRAME STRUCTURES
USING RANDOM FIELDS THEORY

*H. Ukon** and *E. H. Vanmarcke***

*   Information Processing Center, KAJIMA Corporation, Tokyo, JAPAN
**   Dept. of Civil Engineering and Operations Research, Princeton University, N.
     J., USA

ABSTRACT
A dynamic reliability analysis system of arbitrary shaped 3-
dimensional frame structures is presented and its use is illustrated
in bridge seismic safety assessment.   The auto- and cross-spectral
density function, including the effect of kinematic interaction, are
computed by considering the local earthquake ground motion as a
limited-duration segment of a homogeneous random field.   The
spectral density function of response is computed by the frequency
response analysis of a multi-point input system. Reliability is
calculated as a first-passage probability, and response nonstationarity is assessed by assuming a time-dependent pseudo-
damping parameter.

KEYWORDS
*Random Fields Theory ; Frequency Response Analysis ; General-
purpose Program ; 3D Frame Structure ; First-Passage Probability ;
Non-stationarity ; Reliability*

1. INTRODUCTION

In assessing seismic behavior of structures it is rational and efficient to use
stochastic analysis methods to account for unpredictable details in earthquake ground
motion.   Structures occupying a relatively large area in space such as bridges with
long spans are affected by variability in space as well as in time.   In this context, a
number of deterministic analyses of seismic behavior of structures have been done
considering the phase difference and the frequency characteristics of seismic input.

Reliability analysis based on classical random vibration (Refs. [1,2,3,4]) considers
the structure to be known and the excitation to be a stationary random process.   In
this study, a dynamic reliability analysis method of arbitrary shaped 3-dimensional

frame structures is presented and applied to seismic safety assessment of large bridges. The outline of the study is as follows. The auto- and cross-spectral density function including the effect of kinematic interaction are computed first considering the earthquake motion as a partial realization of a homogeneous random field, applying random field theory proposed by Vanmarcke [1]. The spectral density function of the response is computed next. The analytical model is a deterministic linear 3-dimensional framework; response characteristics are computed by means of frequency response analysis of a multi-point input system. Reliability is evaluated as a first-passage probability; non-stationary dynamic reliability is also computed based on the assumption of quasi-non-stationarity of the response. Finally, numerical examples of a 4-span 3-dimensional girder bridge are presented.

2. SUMMARY OF ANALYSIS

(1) Spectral density function of input [1]

Assuming multiple support input, the equivalent stationary stochastic input (during the strong phase of the earthquake) is expressed as a matrix $[G]$ with components $G_{\Delta_{ii}}(\omega)$ as follows (Refer Fig.1):

$$[G] = [G_{\Delta ij}(\omega)] = [\gamma_{ij}(\Delta,\omega)G_{ij}(\omega)\,exp(-i\omega\theta_{ij})] \quad ---(1)$$

where $\gamma_{ij}(\Delta,\omega)$: $i=j$ the frequency - dependent variance function specified by the local averaging area Δ

$i \neq j$ the frequency - dependent cross variance function specified by the interval Δ

$G_{ij}(\omega)$: $i=j$ the auto - spectral function

$i \neq j$ the cross - spectral function

θ_{ij} : $i \neq j$ the phase difference ($=0$ when $i=j$)

Fig.1 Defintion of local averaging area and interval

In case $i=j$, denoting the rectangular averaging area by $D = U_1 U_2$, the s.d.f. of the original random field (no local averaging) by $G(\omega)$, the s.d.f. of local average, $G_{\Delta ii}(\omega)$, is obtained as follows:

$$G_{\Delta ii}(\omega) = \gamma_{ij}(D,\omega)G(\omega) = \gamma_\omega(U_1,U_2)G(\omega) \quad --- (2)$$

Frequency-dependent 2D variance function $\gamma_\omega(U_1,U_2)$ is given as follows:

$$\gamma_\omega(U_1,U_2) = \gamma_\omega(U_1)\gamma_\omega(U_2|U_1) = \left\{1+\left(\frac{U_1}{\theta_\omega^{(1)}}\right)^2\right\}^{-\frac{1}{2}}\left\{1+\left(\frac{U_2}{\theta_{\omega U_1}^{(2)}}\right)^2\right\}^{-\frac{1}{2}} \quad --- (3)$$

$$\theta_\omega^{(1)} = C_1\theta^{(1)}\left\{1+\left(\frac{\omega}{\Omega_1^{(1)}}\right)^2\right\}^{-\frac{1}{2}}, \quad \theta_{\omega U_1}^{(2)} = C_2\theta_{U_1}^{(2)}\left\{1+\left(\frac{\omega}{\Omega_1^{(2)}}\right)^2\right\}^{-\frac{1}{2}} \quad --- (4)$$

where $\qquad \gamma_\omega(U_1)$: 1D variance function,

$\gamma_\omega(U_2|U_1)$: conditional 1D variance function,

$C_1 = a^{(1)}/\theta^{(1)}\theta^t$,

$C_2 = a^{(2)}/\theta^{(2)}\theta^t$,

$a^{(1)}$: the correlation parameter in the plane formed by the axis in the direction of u_1 and the time axis,

$a^{(2)}$: the correlation parameter in the plane formed by the axis in the direction of u_2 and the time axis,

$\theta^{(1)}$: the scale of fluctuation in the direction of u_1,

$\theta^{(2)}$: the scale of fluctuation in the direction of u_2,

θ^t : the scale of fluctuation in the direction of the time axis,

$\Omega_1^{(1)}$: the corner frequency in the direction of u_1,

$\Omega_1^{(2)}$: the corner frequency in the direction of u_2,

and $\qquad \theta^{(2)}{}_{U_1}$: the conditional scale of fluctuation in the direction of u_2

In case $i \neq j$, the cross spectral density function of local average $G_{\Delta ij}(\omega)$ is obtained as follows (Refer Fig.1):

$$G_{\Delta ij}(\omega) = \frac{G(\omega)}{4DD'} \sum_{k=0}^{3} \sum_{l=0}^{3} (-1)^k (-1)^l U_{1k} U_{2l} \gamma_\omega(U_{1k}, U_{2l}) \qquad ---\ (5)$$

where $\qquad D, D'$: the local averaging areas,

U_{1k}, U_{2l} : the intervals between the two areas ($k=0,1,2,3$, $l=0,1,2,3$),

$G(\omega)$: the s.d.f. of the original random field,

and $\qquad \gamma_\omega(U_{1k}, U_{2l})$: the frequency-dependent variance function derived from the intervals U_{1k} and U_{2l}.

(2) Frequency response analysis of a multi-point input system

The equation of motion of a multi-point input system is given as follows;

$$\begin{bmatrix} M_{AA} & M_{AB} \\ M_{BA} & M_{BB} \end{bmatrix} \begin{Bmatrix} Y_A \\ Y_B \end{Bmatrix} + \begin{bmatrix} K_{AA}^* & K_{AB}^* \\ K_{BA}^* & K_{BB}^* \end{bmatrix} \begin{Bmatrix} Y_A \\ Y_B \end{Bmatrix} = \begin{Bmatrix} 0 \\ P_B \end{Bmatrix} \qquad ---\ (6)$$

where $\qquad [M]$: the mass matrix,

$[K^*]$: the complex stiffness matrix,

$\{Y_A\}$: the absolute displacement vector of moving points,

$\{Y_B\}$: the absolute displacement vector of fixed points,

$\{P_B\}$: the reaction force vector at fixed points.

From the upper part of eq.(6), we obtain frequency transfer function $\{H_i\}$ assuming the unit harmonic input at input point i as follows:

$$\{H_i(\omega)\} = \frac{-(-\omega^2[M_{AB}] + [K_{AB}^*])\{V\}_i}{-\omega^2[M_{AA}] + [K_{AA}^*]} \qquad ---\ (7)$$

where $\qquad \{V_i\}$: the indicating vector of seismic input

Denoting the total degrees of freedom of the system by N, the frequency transfer function is finally obtained in the following form.

$$[H] = [\{H_1\} \{H_2\} --- \{H_N\}] = [H_{ij}(\omega)] \qquad\qquad --- \quad (8)$$

Denoting the total number of input points by n, the response of point P $Y_p(t)$ is

$$Y_p(t) = \sum_{i=1}^{n} \left(\int_0^\infty h_{ip}(\tau) X_i(t-\tau) d\tau \right) \qquad\qquad --- \quad (9)$$

where h_{ip} : the unit impulse function of point P subjected to i-supporting point input

X_i : an earthquake input

Fourier transformation of eq.(9) yields

$$\{F_y\} = [H]\{F_x\}, \qquad\qquad --- \quad (10)$$

where $\{F_x\}$ and $\{F_y\}$ are the Fourier Transformation of input and response, respectively. The cross spectral matrix of response $[G_y]$ and the (i,j) component of the matrix are, respectively:

$$[G_y] = \frac{\{F_y^*\}\{F_y\}}{T} = [H^*][G_x][H], \quad G_{yij} = \sum_{k=1}^{n} \sum_{l=1}^{n} H_{ik}^* G_{xkl} H_{lj} \qquad --- \quad (11)$$

(3) Reliability evaluation

The first passage probability is one of the performance measures that focuses on the maximum value of a random process. The dynamic reliability $L_b(t_0)$ is defined by the probability that a random process $X(t)$ does not cross a given level b within a given time period $[0,t_0]$. Assuming Gaussian excitation and accounting for clumping of b-level crossings, $L_b(t_0)$ may be approximated as follows [5]:

$$L_b(t_0) = exp\left[-2v_b^+ t_0 \frac{1 - exp\left\{ -\sqrt{\pi/2}\, \delta r \right\}}{1 - exp\left(-r^2/2 \right)} \right] \qquad --- \quad (12)$$

where v_b^+ : the mean occurrence rate of upcrossings of a given level b,

r : ratio of a level b and the standard deviation σ,

and δ : δ-factor (bandwidth measure) of $X(t)$.

Non-stationarity is taken into account by the assumption of quasi-non-stationarity, i.e., non-stationary response to suddenly applied stationary input; it can be assessed using a fictitious damping factor ζ_t which depends on the true viscous damping factor ζ and the natural frequency ω_n [6]:

$$\zeta_t = \frac{\zeta}{1 - exp(-2\zeta\omega_n t)} \qquad\qquad --- \quad (13)$$

For $t \rightarrow \infty$, $\zeta_t \rightarrow \zeta$, and $\zeta_t = (2\omega_n t)^{-1}$ when $\zeta = 0$.

3. NUMERICAL EXAMPLE

Numerical examples of a 4-span girder bridge shown in Fig.2 are presented in this section.

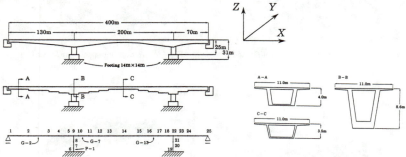

Fig.2 Analytical model and sections

(1) Analysis assumptions

The two supporting points (node 6 and 19) are fixed and the others (node 1 and node 25) are only vertically supported. The smoothed spectral density function for the El-Centro(N-S) earthquake, with Parzen's lag window of 0.5 Hz, is used as the basis for the temporal variation of the stochastic input. The frequency-dependent correlation function obtained by Harichandran and Vanmarcke [7] is used on the two space axes (v_1, v_2), and is assumed to be isotropic.

$$\rho_\omega(v_1, v_2) = \rho_\omega(v) = A exp(-c_1 v) + (1-A)exp(-c_2 v), \ v = (v_1^2 + v_2^2)^{1/2} \qquad --- \quad (18)$$

The corresponding wave number spectrum is

$$s(k_1, k_2) = \frac{A}{2\pi c_1^2}\left\{1 + \left(\frac{k_1}{c_1}\right)^2 + \left(\frac{k_2}{c_1}\right)^2\right\}^{-3/2} + \frac{1-A}{2\pi c_2^2}\left\{1 + \left(\frac{k_1}{c_2}\right)^2 + \left(\frac{k_2}{c_2}\right)^2\right\}^{-3/2} \qquad --- \quad (19)$$

Fig.3 shows eq.(18) and Fig.4 shows eq.(19) at frequency 2.5Hz. The same values of the parameters (c_1 and c_2, frequency-dependent) are used as in Ref. [7].

Fig.5 shows the spectrum of the input field combining the spectral characteristics in time (frequency) and space (wave number); the spectrum is smoothed by Parzen's lag window with 0.5 Hz bandwidth. Fig.6 shows the variance function obtained from eq.(3).

(2) Transfer function

The frequency transfer function $H_{6,i}(\omega)$ of the absolute acceleration in the Y-direction subjected to the unit harmonic acceleration point input at node 6 are shown in Fig.7 to Fig.8. In the same manner, the frequency transfer function $H_{1,i}(\omega)$, $H_{19,i}(\omega)$, $H_{25,i}(\omega)$, subjected to the point input at nodes 1, 19, 25 are computed. When employing a lumped mass matrix system, when the input is spatially uniform, their summation $H_{1,i}(\omega) + H_{6,i}(\omega) + H_{19,i}(\omega) + H_{25,i}(\omega)$ becomes equal to the ordinary frequency transfer function $H(\omega)$ of node i.

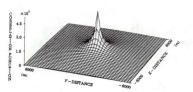

Fig.3 Spatial correlation function(f=2.5Hz)

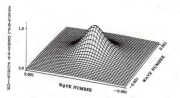

Fig.4 Wave number spectrum(f=2.5Hz)

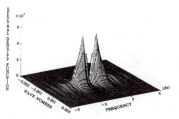

Fig.5 Frequency-wave number spectrum

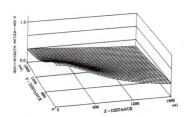

Fig.6 Variance function(f=2.5Hz)

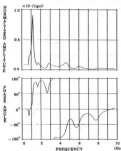

Fig.7 Transfer func.($H_{6,3}(\omega)$)

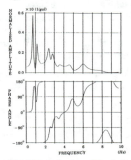

Fig.8 Transfer func.($H_{6,14}(\omega)$)

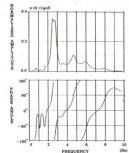

Fig.9 Transfer func.($H_{6,24}(\omega)$)

(3) RMS value of response

Examples of the response s.d.f. are shown in Fig.10 and Fig.11. Fig.10 shows the s.d.f. of the absolute acceleration response node 3 in the Y-direction subjected to Y input. Fig.11 shows the s.d.f. of stress σ_{x1} at (1,1) in the normal sectional coordinate parallel to the (y-z) plane in the local coordinate in member G-2. Table 1 summarizes RMS(Root-Mean-Square) responses to X-input, Y-input and Z-input. $M_{y,z}$ means M_y or M_z, whichever is not equal to 0.

"UNIFORM" in Table.1 is the result based on uniform input, "DRESS" is by a correlated input and "TRAVEL" is by a correlated input considering phase difference effect (V_s=1000m/s).

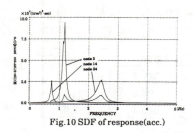

Fig.10 SDF of response(acc.)

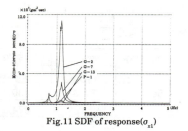

Fig.11 SDF of response(σ_{x1})

Table 1 RMS of responses

INPUT	RESPONSE	node 3(gal)	node 14(gal)	$M_{y,z}$(G-2)(t·m)	σ_{x1}(G-2)(t/m^2)
X-input	UNIFORM	69.67	65.68	1128.17	19.76
	DRESS	66.49	61.46	1542.30	26.05
	TRAVEL	42.26	34.90	2895.57	45.61
Y-input	UNIFORM	128.25	95.35	5282.52	154.67
	DRESS	127.80	93.86	5260.94	154.04
	TRAVEL	101.38	77.78	4100.15	120.05
Z-input	UNIFORM	91.82	113.39	1833.81	28.68
	DRESS	89.12	107.77	1828.13	28.60
	TRAVEL	83.31	105.03	1328.13	21.24

(3) First passage probability

Fig.12 shows quasi-non-stationary evolutionary transfer function at node 3 in the Y-direction subjected to Y-input. The damping value $\zeta = 0.05$ is assumed. Fig.13 shows the evolution in time of the first passage probability of node 3 acceleration response. Table 2 summarizes acceleration level which first passage probability $L_b(t_0)$ becomes some specified value at time = 30.0(sec.), based on correlated input(DRESS Y-input). "STATE" in Table 2 is the result of stationary analysis and "QUASI" is the result of quasi-non-stationary analysis.

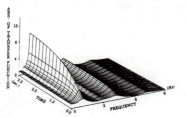

Fig.12 Quasi-non-stationary transfer function

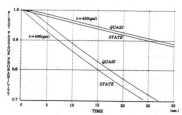

Fig.13 Evolving first passage probability

Table 2 Response level for specified first passage probability(gal, $t_0 = 30$sec)

$L_b(t_0)$	NODE	3	14	24
90.0(%)	STATE	457.0	349.0	411.7
	QUASI	454.4	347.4	410.1
50.0(%)	QUASI	380.6	296.3	347.0
	STATE	374.5	292.4	343.7

4.CONCLUSION
The analytical method proposed in this paper has the following features:
(1) Structural safety is estimated by dynamic reliability analysis; only the statistical description of an ensemble of input motions is used, in terms of the frequency wave number spectrum.
(2) The evaluation of an input field, including the effect of the kinematic interaction and support geometry, is naturally and rationally performed considering the earthquake motion as a space-time random field; it identifies clearly and separately the contribution to the total response from each support point.
(3) Response is computed by the frequency response analysis of a multi-point input system. Reliability estimation is done by means of quasi-non-stationary random vibration analysis.

Variance, spatial correlation and phase difference in seismic input all influence seismic response, as the numerical study illustrates. Which effect dominates depends on the foundation, the earthquake and the structure itself. Hence, this kind of extensive analysis is quite necessary in comprehensive seismic safety evaluation of structures.

ACKNOWLEDGEMENT
Part of this work was accomplished in the cooperative research project between Princeton University and Kajima Corporation.

REFERENCE
[1] Vanmarcke,E.H. : "Random Fields : Analysis and Synthesis", the MIT press, Cambridge, Mass. and London, England, 1983
[2] Lin,Y.K. : "Probabilistic Theory of Structural Dynamics", McGraw-Hill, New York, 1967
[3] Crandoll,S.H.and Mark,W.D. : "Random Vibration in Mechanical Systems", Academic Press,New York, 1963
[4] Nigam,N.C. : "Introduction to Random Vibrations", the MIT Press, Cambridge, Mass. and London, England, 1983
[5] Corotis,R.B.,Vanmarcke,E.H. and Cornell,C.A. : "First Passage of Non-stationary Random Process", EM. Div. ASCE, April, 1972
[6] Vanmarcke,E.H. : "Structural Response to Earthquakes", Ch.8 in Seismic Risk and Decisions, Edited by C.Lomnitz and E.Rosenblueth, Elsevier Publ. Co., 1976
[7] Harichandran,S.,Vanmarcke,E.H. : "Stochastic Variation of Earthquake Ground Motion in Space and Time", EM. Div. ASCE, Feb., 1986

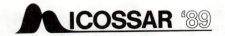
RELIABILITY STUDY OF FERROCEMENT SLABS UNDER CYCLIC LOADING

Ser-Tong Quek[*], Palaniappa Paramasivam[**] and Seng-Lip Lee[***]

[*] Lecturer, Dept. of Civil Engineering, National University
of Singapore, Kent Ridge Crescent, Singapore 0511.
[**] Associate Professsor, Dept. of Civil Engineering, National
Univ. of Singapore, Kent Ridge Crescent, Singapore 0511.
[***] Professor, Dept. of Civil Engineering, National University
of Singapore, Kent Ridge Crescent, Singapore 0511.

ABSTRACT

A simple probabilistic approach to fatigue strength studies
for ferrocement slabs is presented. A preliminary
experimental program to obtain the relevant statistics was
conducted and the laboratory results summarized. Unlike
most fatigue strength studies where the specimens are
cyclically loaded to failure, the approach employed here was
to obtain the residual strength of the specimen after it has
been subjected to some predetermined number of cycles of
loading, n. Probability of failure curves against specified
cracking and ultimate strength conditioned on different
values of n are presented. Within the load range
considered, the results show that ferrocement possesses good
fatigue properties.

KEYWORDS

reliability; fatigue; residual strength; ultimate; cracking;
uncertainty analysis; cyclic;

1. Introduction

Ferrocement elements are currently being proposed for use as secondary
roofing in Singapore. Due to the nature of the envinronmental loading such
as changes in temperature and wetting and drying of the exposed surfaces, the
fatigue strength of such elements subjected to the working condition has to
be evaluated. This will facilitate the proper design of the elements and the
specification of a replacement period, if necessary.

Fatigue of ferrocement slabs subjected to bending have been studied by
various researchers in which the conventional S-n relationships are obtained
for different percentages and types of reinforcement used [1-5]. However,

due to the inherent variability of the materials used, the fabrication process and the loadings, the recommended strength and service life cannot be accurately specified. Such uncertainties should realistically be accounted for through proper quantification by means of a probabilistic approach [6].

In this paper, the reliability approach to fatigue strength studies for ferrocement slabs is presented. The experimental program to supplement the theory is carried out. Unlike the conventional approach of obtaining the stress-life (S-n) curves in fatigue studies, the residual strength-life-stress curves are proposed.

2. Practical Considerations

In conventional fatigue strength studies, specimens are cyclically loaded to failure at different stress levels, S, from which the number of cycles to failure, n, are obtained. This approach has the advantage of compactness in that one curve is obtained for each class of material and type of loading. However, for most materials, n can be very large unless the stress levels are extremely high. At most working stress levels, n is of the order of millions of cycles. The experimental program to obtain the S-n curves can be prohibitively time-consuming and costly, especially at the working stress levels which is the primary range of interest.

As an alternative to the S-n curves, the residual strength of the element, R, can be obtained after it has been subjected to a predetermined number of cycles, n, at a given stress level, S (without having to cycle the element to failure). The values of n to be tested should correspond to values close to the intended working life of the element. Similarly, S should correspond to those close to the working load that the element would be subjected to. The only disadvantage to this approach is that an additional variable, R, has to be considered. However, it provides the flexibility of changing the failure criterion through a change in the allowable R value in design.

3. Experimental Program

The slab proposed for use as secondary roofing element is shown in Fig. 1, with the nominal values of the dimensions and properties tabulated in Table 1. Adopting the rectangular stress block approach as in reinforced concrete design, the computed nominal static cracking and ultimate load are R_c=1.26 kN and R_u=2.52 kN, respectively. The elements were subjected to sinusoidal loading, with a range from 0 to Q_m kN, and tested to some selected n cycles after which the residual strength are obtained. Three value of Q_m are used, namely, 0.75, 1.05 and 1.30 kN. The values of n selected are 3,000, 10,000, 30,000 and 100,000 cycles. Since the temperature is maximum during the day time and minimum during the night, one cycle would thus correspond to a time period of one day. The values of n would therefore correspond to a life of approximately 8, 27, 82 and 273 years, respectively. Three properties are monitored, namely, the first-crack strength, R_c, the ultimate strength, R_u, and the modulus of elasticity within the working range, E. For proper control, the elements are cast in four batches of 18 specimens each, one-third of which are used as control specimens to determine the actual cracking and ultimate strength of the uncycled slabs under static load test. The controlled specimens are tested at the beginning and at the

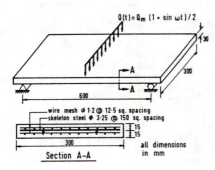

$Q(t) = Q_m (1 + \sin \omega t)/2$

wire mesh ∅ 1·2 ⊕ 12·5 sq. spacing
skeleton steel ∅ 3·25 ⊕ 150 sq. spacing

all dimensions in mm

Section A-A

Fig. 1 - Test Specimen

Table 1 - Ferrocement Data

Item	Nominal Value
Wire Mesh:	
Diameter	1.2 mm
Ultimate Strength	330 MPa
Modulus of Elasticity	180 000 MPa
Volume fraction	0.012
Skeletal Steel:	
Diameter	3.25 mm
Ultimate Strength	660 MPa
Modulus of Elasticity	200 000 MPa
Volume fracton	0.0014
Mortar:	
Cylinder Strength	45 MPa
Tensile Strength	4.5 MPa
Sand: cement ratio	2:1
Water: cement ratio	0.47:1

end of each batch of specimens in the experiment. The test procedure of each slab under cyclic load is as follows. First the specimen is loaded statically up to 1 kN, from which its modulus of elasticity at n=0, E_o, is obtained. The specimen is next subjected to n cycles of loading, at a rate of 5 Hertz. This rate is chosen since previous fatigue studies (e.g. [3]) have shown that for reinforced concrete, a loading rate within the range of 1-10 Hz does not affect the fatigue strength significantly. At the end of the cyclic loading, the specimen is immediately tested for each of the three properties, R_{cn}, R_{un} and E_n by loading the specimen to failure under normal static load test condition.

As a first-order study, the deterioration in the mean values of R_{cn}, R_{un} and E_n with n is studied. Therefore, for each n and Q_m, three specimens are used from which an average value of each of the three parameters under study are estimated. In order to capture the variation of the results due to the inherent variability of the materials used, fabrication process, and loadings, many specimens have to be tested. Due to time and cost constraint, two sets of nine specimens each are tested, at Q_m=1.05 kN and n=30,000 and 100,000 cycles respectively.

4. Experimental Results

The results obtained are plotted in Figs. 2-4. Although the nominal cracking load is 1.26 kN, the values obtained from computation using the experimental cube strength and the actual dimensions of each specimen is much higher and agree reasonably well with the values obtained through testing the control specimens. As such all

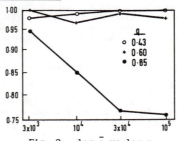

Fig. 2 - log e vs log n

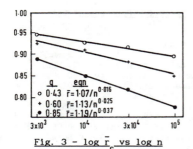

Fig. 3 - log $\bar{r}_c$ vs log n

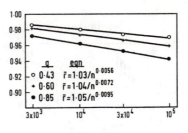

Fig. 4 - log $\bar{r}_u$ vs log n

three load levels are below the actual cracking load. The estimated mean values of $q=(Q_m/R_{co})$ are 0.43, 0.60, and 0.85.

From Fig. 2, it can be seen that there is no significant deterioration in the mean value of $e(=E_n/E_o)$ with n even up to about 60% of the cracking load. Thus, one can assume linearity and the superposition principle may be applicable, justifying the use of Miner's rule when random load range is considered. At the 85% level, minor deterioration is observed in some cases.

In Figs. 3 and 4, R_{cn} and R_{un} are normalized as $r_c(=R_{cn}/R_{co})$ and $r_u(=R_{un}/R_{uo})$. The following considerations are taken for obtaining R_{co} and R_{uo}. Each specimen after failure is dissected from which the relevant dimensions are measured and together with the experimental cube strength, the cracking and ultimate strength of the virgin specimens are computed. This approach, although conceptually attractive, is not feasible here since the value of q during the experiment will vary between specimens and the normalized results may not always appear consistent. Therefore, the results for each batch of specimens including the R_{co} and R_{uo} values obtained experimentally through the control specimens are also plotted against n and a line of best fit obtained. The overall *mean* of R_{co} and R_{uo} for each batch of specimens may then be estimated and used for normalization purposes. Figs. 3 and 4 show the mean deterioration of r_c and r_u (denoted by $\bar{r}_c$ and $\bar{r}_u$, respectively) with n. In view of the time span considered, the results indicate the good fatigue resistance of the specimen under study. The deterioration in the cracking strength is comparatively more significant than that of the ultimate strength. There is no necessity to extrapolate the curve to higher values of n as it is unlikely that the actual slabs in practice will be used longer than the current lease life of the building of 99 years.

The inherent variability in R_c and R_u are estimated from the statistical fitting of the results of specimens cycled with $Q_m=1.05$ kN at n=30,000 and 100,000 cycles, as shown in Figs. 5 and 6. From the four sets of data plotted, the normal distribution may be assumed for R_c and R_u.

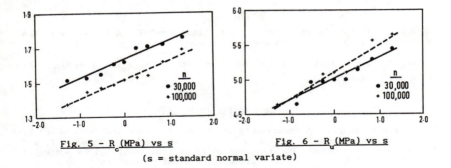

Fig. 5 - R_c(MPa) vs s   Fig. 6 - R_u(MPa) vs s

(s = standard normal variate)

5. Reliability Considerations

When a specimen has been cyclically loaded to n cycles with a load range ratio of q, the residual strength ratio (which may be used to denote either the cracking or the ultimate strength ratio), r, is not a constant due to the existence of uncertainties in the materials, fabrication and loading processes. It should be treated as a random variable with probability distribution function, $F_r(r|q,n)$. The probability, p_f, that a slab will have a residual strength below some allowable value, r_a, after it has been cyclically stressed from 0 to q for n number of cycles is thus given by

$$P_{f|q,n} = P(r<r_a|q,n) = F_r(r_a|q,n) \qquad (1)$$

If $F_r(r|q,n)$ can be approximated by the normal distribution, with a mean of $\mu_{r|q,n}$ and standard deviation, $\sigma_{r|q,n}$, then the allowable strength is related to the probability of the slab not satisfying this criterion by

$$r_a = \mu_{r|q,n} + \sigma_{r|q,n} * \Phi^{-1}(p_{f|q,n}) \qquad (2)$$

where Φ^{-1} is the usual notation for the inverse cumulative normal distribution function.

Throughout the intended life of the slab, the temperature is not constant but of random variation. By measuring the actual temperature variation of some typical roof slabs over a period of time, and looking at past years of temperature variation of the location, a probability density function for the stress ratio may be fitted, denoted as $f_q(q)$. Assuming that within the working range, Miner's rule of linear cumulative fatique damage is applicable, Eq. (1) may be modified to account for stress variations throughout the life of the element as

$$\int_{all\ q} F_r(r_a|q,n).f_q(q).dq = P_{f|n} \qquad (3)$$

$$\sum_1 F_r(r_a | q_i, n) \cdot P_q(q_i) = P_{f|n} \qquad (4)$$

where Eq. 3 is for the case where q is continuous and Eq. 4 is used when the variation of q is divided into discrete ranges.

Hence, for any given two of the three variables, r_a, n and $P_{f|n}$, the third may be determined.

6. Uncertainty Analysis

The main sources of uncertainty in fatigue analysis are contributed by the actual environmental loadings, q, the predicted residual strength, r, and the measure of life, n. For random load range, the variation is accounted for in view of the fact that a probability density function is used as a weighting function. The uncertainty in the value of n may be assumed as negligible, since a cycle of significant stress range occurs once a day with negligible variation over the life span of the element in actual usage.

The uncertainty of r comprises of:
(a) its inherent variability. For a given n and q, the inherent random nature of r is approximately quantified by the standard deviation of r obtained from many samples. An average value of its coefficient of variation (cov) can therefore be obtained, denoted as δ_r. From Figs. 5 and 6, $\delta_r \approx$ 0.07.
(b) uncertainty in $\bar{r}$. From regression analysis using the experimental data, we obtained the relationship $\bar{r} = c\, n^\alpha$, where $\bar{r}$ is the mean residual strength, n is the number of cycles and c and α are constants. In the experiment, n can be accurately controlled. However, the imposed load ratio, q, varies between specimens and its cov have to be estimated, say Ω_q, which arises mainly from two sources. First, there is a slight variation around the value set in the cyclic test machine during the testing of each specimen. By monitoring the range of variation, the contribution is estimated to be about 0.02. Secondly, due to the variability of the strength of each specimen, each regressed line in Figs. 3 and 4 is for an overall mean value of q. However, each specimen will actually experience a value different from this and the possible variation is best estimated from the variation of the cracking load ratio computed from the results of the controlled specimens from each batch. The cov is estimated to be slightly less than 0.06. The overall value of $\Omega_q = 0.06$ is obtained. For each n, a $\bar{r}$ value is obtained. This $\bar{r}$ is a sample mean and will therefore have some variation, which can be estimated from the data. By taking a weighted average of the variance of the mean over the regressed line, an approximate value of $\delta_{\bar{r}}$ (average) of 0.04 is obtained.

The total uncertainty in r to be used in Eqs. 1-4 is thus
$$\Omega_r = \sqrt{\Omega_q^2 + \delta_r^2 + \delta_{\bar{r}}^2} \qquad (5)$$
which is estimated on the average to be approximately 0.10 using the experimental data.

7. Numerical Example

By monitoring the temperature difference across the thickness of a specimen placed on the roof of a typical building, it is observed that a maximum value of 10°C during a 24 hours cycle is not unusual. During the hot season, a maximum difference of 12°C or more may be obtained. The temperature differences of 10°C and 12°C correspond to q values of approximately 0.4 and 0.5 respectively. Hence, the relative frequency of the stress range shown in Table 2 is used in Eq. 4 to compute the probability of failure with respect to different allowable value of r, r_a, for some specified n. The results are plotted in Figs. 7 and 8. From the results, it can be observed that the probability of failure does not decrease significantly with n. This is to be expected since the deterioration rate of the strength within the load range considered as observed in the experiment is low. However, the results also indicate what the maximum allowable r

Table 2 - Load Range

q	relative frequency
< 0.45	0.86
0.45 - 0.65	0.12
0.65 - 0.90	0.02

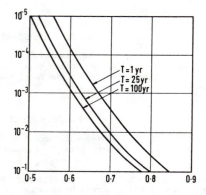

Fig. 7 - p_f (cracking) vs r_a (cracking)

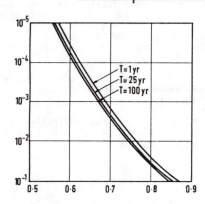

Fig. 8 - p_f (ultimate) vs r_a (ultimate)

value should be if a target reliability is to be achieved throughout the life
of the member, for the type of random load range considered. For example, if
a minimum p_f of 10^{-3} is to be achieved for a specimen to be used for 25
years, the design cracking strength should be factored by 0.63 and the design
ultimate strength should be factored by 0.68.

8. Concluding Remarks

The experimental results and the method used for a probabilistic study
is presented. The results in this paper show that ferrocement has good
fatigue properties within the stress range considered and is therefore an
attractive secondary roofing material. It will be of interest to conduct
similar experiments on specimens cyclic at the non-linear range for which
there may be practical applications, for example, in earthquake-resistant
design. A modification of the above method may be necessary to account for
non-linearity.

9. Acknowledgements

The authors acknowledge the assistance of our laboratory staff, Messrs
B.O. Ang, H.B. Lim and W.M. Ow in the experimental work. This study is part
of an on-going programme under the National University of Singapore research
grant RP880623.

10. References

[1] PARAMASIVAM, P., DAS GUPTA, N.C. and LEE, S.L., "Fatigue Behaviour of
    Ferrocement Slabs", Journal of Ferrocement, Vol. 11, No. 1, pp. 1-10,
    1981.

[2] PICARD, A. and LACHANCE, L., "Preliminary Fatigue Tests on Ferrocement
    Plates", Cement and Concrete Research, Vol. 4, pp. 967-968, 1974.

[3] MCKINNON, E.A. and SIMPSON, M.G., "Fatigue of Ferrocement", Journal of
    Testing and Evaluation, Vol. 3, No. 5, pp. 359-363, 1975.

[4] KARASUDHI, P., MATHEW, A.G. and NIMITYONGSKUL, P., "Fatigue of
    Ferrocement in Flexure", Journal of Ferrocement, Vol. 7, No. 2, pp.
    80-95, 1977.

[5] BALAGURU, P.N., NAAMAN, A.E. and SHAH, S.P., "Fatigue Behaviour and
    Design of Ferrocement Beams", Journal of Structural Division, ASCE, Vol.
    105, ST7, pp. 1333-1346, 1979.

[6] ANG, A.H-S. AND MUNSE, W.H., "Practical Reliability Basis for Structural
    Fatigue", Preprint 2494, ASCE National Structural Engineering Conference,
    April 1975.

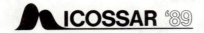
STRUCTURAL SAFETY AND SATISFICING

Colin B. Brown

Department of Civil Engineering, University of
Washington, Seattle, Washington, U.S.A.   98195

Abstract

The mathematical program of a cost extremum constrained by safety, functional and other statements is a comfortable image of design.   In this
paper the various components of this program are
examined in the light of real events.   The conclusion is that the hope of optimizing as a structural engineering design paradigm must be replaced
by a softer viewpoint, such as satisficing, if
realistic events are not to be sacrificed.   The
consequences of such a satisficing approach on
safety measures are considered.

KEYWORDS

Design, Fuzzy Programming, Optimizing, Programs,
Safety, Satisficing.

1.   INTRODUCTION

The mathematical program

MIN   Structural Costs

Subject to: force safety constraints
            displacement constraints
            stability safety constraints
            frequency constraints

has appeal as a model of structural design procedures. Certainly force and displacement constraints
involve static structural analysis in some form,

$$\underline{K} \cdot \underline{x} = \underline{F} \tag{1}$$

where $\underline{K}$ is a stiffness matrix, $\underline{x}$ a kinematic vector
and $\underline{F}$ a force vector. The other two constraints
employ an eigen equation

$$(\underline{K} - \lambda \underline{M}) \underline{x} = 0 \tag{2}$$

where $\underline{M}$ is either the geometric stiffness or mass
matrix. These equations, (1) and (2), are the backbone of structural analysis and therefore seem
entirely cogent to the design procedure. Additionally, a measure of safety and uncertainty must creep
into the program in both the objective function and
the constraints. This again is an area of structural
research and certainly within the normal parlance of
design.

This paper will examine the mathematical program
postulated in the light of real events in the life of
a structure. The consequence of the study will be to
suggest that the program is not appropriate and that
some other, softer, teleological approach must be
offered.

2. OBJECTIVE FUNCTION

The program cited mentioned costs and not weight as
the objective function. Indeed it is costs that
structural engineers seek to minimize and costs are
seldom synonymous with weight. In the structural
design of aircraft the specification calls for a payload of m* and a kinematic performance of a. The
design focuses upon a structure of weight m which,
together with the payload, can be moved as specified
by thrust units of output F. Thus, Newton's second
law of a form F = k (m+m*) a, where k is a gravity
constant, controls the design. Here to obtain specified motion and payload at minimum cost is the same
as producing a minimum weight with associated minimum
thrust. In civil engineering design, the approach is
of statics, and the structural weight appears in F
and not in the inertia effects. The provision of
less weight in this case in no way requires that
costs reduce.

A consideration of costs must involve more than the initial costs C_I; the future costs associated with potential collapse must be included together with the chance of expensive litigation if errors exist. The initial costs will be constrained by the safety and other requirements, and so the safety level as an intended probability, p, will only appear implicitly. However, the value p must be explicit in the rebuilding costs C_R and the chance of errors, P, will appear in C_R and C_L, the litigation expenses. Thus, the objective function will be over the decision variables, $\underline{x}$, as

$$\text{Min}_x \, [C_I(x) + (p+P) \, C_R + P \cdot C_L] \tag{3}$$

and the value of C_L will depend upon P.

The uncertainties, p and P, are the actual intended probabilities of failure and error leading to failure. Only in this way can the objective function of Equation (3) reveal the real design costs. The costs of C_R and C_L will have different values to the actors in the design sequence. To some, they are premiums on exclusion insurance, to others they are actual costs which would have to be met. Owners who sell structures can instruct that the design not include the costs associated with C_R and C_L.

3. CONSTRAINTS

The structural constraints upon the objective function fall into functional and safety types. In the case of functional constraints, the costs of violation, C_L and C_R, may be much smaller than the costs associated with absence of safety. On the other hand, the chances of the display of functional distress are much greater. In the case of functional design, the loading state is very likely and the return period for occurrence short (certainly of the order of the working cycle of the structure; usually daily). Design for safety is concerned with gravity loads with similarly short return periods and extreme lateral loads where the return periods are much the same as the anticipated structural life. These comments suggest that the constraints could be better arranged according to the return period, and hence probability of violation, rather than the force, displacement, frequency and stability separations. Then the program would be concerned with two types of uncertainty measures--one associated with short

(possibly daily) return period occurrences and the
other with the performance during events with return
periods much like the anticipated life of the structure. The probabilities of violation for the second
case are p and P as used in the objective function of
Equation (3). The probabilities of violation for the
short return period cases may well be different.
This difference can be considered for the safety
constraints alone. For gravity effects of load,
these probabilities are p and P; for lateral effects
they are p and P. The costs associated with gravity
effects may be different from lateral ones. Hence
the products in Equation (3) have to be considered as
a mini-max situation. Thus,

$$\text{Min}_x \text{ Max } [C_I(x) + (p + P) \, c_R^L + P \, c_L^L; \qquad (4)$$
$$C_I(x) + (p + P) \, c_R^S + P \, c_L^S]$$

may be a more appropriate objective function where
the superscripts S and L refer to gravity and lateral
loads associated with <u>short</u> and <u>long</u> return periods.
A similar separation of functional constraints may be
applied to the mini-max objective function of
Equation 4.

The various discussed uncertainties in the objective
function are for actual or objective probabilities
rather than notional ones. However, design procedures tend to examine safety in terms of probabilities which refer to a reliability index. Unless the
distributions of probabilities are known, the various
values of p, P, p and P in the constraints are surro-
gates for the actual values used in the objective
functions. This conflict ensures that the proposed
mathematical programs for optimization become design
algorithms. However, the sharp answers provided by
the schemes suggest the best practical solution.

4.   SATISFICING

The difficulties in relating optimization schemes to
real events are concerned with the difficulties in
the assignment of the same probabilities to the
objective function and to the constraints. This may
not be critical if the region around the optimum is
not sharp and it is not crucial from a cost viewpoint
to select x variables conditional on prescribed probabilities <u>assigned</u> to provide the exact optimum.
Such a solution is robust and not sensitive to the

exactness of information available or actions
involved. The obtaining of more realistic information, the formulation of all the alternatives and
the provision of objective probabilities in both the
objectives and constraints is, at the best, costly
and, more likely, will prove elusive. The complete
information required for optimization is, in fact, in
conflict with the engineer's intention of solving
real world problems. Simon [1] urged decision makers
to seek solutions which are possible and acceptable
rather than the best available--the process he termed
satisficing.

In a structural design decision scheme based upon
satisficing, the formalities of optimization, with
the advantage of extensive mathematical under-
pinnings, can be retained without the necessity of
claiming a uniqueness to a valid solution.

A satisficing program could read as:

> MIN   Structural Costs
>
> Subject to: about satisfying safety and
>   functional constraints.

This would provide for structural costs as stated in
Equations (3) or (4) with prescribed and intended
values of p, P, p and P. The constraints would be in
the same mathematical forms as optimizing, but would
involve a vagueness about the occurrence of the probabilities in the objective function. In this way,
satisficing would change to

> MIN   Structural costs from amongst a
>   reasonable set of alternatives
>   for prescribed and intended
>   probabilities (Pp) of failure due
>   to random mishaps and to errors.
>
> Subject to: safety and functional constraints
>   where the prescribed probabili-
>   ties are imprecise and surrogates
>   for Pp.

This program is in fact the subject of the literature
on fuzzy programming. Here the objective is precisely understood and the constraints are fuzzy [2].

5.   FORMALITIES

The move from an optimizing to a satisficing program can be obtained by a formal argument. An optimization program,

$$\underset{\underline{x}}{\text{MIN}} \quad z\ (\underline{x})$$

$$\text{Subject to: } f\ (\underline{x}) \geq \underline{G} \tag{5}$$

$$\underline{x} \geq \underline{o}$$

provides an objective function Z which depends upon the elements of the vector $\underline{x}$. The search for elements of $\underline{x}$ that minimize Z is constrained by the various inequalities.

The simplest form of (5) is linear for both the objective function and the constraints. Then the linear program is

$$\underset{\underline{x}}{\text{MIN}} \quad (Z = \underline{c}^T \underline{x})$$

$$\text{Subject to: } \underline{A}\ (\underline{x}) \geq \underline{G} \tag{6}$$

$$\underline{x} \geq o$$

The objective function and the constraints can be fuzzified. For instance the elements of $\underline{G}$, namely G_i, can be replaced by a fuzzy mean with connotation of "small, but not too small." The objective function can also be softened. This is reasonable when multiple objectives exist. These can then be dealt with as goals in the form.

$$\underline{c}^T \underline{x} \leq \underline{z}^1 \tag{7}$$

This approach is still crisp but can be interpreted as fuzzy objectives. The constraints now appear as the inequalities

$$\underline{k}\ \underline{x} \geq \underline{h} \tag{8}$$

where $\quad \underline{k} = [-\underline{c}^T \underline{A}]^T \tag{9}$

and $\quad \underline{h} = [-\underline{z}^1\ \underline{G}]^T \tag{10}$

The crisp form of the ith constraint is

$$\sum_j k_{ij}\ x_j \geq h_i \tag{11}$$

which can be softened to

$$\sum_j k_{ij} x_j \geq h_i - S_i \qquad (12)$$

where S_i is the slack variable and may have fuzzy form as

$$\mu_i | S_i$$

One form for the supports, μ, is

$\mu = 1$ if (11) applies
$\mu = 0$ if (12) does not apply

$$\mu = \sum_j \left(\frac{k_{ij} x_j - h_i + S_i}{S_i}\right) \text{ if the form}$$

is between (11) and (12).

The optimizing problem of (6) has now been changed to determining, for any vector $\underline{x}$, the maximum of the minimum supports among these fuzzy inequalities. In this way, the minimum supports are the sequence (a_i) and the maximum among these is the "optimum" decision variable.

More formally, the program is

MAX (a_i)

$$\text{Subject to: } a_i \leq \sum_j \left(\frac{k_{ij} - h_i + S_i}{S_i}\right); \qquad (13)$$

$$0 \leq a_i \leq 1;$$

$$\underline{x} \geq 0; \ i = 1, 2 \ldots$$

In this way, the a_i provides a support for "optimal" solution $\underline{x}$, where optimal is interpreted as having the greatest support.

When the objective function is crisp, then we have a mixed program between (6) and (13); namely,

$$\text{MIN } (z^1 = \sum_j c_j x_j)$$

$$\text{Subject to: } a_i \leq \sum_j \left(\frac{k_{ij} - h_i + s_i}{s_i} \right); \quad (14)$$

$$a_i \geq a_L; \; x_j \geq 0, \; \begin{array}{l} i = 1 \text{ to } m \\ j = 1 \text{ to } n \end{array}$$

In this case, a_L, is the minimum support for barely acceptable design.

These approaches whereby the optimization problems of (5) and (6) are softened to the fuzzy problems of (13) and (14) are developed in the book by Zimmermann [2].

6. CONCLUSIONS

The effort in this paper has been to demonstrate the value of the optimization approach to structural design and then to show the difficulties in attaining the identical meanings to probabilities in the objective function and in the constraints. Rather than abandon the advantages of the mathematical programming methodology of optimization, it is suggested that the problem can be restated as that of satisficing where the well developed methods of fuzzy mathematical programming can be employed. Thus, the advantages of a method which is mathematically tractable and, at the same time, physically realistic can be utilized.

Acknowledgment: The work was supported by the National Science Foundation (ECE 8518155).

References:

[1]   SIMON, H.A.: Administrative Behavior, 2d
        Edition, Free Press, New York, 1957.

[2]   ZIMMERMANN, H.J., Fuzzy Sets, Decision Making
        and Expert Systems, Kluwer Academic
        Publishers, Boston, 1987.

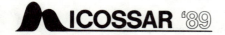

SAFETY ASSESSMENT FOR DEBRIS BASINS

Bilal M. Ayyub* and Richard H. McCuen**

* Associate Professor, Depart. of Civil Eng.,
   University Maryland, College Park, MD 20742, USA
** Professor, Depart. of Civil Eng., University of
   Maryland, College Park, MD 20742, USA

ABSTRACT

Debris basins are engineering control structures
used in locations where debris flows occur.
Knowledge of the risk of failure as a function of
important design variables can improve decision-
making and minimize the hazard associated with
debris flows. The risk was computed for four
policy elements: rainfall frequency, interval
between significant watershed burn, construction
and dredging accuracy, and regularity of
maintenance of debris basins. The burn interval
and rainfall magnitude are the two most important
variables associated with failure risk, with the
risk varying from less than one percent to as much
as 65 percent.

KEYWORDS

Debris flow; Failure probability; Risk assessment;
Uncertainty

1. INTRODUCTION

Debris flows are movements of large soil masses through defined
channel systems. They represent a significant hazard in many
parts of the world. A debris basin is a storage structure used to
contain the debris [1]. These basins are usually located at the
mouths of steeply sloped canyons. While debris flows are a
continual problem, there have been very few systematic efforts
made to compile data on the volumetric characteristics of debris
flows [2]. Thus, accurate design methods are rarely available.
Where data are available, the records are usually short and, thus,
large sampling variation is expected.

Hydrometeorological variability is a primary source of the year-
to-year variation in the magnitude of debris flows. Factors that
are associated with hydrometeorological conditions and that affect
the variability of debris flow volumes include the antecedent
rainfall volume prior to a destabilizing rainfall event, the

intensity and duration of the rainfall event, the occurrence of
lightning that causes extensive forest fires over the watershed,
and surface erosion that occurs during minor storms and is
temporarily stored in the channel system; this inter-storm surface
erosion often becomes part of the interstitial mud of the debris
flow. In addition to the variations caused by the
hydrometeorological factors, watershed and soil characteristics
are important, including the watershed slope, land cover, and both
the particle size distribution and angle of repose of the debris
material.

In designing a debris basin, the volume of storage is a primary
design variable. The volume of storage required for control is
directly related to the hazard presented by the debris flow and
the potential for failure of the basin. Thus, maintenance and
inspection of the basin to ensure adequate storage for the control
of debris flows is essential to maintain acceptable levels of
failure risk. In addition to debris deposited during major storm
events, eroded material continually enters the basin during minor
storms; therefore, the basin must be dredged, usually on an
irregular as-needed maintenance schedule. The potential for
failure of the basin depends on the accuracy of both the design
and the dredging. The risk of failure will increase if the basin
volume after either construction or dredging is less than that
specified in the design.

The objective of this study is to estimate the probability of
failure of debris basins as a function of variables that
contribute to the variations in the supply of and demand for basin
storage. The probabilities of failure can provide useful
information to policy makers and design engineers about the
optimum design. Also, economists could use such probabilities to
evaluate the benefits and costs of alternative designs.

2. RISK ASSESSMENT: MATHEMATICAL DEVELOPMENT

The performance function that expresses the relationship
betweenthe design volume of a debris basin and the volume of a
debris flow can beexpresses by the following equation:

$$Z = g(X_1, X_2, \ldots, X_n) \tag{1}$$

in which the X_i, i=1,...,n, are design variables, with g(.) being the
functional relationship between the design random variables and
failure. The performance function can be defined such that the
limit state, or failure surface, is given by Z=0. The failure
event is defined where Z<0, with the survival event defined as the
space where Z>0. Thus, the probability of failure can be evaluated
by the following integral:

$$P_f = \int\int \cdots \int f_X(X_1, \ldots, X_n) \, dx_1 \, dx_2 \ldots dx_n \tag{2}$$

where f_X is the joint density function of $X_1, X_2, \ldots, X_n$ and the
integration is performed over the region where Z<0. Because each
of the design variables has a unique distribution and they

interact, the integral of Eq. 2 cannot be easily evaluated. A probabilistic modeling approach of Monte Carlo computer simulation with Variance Reduction Techniques (VRT) can be used to estimate the probability of failure.

2.1 Conditional Expectation VRT

The performance function of a fundamental risk assessment case is given by

$$Z = R-L \tag{3}$$

where R is a function of the structural strength or resistance, and L is a function of the corresponding load effect. Therefore, the probability of failure, P_f, is given by

$$P_f = P(Z<0) = P(R<L) \tag{4}$$

For a randomly generated value of L (or R), say l_i (or r_i), the probability of failure is given by

$$P_{f_i} = P(R<l_i) = F_R(l_i) \tag{5a}$$

or $$= P(L>r_i) = 1-F_L(r_i) \tag{5b}$$

where F_R and F_L are the cumulative distribution functions of R and L, respectively. Thus, for N simulation cycles, the mean value of the probability of failure is given by the following equation:

$$\bar{P}_f = (\Sigma\, P_{f_i})/N \tag{6}$$

The variance (Var) and the coefficient of variation (COV) of the estimated probability of failure are given by

$$Var(\bar{P}_f) = Var(P_f)/N$$

$$= [\Sigma\, (P_{f_i}-\bar{P}_f)^2]/(N(N-1)) \tag{7}$$

$$COV(\bar{P}_f) = [Var(\bar{P}_f)]^{1/2}/\bar{P}_f \tag{8}$$

The randomly generated variables should be of least variabilities and the resulting conditional expectation need to be evaluated by some known expression.

The concept and the steps involved are further explained by Ayyub and Haldar [3] and White and Ayyub [4]. According to this method, the variance of the estimated quantity is reduced by removing the variability of the control variable on which conditioning was not done.

2.2 Antithetic Variates VRT

In this method, a negative correlation between different cycles of simulation is induced in order to decrease the variance of the

estimated mean value. If U is a random number uniformly
distributed between 0 and 1 and is used in a computer run to
determine the probability of failure $P_{f_i}{}^{(1)}$, the 1-U should be used
in another run to determine the probability of failure $P_{f_i}{}^{(2)}$.
Therefore, the probability of failure at the ith simulation cycle
is given by

$$P_{f_i} = (P_{f_i}{}^{(1)}+P_{f_i}{}^{(2)})/2 \qquad\qquad (9)$$

The Antithetic Variates VRT is described in details by Ayyub and
Haldar [3] and White and Ayyub [4].

3. RISK ASSESSMENT: DEBRIS BASIN FAILURE

3.1 Factors Affecting Design Risk

Policies intended to control debris flows with debris basins
should address four primary elements: The magnitude and frequency
of precipitation, the frequency of forest fires, the loss of
storage in the basin due to small volumes of debris that
accumulate between major debris-generating storm events, and the
accuracy of excavation during construction and dredging. Debris
flows most often occur when intense rainfalls follow extended
periods of rainfall that saturate steeply sloped portions of the
watershed. While short duration rainfall intensities are used as
input for waterflood estimation methods, longer duration rainfall
volumes are better indicators of debris flows because they reflect
both the antecedent rain and the rain that generates the debris
flow.

In addition to the frequency of rainfall, the frequency of
watershed-scale forest fires is an important element of a design
policy. Forest fires destroy the natural vegetation, thus
exposing large surface areas to the kinetic energy of the
raindrops and the erosive energy of the resulting surface runoff.
Furthermore, the fire sears the surface of the watershed, which
reduces infiltration and increases runoff velocities. Soil
moisture is retained in the soil matrix when there are no trees to
transpire the water, thus increasing both the stress placed on the
failure plane and the potential for debris slides. As the
frequency of forest fires increase, the volume of debris flows is
expected to increase. Therefore, the design model should include
a variable to reflect the expected time interval between forest
fires.

In between the major debris-producing storms, minor storms can
generate significant volumes of sediment that collect in the
drainage system of the watershed, as well as in the debris basin.
The amount of such debris in the basin at the time of occurrence
of a debris-producing storm affects the failure rate of the basin.
Therefore, a debris management policy should include a policy
element that requires inter-storm dredging of sediment that
accumulates in debris basins. Dredging should take place just
before the season when most debris flows occur and when the
storage taken up by inter-storm sediment accumulation exceeds a
certain percentage of the design volume.

Construction accuracy is the fourth factor that may influence the risk of failure of a debris basin. When the as-built volume of the basin is less than the volume specified by the designer, then the risk of failure increases. Given the value of land, there is a natural desire to minimize the area devoted to the debris basin. Thus, inspection of the debris basin to ensure that the as-built volume and the volume after dredging is at least equal to the design volume should be included as part of every debris management policy.

3.2 Formulation of a Debris Model for Risk Assessment

Given the importance of these four factors, a model that allows for the design uncertainty of these factors was formulated. The central part of the model is an empirical formula that relates the volume of debris flow (D_y, in cubic yards) to the 72-hour rainfall depth (P, in inches), the drainage area (A, in square miles), and the time interval between forest fires (t, in years). Data for debris basins in the Los Angeles area were analyzed, with the following result:

$$D_y = 2750 \ P^{0.75} \ A^{1.25} \ [1 + 80e^{-0.62A-0.537t}]^{0.5} \qquad (10)$$

Equation 10 is used as the base model for estimating both the supply and demand functions of Eq. 3. For the analysis of risk of debris basin failure, the demand function reflects the variation in debris flow volumes that results from the natural uncertainty in precipitation and forest fires. A drainage area of one square mile is assumed; since area was considered to be a constant in the estimation of risk, the assumption has no bearing on the assessments of failure. The precipitation is assumed to follow a log-extreme value distribution with a mean and coefficient of variation of 4.5 and 0.444, respectively. The time between forest fires is assumed to follow a log-normal distribution with a mean value and coefficient of variation of 8.0 and 1.375, respectively. Equation 10 is also used to compute the supply function of Eq. 3. In this case, the supply defines the design volume of storage that is available for a debris event. Thus, the base design with Eq. 10 reflects the volume required by the debris management policy, with the policy specifying both a design precipitation depth P and a design fire interval t. In quantifying the supply function, three design precipitation depths P are evaluated, the 2-, 10-, and 100-year events. Since most of the data used to calibrate Eq. 10 had burn intervals of less than 25 years, four burn intervals t are evaluated, 2, 5, 10, and 25 year intervals. Variation in the in-place volume associated with the frequency of dredging can be represented by an exponential distribution; three policy statements are considered, with variation in the allowable inter-storm accumulations of 0, 10, and 25 percent of the design volume. At these percentage increases, of course, the supply of storage for major debris-generating storms decreases. Finally, construction accuracy was assumed to be normally distributed and the risk evaluated for variations of 2 and 5 percent, which reflect the expected construction accuracy for cohesive and noncohesive soils, respectively.

3.3 Failure Assessment

The limit state equation for the purpose of failure assessment is
given in the form of Eq. 3 as follows:

$$Z = V_i - \log_{10}(U) \, k \, V_D - 2750 \, P^{0.75} \, A^{1.25} \, [1 + 80e^{-0.62A-0.537t}]^{0.5} \qquad (11)$$

where V_i is the initial provided volume, V_D is the design volume, k
is the needed fraction for dredging, and U is a uniform random
variable (random number). The failure event is defined as the
event during which the demand for storage exceeds the volume
supplied by the in-place design. For example, using the following
mean values for P, A and t of 13.7, 1. and 2, respectively, V_D can
be determined to be 7760. For a dredging policy factor k = 0.001
($\approx$0) and a coefficient of variation (COV) for V_i of 0.02, the
following simulated probabilities of failure can be generated
using an increasing number of simulation cycles, and P and U as the
control random variable in the Conditional Expectation and
Antithetic Variates VRT. The results are summarized in Table 1.

Table 1. Failure Probabilities using the Simulation Algorithm

Simulat. Cycles	U Control Variable		P Control Variable	
	P_f	COV(P_f)	P_f	COV(P_f)
10	1.000×10^{-37}	0.333	3.838×10^{-3}	0.202
1010	3.960×10^{-3}	0.352	4.103×10^{-3}	0.045
4010	3.990×10^{-3}	0.176	----	---
7010	3.923×10^{-3}	0.134	----	---

It is evident from Table 1 that the variance reduction techniques
in Monte Carlo simulation converge to the correct P_f with
relatively small number of simulation cycles, in this case about
1000 cycles. Using either P or U as a control variable in the
simulation process results in the same estimated P_f. However, P as
a control variable is a better choice since it has a larger COV.
By not randomly generating the control variable P, the COV(P_f) can
be further reduced and convergence can be expedited. The
simulation-based risk assessment algorithm was executed for the
conditions described in the previous section, with two levels for
the construction accuracy and three policy levels each for
precipitation P, burn interval t, and dredging of inter-storm
accumulation. Figure 1 shows the failure surface, which gives the
probability of failure of a debris basin for designs based on
return periods of from 2 to 100 years for the design precipitation
and 2 to 25 years for the design burn interval. There is a
noticeable interaction between the two variables. The risk of
failure is greatest, approximately 67 percent, for a basin that is
only designed to control the 2-year precipitation and on an
infrequent burn interval of 25 years; such a design would have a
very small volume of storage.

Debris basins must be maintained since sediment accumulates in the
basins during minor, non-debris flow storm events. If basins are
not properly maintained by dredging the accumulated sediment, then

the storage specified by the design engineer will not be available
during a debris-producing storm event. Thus, the probability of
failure is expected to increase as the time interval between
dredging increases. Three policy conditions were evaluated, each
representing a different fraction of storage occupied by sediment
accumulation permitted prior to dredging. Specifically, fractions
of 0, 0.1, and 0.25 are considered, with a fraction (k) of 0
indicating a policy that requires the dredging of all sediment or
debris immediately after it has been deposited. This may be
considered impractical since it would require continuous
monitoring. Thus, the other levels studied reflect varying levels
of practicality and the availability of public funds for
maintenance. Table 2 shows the effect. For a policy that allows
as much as 25 percent of the basin volume to be taken up by
sediment deposition from minor storms, the risk of failure
increases about 2 times the risk where continual maintenance is
provided. This risk is sufficient to warrant recognition of the
need for all policies to provide for both maintenance between
debris-generating storms and monitoring of sediment accumulation
during these periods. The policy should specify a value of k that
is reasonable from the standpoint of the availability of
maintenance resources and the cost associated with the risk of
failure.

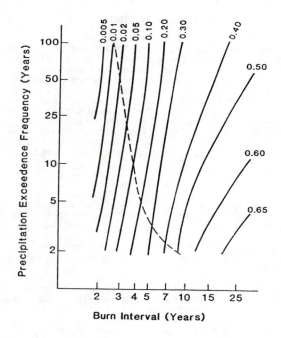

Figure 1.   Probabilities of Debris Basin Failure

Table 2. Failure Probabilities for Alternative Dredging Policies
(k) and Policy Exceedence Frequencies (T) of Precipitation,
with a Policy Burn Interval of Two Years.

T in Years

k	2	10	100
0.00	0.053	0.010	0.0041
0.10	0.064	0.012	0.0050
0.25	0.093	0.021	0.0083

The fourth factor included in the model is the construction and
dredging accuracy. This has assumed to be normally distributed,
which reflects the possibility that the basin may have either a
larger or smaller constructed volume than specified in the design.
A larger in-place volume would reduce the risk of failure. The
resulting estimates of failure risk indicate that the construction
accuracy has little effect on the overall risk of failure, with a
maximum variation of about 3 percent. Thus, design risk is
relatively insensitive to construction accuracy as long as the
construction practice including inspection assures that the volume
is within 5 percent of the design volume.

CONCLUSIONS

The size of a debris basin can be designed by balancing the costs
associated with construction, operation, and maintenance with the
benefits provided through the prevention of debris flows from
damaging downstream public facilities and the loss of life. Safer
and more rational designs result when design practices account for
the risk of failure. The risk of debris basin failure associated
with four controlling factors was studied. These factors are the
return period of the rainfall, the time interval between
substantial burns, and construction and maintenance practices. An
evaluation of the risk of failure of debris basins was made using
the Conditional Expectation VAriance Reduction Technique with a
debris flow model developed from data for the southern California
area.

REFERENCES

[1] Hollingsworth, K. and G.S. Kovacs, "Soil Slumps and Debris
     Flows: Prediction and Protection," Bulletin of the Association
     of Engineering Geologists, Vol. XVIII, No. 1, pp. 17-28, 1981.

[2] Johnson, P.A., R.H. McCuen, and T.V. Hromadka, "Magnitude and
     Frequency of Debris Flows," J. of Hydraulic Engineering, ASCE,
     (underreview), 1988.

[3] Ayyub, B.M. and A. Haldar, "Practical Structural Reliability
     Techniques," J. of Structural Engineering, ASCE, Vol. 110, No.
     8, pp. 1707-1724, August 1984.

[4] White, G.J. and B.M. Ayyub "Reliability Methods for Ship
     Structures," Naval Engineers Journal, ASNE, Vol. 97, No. 4, pp.
     186-96, May 1985.

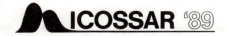
Development of a Deep Knowledge Based Expert System for Structural Diagnostics

A. Jovanović [1], A.C. Lucia [2], A. Servida [2,3], G. Volta [2]

[1] - MPA - Univ. of Stuttgart, Pfaffenwaldring 32, 7000 Stuttgart 80, FR Germany
[2] - Commission of the European Communities, JRC Ispra, 21020 Ispra (Va), Italy
[3] - CEC, JRC Ispra, Italy and JRC Petten, PO Box 2, 1755ZG Petten, The Netherlands

Abstract

The paper tackles development of a structural reliability expert system, the basic goal of which is to improve current possibilities for structural safety and reliability analysis of critical pressurized components (e.g. vessels, piping, feedwater tanks, etc.) of large power and process plants, by applying the deep knowledge modelling and by coupling symbolic analysis with the numerical one. The system should enable quicker assessment and use of non-numerical (qualitative), incomplete, imprecise and/or uncertain data, unconsidered in present methodologies, which are mostly numerical. Incorporation of the deep knowledge leads to better handling of unexpected events. Architecure (modular), paradigms and description of the modules of the system are given in the paper, as well as the considerations regarding the use of the deep knowledge and coupling of symbolic and numerical analyses.

Introduction

Many power and process plants face nowadays the problem of unplanned and costly outages due to structural failures of pressurized components. The expert system DSN (*Deep knowledge based expert system using coupled Symbolic-Numerical analysis for structural diagnostics*), described in this paper, has been designed as a decision aid to plant engineers. The principal task of the system, based on the structural diagnostics, is to indicate the optimal patterns of the future plant/component operation. To perform this task, it must take into account all the information available, i.e. both the standard, numerical, data (usually design data and resumed nondestructive examination (NDE) data), and the relevant information contained in the humans' experience (e.g. experience with similar plants/components, experience

from the material/component fabrication, independent expert opinion, etc.).
This information is mainly qualitative and, very often, incomplete, uncertain
and/or imprecise. The system must also explain its outcomes. Development
of the system [1] resulted mainly from co-ordination of the research programs
of MPA Stuttgart and JRC Ispra. Although the proposal as such has not been
realized, many of its elements have been incorporated in the derived expert
systems, e.g. in the leak-before-break expert system of MPA [2]. The DSN
objectives are described in detail in [1] and summarized in Fig.1.

APPLICATION DOMAIN	A I
Area: Flaw and damage assessment in pressurized components **Conditions:** mechanical and thermal loads, static, cyclic, under normal operation and transients **Geometries:** press. vessels, feedwater-tanks, large piping, nozzles **Result:** Safety and reliability diagnostics, analysis and life time prediction **Knowledge Base:** - General Facts - Facts being part of the current physical situation (e.g.: dimension, position and nature of those defects, environment, loads, material, etc.) - Rules: e.g. those regarding crack behaviour, NDE result interpretation, corrosion water/steel, material technology, Designer's/Operator's experience, etc.	- **Methods:** for treatment of imprecisions and uncertainties - **Non-standard logic** - **Qualitative modelling** (equations and constraints) - **Advanced logic,** e.g. for distinguishing between facts which are always true and facts which are true in the concrete situation only (modal logic), or making conclusions out of negative statements (negative inference), etc. - **New Rules Creation** by learning from application cases - **Deep Knowledge representation** methodologies

DSN Expert System Based Structural Safety Analysis

Fig.1 - The twofold objectives of DSN

Engineering Knowledge in Structural Diagnosis

Structural diagnostics, and structural safety and reliability assessment
are essential for the decision regarding the component life prediction and possible extension. The diagnosis and the assessment capture the possibility of a
structural failure of the component, which, in most of the practical cases, could
lead to leaks and/or catastrophic breaks on/of pressurized components, i.e. to
hazards related to possible contamination of environment, release of the stored
energy, public risks, etc. Excluding of a gross structural failure is primarily
a safety concern, while reduction, prediction and successful management of
minor possible structural failures are mainly reliability, safety and economy

concerns.

The assessment is usually couched in terms of:

- failure risk, as probability of a structural failure, given the loading conditions (normal operation and hypothesized transient and/or accident conditions), and
- remnant (residual) life time estimation, as a function of experienced and/or expected loading conditions.

Both parts of the assessment are basis for the component and/or plant life extension. The assessment is necessarily based on structural diagnostics.

Current standards of material technology, design, manufacturing, inspection and licensing, provide already a high level of reliability of pressurized components. Failure rates by conventional plant components are under the level of 10^{-4} failures/component-year, and approximately 10 to 100 times lower by nuclear and recently constructed non-nuclear components. However, a possibility of failure cannot be completely excluded, as possible causes of structural failures are mostly out of the scope of applicable standards, or under the threshold of analytical modelling built into the procedures and standards (e.g. local damage phenomena, effects of human errors in manufacturing, control and operation, etc.). During the component operation, the structural state of component changes continuously, while the complex space of all significant state variables is only roughly defined and described by standard procedures, based primarily on the "lower bound", conservative assumptions.

An overall, stated and widely accepted technology for accurate and reliable prediction of life time, in terms of a best estimate, does not exist yet. Existing procedures are mainly conservative, entailing wide uncertainty margins. This is true in particular for the two phenomena usually leading to possible structural failures, namely for

- the crack behaviour modelling, and
- the general damage accumulation modelling.

Cracks, developed mainly out from initial defects from the manufacturing (e.g. defects in weldments), and/or due to fatigue, can be considered as the principal possible cause of structural failure in pressurised components. Numerical analysis of the crack behaviour and damage accumulation alone, cannot provide a complete decision basis. In fact, the numerical analysis does not include qualitative information, it is unsuitable for screening of possible future loading scenarios, it must be (when detailed) performed by extremely specialized personnel, it gives few explanations, it is usually expensive and time consuming. Therefore, it has to be combined with human expertise, especially at the decision level. This goal can be achieved through application

of an expert system — in this case the DSN expert system.

Deep Knowledge and Coupling Symbolic and Numerical Analysis

The knowledge representation is of a particular importance in the structural safety diagnostics, because it is a very "ill formed" problem from the AI point of view, entailing the use of a "deep knowledge" based system, so we shall briefly recall the notions of the deep and shallow models [3]:

Shallow models: (see Fig.2a) In shallow models ... *the conclusions are drawn directly from observed facts that characterize a situation. (...) They directly encode the heuristics that experts use in performing their reasoning tasks, and are thus relatively easy to build.*

Deep models: (see Fig.2b) *Deep models correspond more closely to the notion of reasoning from first principles. (...) Deep reasoning is, however, bound to be slower and more complex than shallow reasoning in that a more sophisticated control structure is required.* Or, given the two models M and M', it can be stated that *M' is deeper than M if there exists some implicit knowledge in M which is explicitly represented in M'.*

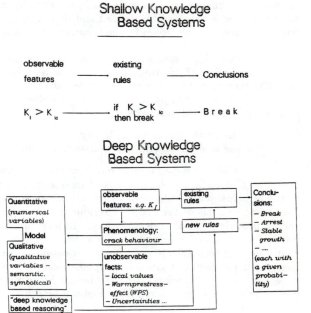

Fig.2 - Shallow (a) and deep knowledge (b) based systems

In the expert systems applied in diagnosis, the deep models offer many significant advantages. when compared to the models based on the shallow and/or "compiled knowledge" only, in spite of some practical advantages of compiled knowledge based models [4]. DSN relies on deep knowledge modelling mainly due to the following two reasons:

- Safety of a pressurized component is mostly affected by unexpected situations (accidents, transients), which, by definition, cannot be fully foreseen in advance. Shallow models are unable to deal with circumstances different from those explicitly anticipated.

- The processes forming the background of the structural safety (e.g. crack behaviour under transient and/or repeated loadings) are providing very poor, inconsistent and uncertain information for the formulation of premises. Reshaping this information is possible only in terms of "checking the compatibility with the first principles", which, again, leads to the use of deep knowledge.

The coupling in the area of structural diagnosis has to be done between the engineering numerical analysis (in limit cases including also the finite element analysis one - excluded, however, from the DSN), and the symbolic analysis used in reasoning [5]. Fuzzy algebra and concept of linguistic variable [6], implement in Lisp based software tools, have been chosen for the symbolic based analysis. The numerical, engineering calculations are in Fortran. The two parts of the analysis interact either within the framework of the expert system shell software [7] (numerical programs included as "methods" in the knowledge base), or the numerics is invoked from Lisp, (Fortran subroutines) [2].

Architecture of the D S N

Principal elements of the DSN architecture are schematically represented in Fig. 3, showing that the system takes information from five sources: from the user, from the "independent expert", from the material data base, from the image knowledge base and from a case history data base. Among the inputs, the most innovative is the Image Knowledge Base, acting as a "bank" of behaviour patterns of the class of components being the target for DSN. Construction of this knowledge base implies development of novel querying, retrieval and processing techniques. In this respect, the major difficulty is the definition of effective measures of similarity and proximity between the stored image patterns and the situation problem to be analyzed. The definition of this measure concerns the general issue of characterizing real entities by analogy. The user's requests provide guidelines for the whole session and they define its goals: structural failure risk assessment, life time prediction, etc. The user is supposed to provide the standard data regarding the component (loads, materials, known state of defects), as well as the evidence (description of the current situation) and the human opinion (e.g. the operator's one). DSN closes the

loop towards the user providing three types of possible outputs: (a) - answer to user's questions, (b) - request for additional information and (c) - input for detailed analysis, consisting of pre-elaborated input data and the solution pattern proposal (e.g. sequence of codes to be used, type of analysis to be performed, etc.).

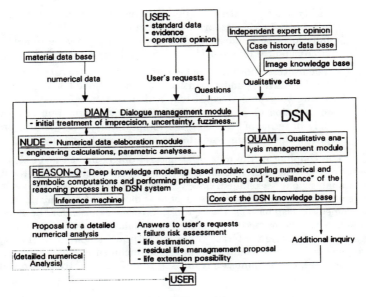

Fig.3 - The DSN architecture

Modules

The dialogue interfacing with the user is obtained through the DIAlogue Management module DIAM. The module will assess, through an interactive dialogue with the user, the correct representation frame, granularity and semantics, for the uncertainty of the information provided as input to the system. Output of this module, to the system, is an internal description of the uncertainty coherent to the internal paradigms of the other modules. The REASON-Q (REASONing in Quick structural diagnostics) module contains the main inference engine and the core of the DSN knowledge base. It behaves as a general controller of the inference process through the different modules of the DSN, looking out for firing the most appropriate metarules, which would allow to achieve the solution by preserving consistency and preventing possible "explosions" in the reasoning process. Two intelligent modules linked to REASON-Q, are NUDE (NUmerical Data Elaboration and analysis) and

QUAM (QUalitative Analysis Module). The capability of coupling will represent the main link with NUDE module, which performs: (a) - standard engineering calculations (stresses, fracture mechanics parameters, etc.) and (b) - parametric calculations. REASON-Q activates also the QUAM module, that carries out the qualitative analysis. QUAM module searches to construct the space of possible states resulting from the situation as described by the user and from the expected and/or hypothesized future loads of the pressurised components ("loading scenarios"). QUAM also makes the first screening of the possible outcomes, i.e. selection of possible solution outcomes eligible in respect to the deep knowledge criteria. The four modules are in a continuous interaction.

Conclusions

Quintessence of the DSN research is the development and application of an expert system for the structural diagnostics, structural safety and reliability assessment and life extension analysis for complex pressurized components. The efficiency of the system, i.e. of the derived systems, is partly tested in the structural reliability experiments at the MPA and JRC, while further verification in practice will be done with collaborating industry partners.

References

[1] Deep knowledge expert system using coupled symbolic numerical analysis for structural diagnostic, *ESPRIT-II Proposal Nr. 2224 (Area II.2.3)*, 1988

[2] JOVANOVIC, A.: Knowledge representation and reasoning in the structural reliability assessment expert systems: Implementation in a Leak-before-Break prototype expert system, accepted for ICOSSAR-89 (R14B-07), San Francisco, 1989

[3] KLEIN, D., FININ, T.: What's in a deep model, Proceedings of the 10th IJCAI-87, vol.1, pp. 559-562, Milano, August 1987

[4] CHANDRASEKARAN, B., MITTAL, S.: Deep versus compiled knowledge approaches to diagnostic problem–solving, *Int.J.Man–Machine Studies*, 1983, vol.19, pp 425-436

[5] KOWALIK, J.S. (Eds.):"Coupling symbolic and numerical computing in expert systems, II, North-Holland, Amsterdam - New York - Oxford - Tokyo, 1988

[6] ZIMMERMANN, H.J.: Fuzzy sets, decision making and expert systems, Kluwer Academic Publishers, Boston - Dodrecht - Lancaster, 1987

[7] KEE^{TM} 3.0 Users' Manual, IntelliCorp, W. Mountain Veiw, CA., 1988

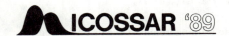

STRUCTURAL RELIABILITY THROUGH MACHINE LEARNING
FROM CASE HISTORIES

John R. Stone and David I. Blockley

Department of Civil Engineering, University of Bristol, BS8 1TR, U.K.

Abstract

The development of a management tool for the control of
structural safety is described. Artificial intelligence techniques
of 'machine learning' are employed to facilitate a method of
learning from past experience. The serial and parallel
approaches of discrimination and connectivity analysis are
employed. A technique for building a hierarchically
structured knowledge base made up from the study of
individual case histories is outlined. Support logic is used to
permit an 'open world' representation of uncertainty.

Keywords: machine learning, knowledge-based systems, case
histories, uncertainty, structural safety

1.Introduction

Engineers are responsible for the safety of the structures they create, and must
respond positively to accidents and failures which occur. The design and construction
process leading to a completed structure embodies the current state of knowledge in
engineering, as expressed in design codes of practice and heuristic methods, and thus
any subsequent failure may be seen as potentially exposing a weakness in that
knowledge. Progress in engineering can be viewed as repeated conjecture and
refutation, with a failure refuting the conjecture of the structure.

A preliminary specification for a safety management system was given by
Blockley [1] at a workshop on modelling human error in structural design and
construction, and identified three main requirements: the examination of case
histories, the need to handle realistically open-world uncertainty, and the use of the
concept of a hierarchically structured knowledge base.

This paper summarises one approach to learning from past events by developing the techniques of 'machine learning' to build a knowledge-based system (KBS) for use in controlling structural safety.

2.Case histories

The study of failures is of interest to social scientists as well as to engineers, since human and organisational factors are commonly significant. Turner [2] has observed that "...many disasters and large-scale accidents display similar features...". It is this regular occurrence of features which is the key to our method of learning.

The first problem to be addressed is that of obtaining accurate details of projects, both failed and successful, past and present. Information is combined from two sources - published reports and recorded interviews. Both have their disadvantages. Written reports commonly follow inquiries into major failures, eg Tacoma Narrows Bridge or Ronan Point. Whilst generally providing a wealth of detail, they represent only one extreme of the continuum of failures. Much more significant in terms of numbers of occurrences are the small to medium failures which may not not undergo detailed and public investigation. The potential information which could be learned from these is lost to the industry. A second approach which has been used with some success by Pidgeon et al.[3] is interviews with key individuals concerned with a project . This enables information to be gleaned from cases which may not otherwise have been reported. However, as Blockley [1] has noted there is a narrow 'window' of time during which suitable case studies are available. This is because recent events may be surrounded with litigation whilst those more than about ten years old are difficult to establish since the memories of participants may begin to fail.

Our method seeks to use details obtained from both the above routes and to combine the information into a knowledge base which may then be consulted by others. To do this, it is necessary to take the case histories, which are in the form of 'stories', and transform them into a structured form suitable for manipulation by computer. The form chosen for this is the event sequence diagram (ESD), as described by Turner [2], which shows the temporal order and relationship of events leading up to a particular outcome.

To develop a useful knowledge base it is necessary to collect a large number of case histories and refine them into a series of event sequence diagrams. This raises the problem of defining a suitable 'vocabulary' for describing events.

It will be shown that the 'learning' method depends on having propositional concepts ('words') which occur in more than one diagram, enabling a 'linkage' to be established. It is therefore proposed to establish an evolutionary 'dictionary' of concepts, sufficiently large to cover the richness of the range of cases held in the knowledge base yet small enough to ensure repeated use.

3. Hierarchies of knowledge

Figure 1 shows a series of ESDs arranged hierarchically. The lowest (deepest) level contains the most detailed information, specific facts from the stories of individual case histories. At the other extreme, the highest (shallowest) level represents the accumulated 'story' of all the case histories in more general terms. This hierarchical structure is useful because it reflects the fact that in some situations very detailed information may be required whilst in other cases general concepts are more meaningful. It enables an appropriate level of problem solving within the knowledge based system. The use of a 'high level' concept, such as 'poor site supervision', may be

sufficient for some purposes whereas a 'low level' one like 'reinforcement starter bars omitted' may be necessary for a more detailed analysis.

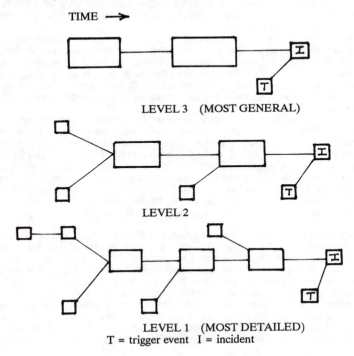

LEVEL 3   (MOST GENERAL)

LEVEL 2

LEVEL 1   (MOST DETAILED)

T = trigger event   I = incident

Fig.1 Hierarchy of event sequence diagrams

4. Uncertainty

A number of different methods have been proposed for handling uncertainty in knowledge based systems. These include certainty factors, fuzzy logic, Bayesian probability and various combinations thereof.

Support logic [4] has been developed to attempt to overcome some of the objections to fuzzy and Bayesian methods. Two values of necessary and possible support give lower and upper bounds respectively on the evidential support for a proposition or event. It is an 'open world' model of uncertainty in the sense that it is possible to represent propositions as true, false or unknown [5].

Each propositional concept in each event sequence diagram input into the KBS is given a support pair which expresses the evidential support for the truth of the proposition within the case history.

5. Machine learning

The topic of machine-learning is an increasingly important area of artificial intelligence research, and the ability to learn from experience would be a useful ingredient of any meaningful KBS.

The method of 'learning' is based around the use of an algorithm developed by Norris [6] to aid clinicians in medical diagnosis . Instead of observed symptoms and sets of possible diseases, concepts in event sequence diagrams and their outcomes are used. Learning is then carried out by induction - that is general rules are proposed from the examination of a number of example cases. There are two phases to the 'learning' process involving algorithms for discrimination and connectivity which will now be described.

Figure 2 shows three imaginary ESDs, in which concept names are represented by numbers. The ESDs are simple 'tree' structures, with each concept being a node of the tree. Each node may in turn be classed as a 'head' or 'tail' node when viewed from another node. For example, in ESD 1, node 10 is a head node of nodes 1, 2 and 3 and a tail node of node 20. In Figure 2 all three ESDs have the same final outcome, 20, and intermediate outcomes 10, 11 and 12. Thus, for example, node 20 might refer to "failure", nodes 10, 11 and 12 to "human error", "limit state" and "random hazard" respectively, and nodes 1, 2 and 3 to "wind loading (code) value found to be too low", "no consideration of progressive collapse in design" and "no consideration of explosive loading in design".

The discrimination algorithm considers each head node in turn and calculates, from the evidence of all the ESDs, which tail node is most indicative of that outcome. For example, in Fig.2, node 9 occurs only as a tail node of node 12 and nowhere else. This suggests that in any other future case the observation of 9 is highly indicative that 12 may occur. Node 4 is a tail node of both 11, in cases 1 and 2, and 10 in case 3. The future occurrence of 4 is therefore evidence for probably 11 but possibly 10.

The connectivity algorithm adopts a parallel approach. Each head node is considered and a search is made for groups of tail nodes which commonly precede it. In Fig.2, nodes 1 and 3 always occur together before node 10. The group of nodes (1,2,3) is also indicative of 10, but less strongly so since the three do not occur in each case.

For a more detailed discussion of these algorithms see Stone et al.[7].

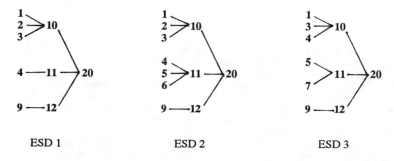

Fig.2 Three imaginary event sequence diagrams

6. Building the knowledge base

The tools of discrimination and connectivity now allow a hierarchical knowledge base to be formed. It is emphasised that the building is done interactively by a professional engineer making decisions based upon information provided by the computer and is not entirely automatic.

A connectivity analysis is performed over all the event sequence diagrams for each head node (ie each final and intermediate outcome). This presents the builder with a set of concepts which frequently occur together, and an associated support pair. Upon studying a set, the builder may decide that they constitute a new, 'higher level' concept, and therefore chooses to rename them as such. For example, if the concepts 'calculation error made in design', 'calculation error missed by checking engineer' and 'wrong drawing issued' are strongly connected, the builder may wish to define the new concept 'poor design office organisation' to represent them. When each connected set has been examined and renamed as appropriate, a new analysis is carried out where the old terms in each event sequence diagram are replaced by the new concepts. Further renaming and substitution is repeated until either no groups are found or until those that are do not suggest new concepts.

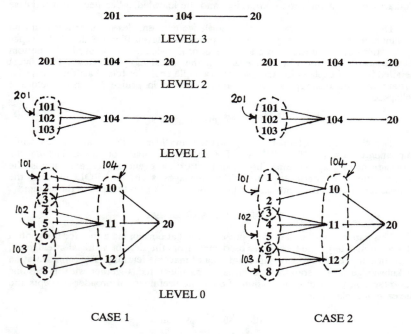

Fig.3 Hierarchical merging of concepts

Figure 3 shows in Venn Diagram form how the concepts are grouped to form new concepts at each hierarchical level. At each stage of the iteration, the builder may merge any of the newly formed ESDs which are considered to be sufficiently similar.

The check for similarity takes two ESDs in turn and proceeds from the final outcome concept to the initial concepts. If the same concepts are present in each ESD, the two can obviously be merged. If the necessary support for any concept present in one ESD and not the other is less than a builder-specified threshold then the two ESDs are sufficiently similar that they can be merged. This avoids the presence of concepts with low support from preventing the merging of ESDs which are broadly similar.

7. Consulting the knowledge base

To fulfill its' desired purpose as a management tool for the control of structural safety, the knowledge base, to be created from the 'learned' case histories as described, must be consulted by a user and offer advice. The typical user envisaged for the system is a project engineer faced with a new or current project and who wishes to assess its safety or proneness to failure. The advice provided by the system is therefore a measure of how closely the details of the new project match those in the knowledge base. The system will therefore request the user to assign supports to those concepts which she/he recognises as applying to the new project. The evidential support for the safety of the project will then be calculated by using a measure of similarity between the proposed project and the knowledge 'learned' from the case histories.

The hierarchical structure of the knowledge base enables the user to ask for an explanation of any answer given at an appropriate level. If more detailed, deeper explanations are required then a lower level of knowledge can be used. The bottom level concepts are the observed events in the actual case histories. At the most detailed level of explanation the user for example could be told that "this particular pattern of events is very similar to that observed in bridge X shortly before it collapsed".

8. Implementation

The 'learning' system is being written in 'Microsoft' C and is currently implemented on an IBM PC AT. Future developments proposed include the provision of of graphics capability in order to draw event sequence diagrams, the consideration of interfaces with other languages such as PROLOG, and the establishment of a sufficiently large number of case histories to input to the system.

9. Discussion and conclusions

This paper describes a current project aimed at developing a tool for the control of structural safety. The techniques used come from the field of artificial intelligence, in particular the increasingly important area of machine learning. The representation of knowledge in a hierarchical form has been illustrated, together with its ensuing advantages in terms of compression of data, the usefulness of broader concepts and the ease of explanation.

10. Acknowledgement

This research is funded by a grant from the U.K. Science and Engineering Research Council.

References

[1] Blockley,D.I.:"An A.I. tool in the control of structural safety ". In A.S.Nowak (Ed.) "Modelling human error in structural design and construction". New York: A.S.C.E., 1986.

[2] Turner,B.A.:"Man made disasters". London: Wykeham, 1978.

[3] Pidgeon,N.F., Blockley,D.I. & Turner,B.A.:"Site investigations - lessons from a late discovery of hazardous waste". The Structural Engineer, Vol.66, No.19, pp311-315, 1988.

[4] Blockley,D.I. & Baldwin,J.F.:"Uncertain inference in knowledge-based systems". Journal of Engineering Mechanics, ASCE, Vol.113, No.4, pp467-481, 1987.

[5] Blockley,D.I.:"Open world problems in structural reliability". Proc. 5th ICOSSAR, San Francisco, 1989.

[6] Norris,D., Pilsworth,B.W. & Baldwin,J.F.:"Medical diagnosis from patient records - a method using fuzzy discrimination and connectivity analyses". Fuzzy Sets and Systems. Vol.23, pp73-87, 1987.

[7] Stone,J.R., Blockley,D.I. & Pilsworth,B.W.:"Towards machine learning from case histories". (in press).

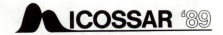
Qualitative Design for Safety and Reliability

Shuichi Fukuda*

*   Associate professor, Osaka University, Welding
    Research Institute, 11-1, Mihogaoka, Ibaraki,
    Osaka, 567, Japan

Abstract

This work aims at developing a fundamental
methodology for the qualitative reliability
design of machines and structures to cope with
the present situation of small production with a
wide variety of kinds in the hope of comple-
menting the present quantitative reliability
methods. This paper points out that if we note
the topological features of products, then we can
feed back the past experience especially of
manufacturing with great ease to the design stage
and improve the reliability. The fundamental
techniques using B-reps and Prolog to extact
features are described with several verifying
examples of a boom, a girder and bridges.

KEYWORDS

Design; feature; topology; B-reps; construction
machine; girder; bridge

1. Introduction

Safety and reliability become more and more important
recently with the rapid transition from mass to small production
and from a limited to a wide varieties. Thus, the number of
identical or similar products is quickly decreasing so that
higher reliability than ever before is required and reliability
is often stressed to distinguish a product from another in order

to add more value to it.

It is well known that shape is a most important factor for the function of a product. But it seems very few works have been carried out to correlate shape with reliability. This is considered because any particular attention to shape has not been necessary up to now.

But the number of production of large machines and structures has been and is very small no matter how the situation of production might change. In such cases, great efforts have been made to utilize the past experience. For example, welds are classified according to their shapes and necessary information to secure a good quality weld is added as attributes to each shape. And these pieces of knowledge are compiled as a standards or guidelines for welding. Thus, the welding engineer can apply the pieces of knowledge from the past experience to the production of large machines or structures which are different from the ones he constructed.

The situations are similar in these large scale machines and structures as well as in small scale ones because their requirements and operating conditions are getting more and more severe. Therefore some new methods are called for to complement the existing ones.

This work is an attempt to establish a methodology for the reliability design with special attention paid to the qualitative aspect, i.e., to the topological features and qualitative attributes attached to them. Thus, what differentiates this work from the conventional CAD is that we examine the topological aspects expressed in symbols, while the conventional CAD system discusses the geometrical features which fundamentally require numerical processing.

The reasons why we paid our attention to topology are

(1) To look at the manufacturing processes in terms of shape, welding may be said to be a technique to combine two topologies into one while cutting breaks one topology into two. And lathing and grinding are more concerned with geometrical manipulation. Thus if we are to consider problems arising from such topology manipulating manufacturing processes, we have to pursue another methodology which can deal with topology more easily.

(2) Mechanical machining can be carried out with high precision because the works remain solid throughout the whole process. But in such processes as welding where phase changes are involved pieces of knowledge on how to secure good quality are assembled more in terms of topologies than geometries.

This work is a preliminary attempt to establish a fundamental methodology for the qualitative reliability design noting the topological features of products.

2. Representation of topology model

As this is a preliminary work, we considered only machines and structures in the form of simple polyedra. But it can be extended to more complicated forms. In this work, B-reps is used to represent topologies. For example, the topology of the structure shown in Fig. 1 is represented using B-reps as in Fig.

2. Thus, each face is defined in terms of more than or equal to
one face_loop(s) and each face_loop is represented by directed
edges with its direction selected so that a right-handed screw
moves toward the exterior of the body. Multiple face_loops where
the number of face_loops is more than one occur when there is a
protrusion or a hole.
        The advantage of using B-reps is that we can easily extend
our topological discussion to geometrical one by adding the vertex
locations.
        We used Prolog because we can express topologies very
concisely and can attach attributes easily. Furthermore, the
inference mechanism embedded in Prolog fits very well for the
purpose. Fig. 3 shows the Prolog expression.

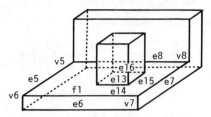

Fig.1 Sample structure

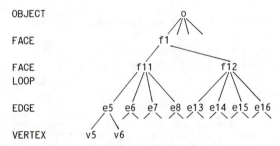

Fig.2 Boundary representation
of the sample structure

```
object(object_name:atom,face_name:list)
face(face_name:atom,face_loop_name:list)
flp(face_loop_name:atom,edge_name:list)
edge(edge_name:atom,initial_vertex:atom,
      terminal_vertex:atom,
      凸凹flag:{1 凸|-1 凹})
vertex(vertex_name:atom)
```

Fig.3 Prolog representation

3. Extraction of topological features

We will describe some of the fundamental techniques to
extract topological features based on such connectivity relations
between face and face, face and edge, etc.

3.1 The determination of concavity-convexity between
the adjacent faces

The most fundamental technique to extract topological
features is to determine whether adjacent faces are connected in
the concave manner or in the convex manner. This can be inferred
mathematically but in this work a flag is prepared for each edge
to show whether it is concave or convex.

3.2 Adjacent faces with the common edge

And edge is always shared by two faces. If we denote the
fact that an edge e is a component of a face f by $e \in f$, the
adjacent faces f_j and f_k which share an edge e_i can be expressed
as follows;

$$F_w(e_i) = \{(f_j, f_k) \mid (e_i \in f_j \wedge e_i \in f_k)\}$$
$$(J \neq k) \qquad \cdots\cdots\cdots\cdots\cdots\cdots (1)$$

3.3 A face adjacent to two faces

The face f_k adjacent to faces f_i and f_j is given as
follows.

$$F_c(f_i, f_j) = \{f_k \mid (e_i \in f_i) \wedge (e_i \in f_k) \wedge$$
$$(e_j \in f_j) \wedge (e_j \in f_k)\}$$
$$(i \neq j \neq k) \qquad \cdots\cdots\cdots\cdots\cdots\cdots (2)$$

3.4 Protrusion

Whether a protrusion exists or not on a certain face f is
determined in the following manner. If there is a protrusion or
a hole, the face f has multiple face loops. If the remaining
elements subsequent to the first one in the face loop list are
concave, then the body which contains this face loop is a
protrusion. A groove or a hole can be detected in the similar
manner.

4. Search by Prolog: an example

For example, the face F adjacent to faces F1 and F2
can be sought by using Prolog in the following manner.

```
common(F1,F2,F):-adjacent(F1,F), adjacent(F2,F).
adjacent(F1,F2):-face edge(F1,E1),!,member(E,E1),
            wing(E,F),member(F2,F),F2/==F1.
```

Other cases can be sought in the similar manner.

5. Sample problems

The present system consists of version 1 and 2. Example 1 and 2 show the cases solved by version 1 and the following 3 examples are solved by version 2.

[Example 1] Let us consider the boom of a construction machine. Suppose we wish to examine whether it is appropriate or not to attach a reinforcement to it in order to increase its rigidity (Fig. 4). The reinforcement can be detected as a protrusion on the upper face of the boom. If the protrusion or reinforcement is joined to the face by welding, then the corner welding is necessary, which has the large possibility of producing high local stress concentration as shown in the figure. If the loading is cyclic and fatigue is expected, these parts become dangerous. Therefore, we have to take such measures as to increase the rigidity by changing the geometry of the boom without attaching a reinforcement or by decreasing the stress concentration using TIG welding if we wish to attach a reinforcement.

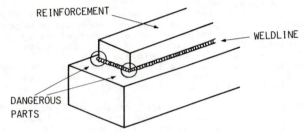

Fig.4 Boom of construction machine

[Example 2] Let us consider the design plans shown in Fig. 5 and 6. If the members are connected in such a manner shown in Fig. 5, the part shown in the figure becomes dangerous because here stress is concentrated and furthermore, three weldlines meet at this point thus creating a very complicated thermal history and starts and stops of weld procedures occur, so that a large deal of deterioration of weld quality is expected. But if we take the design plan of Fig. 6, then such problems can be avoided. Better still, robotization becomes far easier. The difference between Fig. 5 and 6 is how the transverse member is connected. Therefore, if a designer comes up with the Fig. 5 plan, then the computer suggests the change of his plan to the Fig. 6 one by changing the original topology into such that the transverse member becomes a bridge component.

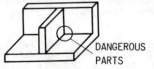
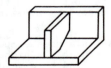

Fig.5 Original plan   Fig.6 Improved plan

Version 2 which can of course solve the above problems was
developed with special attention paid to the problem of
robotization.   The idea is that if we could change the structural
shapes, we could use less intelligent robots.   Nowadays we are
going more and more toward developing a highly intelligent robot
but such a system inevitably complex and has the possibility of
some unexpected factors of failures creep in.   But if we could use
less intelligent robots, we could not only save a large amount of
money but also we could utilize our past experience more
straightforwardly thus we could secure reliability with more ease.
Version 2 was developed chiefly with a bridge structure in mind.
In the case of a bridge, structural members can be classified
into open and closed sections.   So our system first classifies
the target structure into either of the two, applies such
following rules and change the topologies.

[rule 1] If the section is closed, avoid welding inside.   (This is
a problem of workability.)
[rule 2] To produce a good quality weld, horizontal position is
preferred for a welding robot.
[rule 3] If the weld position is the same, it is better if the
number of times of positioning is less and such works which call
for jigs are less.   (It requires a large amount of money to
design, manufacture and place jigs.   And jigs can be more often
that not used for special purposes alone.)
[rule 4] To maximize the arc time for a cycle of a robot motion.
[rule 5] To utilize the standard section as much as possible.   (It
is much easier to secure accuracy than fabrication.)

[Example 3] This is an example of application of rule 1.   The
system first confirms that the section is closed and extracts its
topological feature.   Thus it detects that welding is planned to
be done inside.   So it applies the rule 1 and places the ribs
outside instead of inside as shown in Fig. 7.

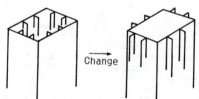

Fig.7 Change of rib locations

[Example 4] This is another example of a closed section.   If we
weld the original design plan in the flat position, we have to
position the work three times and if we weld it in the horizontal
position, we have to position it once.   But if we change the
topology, we can perform horizontal welding without changing its
position (Fig. 8).

[Example 5] This is a most complicated case.   If we wish to put
the quality of weld at the highest priority, we had better weld
the I-girder first and then weld the gusset.   Note that the
welding position is changed from the middle to the bottom flange

in Fig. 9.

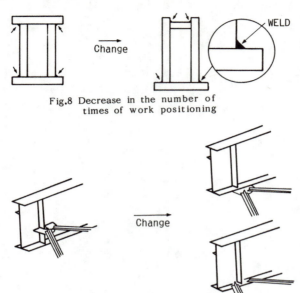

Fig.8 Decrease in the number of
times of work positioning

Fig.9 Improvement of worability

6. Summary

     In welding, there is a concept of welded joints which is none
other than extraction of partial geometry or topology to be exact
from the whole structure. Welding engineers classified welded
joints into T, corner, etc. according to their topologies and
attached necessary attributes such as the appropriate procedures
or conditions to each of them. Thus other engineers can find out
the proper solution using the concept of welded joints. This
concept also plays an important role for statistical analyses
since such classification provides a basis for setting up a
parent population.
     Therefore, what our system performs is a substitution of what
an engineer did by a computer. But up to now such qualitative
examination has been very difficult to be carried out by a
computer. The emerging new computers which provide better
environment for symbolic processing make this task possible at
the practical application level.
     There are two aspects to reliability. One can be discussed
with numbers while another can be more easily discussed with
symbols. This work is a preliminary step toward establishing a
methodology for qualitative reliability engineering which will
complement the present quantitative-oriented one.

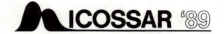
Application of Fuzzy-Bayesian Analysis
to Structural Reliability

Seiichi Itoh[1] and Hiroshi Itagaki[2]

[1] Researcher, Airframe Division, National Aerospace Laboratory
6-13-1 Osawa, Mitaka, Tokyo 181 JAPAN
[2] Prof., Dept. of Naval Architecture and Ocean Engineering,
Yokohama National University
156 Tokiwadai, Hodogaya-ku, Yokohama, Kanagawa 240 JAPAN

Abstract
    The purpose of this study is to develop a fuzzy-Bayesian
reliability approach for the decision of proper inspection
schedule to maintain structural integrity. According to this
approach, a priori uncertain parameters can be estimated from the
fuzzy information.

Introduction
    Somewhat subjective is said to exist in the results of
structural inspection since inspectors are forced to make
decision based on uncertain information and sometimes the
expression is not only subjective but also fuzzy. Therefore,
conventional statistical theories hardly make efficient use of
inspection results. The authors have been treating this problem
from the viewpoint of subjective (Bayesian) reliability theory
and obtained some useful conclusions(Shinozuka,Itagaki and Asada
1981). On applying Bayesian theory, two quite different
approaches are possible. The one is to construct a crisp
mechanical and stochastic model as completely as possible
introducing many uncertain parameters to be estimated. The other
is to use rather simple model and a few uncertainties are
introduced. The former may need data which are as exactly as
possible because the sophistication becomes meaningless,
otherwise. On the contrary the latter is intended to treat the
uncertainties as a whole so that precision of the measurement of
the individual parameter may have little effect on the final
decision. In such a case, the fuzzified Bayes theorem can be a
very powerful tool of analysis.
    The authors try to express the results of inspection in
terms of linguistic variables and apply the fuzzified Bayesian
analysis to obtain the posterior probabilities of the
uncertainties. Numerical examples are presented to compare the
effect of fuzzy information(FI) with that of objective one(OI).
    The Bayes theorem is given by the following equation, if

there are no fuzziness exists.

$$P(\theta|X) = P(X|\theta) \cdot P(\theta)/P(X) \quad \cdots\cdots\cdots\cdots\cdots\cdots\cdots\cdots\cdots\cdots\cdots\cdots \quad (1)$$

In this equation, θ is an unknown parameter, having a certain prior distribution $P(\theta)$. $P(\theta|X)$ is commonly known as a posterior probability of θ after the statistical event X has occurred.

However, the information obtained during inspections is not always well defined. Such information might be, for example, "A small crack is detected", or "A large crack is detected". The above equations cannot be used with such qualitative information. Therefore, eqs.(2) and (3) are used introducing a linguistic variable to express the fuzzy information. In the following, three events are considered, namely, "A small (medium, large) crack was detected". The corresponding membership functions are defined as shown in Fig.1. By definition, the conditional probability of a fuzzy event is given as

$$P(\underline{F}|\theta) = \sum_{x} \mu_F(x) \cdot P(x|\theta) \quad \cdots\cdots\cdots\cdots\cdots (2)$$

and the posterior probability

$$P(\theta|\underline{F}) = \sum_{x} \mu_F(x) \, P(x|\theta) \, P(\theta)/P(\underline{F}) \quad \cdots (3)$$

This is called the fuzzified Bayes theorem(Asai and Negoita 1978).

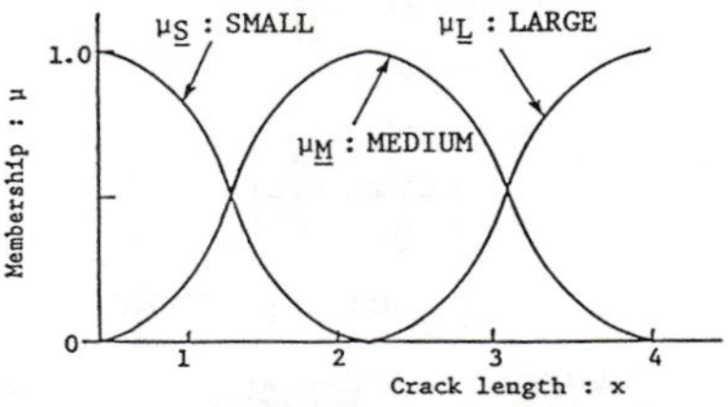

Fig.1 Membership functions for linguistic variables

Structural Inspection

1. Statistical parameters

In this report, attention is focused on fatigue damage. The statistical distribution of the time to crack initiation(TTCI) is assumed to follow a two-parameter Weibull distribution given by eq.(4).

$$f_{t_0}(t) = \alpha/\beta(t/\beta)^{\alpha-1} exp\{-(t/\beta)^{\alpha}\}, \; t \geq 0 \quad \cdots\cdots\cdots\cdots\cdots\cdots\cdots\cdots\cdots\cdots\cdots (4)$$

The size of crack initiates at t_0 is x_0. The probability of detecting a crack of length x is given by

$$\left\{ \begin{array}{l} D(x) = \{(x-x_0)/(x_B-x_A)\}^m \; : \; x_A \leq x \leq x_B, \quad D(x) = 0 \; : \; otherwise \\ \overline{D}(x) = 1 - D(x) \end{array} \right. \quad \cdots\cdots\cdots\cdots\cdots (5)$$

Evaluation of the fatigue crack propagation is done by applying the nonhomogeneous Poisson process model(Wen-Fang Wu 1986). According to this model, the transition probability that a crack of size x_1 at t_1 will propagate to x at t_2 is expressed as

$$q(x:t_2|x_1:t_1) = exp\{-M(x-x_1)\} \cdot \{M(x-x_1)\}^{t_2-t_1-1} \cdot I(x-x_1)/(t_2-t_1-1)! \quad \cdots\cdots\cdots (6)$$

in which $M(x)$ and $I(x)$ are defined by

$$\left\{ \begin{array}{l} M(x) = S \cdot exp\{-D(x_1-x_0)\} \cdot \{1-exp(-D \cdot x)\}, \\ I(x) = d\{M(x)\}/dx \end{array} \right. \quad \cdots\cdots\cdots\cdots\cdots\cdots\cdots\cdots\cdots\cdots (7)$$

where S and D are parameters which give the mean of crack propagation life.

2. Probability of inspection results

It is assumed that the structure consists of only one critical element which undergoes periodic inspections. The probability of outcomes is as follows. The distribution of fatigue crack size just before the i-th inspection is written by

$$h(x:t_i) = \int_{t_{i-1}}^{t_i} q(x:t_i-t|x_0:0) \cdot f_{t_0}(t) dt + \int_{x_0}^{\infty} q(x:t_i|y:t_{i-1}) \cdot \overline{h}(y:t_{i-1}) dy \quad \cdots\cdots\cdots (8)$$

in which $\overline{h}(x:t_{i-1})$ is the distribution of remaining crack size after the (i-1)-th inspection. Using above equations, the probabilities of events at the i-th inspection are

$$P_D(x:t_i) = h(x:t_i) \cdot D(x) \quad\text{...} \quad (9)$$

$$P_{ND}(t_i) = \int_x h(x:t_i) \cdot \overline{D}(x) dx + R_{t0}(t_i), \quad R_{t0}(t_i) = 1 - \int_0^{t_i} f_{t0}(t) dx \quad\text{.........} \quad (10)$$

$$\overline{h}(x:t_i) = h(x:t_i) \cdot \overline{D}(x) \quad\text{...} \quad (11)$$

where $P_D(x:t_i)$ is the probability density that a crack of size x is detected by the i-th inspection for the first time. $P_{ND}(t_i)$ is the probability that no crack is detected throughout inspections.

While these equations are for a crisp event of the detection of a certain size of crack, the probability of a fuzzy event, such as the detection of a small crack, is given

$$P_D(\underline{S}:t_i) = \int_x \mu_{\underline{S}}(x) \cdot h(x:t_i) \cdot D(x) dx \quad\text{...} \quad (12)$$

3. Parameter estimation

When a cracked element n in a structure is detected and replaced s times in the past, the probability of detecting a crack at the i-th inspection is given as,

$$L_n(i|\theta) = P_D(\underline{F}:T_1) \cdot P_D(\underline{F}:T_2-T_1) \cdot ,,, \cdot P_D(\underline{F}:t_i-T_s) \quad\text{............................} \quad (13)$$

where all of the events for crack detection are considered to be fuzzy. If any crack cannot be detected at the i-th inspection, the probability of not detecting a crack under the same inspection history as just mentioned is

$$L_n(i|\theta) = P_D(\underline{F}:T_1) \cdot P_D(\underline{F}:T_2-T_1) \cdot ,,, \cdot P_{ND}(t_i-T_s) \quad\text{............................} \quad (14)$$

For a fleet size N, the probability of outcomes of inspections is

$$L_{SYS}(i|\theta) = \prod_{n=1}^{N} L_n(i|\theta) \quad\text{...} \quad (15)$$

If a prior density function for uncertain parameters can be given, the posterior density function after the i-th inspection is derived from the fuzzified Bayes theorem Eq.(3).

$$g^{(i)}(\theta) = L_{SYS}(i|\theta) \cdot g^{(0)}(\theta) / \int_\theta (\text{Numerator}) d\theta \quad\text{...........................} \quad (16)$$

Numerical Examples and Discussion

The uncertainty on the scale parameter β of the probability density function(PDF) of TTCI and the parameter m of the NDI detectability are considered. Parameter values chosen for the numerical example are shown in Table 1 , in which the time in the model are normalized.

Figure 2 shows the posterior joint PDF of β and m after several inspections. Table 2 shows the coefficient of variation (COV) for the modal values. It can be observed that the concentration of probability at the mode in the joint PDF estimated from the FI is similar to that estimated from the OI. It is necessary to know or measure the exact crack size x for analysis with OI. On the other hand, the crack size can be evaluated in terms of verval expressions for the FI. Therefore, considering the inspection cost, it may be said that there are some cases where the FI is more effective than the precise OI.

Conclusion

Applying the fuzzified Bayes theorem, an example of

formulation to estimate the uncertain parameters is given. It is
demonstrated that the mode estimated from Bayesian and fuzzified
Bayesian analysis are not different from each other.

Acknowledgments
    The authors would
like to thank Prof. M.
Shinozuka, Princeton
University and Dr. H.
Asada, National Aero-
space Laboratory for
their helpful advice.

Table 1 List of parameter
          values

· Fatigue crack initiation	
Shape parameter : α	4.0
Scale parameter : β	
(to be estimated)	
True value (unknown) : β_T	1.0
Assumed range of β	
(Flat prior)	
Lower limit : β_L	0.5
Upper limit : β_U	1.5
· Fatigue crack propagation	
Initial crack length : x_0	0.4
Fail-safe crack length : x_C	4.0
Parameter of Poisson	
process model	
S	66.0
D	0.06
· NDI detectability	
X_A	0.4
X_B	4.5
m	
True value (unknown) : m_T	0.3
Assumed range of m	
(Flat prior)	
Lower limit : m_L	0.1
Upper limit : m_U	0.5
· Fleet size : N	25
· Inspection interval : ΔT_I	0.15
· Design life : T_D	1.5

References

Asai,K. and Negoita,C.V.
  (1978). *Introduction to
  fuzzy set theory.* OHM
  Press, Tokyo.
Shinozuka,M., Itagaki,H.,
  and Asada,H. (1981).
  "Reliability analysis of
  structures with latent
  cracks." *Proc.* the US-
  Japan cooperative
  seminar, Honolulu, 237-
  247.
Wen-Fang Wu. (1986). "On
  the markov approximation
  of fatigue crack growth."
  Probabilistic mechanics,
  1(4), 224-233.

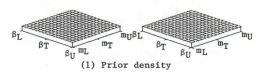

(1) Prior density

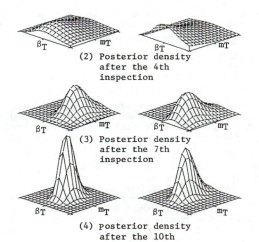

(2) Posterior density
    after the 4th
    inspection

(3) Posterior density
    after the 7th
    inspection

(4) posterior density
    after the 10th
    inspection

2.1 Objective   2.2 Fuzzy
    information   information

Fig.2 Example of prior and posterior joint
       probability densities

Table 2 Statistics of modal values for posterior
         joint probability density

	Inspec.	Mean		Standard d.		COV	
	No.	OI	FI	OI	FI	OI	FI
β	4	1.11	1.12	0.12	0.12	0.10	0.10
	7	1.09	1.09	0.07	0.08	0.07	0.07
	10	1.02	1.01	0.06	0.06	0.05	0.06
m	4	0.28	0.28	0.11	0.11	0.38	0.41
	7	0.28	0.28	0.07	0.09	0.26	0.31
	10	0.31	0.34	0.06	0.07	0.18	0.20

OI : Objective information
FI : Fuzzy information

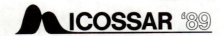
FUZZY RANDOM DYNAMICAL RELIABILITY ANALYSIS
OF ASEISMIC STRUCTURES

Jin-ping OU and Guang-yuan WANG

Research Institute of Structural Science
Harbin Architectural and Civil Engineering Institute
Harbin, China

ABSTRACT

In our previous papers[1, 2], by taking account of the
fuzziness in the earthquake excitation, a fuzzy random model
of earthquake ground motion has been established and a method
of fuzzy random response analysis of aseismic structure has
been presented. In this paper, the fuzziness of the damage
grades is further taken into account and the fuzzy safe criterion is presented. Then, by making use of the results of
fuzzy random seismic response analysis, the fuzzy random
dynamical reliabilities of a SDOF hysteretic structure, and
a MDOF hysteretic structure subjected to earthquake are
researched and their concrete formulas are derived under some
proper assumptions.

KEYWORDS

Fuzzy random response, fuzzy safe criterion, fuzzy random
dynamical reliability.

1. FUZZY SAFE CRITERION OF ASEISMIC STRUCTURES

The dynamical reliability analysis of aseismic structures is the problem to
find the probability of the random response without exceeding the safe boundary
in the duration [0, T] of earthquake. At present, this boundary is considered
as a clear-cut line, which is very unreasonable for many practical problems.
In the earthquake engineering, the damage grades of a structure are often
devided into the following five grades; slight, moderate, severe, destructive
and collapsed. The definitions of these damage grades evidently possess strong
fuzziness and can not use clear-cut values of structural response as the boundaries between these grades. Therefore, if the response X(t) of a structure,
which may be the displacement, stress or cumulative energy response, is taken to
define the damage grades of the structure and the damage grades are divided into

$$\{\underset{\sim}{B_1}, \underset{\sim}{B_2}, \underset{\sim}{B_3}, \underset{\sim}{B_4}, \underset{\sim}{B_5}\} = \{\text{slight, moderate, severe, destructive, collapsed}\} \quad (1)$$

then every damage grade $\underset{\sim}{B_i}$ (i=1, 2, ..., 5) should be a fuzzy subset on the value region $R^1 = (-\infty, \infty)$ of the response X(t). The membership functions of $\underset{\sim}{B_i}$ (i= 1, 2, ..., 5) have the characters as shown in Fig.1.

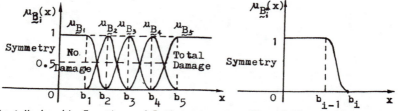

Fig.3 Membership Functions of $\underset{\sim}{B_i}$ (i=1, 2, ..., 5)   Fig.2 Membership Function of $\underset{\sim}{B_i}$

Let $\underset{\sim}{B_i^*}$ represent the fuzzy safe region in which $\underset{\sim}{B_i}$ or more severe damage will not occur for the structure, its membership function curve should have the form as shown in Fig.2 and may be expressed by[3]

$$\mu_{\underset{\sim}{B_i^*}}(x) = \begin{cases} 1 & (x \leqslant b_{i-1}) \\ 0.5\{1 - \sin[(x - b_{i-1})/(b_i - b_{i-1}) - 0.5]\pi\} & (b_{i-1} < x \leqslant b_i) \\ 0 & (x > b_i) \end{cases} \quad (2)$$

Obviously, when the earthquake response X(t) of a structure is fallen into the fuzzy safe region $\underset{\sim}{B_i^*}$, the structure will not suffer $\underset{\sim}{B_i}$ or more severe damage to certain extent. Thus, this safe criterion can be expressed as

$$\underset{\sim}{\Theta} = \{ X(t) \underset{\sim}{\subseteq} \underset{\sim}{B_i^*}, \ 0 \leqslant t \leqslant T \} \tag{3}$$

This is a fuzzy safe criterion, in which the response X(t) may be random or fuzzy random one. The probability that the respose satisfies this criterion is defined as the fuzzy random dynamical reliability of the structure since not only the randomness but also the fuzziness are taken into account.

2. FUZZY RANDOM DYNAMICAL RELIABILITY OF A SDOF HYSTERETIC STRUCTURE

In this section, we consider the SDOF hysteretic structure under earthquake which fuzzy random response has been analyzed in Ref.[1]. The displacement response $Y(\xi, t)$ of the structure is a random process with a fuzzy parameter ξ, in which ξ is a fuzzy variable relative to the fuzzy earthquake intensity and its possibility distribution funtion can be expressed as

$$\pi_{\underset{\sim}{\xi}}(u) = 0.5\sin(1.4427 \ln u - I + 1.178)\pi + 0.5, \qquad u \in [f(I-1), f(I+1)] = [u_1, u_*] \tag{4}$$

where I=1, 2, ..., 12, is the fuzzy intensity ordinal number and f is the relationship function[1].
If the displacement response is taken as the control response of the structure safety, then the fuzzy random dynamical reliability that the structure will not suffer $\underset{\sim}{B_i}$ or more severe damage can be defined as

$$P_*(\xi, \underset{\sim}{B_i}, T) = P\{\underset{\sim}{\Theta}\} = P\{Y(\xi, t) \underset{\sim}{\subseteq} \underset{\sim}{B_i^*}, \ 0 \leqslant t \leqslant T\} \tag{5}$$

For any a F-sample process Y(u, t) of $Y(\xi, t)$, obviously

$$P_*(u, \underset{\sim}{B_i}, T) = P\{Y(u, t) \underset{\sim}{\subseteq} \underset{\sim}{B_i^*}, \ 0 \leqslant t \leqslant T\} = P\{Y_*(u) \underset{\sim}{\subseteq} \underset{\sim}{B_i^*}\} \tag{6}$$

in which

$$Y_* = \max_{t \in T} | Y(u, t) | \tag{7}$$

Thus, according to the probability formula of a fuzzy random event, we have

$$P_{\bullet}(u, \underline{B}_{\iota}, T) = \int_{0}^{b_i} p_{v}(u, y) \mu_{\underline{B}_{i}}(y) dy \qquad (8)$$

in which $p_{v}(u, y)$ is the probability density function of $Y_{\bullet}(u)$. For any a fixed u, by making use of the theory of random dynamical reliability, the probability distribution function of $Y_{\bullet}(u)$ can obtained as

$$P_{v}(u, y) = \exp\{-\sigma_{\dot{v}}/(\pi \sigma_{v})\exp[-y^2/(2\sigma_{v})T]\} \qquad (9)$$

in which σ_{v} and $\sigma_{\dot{v}}$ are respectively the stationary standard variances of $Y(u, t)$ and $\dot{Y}(u, t)$, which have been obtained in Ref.[1]. Substituting Eqs.(2) and the derivative of $P_{v}(u, y)$ for y into Eq.(8), we obtain

$$P_{\bullet}(u, \underline{B}_{\iota}, T) = 0.5 [P_{v}(u, b_{\iota})+P_{v}(u, b_{\iota-\iota})-2P_{v}(u, 0)-D_{\iota}(u)] \qquad (10)$$

in which
$$D_{\iota}(u) = \int_{b_{i-1}}^{b_i} p_{v}(u, y)\sin[(y-b_{\iota-\iota})/(b_{\iota}-b_{\iota-\iota})-0.5] \pi \, dy \qquad (11)$$

Substituting ξ for u in Eq.(10), the fuzzy random dynamical reliability $P_{\bullet}(\xi, \underline{B}_{\iota}, T)$ can be obtained, which is the function of the fuzzy variable ξ. Taking F-expectation for $P_{\bullet}(\xi, \underline{B}_{\iota}, T)$, the corresponding dynamical reliability in nonfuzzy form can be obtained, i.e.

$$P_{\bullet}(\underline{B}_{\iota}, T) = E_{v}[P_{\bullet}(\xi, \underline{B}_{\iota}, T)] \qquad (12)$$

in which $E_{v}[\cdot]$ is the F-expectation which definition can be seen in Ref.[4].

Since it is difficult to obtain the analytical expression of F-expectation $P_{\bullet}(\underline{B}_{\iota}, T)$, by evenly taking J F-samples of ξ in the F-sample set $G=\{u \mid \pi_{\xi}(u)>0\}$, the F-expectation can be approximately obtained as

$$P_{\bullet}(\underline{B}_{\iota}, T) = \sum_{j=1}^{J} P_{\bullet}(u_{\iota}, \underline{B}_{\iota}, T)\pi_{\xi}(u_{\iota}) / \sum_{j=1}^{J} \pi_{\xi}(u_{\iota}) \qquad (13)$$

$P_{\bullet}(\underline{B}_{\iota}, T)$ is a nonfuzzy value, which can be used in the similar way to the ordinary dynamical reliability.

3. FUZZY RANDOM DYNAMICAL RELIABILITY OF A MDOF HYSTERETIC STRUCTURE

In this section, we consider the MDOF hysteretic shear structure under earthquake which fuzzy random response has been analyzed in Ref.[2]. The relative displacement responses of the structure can be expressed as a random process vector with the fuzzy parameter ξ, i.e. $\bar{Y}(\xi, t)=(Y_{\iota}(\xi, T), Y_{\bullet}(\xi, t), \ldots, Y_{\bullet}(\xi, t))^{\tau}$, and their statistical moments have been obtained in Ref.[2].

3.1 Fuzzy Random Dynamical Reliability of the Structural Storeys

If every storey of the shere structure is regarded as the member of the structure, the fuzzy random dynamical reliabilities of the structural storeys are those of the members, which may be used in the probabilitical design of the structural members.

Assuming the relative displacement $Y_{\bullet}(n=1, 2, \ldots, N)$ is taken as the control response of the nth storey safety, the fuzzy safe region $B_{\iota n}$ that the nth storey will not suffer $\underline{B}_{\iota}$ or more severe damage can be determined similar to B_{ι} and its membership function can be also expressed by Eq.(2) only by substituting in for i in Eq.(2). Therefore, the definition, analysis and concrete formula of the fuzzy random dynamical reliability that the nth storey will not suffer B_{ι} or more severe damage is completely similar to the section 2.

3.2 Fuzzy Random Dynamical Reliability of the Structural System

The fuzzy random dynamical reliability of the structural system depends on

the requirement for the system safety. If any storey suffers $\underline{B}_i$ or more severe damage, the system will regarded as suffering this kind of damage. Thus, the fuzzy random dynamical reliability that the system will not suffer $\underline{B}_i$ or more severe damage can be defined as

$$P_s(\xi, \underline{B}_i, T) = P\{\bigcap_{n=1}^{N} [Y_n(\xi, t) \subseteq \underline{B}_{in}^*, 0 \leqslant t \leqslant T)\} \tag{14}$$

In the following, we will find its approximate formulas under some assumptions.
1) The weakest link assumption [5,6]   If it is assumed that the event for every storey to suffer the damage is independent, Eq.(14) will become

$$P_s(\xi, \underline{B}_i, T) = \prod_{n=1}^{N} P_s^{(n)}(\xi, \underline{B}_{in}, T) \tag{15}$$

in which $P_s^{(n)}(\xi, \underline{B}_{in}, T)$ is the fuzzy random dynamical reliability of the nth storey.
2) The weakest element assumption [5]   If in the structure there is such a weakest storey, denoted as n^*th storey, that the n^*th storey will first suffer the damage (i.e. $\underline{B}_i$ or more severe damage), or if any other storey suffer the damage, the n^*th storey must suffer the damage, then refering to Ref.[5] and according to Eq.(14), it can be derived that

$$P_s(\xi, \underline{B}_i, T) = P_s^{(n^*)}(\xi, \underline{B}_{in^*}, T) = \min_{n \in \bar{N}} P_s^{(n)}(\xi, \underline{B}_{in}, T) \tag{16}$$

in which $\bar{N} = \{1, 2, \ldots, N\}$ is the index set of storey numbers.
    After the fuzzy random reliability that of system is obtained by Eq.(15) or (16) according to the different practical cases, the corresponding F-expectation can be obtained by Eq.(13).

CONCLUSIONS

    The damage grades of aseismic structures expressed by the fuzzy subsets on the response region are reasonable and by using them we can well determine the fuzzy safe regions that structural members and system will not suffer a certain grade or more severe damage. Therefore, the fuzzy random dynamical reliability and their F-expectations of a SDOF structure and of the members and system of a MDOF structure subjected to the earthquake corresponding to any fuzzy intensity can be easily obtained by making use of the results of the fuzzy random seismic response analysis and under some proper assumptions. These results offer a good estimation for the safeties of the structures corresponding to the different safe grades.

ACKNOWLEDGMENTS

    This paper is a part of the project supported by the Educational Foundation of Huo Ying-dong.

REFERENCES

[1] Wang Guang-yuan and Ou Jin-ping,"Fuzzy Random Vibration of Structures Subjected to Earthquake", Proc. Int. Symp. on Fuzzy Math. in Earthquake Res. Vol.1, pp.205-216, Seismological Press, 1985.
[2] Wang Guang-yuan and Ou Jin-ping,"Fuzzy Random Vibration of MDOF Hysteretic Systems Subjected to Earthquake", Earthquake Eng. and Struct. Dynamics, Vol.15, No.5, pp.539-548, 1987.
[3] Ou Jin-ping and Wang Guang-yuan,"Dynamical Reliability Analysis of Aseismic Structures Based on Fuzzy Damage Criterion", Earthquake Eng. and Eng. Vibration, Vol.6, No.1, pp.1-11, 1986, (in Chinese).
[4] Wang Guang-yuan and Ou Jin-ping,"Fuzzy Random Vibration of Aseismic Structures", J. of Vibration and Shock, Vol.6, No.2, pp.23-30, 1988, (in Chinese).
[5] Ou Jin-ping,"Dynamical reliability Analysis of MDOF Systems", Earthquake Eng. and Eng. Vibration, Vol.7, No.4, pp.1-12, 1987, (in Chinese).
[6] Moses, M.,"Reliability of Structural System", J. of Struct. Div., ASCE, Vol.100, No.9, pp.1813-1820, 1974.

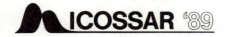
KNOWLEDGE REPRESENTATION AND REASONING IN STRUCTURAL RELIABILITY ASSESSMENT EXPERT SYSTEMS:
Implementation in the Leak-Before-Break Prototype System

A. Jovanović

MPA - Univ. Stuttgart, Pfaffenwaldring 32, 7000 Stuttgart 80, FR Germany

Abstract

Knowledge reperesentation and reasoning, considered of a major importance for building and practical application of the structural reliabilty assessment expert systems, are the issues tackled in the research described in the paper. The concept developed in the research allows succesful introduction and use of standard and/or heuristic engineering knowledge (procedures, code calculations, etc.), of preceding operational experience, of testing results and experimental evidence, etc., into expert systems. The concept has been applied in the prototype expert system for the leak-before-break analysis, where the results of the over 50 large scale MPA-tests have been taken as a basis for knowledge elicitation and verification of the system.

Introduction

Several important subdomains of structural safety and reliability entail strong involvement of heuristics: e.g. interpretation of nondestructive examination and/or material testing results, pressurized thermal shock analysis [1], treatment of uncertainties, corrosion influence assessment, Leak-before-Break (LBB) analysis, etc. Numerical analysis alone, even when very complex, provides just a part of the total information necessary for reaching a valuable expertise, judgement and decision in the application domain. The expertise is necessarily linked to active role of humans, which cannot be avoided. However, current state of the art in the field of knowledge engineering (KE) and expert systems (ES), can offer a substantial complement to the "conventional" decision basis for solving some of the domain-cases listed above. The case tackled here is the LBB-analysis, where the term "leak-before-break" [2] describes behaviour of a pressurized component (pipe, vessel, etc.) during an actual or hypothesized failure caused by fatigue crack growth. Namely, LBB means that, once the crack has grown enough to penetrate the wall, only a leak occurs, leaving, under all possible circumstances, a substantial margin (e.g. in terms of safety and time for the component replacement/repair) between the

leak occurance and final rupture of the component caused by the fact that the crack has reached the critical size ("break"). LBB is essential for inherent safety of pressurized systems and components (e.g. those in power plants), because hazards and technical remedies related to leak usually differ drastically from those of a break.

Design and practical development of the LBB expert system has been strongly related to other research efforts in the field of ES/KE applications in structural safety and integrity assessment, at the State Materials Testing Institute - MPA, Stuttgart. Knowledge elicitation, knowledge representation and reasoning resulted to be very important for this research, so the work on them lead to development and application of a new and original concept.

The Concept

Several methodologies enable numerical analysis of the LBB behaviour (see, for instance, [2]). However, many of them tend to be very conservative and/or with large uncertainty margins. Applied to the same data base as the one used for the development of the LBB expert system, the methods and the data with which they have been applied, resulted in imprecision of -30 to +40%, when predicting the critical bending moment. They have also entailed a relatively high degree of uncertainty in qualitative prediction of test outcomes (break / leak), confirming thus that the numerical methods cannot encompass all the factors relevant for the structural failure, such as toughness, corrosion influence, stiffness of piping, nature of crack, temperature influence, etc.

The adopted concept is based on analysis of analogy between the given case and the cases in the knowledge base (KB). Numerical and non-numerical information from tests and operation (currently about 50 tests [3]) are stored in the KB. Analogy in single factors (temperature, loading conditions, material toughness, etc.) is combined with results of the numerical analysis [2]. Then, the relevant cases (if any!) are extracted form the knowledge base. "Relevant" are either those cases showing very high analogy with the analyzed one, or those being totaly different. The prediction *leak* or *break* is made accordingly.

Knowledge Elicitation and Representation

Three methods have been used in knowledge elicitation: organized inquiry, analysis of test reports and direct interviewing of the domain experts. The inquiry, perfomed among the engineers of MPA and other institutions and companies (46 participants) [4], provided information regarding possibility distributions of numerical and linguistic variables used in the LBB analysis. Analysis of test reports gave the basic information (both numerical and qualitative) about the tests. Interviewing has been done with the MPA experts only, and its results have been used mainly for confirmation/redefinition of the heuristic rules (used in the ES) derived from analysis of reports.

In the Lisp version of the ES, described here, the knowledge base is organized as a series of lists, containing both numerical and qualitative data. Numerical data taken into account are, e.g., geometry of the component, crack

dimensions, working temperature, loads, material characteristics, etc. Qualitative information regards type of the crack (inside/outside, fatigue crack, welding defect,etc.), stiffness of the construction, etc. Rules, in this version of ES, are purely heuristic, e.g. *"when the crack is deep, then the chance of the failure mode leak is high"*. The terms like *deep* and *very high chance* are defined as trapeziodal fuzzy numbers [5], the parameters of which are derived from the inquiry [4]. Technically, the solution is designed as an open architecture, shell free, intelligent module, interacting with the user in the search of the appropriate representation of various pieces of knowledge and in determination of its importance for the final conclusion.

Reasoning by Analogy

The main task of the reasoning process is the intelligent recognition of analogy/similarity between the analyzed case and the cases in the knowledge base. To each piece of knowledge regarding the influencing factors, is attributed a weighting factor, the initial value of which is provided by the domain expert, but the value of which can be reviewed interactively. Analogy between numerical data is defined as a combination of the weighting factors and the algebraic ratio/difference. Obviously, the operating temperature of 120°C is more similar to 150°C than to 30°C . Analogy in qualitative data is more difficult to define. So far, the issue in the LBB expert system is tackled either by direct specification of the granularity of the descriptors (*very corrosive* is more similar to *corrosive*, than to *neutral*), or by assigning membership functions to the descriptors (estimators) [4,5]. Thus, the analogy/similarity can be "quantified" and, later on, compared. Search for analogues is still in the way thath underline all cases in the knowledge base are compared with the actual one. This issue, however, has to be improved further on, as the number of cases in the knowledge base is expected to increase.

Reasoning has two levels: the one of the pre-established production rules and the one of the self-generated and "implicit" rules. Its main result is the diagnosis/prediction of the structural state, in terms of *leak, break, no failure* and *prediction impossible* (not any). In addition, semantic (linguistic) probability estimators (e.g. *very high chance, meaningful chance, it may be,* etc.) are assigned to ES outcomes, accordingly to how strong the analogy between the analyzed and the reference case(s) was. Transition between the semantic and numerical probabilities is obtained by means of fuzzy algebra [5]. Precision of the system can be tuned through interactive definition of the analogy significance limits. Thus, if a high certainty of answers is required only "sure" analogies will be identified, with the consequence that the probability of not finding the significant analogues increases (Tables 1 and 2).

Results: Application and Verification of the Concept

The concept has been first verified on the MPA LBB full-scale tests [3] (internally pressurized pipes, $\phi = 800 \ mm$, exposed to bending and containing a circumferential defect). Test results, versus results of the ES-based analysis

are given in Tables 1 and 2, showing that optimization of the weighting factors
and narrowing of analogy significance limits lead to higher precision of predictions, but contemporary, to reduction of number of cases in which the system
provides an answer (Table 2).

Table 1: LBB expert system results, for
unoptimized weighting factors and the
analogy significance limits 0.3 and 0.7

Table 2: LBB expert system results, for
optimized weighting factors and the
analogy significance limits 0.1 and 0.9

Test Nr.	Test result	ES prediction	Probability descriptor
BVZ110	LEAK	BREAK	meaningf.chance
BVZ116	LEAK	LEAK	meaningf.chance
BVZ130	LEAK	LEAK	meaningf.chance
BVZ140	LEAK	BREAK	meaningf.chance
BVZ150	BREAK	BREAK	meaningf.chance
BVZ040	BREAK	BREAK	meaningf.chance
BVZ050	BREAK	LEAK	it may be
BVS060	BREAK	BREAK	meaningf.chance
BVS070	LEAK	LEAK	meaningf.chance
BVS080	BREAK	LEAK	meaningf.chance
BVS102	BREAK	LEAK	meaningf.chance
BVS050	BREAK	BREAK	meaningf.chance

Nr. of predictions vs. analyzed cases: 12/12

Exact predictions   58%

Test Nr.	Test result	ES prediction	Probability descriptor
BVZ110	LEAK	not any	—
BVZ116	LEAK	LEAK	meaningf.chance
BVZ130	LEAK	not any	—
BVZ140	LEAK	not any	—
BVZ150	BREAK	not any	—
BVZ040	BREAK	BREAK	meaningf.chance
BVZ050	BREAK	not any	—
BVS060	BREAK	BREAK	meaningf.chance
BVS070	LEAK	not any	—
BVS080	BREAK	not any	—
BVS102	BREAK	not any	—
BVS050	BREAK	not any	—

Nr. of predictions vs. analyzed cases: 3/12

Exact predictions   100%

Conclusions

The LBB expert system has been designed and developed as a practice
oriented, KE-based engineering tool. Although simple, its first results appear
to be good, and allow to expect that, with future developments, they will be
further improved

References

[1] OKAMURA,H., YAGAWA, G.: An expert-interactive algorithm of structural
     integrity evaluation and its application to thermal shock, in SMiRT - Advances
     1987, ed. F.H.Wittmann, Balkema, Rotterdam, Boston, 1987
[2] MUNZ, D.: Development of a leak-before-break methodolgy, Ibid
[3] STOPPLER, W., SCHIEDERMEIER, J., HIPPELEIN, K., STURM, D., SHEN,
     S.: Bruchverhalten von Rohren unter Innendruck mit gleichzeitig wirkendem äuß-
     erem Biegemoment, 13th MPA Seminar, Stuttgart, 1987
[4] JOVANOVIC, A., HASSLER, M.: Vorläufige Ergebnisse der Umfrage - Anwen-
     dung der Methoden des Knowledge Engineering in der Struktursicherheitsanalyse,
     Zwischenbericht, MPA Stuttgart, 1989
[5] JOVANOVIC, A., SERVIDA, A., SAUTER, A.: Application of fuzzy algebra for
     treatment of uncertainties in the structural analysis of pressure vessels exposed
     to thermal shock. Proc. 4th SAS-World Conf. - FEMCAD, Vol. 2, Paris, Oct.
     1988

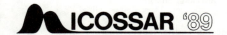 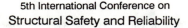

LIFETIME-MAXIMUM LOAD EFFECT FOR HIGHWAY BRIDGES BASED ON
STOCHASTIC COMBINATION OF TYPICAL TRAFFIC LOADINGS

Hiroyuki Kameda* and Masakuni Kubo**

* Professor, Disaster Prevention Research Institute,
  Kyoto University, Gokasho, Uji, Kyoto 611, Japan
** Manager of Design Section, SOGO Engineering Inc., 3-5-9,
  Higashinakajima, Higashiyodogawa-ku, Osaka 533, Japan

ABSTRACT

An efficient and accurate method is proposed for probabilistic
assessment of lifetime maximum load effect on highway bridges
caused by traffic load. Probability distribution of the lifetime maximum girder response is derived in terms of a convolution from the
"main loading" and "subsidiary loadings", the former being the
effect of a heavy vehicle loaded immediately at the evaluated girder
section, and the latter being the effect of a group of vehicles loaded
at other locations along the same lane and neighboring lanes. It is
demonstrated that the present method provides much higher computational efficiency than a recent improved method of direct
Monte-Carlo simulation technique. It is also shown that the present
method can be used easily with analytical load models, which allows
one to develop theoretical insight of the probability distribution of
lifetime maximum load.

KEYWORDS

highway bridges; vehicle loads; lifetime-maximum load; probability
convolution; main loading; subsidiary loading; Monte-Carlo simulation; exponential idealization

1. INTRODUCTION

1.1 Overview

Rational assessment of the effect of vehicle loads is a major issue in the design of
highway bridges [1]-[5]. The lifetime maximum load should be assessed in a reasonable
manner. The vehicle load effect on main girders are observed as combined contributions

from individual vehicles typically observed during traffic congestions, which will constitute a long train of vehicles moving slowly and continuously as illustrated in Fig.-1. In the previous studies, Monte-Carlo simulation technique has been used extensively. Particularly, the simulation technique developed by Fujino et al.[6] is efficient in the sense that it is in principle an application of importance sampling concept.

It should be noted, however, that even efficient improved Monte-Carlo technique requires a tremendous amount of computational effort when lifetime, say 50 years, maximum load is involved. It is clear also that Monte-Carlo technique is an experimental method which often prohibits one to develop a theoretical insight into the obtained results in return to its capability of incorporating realistic complex models. It is, therefore, desirable to develop a method that can take advantage both of the generality of theoretical analysis and the flexibility of Monte-Carlo technique.

The present paper is an output from the efforts made in such a direction. The method developed herein deals with the effect of major heavy vehicles, called "main loading", in terms of a single probability distribution, and the effects of other vehicles, called "subsidiary loading", which are evaluated from Monte-Carlo simulation. Then the maximum girder response is obtained from the convolution of these two distributions. In this way, high computational efficiency of Monte-Carlo simulation is realized.

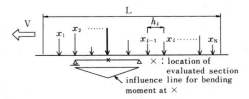

Fig.-1 Schematic Presentation of Vehicle Loads over a Single Span Girder

By applying exponential distribution for idealization, the entire procedure can be formulated analytically in a closed from. This helps greatly develop theoretical understanding of the characteristics of the lifetime maximum load on a sound physical basis.

1.2 Traffic Load Models and Structural Models

The traffic load models used in this study are based on the activities by the Committee on Design Load Development for Hanshin (Osaka-Kobe) Expressways (HDL Committee) [7]-[8]. For the details of the model development, see the indicated references. Fig.-2 outlines the vehicle classifications, the range of vehicle weight for each category, and their proportions relative to the total traffic volume. Table-1 shows the models of the occurrence of traffic congestions.

Single girder bridges with cross section shown in Fig.-3 are dealt with throughout this study. It is assumed that the lateral load distribution for each plate-girder is uniform along the span and is given in Fig.-3. Under the Japanese highway bridge code, the design mid-span bending moment M^* assumes the values indicated in Table-2. In the following, the response to vehicle load, or load effect, will be represented by the mid-span bending moment M_o caused by the vehicle load considered in the analysis as normalized by M^*; i.e., the load effect will be represented by the moment ratio expressed as

$$Y_o = M_o / M^* \qquad (1)$$

Table-1 Modeling of Traffic Congestions

	traffic condition	
	case-I	case-II
type of traffic congestion	natural	accidental
occurrence rate No (1/month)	50	5
total length L (km)	50	1
vehicle speed V (km/hr)	20	1
traffic volume in a congestion N	3,636	92

Table-2 Design Bending Moment M^* under the Japanese Highway Bridge Code

	Girder No.	span-length l (m)			
		20	40	60	80
M^* (tf/m)	G_0	-	286.0	-	-
	G_1	163.8	439.6	823.6	1314.1
	G_2	-	409.9	-	-
	G_3	-	423.7	-	-

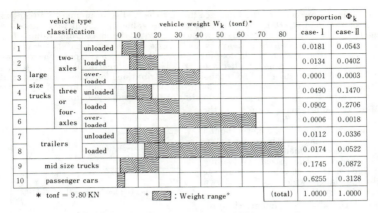

Fig.-2 Vehicle Load Models (See Ref.[7]-[8] for details)

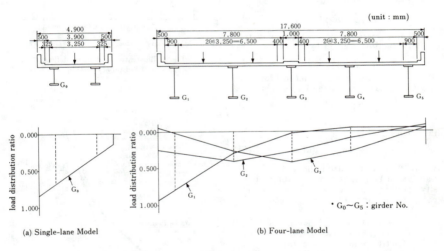

Fig.-3 Structural Model of Girder Section

2. MAXIMUM GIRDER RESPONSE BY CONVOLUTION OF MAIN LOADING AND SUBSIDIARY LOADING

2.1 Theoretical Basis

Let X_i represent the vehicle weights acting on a bridge. Under this condition, the moment ratio Y_o is given by

$$Y_o = \sum_{(i)} \xi_i X_i / M^* \tag{2}$$

in which ξ_i is the ordinate of the influence line for the mid-span moment corresponding to position of the load X_i. It is clear that an extreme value of Y_o will take place when an extremely heavy vehicle is loaded at a position with the maximum value of ξ_i. This corresponds to a case where a heavy vehicle rests at mid-span as shown in Fig.-4(a). Herein this state is called "main loading". The other vehicles expect the one used for main loading will have smaller effects on Y_o. They are illustrated in Fig.-4(b), and is called "subsidiary loading". The probability distribution for these loading states as sketched in Fig.-4 are represented by the probability density functions of Y_o corresponding to the respective states; $g_{kx}(y)$: main loading, $h_k(y)$: subsidiary loading.

Here the subscript k stands for the vehicle category used for main loading. The notation x denotes the lower bound X_c on main loading and is to be determined for the convenience of analysis. It is also noted that the subsidiary loading can include a train of vehicles along the neighboring lanes which is not shown explicitly in Fig.-4(b).

From the above argument, the probability density function of Y_o at an arbitrary instant is represented by the following convolution integral.

$$f_{kx}(y) = \int_0^\infty g_{kx}(y-t)\, h_k(t)\, dt \qquad (3)$$

The corresponding cumulative distribution function is given by

$$F_{kx}(y) = \int_0^y f_{kx}(u)\, du \qquad (4)$$

Then the distribution function for the monthly maximum value of Y_o yields

$$F_z(y) = \prod_{(k)} [\, F_{kx}(y)\,]^{n_{kx}} \qquad (5)$$

in which

$$n_{kx} = N_o\, N\, \Phi_k \int_z^\infty q_k(u)\, du \qquad (6)$$

where N_o, N and Φ_k are indicated in Fig.-2 and Table-1. $q_k(u)$ is the probability density function of the vehicle weight for category k, and has been specified in Ref.[7].

Whereas the probability distribution $g_{kx}(y)$ can be derived easily from $q_k(y)$, the only effective method to obtain $h_k(y)$ is Monte-Carlo simulation. However, it is noted that the major contribution from $h_k(y)$ to $f_{kx}(y)$ is small, and its tail characteristics do not influence the result. This allows one to evaluate $h_k(y)$ from a relatively small number of simulations, which saves much of computational efforts with a satisfactory accuracy. The values for $g_{kx}(y)$ and $h_k(y)$ have been obtained, but omitted in this paper due to the limitation of space.

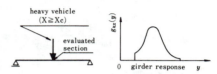

(a) Main Loading by a Heavy Vehicle

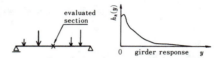

(b) Subsidiary Loading by a Random Vehicle Train

Fig.-4 Illustration of Main Loading and Subsidiary Loading

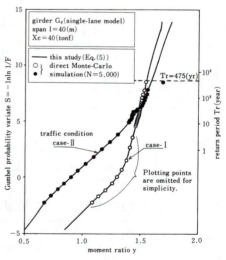

Fig.-5 Probability Distribution of Monthly Maximum Load Effect (Sigle-lane Model)

2.2 Girder Response for a Single-lane Model

Fig.-5 shows the probability distribution of monthly maximum girder response for a case of a single-lane model.

The analytical value from Eq.(5) is plotted along with the results of direct Monte-Carlo simulation. Observe that the analytical results and simulated results generally agree very well. It should be noted, however, that the number of Monte-Carlo simulations N=5000 used for this case is not yet enough when a return period of Tr=475 years is discussed. This corresponds to a case of 10% exceedance probability in 50 year lifetime, which is a standard target risk level in many of design evaluation. In contrast, the analytical results are valid for much larger values of return period, for example, over 1000 year lifetime.

Fig.-6 shows the computer cpu time required on a digital computer with 1 mips. Considering the above discussion claiming that direct simulation requires the number of repetitions of more than 5000, one can see that the method of this study has a high computational efficiency.

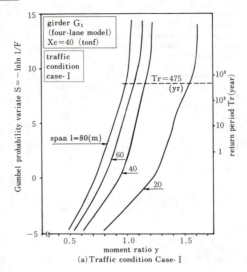

(a) Traffic condition Case- I

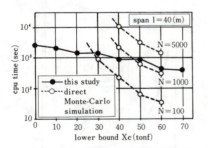

Fig.-6 Comparison of Computing Time (ACOS 430-20)

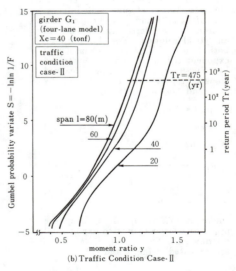

(b) Traffic Condition Case- II

Fig.-7 Probability Distribution of Monthly Maximum Load Effect (Four-lane Model) (1)

2.3 Girder Response for a Four-lane Model

Fig.-7 and -8 show the probability distribution of monthly maximum

girder response for a case of a four-lane model. Observe that the load values for a given return period Tr vary with the span length l = 20-80 (m), load case-I or case-II, or location of the plate girder G_1, G_2 or G_3. As the moment ratio y is being evaluated by Eq.(1) with respect to the design load under the Japanese highway bridge code, these results demonstrate that the current codes do not provide uniform safety as long as the live load specifications are concerned. The bridges with short spans sustain heavier load compared to bridges with long spans, because the main loading by a heavy vehicle has large effect to the maximum girder response.

3. ANALYTICAL EVALUATION BY IDEALIZED EXPONENTIAL LOAD MODELS

3.1 Idealization of Main Loading and Subsidiary Loadings by Exponential Distribution

The numerical results for $F_x(y)$ by Eq.(5) in the previous chapter was obtained through numerical integration of Eqs.(3) and (4). This chapter is intended to derive its closed-form analytical solution. For this purpose, both $g_{kx}(y)$ and $h_k(y)$ are idealized with exponential distributions as illustrated by Fig.-9 and -10. The idealized models, represented respectively, by $f_1(y_1)$ and $f_2(y_2)$ are given by

$$f_1(y_1) = A_1 e^{-a_1(y_1-b_0)} \; ; \; b_0 \leq y_1 \leq b_1 \quad (7)$$

$$f_2(y_2) = A_2 e^{-a_2 y_2} \quad ; \; 0 \leq y_2 \leq b_2 \quad (8)$$

in which

$$A_1 = \frac{a_1}{1-e^{-a_1(b_1-b_0)}} \quad (9)$$

$$A_2 = \frac{a_2}{1-e^{-a_2 b_2}} \quad (10)$$

where $a_1 = 7.640$, $a_2 = 36.0$. They are upper and lower truncated exponential distributions. The bounds b_0, b_1 and b_2 are indicated in Fig.-9 and -10.

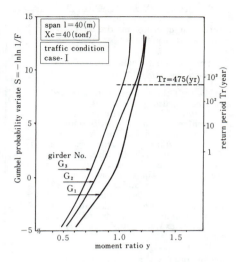

Fig.-8 Probability Distribution of Monthly Maximum Load Effect (Four-lane Model) (2)

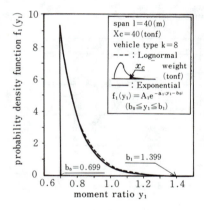

Fig.-9 Idealization of Load Model by Exponential Distribution (Main Loading) (1)

3.2 Girder Response by Single-fold Convolution

The convolution integral corresponding to Eq.(3) is then performed analytically, and its closed-form solution are represented by

$$f_0(y) = \int_0^\infty f_1(y-t) \, f_2(t) \, dt \qquad (11)$$

(1) $y < b_0$, $y \geq b_1+b_2$; $\qquad (12)$

$$f_0(y) = 0$$

(2) $b_0 \leq y < b_0+b_2$;

$$f_0(y) = \frac{A_1 A_2}{a_1-a_2} \left[e^{-a_2(y-b_0)} - e^{-a_1(y-b_0)} \right]$$

(3) $b_0+b_2 \leq y < b_1$;

$$f_0(y) = \frac{A_1 A_2}{a_1-a_2} \left[e^{-a_1(y-b_0-b_2)-a_2 b_2} - e^{-a_1(y-b_0)} \right]$$

(4) $b_1 \leq y < b_1+b_2$;

$$f_0(y) = \frac{A_1 A_2}{a_1-a_2} \left[e^{-a_1(y-b_0-b_2)-a_2 b_2} - e^{-a_2(y-b_1)-a_1(b_1-b_0)} \right]$$

Using these results for $f_{kz}(u)$ in Eq.(4), we can again obtain the probability distribution of the maximum girder response.

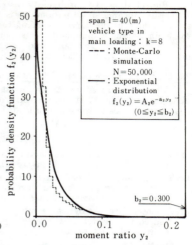

Fig.-10 Idealization of Load Model by Exponential Distribution (Subsidiary Loading) (2)

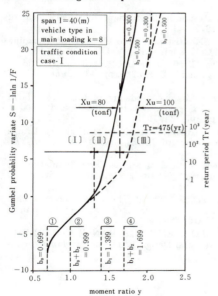

Fig.-11 Monthly Maximum Load Effect by Idealized Exponential Distributions (Single-Fold Convolution)

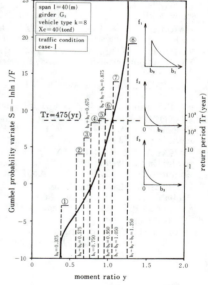

Fig.-12 Theoretical Estimation of Monthly Maximum Value by Convolution of three Exponential Distributions

Fig.-11 shows the probability distribution of the monthly maximum girder response. As the significance of all parameters is clear in this case, as shown in the figure, the physical meaning of the curves changing their slopes can be discussed. The region [I] is where the main loading is dominant, the region [II] is where the main loading and the subsidiary loading are both significant, and the region [III] is where the upper bound loads are involved. This type of observation is a great advantage of performing theoretical analysis.

3.3 Girder Response by Multi-fold Convolution

The same procedure can be repeated by introducing additional distribution of subsidiary loading. Fig.-12 shows a result from two-fold convolution, in which the distribution $f_3(y_3)$ corresponds to the effect of loads in the neighboring lanes.

4. CONCLUSIONS

(1) An efficient and accurate method was developed for probabilistic assessment of maximum girder response of highway bridges caused by traffic load.

(2) The method developed herein uses a concept of "main loading" and "subsidiary loading" and their combination through probabilistic convolution. It is particularly useful in reducing the computational efforts in Monte-Calro simulation.

(3) A closed-form analytical solution was derived by using idealized exponential load models. It is useful for the development of a theoretical interpretation of the load distribution characteristics.

(4) On this basis, a critical appraisal of the current Japanese design code was also made from a viewpoint of uniform load safety.

REFERENCES

[1] LENNARD,D.R.: "A Traffic Loading and Its Use in the Fatigue Life Assessment of Highway Bridges", TRRL Report, LR252, 1972

[2] HARMAN,D.J. and DAVENPORT,A.G.: "A Statistical Approach to Traffic Loading on Highway Bridges", Canadian Journal of Civil Engineering, Vol.6, 1979

[3] FUJINO,Y., ITO,M. and ENDO,M.: "Design Traffic Live Load for Highway Bridges Based on Computer-Simulation", Proceedings of JSCE, No.286, pp.1-13, June 1979 (in Japanese)

[4] DITLEVESEN,O. and MADSEN,H.O.: "Probabilistic Modeling of Man-made Load Processes and Their Individual and Combined Effects", Structural Safety and Reliability, Proceedings of ICOSSAR'81, Elsevier, pp.103-134, 1981

[5] SHINOZUKA,M., MATSUMURA,S. and KUBO,M.: "Analysis of Highway Bridge Response to Stochastic Live Loads", Proceedings of JSCE, No.344/I-1, pp.367-376, April 1984 (in Japanese)

[6] TAKADA,K and FUJINO,Y.: "An Efficient Method of Monte-Carlo Simulation for Evaluating Extremes of Traffic Live Load on Highway Bridges", Journal of Structural Engineering, JSCE, Vol.32A, pp.551-559, Mar. 1986

[7] KONISHI,I., KAMEDA,H. and MATSUMOTO,T.: "Traffic Load Measurement and Probabilistic Modeling for Structural Design of Urban Expressways", Structural Safety and Reliability, Proceedings of ICOSSAR'85, June 1985

[8] COMMITTEE ON DESIGN LOAD DEVELOPMENT FOR HANSHIN EXPRESS-WAYS (HDLC): "Investigation on the Structural Design Load for Hanshin Expressways", Final Report, Dec. 1986 (in Japanese)

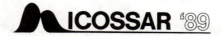
THE N-YEAR MAXIMUM MEAN VALUES AND THE COEFFICIENTS OF VARIATION OF LOADS SUBJECTED TO PORT AND OFFSHORE STRUCTURES

S. SHIRAISHI* and S. UEDA*

* Structural Engineering Division, Port and Harbour Research Institute, Ministry of Transport, Yokosuka, Kanagawa, Japan

ABSTRACT

It is necessary to know mean values and coefficients of variation of loads in order to verify the safety of structures. In this paper, the N-year maximum mean values and the coefficients of variation of wind loads, seismic loads and wave loads are calculated by means of observed data of wind speed, magnitude of earthquake and wave height in Japan. Furthermore, load factors are calculated by means of the N-year maximum mean values and the coefficients of variation of loads. And regional variations of the N-year maximum mean values and the coefficients of variation of loads are examined.

KEYWORDS

Load; N-year Maximum Mean Values and Coefficients of Variation; Load Factor; Port Structures ; Offshore Structures.

1.INTRODUCTION

It is necessary to know the N-year maximum **mean values** (hereafter denoted **MEANs**) and the **coefficients of variation** (hereafter denoted **COVs**) of loads in order to verify the safety of structures. In this paper, the N-year maximum MEANs and COVs of wind loads, seismic loads and wave loads are calculated by means of observed data of wind speed, magnitude of earthquake and wave height in Japan[1].

The N-year maximum **probability distribution functions** (hereafter denoted **PDFs**) are obtained fitting to one of three types of asymptotic distribution forms of extreme values which PDFs are the Type I , the Type II and the Type III asymptotic distribution

forms. As for the Type III asymptotic distribution forms, the parameters A_3, B_3 are fitted to each shape parameter k 0.75, 0.85, 1.0, 1.1, 1.25, 1.5, 2.0, respectively[2].

The N-year maximum MEANs and COVs are calculated by means of the N-year maximum PDFs which are obtained from the best-fitting yearly maximum PDFs to observed samples. Ravindra et al.[3] obtained the N-year maximum MEANs and COVs of wind loads by means of the Type II asymptotic distribution form. As for the Type III asymptotic distribution forms, the N-year maximum MEANs and COVs cannot be written with functions. Then, in this paper the N-year maximum MEANs and COVs are calculated both by means of binomial expression and numerical integration.

Furthermore, regional variations of the N-year maximum MEANs and COVs of loads are examined. And load factors for wind loads, seismic loads and wave loads are calculated by means of the N-year maximum MEANs and COVs of loads.

2.METHOD OF CALCULATION

In order to evaluate N-year maximum MEANs and COVs of loads, it is necessary to obtain data in long term, but it is usually difficult to obtain such kind of data. In this paper, the N-year maximum MEANs and COVs are calculated by means of the N-year maximum PDFs of loads which are approximately defined from the yearly maximum PDFs of loads. In this paper, following three types of asymptotic distribution forms are examined.

Type I   : $P(x) = \exp[-\exp(-(x-B_1)/A_1)]$ (1)

Type II   : $P(x) = \exp[-(A_2/x)^k]$ (2)

Type III : $P(x) = 1-\exp[-((x-B_3)/A_3)^k]$ (3)

in which P(x) is PDF of loads and A_i, B_i are parameters where subscripts i=1,2,3 correspond to the Type I , the Type II and the Type III asymptotic distribution form, respectively.

The N-year maximum PDF $P_N(x)$ is approximately defined from the yearly maximum PDF P(x) as follows.

$$P_N(x) = [P(x)]^N \qquad\qquad (4)$$

Type I   : $P_N(X) = \exp[-\exp(-(x-B_1-A_1\ln N)/A_1)]$ (5)

Type II   : $P_N(x) = \exp[-(A_2 N^{1/k}/x)^k]$ (6)

Type III : $P_N(x) = [1-\exp(-((x-B_3)/A_3)^k)]^N$ (7)

As for the Type I and the Type II asymptotic distribution form, the N-year maximum MEANs x_{NM} and COVs V_N are written with functions.

Type I   : $x_{NM} = (B_1 + \gamma A_1) + A_1 \ln N$ (8)

              $V_N = \pi A_1 / (\sqrt{6}\ x_{NM})$ (9)

Type II   : $x_{NM} = A_2\ \Gamma(1-1/k) N^{1/k}$ (10)

$$V_N = [\ \Gamma(1-2/k)-(\ \Gamma(1-1/k))^2]^{1/2}/\ \Gamma(1-1/k) \qquad (11)$$

in which γ is the Euler's constant($\gamma=0.5772$), π is the circular constant and $\Gamma(\ )$ is the Gamma function.

As for the Type III asymptotic form, the N-year maximum MEANs and COVs cannot be written with functions. In this paper, the N-year maximum MEANs and COVs are calculated by means of both **the binomial expression** and **the numerical integration of the N-year maximum PDF.**

Method by the binomial expression

Eq.(7) is expanded into the binomial expression as follows.

$$P_N(x)=1+ \sum[_NC_i(-1)^i\exp(-i((x-B_3)/A_3)^k)] \qquad (12)$$

in which $_NC_i$ is the binomial coefficient and i is the imaginary unit.

Thus, the N-year maximum MEANs and variances are written as follows.

$$x_{NM}= \sum[_NC_i(-1)^{i+1} (A_3i^{-1/k} \Gamma(1+1/k)+B_3)] \qquad (13)$$

$$\delta_N^2= \sum[_NC_i(-1)^{i+1} [A_3^2i^{-2/k} \Gamma(1+2/k) + 2A_3B_3i^{-1/k} \Gamma(1+1/k)$$
$$+ B_3^2]] - x_{NM}^2 \qquad (14)$$

Method by the numerical integration of PDF

The MEANs and **standard deviations** (hereafter denoted SDs) are calculated by means of the numerical integration with following four models of integration step Δp_i. Through the comparison of the results each other, a suitable integration step is selected.
(a) $\Delta p_i=0.01$ (b) $\Delta p_i=0.001$
(c) $\Delta p_i=0.0001$ at $p_N(x)=0.0001$ to 0.001 and $p_N(x)=0.999$ to 0.9999
   $\Delta p_i=0.001$ at $p_N(x)=0.001$ to 0.01 and $p_N(x)=0.99$ to 999
   $\Delta p_i=0.01$ at $p_N(x)=0.01$ to 0.99
(d) $\Delta p_i=0.00001$ at $p_N(x)=0.00001$ to 0.0001 and
   $p_N(x)=0.9999$ to 0.99999
   Δp_i are same as at $p_N(x)=0.0001$ to 0.9999

At first, models of integration step are examined for the Type I asymptotic distribution form of the wind pressures ($A_1=249.2$ N/m^2, $B_1=418.4$N/m^2) which N-year maximum MEANs and SDs are able to calculate by means of functions (Eqs.(8) and (9)). The results are tabulated in **Table-1.** The results calculated with the integration step of the model(c) and (d) agree with the results by means of functions, the model (c), then, is selected for a integration step of the numerical integration.
   Next, regarding to the Type III asymptotic distribution form ($k=0.85$,$A_3=256.4$N/m^2, $B_3=280.5$N/m^2) , the N-year maximum MEANs and COVs of the wind pressures are calculated by means of both the binomial expression and the numerical integration. The results are tabulated in **Table-2.** The results calculated by means of the binomial expression agree with the results by means of the numerical integration for those 10, 20, 30, 40 and 50 years. But,

the results for 60 years which are calculated by means of the
binomial expression do not seem accurate comparing those by means
of the numerical integration. This causes according to the error
in the numerical calculation of the binomial expression in Eqs.
(13) and (14). In this paper, regarding to the Type III asymptotic
distribution forms, the numerical integration is adopted for the
calculation of the N-year maximum MEANs and COVs.

**Table-1 The N-year maximum MEANs and SDs calculated by means
of the numerical integration and functions**

N-year	MEAN (N/m^2)					SD (N/m^2)				
	F.	(a)	(b)	(c)	(d)	F.	(a)	(b)	(c)	(d)
10	1136	1107	1133	1136	1136	320	289	314	319	320
20	1309	1276	1305	1308	1309	320	289	314	319	320
30	1410	1375	1406	1410	1410	320	289	314	319	320
40	1482	1445	1477	1481	1482	320	289	314	319	320
50	1537	1500	1533	1537	1537	320	289	314	319	320

F.:Result by means of functions

**Table-2 The N-year maximum MEANs and SDs calculated by means of
the numerical integration and the binomial expression**

N-year	MEAN (N/m^2)					SD (N/m^2)				
	B.E.	(a)	(b)	(c)	(d)	B.E.	(a)	(b)	(c)	(d)
10	1204	1167	1199	1204	1204	468	416	457	467	469
20	1450	1409	1445	1450	1451	490	437	479	489	491
30	1601	1557	1596	1601	1601	501	447	490	500	501
40	1710	1664	1705	1710	1710	507	454	497	506	508
50	1796	1748	1790	1796	1796	513	458	501	511	513
60	1284	1817	1861	1867	1867	1522	462	505	515	517

B.E.:Result by means of the binomial expression

3.RESULT OF CALCULATION

(1) Wind Loads
    The N-year maximum MEANs and COVs of the wind pressures are
calculated by means of the wind speed data observed at 131
observatory points during 1929 to 1983. Followings are the
procedures of the calculation.
 a) Modify the observed wind speed to the value at height of 10m.
 b) Calculate the yearly maximum wind pressure.
 c) Determine the best-fitting PDF.
 d) Calculate the N-year maximum MEANs and SDs of the wind
    pressures.
    In the calculation, the effect of variation of wind drag
coefficient and the vertical distribution of wind speed are not
considered as the causes of variation of wind loads.
    Figure-1 shows the 50-year maximum COVs of the wind pressures
for each Type of asymptotic distribution form. Closed circles are
denoted the 50-year maximum COVs of the wind pressures for each
Type of asymptotic distribution form. And open circles are
denoted the 50-year maximum COVs of the wind pressures for the

Type I and the Type III asymptotic distribution forms. Regarding to the Type III asymptotic distribution form, the 50-year maximum COVs of the wind pressures are calculated in the range of 0.30 to 0.35 when the shape parameter k is 0.75, while, those are calculated in the range of 0.08 to 0.12 when k is 2.0. As for the Type II asymptotic distribution form, the 50-year maximum COVs of the wind pressures are greater than those for the Type I and the Type III asymptotic distribution forms. And for some

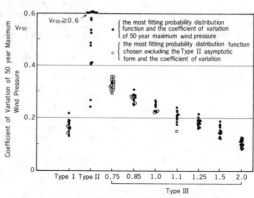

Fig.-1 Best-fitting PDFs and COVs

observatory points the COVs are greater than 0.6 which seem unreasonably large. Regarding to these observatory points, the Type III asymptotic distribution form (k=0.75) are fitted and the 50-year maximum COVs are calculated in the range of 0.28 to 0.36 which seem reasonable. Accordingly, the Type II asymptotic distribution form is not adopted for the best-fitting distributions for N-year maximum PDF.

Figure-2 shows the relation between the 50-year maximum MEANs and the sums of the MEANs and SDs of wind pressures vs. to the wind pressures for 50 years return period. Closed circles and open circles show the 50-year maximum MEANs and the sum of MEANs and SDs of the wind pressures, respectively. The ratios of the values

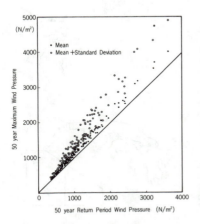

Fig.-2 50-year Maximum
       Wind Pressures

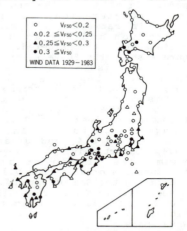

Fig.-3 50-year Maximum COVs
       of Wind Pressures

for 50 years return period vs. to the 50-year maximum MEANs and the sum of MEANs and SDs of the wind pressures are in the range of 1:1.05 to 1:1.2 and 1:1.2 to 1:1.5, respectively. The MEANs and SDs vary with observatory points, then, regional variations of the N-year maximum COVs of the wind pressures are examined.

Figure-3 shows the 50-year maximum COVs of wind pressures which are denoted V_{F50}. Open circles, open and closed triangles and closed circles show the observatory points where V_{F50}s are corresponding to less than 0.2, in the range of 0.2 to 0.25, in the range of 0.25 to 0.3, and greater than 0.3, respectively.

(2) Seismic Loads

The N-year maximum MEANs and COVs of the seismic coefficients are calculated from the data of the magnitudes of earthquake for 190 points during 1885 to 1981. Followings are the procedures of the calculation.

a) Calculate the seismic coefficients from maximum accelerations by means of the approximate formula[4].

b) Determine the best-fitting PDF.

c) Calculate the N-year maximum MEANs and SDs of the seismic coefficient.

In the calculation, the variation both of the relationship between the magnitude of earthquake and the maximum acceleration, and the relationship between the maximum acceleration and the seismic coefficient are not considered.

Figure-4 shows the 50-year maximum COVs of the seismic coefficients which are denoted V_{kh50}. Open circles, open and closed triangles and closed circles show the locations where V_{kh50}s are corresponding to less than 0.2, in the range of 0.2 to 0.3, in the range of 0.3 to 0.4, and greater than 0.4, respectively. The N-year maximum COVs of the seismic coefficients are greater·than 0.3 in the area along the Sea of Japan. And the N-year maximum COVs are smaller than 0.2 in the area along the Pacific Ocean.

(3) Wave Loads

The N-year maximum MEANs and COVs of the significant wave heights are

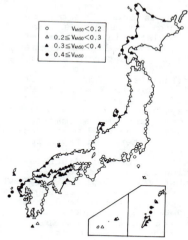

Fig.-4 50-year Maximum COVs of Seismic Coefficients

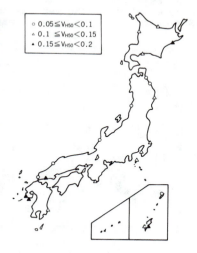

Fig.-5 50-year Maximum COVs of Significant Wave Heights

calculated by means of wave data obtained at 23 observatory points
during 1972 to 1981. Followings are the procedures of the
calculation.
 a) Determine the best-fitting PDF.
 b) Calculate the N-year maximum MEANs and SDs of the significant
wave heights.

In the calculation, the effects of
wave period, wave deformation in
shallow water, deformation by wave
breaking are not considered as the
factors of variation of loads.

Figure-5 shows the 50-year maximum
COVs of the significant wave
heights which are denoted V_{H50}. Open
circles, open and closed triangles
show locations where V_{H50}s are
corresponding to less than 0.1, in
the range of 0.1 to 0.15, and
greater than 0.15, respectively. The
N-year maximum COVs of the
significant wave heights on the
coast along the Sea of Japan and in
the northern part of Japan along the
Pacific Ocean are smaller than the
values in the southern area along
the Pacific Ocean. But the
deviations of the values are not so
large as the values of wind loads
and seismic loads.
(4) Load factors

The load factors for wind loads,
seismic loads and wave loads are
calculated with the N-year maximum
MEANs and COVs of loads. Followings
are the procedures of the
calculation.
 a) Calculate the load factor at
each observatory point by Eq.(15).

$$\gamma_S = (S_M/S^*)\exp(\alpha_1\alpha_2\beta V_S) \qquad (15)$$

in which S_M is a MEAN of the loads,
S^* is a characteristic value of the
loads, β is a safety index, α_1, α_2 are
linearized coefficients (here α_1, α_2=
0.75) and V_S is a COV of the loads.
 b) Define the boundary of the area
where the deviation of load factors
are not so large each other.

Figures-6 and -7 show the load
factors for wind loads and seismic
loads at each observatory point. In
the calculation, safety indices
assumed to 3.0. The regional variations of load factors for wind
loads and seismic loads are found as
shown in **Figs.-8 and -9.**

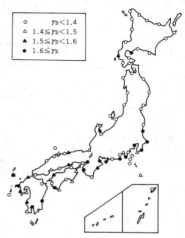

circle: $\gamma s < 1.4$
△ $1.4 \leqq \gamma s < 1.5$
▲ $1.5 \leqq \gamma s < 1.6$
● $1.6 \leqq \gamma s$

**Fig.-6 Load Factor
for Wind Loads**

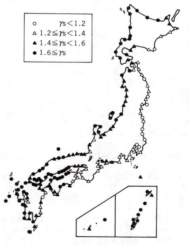

circle: $\gamma s < 1.2$
△ $1.2 \leqq \gamma s < 1.4$
▲ $1.4 \leqq \gamma s < 1.6$
● $1.6 \leqq \gamma s$

**Fig.-7 Load Factor
for Seismic Loads**

4.CONCLUSION

In this research, followings are concluded.
1) Regarding to the Type III asymptotic distribution forms, the N-year maximum MEANs and SDs cannot be calculated by means of functions, the N-year maximum MEANs and COVs are calculated by means of both the binomial expression and the numerical integration. It was found that the N-year maximum MEANs and COVs are calculated by means of the numerical integration seem appropriate.
2) The N-year maximum COVs of wind loads, seismic loads and wave loads vary with locations and periods. The 50-year maximum COVs of the loads are in the range of 0.08 to 0.35 for the wind pressures, in the range of 0.05 to 0.55 for the seismic coefficients and in the range of 0.05 to 0.20 for the significant wave heights. And regional variations of the N-year maximum COVs of loads are found.
3) The load factors for wind loads, seismic loads and wave loads are calculated at each observatory point and the area with the N-year maximum MEANs and COVs of those loads and the safety indices of the structures. Regional variations of the load factors are found.

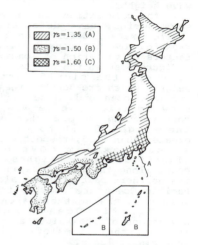

Fig.-8 Regional Variation of Load Factor for Wind Loads

REFERENCES

[1] SHIRAISHI,S. and S.UEDA : "Study on the Method of Verification of Structural Safety of Port and Offshore Structures ", Report of the Port and Harbour Research Institute (PHRI), Vol.26-2, pp.493-576, 1987
[2] Petruaskas,C. and P.M.Aagaard : "Extrapolation of Historical Storm Data for estimating design Wave Heights", Proc. of 2nd Offshore Technology Conference, pp.I-409-428, 1970
[3] Ravindra,M.K. et al.: "Wind and Snow Load Factors for Use in LRFD", Proc. of ASCE, Vol.104, No.ST9, pp.1443-1457, 1978
[4] NODA,S. et al. : "Relation between Seismic Coefficient and Ground Acceleration for Gravity Quaywall", Report of the PHRI,Vol.14 -4, pp.67-111, 1975

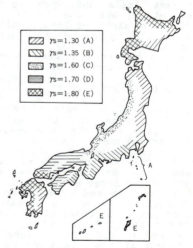

Fig.-9 Regional Variation of Load Factor for Seismic Loads

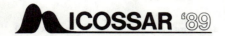
EXTRAORDINARY LIVE LOAD MODEL IN RETAIL PREMISES

Jun Kanda* and Kazushige Yamamura**

* Dept. of Architecture, University of Tokyo,
  7-3-1 Hongo, Bunkyo-ku, Tokyo, 113 Japan
** Dept. of Architecture, Tokyo Metropolitan University,
  2-1-1 Fukazawa, Setagaya-ku, Tokyo, 158 Japan

ABSTRACT

Based on a live load survey on six supermarkets, the probabilistic nature of sustained live loads and simulated extraordinary live loads were examined. The simulation was performed to represent personnel concentration, i.e. crowding in emergency situation with the personnel density as a parameter. An empirical extreme value distribution with upper and lower bound limits was successfully applied to match the cumulative probability plots of simulated results. The reliability index was computed to discuss the safety level of the current Japanese design practice.

KEYWORDS

Live load; gamma distribution; extreme value distribution; equivalent uniformly distributed load; reliability index

1. Introduction

Possible occurrences of live load extremes are considered to be caused either by the furniture concentration or the personnel concentration. The situation of concentration occurrences is postulated to create realistic probabilistic models. Furniture concentration simulation was adopted for a probabilistic model for live load extremes in office buildings[1]. For such occupancy types as retail premises, conference halls and auditoriums, crowding situations are considered to be dominant causes for live load extremes. It is the purpose of this paper to provide a

probabilistic model for the extraordinary live load in retail premises as a consequential estimation based on surveyed sustained live load distributions. Crowding situations are introduced parametrically but in a fairly simple manner in comparison with existing models[2].

2. Sustained Live Load Statistics

A live load survey was carried out for six supermarkets with total area of 2837 m^2 in August 1986 and January in 1987 in Tokyo and Atsugi, Japan. The number of samples may not be sufficient to discuss the adequacy of codification purposes, however, certainly provide sufficient information on the probabilistic nature of live load distributions in retail premises.

Data were stored in a computer file in terms of a room plan with specified shape and weight of shelves or racks with goods as an arbitrary point-in-time situation. Table 1 shows statistics of data by averaging live load intensities for square units with a size varying from 1 m to 17 m. Each shelf or unit rack load was regarded as uniformly distributed in the corresponding projected area.

TABLE 1. Statistics of sustained live load

AREA (m2)			NO. OF DATA	MEAN (Pa)	STDV (Pa)	COV (%)
1 x 1 =	1		10615	1297	834	64
2 x 2 =	4		2421	1262	527	42
3 x 3 =	9		1084	1327	417	31
4 x 4 =	16		566	1297	392	30
5 x 5 =	25		311	1321	343	26
6 x 6 =	36		229	1303	327	25
7 x 7 =	49		156	1333	296	22
8 x 8 =	64		101	1297	275	21
9 x 9 =	81		87	1312	261	20
10 x 10 =	100		61	1327	234	18
11 x 11 =	121		54	1316	255	19
12 x 12 =	144		42	1305	221	17
13 x 13 =	169		32	1321	188	14
14 x 14 =	196		22	1291	217	17
15 x 15 =	225		18	1277	193	15
16 x 16 =	256		17	1283	186	15
17 x 17 =	289		16	1269	181	14

Personnel loads are postulated to be 700 Pa which corresponds to one person /m^2 , and uniformly distributed in the area not occupied by goods display(vacant area), commonly for all cases.

The mean value was approximately 1300 Pa irrespective of unit area and the coefficient of variation (c.o.v.) asymptotically decreases with the area from 64% for 1 m^2 to 15% for 200 m^2 or over. Mean values are somehow greater than those of some past live load survey data in U.K.[3], mostly because the difference in display systems of goods and partly because refrigerating installations were all excluded in the past survey[3]. It is worthy of note that a similar c.o.v. value, in the order of 100%, for 1 m^2 unit was reported for different occupancy types such as office buildings[1] and also residential flats[4]. A general tendency of c.o.v. variation with the area was comparable to the survey in U.K.[3].

Frequency distributions of sustained live load intensity for typical unit areas are shown in Fig. 1. The distribution can be

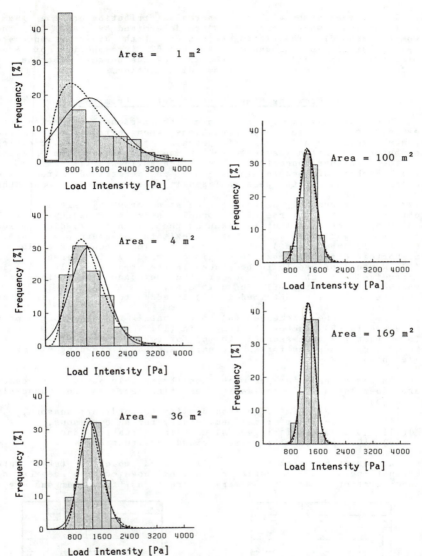

Fig. 1   Frequency distribution of sustained live load intensity
for various unit areas, where
———— moment fit curve of Normal distribution
----- moment fit curve of Gamma   distribution

well represented by either the normal distribution or the gamma
distribution, whose parameters are determined by means of the
moment method. The fitting errors for both distributions are
fairly small, say around or less than 5% irrespective of area
except for 1 m² and 4 m² in the case of the normal distribution
and 1 m² in the case of the gamma distribution.

3. Crowding Simulation in Retail Premises

Since it is very difficult to collect personnel gathering data
especially in emergency situations when extraordinary load
concentration could be anticipated, a fairly simple crowding
computer simulation was performed in the following scheme similar
to the furniture concentration in office buildings[1];
(1) Number of people in a shop is determined from the
multiplication of the personnel density before crowding and the
vacant area.
(2) Unit personnel occupying area at a crowding situation is
obtained as the inverse of personnel density of crowding.
(3) One corner of a shop is taken as the origin of crowding which
can be regarded as a hypothetical exit for an emergency escape.
(4) A unit area (1x1m²) with a vacant portion, whose center is in
a shortest distance from the origin of crowding, is chosen.
(5) Number of people to be placed in the area chosen in (4) is
determined so that the sum of the personnel occupying area
reaches the vacant portion of unit area.
(6) Personnel load obtained in (5) is added to the load in that
unit area.
(7) Repeat the processes (4) to (6) until the sum of number
obtained in (5) reaches the number in (1).
(8) Repeat the processes (3) to (6) for four corners.
Schematic room plans for an ordinary case and a crowding situation
are shown in Fig. 2.

As basic cases for personnel density in this study, four cases
are postulated by considering existing surveys for assembly
loads[5] as,
(i)   0.3 person/m² : maximum density for ordinary seasons,
(ii)   1.0 person/m² : maximum density for sale seasons,
(iii) 5.0 person/m² : general crowding situation, and
(iv) 10.0 person/m² : extreme crowding situation.

An existing probabilistic live load model[2] takes into
account personnel gathering situations, however, the degree of
concentration seems to be very moderate as for emergency-type

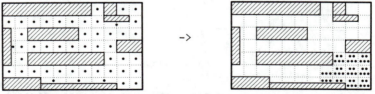

Fig. 2 Diagrammatical explanation of personnel concentration

situations considering parameter values adopted. Four typical
crowding cases are chosen as;
   (a) 0.3 -> 5.0 person/m^2,   (b) 0.3 -> 10.0 person/m^2,
   (c) 1.0 -> 5.0 person/m^2, and (d) 1.0 -> 10.0 person/m^2.
Cases (a) and (b) are concentration situations with the density
(iii) and (iv) for ordinary seasons of density (i). Cases (c) and
(d) are those for sale seasons, which are less likely to occur
than cases (a) and (b).

4. Equivalent Uniformly Distributed Loads

The equivalent uniformly distributed loads (EUDL) were
computed for slabs and girders in a similar manner to the previous
study[1]. Load effects were examined in terms of the maximum end
bending moments $Mx1$, $My1$ and the maximum center bending moments
$Mx2$, $My2$ and the maximum shear forces Qx, Qy for slabs with fixed
ends where subscripts x and y denote shorter and longer span
direction respectively and the larger end bending moment $M1$, the
center bending moment $M2$ and the maximum shear force Q for girders
with fixed ends. Results are listed in Table 2 for (a) slabs and
(b) girders for the sustained load model and the extreme live load
model cases (a) and (d). The mean values for case (a) is less
since the lower personnel density was adopted. Considerable
increase of c.o.v. can be pointed out due to the crowding
simulation especially for case (d).

The probabilistic nature was examined and cumulative
distributions of slab $Mx1$ EUDL were shown on the Type I extreme
value probability paper as an typical example in Fig. 3 for four

TABLE 2 (a) Statistics of EUDL for slabs (Mean Area = 24.5 m^2)

		Sustained Load			Extreme Load					
					Case (a)			Case (d)		
		MEAN (Pa)	STDV (Pa)	COV (%)	MEAN (Pa)	STDV (Pa)	COV (%)	MEAN (Pa)	STDV (Pa)	COV (%)
Load Intensity		1328	331	25	990	552	56	1293	1156	90
EUDL	Mx1	1482	508	34	1171	692	59	1596	1501	94
	Mx2	1328	481	36	1042	692	66	1460	1518	104
	My1	1356	424	31	1084	702	65	1527	1524	100
	My2	1777	993	56	1521	1106	73	2018	1845	92
	Qx	1765	679	39	1438	755	53	1906	1572	82
	Qy	1550	531	34	1302	719	55	1759	1522	87

TABLE 2 (b) Statistics of EUDL for girders (Mean Area = 48.1 m^2)

		Sustained Load			Extreme Load					
					Case (a)			Case (d)		
		MEAN (Pa)	STDV (Pa)	COV (%)	MEAN (Pa)	STDV (Pa)	COV (%)	MEAN (Pa)	STDV (Pa)	COV (%)
Load Intensity		1317	290	22	984	464	47	1283	1057	82
EUDL	M1	1291	385	30	958	578	60	1314	1295	99
	M2	1245	382	31	897	561	63	1247	1295	104
	Q	1326	383	29	1011	587	58	1370	1291	94

TABLE   3.   Examples of percentile values
                    for extraordinary live load models (unit:Pa)

Case	slab Mx1			girder M1		
	95.0%	99.0%	99.9%	95.0%	99.0%	99.9%
(a)	2584	3339	3379	2139	2971	3622
(b)	2934	4590	6120	2135	3448	5211
(c)	3495	3839	3913	3287	3793	3909
(d)	4911	6402	6817	4271	6112	6774

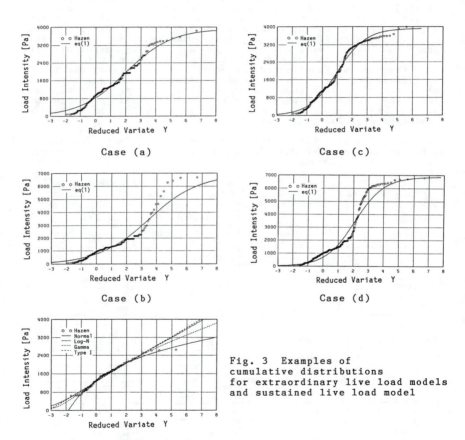

Case (a)   Case (c)

Case (b)   Case (d)

Fig. 3 Examples of
cumulative distributions
for extraordinary live load models
and sustained live load model

Sustained live load

extreme load models (a), (b), (c) and (d), and sustained load
model. The existence of upper bound limit is clearly observed in
Fig. 3 (a) to (d). This was caused by a very high load intensity
in some unit areas due to an extreme crowding situation, which did
not appear in the furniture concentration in office buildings[1].
An empirical extreme value distribution defined in eq(1) [6] was
introduced as a possible best fit cumulative distribution
function.

$$F_x(x) = \exp\left(- \left\{\frac{w-x}{ux}\right\}^\gamma \right) \qquad 0 < x < w \qquad (1)$$

where the upper bound limit w = 4.0 KPa for cases (a) and (c), and
w = 7.0 KPa for cases (b) and (d), and parameters u,γ were
obtained by fitting to the Hazen plots on the Type I probability
paper.

Summarized results for percentile values are listed in Table 3.
99 percentile values for E.U.D.L. of average-sized slab dominant
fixed end bending moment for four cases are (a) 3.3 KPa, (b) 4.6
KPa, (c) 3.8 KPa and (d) 6.4 KPa respectively. Because of the
degree of concentration, the order of 95 percentile values for
cases (b) and (c) is reverse and the difference between 99.9
percentile values for cases (a) (c) and (b) (d) becomes very
significant. When we discuss the reliability of load evaluation in
exceptionally rare cases, it is necessary to collect more actual
data for crowding situations as rare events.

5. Safety Evaluation for Extraordinary Live Load Model

The safety level of the current Japanese design practice for
live load in retail premises is discussed in terms of the safety
index β of A.F.O.S.M. Typical dead load and specified live load
values were taken as 3.2 KPa and 3.0 KPa for slabs and 6.0 KPa and
2.4 KPa for girders. The ratio of the mean failure strength to
the allowable design strength was assumed to be 1.65, which
corresponds to the ratio of the mean yield stress of steel to the
longterm allowable stress of steel. The probability distributions
of failure strength and dead load were assumed to be log-normal
and normal with the c.o.v. of 20% and 10% respectively. Then the

TABLE 4. Comparison of reliability index β
for extraordinary live load models with one concentration, where
eq(1) is used for supermarkets and the Type I for office

| | | Crowding in Supermarket | | | | Furniture Concentration in Office |
		Case(a)	Case(b)	Case(c)	Case(d)	
Slab	Mx1	3.05 (2.97)*	2.61 (2.61)	2.36 (2.47)	1.92 (2.05)	3.00
	Mx2	3.09 (2.98)	2.56 (2.55)	2.42 (2.47)	1.92 (2.08)	3.14
	My1	3.07 (2.34)	2.64 (2.67)	2.47 (2.43)	1.88 (2.06)	2.94
	My2	2.15 (2.34)	2.09 (2.18)	1.77 (2.14)	1.60 (1.79)	2.34
	Qx	2.86 (2.80)	2.47 (2.48)	2.36 (2.40)	1.82 (1.93)	2.61
	Qy	2.97 (2.89)	2.57 (2.60)	2.35 (2.42)	1.86 (2.00)	2.65
Girder	M1	2.92 (2.89)	2.88 (2.89)	2.41 (2.52)	2.10 (2.25)	2.74
	M2	2.97 (2.93)	2.94 (2.93)	2.42 (2.53)	2.11 (2.27)	2.79
	Q	2.87 (2.85)	2.83 (2.85)	2.41 (2.51)	2.09 (2.24)	2.70

* ( ) : based on log-normal distribution

reliability index was computed by the iterative way[7] modifying a
published program[8].

Results are listed in Table 4 where the probability
distribution for live load were represented by eq(1) and the log-
normal with corresponding 1st and 2nd moments. Difference in β
value due to the choice of the probability distribution for
extraordinary live load model was not very significant for this
level of safety. The reliability index was also computed for the
extreme load model in office buildings for the comparison, where
the Type I distribution was adopted[1]. If the case (d) is
reasonably accepted as a typical crowding situation in
supermarket, the safety level is considerably lower than the
furniture concentration in office buildings. However it should be
noted that these values are obtained for the limit state of
steel yielding and not for the ultimate limit, and at least the
level of safety at present has not been criticized seriously.

6. Conclusions

A live load survey was carried out for six supermarkets and
statistical data was presented. Crowding simulation was proposed
to create an extraordinary live load model for retail premises.
Numerical examples were examined to discuss the safety level of
the current design practice in Japan. The usefulness of the
empirical extreme value distribution with upper and lower limits
was confirmed.

Acknowledgements

Authors are indebted to Prof. Yukio Tamura, Tokyo Institute of
Polytechnics for his help in surveying and data processing.

References

[1] KANDA, J. and KINOSHITA, K.:"A probabilistic model for live
load extremes in office buildings", Proc. 4th Int. Conf. Str.
Safety and Reliability, pp.II-287-296,1985.
[2] HARRIS, M. E., COROTIS, R. B. and BOVA, C. J.:"Area-dependent
processes for structural live loads", Proc. ASCE, J. Str. vol.107
pp857-872,1981.
[3] MITCHELL, G.R. and WOODGATE, R. W.:"Floor loadings in retail
premises - the results of a survey", C.P.25/71, Build. Res. St.,
Garston, U.K., 1971.
[4] YAMAMURA, K and KANDA, J.:"Stochastic analysis of live load
for apartment buildings", Sum. Tech. Pap. Annu. Meet. Archit.
Inst. Japan Str.I pp 37-38,1986 (in Japanese).
[5] SENTLER, L.:"Live Load surveys - a review with discussions",
Rep.78 Div. Build. Tech., Lund Inst. Tech., Sweden, 1976.
[6] KANDA. J.:"A new extreme value distribution with lower and
upper limits for earthquake motions and wind speeds", Theor. Appl.
Mech. vol.31, Univ. Tokyo Press,pp 351-360,1981.
[7] ANG.A.H-S. and TANG, W.H.:"Probabilistic Concepts in
engineering planning and design" vol.II John Wiley and Sons,
pp.334-382, 1984.
[8] HOSHIYA, M. and ISHII, K.:"Reliability design of structures",
Kajima Pub. pp.198-208, 1986, (in Japanese).

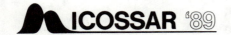
AN INVESTIGATION OF LIVE LOADS ON CONCRETE STRUCTURES
DURING CONSTRUCTION

Saeed Karshenas[*] and H. Ayoub[**]

* Assistant Professor, Dept. of Civil Engineering,
   Marquette University, Milwaukee, WI
** Graduate Student of Construction Engineering
   Marquette University, Milwaukee, WI

ABSTRACT

Slab formwork live loads include the
weight of equipment, construction material
storage loads, and impact loads when placing the concrete. In this paper material
and equipment loads are investigated.
Data on equipment and material storage
loads are collected by surveying twenty
concrete buildings under construction.
Direct weighing method and inventory technique were used to collect material and
equipment load data. The collected load
data are used to develop probability
models representing slab formwork loads.

KEYWORDS

Construction, live load, concrete, probability, formwork

1. INTRODUCTION

Failure of concrete structures during construction
constitutes a significant percentage of construction
failures. Formwork failure is among the major causes of
concrete construction failure. (In this paper the term
formwork is taken in its broadest sense to mean the
total system of support for the freshly placed concrete.)

One of the causes of slab formwork failures is overloading. The major slab formwork loads include dead, live,
equipment impact, and wind loads. In this paper formwork live loads are investigated. Concrete formwork
live loads include construction material storage loads,
weight of equipment and workers, and impact loads when
placing the concrete. Although the construction live
load imposed on a temporary support system is transient
in nature, this short-term peak load may be critical to
the safety of the overall building.

Developing live load models for slab formwork requires a data bank of actual formwork live loads. The
required data for this study are obtained from twenty
different concrete building projects. These buildings
are located in the cities of Chicago, Milwaukee, Minneapolis, Atlanta, Orlando, and Tampa.

2. DATA COLLECTED

Data on formwork live loads during various
construction activities have been collected and are
being analyzed. This paper presents a statistical
analysis of the live loads on slab formworks after the
slab is poured and preparations for construction of the
next floor is underway.

In multistory concrete construction as soon as a
section of a slab is poured and the concrete is hard
enough to walk on, material and equipment required for
construction of the rest of the slab and erection of the
formwork of the next level are lifted and stored on the
partially cured slab. These material and equipment
usually include large tool boxes, bundles of reinforcement, stacks of shores, wall and column forms, air com-
pressors, portable toilets, dumpsters, etc. These objects are usually stored on a small area to keep the
rest of the slab area unobstructed for various construction activities. To develop a probability model for
representing live loads on a newly poured slab, a large
sample of the weights of the objects on concrete slabs
during construction is required.

In structural design, it is not the load itself but
different load effects such as axial load, shear, and
bending moment which are used for the design of a member.
It is, therefore, necessary to transform the load into
the load effect in the process of design. For reliability analysis, the statistical moments of load effects
may be calculated analytically by using influence surfaces (Pier, 1971). However, application of the influence surface method requires the knowledge of spatial

correlation (Corotis, 1968, 1979) of construction live loads. For determining spatial dependence or independence of live loads, in addition to the weight of the objects on a slab, their sizes and locations on the plan of the slab should also be recorded.

To collect the above-mentioned data, in each project site that was visited, the survey crew first prepared a sketch of the building floor plan. On this plan, the area of the floor that at the time of the survey was serving as storage area for material and equipment was identified. For each object on the loaded floor area, data on the weight, size, contact area with the slab, and location relative to a coordinate system were recorded. Later in the office, this information were used to draw a plan of the loaded area with all objects shown to the scale at their original locations.

3. DATA COLLECTION TECHIQUES

Direct weighing and inventory techniques (Culver, 1975, 1976) were used for determining the weight of various objects. In the inventory technique, instead of directly weighing an object, observable physical characteristics from which weight information could be obtained are collected for each load item. For example, the weight of a bundle of reinforcing bars can be calculated if the number, length, and size of the bars are known. In case of equipment, weight could be determined from the manufacturer's catalog if the type, model, and manufacturer's name are known.

If the weight of an object could not be determined by the inventory technique, the direct weighing method was used. For direct weighing of an object the contractors crane at the site was used. By attaching a scale to the sling of the crane and lifting the object by the scale's hook, the weight of the object was determined. Data from some projects could not be included because the contractor did not allow the survey crew to use the crane for direct weighing.

4. DATA ANALYSIS

To analyze the collected data, the surveyed floor areas were divided into successive sets of rectangular grids of various areas and aspect ratios. To do this, a piece of tracing paper large enough to cover the loaded area of each building floor surveyed was divided into 0.093 m^2 grids (1 ft^2). The tracing paper was then placed over the plan of the loaded area of each surveyed floor and the loads within each grid were recorded.

The live load intensities for grids of 2.3 m^2 (25 ft^2) were obtained by adding the live loads of 25 adjacent grids of 0.093 square meters. Similarly, the live loads over grids of 5.94 m^2 (64 ft^2), 9.3 m^2 (100 ft^2), 13.36 m^2 (144 ft^2), 20.9 m^2 (225 ft^2), and 37.1 m^2 (400 ft^2) were obtained and the corresponding histograms were plotted. The unit loads were calculated for grid areas up to 37.1 m^2 (400 ft^2) due to the fact that the influence areas of various formwork load effects are usually within this range.

To investigate the effect of the grid shape on the unit load histogram, 99% probable loading intensities for various aspect ratios (grid width/grid length) were determined from observed load data. Table 1 shows 99% probable load intensities for four grid sizes and three different aspect ratios. The figures in Table 1 show that there is no consistent relationship between the 99% probable loading intensity and the shape of the grid used. Therefore, the data analysis is performed for square grids only.

Table 1.   Observed 99% Probable Level of Load Intensity for Various Grid Shapes (KN/m^2)

Bay Area (m^2)	Aspect Ratio (Width/Length)		
	0.25	1	4
0.36	4.74	5.89	5.03
3.3	3.08	2.475	2.92
9.3	1.992	2.226	1.752
37.1	1.245	1.269	1.168

Frequency diagrams of load intensities for various grid sizes are prepared. While space does not permit presenting all graphical results, a typical histrogram is shown in Figure 1.

The shape of the frequency diagram in Fig. 1 suggests a mixed distribution for the observed data, a discrete part at zero load intensity and a continuous part for load intensities greater than zero. Several theoretical distributions were fitted to the nonzero portion of the observed data. Table 2 shows the Kolmogorov-Smirnov statistics for lognormal, exponential, gamma, and Weibul distributions. Although gamma, exponential and Weibul distribution all provide reasonably good fit to the observed data, Weibul distribution gives the

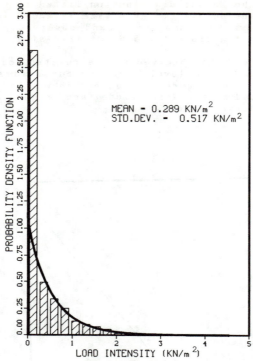

Figure 1. - Histogram and the fitted
density function for 5.95 m² grid area.

Table 2. Kolmogorov-Smirnov Statistics
for Various Distributions

Grid size m²	K-S Statistics			
	Lognormal	Exponential	Gamma	Weibul
2.32	0.0589	0.0237	0.0352	0.0327
5.95	0.132	0.0262	0.0231	0.0225
9.25	0.112	0.0423	0.0568	0.0532
13.36	0.112	0.0268	0.0445	0.0399
20.9	0.117	0.0401	0.0269	0.0262
37.16	0.151	0.109	0.115	0.0676

best fit. The Weibul distribution fitted to the observed live loads on 5.95 m^2 grid size is shown in Fig. 1.
Figure 2 shows the theoretical and observed cumulative
distribution functions for 5.95 m^2 grid size.

In Table 3 the observed and calculated loads at 95%
and 99% probability levels for various grid sizes are
compared. As the figures in Table 3 show, the load values calculated by the Weibul model are in good agreement
with the observed values.

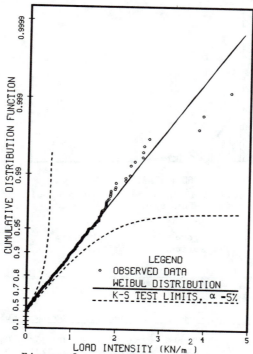

Figure 2. - The observed and theoretical
cumulative distribution functions for
5.95m^2 grid area.

Table 3. Comparison of Observed and Calculated
          Loads for Various Grid Sizes

Grid size m^2	95% Load, KN/m^2 observed	Weibul	99% Load, KN/m^2 observed	Weibul
2.32	1.671	1.637	2.968	2.959
5.95	1.376	1.393	2.25	2.331
9.25	1.297	1.307	2.226	2.173
13.36	1.12	1.264	2.059	2.092
20.9	1.01	1	1.537	1.57
37.16	0.91	0.91	1.269	1.326

5. SUMMARY AND CONCLUSIONS

A summary of the results of a statistical analysis of live load data from twenty concrete buildings under construction are presented. The analyzed data were collected from the surface of newly poured slabs supported by formwork.

The observed data were compared with several theoretical probability distributions. Visual and quantitative goodness-of-fit tests show that a mixed distribution best describes the data. The mixed distribution consisted of a discrete part at zero load intensity and a Weibul distribution for nonzero load intensities. The observed and calculated loads at 90%, 95%, and 99% probability levels are in good agreement.

Effort for calculating an equivalent distributed load for formwork design is currently underway.

6. REFERENCES

1. Peir, J.C., and Cornell, C.A., "A Stochastic Live Load Model for Buildings," Research Report R711-35, Dept. of Civil Engineering, M.I.T., September 1971.

2. Corotis, R.B., and Jaria, V.A., "Stochastic Nature of Building Live Loads," ASCE Journal of Structural Div., March 1979.

3. Corotis, R.B., "Statistical Measurement and Prediction of Building Design Loads," M.S. Thesis, Dept. of Civil Engineering, M.I.T., Sept. 1968.

4.   Culver, C.G., and Kushnar, J., "A Program for Sur-
      vey of Fire and Live Loads in Office Buildings,"
      Technical Note 858, National Bureau of Standards,
      Washington, DC, 1975.

5.   Culver, C.G., "Survey Results for Fire Loads and
      Live Loads in Office Buildings," Building Science
      Series 85, National Bureau of Standards, Washing-
      ton, DC, 1976.

7.   ACKNOWLEDGEMENT

This research is supported by the National Science
Foundation, Grant No. MSM-8617807.

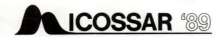
NONLINEAR LOAD EXCEEDANCES WITH CORRELATED
SUSTAINED LOAD PROCESS

Karen C. Chou

Assoc. Prof., Dept of Civil & Envir. Engrg., Syracuse Univ., Syracuse, NY 13244-1190, U.S.A.

ABSTRACT

Past studies on nonlinear structural response of a
single member only concerned with independent,
identically distributed load magnitudes. When the
structural member is subjected to cumulative loads
such as a column at the lower level of a multi-
stories building, the cumulative load magnitudes
will no longer be independent even though the
individual loads may be independent. In this
paper, a discussion on the analysis of nonlinear
load exceedances of a single structural member
subjected to correlated sustained load process is
presented.

KEYWORDS

Nonlinear Response, Stochastic Processes, Live
Loads, Buildings, Correlated Loads, Load Exceedances, Probability Theory, Structures.

1. INTRODUCTION

In the past two decades, extensive studies have been performed on the reliability of single structural member. In the
earlier stage of the studies, focuses were on the stochastic
modelling of load processes (e.g., [1]) and on the linear load
effect analysis (e.g., [2,3]). More recently, studies have
extended to include the nonlinear structural response of a single
member subjected to live load processes [4-7]. Since the studies
in load processes only described the behavior of an individual
occupant, the load processes consist of Poisson arrivals of
occupants and independent, identically distributed (iid) load
magnitudes [1]. The reliability analyses performed using this

load model are, therefore, applicable only if one occupant is
contributing to the response of the structural member. However,
if one considers a cumulative load effect on a structural member
from a number of occupants contributing identical type of loads
(e.g., all have sustained loads), then the total load magnitudes
will not be independent. As an example, consider a column at the
lower level of a multi-stories building. For simplicity, assume
the floor loads share a common supporting column. This lower
level column would then have to support, partially, all the floor
loads above it. If one of the occupant is replaced within these
floors, the cumulative load magnitude applied on this lower level
column will change. Although the load magnitudes of individual
occupant is independent with each tenant change, the cumulative
load magnitude applied on this lower level column is not. Thus
it is desirable to further extend the reliability analysis to
include correlated load magnitudes.

    In the case of linear load effects, the analysis would be
simple because one can utilize the principle of superposition to
obtain the cumulative effects. However, in the case of nonlinear
response, the analysis would not be quite as simple since nonlinear response is load path dependent [4]. In this paper the
reliability analysis for single structural members subjected to
correlated live load processes with nonlinear response is discussed. A one-step correlated magnitude is assumed in the analysis. This assumption is of practical interest especially when
the number of occupants considered is small. Therefore the
cumulative load magnitudes are correlated more significantly
between any two magnitudes differed by one tenant change than
with those affected by two or more tenant changes. In addition,
it is assumed that the load arrivals of individual tenants is
Poisson distributed and the material follows a bilinear load-
deformation relationship with unloading parallel to the initial
elastic range [4].

2. NONLINEAR LOAD EXCEEDANCES

Independent, Identically Distributed load Magnitudes

    In the studies of load exceedances, an exceedance is defined
as the occurrence of a load whose magnitude is greater than a
certain threshold level. For linear analysis, the threshold
level remains constant throughout the analysis. In the nonlinear
analysis, exceedances based on threshold deformation are load
history dependent. The threshold force corresponding to the
predetermined threshold deformation changes with time [4]. Thus,
the exceedance rate is no longer constant. It is a process which
can be found via the conditional exceedance rate [7]. Since
sustained live loads follow a Poisson process, the conditional
exceedance rate $v(t)$ given that the effective threshold A_t at
time t is α_t becomes

$$v(t) = v_S G_S(\alpha_t) \tag{1}$$

in which v_S = the arrival rate of the load process, and $G_S(\cdot)$ =
the complementary cumulative distribution function (cdf) of the
load magnitude S. The unconditional exceedance rate at time t
becomes

$$v(t,A) = \int_{\alpha_t} v_S G_S(\alpha_t) f_{A_t}(\alpha_t) \, d\alpha_t \tag{2}$$

in which

$$f_{A_t}(\alpha_t) = \frac{\gamma}{\gamma-1} v_S t f_S(\xi) \exp\left[-v_S t G_S(\xi)\right] \tag{3}$$

in which γ = coefficient of reduced stiffness, $\xi = (\gamma\alpha_t-\alpha_0)/(\gamma-1)$ Since the load exceedance process N_t is in the family of Poisson process [4,7], it is possible to find the expected value of N_t once the exceedance rate is known.

Correlated Load Magnitudes

Consider a load process X which is a cumulation of m load processes. For instance, a load process of a column supporting m sustained loads. Let X_i be the ith cumulation of m sustained loads, i.e., $X_i = S_{i_1} + S_{i_2} + S_{i_3} + \ldots + S_{i_m}$. Since sustained load S is a Poisson process, the load process X also has Poisson arrivals with rate v_X. However, the probability density function (pdf) of X, f_X, is no longer iid as in the case of f_S. Therefore Eqs. 2 and 3 would not be applicable here. In order to incorporate the time dependent load magnitude X, it is, perhaps, easier to discretize the load process, and to derive the load exceedance process via its conditional situation.

For linear response where the threshold α is constant, the expected number of exceedances Y can be found as

$$E[Y] = \sum_{k=1}^{\infty} P_K(k) \sum_{i=1}^{k} G_{X_i}(\alpha) \tag{4}$$

in which K = the number of load arrivals (or changes) of X in time t, and

$$P_K(k) = \frac{(v_X t)^k \exp(-v_X t)}{k!} \tag{5}$$

As discussed in previous section, the threshold level is not constant for nonlinear response. Therefore Eq. 4 can be considered as a conditional expectation of the number of nonlinear load exceedances N given the effective threshold level for X_i is α_{i-1} (since the (i-1)th load determines the effective threshold level for the ith load),

$$E\left[N \text{ given } A_{i-1} = \alpha_{i-1}\right] = \sum_{k=1}^{\infty} P_K(k) \sum_{i=1}^{k} G_{X_i}(\alpha_{i-1}) \tag{6}$$

The unconditional expectation of N becomes

$$E[N] = \sum_{k=1}^{\infty} P_K(k) \left\{ \sum_{i=1}^{k} \int_{\alpha_{i-1}} G_{X_i}(\alpha_{i-1}) f_{A_{i-1}}(\alpha_{i-1}) \, d\alpha_{i-1} \right\} \tag{7}$$

in which

$$f_{A_{i-1}}(\alpha_{i-1}) = \frac{\gamma}{1-\gamma} \, f_{X_{i-1}}(\xi_{i-1}) \quad ; \quad \xi_{i-1} = \frac{\gamma \alpha_{i-1} - \alpha_0}{\gamma - 1} \tag{8}$$

In the derivation, no specific constraint is placed on the distribution of X_i except that the load magnitudes are assumed to be non-negative. For the example of a lower level column discussed in the Introduction, $f_{X_{i-1}}$ and f_{X_i} would be correlated probability density functions. The exact pdf will depend on the pdf of X's. For instance, if X_0 is Gaussian distributed, then X_1 has a conditional normal distribution conditioned on X_0. Similarly, X_2 is also conditional normally distributed, but is conditioned on X_1, and so on until X_k. If X_0 is not a Gaussian variate, then it would be a bit complicated in determining the conditional pdf for X_1 and susequent X_i's.

3. DISCUSSION

The concept of determining the expected number of nonlinear load exceedances with correlated sustained load process is presented. Although no verification is presented in the paper (it is currently being performed), the logic of the derivation appears to be meritorious. The results of this study will enhance one's ability to assess structural components that are subjected to loads from various contributions such as lower level column described in the Introduction. Although the analysis presented here was aimed at correlated load magnitudes, it was not one of the assumptions used in the derivation. The analysis is valid for X_i's that are correlated as well as variables with totally different probability models.

4. REFERENCES

[1] HARRIS, M.E., COROTIS, R.B. and BOVA, C.J.:"Area-Dependent Processes for Structural Live Loads", Journal of the Structural Division, ASCE, Vol. 107, No. ST5, pp.857-872, 1981.

[2] WEN, Y-K.:"Stochastic Dependencies in Load Combination", ICOSSAR'81, Trondheim, Norway, June 23-25, 1981.

[3] COROTIS, R.B. and TSAY, W-Y.:"Probabilistic Load Duration Model for Live Loads", Journal of Structural Engineering, ASCE, Vol.109, No.4, pp.859-874, 1983.

[4] CHOU, K.C., COROTIS, R.B. and KARR, A.F.:"Nonlinear Response to Sustained Load Processes", Journal of Structural Engineering, ASCE, Vol.111, No.1, pp.142-157, 1985.

[5] CHOU, K.C.:"Probabilistic Analysis of Nonlinear Response to Sustained Load Processes", Structural Safety, Vol.4, No.1, pp.1-13, 1986.

[6] CHOU, K.C. and THAYAPARAN, P.A.:"Nonlinear Structural Response to Live Load Processes", Journal of Structural Engineering, ASCE, Vol.114, No.5, pp.1135-1151, 1988.

[7] FLEISCHMAN, W.M. and CHOU, K.C.:"On Computing the PMF of the Number of Nonlinear Load Exceedances", Proceeding of the Symposium on Reliability-Based Design in Civil Engrg., Lausanne, Switzerland, pp.49-56, July 7-9, 1988.

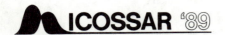
EXTREME VALUE DISTRIBUTION FOR STOCHASTIC PROCESSES
SUBJECT TO A FULL LOAD CLIMATE

J.B. Mathisen

A.S Veritas Research, P.O. Box 300, N-1322 Høvik, Norway

ABSTRACT

Reliability analysis of structures subject to environmental loads generally involves load effects which are properly treated as stochastic processes, and requires determination of the extreme value distribution due to these loads. For ships and offshore structures, the extreme value distribution is often obtained via combination of the distribution of the load effect for a set of environmental states, together with the probability of occurrence of those states. This procedure becomes cumbersome if the set of environmental states is large, or if the load effect requires much effort for each environmental state. An approximation for the extreme value distribution is presented, which only requires results from a limited number of environmental states. Bounds are included with the probability estimate. Results are obtained for an idealised offshore response to combined wind and waves.

KEYWORDS

stochastic processes, wave loads, wind loads, offshore structures

1. INTRODUCTION

Long term distributions of wave-induced loads acting on ships have been available as routine calculations for about two decades. Early descriptions of this type of calculation procedure are given by Jasper [1], and Bennet et al. [2]. The "long term" designation emphasizes that these marginal distributions correspond to long time periods, of the order of several years, as opposed to "short term" conditions of the order of one hour, for which response calculations can be based on an assumption of stationarity. Extreme value distributions are obtainable from the long term distributions.

In essence, the typical long term distribution procedure utilises a straight-forward calculation of response to Gaussian wave excitation under stationary environmental conditions, and then combines the conditional response distributions with the probability of occurrence of the environmental conditions. The environmental conditions are usually described by three random variables; namely the significant wave height, the wave zero-up-crossing period, and the direction of the waves relative to the ship. The range of environmental conditions is discretised into a set of environmental states, typically involving 20 significant wave heights, 20 wave zero-up-crossing periods, and 8 wave directions. Hence a total of 3200 discrete environmental states might be considered. Response calculations are required for each of these environmental states. This task might seem daunting, but actually only requires a moderate amount of computational effort when the response is linear.

Bitner-Gregersen and Haver [3] have recently developed a more detailed statistical description of the offshore environment, including wind and current speeds jointly with wave variables. Their

environmental model is intended to allow a more rational combination of the load effects due to these three environmental processes, with particular reference to jacket-type offshore structures. If the above procedure to calculate long term response distributions is generalised to include wind and current effects, then the computational effort required to consider the full range of environmental states is increased by one or two orders of magnitude. This quickly becomes impractical, and some means of making these calculations more efficient is required. In the present context, it is appropriate to focus on high response levels, and the corresponding low probabilities of exceedence. Experience shows that only a portion of the entire set of environmental states contribute significantly to the probabilities of exceedence at high response levels. Thus, some means of including only these "significant" environmental states in the computation is required. First or second order reliability methods, as described by Madsen, Krenk and Lind [4] can be applied to this problem. These methods essentially calculate a probability integral by analytical approximation in the vicinity of the point which contributes most to the integral. Although these methods are very efficient, they introduce continuity requirements which may not always be fulfilled. In this paper, a discrete search and summation procedure is followed, which parallels the reliability methods without placing the same continuity requirements, and which may also be seen as a generalisation of the method applied to ship response.

2. THEORY

It is assumed that the environmental conditions to which a structure is subjected may be described by a procession of piecewise stationary environmental states. A random vector Ψ is introduced to describe each stationary environmental state. The components of Ψ might typically include the significant wave height $\Psi_1=H_s$, the peak wave period $\Psi_2=T_p$, the wave direction $\Psi_3=\Theta_w$, etc. It is assumed that a joint distribution $F_\Psi(\psi)$ may be determined for these environmental variables, at the site under consideration. The domain of environmental conditions is divided into an exhaustive and disjoint set of m discrete environmental states, with the centre points of these states at ψ_i, $i=1,2,\ldots,m$, and the probability of occurrence of each state denoted by $p(\psi_i)$.

Given a stationary environmental state, then it is assumed that the conditional distribution $F_{Q|\Psi}(q|\psi)$ of individual response maxima Q to the environmental loads within that environmental state may be determined, together with the up-crossing rate $\nu(\psi)$ of the response through the mean level.

Finally, it is assumed that the response distribution in one environmental state is statistically independent of the response in any preceding state. Under these conditions, the theorem of total probabilities may be invoked to obtain the long term (marginal) response distribution as

$$F_Q(q) = \sum_{i=1}^m F_{Q|\Psi}(q|\psi_i)\, p(\psi_i)\, \nu(\psi_i) / \bar\nu \qquad (1)$$

where $\bar\nu$ is the long term mean up-crossing rate of the response, given by $\bar\nu = \sum_{i=1}^m \nu(\psi_i)\, p(\psi_i)$.

The ratio of up-crossing rates included in equation (1) is required to convert from the occurrence frequency of environmental states to the occurrence frequency of response maxima; cf. Battjes [5].

The expected number of response maxima during a time duration d is $n=\bar\nu d$. If the response maxima may be assumed independent, then the distribution $F_R(r)$ of extreme values R of the response maxima Q may be written

$$F_R(r) = [\, F_Q(r)\,]^n \qquad (2)$$

The assumption of independent maxima is questionable, particularly if the response process is narrow-banded. However, it is usually accepted to lead to a useful approximation at low probabilities of exceedence. A more theoretically satisfying description can be made by use of stochastic processes.

With the above formulation, the basic problem is the determination of the long term distribution $F_Q(q)$, and this is addressed in the following. Consider an expression for the probability of exceedence corresponding to the cumulative probability in equation (1). Suppose that the values of the summand in this equation may be determined in decreasing order of magnitude, and consider an estimate for the probability of exceedence obtained by taking the k largest summands

$$[1 - F_Q(q)]_{(k)} = \sum_{j=1}^k [1 - F_{Q|\Psi}(q|\psi_{\kappa(j)})]\, p(\psi_{\kappa(j)})\, \nu(\psi_{\kappa(j)}) / \bar\nu \qquad (3)$$

where $\kappa(j)$, $j=1,2,\ldots,k$ provides the index values of the leading summands in terms of the numbering scheme adopted for the discretisation of the environment; i.e. $\psi_{\kappa(1)}$ is the environmental state which contributes most to the response probability of exceedence. All the $m-k$ terms which are not included in the partial summation will contribute less than the last included term. Hence, the probability of exceedence is bounded as follows

$$[1 - F_Q(q)]_{(k)} \leq [1 - F_Q(q)] \leq [1 - F_Q(q)]_{(k)} + (m - k)\,[1 - F_{Q|\Psi}(q\,|\psi_{\kappa(k)})]\,p(\psi_{\kappa(k)})\,\nu(\psi_{\kappa(k)})/\bar{\nu} \qquad (4)$$

To implement a calculation of the bounds given in equation (4), algorithms are required to find the discrete environmental state giving the largest contribution to the response probability of exceedence, and the subsequent next largest contributions. In order to ease this task, the domain of environmental conditions is discretised in such a way that the centre points of any pair of adjacent cells differ only by changes in one component of the environmental vector ψ. The cell giving the largest contribution is basically found by moving stepwise between adjacent cells until a maximum is located. This cell is then included as the first contribution to the sum in equation (3). A list of perimeter cells adjacent to the included cell is established. The next included cell is the cell on the perimeter list with the largest contribution. The perimeter list is subsequently updated to include adjacent cells to the last included cell, which are not already included in the partial sum. This process may then be continued until the bounds in equation (4) have converged sufficiently. Circular variables, such as wave direction, can easily be implemented in this scheme by making 0° and 360° adjacent values.

3. TRIAL APPLICATION

A preliminary version of the joint environmental model developed by Bitner-Gregersen and Haver [3] has been utilised, including waves and wind, but initially omitting ocean current effects. The joint probability density function for the environmental random variables is here written

$$f_{\Psi}(\psi) = f_{\Psi_1|\Psi_3}(\psi_1|\psi_3)\;f_{\Psi_2|\Psi_1\Psi_3}(\psi_2|\psi_1,\psi_3)\;f_{\Psi_3}(\psi_3)\;f_{\Psi_4|\Psi_1\Psi_3}(\psi_4|\psi_1,\psi_3) \qquad (5)$$

where Weibull distributions are utilised for the significant wave height Ψ_1 and for the one-hour average wind speed Ψ_4 at 10 m above sea level. The peak period of the wave spectrum Ψ_2 has a log-normal distribution. Wind and wave effects are assumed co-directional in this model, with all the other variables conditional on the wave direction Ψ_3, which is, itself, discretised into 8 compass directions, with empirical probabilities of occurrence. The component distribution functions and numerical values of the distribution parameters are omitted here to save space, but may be found in reference [3], for an offshore site at Haltenbanken (65° 2' N 7° 33' E) off the North-West coast of Norway.

A very simple response model is adopted to ease experimentation with the long term calculation procedure. The total response X is taken as the sum of a wind load X_v and a wave load X_w, where the wind load is a random variable and the wave load is a stochastic process. The mean wind load under stationary environmental conditions is obtained as a drag load of the form $\bar{x}_v(\psi) = g(\psi_3)\,\psi_4^2$, where the drag load coefficient $g(\psi_3)$ varies with direction, and is given in Table 1. The standard deviation of the wave load under stationary conditions is written as

$$\sigma_w(\psi) = \psi_1 h_1(\psi_3)\,/\,\{\;\sqrt{8}\,[(1 - h_3\omega_p)^2 + h_2\omega_p]\;\} \qquad (6)$$

where the variation of the wave load with direction is arranged through the coefficient $h_1(\psi_3)$, which is also given in Table 1. This wave load is intended to be mildly resonant in nature, as determined by coefficients $h_2=0.5$ and $h_3=1.0$, giving a response peak when the peak wave period $\psi_2 = 2\pi/\omega_p = 6.3$s. Under stationary conditions, the maxima Q of the total response X are taken to be distributed according to a non-central Rayleigh distribution function

$$F_{Q|\Psi}(q\,|\psi) = 1 - \exp[-(q - \bar{x}_v(\psi))^2\,/\,(2\,\sigma_w^2(\psi))]\;;\quad q \geq \bar{x}_v(\psi) \geq 0 \qquad (7)$$

The up-crossing rate of the total response through the mean level, is taken as $\nu(\psi) = 1.4/\psi_2$.

Some experimentation has been carried out with the range of the environmental variables, and with the fineness of the discretisation. Relatively high values for the upper levels of wave heights, wave periods and wind speeds have been utilised, to ensure that no relevant environmental states are omitted from the summation. The present example converged adequately when 25 equi-spaced intervals were used for both wave height and wave period, 20 intervals for wind speed, and the number of directions was fixed at 8 sectors. The total number of environmental states was $m=100000$ with this discretisation. The resulting probabilities had converged to within about 20% of their final values after roughly $k=1000$ environmental states had been included in the summation. By only carrying out a partial

summation, the computational effort has been reduced by a factor of 100. Typical results are shown in figure 1. The upper bound shown on this figure corresponds to 5 times the probability of exceedence, and was used to truncate the summation at this level of accuracy, since little change in the estimated probabilities occurred if the summation was continued to include additional contributions.

Direction ψ_3	Wave load coeff. $h_1(\psi_3)$	Wind load coeff. $g(\psi_3)$
N	1.00	0.006
NE	0.85	0.005
E	0.70	0.004
SE	0.85	0.005
S	1.00	0.006
SW	0.85	0.005
W	0.70	0.004
NW	0.85	0.005

Table 1   Wave and wind load coefficients.

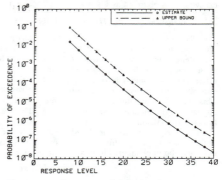

Fig.1   Long term distribution for
               combined wind and wave load.

4. CONCLUSION

A generalised discrete summation technique has been developed, based on the standard calculational procedures used in the determination of long term distributions of wave-induced loads, but providing higher computational efficiency by considering only the environmental states which contribute significantly to the probability of exceedence. Bounds are available for the estimated probability, but the accuracy of the method appears to be considerably better than indicated by the upper bound.

The method should be compared to asymptotic results based on first or second order reliability methods. It appears likely that the asymptotic methods will be more computationally efficient. However, the present method may be useful as a check on asymptotic results, and when the continuity requirements of the asymptotic methods are not met.

5. ACKNOWLEDGEMENT

This paper has been prepared within the research program "Reliability of Marine Structures," which is sponsored by Veritas Research, Saga Petroleum, Statoil, and Conoco. The opinions expressed herein are those of the author and should not be construed as reflecting the views of these companies.

6. REFERENCES

[1]   Jasper,N.H.: "Statistical Distribution Patterns of Ocean Waves and of Wave-Induced Ship Stresses and Motions, with Engineering Applications," Trans. SNAME, Vol.64, 1956.

[2]   Bennet,R., Ivarson,A., Nordenstrøm,N.: "Results from Full Scale Measurements and Predictions of Wave Bending Moments Acting on a Ship," Swedish Shpbldg. Rsch. Fnd. Rep. No.32, 1962.

[3]   Bitner-Gregersen,E.M., Haver,S.: "Joint Long Term Description of Environmental Parameters for Structural Response Calculation," 2nd Int. Workshop on Wave Hindcasting and Forecasting, Vancouver, 1989.

[4]   Madsen,H.O., Krenk,S. Lind,N.C.: "Methods of Structural Safety," Prentice-Hall, Englewood Cliffs, New Jersey, 1986.

[5]   Battjes,J.A.: "Long-Term Wave Height Distributions at Seven Stations around the British Isles," Deutsche Hydrographische Zeitschrift, Vol.25, No.4, 1972.

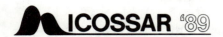
THE EXTREMES OF COMBINATIONS OF ENVIRONMENTAL LOADS

Marc A. Maes

Dept. of Mathematics and Statistics,
Queen's University, Kingston, Canada, K7L 3N6.
(formerly with Det Norske Veritas (Canada) Ltd)

ABSTRACT

The objectives and the general approach of an offshore environmental data and extremal analysis are described. The intention is to develop long-term design values of wave, wind and current processes which, together, generate extreme load effects in offshore structures. The analysis procedure is to take into account factors such as nonstationarity, auto- and crosscorrelation, nonlinearities and uncertainties in the load modeling, and arbitrary combinations of processes. A preliminary data inspection shows which analysis features and techniques are appropriate.

KEYWORDS

Extreme Values; Environmental Loads; Load Combinations; Uncertainties; Discrete Time Series Analysis; Design Criteria.

1. Introduction

The development of optimal design criteria for long-term wave, wind and current loading on offshore installations has recently received considerable attention [1,2]. Environmental variables with large return periods need to be determined from a detailed analysis of the extremes of the random load processes generated by the environment. The most widely used practical approach for the long term is to either fit an extreme value distribution to observed monthly or annual extreme winds, waves or currents - on the grounds that these data are, themselves, the largest of very many individual values - , or else to derive an extreme value distribution analytically from a marginal point-in-time distribution fitted to the bulk of data. In both cases, the r-year return value is then determined as a percentile from the extreme value distribution and is used as a parameter in design criteria.

There are several shortcomings to this approach. First, notwithstanding the fact that the procedure may provide suitable estimates of the extreme *environmental* parameters, say wind speed or significant wave height, it does not necessarily yield good estimates of the extreme *load effects* experienced by a structural member. Not only are there many nonlinearities involved in the procedure of transforming the former into the latter (e.g. wave load theories), but in addition, the consideration of unavoidable - and often sizable - modeling uncertainties influences the overall probability distribution of the extremes [2].

Second, the empirical extreme values approach does not lend itself to the analysis of the following critical question : which *combined* environmental conditions need to be considered in design ? This problem cannot be investigated if only the extremes of the contributing processes are analyzed instead of (1) their joint behaviour in time, and (2) the extreme values of the combined (total) load effect process generated by the environmental agents. Indeed, any suitable approach should be founded upon the reasonable assumption that the exceedance probability of any extreme load combination must be less than, and ideally (to spread design risks uniformly) as close as possible to the exceedance probability of the extreme loads acting in isolation.

Research currently in progress at Queen's in Kingston focuses on appropriate methods of analysis of long-term environmental data with the objectives (a) to account for uncertainties associated with factors other than the random environment itself, (b) to determine the environmental conditions which generate extreme load effects, rather than simply determining the extremes of the environmental parameters, and (c) to study and to transform into usable design criteria, the *combinations* of load effect processes generated by the "total" environment. In the present paper certain results from a preliminary investigation of long-term drill rig data achieved by the Canadian Climate Centre of Atmospheric Environment Services Canada, are used to challenge and/or investigate some of the critical features of suitable long-term extreme value analysis procedures.

2. Preliminary Data Inspection

It is instructive to start a preliminary analysis by simply plotting long-term observations of each environmental variable as a function of time. An inspection of the plots allows several preliminary conclusions to be drawn :

(1) The stochastic processes of the long-term environmental parameters, $X(t)$, are clearly not stationary. The nonstationarity of the mean is particularly evident when smoothing sequences are applied [3] such that seasonality emerges.

(2) Sequences of observations are correlated. They cannot be treated as stochastically independent.

(3) The long-term environmental processes are not Gaussian, even after removal of nonstationarity. Plots show that there is a clear asymmetry in favour of the higher $X(t)$ values.

(4) The processes are to be analyzed as discrete-parameter continuous processes; there is little advantage in working with correspondent continuous time series : this could possibly facilitate mid-range prediction but it would likely worsen the tail behaviour, which is of critical interest.

(5) Preliminary calculations suggest that the means and the stan-

dard deviations of $X(t)$ are roughly proportional. In other words, the random variability during a given season seems to be scaled in accordance with the seasonal mean itself. This would indicate that a logarithmic transformation of $X(t)$ is appropriate.

The findings under (1) and (2) warn against the unbridled use of an asymptotic extreme-value distribution for the extremes of an environmental process if its parameters are derived from observations that are assumed to be stationary and independent. Appreciably, the effects of nonstationarity and dependence are even more pronounced when extremes of combinations are studied.

3. Approach and Analysis

Mathematical results in the field of extremes of dependent random sequences are well described by several authors [4,5]. The associated techniques are mostly extensions of the classical asymptotic theory of extremes of independent and identically distributed random variables (iid) [6] under certain well defined, restrictive conditions. Suppose that it is decided to model nonstationarity of the mean only, and that the autocorrelation structure of the stationary difference process is such that correlation decays weakly with increasing distance between points in time, according to the criteria set out in [5,7], then certain extreme value results are applicable depending on the type of probability function chosen for the stationary difference process and the degree of nonstationarity of the mean. Specifically, Weibull, gamma [8] and, most prominently, normal [9] models have been developed for similar procedures in the field of air quality analysis and are being explored for the present application.

If the discrete long-term sample functions of waves, winds and currents recorded over a period of time of 1 to several years can be considered sufficiently representative of future weather patterns, then it is reasonable to assume that the distribution of the maximum total load effect generated by these environmental agents can be used as the basis for formulating extreme design criteria. For clarity, consider long-term stochastic processes **X** and **Y** of just two environmental variables X and Y, say wind speed and significant wave height, then a key objective of the research is to study the extreme value distributions of the following 3 random variables :

$$Z_1 = max \; (a_1 X + a_2 Y) \tag{1}$$

$$Z_2 = max \; (X^\alpha) \tag{2}$$

$$Z_3 = max \; (A.X) \tag{3}$$

as a function of the extreme values of max **X** and max **Y** themselves. The equations are sufficiently general to cover a wide range of combinations and transformations, including uncertainties represented by random variables or stochastic processes A. As an example of the type of result that should be produced in the course of the current research, consider Fig 1(a) which refers to equation (2) for Z_2 : δ_X is the coefficient of variation (cov) of the stationary difference process which is assumed to be lognormal with zero mean and stationary covariance matrix; the one-hundred year return period value of X, raised to the power α, is compared with the same value for X^α, i.e. the true extreme value. Fig 1(b) is a

similar plot of the extreme of $A.X$, where A is a random variable with cov δ_A, representing a model uncertainty. In both cases, mild nonstationarity is included using a four-term trigonometric polynomial.

Acknowledgements

Financial assistance from Queen's University and the National Science and Engineering Research Council of Canada is appreciated. Data access was kindly provided by Mr. V. Swail of the Canadian Climate Centre, A.E.S., Downsview, Canada.

References

[1] CANADIAN STANDARDS ASSOCIATION :"Code for the Design, Construction and Installation of Fixed Offshore Structures", Preliminary Standards S471,472,473,474 and 475, 1988 al 1989.

[2] MAES, M.A.:"A Study of a Calibration of the New CSA Code for Fixed Offshore Structures", Techn. Rep. No.7, 9 & 10, Environmental Protection Branch, COGLA, 1986.

[3] TUKEY, J.W.:"Exploratory Data Analysis", Addison - Wesley Publ. Cy. Inc., Reading, Mass., 1977.

[4] CRAMER, H. and LEADBETTER, M.G.:"Stationary and Related Stochastic Processes", Wiley, New York, 1967.

[5] LEADBETTER, M.R., LINDGREN, G., HOLGER, R.:"Extremes and Related Properties of Random Sequences and Processes", Springer Verlag, New York, 1983.

[6] GUMBEL, E.J.:"Statistics of Extremes", Columbia U.P., New York, 1958.

[7] BERMAN, S.M.:"Limit Theorems for the Maximum Term in Stationary Sequences", Ann. Math. Statist., 35, 502-516, 1964.

[8] BARLOW, R.E.:"Averaging Time and Maxima for Air Pollution Concentrations", Proc. 6th Berkeley Symp. on Math. Statistics and Probability, Vol.VI, pp. 433-442, 1972.

[9] HOROWITZ, J.:"Extreme Values from a Nonstationary Stochastic Process : An Application to Air Quality Analysis", Technometrics, 22, pp. 469-478, 1980.

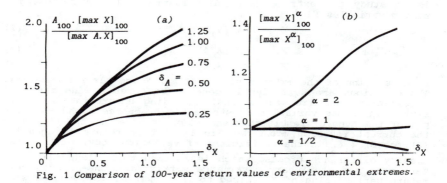

Fig. 1 *Comparison of 100-year return values of environmental extremes.*

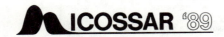 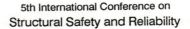
PROBABILISTIC MODELS FOR LOADS DUE TO BULK SOLIDS

Lam Pham

CSIRO Division of Building, Construction and Engineering
Graham Road, Highett, Victoria 3190, Australia

ABSTRACT

This paper presents two probabilistic models for loads due to bulk solids.
One model seeks to obtain an accurate presentation of the loads; the model
is used as a basis for a systematic treatment of available experimental data.
The other model seeks to account for various components of uncertainties
in the estimates of the loads; the model is used as a basis for the derivation
of a load specificaton.

KEYWORDS

Bulk solids, loads, probabilistic method, structural reliability.

1. INTRODUCTION

Although the variable nature of loads in bins and silos due to bulk solids is widely acknowledged
and reported in literature, there has been no systematic numerical assessment of this variability. The
sources of this variability are many and complex, but can be broadly classified into two groups: (a)
variability caused by the inherent stochastic nature of the loads, and (b) variability caused by the
uncertainties introduced in the process of assessing the loads for design. Probabilistic modelling
provides a rational framework for the assessment of both sources of variability. Two types of
probabilistic models are relevant to current needs: (a) a scientific model seeking to represent the
loads due to bulk solids in the most accurate manner taking account of their inherent variable
characteristics, and (b) a code model seeking to account for various components of uncertainties in
the load assessment process. The first is used to provide a framework for the systematic treatment
of available experimental data, while the second is used in the derivation of a load specification.
For simplicity, only normal wall pressure due to bulk solids is discussed in this paper.

2. PROBABILISTIC MODELS

2.1 Scientific Model

Preliminary statistical analysis of one set of experimental data [1] has shown that the total normal
wall load is the same during discharge as during storing and that the problem of discharging
pressure is a problem of load redistribution. A common model for both storing and discharging
pressure is therefore both possible and desirable.

For an axisymmetric container, the normal wall pressure p at a point (x,θ) (Fig. 1) is given by

$$p = p_1(x,H) + p_2(x,\theta,H,\tau) + p_3(x,\theta,H,\tau) \qquad (1)$$

where H is the surface height of the bulk solid mass; τ is the time sequence of storing and discharge; p_1 is the axisymmetric component of wall pressure representing the mean pressure at a given level; p_2 is the non-axisymmetric component of wall pressure representing the systematic variation caused by features such as eccentric filling and discharge; and p_3 is the random component taking into account influences such as wall imperfection, temperature changes, moisture movement, and time-dependent material properties (consolidation, creep, etc.). These influences are in principle not random but shall be treated as such because their precise influence on the loads are not known at present.

The above model has been used together with an harmonic analysis of five sets of experimental data on the same barley silo with the same eccentric filling and discharging conditions [2]. The results are presented in Fig. 2 together with the following comments:

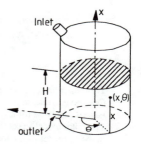

(a) The axisymmetric component p_1 is approximately the same for both storing and discharging conditions and can be represented by Janssen-type pressure distribution, i.e.

$$p_1 = (\gamma D/4\mu) [1 - \exp(-\mu K_j.4(H-x)/D)]$$

with γ = unit weight = 7.9 kN/m^3 (barley),
      μ = 'fitted' wall friction coefficient = 0.4, and
      K_j = 'fitted' ratio of vertical to horizontal pressure
          = 0.35.

Figure 1. Notation and coordinate system for axisymmetric bin

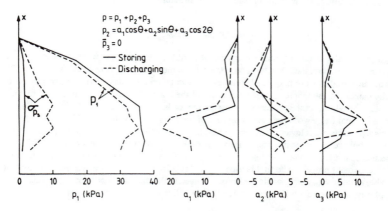

Figure 2. Components of normal wall pressures due to bulk solids

(b) The non-axisymmetric component p_2 can be represented by:

$$p_2 = a_1\cos\theta + a_2\sin\theta + a_3\cos2\theta$$

with the values of a_1, a_2 and a_3 as given in Fig. 2. The values of a_1, a_2 and a_3 are generally much larger for discharging conditions than those for storing conditions. The terms a_1 and a_3 represent the systematic variations caused by the eccentricities of filling and discharge. The presence of a non-zero value for a_2 is puzzling and can only be attributed to the peculiarities of the test silo.

(c) The random variation component p_3 can be represented by a zero mean distribution with standard deviation σ_{p3} as given in Fig. 2. Processing of further experimental data is needed before p_2 and p_3 can be expressed analytically.

2.2 Code Model

At present all bulk solid loading codes specify axisymmetric peak pressure envelopes as the basis for design [3-5]. The aim of the code model is to assess the variability associated with the determination of this peak pressure envelope. The normal wall peak pressure envelope can be represented by

$$p = kp_J \tag{2}$$

where p_J is the corresponding Janssen pressure, a random variable which is a function of the container geometry and bulk solid properties; and $k = k_1 k_2 \ldots k_N$ where k_1 is a factor to account for the variable filling or discharging condition, k_2 a factor for the effect of variable temperature, k_3 for moisture content, k_4 for wall flexibility, etc.

Unfortunately, very little reliable experimental data are available for many of the above effects, and only two factors can be taken into account quantitatively: (a) variability of properties of bulk solids, and (b) variability associated with centric filling and discharging conditions. Due to space limitations, the influences of these factors can only be discussed in general terms.

2.2.1 Properties of bulk solids

There has been little published data on the variability of properties of bulk solids until recently [6]. For the code model, this variability is reflected in the variability of the Janssen pressure. Simulation studies [7] have shown that the variability of the Janssen pressure is dominated by the variability of the internal friction property of the material for squat containers, while for deep containers the dominating factor is the variability of wall friction. Variability caused by the variation in bulk solid properties is a significant but not a dominating component of the total variability. The coefficient of variation of properties of bulk solids is of the order of 10-20% and the resulting coefficient of variation of Janssen pressure is about 25%.

2.2.2 Factors for storing and discharging pressures

An examination of published experimental data [8] shows that the mean observed axisymmetric component of pressure under storing is about 1.08 times Janssen's pressure with a coefficient of variation of 18%. The unsymmetrical component of normal wall pressure however can be quite

large in some experiments. This is caused by the eccentricity of filling conditions, wall imperfection and the anisotropic characteristics of the bulk solids. Under centric discharge, the observed mean peak normal pressure is about 1.6 times the static Janssen pressure with a coefficient of variation of 33%. As seen in the scientific model, a large non-axisymmetric component of normal wall pressure could be present due to the eccentricity of the discharge.

2.2.3   Application

Based on the above consideration, it can be shown that the current Australian specified normal wall pressure values [5] (the most severe in the world) are about the 90-percentile values of the peak loads [7].

3. CONCLUSIONS

Two probabilistic models for normal wall pressure due to bulk solids have been presented. Their use as a basis for a systematic treatment of experimental data and as a method for assessing the probability levels of current specified loads are demonstrated.

4. REFERENCES

[1] PHAM, L, NIELSEN, J. and MUNCH-ANDERSEN, J.: 'Statistical characteristics of silo pressure due to bulk solids', Proc. Second International Conference on Bulk Materials Storage, Handling and Transportation, Wollongong, Australia, pp. 132-136, July 1986.

[2] OOI, J.Y, PHAM, L. and ROTTER, J.M.: 'Systematic and random features of measured pressures on silo walls', Proc. Third International Conference on Bulk Materials Storage, Handling and Transportation, Newcastle, Australia, June 1989.

[3] DIN: 'Design loads for buildings: loads on silos', German Standard Sheet 6, February 1984.

[4] ACI: 'Recommended practice for design and construction of concrete bins, silos and bunkers for storing granular materials', ACI 313-77, American Concrete Institute, 1977.

[5] THE INSTITUTION OF ENGINEERS, AUSTRALIA: 'Guidelines for the assessment of loads in bulk solid containers', I.E.Aust, Canberra, 1986.

[6] PHAM, L.: 'Reliability of mass flow prediction', Proc. Second International Conference on Bulk Materials Storage, Handling and Transportation, Wollongong, Australia, pp. 237-240, July 1986.

[7] PHAM, L., ROTTER, J.M., and GORENC, B.E.: 'The use of probability methods in the derivation of bin loads for structural design', Proc. Second International Conference on Bulk Materials Storage, Handling and Transportation, Wollongong, Australia, pp. 104-109, July 1986.

[8] PHAM, L.: 'Variability of bin loads due to bulk solids for structural design', *I.E.Aust., Civil Engg Trans*, Vol. CE27, No. 1, pp. 73-78, February 1985.

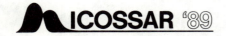
LOAD MODELLING FOR SINGLE STORY STEEL BUILDINGS

H. Pasternak

Institute of Steel Structures, Technical University
Carolo-Wilhelmina, Braunschweig, Germany

ABSTRACT

Realistic load assumptions for single story buildings are thus far complicated by the insufficient
state of the modelling of individual loads as random variables or stochastic processes and the
works on a procedure which takes the diverse load
models into consideration and nonetheless combines
individual loads in a practicable way. The Braunschweig research program comprises, above all, the
compilation of a universal program for load combination by utilizing the point-crossing method, the
modelling of crane loads as well as the integration of known models for other actions. Experiences and initial results will be reported.

KEYWORDS

Single story buildings; load models; load combination; point-crossing method.

1. Introduction

Single story steel buildings are among the typical products of
the steel structure industry; e.g. in Germany approximately 40
per cent of these buildings are constructed of steel.

Characteristic for this category of structures is a large number
of actions having stochastic properties, moreover its probabilities of occurence may be quite variable and its ordinates vari-
ously distributed. In addition to this is the partially spatial
structure of the applied loads. Fig. 1 verifies the random character of internal forces (long-term measurements) for a chosen
cross brace of an erection bay column (N - normal distribution
approximation). The evaluation of the cases involving damages also shows: with current load assumptions, in general, sufficient

safety is attained in the ultimate limit states. The situation in
the serviceability limit states, however, frequently unsatisfactory, they are often in accordance with, neither qualitatively
nor quantitatively, the real utilization. Improved load assumptions also belong to an improvement of this state.

2. Load Modelling and Combination

Load models and a procedure that allows the combination of any
load model (random variable and random process) should take the
facts mentioned above into account but also remain comprehensible
with respect to the re-
quired data. The latter
is especially important
because of the mass cha-
racter of single story
steel buildings. They
should also give suffi-
ciently accurate results
for the safety analysis
according to level III
and should be suitable
for the partial safety as
well as load combination
factor calibration for
today's code purposes. On
the basis of these consi-
derations, a combination
of classic reliability
analysis methods with the
point-crossing method was
chosen.

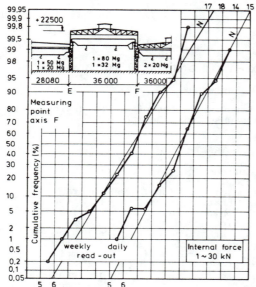

Fig.1 Example of a fre-
quency distribution

Within the framework of
the research project, the
available informations on
the modelling of the ac-
ting loads were first gathered. Using selected data, it was begun
with the verification of known models and, if required, new models were proposed. For all load models, the PDF and, if applicable, the upcrossing rate will be compiled into a process base.
For the sake of simplicity it is assumed that all actions are
stationary and completely correlated or not correlated and that
all load influence coefficients are scalar. The following were
analysed to now: the dead load of supporting and non-supporting
members, out-of-plumb, as well as snow and wind loads, vertical
and horizontal traveling crane loads and load caused by material
storing processes; later gradually will be introduced: ground
settlement, prestressing, temperature load, customary operational
impact and paraseismic loads. Current model recommendations and
specifications as to data availability are collected in Fig. 2.

Exemplary data concerning modelling of crane loads. Along with
the registration of deterministic values, the characteristics of

the respective transport process as typical input data were to be considered, i.e. typical PDFs of the imposed load, PDFs of the truck position (uniform or one-sided running), the frequency with which a column (as a result of crane operation) is loaded. Crane loads are location dependent. It is a part of actual practice to investigate plane frames; with regard to a yet surveyable reliability analysis, the continuation of this practice is sensible.

Load description		Model		Distribution of amplitude	Data
dead load			random variable	normal	available
out-of-plumb			random variable	normal	available
ground settlement			random variable	gamma	incom-plete
prestressing			random variable	normal	insuffi-cient
snow load			triangle or exponentially declining renewal pulse process	lognormal	available
wind load			continuous process	Weibull	available
temperature load			2 deterministic sinusoidal curves + Gaussian process	normal	available
vertical crane load			rectangular renewal pulse process	different	incom-plete
horizontal crane load			2 filtered Poisson processes	different	incom-plete
load caused by material storage			rectangular renewal pulse process	gamma	insuffi-cient
impact & paraseismic load			filtered Poisson process or triangle renew. pulse process	lognormal or gamma	insuffi-cient

Fig.2 Proposed load models

The problem of location dependent actions for structures can, at least approximately, be solved, in those for location dependent influence factors c(x) conservative approximations with scalar values c can be found. From this follows: Vertical crane loads can be figured as a rectangular or trapezoidal renewal pulse pro-

cess. The PDF of the load amplitude may have symmetrical or asymmetrical distribution and a spike at zero corresponding to the "off"-times. Additionally (sensible only in the case of great inaccuracy), the influence of crane runway imperfections on the vertical load may be approximated by a Gaussian process with "off"-times. The model for horizontal loads includes loads caused by oblique position of the traveling crane (frequently useful model: stationary Gaussian process with "off"-times and truck braking (frequently useful model: filtered Poisson process with normal distributed amplitude). For better simplification the load from crane oblique position can also be regarded as a filtered Poisson process. The correlation between vertical and horizontal loads decreases with increasing inclination of the crane wheels and increasing imperfections of the crane runway in the horizontal plane; in many cases it can be neglected.

The calculation of the upcrossing rate for n time dependent load components demands, as is known, the multiple convolution of n-1 PDFs and an upcrossing rate. After addition of the part of time invariant loads, the upcrossing probability of a given threshold by the defined load combination is known (this procedure is sensible in regard to the ultimate limit states in deterministic codes and serviceability limit states). When considering the resistance as a random variable, a further integration with its PDF gives the failure probability of the investigated (sub)structure.

3. Computer Program

After a few trials, a computer program is now being developed, using the above described procedure. For each of the time dependent loads a process can be chosen, for which the necessary characteristics of the PDF of the load amplitude, the upcrossing rate of the load process and the influence factor are to be entered. For every time invariant load, the characteristics of its PDF and the influence coefficient are to be entered. In order to simplify the load combination procedure for the user it is expedient that the program sorts elementary input data to a a-priori load assumption (eg., with the input of the truss type, the imposed load and the span width one obtains the PDF of the dead load). Alternatively, the load data can be directly input. The convolution operations have been performed using a FFT algorithm. As output, one can obtain either the probability of the upcrossing of defined thresholds by the load combination, or with input PDF of the resistance the mentioned failure probability.

4. Conclusions

The introduced procedure was successfully employed during the investigation of crane loads. At this time the combination of the horizontal and vertical loads of two cranes on one craneway as well as in neighboring bays is being extensively analysed. The future aim is to investigate structures under all relevant loads. The integration of current knowledge about individual loads into simple models for the load combination has already appeared as one difficulty.

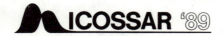
STATISTICAL ANALYSIS OF AUSTRALIAN GROUND SNOW LOADS

Lam Pham

CSIRO Division of Building, Construction and Engineering
Graham Road, Highett, Victoria 3190, Australia

ABSTRACT

This paper presents a statistical analysis of ground snow data from the alpine areas of south-eastern Australia (above 1200 m). The characteristic values of ground snow load S_G (in kPa) and ground snow depth D (in metres) can be represented by $S_G = 3.74\,D^{1.20}$. The mean relation between S_G (in kPa) and site elevation H (in metres) is $S_G = (H/1000)^{4.4}$. Terrain has considerable effects on ground snow distribution; the main factors are wind exposure, sun exposure, topography and distance from snow-producing ridges. A system of terrain classification has been devised to improve the specification of ground snow loads.

KEYWORDS

Ground snow, loads, snow loads, statistical analysis, terrain classification.

1. INTRODUCTION

This paper is concerned with the statistical analysis of snow data from the alpine areas of south-eastern Australia (above 1200 m). These areas are shown in Fig. 1. In these areas, the local variation is large and contour mapping is not meaningful. In most countries, the snow loads for similar locations are site specific. Unfortunately for Australia, locations where snow data are available are not those where developments occur; thus it is necessary to develop a general procedure to estimate snow loads for any site. In this paper, Australian ground snow load data are first analysed, the effects of terrain on the distribution of ground snow loads are then discussed, and finally the proposed Australian method of specifying ground snow loads for the alpine area is described.

2. AUSTRALIAN GROUND SNOW LOAD DATA

2.1 Characteristic Values

Annual maximum values of snow depths and loads have been extracted from records of snow data from the Snowy Mountain Hydro Electric Authority, the State Electricity Commission (Victoria) and the Road Construction Authority (Victoria). Characteristic values are defined as the annual maximum values with a five percentile chance of exceedance in any one year. They have been computed from the means and coefficients of variation of the annual maximum values using an extreme value distribution Type I [1] and alternatively a log-normal distribution [2] with negligible differences.

2.2 Relation Between Characteristic Values of Ground Snow Loads and Ground Snow Depths

It is necessary to establish this relationship because, while ground snow depths data are available for all sites, ground snow load data are only available for a limited number of sites.

The characteristic values of ground snow load S_G (in kPa) and ground snow depth D (in metres) are well correlated and can be represented by

$$S_G = 3.74 \, D^{1.20} \qquad (1)$$

with a coefficient of determination $R^2 = 0.967$ (Fig. 2). Australian snow is considerably wetter than other countries'; the average 'design' density of snow is 42% of water with a coefficient of variation of 18% compared with a European average density of less than 30% [1].

2.3 Relation Between Characteristic Ground Snow Loads and Elevations

Index snow courses are locations that the Snowy Hydro Electric Authority has been using as major reference points for its hydrological estimates. There is a good correlation between the values of characteristic ground snow loads and elevations for the index snow courses (marked in Fig. 3 as ■). The relation between the values of characteristic ground snow loads S_G (in kPa) and elevation H (in metres) for the index snow courses is given by

$$S_G = (H/1000)^{4.4} \qquad (2)$$

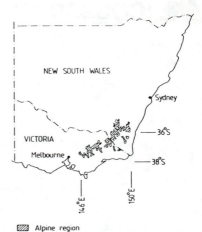

Figure 1. Map of south-eastern Australia showing alpine regions under consideration (above 1200 m)

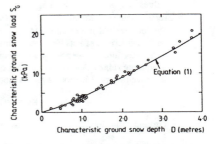

Figure 2. Relation between characteristic values of ground snow loads and ground snow depths

Applied to all available data, the measured characteristic snow loads can be from 0.3 to 1.7 times the value predicted by equation (2) with a coefficient of variation of about 45%. Equation (2) is therefore not accurate enough on its own to be used for ground snow specification, and the effects of terrain on ground snow distribution have to be studied.

3. EFFECTS OF TERRAIN ON GROUND SNOW DISTRIBUTION

3.1 Factors Affecting Ground Snow Distribution

In alpine areas of mainland Australia, years with deep snow accumulation have been known to occur with strong and persistent westerly wind. In strong westerly airflows, the mountains produce ascending air to their west, snow forms in the ascending current, carries across the barrier and

deposits on the eastern side. From an examination of the sites with ground snow much greater or much smaller than that predicted by equation (2), the following interrelated factors are known to have effects on the accumulated amount of snow on the ground: (a) wind exposure, (b) sun exposure, (c) topography, and (d) distance from snow-producing ridges.

3.2 Terrain Classification

A system of terrain classification has been devised with the specified snow load 30% larger than that given by equation (2) (Terrain Classification 1), equal to that given by equation (2) (Terrain Classification 2), and 30% smaller than that given by equation (2) (Terrain Classification 3). The classification of an area is given by a numerical method obtained from an assessment of wind exposure, topography and distance from snow-producing ridges in accordance with the following points scales.

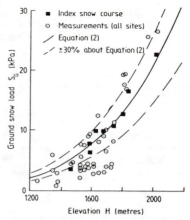

Figure 3. Relation between characteristic ground snow load and elevation

(a) Wind exposure: 1 = exposed, 2 = average, 3 = sheltered.
(b) Topography: 1 = ridge, 2 = valley and slope except those facing eastward, 3 = slope facing eastward.
(c) Distance from snow-producing ridge: 1 = distance greater than 5 km, 2 = distance between 5 km and 0.5 km, 3 = distance less than 0.5 km.

From the above, a terrain classification can be established by totalling the points as follows:

Terrain classification	Number of points
3	≤ 5
2	$>5 <7$
1	≥ 7

The result of the application of the above system to all sites with known snow measurements are presented in Table 1, from which it can be seen that the system does provide an improvement in the specification of ground snow loads.

4. CONCLUSION

A statistical analysis of Australian snow load data has been presented. The relationship between characteristic values of ground snow loads and ground snow depths has been developed. On average, Australian snow contains 40% more water than European snow. Characteristic ground snow load is strongly correlated with elevation, however the effect of terrain on the ground snow distribution is considerable. A system of terrain classification has been proposed which will considerably improve the snow load specification.

Table 1

Location	No. of sites	Measurement/specification			
		Mean	Cov	Max.	Min.
Snowy Mountains – Terrain Class 1	14	1.0	0.15	1.27	0.93
2	13	0.9	0.25	1.21	0.42
3	26	0.8	0.40	1.38	0.47
Victorian Alpine – all terrain classes	7	1.1	0.15	1.26	0.79
All sites without terrain classification	60	0.85	0.45	1.70	0.30

5. ACKNOWLEDGMENT

The author wishes to thank Mr J. Grimstead of the Snowy Mountain Hydro Electric Authority for supplying the data on Snowy Mountain sites and Mr J. Colquhuon of the Australian Bureau of Meteorology for information regarding snow formation.

5. REFERENCES

[1] INTERNATIONAL STANDARD ORGANISATION: 'Snow loads on roofs', ISO 4355, 1981.
[2] ELLINGWOOD, B. and REDFIELD, R.: 'Ground snow load for structural design', *Journal of Struct. Eng., ASCE*, Vol.109, No.4, pp.950-964, April 1983.
[3] STANDARDS ASSOCIATION OF AUSTRALIA: Draft Australian Standard AS 1170.3, 'Minimum design loads on structures – Part 3: Snow loads', DR88133, August 1988.

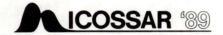
STRUCTURAL RELIABILITY OF WOOD BEAM–COLUMNS

Ricardo O. Foschi* and Felix Z. Yao**

*Professor and **Research Engineer, Dept. of Civil Engineering
University of British Columbia, Vancouver, B.C. Canada

ABSTRACT

A formulation for reliability–based design procedures of wood columns and beam columns is presented. The load carrying capacity of these structural members is obtained using a non–linear finite element model. The adopted structural analysis method is applicable over the full range of slenderness ratios. A limit state design equation is proposed which includes specified compression and bending strengths, resistance factors and an adjustment for slenderness. Reliability calculations are carried out using *FORM* procedures.

KEYWORDS

Reliability–based design; columns; beam–columns; finite elements; code calibration.

1. INTRODUCTION

A typical beam column, under an eccentric axial load P (with an eccentricity e) and a uniformly distributed transverse load Q, is illustrated in Fig.1. In the analysis which follows it is assumed that the beam–column is laterally restrained. The mode of failure of such a member depends on its slenderness ratio, λ. For beam–columns of rectangular cross–section $\lambda = L/H$, which is the ratio of span to section depth in the direction of bending. The relationship between the mode of failure and the slenderness ratio will first be discussed in the context of simple columns. For columns with low slenderness, the load–carrying capacity is controlled by the compression strength. At the other extreme, the capacity of high slenderness columns approaches that given by elastic Euler buckling and is controlled by stiffness. The capacity of columns of intermediate slenderness is more difficult to estimate, as their behavior is now determined by the material nonlinearities in tension and compression. Traditionally, design

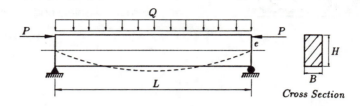

Figure 1: Typical Layout of a Beam–Column.

codes have distinguished between "short", "intermediate" and "long" columns, using Euler's equation for high λ–values, straight compression for low λ–values and a suitable interpolating equation for intermediate λ–values.

The reliability of a column is of course influenced by the randomness in the loads and the material properties, plus the uncertainties in load eccentricities and column out–of–straightness. For wood, the randomness in the distribution of material properties along the length of the column may also be an important factor. In general, if a procedure to calculate the load–carrying capacity P_{max} of a column is available, when values for all the intervening variables are given, the reliability may be estimated by considering the performance function

$$G = P_{max} - P = P_{max} - (P_d + P_l) \tag{1}$$

where P_d and P_l refer, respectively, to the dead and live load components of the total axial load P.

For a beam–column, when in addition to P there is a transverse load Q, the safe domain and failure domain will be separated by an interaction curve as schematically shown in Fig.2. In this case, the performance function G may be written as the difference between the lengths of the vectors OB and OA (see Fig.2), or

$$G = [P_{max}^2 + Q_{max}^2 L^2] - [(P_d + P_l)^2 + (Q_d + Q_l)^2 L^2] \tag{2}$$

where P_{max} and Q_{max} satisfy the interaction equation and correspond to point B in Fig.2. In Eq.(2) the components Q_d and Q_l correspond to the dead and live components of the transverse load Q.

In general, explicit expressions are not available for the evaluation of P_{max} and Q_{max}. This paper outlines a procedure for the calculation of P_{max} and/or Q_{max}, based on a non–linear finite element analysis of the column or beam–column. The reliability of these structural members is then obtained using $FORM$, with Eqs. (1) or (2) describing the limit state condition. These results are used to calibrate a limit state design equation, which is valid for all slenderness ratios.

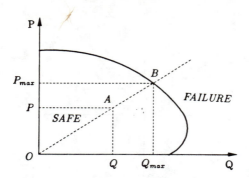

Figure 2: $P - Q$ Failure Interaction Relationship.

2. STRUCTURAL ANALYSIS

The structural capacity of a beam–column is obtained via the finite element method. Figure 3 illustrates the discretization of a beam–column using one–dimensional beam finite elements. These have two nodes, each with four associated degrees–of–freedom: u, du/dx, w and dw/dx. The displacements u and w are thus interpolated within an element using cubic polynomials. With the assumption of plane sections remaining plane, the strain ϵ at a distance z from the centroidal axis, is given by

$$\epsilon = \frac{du}{dx} + \frac{1}{2}\left(\frac{dw}{dx}\right)^2 - z\frac{d^2w}{dx^2} \tag{3}$$

The assumed non–linear stress–strain relationship is shown in Fig.4. Tensile stresses obey a linear elastic law, with modulus E, up to a maximum stress F_t at which point brittle failure occurs. Compression stresses obey a linear elastic law, with modulus E, up to a maximum stress F_c, followed by a linear softening branch with modulus mE ($m < 0$).

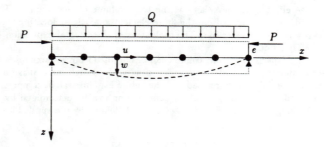

Figure 3: Finite Element Discretization of the Beam–Column.

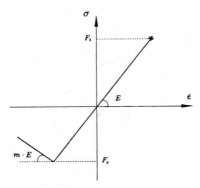

Figure 4: Stress–Strain Relationship.

The assumptions of large deformations, as expressed by Eq.(3), plus nonlinear material behavior in compression, result in a nonlinear problem which can be solved by standard iterative techniques like, for example, the Newton–Raphson procedure [4]. Failure of this method to converge to a solution implies that the applied load exceeds an ultimate load controlled by compression. On the other hand, convergence to a solution does not always means survival: failure by tensile stresses would follow if these satisfy the criterion for brittle fracture. Such brittle behavior implies strength dependency on specimen size and stress distribution. The brittle fracture criterion used [1], links the strength of the column during bending to the strength that it would have if tested in pure tension. Thus, if σ_{max} is the largest tensile stress produced by bending, and σ are the bending tensile stresses distributed over the volume V^+, failure will occur when

$$\sigma_{max}^k \int_{V^+} \left(\frac{\sigma}{\sigma_{max}} \right)^k dV \geq F_t^k V \tag{4}$$

where k is the "size effect" parameter and V the total volume of the column. The actual volume V^+ in tension is determined from the finite element analysis by locating the position of the neutral axis at each cross–section. The evaluation of the integral expression in Eq.(4) is performed using Gauss quadrature.

The load–carrying capacity P_{max} of a column can be obtained from a search following finite element analyses at different applied load levels. Initially, the search interval is bounded by zero and the lower of the Euler buckling load and the failure load in straight compression. The load–carrying capacities P_{max} and Q_{max} of a beam–column can be determined by a similar incremental search along the vector direction OA of Fig.2, defined by the applied loads P and Q.

3. RELIABILITY ANALYSIS

Reliability analysis was carried out using the performance functions given by Eqs.(1) and (2). The gradient of Eqs.(1) or (2), required by *FORM*, was obtained by numerical differentiation using the values P_{max} and/or Q_{max} computed via the finite element analysis. The

capacities P_{max} and/or Q_{max} were assumed to depend on the following random variables: the strengths F_c and F_t, the modulus of elasticity E, the axial end eccentricity e and the dead and live load components. All other variables (m, k, B, H, L) were assumed deterministic. The random material variables were assumed to be distributed according to 2–Parameter Weibulls. A positive correlation was assumed between F_c and E and F_t and E. For both cases, the correlation coefficient was taken to be 0.60. These correlated variables were treated using the procedure proposed by Der Kiureghian [2]. The eccentricity e was assumed normally distributed, with a mean of $0.05H$ and a coefficient of variation of 20%. The live loads considered were produced by snow. These were taken to be the maxima occurring in a 30–year design life, and were represented by Extreme Type I distributions. The dead load component was assumed normally distributed.

The proposed limit state equation has the form

$$\left(\frac{\alpha_d P_{dn} + \alpha_l P_{ln}}{\phi_c F_{cn} A K}\right) + \left(\frac{\alpha_d E(Q_{dn}, P_{dn}) + \alpha_l E(Q_{ln}, P_{ln})}{\phi_b F_{bn} S}\right) = 1.0 \tag{5}$$

where (P_{dn}, P_{ln}) and (Q_{dn}, Q_{ln}) are, respectively, the dead and live load components of the nominal (design) loads P_n and Q_n. The nominal dead load component is usually the mean value, while the nominal live load is a load with low exceedance probability. In this work, the nominal live load equals the 30–year return snow load. The bending effects $E(Q_{dn}, P_{dn})$ and $E(Q_{ln}, P_{ln})$ are un–amplified maximum bending moments produced by the applied loads. For the type of beam–column shown in Fig. 1

$$E(Q_{dn}, P_{dn}) = -P_{dn} \cdot e_n + \left(\frac{Q_{dn} \cdot L^2}{8}\right) \tag{6}$$

where e_n is a nominal end eccentricity. Also in Eq.(5), α_d and α_l are the load factors associated with the dead and live loads (in Canadian Limit State Design codes $\alpha_d = 1.25$ and $\alpha_l = 1.50$); ϕ_c and ϕ_b are resistance factors associated, respectively, with the compression and bending capacities; F_{cn} is a nominal compression strength and F_{bn} the corresponding nominal bending strength; A is the cross–sectional area and S is the section modulus. Finally, K is a slenderness adjustment factor which is function of the slenderness ratio λ. The form of K must be as shown in Fig.5, starting from 1.0 at $\lambda = 0$ and approximating Euler's equation for high values of λ. In the transition region, corresponding to intermediate slendernesses, the values of K are usually interpolated using empirical results. In the reliability context, the "goodness" of the interpolation should be judged in terms of the variation in the reliability index β as the slenderness ratio is changed. In this work, the following expression for K was used [3]

$$K(\lambda) = \left(1.0 + \frac{F_{cn}\lambda^3}{N E_n}\right)^{-1} \tag{7}$$

where E_n is a nominal modulus of elasticity, and N is a calibration parameter. The value of N can be obtained by calibrating Eq.(7) to Euler's equation for a high value of λ. If F_{cn} and E_n are 5th–percentile values, ratios of E_n/F_{cn} between 300 and 500 are common for wood. Accordingly, values of N between 34 and 49 would be required to calibrate Eq.(7) to Euler's equation in the range $\lambda = 30$ to 50. The value $N = 35$ was chosen as a conservative value.

Defining the ratios $\gamma_p = P_{dn}/P_{ln}$, $\gamma_q = Q_{dn}/Q_{ln}$ and $\varepsilon = (Q_{ln} \cdot L)/P_{ln}$ permits Eq.(5) to be rewritten as

$$P_{ln} = \left\{\left(\frac{1.25\gamma_p + 1.50}{\phi_c F_{cn} A K}\right) + \left(\frac{1.25(-e_n\gamma_p + \varepsilon\gamma_q L/8) + 1.50(-e_n + \varepsilon L/8)}{\phi_b F_{bn} S}\right)\right\}^{-1} \tag{8}$$

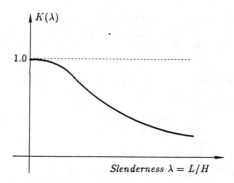

Figure 5: Relationship between K and the Slenderness Ratio λ.

Note, the quantity ε represents the ratio of the transverse live load to the axial live load. The applied loads P and Q can also be written as

$$P = P_{ln}[\gamma_p d_p + l_p]; \qquad Q = P_{ln} \cdot (\frac{\varepsilon}{L})[\gamma_q d_q + l_q] \qquad (9)$$

where $d_p = P_d/P_{dn}$, $l_p = P_l/P_{ln}$, $d_q = Q_d/Q_{dn}$ and $l_q = Q_l/Q_{ln}$. Using Eqs.(8) and (9), the performance function given by Eq.(2) and the associated reliability can be evaluated for different factors ϕ_c and ϕ_b, ratios ε, γ_p and γ_q and slenderness ratio λ.

The reliability analysis results presented herein correspond to one wood species classification (Spruce–Pine–Fir, Select Structural), one cross–section (38 mm x 184 mm), with material properties as specified in Table 1. The snow load cases for three locations in Canada were considered. The live load components l_p and l_q were taken to be the product $l_G \cdot r$, where l_G is the normalized load based on ground snow and r is an adjustment for snow on roofs. Statistics for l_G and r are shown in Table 2. Table 3 gives the reliability indeces for columns, obtained for different performance factors ϕ_c, at different γ_p and slenderness λ. The reliability analysis results pertaining to beam–columns are presented in Fig.6. These results are for the Ottawa snow load case, using $\phi_c = 0.80$, $\phi_b = 0.90$, $\gamma_p = 0.25$ and $\gamma_q = 1.00$. This figure shows the variation in β with ε when $\lambda = 10$ and $\lambda = 40$.

Table 1: Material Parameters.

Parameter	2–P Weibull Distribution	
	Mean (MPa)	Std. Dev. (MPa)
E	1.01×10^4	1.92×10^3
F_t	18.19	3.22
F_c	26.54	3.99

Table 2: Snow Load Statistics.

City	l_G, Extreme Type I		r, Lognormal	
	Mean	Std. Dev.	Mean	Std. Dev.
Arvida	1.08	0.18	0.60	0.27
Saskatoon	1.10	0.21	0.60	0.27
Ottawa	1.11	0.23	0.60	0.27

Table 3: Reliability index β for Columns.

Location	γ_p	λ	$\phi_c = 0.4$	$\phi_c = 0.6$	$\phi_c = 0.8$
	0.25	10	4.11	3.23	2.73
		20	4.05	3.34	2.81
		40	4.16	3.57	3.10
Arvida	1.00	10	3.23	3.59	2.89
		20	4.23	3.47	2.90
		40	4.20	3.61	3.14
	4.00	10	4.24	3.42	2.68
		20	3.95	3.22	2.67
		40	4.04	3.42	2.90
	0.25	10	4.04	3.27	2.68
		20	4.01	3.27	2.71
		40	4.12	3.62	3.06
Saskatoon	1.00	10	4.27	3.55	2.82
		20	4.12	3.46	2.86
		40	4.19	3.59	3.13
	4.00	10	4.24	3.42	2.69
		20	4.03	3.22	2.66
		40	4.04	3.41	2.90

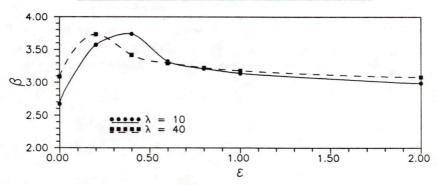

Figure 6: $\varepsilon - \beta$ Relationship for Beam–Columns.

It is of interest to note that the search for ultimate loads, with a finite element analysis at each step, plus the iterative *FORM* calculation were carried out in a Sun Microsystems 4/260 minicomputer, with an average time of 20 CPU minutes per complete analysis (i.e. one β–value in Table 3).

4. DISCUSSION AND CONCLUSIONS

Table 3 shows, that, for a given ϕ_c, the adopted slenderness adjustment factor $K(\lambda)$ permits the maintenance of sufficiently uniform reliability over a range of loading conditions, dead/live load ratios and slendernesses. Furthermore, Fig.6 shows that the format of the limit state design equation (which is essentially a linear interaction equation between compression and bending) is a conservative format, in that the minimum β–value obtained is for the case $\varepsilon = 0$. Ideally, it would be desirable to use a format that would result in less variation of β with ε. However, this is difficult to achieve while maintaining relative simplicity in the limit state checking equation. The actual behavior is quite complicated. For example, the shape of the curves in Fig. 6 result from the interaction of two processes. At low ε, failure is influenced more by compression behavior due to low slenderness; while at high ε, failure is influenced more by bending. Overall, it may be concluded that the proposed limit state design equation is an adequate and relatively simple compromise, leading to safe designs for all combinations of axial and transverse loads.

Finally, this study shows that *FORM* procedures can be efficiently coupled with nonlinear structural analyses, while maintaining the simplicity of the reliability algorithm.

REFERENCES

[1] BOLOTIN, V.V.: " *Statistical Methods in Structural Mechanics* ", Holden Day, Inc., San Francisco, California, Chapter 3,1969

[2] DER KIUREGHIAN, A.: " Structural Reliability Under Incomplete Probability Information ", *Journal of Engineering Mechanics*, ASCE, Vol.112, No.1, January, 1986

[3] JOHNS, K.C.: " Semi–Empirical Column and Beam Instability Formulae ", *Rapport Technique KJ.03.88, Departement de Génie Civil, Université de Sherbrooke*, June, 1988

[4] KOKA, N.E.: " Laterally Loaded Wood Compression Members: Finite Element and Reliability Analysis ", *The Thesis for the Degree of Master of Applied Science*, University of British Columbia, October, 1987

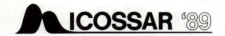
PROBABILISTIC APPROACH TO STRESS DISTRIBUTIONS
FOR ORTHOTROPIC WOOD PLATES

Akira Nishitani* and Gengo Matsui*

*   Dept. of Architecture, Waseda University
    Shinjuku-ku, Tokyo, Japan

ABSTRACT

Probabilistic approach to the stress distributions of an orthotropic wood plate is discussed.
The stress distributions for the orthotropic elastic media are fully dependent of the material properties. Hence taking into account uncertainty of
the elastic coefficients leads to variations of
the stresses. Utilizing the Monte Carlo simulation and second order approximation techniques,
probability-based evaluations of the orthotropic
stresses are discussed.

KEYWORDS

Orthotropic medium; elastic modulus; stress distribution; Monte Carlo method; second order
approximation method

1. INTRODUCTION

For homogeneous isotropic media the stress distributions are
independent of any elastic coefficients such as Young's moduli,
shear moduli and Poisson's ratios; that is, there is no need of
direct involvement of these elastic coefficients in solving the
differential equations with respect to the stress functions [5].
On the other hand, the stress distributions for the orthotropic media are fully dependent of the elastic coefficients [4].
This fact makes the orthotropic stress analyses either quite hard
or troublesome. Hence uncertainty of these elastic coefficients
leads to variability of the stress distributions.
Among various kinds of structural materials, wood is a typical orthotropic medium, since Young's modulus for the direction

parallel to the fibers is different from that for the direction
perpendicular to the fibers. Considering the fact that wood is
a natural material, elastic property variation or uncertainty
should be taken into account for the stress analyses.

Based on the two-dimensional orthotropic elastic analysis
under the plane stress condition, which was developed by the
authors [1], this paper attempts a probabilistic approach to the
orthotropic stresses for an wood plate subject to a concentrated
load at the end.

Sensitivity investigation of the elastic coefficients concluded that two Young's moduli both for the directions parallel
and perpendicular to the wood fibers and the shear modulus make
significant contributions to the stress distributions [3]. In
the sense of sensitivity Poisson's ratios are much less significant [3]. Therefore, the probabilistic analysis in this study
treats two Young's moduli and the shear modulus as stochastic,
while Poisson's ratios are regarded as deterministic. For the
analysis, two kinds of techniques are employed : Monte Carlo
method and second order approximation method.

2. ANALYTICAL SOLUTIONS FOR ORTHOTROPIC WOOD STRESSES

In this chapter analytical solutions of the orthotropic
stresses are demonstrated for a two-dimensional wood plate with
a concentrated load at the end as depicted in Fig.1.

In Fig.1, c represents a distance between the center of the
plate and the load point at the end and
l is a half of the width of the plate.

First of all, the following
constants, e and g, are introduced:

$$e = \sqrt{E_x/E_y} \qquad (1)$$

$$g = (\sqrt{E_x/E_y}/G_{xy}) - 2(1 + \sqrt{v_{xy}\, v_{yx}}) \qquad (2)$$

in which E_y = Young's modulus in the
direction parallel to the fibers,
E_x = Young's modulus in the direction
perpendicular to the
fibers, G_{xy} =
shear modulus, v_{xy} = Poisson's ratio
with respect to the x-axis direction
due to the normal stress in the y-axis
direction, v_{yx} = Poisson's ratio with
respect to the the y-axis direction
due to the normal stress in the x-axis
direction.

Following the introduction of e
and g, P_1 and P_2 are defined as

$$P_1 = \sqrt{n + \sqrt{n^2 - m}} \qquad (3)$$

$$P_2 = \sqrt{n - \sqrt{n^2 - m}} \qquad (4)$$

with

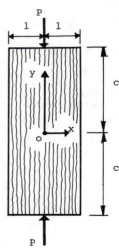

Fig.1 Orthotropic
Wood Plate

$$m = e^2 \; ; \; n = e(2+g)/2 \qquad (5)$$

In association with the coefficients P_1 and P_2, the Fourier series expansion technique provides the analytical solutions both for the normal stress parallel to the loading direction, σ_{yy}, and the shear stress, σ_{xy}, under the assumption of $c \gg 1$:

$$\sigma_{yy} = -(P/2l)-(P/l)\sum_{k=1}^{3} [P_1/(P_1-P_2)] \cdot$$

$$\exp[\frac{k\pi}{l}P_2(y-c)]-(P_2/P_1)\exp[\frac{k\pi}{l}P_1(y-c)] \cos\frac{k\pi}{l}x \quad (6)$$

$$\sigma_{xy} = (P/l)\sum_{k=1}^{3} [P_1P_2/(P_1-P_2)] \cdot$$

$$\exp[\frac{k\pi}{l}P_2(y-c)]-\exp[\frac{k\pi}{l}P_1(y-c)] \sin\frac{k\pi}{l}x \qquad (7)$$

These analytical solutions thus estimated were proved to be accurate by comparing with the experimental results [1]. The finite element results were also in good agreement with the theoretical stresses [2]. Sensitivity studies to the orthotropic stresses conclusively showed that the value of e and the first term value in g (Eq.2) are of significance; that is, three sorts of elastic moduli, E_x, E_y and G_{xy}, are much more significant than Poisson's ratios, v_{xy} and v_{yx}. Consequently, in this paper, these three moduli are treated as stochastic in developing a probabilistic approach to the orthotropic wood stress distributions.

3. MONTE CARLO FORMULATION

The preceding chapter was devoted to present the analytical solutions for the orthotropic stresses. It has been mentioned that the stresses fully depend on those constants which are determined by E_x, E_y, G_{xy}, v_{xy} and v_{yx}. In this study, however, only E_x, E_y and G_{xy} are regarded as stochastic since they are of great significance among all the elastic coefficients necessary for the two-dimensional plane stress problems.

In considering the variations of E_x, E_y and G_{xy}, the following forms are introduced:

$$E_x = E_{xo}(1+A_1) \; (8); \quad E_y = E_{yo}(1+A_2) \; (9); \quad G_{xy} = G_{xyo}(1+A_3) \; (10)$$

in which A_1, A_2 and A_3 are assumed to be Gaussian random variables with zero-means and hence E_{xo}, E_{yo} and G_{xyo} are the mean values of E_x, E_y and G_{xy}, respectively.

In order to conduct the Monte Carlo simulation, the standard deviations of A_1, A_2 and A_3 i.e. the coefficients of variation for E_x, E_y and G_{xy} are required to be designated. Once they have been specified, E_x, E_y and G_{xy} are determined by the simulated values of A_1, A_2, A_3, then providing P_1 and P_2. With P_1 and P_2 obtained, σ_{yy} and σ_{xy} are calculated by means of Eqs.6,7. Repetition of this procedure makes the necessary statistic data available.

4. FORMULATION OF SECOND ORDER APPROXIMATION

Formulation of approximation method involving terms up to second order is developed in this chapter.

The second order approximation method requires the first- and second-order partial derivatives with respect to the random variables.

However, the analytical solutions of the orthotropic stresses given by Eqs.6,7 are of too complicated forms to directly differentiate with respect to the random variables A_1, A_2 and A_3, which were introduced in Eqs.8-10. In spite of direct differentiation, two steps are employed as in the following.

Since the elastic moduli have been already assumed to be presented by Eqs.8-10, the coefficients P_1 and P_2 are not constant any longer. First step begins with expanding them in the second order forms with respect to the non-dimensional random parameters A_1, A_2 and A_3:

$$P_1 = (P_1) + \sum_{m=1}^{3} (P_{1,m})A_m + 0.5\sum_{m=1}^{3}\sum_{n=1}^{3} (P_{1,mn})A_mA_n \tag{11}$$

$$P_2 = (P_2) + \sum_{m=1}^{3} (P_{2,m})A_m + 0.5\sum_{m=1}^{3}\sum_{n=1}^{3} (P_{2,mn})A_mA_n \tag{12}$$

or, more compactly with the summation convention rule utilized,

$$P_i = (P_i) + (P_{i,m})A_m + 0.5(P_{i,mn})A_mA_n \quad (i=1,2;m,n=1,2,3) \tag{13}$$

in which $P_{i,m}=\partial p_i/\partial A_m$, $P_{i,mn}=\partial p_i^2/\partial A_m\partial A_n$ and ( ) denotes the evaluation at $A_1=A_2=A_3=0$.

For simplicity let us assume that A_1, A_2 and A_3 are mutually independent. Then the mean and standard deviation of P_i can be estimated:

$$E[P_i]=(P_i) + 0.5(P_{i,m})\sigma[A_m]^2 \quad (i=1,2;m=1,2,3) \tag{14}$$

$$\sigma[P_i]= \sqrt{(P_{i,m})\sigma[A_m]^2+0.5\ (P_{i,mn})(P_{i,mn})\sigma[A_m]^2\sigma[A_n]^2}$$
$$(i=1,2;m,n=1,2,3) \tag{15}$$

in which the notations E[ ] and σ[ ] denote the expected value and standard deviation, respectively. Equation 15 is obtained by utilizing the relationship for Gaussian variables : $E[A_kA_lA_mA_n]$ = $E[A_kA_l]E[A_mA_n]+E[A_kA_m]E[A_lA_n]+E[A_kA_n]E[A_lA_m]$.

The expected value of the product of $P_1-E[P_1]$ and $P_2-E[P_2]$ can be also estimated,

$$E[(P_1-E[P_1])(P_2-E[P_2])]=$$
$$(P_{1,m})(P_{2,m})\sigma[A_m]^2+0.5(P_{1,mn})(P_{2,mn})\sigma[A_m]^2\sigma[A_n]^2 \tag{16}$$

From the above formulation $E[P_1]$, $E[P_2]$, $\sigma[P_1]$, $\sigma[P_2]$ and $E[(P_1-E[P_1])(P_2-E[P_2])]$ are available, which are necessary to

estimate the mean values and standard deviations of the orthotropic stresses.

To start the second step, P_i is assumed to be as in the
following:

$$P_i = E[P_i](1+Q_i) \qquad\qquad (i=1,2) \qquad\qquad (17)$$

Accordingly, $E[Q_i]=0$, $\sigma[Q_i]=$coefficient of variation for P_i,
and $E[Q_iQ_j]=E[(P_i-E[P_i])(P_j-E[P_j])]/E[P_i]E[P_j]$.

Using Q_1 and Q_2, the stresses σ_{yy} and σ_{xy} are similarly
expanded in the following manner:

$$\sigma_{yy}=(\sigma_{yy})+(\sigma_{yy},i)Q_i+0.5(\sigma_{yy},ij)Q_iQ_j \qquad (i,j=1,2) \qquad (18)$$

$$\sigma_{xy}=(\sigma_{xy})+(\sigma_{xy},i)Q_i+0.5(\sigma_{xy},ij)Q_iQ_j \qquad (i,j=1,2) \qquad (19)$$

in which repeated subscripts indicate summation, $\sigma_{yy},i=\partial\sigma_{yy}/\partial Q_i$,
$\sigma_{yy},ij=\partial^2\sigma_{yy}/\partial Q_i\partial Q_j$,etc., and ( ) means the evaluation at $Q_i=Q_j=0$.

The means and standard deviations are estimated as

$$E[\sigma_{yy}]=(\sigma_{yy})+0.5(\sigma_{yy},ij)E[Q_iQ_j] \qquad (i,j=1,2) \qquad (20)$$

$$E[\sigma_{xy}]=(\sigma_{xy})+0.5(\sigma_{xy},ij)E[Q_iQ_j] \qquad (i,j=1,2) \qquad (21)$$

and

$$\sigma[\sigma_{yy}]= \sqrt{(\sigma_{yy},ij)E[Q_iQ_j]-0.25(\sigma_{yy},ij)(\sigma_{yy},kl)E[Q_iQ_j]E[Q_kQ_l]}$$

$$(i,j,k,l=1,2) \qquad (22)$$

$$\sigma[\sigma_{xy}]= \sqrt{(\sigma_{xy},ij)E[Q_iQ_j]-0.25(\sigma_{xy},ij)(\sigma_{xy},kl)E[Q_iQ_j]E[Q_kQ_l]}$$

$$(i,j,k,l=1,2) \qquad (23)$$

Equations 22 and 23 are obtained by neglecting $E[Q_iQ_jQ_k]$ and
$E[Q_iQ_jQ_kQ_l]$.

5. NUMERICAL EXAMPLES

Numerical examples are shown in this chapter.

In the beginning, it is assumed that E_{xo} = 17,800 kgf/cm²
(1,740 MPa), E_{yo} = 204,000 kgf/cm² (20,000 MPa), G_{xyo} = 9,100
kgf/cm² (893 MPa), v_{xy} = 0.15 and v_{yx} = .013. These values have
been based on the experiments conducted by the authors [1].

Before going through the probabilistic calculations, the
deterministic stress distributions based on Eqs.6,7 are shown in
Fig. 2, in which there is no involvement of the random variables
A_1, A_2 and A_3. Figure 2 enables us to recognize how a concentrated load transfers through the wood plate with the material
properties above specified.

In performing the probability-based evaluation it is needed
to specify the standard deviation of the three independent random
variables A_1, A_2 and A_3. Consider two cases : $\sigma[A_1]=\sigma[A_2]=\sigma[A_3]=$
0.1 and 0.2.

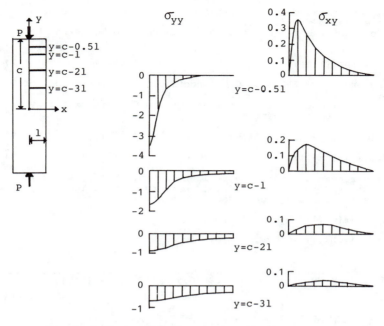

Fig.2 Orthotropic Stress Distributions (unit : P/l)

Along a line which is given by y=constant, the maximum
normal stress σ_{yy} is recognized at the center i.e. x=0. In the
sense of stress concentration the values of σ_{yy} at x=0 are of
significance. Division of these values by 0.5P/l leads to the
stress concentration factors. Tables 1 and 2 show the means and
standard deviations of the maximum σ_{yy} values based on both of

Table 1. Means and Standard Deviations of
σ_{yy} at x=0 for $\sigma[A_i]$=0.1

$\frac{y-c}{l}$	M.C. Mean (S.D.) [P/l]	S.O.A. Mean (S.D.) [P/l]
-0.5	-3.43 (.199)	-3.46 (.204)
-1.0	-1.73 (.099)	-1.74 (.101)
-2.0	-0.92 (.047)	-0.92 (.048)
-3.0	-0.68 (.027)	-0.69 (.028)

Table 2. Means and Standard Deviations of
σ_{yy} at x=0 for $\sigma[A_i]=0.2$

$\frac{y-c}{l}$	M.C.	S.O.A.
	Mean (S.D.) [P/l]	Mean (S.D.) [P/l]
-0.5	-3.46 (.423)	-3.53 (.406)
-1.0	-1.74 (.211)	-1.78 (.202)
-2.0	-0.92 (.099)	-0.94 (.095)
-3.0	-0.69 (.057)	-0.70 (.055)

the Monte Carlo (M.C.) and second order approximation (S.O.A.)
methods for $\sigma[A_i]=0.1$ and $\sigma[A_i]=0.2$, respectively. For Monte
Carlo simulation, 1000 sets of sample solutions are calculated.
It is found that the results from M.C. and S.O.A. are generally
in good agreement and that there exist nontrivial variations in
σ_{yy} at the vicinity of the loading point.
     Tables 3 and 4 compare the means and standard deviations of

Table 3. Means and Standard Deviations of σ_{xy}
along (y-c)/l=-0.5 for $\sigma[A_i]=0.1$

x[l]	M.C.	S.O.A.
	Mean (S.D.) [P/l]	Mean (S.D.) [P/l]
0.0	.000 (.000)	.000 (.000)
0.1	.358 (.004)	.357 (.004)
0.2	.281 (.014)	.278 (.014)
0.3	.193 (.012)	.191 (.012)
0.4	.132 (.008)	.131 (.008)
0.5	.092 (.006)	.091 (.006)

Table 4. Means and Standard Deviations of σ_{xy}
along (y-c)/l=-0.5 for $\sigma[A_i]=0.2$

x[l]	M.C.	S.O.A.
	Mean (S.D.) [P/l]	Mean (S.D.) [P/l]
0.0	.000 (.000)	.000 (.000)
0.1	.356 (.009)	.355 (.007)
0.2	.280 (.028)	.276 (.028)
0.3	.193 (.025)	.189 (.024)
0.4	.133 (.018)	.130 (.017)
0.5	.092 (.013)	.090 (.012)

the shear stresses σxy at $(y-c)/l=-0.5$ for $\sigma[A_i]=0.1$ and 0.2, respectively. They demonstrate only the values at $x/l=0.1$, 0.2, 0.3, 0.4 and 0.5 because the shear stresses for $x/l>0.5$ are less significant. Similar results have been produced by either M.C. or S.O.A. Nevertheless, the variations are not so significant as far as the maximum shear stresses are concerned.

6. CONCLUDING REMARKS

In the sense of material property variations wood is one of the most significant structural materials. Probabilistic treatment of the orthotropic wood stresses is necessary in performing an elastic analysis.

Probabilistic formulations based on the Monte Carlo and second order approximation techniques are presented for the orthotropic stress distributions of an wood plate with a concentrated load at the end.

The resulting means and standard deviations from these two techniques are quite similar. Especially for $E[A_i]=0.1$, the two-step second order approximation method developed in this paper produces almost same results as the Monte Carlo solutions. This approximation method is expected to be very useful and efficient in the case of small variability of elastic moduli.

REFERRENCES

[1] Matsui,G., and Nishitani,A., "Elastic Analysis of Stress Distributions for Orthotropic Woods with Concentrated Forces," J.Structural and Construction Engineering, Trans. of Architectural Institute of Japan, No.362, 1986 ,pp.116-122
[2] Matsui,G., and Nishitani,A., "Finite Element Analysis of Stresses for Orthotropic Woods," Bulletin of Science and Engineering Research Laboratory, Waseda University, No.118, 1988, pp.32-39
[3] Nishitani,A., Orthotropic Analysis of Bolt-Jointed Woods, Doctoral dissertation, Waseda University, 1987
[4] Sokolnikoff,I.S., Mathematical Theory of Elasticity (2nd Ed.), McGraw-Hill, New York, NY, 1956
[5] Timoshenko,S.P., and Goodier,J.N., Theory of Elasticity (3rd Ed.), McGraw-Hill, New York, NY, 1951

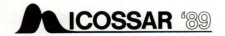
DEVELOPMENT OF RELIABILITY-BASED INTEGRATED DESIGN-CONSTRUCTION
QUALITY ASSURANCE SYSTEM OF PENSTOCK

Yoshimitsu Ichimasu*, Takayuki Nakamura*
Hirokazu Tanaka*, Masaaki Yamamoto**, Yasuaki Shimizu**
and Harumitsu Kemboh**

* Construction Dept., Tokyo Electric Power Co., Inc.
  1-1-3, Uchisaiwai-cho, Chiyoda-ku, Tokyo 100, Japan
** Civil Engineering Design Division, Kajima Corporation,
  6-5-30, Akasaka, Minato-ku, Tokyo 107, Japan

ABSTRACT

A more rational design-construction quality assurance system is proposed by applying the concepts of
the reliability-based design methods. By applying
this system, the design level of safety of a structure can be assured during and after the construction. The proposed system is applied to a penstock
project of the actual pumped storage hydroelectric
power plant to demonstrate the effectiveness of this
system.

KEYWORDS

Safety assurance, reliability, sensitivity analysis,
safety control, reliability-based design, penstock

1. INTRODUCTION

In these days, the design levels of structural safety can be
evaluated more rationally by using the reliability-based design
methods, whereas the levels of safety of structures during and
after the construction are not clearly examined. By applying the
concepts of the reliability-based design methods to the quality
control procedure during construction, we propose a rational
quality assurance system which can assure the levels of safety of
structures during and after the construction.

The concepts of the proposed quality assurance system is
presented. Also, by applying this system to the penstock project

of the actual pumped storage hydroelectric power plant, a numerical example is performed.

2. CONCEPTS OF QUALITY ASSURANCE SYSTEM

The proposed quality assurance system is composed of three subsystems (Fig. 1):

    a.   Design requirements establishing stage
    b.   Design stage
    c.   Construction control stage

The concepts of these subsystems are described in the following subsections.

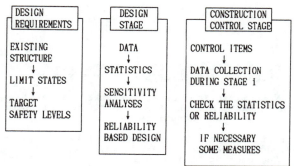

Figure 1 CONCEPTS OF SAFETY ASSURANCE SYSTEM

2.1 Design Requirements Establishing Stage

Based on the design and construction experience of existing structures, the performance functions corresponding to limit states are defined and the statistics (e.g. mean values, standard deviations) of the design parameters of these performance functions are obtained. Then, by evaluating the reliability of existing structures, the target reliability (or the target reliability index) is determined.

2.2 Design Stage

The statistics of design parameters composing the performance functions are set up from the data of measurements and experiments. Then the sensitivity analyses are performed to realize the sensitivities of the design parameters to the safety level of the structure: for example, the sensitivities to the mean value of the performance function, the sensitivities to the variance of the performance function, and the sensitivites to the reliability. Based on these sensitivites, the design alternatives are established effectively.

The structure is designed to satisfy the condition that the reliability of the structure should not be less than the target reliability.

2.3 Construction Control Stage

The important design parameters are selected as the construction control parameters based on the results of the sensitivity analyses conducted at the design stage. If these parameters are difficult to be measured and controled directly, other parameters which are correlated to the design parameters are used as the construction control parameters. In such a case, the correlation between the design parameters and the construction control parameters should be adequately examined in advance.

The actual construction of the structure is divided into several stages and the construction control parameters are different among these construction stages. For each construction stage, the statistics of the design parameters are estimated based on the data measured during the construction. If these statistics are in the safe zone in comparison with those assumed in the design stage, then move on to the next construction stage. In the contrary case, the reliability is reexamined using the measured data statistics. If the reliability level is satisfactory, then proceed on to the subsequent construction stage, and if not, provide necessary measures.

By repeating these steps, it is possible to assure the level of safety of the structure as provided at the design stage.

3. NUMERICAL EXAMPLE

In these days, design and construction of penstocks become more complicated because of high water pressure and lack of proper construction sites for pumped storage hydroelectric power plants. Penstocks constructed in rock deposites are designed as composite structures of steel pipes, concrete rings and surrounding rocks. The uncertainties of these structural elements' properties require a more sophisticated design and construction control approach for penstock projects.

The following is an example of how the integrated design-construction quality assurance system has been applied to an actual penstock project.

3.1 Design Requirements Establishing Stage

(1) Design and construction experience of existing structures

As a result of the studies made on the engineering standards [2,3] and the documents of design and construction control of the penstock project of Imaichi Hydro-electric Power Plant, the findings are as follows:

a. Design of the existing structures

The penstock is composed of a steel pipe, filled concrete and surrounding base rock as shown in Figs. 2 and 3. Also, drain pipes are laid in the filled concrete area for the purpose of reducing outer water pressure.

Acting on the pen-
stock are the following
three major loads:

. Inner water pressure
  due to static water
  pressure, water hammer
  and surging.
. Outer water pressure
  due to the underground
  water.
. Difference in temper-
  atures between the
  water and the steel
  pipe (the base rock).

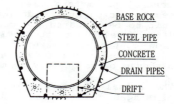

Figure 2   SECTION OF PENSTOCK

The primary limit states are tension failure of the
steel pipe due to the inner water pressure and buckling of
the steel pipe due to the outer water pressure.

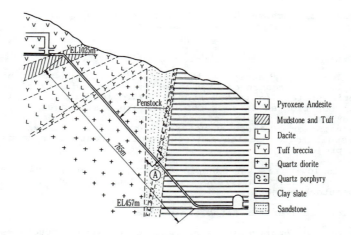

Figure 3   PENSTOCK AT IMAICHI PROJECT

b. Construction procedure

The construction is executed in the following sequences:

1) Drift excavation   2) Widening cut work
3) Steel pipe setting and concrete filling

The drift excavation is executed from the bottom toward the top. Then the widening cut work is executed from the top toward the bottom. Steel pipe setting and concrete filling is executed from the bottom toward the top. The steel pipe is set per each construction block and is welded to the steel pipe of the preceding block. After laying the drain pipes, concrete is filled in the clearance between the steel pipe and the base rock.

(2) Performance function

The tension failure and the buckling of the steel pipe are the primary limit states. In the design of existing structures, the thickness of the pipe is mainly determined by the tension failure due to the inner water pressure. Accordingly, in this report, the tension failure of the steel pipe is studied as an example.

The performance function of tension failure is shown in equation (1), where the base rock load bearing ratio is given by equation (2).

$$Z = \eta \cdot \sigma_y - \sigma_\theta = \eta \cdot \sigma_y - \frac{(P - P_0)\, D}{2t}\, (1 - \lambda) - \sigma_0 \qquad (1)$$

$$\lambda = \cfrac{1 - \cfrac{Es}{P} \cdot \alpha_s \cdot \Delta T \cdot \cfrac{2t}{D}}{1 + (1 + \beta c) \cdot \cfrac{Es}{Ec} \cdot \cfrac{2t}{D}\ln\cfrac{DR}{D} + \cfrac{Es}{Dg} \cdot \cfrac{mg+1}{mg} \cdot \cfrac{2t}{D}} \qquad (2)$$

σ_θ: Tensile stress of steel.

λ : Base rock bearing ratio.

P_0: Inner pressure corresponding to σ_0.

σ_0: Pipe stress when the amount of deformation towards the radius direction reached the initial clearance amount.

For other symbols, see Table 1.

(3) Statistics of design parameters

Tables 1 and 2 show the statistics of the design parameters of the penstock project of Imaichi Hydroelectric Power Plant.

(4) Target reliability

The reliability indices evaluated for the design sections of the Imaichi penstock are shown in Figure 4. The target reliability $\beta_T = 6.8$ is determined by taking the minimum value.

Table 1   EXAMPLE OF DESIGN PARAMETERS

Inner Water Pressure		P	6.25 MPa
Temperature difference		ΔT	20 °C
Steel (HT80)	Diameter	D	5.50 m
	Thickness	t	37 mm
	Reduction Factor for Welding	η	0.95
	Yield Stress	σy	790.4 MPa
	Young's Modulus	Es	206 GPa
Concrete	Plastic Deformation Factor	βc	0
	Young's Modulus	Ec	23 GPa
Rock (Sand Stone)	Diameter	DR	7.02 m
	Deformation Modulus	Dg	8.53 GPa
	Poisson Number	mg	4

Table 2   EXAMPLES OF STATISTICS

Parameters	Mean	Stand.Dev.	C.O.V
σy	790.4 MPa	20.5 MPa	0.03
Dg	8.53 GPa	7.16 GPa	0.8
DR	7.02 m	0.24 m	0.03
Ec	23.0 GPa	9.81 GPa	0.04

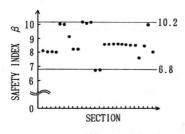

Figure 4   RELIABILITY OF
            EXISTING PENSTOCK

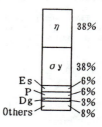

Figure 5   EXAMPLE OF
            SENSITIVITY ANALYSIS

3.2   Design Stage

(1)   Sensitivity analysis

　　　　At the section A shown in Figure 3, the sensitivities to
the mean value of performance function given by Equation (3)
are obtained as shown in Figure 5.

$$\text{Sensitivity of } X_i = \frac{(\partial Z / \partial X_i)_{\underset{\sim}{\overline{X}}} \cdot \overline{X_i}}{\sum_{j=1}^{n} (\partial Z / \partial X_j)_{\underset{\sim}{\overline{X}}} \cdot X_j} \times 100\% \qquad (3)$$

where,   Z: Performance function
        Xi, Xj: Design parameters
        $\overline{X}i, \overline{X}j$:   Mean values of Xi, Xj

(2) Design of cross section

Assume that the statistics are obtained for four design parameters as shown in Table 2: yield stress of steel σ_y, deformation modulus of rock Dg, excavation diameter of cross section D_R, and Young's modulus of concrete E_c.

At the Section A, the reliability index is ß = 6.88 when the thickness of the steel pipe is t = 28mm (HT 80). This satisfies the target reliability index β_T = 6.8. Therefore, the thickness of the steel pipe at the Section A is designed to be 28mm.

Regarding the other design sections, the thicknesses of the steel pipes are determined in the same way.

3.3 Construction Control Stage

(1) Selection of important control parameters

In this case, the yield stress of steel σ_y, the deformation modulus of rock Dg, and the excavation diameter D_R are selected as the important control parameters, because these parameters have considerable effects on the safety of the structure (Fig. 5) and can be monitored during the construction.

(2) Execution of construction control

Suppose the construction stage that the widening cut is completed and that the statistics of the yield stress of steel and the excavation diameter are examined as almost the same as the design statistics. Then, assume that the measured deformation modulus of the rock surrounding the section A is realized that the standard deviation is nearly the same as the design value, whereas the mean value (7.8GPa) is less than the design value (8.5GPa).

If the statistics of the deformation modulus of the rock measured during the construction is in the safe zone compared to the design statistics, the construction may be advanced to the succeeding stage. However, in this case, the mean value of the deformation modulus of the rock measured during the construction is in the unsafe zone. Then, the reliability is calculated using the new statistics with the result of ß = 6.2 which is less than the target reliability of β_T = 6.8. Therefore, some corrective measures become necessary.

Among several corrective measures, the improvement of the base rock by the high pressure grouting is considered to be superior because of its workability and effectiveness on safety. Also, the extent of its effect has been confirmed by preliminary experiments. Therefore, it is decided to improve the base rock with the high pressure grouting to meet the grade of Dg = 8.5 GPa. After executing this improvement, the safety of the structure is confirmed by measuring the deformation modulus of rock.

Repeating these steps at each construction stage, the safety of the structure is confirmed finally.

4. CONCLUDING REMARKS

By applying the concepts of the reliability-based design methods to the construction control procedure, a system, which assures the safety of the structure during and after the construction, is proposed in this report. Furthermore, taking the penstock project of the actual pumped storage hydroelectric power plant as an example, the effectiveness of the proposed system is demonstrated.

In the presented example, it is assumed that plentiful data can be available at the construction control stage to evaluate the reliability with sufficient accuracy. However, for design parameters of which adequate data are difficult to be obtained and which have less influence on the safety of the structure, a more simple construction control procedure should be useful. Therefore, the study efforts will be required to establish a further realistic quality assurance system, for example, by applying the concepts of sampling inspection.

REFERENCES

[1] Ang. A.H-S. and Tang, W.H.: Probability Concepts in Engineering Planning and Design Volume II, John Wiley & Sons, Inc., 1984.

[2] Hydraulic Gate and Penstock Association (Japan): Technical Standards for Gate and Penstocks, 1981.

[3] Hydraulic Gate and Penstock Association (Japan): Technical Standards for Gate and Penstocks - supplements -, 1974.

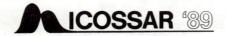

PROBABILISTIC AND RELIABILITY ANALYSIS OF ASPHALT
CONCRETE AIRFIELD PAVEMENTS

Yu T. Chou[1], Member ASCE

Abstract

A procedure was developed to analyze layered
elastic asphaltic concrete airfield pavement systems in
terms of probability and reliability. Rosenblueth
method is used to estimate the expected value and variance of the strains (dependent parameters) based on the
input mean values of independent parameters, i.e., aircraft load, layer thicknesses, and material moduli.
The relationships between the reliability level and the
allowable strain repetition of the designed system provide a decision-making tool for engineers to design
pavements at desired reliability level. The performance of a two-layer asphaltic concrete pavement is
sensitive, in the descending order, to variations of
the input parameters concrete thickness, gear load,
asphaltic concrete modulus, and subgrade modulus.

Introduction

The design of asphaltic concrete airfield pavements in the US Army Corps of Engineers (USACE) is
deterministic. The effect of material variability on
pavement performance is considered in the designer's
selection of the subgrade stiffness value, and the
design safety factor is implicitly contained within
construction specifications such as compaction
requirements. A design procedure, expressed in probabilistic and reliability terms is presented in this
paper. The designer can select the pavement thickness
and in some cases develop an overlay design scheme
based on the desired reliability level. By using the
procedure, the partial effect of the variability of
each design parameter on pavement performance can also

[1]US Army Engineer Waterways Experiment Station
Vicksburg, MS 39180

be investigated, and its effects on the final design can be quantified. Emphasis can be placed on the crucial parameters to be tightly controlled in the construction phases.

Probabilistic and Reliability Approach

A two-layer flexible pavement is a pavement structure composed entirely of asphalt concrete on the natural or prepared subgrade. The pavement failed by the traffic load is due to fatigue cracking of the asphaltic concrete or rutting in the subgrade.

In analyzing a pavement structure in probabilistic and reliability terms, the expected value and variance of a function (such as the computed stresses, strain, or load repetition) should first be determined, and the reliability of the design can then be evaluated. Based on Taylor series expansion, the variance of a variable x is

$$\sigma_x^2 = E[x^2] - [E(x)]^2 \tag{1}$$

and the variance of a function $f(x)$ is

$$V[f(x)] = E[f^2(x)] - E[f(x)]^2 \tag{2}$$

Rosenblueth (1) used the point estimates of the function to determine the variance and the expressions for the expected values are:

$$E[\varepsilon^N] = \frac{1}{2}\left(\varepsilon_+^N + \varepsilon_-^N\right) \quad \text{for one variable} \tag{3}$$

$$E[\varepsilon^N] = \frac{1}{2^2}\left(\varepsilon_{++}^N + \varepsilon_{+-}^N + \varepsilon_{-+}^N + \varepsilon_{--}^N\right) \text{ for two variables} \tag{4}$$

$$E[\varepsilon^N] = \frac{1}{2^3}\left(\varepsilon_{+++}^N + \varepsilon_{++-}^N + \varepsilon_{+--}^N + \varepsilon_{+-+}^N + \varepsilon_{-++}^N + \varepsilon_{-+-}^N \right.$$
$$\left. + \varepsilon_{--+}^N + \varepsilon_{---}^N\right) \quad \text{for three variables} \tag{5}$$

$$E[\varepsilon^N] = \frac{1}{2^4}\left(\varepsilon_{++++}^N + \varepsilon_{+++-}^N + \varepsilon_{++-+}^N + \varepsilon_{++--}^N + \varepsilon_{+-++}^N \right.$$
$$+ \varepsilon_{+-+-}^N + \varepsilon_{+--+}^N + \varepsilon_{+---}^N + \varepsilon_{-+++}^N + \varepsilon_{-++-}^N$$
$$\tag{6}$$
$$+ \varepsilon_{-+-+}^N + \varepsilon_{-+--}^N + \varepsilon_{--++}^N + \varepsilon_{--+-}^N$$
$$\left. + \varepsilon_{---+}^N + \varepsilon_{----}^N\right) \quad \text{for four variables}$$

$$E[\varepsilon^N] = \frac{1}{2^M} \left(\varepsilon^N_{\underbrace{+++++++}_{M}} + \cdots + \varepsilon^N_{\underbrace{------}_{M}} \right) \qquad (7)$$

for M variables

Note that the number of total terms to calculate the expected value of a function ε which has M variables is 2^M, and N has a value of either 1 or 2 as shown in Equation 2, i.e., ε is represented by $f(x)$, and N is the power of the function.

The independent parameters for an asphaltic concrete pavement are the wheel load P, the elastic modulus and thickness of the asphalt layer E_1 and h_1, respectively, and the elastic modulus of the subgrade E_2. For a four-parameter problem, Equation 6 is used to determine the expected value of the strain ε. The variance of ε is computed using Equation 2.

Two failure criteria are used. The criteria for allowable strain repetitions N for the asphaltic concrete can be mathematically expressed as

$$N_{Allowable \atop (AC)} = 10^A \qquad (8)$$

where

$$A = -5 \, Log_{10} \, \varepsilon - 2.665 \, Log_{10} \, (E_{AC}) + 2.68$$

ε = maximum horizontal tensile strain at the bottom of the asphaltic concrete layer
E_{AC} = elastic modulus of the asphaltic concrete, psi

The criteria for allowable strain repetition N for the subgrade can be expressed as

$$N_{Allowable \, (subgrade)} = 200,000 \left(\frac{A}{\varepsilon_{subg}} \right)^B \qquad (9)$$

where
$$A = 0.000247 + 0.000245 \, Log_{10}(E_{subg})$$

E_{subg} = subgrade modulus, psi

ε_{subg} = subgrade strain, dimensionless

$$B = 0.0658 \, (E_{subg})^{0.559}$$

The expected strain value ε (in each term) in Equations 3 to 7 is computed using the layered elastic method. With the strain value ε assumed normally distributed, the number of strain repetitions corresponding to $\varepsilon + \varepsilon\sigma_\varepsilon$ (or $\varepsilon[1 + C \cdot CV(\varepsilon)]$) can be determined from Equations 8 and 9, and the probability of $\varepsilon \leqq \varepsilon[1 + C \cdot CV(\varepsilon)]$ is taken from the normal distribution and C is the selected number varying from -3 to +3. The computations of the reliabilities and allowable strain repetitions are conducted in a computer program called RELIBISA. The program logic of RELIBISA is presented in Reference 2.

Analysis of Asphaltic Concrete Pavements

For asphaltic concrete pavements, the critical periods for both the asphaltic concrete and the subgrade are in the warmer summer months. A concrete modulus value of 250,000 psi and a Poisson's ratio of 0.5 are used in the analysis for both strain criteria. For subgrade soil, a modulus of 9,000 psi and a Poisson's ratio of 0.45 are used in the computation. A B-747 aircraft was used in the computation.

Figure 1 shows the relationships between reliability level and strain repetition for pavement with varying thicknesses. The CV's , P , E_1 , h_1 , and E_2 for the input parameters are 0.1, 0.15, 0.1, and 0.25, respectively. The significance of the curves is that for a given concrete thickness, the allowable strain repetition to failure is presented at its reliability level. In the figure, the flatter the slope of the curves, the greater are the uncertainties involved in the design, and the steeper the slopes of the curves, the lesser are the uncertainties involved in the design.

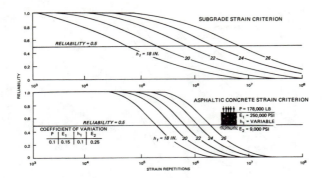

Figure 1. Relationships between
reliability and strain repetitions for asphaltic concrete
pavements with varying CV

For a given design strain repetition, the relationships between reliability and concrete thickness
can also be obtained from Figure 1. Engineers can
choose the concrete thickness suitable for the selected
reliability level of the design. This point can best
be explained from the curves presented in Figure 2
which is plotted from Figure 1 showing the relationships between the strain repetition and concrete
thickness at different reliability levels. The relationships in Figure 2 can be helpful to designers in
selecting the allowable strain repetitions of a given
pavement section for a desired reliability level or to
vary the concrete thickness that will be suitable for a
specific design performance level of the pavement for a
given reliability level. The slope of the curves shown
in Figure 2 is the rate of increase of allowable strain
repetition due to the increase of pavement thickness
for a given reliability level. In Figure 2 the slopes
of the curves for the subgrade strain criterion are
much steeper than those for the bituminous concrete
strain criterion, indicating that at a given reliability level of the design, increasing concrete
thickness would increase the allowable strain repetition (or pavement service life) with respect to the
subgrade failure more rapidly than would for the
fatigue cracking failure of the asphaltic concrete.

For a good pavement design, it is ideal to have
the pavement structure failed in fatigue cracking and
subgrade failure at nearly the same traffic level as
well as the same reliability level. An optimum

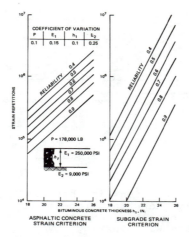

Figure 2. Relationships between
pavement thickness and strain
repetition (from Figure 1) of
asphaltic concrete

asphaltic concrete thickness may be selected using the
relationships similar to those plotted in Figure 2
based on actual subgrade modulus value.

       The results presented in Figure 1 assume that
all of the four design input parameters have variations. To study the effect of each individual parameter on pavement performance, computations were made
to vary only one parameter each time while the variations of the other three parameters were zero. The
results are plotted in Figure 3 for the CV of 0.1.
Figure 3 shows that for both failure criteria, the
pavement performance (allowable strain repetition) is
least sensitive to the variation of the subgrade
modulus and followed by the variation of the bituminous
concrete modulus. The pavement performance is most
sensitive to the variation of the bituminous concrete
thickness and is followed by the aircraft gear load.

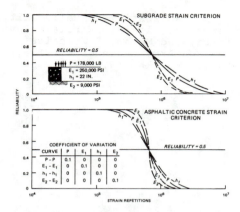

Figure 3.   Relationships between
reliability and allowable strain
repetition for concrete pave-
        ments varying CV

Conclusions

        The relationships between reliability and strain
repetition developed for asphaltic concrete pavements
(Figures 1 to 3) can be used to optimize the design.
The performance of an asphaltic concrete pavement is
sensitive to variations of the input parameters (in the
descending order) concrete thickness, gear load,
asphaltic concrete modulus, and subgrade modulus for
both asphaltic concrete and subgrade strain criteria
(Figure 3).

Acknowledgements

        The work described in this paper was funded by
the U.S. Army. The views of the author presented
herein do not purport to reflect the position of the
Dapartment of the Army. This paper is published with
the permission of the Chief of Engineers.

References

[1]   ROSENBLUETH, EMILIO.:   "Point Estimates for
        Probability Moments," Proceedings, National
        Academy of Sciences, USA, Mathematics, Vol 72,
        No. 10, pp 3812-3814, 1975 (Oct).

[2]   CHOU, Y. T.: "Probabilistic and Reliability
        Design Procedures for Flexible Airfield Pave-
        ments, Elastic Layered Method," Technical
        Report GL-87-24, US Army Engineer Waterways
        Experiment Station, Vicksburg, Miss.

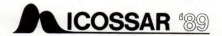
OPTIMIZATION OF STRUCTURES UNDER STOCHASTIC LOADS

S. H. Kim* and Y. K. Wen**

*   Korea Institute of Construction Technology,
    Yeongdungpo-Gu, Seoul, 150, R.O.K.
** Department of Civil Engineering, University of
    Illinois at Urbana-Champaign, Urbana, IL 61801, USA

ABSTRACT

A method for design under multiple stochastic loads
based on optimization is developed. The objective
function such as cost is minimized under the
constraint that the probabilities of various limit
states being reached are within allowable limits.
The loads are treated as random processes and their
combined effects evaluated by the load coincidence
method.   The limit states considered are either
yielding or first plastic hinge at the member level
or plastic collapse at the system level. A penalty
function approach and a Sequential Unconstrained
Minimization method are used.   To expedite the
analysis, a LRFD format is developed and used. The
dependence of the resultant optimal design on the
constraints used, e.g. at the member or system
level is investigated. The proposed approach to the
problem is shown to be a viable method and the
design using a bi-level (member and system)
reliability constraints is shown to yield more risk
consistent results.

KEYWORDS

Optimization; load combination; random process;
LRFD; nonlinear programming.

1. INTRODUCTION

In most reliability-based structural optimization studies, the loadings have been idealized as time invariant, i.e. as random variables. The loadings, however, are known to fluctuate in both time and space. Also, in the course of the structure's life they may occur simultaneously causing serious consequences, even though the chances of such occurrences are small.

The objective of this study is to develop a method for design under multiple time varying loads based on optimization. The loadings are treated as random processes with proper consideration of the variabilities of the loadings such as random occurrence time, random intensity and duration in each occurrence. The probabilities of the individual and combined load effects on the structure are evaluated by the previously developed and proven load coincidence method. The structures considered are ductile frames and the structural limit states considered include first yield, first plastic hinge formed at the member level, and the system plastic collapse. A penalty function approach in conjunction with a Sequential Unconstrained Minimization (SUMT) method is used to find the optimal design structural parameters.

The method developed is used to examine: (1) the consistence of the reliability of the resultant design based on member-level or system-level reliability constraints, and (2) the risk consistency of current load combination formula and possible improvements. The methodology and results are outlined in the following, details are available in Kim and Wen [1].

2. OPTIMIZATION FORMULATION

Reliability-based optimization can be formulated according to whether constraints are used and how costs and reliability of the structure are considered. The commonly used optimization procedures include:

(1) minimizing the initial cost (or weight) subject to the constraint of a prescribed structural reliability, or
(2) minimizing the probability of failure subject to a prescribed initial cost, or
(3) minimizing the total expected cost including the initial cost and expected cost of failure.

Other combinations of objective function and constraints have also been suggested which give more freedom in the description of the problem [2].

An important consideration in the formulation is the selection of the structural limit state and the corresponding target reliability or target cost limit. The limit state may be either the ultimate failure (collapse) or the unserviceability, of a particular structural member or the structural system as a whole. Depending on the limit state considered, the target reliability is chosen accordingly; e.g., a lower risk level for the ultimate limit state and a moderate risk level for the serviceability limit state.

It is well known that the establishment of an acceptable target reliability at either the member or system level is not an easy task. The same is true of the estimation of the costs or benefits involved; i.e., the cost of construction, the cost of consequence of adverse effect caused by loadings including damage and collapse, and the consideration of the discount factor for future loss or benefit in terms of present value, e.t.c. None of these are easy to determine and almost all are subject to uncertainties.

Alternatively, the reliability level implied in the current building code of practice, which has been continually modified as experience and knowledge on building and loading accumulate, may serve as a useful guidance for selecting the target reliability. Current codes are mainly concerned with safety checking of structural elements such as beams, columns, and connections. Design according to current code procedure may or may not yield a satisfactory system reliability. Conversely, a design satisfying only the system reliability may not meet all the code requirements for the elements. Therefore, an optimization with a consideration of both the system and member failures may produce a more balanced design. With the above objectives in mind, the following formulation is used in this study: to determine the structural parameters $\underset{\sim}{x}$ which

minimizes $\quad f(\underset{\sim}{x}) = w(\underset{\sim}{x}) + CP_s(\underset{\sim}{x})$

subject to $\quad P[h_i(\underset{\sim}{x}) \geq o] \leq P_i \qquad$ for i=1 to n

(1)

in which

$\qquad f(\underset{\sim}{x})$ = objective function
$\qquad w(\underset{\sim}{x})$ = weight or cost of construction
$\qquad C$ = cost due to system failure
$\qquad P_s$ = probability of system failure
$\qquad h_i(\underset{\sim}{x})$ = i-th element limit state function
$\qquad P_i$ = allowable probability of failure of the i-th element
$\qquad n$ = number of constraints

In this formulation, the design is determined from within a feasible region of member-level reliability constraints and is optimal in terms of system reliability. This may be called a bi-level formulation; it is in fact a combination of the formulations (1) and (3) mentioned in the foregoing. The formulation (1), however, is also used in this study with the reliability constraints at either the system or member level to investigate the sensitivity of design to formulation.

3. LOAD MODELING AND RELIABILITY ANALYSIS

The random fluctuations of the loadings considered in this study, such as live load, wind load, and earthquake load are modeled by pulse processes and intermittent continuous random processes [3]. Details of these load process models and their applications to structural reliability can be found in Ref. 3. The method based on a load coincidence consideration [3] is used to evaluate the probability of combined load effect in this study.

Structural system failure consists of a large number of possible failure modes. To evaluate the probability of individual failure mode, the well known first order method [4] is used. Since the modes are generally correlated, the evaluation of system relia- bility is a complex problem. Approximations such as those via a bounding technique [5] can be used. However, the bounds may become very wide when the number of failure mode is large which would hamper the mathematical programming required in the optimization process. In this study an approximate point estimate method is developed for this purpose using three-mode intersections. Details of this method is available in Ref. 1.

The probability terms required in the optimization formu- lation, i.e. the system failure probability P_s (x) and member failure probabilities $P[h_i(x) > 0]$, i=1,n over the period of time t are given by the generic solution:

$$P(t) = 1 - [1-P(0)] \exp(-\nu t) \tag{2}$$

in which $P(0)$ = probability of failure at t=0
ν = mean rate of failure

$$= \sum_{i=1}^{m} \lambda_i P_i + \sum_{i=1}^{m-1} \sum_{j=i+1}^{m} \lambda_{ij} P_{ij} + \sum_{i=1}^{m-2} \sum_{j=i+1}^{m-1} \sum_{k=j+1}^{m}$$

$$\lambda_{ijk} P_{ijk} + \cdots$$

where

λ_i = mean rate of occurrence of load $S_i(t)$
λ_{ij}, λ_{ijk} = mean rate of joint occurrence (coincidence) of load $S_i(t)$ and $S_j(t)$; and load $S_i(t)$, $S_j(t)$ and $S_k(t)$, respectively.
P_i = conditional probability of failure given the occurrence of $S_i(t)$
P_{ij}, P_{ijk} = conditional probability of failure given the coincidence of load $S_i(t)$ and $S_j(t)$; and load $S_i(t)$, $S_j(t)$ and $S_k(t)$, respectively.

4.   OPTIMIZATION ALGORITHM

The optimization problem formulated above involves objective functions and constraint equations that are generally nonlinear. Therefore nonlinear programming methods are required to find the design variables. The interior penalty function approach is most appropriate for the problem at hand because all intermediate steps are in the feasible region and the algorithms for the unconstrained minimization are well developed and generally reliable. Thus the Sequential Unconstrained Minimization Technique (SUMT) in conjunction with the interior penalty function method [6] is used. The sequential nature of the method allows a gradual approach to the criticality of the constraints.

As is usually the case in the problem with probabilistic constraints, the derivatives are difficult to obtain analytically. Therefore, the variable metric method [6] is used for determining the search direction in the unconstrained minimization.

Once the search direction is determined a one-dimensional minimization is necessary. Since a considerable portion of the computation is spent on the one-dimensional search, an efficient technique is needed to find the minimum point. Various methods were tested, it is found that a method based on an inverse B-spline curve fitting technique [7] is most efficient compared with other methods such as Golden section and polynomial fitting.

5. LRFD FORMAT FOR OPTIMIZATION-BASED DESIGN

The reliability-based optimization as outlined in the fore-going, although being a rational design procedure, is limited by the repeated reliability analysis required which is costly when applied to large and complex systems. The numerical effort generally increases geometrically with the number of the design variables. This difficulty may be alleviated by the use of a load and resistance factor design (LRFD) format which ensures, at least approximately, the required reliability under multiple loads by a set of load and resistance factor equations that the design has to satisfy.

The LRFD format recommended in the ANSI 58.1-1982 is based on the requirement that the design satisfies a set of load combination equations. Each equation is derived from a target reliability index β_T: β_T=3.0 for the combination involving only gravity loads; 2.5 for combination involving wind load; and 1.75 for combination involving earthquake load. These reliability levels have been shown to be consistent with current practice. The implied lifetime reliability of such a design under all loads, however, is not defined; and may not be at the intended levels for different design situations, e.g., load ratios (for details, see Ref. 1). Therefore, an alternative format needs to be developed for the optimization analysis.

To correct for the inconsistency, it is proposed to add a new checking equation considering all loads; i.e., the resistance is required to satisfy the following

$$P[R < D + L_S + L_T + W + E] \leq \Phi(-\beta_T)$$
$$P[R < D + L_S + L_T] \leq \Phi(-3.0)$$
$$P[R < D + L_S + W] \leq \Phi(-2.5)$$
$$P[R < D + L_S + E] \leq \Phi(-1.75) \qquad (3)$$

in which β_T is the target lifetime reliability index. β_T is taken as the larger value between the implied lifetime reliability in current design and the linear interpolation value based on different design load ratios [1].The load and resistance factors are then determined by minimizing the difference between the resistance according to the load and resistance factor equation and that which satisfied Eq. 3 for all the design situations considered [8]. The resultant LRFD equations are

$$0.9R > 1.1D_n + 0.87L_n + 1.02W_n + 0.64E_n$$
$$0.9R > 1.1D_n + 1.66L_n$$
$$0.9R > 1.1D_n + 0.62L_n + 1.5W_n$$
$$0.9R > 1.1D_n + 0.07L_n + 2.2E_n \qquad (4)$$

in which the resistance factor is fixed at 0.9, the dead load factor at 1.1, and the subscript n refers to nominal load.

The lifetime reliability achieved using Eq. 4 is found to be much closer to the target reliability with a deviation generally less than 5%.

In this study, to ensure the reliability of structural members for large systems, the required LRFD formats are determined as illustrated above corresponding to a prescribed target reliability under all loads and used as constraints in the optimization.

6. APPLICATIONS

Numerical examples are carried out to illustrate the methodology and investigate the dependence of the design on optimization formulation and risk consistency of the resultant optimal design. Details are available in Ref. 1.

The results generally indicate that the member reliability constraints plays an important role of achieving more risk consistent design, perhaps at a slightly higher cost. Since the member reliability is comparatively much easier to evaluate, it may provide a convenient means to regulate the system reliability in a optimization-based design. This remains to be verified for systems of more complexity.

The method is applied to the design of a two-bay two-story frame shown in Fig. 1 using W shape steel members. The assumed critical sections are indicated in the figure. The loadings include dead, sustained live, transient live, wind and earthquake loads with the statistics comparable to those in Ref. 8. It is a problem of 6 variables and large number of constraints. A target reliability index of = 2.00 is assigned for all members. Additional approximations are necessary in the conversion from the available discrete member sizes to continuous member properties to be used in the optimization algorithms. The load and resistance factor design (LRFD) type format proposed in the foregoing together with a grouping of members is used to expedite the reliability evaluation. Details are available in [1]. The proposed method is found to be quite efficient in reaching the optimal design. The results are shown in Table 1. It is seen that the resultant design has reliability indices close to the target value at all the critical sections which serve as active constraints.

7. CONCLUSIONS

Numerical results on design of simple as well as reasonably realistic structural frames show that:

(1) The proposed method provides a viable procedure for optimal design under time varying random loads.
(2) The selection of optimization formulation is an important consideration; it has significantly effect on the resultant design.

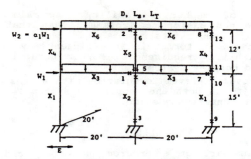

Fig. 1 2-Bay 2-Story Frame Under Time Varying Loads

Table 1 Optimum design using AISC W shape sections

	1	2	3	4	5	6
Designation	W 30x116	W 36x135	W 30x108	W 27x 94	W 30x 99	W 27x 94
Area (A)	34.2	39.7	31.7	27.7	29.1	27.7
Plastic Section Modulus	378	509	346	278	312	278
Slenderness Ratio of Column (H/r)	15.0	12.9	—	13.2	12.3	—

	Critical Section Number		Related Design Variable		Lifetime Reliability (β)	
Active Constraints	2		X_6		2.377	
	3		X_2		2.092	
	6		X_5		2.062	
	7		X_3		2.101	
	9		X_1		2.100	
	12		X_4		2.304	
Inactive Constraints	1		X_3		2.126	
	4		X_2		2.463	
	5		X_5		2.256	
	8		X_6		2.364	
	10		X_1		2.575	
	11		X_4		3.646	

(3) The use of member reliability constraints is more effective and easier to implement than that of system reliability constraints.

(4) Use of the LRFD format is an efficient way of enforcing the reliability requirements for structural members of large and complex systems.

(5) The proposed bi-level formulation with a consideration of reliability at both the member and system levels gives a more risk-consistent design.

ACKNOWLEDGMENT

This study is part of a program on safety and design of structures under multiple hazards supported by the National Science Foundation under grants CEE 82-07590 and DFR 84-14284. The supports are gratefully acknowledged.

REFERENCES

1. S. H. Kim and Y. K. Wen, "Reliability-based Structural Optimization Under Stochastic Time Varying Loads," Civil Engineering Studies SRS No. 533, University of Illinois, Urbana, IL, February 1987.

2. D. M. Frangopol, "Structural Optimization Using Reliability Concepts Design," _Journal of Structural Engineering_, ASCE, Vol. 111, No. 11, November 1985.

3. H. T. Pearce and Y. K. Wen, "Stochastic Combination of Load Effects," _Journal of Structural Engineering_, ASCE, July 1984, pp. 1613-1629.

4. H. O. Madsen, S. Krenk, and N. C. Lind, "Methods of Structural Safety," Prentice-Hall, Engelwood Cliffs, NJ, 1986.

5. R. M. Bennett and A. H-S. Ang, "Formulation of Structural System Reliability," _Journal of Engineering Mechanics_, ASCE, Vol. 112, pp. 1135-1151.

6. S. S. Rao, _Optimization Theory and Applications_, Halsted Press, John Wiley & Sons, New York, NY, 1979.

7. W. J. Gordon and R. F. Riesenfeld, "B-Spline Curve and Surfaces," _Computer-Aided Geometric Design_, Barnhill, R. E. and Riesenfeld, R. F., Editors, Academic Press, New York, NY, 1974.

8. B. Ellingwood, T. V. Galambos, J. G. MacGregor, and C. A. Cornell, "Development of a Probability-Based Load Criterion for American National Standard A58," National Bureau of Standards, NBS Special Publication 577, Washington, D.C., June 1980.

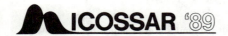
LIMIT STATES RELIABILITY INTERACTION IN
OPTIMUM DESIGN OF STRUCTURAL SYSTEMS

Dan M. Frangopol * and Gongkang Fu **

* Dept. of Civil Eng., University of Colorado, Boulder, CO 80309-0428, USA.
** Dept. of Civil Eng., Case Western Reserve Univ.,Cleveland, OH 44105, U.S.A.

ABSTRACT

A multiobjective optimization approach to deal with structural reliability-based design under multiple limit states is developed. This approach resolves the limit states reliability interaction problem. In fact, keeping track of the reliability level of each limit state of interest becomes an intrinsic part of the optimization method used to find the objective solution set. It is concluded that the proposed approach produces a more balanced optimum solution.

KEYWORDS

Decision space; design criteria; frame design; limit states; multiobjective optimization; Pareto solution set; structural optimization; structural reliability; system reliability.

1. INTRODUCTION

Between the late 1960's and the early 1980's, attempts to optimize structures using the reliability-based design philosophy were made in relation to a single limit state. The objective was to minimize the total structural weight or total cost subject to an allowable value of the probability of occurrence of this limit state. In almost all these attempts the limit state considered was the ultimate state of plastic collapse. A major drawback to such an approach is that design optimization decisions at this limit state must be made in the absence of information as to their consequences at other limit states of concern. In this context, it was shown by Frangopol and Nakib [1,2] that the design satisfying the ultimate state of plastic collapse requirement may not satisfy the serviceability requirement and vice-versa [1,2].

Subsequently, the need for including multiple limit states in reliability-based structural optimization was recognized and dealt with successfully by Frangopol [3], Thoft-Christensen and Murotsu [4], and Kim and Wen [5], among others. However, these studies were formulated within the simplifed framework of multiple system and/or member performance constraints where the interaction among reliability levels of various limit states cannot be considered.

This paper is a sequel to Fu and Frangopol [6,7] and Frangopol and Fu [8]. It proposes a multiobjective optimization approach to deal with structural reliability-based design under multiple limit states. This approach resolves the limit states reliability interaction problem because keeping track of the reliability level of each limit state of interest becomes an intrinsic part of the optimization method used to find the objective solution set. It is concluded that the proposed approach produces a more balanced optimum solution.

2. MULTIOBJECTIVE FORMULATION
As previously indicated, in order to reach more significant levels of reliability-based structural optimization it is necessary to consider (a) all levels of failure including serviceability and ultimate limit states, and (b) the interaction among reliability levels of various limit states. These two aspects will be considered simultaneously by using a multiobjective optimization formulation. After the general formulation is presented a specific formulation for structures is proposed.

2.1 General formulation
The general formulation of a multiobjective optimization problem is stated as:

$$find \quad \mathbf{x} \; \epsilon \; \Omega \tag{1}$$

$$which \; minimizes \quad \mathbf{f}\,(\mathbf{x}) \tag{2}$$

where $\mathbf{x}$ and $\mathbf{f}$ are design variable and objective vectors, respectively, and Ω is the feasible space determined by a set of constraints of the form

$$h_1(\mathbf{x}) \; \leq \; 0 \tag{3}$$

$$h_2(\mathbf{x}) = 0 \tag{4}$$

The objective vector (2) and the design variable vector (1) are given by:

$$\mathbf{f}\,(\mathbf{x}) \; = \; (f_1\,(\mathbf{x}),\; f_2(\mathbf{x}),\; \cdots,\; f_i(\mathbf{x}),\; \cdots,\; f_m(\mathbf{x}))^t \tag{5}$$

$$\mathbf{x} \; = \; (x_1, x_2, \cdots, x_j, \cdots, x_n)^t \tag{6}$$

where each of the m components $f_i\,(\mathbf{x})$ of the vector $\mathbf{f}$ is an individual objective, and each of the n components x_j of the vector $\mathbf{x}$ is a design variable that is varied by the optimization procedure.

2.2 Specific formulation for structures

The multiobjective structural optimization problem considered in this study is formulated as follows

$$min[V(\mathbf{x}),\ P_{f(COL)}(\mathbf{x}),\ P_{f(YLD)}(\mathbf{x}),\ P_{f(DFM)}(\mathbf{x})] \qquad (7)$$

$$for\ \ \mathbf{x}\ \epsilon\ \Omega = \{\mathbf{x}|h_1\ (\mathbf{x})\ \leq\ 0\ ;\ h_2\ (\mathbf{x})\ =\ 0\} \qquad (8)$$

where $\mathbf{x}$ is the design variable vector to be determined (i.e., cross-sectional areas of structural members); V, $P_{f(COL)}$, $P_{f(YLD)}$, and $P_{f(DFM)}$ are the four objectives representing the total material volume of the structure, the probability of collapse, the probability of first yielding, and the probability of excessive elastic deformation, respectively; and $h_1(\mathbf{x}) \leq 0$ and $h_2(\mathbf{x}) = 0$ are constraints on the design variable vector $\mathbf{x}$ and serve to define the feasible solution set Ω.

The three system failure probabilities in (7) are computed as follows:

$$P_{f(COL)} = Prob\ [any\ g_{ci} \leq 0] \qquad (9)$$

$$P_{f(YLD)} = Prob\ [any\ g_{yk} \leq 0] \qquad (10)$$

$$P_{f(DFM)} = Prob\ [any\ g_{dj} \leq 0] \qquad (11)$$

where $g_{ci} \leq 0$, $g_{yk} \leq 0$ and $g_{dj} \leq 0$ are failure mode expressions with regard to collapse, yielding and excessive elastic deformation, respectively.

3. SOLUTION SEARCHING

As Duckstein [9] shows, for the general multiobjective problem (1)-(2), "one cannot speak of an optimum solution point but of a 'satisficing' solution". This solution is defined as a vector $\mathbf{x}^0$ which belongs to the feasible space Ω and under which none of the objectives $f_i(\mathbf{x})$ can be further reduced without increasing at least one other objective, since the objectives are *interacting*. In general, a set of such 'satisficing' solutions exists and is called a Pareto-optimum or non-dominated solution set. The objectives evaluated at those Pareto solutions are referred to as Pareto optimal objectives thereafter.

A strategy to find the Pareto solution set to a structural optimization problem similar to that given in (7)-(8) was reported in [8]. This strategy consists of three steps as follows: (a) choosing ranges on upper limits of system failure probabilities to be considered; (b) solving problems of biobjective optimization decomposed from the original problem; and (c) solving the original multiobjective problem by using the ε-constraint method. This three-step solution searching strategy keeps track of both the total volume (i.e., weight) level and the reliability level of each of the limit states of interest in the objective vector (7). Therefore, it resolves the limit states interaction problem.

4.   FRAME OPTIMIZATION EXAMPLE

Consider the simple nondeterministic steel portal frame shown in Fig. 1. It has seven critical sections with random plastic bending strengths $(M_1, M_2 \ldots, M_7)$ and is acted on by two random loads S_1 and S_2. The random plastic moments M_i (in kNcm-units) can be expressed in terms of deterministic cross sectional area A_i (in cm²-units) and random yielding stress σ_i (in kN/cm²-units), as follows [10]:

$$M_i = 1.670 A_i^{3/2} \sigma_i \qquad (i = 1, 2, \cdots, 7) \qquad (12)$$

Considering the value 1.15 for the shape factor, the elastic capacities of the critical sections are

$$M_{i,el} = 1.452 A_i^{3/2} \sigma_i \qquad (i = 1, 2, \cdots, 7) \qquad (13)$$

All the random variables are assumed to be normal distributed and the statistical information is as follows: mean loads $[\bar{S}_1 = 1 \text{ kN}, \bar{S}_2 = 2.5 \text{ kN}]$, coefficients of variation of loads $[V(S_1) = 0.15, V(S_2) = 0.20]$, coefficient of correlation among loads $[\rho(S_1, S_2) = 0]$, mean yielding stress $[\bar{\sigma}_i = 14 \text{kN/cm}^2]$, coefficient of variation of yielding stress $[V(\sigma_i) = 0.10]$, and coefficients of correlation among plastic (or elastic) moment capacities [perfect within beam correlation: $\rho(M_3, M_4) = \rho(M_4, M_5) = \rho(M_3, M_5) = 1$; perfect within column correlation: $\rho(M_1, M_2) = \rho(M_6, M_7) = 1$; and zero otherwise]. The modulus of elasticity is taken as E = 20,000 kN/cm².

The expressions of the ten failure modes for system collapse g_{ci}, seven failure modes for yielding occurrence g_{yk}, and two failure modes for excessive deformation g_{dj} [i.e., excessive vertical deflection at node 3: $\Delta_v \geq 2cm$; and excessive drift at node 4: $\Delta_h \geq 2cm$], are given in [11]. The three system failure probabilities (9),(10) and (11) are evaluated by Ditlevsen's upper bound method [12]. The design variable vector **x** has two deterministic components A_1 (column cross-sectional area) and A_2 (beam cross-sectional area).

In order to investigate the interaction of various objective functions, the four-objective optimization (7) is decomposed into three biobjective subproblems

$$min[V(\mathbf{x}), P_{f(COL)}(\mathbf{x})] \qquad (14)$$

$$min[V(\mathbf{x}), P_{f(YLD)}(\mathbf{x})] \qquad (15)$$

$$min[V(\mathbf{x}), P_{f(DFM)}(\mathbf{x})] \qquad (16)$$

by considering only one system failure probability along with the total volume V. The corresponding Pareto optimal biobjectives (14), (15) and (16) are shown in Fig. 2. For the biobjective optimization (14), for instance, it is found that the volume has to exceed 10883 cm^3 if the probability of system collapse is required to be lower than 10^{-5}. The optimal solutions (A_1, A_2) associated with formulations (14), (15) and (16) are shown in Figs. 3(a), 3(b) and 3(c), respectively. It is observed that stronger columns $(A_1 > A_2)$ are needed in order to maintain the required reliability

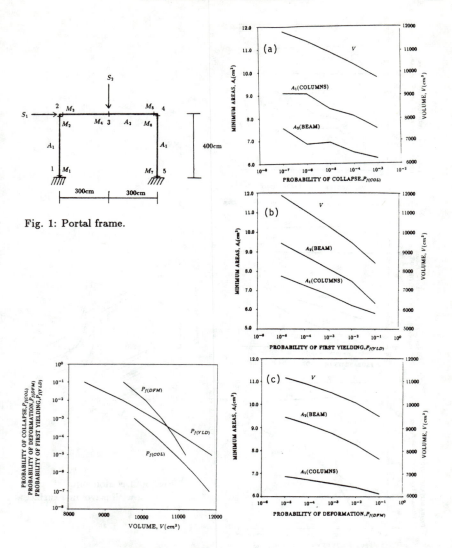

Fig. 1: Portal frame.

Fig. 2: Biobjective optimizations (14), (15) and (16).

Fig. 3: Biobjective optimization (14), (15) and (16) and associated solutions (a), (b) and (c), respectively.

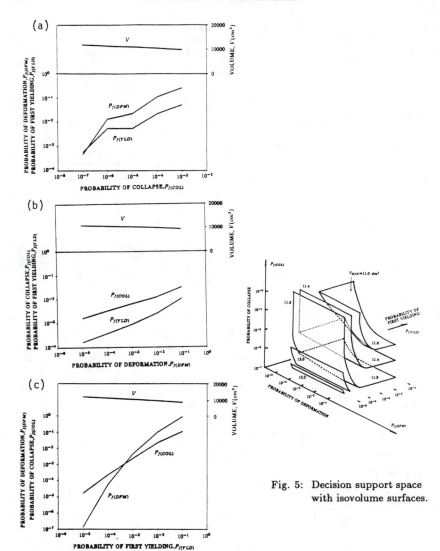

Fig. 5:   Decision support space
         with isovolume surfaces.

Fig. 4:   Biobjective optimizations (14), (15) and
         (16) and associated system unreliability
         levels (a), (b) and (c), respectively.

level with respect to plastic collapse, and stronger beams $(A_2 > A_1)$ are needed in order to maintain the required reliability level with respect to both first yielding and deformation. Given the optimal objectives (14), (15) and (16) the associated system failure probabilities of first yielding and deformation, collapse and deformation, and collapse and first yielding are shown in Figs. 4(a), 4(b) and 4(c), respectively. These figures indicate that some limit state reliability levels are out of control (very large failure probabilities, e.g. $> 10^{-2}$) if they are not included in the optimization process.

Fig. 5 presents the decision support unreliability space (i.e., Pareto objective set) for the multiobjective optimization problem (7) associated with the frame in Fig. 1. It shows a group of five isovolume surfaces in the unreliability space defined by the three system failure probabilities. This decision space represents interactions and trade-offs among the four objectives and provides options for final decisions. It is important to notice that some of the objectives are not critical under certain circumstances.

5. CONCLUSIONS

Based on the results of this study and on the computational experience of the authors [11], the following conclusions are reached: (1) For a reliability-based structural optimization, a formulation with consideration of multiple limit states and other criteria (i.e., weight, cost) produces a more balanced design; and (2) The identification of the decision support region in which limit states reliability interaction exists saves significant computational effort without compromising on accuracy.

6. ACKNOWLEDGMENT

Support of this study by the National Science Foundation through Grant No. MSM-8618108 is gratefully acknowledged.

References

[1] FRANGOPOL, D.M. and NAKIB, R.: "Isosafety loading functions in system reliability analysis", *Comput. Struct.*, Vol. 24, No. 3, 1986, pp. 425-436.

[2] FRANGOPOL, D.M. and NAKIB, R.: "Reliability of structural systems with multiple limit states", in *Materials and Member Behavior* (Ed. D.S. Ellifritt), ASCE, New York, 1987, pp. 638-646.

[3] FRANGOPOL, D.M.: "Structural optimization using reliability concepts", *J. Struct. Engrg.*, Vol. 111, No. 11, 1985, pp. 2288-2301.

[4] THOFT-CHRISTENSEN, M. and MUROTSU, Y: *Application of Structural Systems Reliability Theory*, Springer-Verlag, 1986.

[5] KIM, S.H. and WEN, Y.K.: "Reliability-based structural optimization under stochastic time varying loads", Str. Res. Series No. 533, Univ. of Illinois at Urbana-Champaign, Urbana, Illinois, 1987.

[6] FU, G. and FRANGOPOL, D.M.: "Multicriterion reliability optimization of structural systems", in *Probabilistic Methods in Civil Engineering* (Ed. P.D. Spanos), ASCE, New York, 1988, pp. 177-180.

[7] FU, G. and FRANGOPOL, D.M.: "System reliability and redundancy in a multi-objective optimization framework", in *New Directions in Structural System Reliability* (Ed. D.M. Frangopol), University of Colorado, Boulder, 1989, pp. 147-157.

[8] FRANGOPOL, D.M. and FU, G.: "Optimization of structural systems under reserve and residual reliability requirements", in *Reliability and Optimization of Structural Systems 2* (Ed. P. Thoft-Christensen), Lecture Notes in Engineering, Springer-Verlag, New York, 1989 (in print).

[9] DUCKSTEIN, L.: "Multiobjective optimization in structural design: The model choice problem", in *New Directions in Optimum Structural Design* (Eds. E. Atrek, R.H. Gallagher, K.M. Ragsdell, O.C. Zienkiewicz), John Wiley & Sons, 1984, pp. 459-481.

[10] KOSKI, J.: "Multicriterion optimization in structural design", in *New Directions in Optimum Structural Design* (Eds. E. Atrek, R.H. Gallaher, K.M. Ragsdell, O.C. Zienkiewicz), John Wiley & Sons, 1984, pp. 483-503.

[11] FU, G. and FRANGOPOL, D.M.: "Reliability-based multiobjective structural optimization: Applications to frame systems", Struct. Res. Ser. No. 88-01, Dept. of Civ. Engrg., Univ. of Colorado, Boulder, 1988.

[12] DITLEVSEN, O.: "Narrow reliability bounds for structural systems", J. Struct. Mech., Vol. 7, 1979, 453-472.

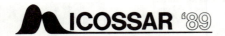
STRUCTURAL OPTIMIZATION BASED ON
COMPONENT-LEVEL RELIABILITIES

Sankaran Mahadevan* and Achintya Haldar**

*   Assistant Professor, Department of Civil and Environmental Engi-
     neering, Vanderbilt University, Nashville, TN 37235
** Professor, Department of Civil Engineering and Engineering Mechan-
     ics, University of Arizona, Tucson, AZ 85721

ABSTRACT

A minimum-weight optimum design procedure is proposed including
the reliabilities of various elements in a structure as constraints. The
reliability indices corresponding to the various limit states are computed
using the Stochastic Finite Element Method, and are required to be
within a desired narrow range. A derivative-free constrained
optimization algorithm with variable discrete step sizes is used to obtain
the optimum design. Both serviceability and ultimate limit states can be
incorporated; also, different levels of risk can be assigned to different
limit states indicating their relative importance. The procedure can
include system reliability constraints, has options for the addition or
deletion of constraints, and can use design groups of similar members
for computational efficiency.

KEYWORDS

Finite elements; limit states; optimization; probability theory; reliability;
safety; structural engineering.

1. INTRODUCTION

Reliability theory is being increasingly incorporated in structural optimization studies in
recent years, due to recognition of the fact that a design has to ensure both reliability of performance and economy to be most desirable from the designer's point of view. Due to the uncertainties in many of the structural parameters, such as loads, geometry, material properties and boundary conditions, reliability of performance cannot be measured without using rational methods of
probabilistic structural analysis. Thus several procedures for reliability-based structural

optimization have been developed in the past.

Most of these procedures make use of the concepts and methods of system reliability. An overall failure probability, or system reliability is imposed as the only behavior constraint in reliability-based optimization. While this approach ensures adequate overall reliability and reduces the number of constraints, it results in non-uniform risk among the various elements of a structure; in general, the control of the designer on reliability at the element-level is lost. Furthermore, in practical design, the members are proportioned according to code specifications that concentrate on the limit states at the member-level. Therefore a more complete approach is required in structural optimization that includes the consideration of both component-level and system-level reliabilities. Such an optimization procedure is proposed in this paper and illustrated with the help of a numerical example.

2. PROPOSED METHODOLOGY

A trial structure designed according to a simplified deterministic procedure is used as the starting point in the proposed algorithm. Given a probabilistic description of the basic parameters of the structure, the reliability indices corresponding to various performance criteria of strength and serviceability are computed at each iteration. The probabilistic information used relates to the means, the coefficients of variation and the probability distribution functions of the parameters. The design variables used here are the cross-sectional properties of the members. Other design variables such as material properties, structural geometry and topology are assumed to be already determined. The improvement in the design is achieved by iterating with the design variables so that all the reliability indices are within a desired, narrow range. For practical application of this strategy to complicated structures, two important problems must be addressed: (i) computation of the reliability indices corresponding to each limit state for every component of the structure, and (ii) formulation of an efficient iterative algorithm that uses these quantities to achieve an optimum design. These two tasks are accomplished as described below.

2.1 SFEM-Based Computation of the Reliability Indices

It is easier to compute the reliability indices as well as the system reliability of simple structures which have closed-form solutions. However, most practical structures have complicated configurations, and their response can only be expressed through a numerical algorithm, such as a finite element procedure. For such structures, the Stochastic Finite Element Method (SFEM) developed in recent years [1] is able to perform the reliability analysis as described below.

In the Advanced First Order Second Moment (AFOSM) method of reliability analysis, a reliability index β is defined as the minimum distance from the origin to the limit state surface given by $G(Y)=0$, where Y is the vector of basic random variables of the structure transformed to the space of reduced variables. The probability of failure is then given by $p_f = \phi(-\beta)$ where ϕ is the cumulative distribution function for a standard normal variable.

The reliability index β can be expressed as $\beta = (y^{*T} y^*)^{1/2}$, where y^* is the point of minimum distance from the origin on the limit state surface. The starting point for an algorithm that searches for y^* may not, in general, be on the limit state surface. Rackwitz and Fiessler [2] proposed a numerical iterative scheme to evaluate y^* by establishing a sequence of linearization points $y_i, y_{i+1}, \ldots\ldots$, according to the rule

$$y_{i+1} = [y_i^T \alpha_i + \frac{G(y_i)}{\nabla G(y_i)}] \alpha_i \qquad (1)$$

where $\nabla G(y) = [\delta G(y)/\delta y_1, \ldots\ldots, \delta G(y)/\delta y_n]^T$ is the gradient vector of the performance function

and $\alpha_i = -\nabla G(y_i)/|\nabla G(y_i)|$ is the unit vector normal to the limit state surface away from the origin. In many cases, the performance function may not be available in analytical form. Der Kiureghian and Ke [3] used a linear transformation from X to Y and obtained an expression for $\nabla G(y)$ using the chain rule of differentiation. Furthermore, the differentiation of the finite element equations after an appropriate transformation from the displacement to a general response makes it possible to compute $\nabla G(y)$ numerically. The Stochastic Finite Element Method enables this by computing and assembling the partial derivative matrices of each quantity computed along the deterministic analysis with respect to the basic random variables.

2.2 Optimization Algorithm

Any optimization procedure has three aspects: objective function, constraints, and the algorithm to search for the optimum solution. A simple and convenient objective function, the minimization of the total weight, is selected as the objective function. The element reliability constraints are written as

$$\beta_i^l \le \beta_i \le \beta_i^u , \quad i = 1,2,...,m \tag{2}$$

where the lower bound β_i^l specifies the minimum required safety level for the ith limit state, while the upper bound β_i^u indicates the desired range of β_i, and m is the number of limit states. The optimum design is said to be reached if all the β_i values fall within the desired range. The feasible region for the design is defined by the lower bound, indicating the acceptable level of risk for each limit state. Reliability-based design formats such as LRFD are derived based upon this idea of acceptable risk levels; the load and resistance factors correspond to prespecified target values of β. Thus one may select the lower bounds same as the target β values used in reliability-based design codes. The upper bounds of β are established such that the β values of different elements fall within a narrow range to assure uniformity in the risk levels. Alternatively, the constraints may simply require the satisfaction of the performance function at the nominal values, as in the case of code-specified serviceability criteria. Such constraints may be written as

$$g_j \ge 0.0 , \quad j = 1,2,...,l \tag{3}$$

where g_j is the performance function for the j such limit state and l is the number of such limit states. The proposed algorithm has the flexibility to handle both types of constraints. Referring to Eq. (2), it can be seen that it is also possible to specify different desired risk levels for different limit states, thus accounting for the fact that all the limit states may not have equal importance.

The system reliability constraint may be written as

$$p_f \le p_f^0 \tag{4}$$

where p_f is the overall failure probability of the structural system, which is required to be less than an acceptable value p_f^0.

A simple, derivative-free search procedure is used to find the optimum design. A 'safe' trial structure (i.e., a design that satisfies the lower bounds of the reliability constraints), selected using an approximate deterministic procedure, is used as the starting point. The algorithm then achieves uniform risk within the feasible region, by searching only in the directions of reducing β values. This means that the algorithm needs to examine only those configurations whose member sizes are less than those of the trial structure. Any movement produces a reduction in weight. If the new design still satisfies the lower bounds of Eqs. (2)-(4), it is accepted as a success; otherwise it is rejected as a failure and the step size is halved in that direction until no significant improvement is possible.

The convergence of the algorithm is accelerated by using discrete step sizes which are determined by different ranges in the values of $(\beta_i - \beta_i^l)$ at any iteration. For example, one may choose step sizes as

$$\Delta = 0.3 \quad \text{for} \quad \beta_i - \beta_i^l \geq 2.0$$

$$= 0.2 \quad \text{for} \quad 1.0 \leq \beta_i - \beta_i^l < 2.0 \qquad (5)$$

$$= 0.1 \quad \text{for} \quad 0.25 < \beta_i - \beta_i^l < 1.0$$

Such a method is easy and fast to implement; even though it is an approximate rule, it is sufficient since the purpose of the algorithm is not to find an absolute optimum but only to ensure that all the β_i values are within a desired range. Furthermore, it also allows the use of different step sizes in different directions. The search is stopped when either all the β_i's are within the desired range or when the smallest step size in every coordinate direction is smaller than a prescribed tolerance level.

If the trial structure selected by using approximate deterministic methods is found to be infeasible, then this design should first be made safe or feasible before starting the weight-minimization algorithm. This improvement can be achieved by simply reversing the search directions and using only the lower bounds of the constraints.

3. NUMERICAL EXAMPLE

A steel portal frame, shown in Fig. 1, is subjected to a lateral load H and a vertical load V. There are ten random variables in this structure, whose statistical description is given in Table 1. The design variables are the area of cross-section A, moment of inertia I, and the plastic

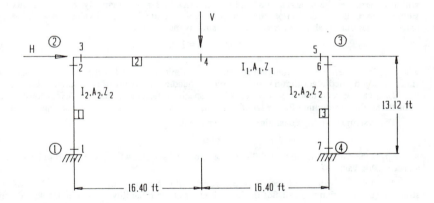

Fig. 1.   Numerical Example—Steel Portal Frame

section modulus Z of the various members. The two columns have identical cross-sections. In the design, the beam and the columns are required to satisfy two types of strength limit states: (i)

Table 1. Description of Basic Random Variables

No.	Symbol	Units	Mean	Coefficient of Variation	Type of Distribution
1	H	kips	4.50	0.15	Type I
2	V	kips	18.00	0.20	Lognormal
3	E	ksi	29000.00	0.10	Lognormal
4	F_y	ksi	36.00	0.10	Lognormal
5	I_1	in^4	740.00	0.10	Lognormal
6	A_1	in^2	132.00	0.10	Lognormal
7	Z_1	in^3	25.60	0.10	Lognormal
8	I_2	in^4	285.00	0.10	Lognormal
9	A_2	in^2	40.30	0.10	Lognormal
10	Z_2	in^3	51.20	0.10	Lognormal

combined axial compression and bending, and (ii) pure bending. The performance functions for the first limit state are written as [4]

$$g_1 = 1.0 - \frac{P}{P_u} - \frac{C_m M}{M_p (1 - P/P_E)} \tag{6}$$

$$g_2 = 1.0 - \frac{P}{P_y} - \frac{M}{1.18 M_p}. \tag{7}$$

Eq. (6) provides an overall stability criterion for the beam-column, whereas Eq. (7) provides a local strength requirement. In these equations, P is the applied axial load on the member, M is the applied bending moment, P_u is the ultimate axial load that can be supported by the member when no moment is applied, P_E is the Euler buckling load in the plane of M, $P_y = A F_y$, where F_y is the yield strength, $M_p = Z F_y$ is the plastic moment capacity and C_m is as defined in the AISC LRFD Specifications [5]. For all three members in the frame, $C_m = 0.85$ is used.

The performance function for the second strength limit state (pure bending) is written as

$$g_3 = 1.0 - \frac{M}{M_u} \tag{8}$$

where M is the applied moment, and M_u is the flexural strength of the member. The flexural strength depends on the range of behavior of the member, i.e., whether it is plastic, inelastic, or elastic. In this example, it is assumed that $M_u = M_p$, the plastic moment capacity of the member. The reliability constraints corresponding to both performance criteria for all the three members are given by

$$3.0 \leq (\beta_1 \text{ or } \beta_2) \leq 3.25 \tag{9}$$

The design is also required to satisfy two serviceability constraints. The limiting vertical deflection at the midspan of the beam = span/240, while the limiting side-sway at the top of the frame = height/400. In the present example, it is required for the sake of illustration that the serviceability limits be satisfied at the mean values of the random variables; thus no reliability ranges are defined. Therefore these two constraints are written as

$$g_4 = 1.0 - \frac{\text{midspan deflection of beam}}{\text{span}/240} \geq 0.0 \tag{10}$$

$$g_5 = 1.0 - \frac{\text{sideway at the top}}{\text{height}/400} \geq 0.0 \tag{11}$$

The strength constraints need to be checked only at the critical section of each member. In the case of the two columns which have identical cross-sections, only that column which carries more load needs to be checked. Furthermore, of the three performance criteria given by Eqs. (6), (7), and (8) for each member, the one having the lowest reliability is critical; therefore, only this critical limit state needs to be checked for reliability after the first iteration in the optimization algorithm. In this example, the above arguments result in four constraints, namely, Eq. (6) for member 2 at midspan, Eq. (8) for member 3 at node 3, and the two serviceability constraints. Such considerations become advantageous in the design of large structures that have several members with identical cross-sections forming one design group, and with several performance criteria, thus reducing the computational effort.

The computations are further reduced by expressing the area and the moment of inertia in terms of the plastic section modulus with the help of regression analysis over a range of I-sections used, as [6]

$$A = -6.248 + 1.211\, Z^{2/3} \tag{12}$$

$$I = 20.36 + 22.52\, A + 0.22\, A^2 \tag{13}$$

This results in the reduction of the number of design variables to only two in the present problem. The step sizes for the optimization algorithm are as shown in Eq.(5). Mahadevan and Haldar [6] discussed elsewhere in detail the method to implement these requirements in reliability-based optimization.

3.1 Results and Observations

Table 2 traces the steps in the proposed reliability-based weight-minimization procedure for designing the portal frame. The use of discrete step sizes which are different in different directions is observed to be a fast means of achieving the optimum design, since it is reasonable that a member with a higher reliability should have a greater reduction in its section properties than a member with a lower reliability. The provision to halve the step sizes in the event of an infeasible move further improves the efficiency of the algorithm in the neighborhood of the optimum solution. While the uniformity in the risk levels achieved is apparent, it is also possible to include system reliability constraints in the proposed algorithm, as described in [6]. However, the difficulty in the computation of system reliability for realistic framed structures with complicated configurations by combining different limit states inhibits its use in practical applications of reliability-based structural optimization at present.

The proposed method, which includes information about the probability distribution functions and the statistical correlations among the random parameters, is a significant improvement over the simple Level I fixing of the load and resistance factors by design codes. Furthermore, the design of the members in this method is directly based on their reliabilities in the actual configuration, instead of on code-specified safety factors that are either empirical or are derived based on the reliability analysis of isolated, simple structural elements [1].

Table 2. Steps in minimum weight design

1 Iteration No.		2 Section Properties		3 β and g values	4 Feasibility of design
		Beam	Columns		
1 (Trial Structure)	Z A I	132.0 25.6 740.0	51.2 10.3 285.0	$\beta_1 = 4.80$ $\beta_2 = 4.66$ $g_4 = 0.67$ $g_5 = 0.68$	Yes
		$1.0 < \beta_1 - \beta_1^l < 2.0 \rightarrow \Delta_1 = 0.2$			
2 ($\Delta_1 = 0.2$)	Z A I	93.0 18.6 515.2	Same as above	$\beta_1 = 3.23$ $\beta_2 = 4.21$ $g_4 = 0.58$ $g_5 = 0.65$	Yes
		$1.0 < \beta_2 - \beta_2^l < 2.0 \rightarrow \Delta_2 = 0.2$			
3 ($\Delta_2 = 0.2$)	Z A I	Same as above	41.0 8.1 218.2	$\beta_1 = 3.24$ $\beta_2 = 3.50$ $g_4 = 0.54$ $g_5 = 0.57$	Yes
		$0.25 < \beta_2 - \beta_2^l < 1.0 \rightarrow \Delta_2 = 0.1$			
4 ($\Delta_2 = 0.1$)	Z A I	Same as above	36.9 7.2 193.0	$\beta_1 = 3.25$ $\beta_2 = 3.16$ $g_4 = 0.52$ $g_5 = 0.53$	Yes

$$0.0 < \beta_1 - \beta_1^l \leq 0.25 \qquad 0.0 < \beta_2 - \beta_2^l < 0.25$$

$\rightarrow$ stop search in $\rightarrow$ stop search in
this direction   this direction

Note: Units for Z, A, and I are in in^3, in^2, and ^{4}, respectively.

4. CONCLUSIONS

A procedure for structural optimization that includes the reliabilities of individual components has been presented in this paper. The method has an advantage over other methods of reliability-based optimization in that it is simple to formulate and implement, and can consider both system and element reliabilities. The attention to element reliabilities is similar to the practical approach used in design offices. Such consideration also results in uniform distribution of risk in the structure. The optimization algorithm has the facility to handle different requirements of different limit states easily, and can be extended to more limit states including those related to system reliability. The use of variable discrete step sizes makes the algorithm easy and fast; its efficiency is also improved by the use of strategies to reduce the constraints and the design variables, making the method attractive for practical application.

5. ACKNOWLEDGEMENTS

This paper is based upon work partly supported by the National Science Foundation under Grants No. MSM-8352396, MSM-8544166, MSM-8746111, MSM-8842373, and MSM-8896267. Financial support received from the American Institute of Steel Construction, Inc., Chicago, is also appreciated. Any opinions, findings, and conclusions or recommendations expressed in this publication are those of the authors and do not necessarily reflect the views of the sponsors.

6. REFERENCES

1. Mahadevan, S., "Stochastic finite element-based structural reliability analysis and optimization," Ph.D. Thesis, Georgia Institute of Technology, Atlanta, Georgia, 1988.
2. Rackwitz, R. and Fiessler, B., "Structural reliability under combined random load sequences," Computers and Structures, Vol. 9, pp. 489-494, 1978.
3. Der Kiureghian, A. and Ke, J.B., "Finite element-based reliability analysis of framed structures," Proceedings, 4th International Conference on Structural Safety and Reliability, ICOSSAR'85, Kobe, Japan, Vol. 1, pp. 395-404, 1985.
4. Bjorhovde, R., Galambos, T.V., and Ravindra, M.K., "LRFD criteria for steel beam-columns," Journal of the Structural Division, ASCE, Vol. 104, No. ST9, pp. 1371-1388, 1978.
5. American Institute of Steel Construction, "Manual of steel construction: load and resistance factor design," Chicago, Illinois, 1986.
6. Mahadevan, S. and Haldar, A., "Efficient algorithm for stochastic structural optimization," Journal of Structural Engineering, ASCE, July 1989.

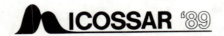
OPTIMUM STUDIES OF COEFFICIENT VARIATION
OF UBC SEISMIC FORCES

Franklin Y. Cheng* and Chein-Chi Chang**
*Curators' Professor of Civil Engineering
University of Missouri-Rolla, Rolla MO 65401
**Former Graduate Assistant, Department of Civil Engineering, University of
Missouri-Rolla
Rolla, MO 65401 U.S.A.

ABSTRACT

This paper presents an optimization algorithm for finding optimal parameters of a nondeterministic structure
subjected to UBC seismic load for both normal and lognormal
distribution models.   The optimization algorithm is based
on the first order approximation of reliability analysis;
construction cost and expected failure cost; and a variety
of constraints such as displacements of a system, yielding
and buckling of constituent members. Several numerical
examples are given which show that the optimum solutions
are very sensitive to the high variation of UBC seismic
load and high design reliability criteria; and   the
lognormal distribution demands higher optimum solution than
the normal.

Keywords: design, earthquake, failure, lognormal, normal,
           optimization, random, steel, structure, UBC.

1. INTRODUCTION

The advent of computer technology and the development of optimization
algorithms have led the earthquake resistant design to a new direction of
seismic design. In the past, the studies on optimum seismic structures are
assumed to be deterministic based on various building codes. The design
results include the effectiveness of various bracing systems, the effect of the
$P - \Delta$ forces and vertical excitations on the design, and the stiffness
requirements for various constraints of allowable story drifts, displacements,
stresses, and frequencies [1, 2, 3]. In this paper a probabilistic approach
based on reliability concept is employed where the resistances and responses
are assumed to be random. The probability of failure (opposite meaning of
reliabilities) is computed on the basis of the responses being larger than
resistances.   A shear building of steel is used to study the optimum solutions
subjected to UBC seismic load for various estimations of nonstructural cost
and expected failure cost with emphasis on the sensitivity of the coefficient
of variation of the seismic forces.

RELIABILITY ANALYSIS

The probability of failure can be expressed as

$$P_f = P(R<S) = \int_o^\infty F_R(r)f_S(r)dr \tag{1}$$

where R = resistance, S = response, r = random parameters, F_R = cumulative probability distribution function for a resistance, and f_S = probability density function for a response. If the probability of failure is assumed to be a normal distribution, they can be expressed as

$$P_f = 1 - \Phi(\beta) \tag{2}$$

where $\Phi(\ )$ = standard normal probability distribution and β = safety factor.

Due to the relationship between response and resistance, the safety factor may be a normal or lognormal distribution:

normal

$$\beta = \frac{\bar{R} - \bar{S}}{\sqrt{\sigma_R^2 + \sigma_S^2}} \tag{3}$$

lognormal

$$\beta = \ln \frac{\left[(\bar{R} - \bar{S}) \sqrt{\dfrac{1 + V_S^2}{1 + V_R^2}} \right]}{\sqrt{\ln[(1 + V_R^2)(1 + V_S^2)]}} \tag{4}$$

where $\bar{R}$, $\bar{S}$ = means of resistance and response, respectively; σ_R^2, σ_S^2 = variances of resistance and response, respectively; and V_R, V_S = coefficients of variation of resistance and response, respectively.

Because the functional relationship between the response (or resistance) and the parameters is difficult to be determined, a first-order approximation is adopted to find mean and variance of response (resistance) based on mean parameter values, i.e.,

$$\bar{S}(r) = S(\bar{r}) \ , \quad \bar{R}(r') = R(\bar{r}') \tag{5a,b}$$

$$\sigma_S^2(r) = \sum_{ij} (\frac{\partial S}{\partial r_i})_{\bar{r}}(\frac{\partial S}{\partial r_j})_{\bar{r}}\rho_{r_i r_j} V_{r_i} V_{r_j} \bar{r}_i \bar{r}_j \tag{6}$$

$$\sigma_R^2(r) = \sum_{i'j'} (\frac{\partial R}{\partial r_i'})_{\bar{r}'}(\frac{\partial R}{\partial r_j'})_{\bar{r}'}\rho_{r_i' r_j'} V_{r_i'} V_{r_j'} \bar{r}_i' \bar{r}_j' \tag{7}$$

where r (r') = random parameters of response (resistance), $\rho_{r_i r_j}(\rho_{r_i' r_j'})$, $V_{r_i}(V_{r_i'})$, $V_{r_j}(V_{r_j'})$, $\bar{r}_i(\bar{r}_i'), \bar{r}_j(\bar{r}_j')$ = the correlation coefficients, coefficients of variation, and mean values of ith and jth random parameters of the responses (resistances), respectively [4, 5, 6].

3. UBC SEISMIC LOAD AND LOAD EFFECTS

UBC Seismic Load. -- The earthquake load expression in the Uniform Building Code (UBC) [7] is given by the ground base shear force, and then distributed to the desired structural levels. The determination of base shear force can be shown in the following equation

E = Z I K C S W   (8)

where E = the shear force at the base; Z = numerical coefficient depends on the
zone that the structure is located, for zone 1,2,3,4; Z = 3/16, 3/8, 3/4, 1
respectively; I = occupancy importance factor; K = numerical coefficient; C =
$1./(15\sqrt{T})$, and the value need not exceed 0.12; T = structural fundamental
period; S = numerical coefficient for site-structure resonance, when
characteristic site period is not properly established, the value of S shall be
1.5; W = the total dead load.

The statistics of the UBC seismic load may be determined as follows:

mean $\bar{E} = \bar{Z}\ \bar{I}\ \bar{K}\ \bar{C}\ \bar{S}\ \bar{W}$ (9)

where the bar over the parameters indicates the mean parameter values.

The coefficient of variation of earthquake may be assumed to be a Type II
extreme value distribution [4] which is given by distribution form as

$$E'(e_1) = e^{(-\frac{e_1}{u_1})^{-k_1}}$$ (10)

where E' = earthquake random variable, e_1 = a random value of E', and u_1, k_1
are parameters which are determined from seismological data survey.

For this type of distribution, the coefficient of variation of earthquake,
$V_{E'}$ is

$$V_E = \sqrt{\frac{\Gamma(1-\frac{2}{k_1})}{\Gamma^2(1-\frac{1}{k_1})} - 1}$$ (11)

where $\Gamma(\ )$ is a Gamma function.

Although due to the highly complicated and unpredictable phenomenon of
earthquake, the exact value of k_1 is impossible to be determined, many values
of k_1 and coefficients of variation of earthquake were proposed in the past
decade. Among them the value of k_1 is 2.3, which corresponds to coefficient of
variation, 1.38, reported in National Bureau of Standards Special Publication
No. 577 and is used for some seismic building designs. The value of k_1, 2.7,
which corresponds to coefficient of variation, 0.85, is recommended for the
nuclear power plant design [8].

4. OPTIMIZATION FORMULATIONS

For a structural problem the optimum design can be described by:

minimizing   objective function
subject to   constraints

OBJECTIVE FUNCTION. The objective function of structural design problem
may be weight and cost which are described as follows:

Weight. -- Weight (W) is the constituents of structural member weights and
can be expressed as

$$W = \sum_i r_{di} l_i A_i \tag{12}$$

where r_{di}, l_i, A_i = the mass density, length, and area of a member.

Total Structural Cost. -- Total structural cost (C_T) may include two parts: initial construction cost (C_I) and expected future failure losses ($L_f P_{fT}$); i.e.

$$C_T = C_I + L_f P_{fT} \tag{13}$$

where L_f = expected failure cost; P_{fT} = system failure probability.

Initial construction cost (C_I) comprises of structural members cost and nonstructural members cost. Expected failure cost (L_f) is the total loss incurred in a structural failure state. This includes additional replacement cost, damage to property, liability due to death and injury, business interruption.

Although initial construction cost and expected failure cost can be classified into many items. However, these quantities are difficult to be estimated. Therefore the ratio of initial cost to member cost (C_{in}) and the ratio of future failure cost to initial cost (C_{VL}) are used in the cost function to represent the various magnitudes of initial construction cost and future failure cost. Based on two above coefficients the initial cost and failure cost function can be calculated as follows:

$$C_I = C_{in} C_u \sum_i l_i A_i \tag{14}$$

$$L_f = C_{VL} C_I \tag{15}$$

where C_u is the unit steel volume cost. Through these two coefficients, the influences of nonstructural cost and expected failure cost in cost optimum design problem will be observed.

CONSTRAINTS. In reliability design the constraints may be the failures of displacements, beams, and columns. The fail-ures can be expressed in terms of safety factor or probability of failure. The structural resistances are allowable dis-placement for displacement failure; the yielding moment for beam failure; the combined stress of axial force and bending moments for column failure [4]. The failure modes could be due to individual components of members or due to structural displacements.

5. NUMERICAL EXAMPLES AND OBSERVATIONS

To illustrate the applicability of the proposed optimum design, a ten-story steel shear building shown in Figure 1 is implemented based on a computer program developed at UMR. The results yield a great deal of information, such as optimum cost, optimum weight, member sizes, and others. This paper only shows the effect of coefficient of variation of UBC seismic load on the optimum solutions of structural weight of the system and the moments of inertia of individual members. The notations in the Figs. 1-5 are: the numbers represent the degree of freedom and members' number; N and LN designate normal and lognormal distribution, respectively; and P_{f0} represents allowable failure probability. The parameters used in examples are: allowable mean displacement = 0.005 times corresponding the story height, allowable variance of displacement = 0.0, mean yielding strength = 36 ksi (2.4804×10^5 kPa), mean

elastic modulus = 30000 ksi (6.201 x 10^2 kPa), coefficient of variation of elastic modulus = 0.06, coefficient of variation for moment of inertia = 0.05, V_{My} = 0.12, V_{Mcr} = 0.2, C_m = 1.0. The parameters for UBC seismic load are Z = 1.0, I = 1.0, K = 1.0, S = 1.5, h_n = 27 ft. (822.96 cm), D_n = 30 ft (915.4 cm).

Since the exact value of coefficient of variation of UBC seismic load, V_E is still unknown due to complicated phenomenon of earthquake, we let V_E vary from 0 to 1.38 and then observe the influence of the coefficient on the design results for normal and lognormal distribution with one low and one high allowable reliability levels.

The optimum weights of the reliability-based design with normal and lognormal distributions are higher than those of deterministic design. The weight percentage increases are from 7.4% to 90.85% for V_E = 0 to V_E = 1.38 at P_{f0} = 10^{-1} when the normal distribution is used; and 7.57% to 63.13% when lognormal distribution is used. The weight percentage increases are from 29.76% to 393.3% for V_E = 0 to V_E = 1.38 at P_{f0} = 10^{-7} with normal distribution; and 44.13% to 2976.9% with lognormal distribution. Therefore the optimum weight and moments of inertia due to change of variation of earthquake at high reliability are faster than those at low reliability with the normal distribution. In particular the increases are much faster for the lognormal distribution and at the high variation of earthquake.

It is apparent that the optimum solutions are sensitive to the coefficients of variation of UBC seismic load and reliability levels. And both the normal and lognormal distributions yield similar observations. The lognormal distribution, however, demands much heavier structural design than the normal distribution, at high reliability and high variation.

6. ACKNOWLEDGEMENTS

This work presented herein is a partial result from the project supported by the National Science Foundation under Grant No. CEE8403875. The financial support is gratefully acknowledged.

7. REFERENCES

1. Cheng, F.Y. and Juang, D.S., "Assessment of Various Code Provisions Based on Optimum Design of Steel Structures," International Journal of Earthquake Engineering and Structural Dynamics, Vol. 16, pp. 45-61, 1988.

2. Cheng, F.Y. and Juang, D.S., "Recursive Optimization for Seismic Steel Frames," Journal of Structural Engineering, ASCE, Vol. 115, pp. 445-466, 1989.

3. Cheng, F.Y. and Truman, K.Z., Optimum Design of Reinforced Concrete and Steel 3-D Static and Seismic Building Systems with Assessment of ATC-03. Final Report for the National Science Foundation. Available at the U.S. Department of Commerce, NTIS access No. PB 87-168564/AS (414 pages), 1985.

4. Cheng, F.Y., and Chang, C.C., "Optimum Design of Steel Buildings with Consideration of Reliability," Proceedings of the 4th International Conference on Structural Safety and Reliability, Kobe, Japan, Vol. 3, pp. 81-88, 1985.

5. Cornell, C.A., "Engineering Seismic Risk Analysis," Bulletin of the Seismological Society of America, Vol. 58, pp. 1583-1600, 1968.

6. Ellingwood, B., and Ang, A.H.-S., A Probabilistic Study of Safety Criteria for Design. Structural Res. Ser. No. 387, U. of Ill., 1972.

7. International Conference of Building Officials, Uniform Building Code, 1985 ed., Whittier, C.A., 1985.

8. Hwang, H., Kagami, S., Reich, M., Ellingwood, B., Shinozuka, and Kao, C.S., Probability Based Load Combination Criteria for Design of Concrete Containment Structures, U.S. Regulatory Commission Report NUREG/CR-3876, Washington D.C., March 1985.

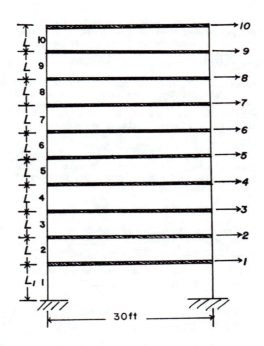

Fig. 1 - 10-Story Shear Building Structure
(L$_1$ = 15 ft, L = 12 ft, 1 ft = 30.48 cm)

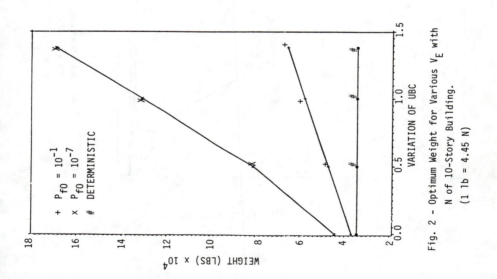

Fig. 3 - $I_1 - I_5$ for Various V_E with N of 10-Story Building.
($1 \text{ in}^4 = 41.62 \text{ cm}^4$)

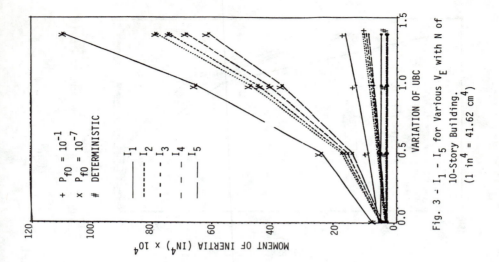

Fig. 2 - Optimum Weight for Various V_E with N of 10-Story Building.
($1 \text{ lb} = 4.45 \text{ N}$)

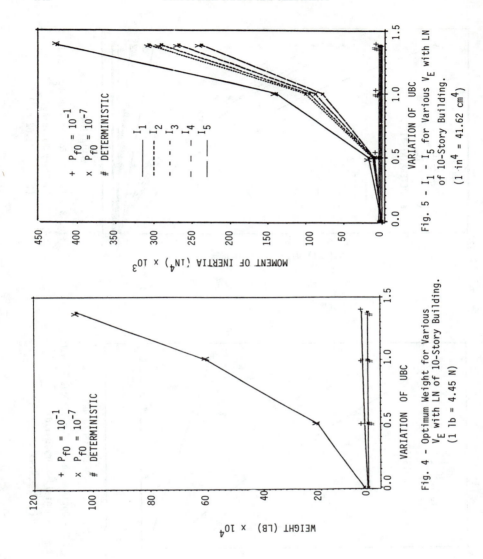

Fig. 5 - I_1 - I_5 for Various V_E with LN
of 10-Story Building.
($1\ in^4 = 41.62\ cm^4$)

Fig. 4 - Optimum Weight for Various
V_E with LN of 10-Story Building.
($1\ lb = 4.45\ N$)

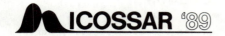
EVALUATION OF LIFETIME RISK OF STRUCTURES
-RECENT ADVANCES OF STRUCTURAL RELIABILITY IN JAPAN-

Naruhito Shiraishi and Hitoshi Furuta

Department of Civil Engineering, Kyoto University, Kyoto
606, Japan

Abstract

The Task Committee on Structural Safety and
Reliability of the Japan Society of Civil Engineers
has made survey of recent advances in the field of
structural reliability. Consequently, it reached a
conclusion that the evaluation of lifetime risk is
quite important to ensure a satisfactory safety or
serviceability level through the life of structures.
The committee published the final report which
includes the results obtained through its research
activity during the last four years. In this paper,
the outline of the report will be introduced with
some discussion on each current issue.

keywords

Damage assessment; design code; human error;
lifetime risk; structural reliability.

1.Introduction

The Task Committee on Structural Safety and Reliability of
the Japan Society of Civil Engineers has made survey of recent
advances of research activities in the field of structural
reliability. Consequently, it reached a conclusion that the
evaluation of lifetime risk is quite important to ensure a
satisfactory safety or serviceability level through the life of
structures. The term of "lifetime risk" means various kinds of
natural and man-made risk which may suffer in the constitutive
process of structures; planning, designing, fabrication and
maintenance. The committee published the final report [1] which

includes the results obtained through its research activity during the last four years. In this paper, the outline of the report will be introduced with some discussion on each current issue.

This research report consists of seven chapters and five appendices. Chapter 1 is the introduction which describes the background and particular characteristics of structural reliability problems in civil engineering structures and the significance of the reliability-based design method. Chapter 2 is concerned with the concept of structural safety; the definition of safety and reliability and analytical method of reliability assessment. Chapter 3 provides the evaluation methods of structural reliability at the various stages of constructing structures. The reliability analyses are explained by paying attention to difference among the planning, designing and constructing processes.

In chapter 4, emphasis is placed on the diagnosis of existing structures whose importance has been well recognized recently. Many examples of structural damage are collected and analyzed, which occurred in both steel and concrete structures. Several testing techniques available for the structural diagnosis are introduced, which include non-destructive testing, field testing and laboratory testing. Some damage measures proposed so far [2] are reviewed and discussed. Moreover, new rating methods are presented, which are based on the multi-criteria analysis and the technique of knowledge-based expert system. Chapter 5 is devoted to the treatment of labor accidents which is discussed from the standpoint of structural reliability.

In chapter 6, human errors (HE) are summarized and classified into several groups; HE in designing, HE during construction and HE in use. It is concluded that education is important for diminishing the potential risk of human errors which sometimes cause a great loss. In chapter 7, several guidelines and design codes are reviewed and compared from the viewpoint of structural safety, which have been used in the design of bridge, port, dam, tunnel, river, lifeline, tank and marine structures. Also, the difference of reliability concept is made clear by comparing with those of the naval, aeronautical, architectural and nuclear engineerings. In the appendices, fundamental bases of the reliability theory are given with emphasis on the system reliability theory, time-dependent theory, application of fuzzy set theory and risk assessment.

2.Uncertainty and Structural Safety

Problems of structural safety have become important due to the existence of uncertainty inherently involved in structural engineering. While our lives are closely related to surrounding natural and man-made environments, we do not have complete information about the environments whose characteristics are not certain or definite. Uncertainties can be classified into seven categories; 1)randomness, 2)modeling error, 3)imperfect information, 4)ambiguity, 5)generality of concept, 6)variation of value and 7)unknown factor. To deal with these uncertainties, probabilistic model, fuzzy model and game model are proposed.

In the decision making under such uncertain environments, the uncertainties to be considered should be analyzed paying attention to their characteristics such as the occurrence time, place and cause. Every uncertainties are checked whether they can be

eliminated and further whether they can be quantified. Based on
the results, the feasibility of several alternative strategies are
examined and determined through a cost effective analysis. To
realize the strategy chosen, the following methods are, for
example, enumerated; 1)all population test, 2)cause and effect
analysis, 3)safety margin method, 4)reliability-based design
method, 5)min and max method, 6)redundant system method, 7)dynamic
programming method, 8)observation-based construction method,
9)provisional method and 10)evasion method.

3.Diagnosis of Existing Structures

Diagnosis of existing structures is becoming important to
establish a rational maintenance or repair program [3,4]. As
representative damage causes, fatigue, delayed failure and
corrosion are considered for steel structures, and in addition to
them, cavitation, chemical reaction and alkali aggregate reaction
are considered for concrete structures. In the research report of
the Task Committee on Structural Safety and Reliability, lots of
damage examples are collected and analyzed. Several testing
techniques available for the structural diagnosis are examined,
which include non-destructive testing, field testing and
laboratory testing. Several damage measures proposed so far are
reviewed and discussed [5].

To perform the structural diagnosis comprehensively, new
diagnosis methods are presented, which are based on the multi-
criteria analysis and the technique of knowledge-based expert
system. In actual maintenance work, it is necessary to determine
the rank of damage for underlying structures. The multi-criteria
analysis [6,7] is introduced into the assessment of structural
damage. The multi-criteria analysis is one of the comprehensive
evaluation systems of alternatives and can be used to deal with
the rated values of damage factors in the more straight and
positive way. Moreover, if the structures under consideration
increase in number, the new data can be used effectively to lead
to a reasonable estimation.

In order to account for the difference of structural types
and surroundings, a pair of alternative indices, "integrity rate"
and "damage rate", is presented as a measure of integrity and
damage states [8]. The integrity rate C_i and damage rate D_i are
defined by use of the concordance analysis [6], which is a
representative of multi-criteria analysis. C_i and D_i stand for the
relative integrity and damage rates among all the investigated
structures.

So far daily maintenance work has been performed on the
basis of intuition and engineering judgment of experienced
engineers. In order to make effective use of knowledge and
experience of experts, many rule-based systems have been
developed in various fields of science, medicine and engineering
[10,11]. A rule-based system is developed here to assess damage
state of bridge structure. The focus is put on the reinforced
concrete deck, because its failure has been occasionally reported
in Japan. The present system consists of main three parts;
interpreter, data-base and rule-base. All rules involved are
described through production rules with certainty factors [9,10].

In this system [13], the information obtained by visual
inspection is used as the input data. Supposing that the
inspection results regarding crack are input into the system,

rules concerning their damage degree, damage cause and damage expanding speed are invoked to provide a solution for the damage assessment. The total number of available rules is 848; 92 rules for damage degree, 258 rules for occurrence time of damage, 365 rules for damage cause, 65 rules for damage expanding speed, 9 rules for description of damage state, 30 rules for damage pattern, and 29 rules for proceeding pattern of damage.

4.Labor Accidents

Every year a lot of casualties take place in the construction industry. Recently labor accidents associated with construction work account for about one third of all occupational accidents and represent nearly a half of the number of deaths for all industry in Japan [14]. This means that the labor accident analysis is one of important issues in the field of structural reliability.

The research report presents 1)definition of labor accident, 2)measures of labor accidents, 3)labor accidents analyses and 4)guideline of the safety assessment for construction work.

In the analysis of labor accidents, it is necessary to clarify the causes of accidents and to develop an appropriate causal model. Causal tree analysis is one of useful methods for labor accident analysis. In order to describe the process leading to an accident graphically, a causal tree diagram is drawn out. The causal tree diagram is made up by indicating events which contributed the occurrence of an accident, and by connecting these events in three kinds of connection type: sequence type, disjunction type and conjunction type. As an example, one of these causal tree diagrams is shown in Fig. 1. Then the linkage of events in the causal tree diagram is investigated and factors contributing the occurrence of the accident is picked up. These factors are identified and classified into five categories; 1)I(human, individual) factor, 2)T(task) factor, 3)M(material) factor, 4)E_p(physical environment) factor and 5)E_s(social environment) factor. In accordance with the definition of

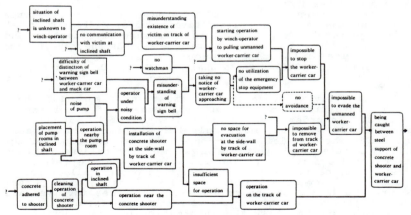

Fig. 1 Causal tree diagram for an accident

classification of these five contributive factors, the frequencies of the factors leading to labor accidents, and the relationships among the factors are investigated. The description of the relationships between the five contributive factors are arranged and illustrated in Table 1.

Table 1 Typical interrelationship between contributive factors to accidents

		Subsequent factor				
		I	T	M	Ep	Es
Previous factor	I	difficulty of communication between individuals (ex. default sign) etc	operation disturbance by improper behavior (ex. unskilled worker) etc	improper use of machinery/tools etc	influence for working condition by behavior of individual etc	disturbance in organization
	T	dangerous works (ex. operation at a high elevated place) etc	inadequate adjustment between concurrent operation etc	improper task assigned to machinery/equipment etc	influence on working condition by inadequate operation mode etc	confusion of working crew due to changing operation mode etc
	M	insufficient protection of worker from machinery etc	disturbance in operation by using improper tool/equipment etc	mismatching use of machinery and material etc	influence on working place by using machinery (ex. vibration) etc	confusion of working crew by using new-machinery etc
	Ep	un-hygienic working place etc	disturbance in operation by working condition (ex. inadequate lighting) etc	environmental effect on machinery/equipments (ex. humidity, heat) etc	environmental effect on working place(ex. decrease stability of sloop by rain) etc	delay the work due to climate condition etc
	Es	Inadequate management system in communication etc	Allowing inadequate work mode etc	absence of person in charge of examining machinery etc	work areas left in a disorderly fashion etc	disorganization due to inadequate adjustment between working crews etc

5.Structural Safety and Human Error

The safety of structures through their lifetime depends on how human beings take part at every phase of the lifetime of structures; planning, designing, constructing and maintenance phases. Therefore, in the evaluation of lifetime risk, the human factors, especially, human errors, should be sufficiently studied. The safety of structures decreases or increases due to the existence of human errors. In the ordinary safety analysis, only uncertainties in the loads acting structures and the resistances of structures are taken into account [15]. However, according to investigations of actual failure cases of structures, it has been pointed out that main contributions are human errors.

Basically, human beings are the creature who makes errors. Some people make errors often, while some people do not make errors so often. Moreover, the frequency of errors increases because of a lack of time, overpractice, emotional instability, physical imperfection and so on. This tendency is also true to civil engineers who plan, design, construct and maintain structures.

In the final report, the human error is defined as that if a failure or defect happens and the causes are not uncertainties in loads and resistances but due to the phenomena reflected to the design codes, human errors exist. The definition of a human error is schematically shown in Fig. 2. The abscissa represents the time and the ordinate represents the engineering level. The level of 100 % means that every phenomenon has been elucidated. The phenomena under the curved line in the figure have been made clear while the phenomena above the curve have not been made clear. The phenomena above the step line have not been reflected to design codes and are unknown for the ordinary engineers and impossible to take countermeasures. If a failure happens because of a phenomenon in the area(1), it is considered that a human error occurred. At the present time, if the failure happens because of the phenomenon A, a human error should have occurred; on the

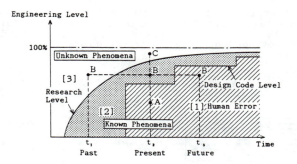

Fig. 2 Definition of human error

contrary, if a failure happens because of the phenomenon B or C, a human error has not occurred. However, judgment of a human error changes as time goes. Taking the phenomenon B as example, if a failure happens because of B at the past time(area(3)) or the present time(area(2)), no human error has happened. However, if the phenomenon B occurred in the future time when the phenomenon was included in the code, it is a human error. Taking the liquefaction as an example, the collapse of bridges because of liquefaction under the Niigata earthquake in 1964 is not human error. After the earthquake, the liquefaction was studied extensively by many researchers and reflected to the pertinent design codes in 1979 and 1980. Therefore, if a failure happens after this time, it is a human error.

6.Structural Safety in Design Codes and Guidelines

In general, there are codes or guidelines for the stages of planning, designing, constructing and using of actual structures. Here, an attempt is made to review how the philosophy of structural safety was introduced in the composition of codes or guidelines. This may aid to bridge the reliability analysis and the practical design and construction works. At first, the thinking way of structural safety in design codes are described comparing those for various kinds of civil engineering structures; harbor structures, dams, river structures, tunnels, lifelines, tanks and marine structures. The comparison is useful for recognizing the difference of design requirements or checking levels for safety peculiar to those structures. Next, an international comparison among design codes proposed in the world is made with emphasis on the design of road bridges. The recognition of different and common points in these codes or specifications is very important to pursue a more rational design format. Investigation and comparison are extended to the design codes used in other fields than civil engineering. As representative structures, building structures(in Japan, the design of buildings is belonging to architectural engineering), aeronautical structures, shipbuilding structures, nuclear power plants and cranes are considered. Table 2 shows the items

regarding safety assessment, in which the circle means that the item is taken into account in the design code. It is seen that in each design code code writers sufficiently recognize the importance of limit states and safety factors, but unexpectedly do not pay attention to problems of durability and human errors.

Table 2 Safety items in various design codes

Item / Structure	Limit State	Safety Factor	Design Life	Safety in Construction	Inspection	Human Error	Safety Index	Economy & Reliability	Unexpected Structure	Future Trend
Road Bridge	O	-	-	-	-	-	-	-	-	O
Railway Br.	O	O	O	-	O	-	-	-	-	O
Port	O	O	O	-	O	-	-	-	O	O
Dam	O	O	-	-	-	O	-	O	-	O
River Struc.	O	-	-	-	-	-	-	O	-	O
Mt. Tunnel	-	O	-	-	-	-	-	-	-	O
Shield Tun.	O	O	-	-	-	-	-	-	-	-
Water Supply	O	-	-	-	-	-	-	-	-	O
Pipeline	O	-	-	-	-	-	-	-	-	O
Tank	O	O	-	O	O	-	-	-	-	-
Marine Struc.	O	O	O	O	O	-	-	-	-	-
Building	O	O	-	-	-	-	O	-	O	O
Airplane	O	O	-	-	O	-	-	-	-	-
Ship	O	-	O	-	O	O	-	-	-	O
Nuclear Pl.	O	-	-	O	O	-	-	-	-	O
Crane	O	O	-	-	O	O	-	-	-	O

7. CONCLUSIONS

In this paper, the outline of the final report of the Task Committee on Structural Safety and Reliability of the Japan Society of Civil Engineers is introduced with some discussions on each current issue. Through the investigation during the last four years, it is concluded that the evaluation of lifetime risk is quite important to ensure a satisfactory safety or serviceability level throughout the entire life of structures. Then, problems with respect to the safety assessment in the constitutive process of structures, diagnosis of existing structures, labor accidents and human errors are considered to require further investigations. To make the theory of structural reliability more useful and practical, it is desirable to examine the possibility of new methods and techniques such as risk assessment, fuzzy set theory [16] and expert systems.

Human error and/or labor accidents are obviously important in assessment of structural reliability. Civil engineers have to bear in mind that human errors may occur at any phase of the lifetime of structures. In order to evaluate the safety of structures and construct the rational structures, civil engineers have to admit existence of human errors, and make the maximum effort to decrease occurrence of human errors as few as possible. To diminish the potential risk of human error, it is inevitable to develop an effective education system by which engineers can understand the characteristic feature of structural safety or structural accident.

Acknowledgment

This paper is a summary of the final report of the Task Committee on Structural Safety and Reliability of the Japan Society of Civil Engineers. The authors are greatly acknowledged

for the supports and contributions of the committee members and writers. Members of the Task Committee are as follows: N. Shiraishi(Chairman), S. Ikeda, N. Inaba, K. Kawashima, H. Kameda, K. Kuroda, T. Kobori, S. Saeki, M. Nakano, Y. Iwasaki, T. Cho, T. Nakayama, A. Hasegawa, N. Sato, S. Hanafusa, Y. Fujino, H. Furuta, M. Hoshiya, C. Miki, and M. Yamamoto

References

1) SHIRAISHI, N.(Ed):Lifetime Risk Assessment of Structures, the Japan Society of Civil Engineers, 1988. (in Japanese)
2) YAO, J. T. P.:Safety and Reliability of Existing Structures, Pitman Advanced Publishing Program, 1985.
3) YAO, J. T. P.:"Probabilistic method for the evaluation of seismic damage of existing structures", Soil Dynamics Earthquake Engineering, Vol. 1, pp.130-135, 1982.
4) PARK, Y. J., ANG, A. H-S. and WEN, Y. K.:"Seismic damage analysis of reinforced concrete buildings", J. of Struc. Eng., ASCE, Vol. 111, pp.740-757, 1985.
5) SHIRAISHI, N. and FURUTA, H.:"Analysis of structural damage", Proc. of International Symposium on Geomechanics, Bridges and Structures, 1986.
6) NIJKAMP, P.:Theory and Application of Environmental Economics, North-Holland, 1977.
7) SHIRAISHI, N. and FURUTA, H.: "Application of multi-criteria analysis to damage assessment of bridges", Proc. of 1st East-Asia Symposium on Structural and Constructing Engineering, 1986.
8) SHIRAISHI, N., FURUTA, H. and SUGIMOTO, M.: "Integrity assessment of bridge structures based on extended multi-criteria Analysis", Proc. of ICOSSAR-4, pp.I505-I509, 1985.
9) FURUTA, H., FU, K. S. and YAO, J. T. P.:"Civil Engineering applications of expert systems", Computer-Aided Design, Vol. 17, pp.410-419, 1985.
10) FURUTA, H., SHIRAISHI, N. and YAO, J. T. P.:"An expert system for evaluation of structural durability", Proc. of 5th OMAE, Vol. 1, pp.11-15, 1985.
11) SHORTLIFFE, E. H.:Computer-Based Medical Consultations:MYCIN, American Elsevier, 1976.
12) SHIRAISHI, N., FURUTA, H., UMANO, M. and KAWAKAMI, K: " An expert system for damage assessment of reinforced concrete bridge deck", 2nd IFSA Congress, Vol. 2, pp.160-163, 1987.
13) SHIRAISHI, N. and FURUTA, H.:"Rule-based system for damage assessment of RC bridge deck, Proc. of International Symposium on Re-Evaluation of Concrete Structures, Denmark, 1988.
14) HANAFUSA, S., SUZUKI, Y. and MAE, I:"Accident analysis and safety assessment for tunneling works", Preliminary Report of IABSE Symposium on Safety and Quality Assurance of Civil Engineering Structures, Tokyo, pp.393-400, 1986.
15) YAMAMOTO, M.:"Human error and its effect on safety of structures", Proc. of Korea-Japan Joint Seminar on Emerging Technology in Structural Engineering and Mechanics, pp.163-171, 1988.
16) YAO, J. T. P. and FURUTA, H.: "Probabilistic treatment of fuzzy events in civil engineering", Jour. of Probabilistic Mechanics, Vol. 1, pp.58-64, 1986.

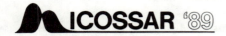
RISK ACCEPTANCE CRITERIA FOR PERFORMANCE–ORIENTED
DESIGN CODES

Stuart G. Reid

School of Civil and Mining Engineering, University of Sydney, NSW,
Australia 2006

Abstract

A general basis for the determination of 'acceptable risks' needs to be
established to facilitate the development of risk–based
performance–oriented design Codes. To determine acceptable risks, it
has been proposed that the methods of probabilistic risk assessment
should be used. The paper briefly reviews the methods of
probabilistic risk assessment, with regard to applications for the
purposes of building regulation. It is concluded that design decisions
should not be based on single–valued measures of risk such as those
contrived according to the methods of probabilistic risk assessment. A
general basis for the development of risk–based design criteria is
outlined.

Keywords

Building Codes; Design Criteria; Cost–Benefit Analysis; Risk
Assessment; Acceptable Risks.

1. Introduction

The development of probabilistic methods for the analysis of structural safety
has led to the development of reliability–based structural design codes. These
codes are limit–states–design codes with partial safety factors to account for the
variability of loads and resistances. The specified values of the partial safety
factors have been chosen to provide consistent reliabilities related to target values
of the nominal probabilities of failure (generally expressed in terms of safety
indices). In most cases, target reliabilities have been chosen to maintain the
structural reliability levels implied by previous code requirements. However, the
currently accepted levels of structural safety and serviceability are not necessarily
the most appropriate, and there is a need to establish a general basis for the
determination of 'acceptable risks' associated with structural failures.

To determine acceptable risks, it has been proposed that the methods of quantitative (probabilistic) risk assessment should be used. The appropriate role of probabilistic risk assessments in the development of risk−based criteria for design is discussed in this paper. Methods of probabilistic risk assessment are briefly reviewed with regard to the general principles, characteristics and limitations of risk assessment procedures, rather than technical details. Conclusions are presented concerning: the state−of−the−art of probabilistic risk assessment; applications of probabilistic risk assessments for the purposes of building regulation; and the development of risk−based performance−oriented design Codes.

2. Risk Assessment

The term 'risk' means different things to different people, and it is not possible to give a precise and concise definition that conveys the full meaning and the connotations of risk. In general terms, risk refers to the danger associated with processes with uncertain outcomes. The nature of risks is extremely complex, and the perception (and acceptability) of risks is affected by many factors related to: hazard characteristics (natural/man−made, avoidable/unavoidable, controllable/ uncontrollable, local/global, continuous/periodic, familiar/unfamiliar, old/new, known/unknown, certain/uncertain, predictable/unpredictable, changing/unchanging, stable [self−limiting]/unstable, etc); exposure characteristics (voluntary/involuntary, compensated/uncompensated, occupational/non−occupational, continuous/periodic/ discrete, controllable/uncontrollable, equitable/unequitable); characteristics of possible consequences (likely/unlikely, minor/major/disastrous/catastrophic, personal/group/ communal/societal, national/international/global, known/unknown, normal/dreadful, familiar/unfamiliar, permanent/temporary, controllable/uncontrollable, reversible/ irreversible, immediate/cumulative/delayed, equitable/inequitable [distribution], etc.); and characteristics of associated benefits (known/unknown, certain/uncertain, essential/non−essential, equitable/inequitable [distribution], etc.).

Risk assessment is a field of common interest to social scientists, physical scientists, statisticians, engineers, economists, technologists and others. However, a general unifying theory of risk assessment has not been developed. Thus risk assessment is a multi−disciplinary conglomerate; it is not a distinct discipline. Accordingly work in the field of risk assessment is characterised by a lack of agreement on fundamental principles, and much of the work is goal−oriented and based on expedient and unjustified assumptions. Reviewers have noted these characteristics and generally concluded that risk assessment is an immature science.

However, it is the writers view that work in the field of risk assessment is essentially dichotomous, concerning either the social sciences, or an immature technology of quantitative (probabilistic) risk assessment. The work of the social scientists concerns the fundamental nature of risks, the development of realistic models of risk assessment based on heuristics, and the role of the risk assessment technology in societal decision−making. The work of the technologists concerns the development of quantitative procedures to expedite risk management (and regulation). The technology of probabilistic risk assessment is not based on the (social) science of risk assessment. Accordingly the technology is immature, and the results are unreliable.

The technology of probabilistic risk assessment is based on the presumption that risks can be represented in terms of probabilities and expected 'costs' of possible outcomes of risk−producing processes. Accordingly, the many non−quantifiable factors relevant to intuitive (heuristic) risk assessment are ignored. Techniques for probabilistic risk assessment include techniques for risk analysis,

risk evaluation and risk management. Probabilistic risk analysis yields estimates of the probabilities of undesireable outcomes associated with hypothetical modes of failure. The estimated probabilities of failure are used as the basis for risk evaluation, using techniques involving risk comparisons (including comparisons with 'acceptable risk' levels), the cost–effectiveness of risk–reduction, and cost–risk–benefit analyses. Corresponding risk management techniques involve risk acceptance, risk reduction, and risk optimisation based on statistical decision theory.

2.1. Risk Comparisons

The risk comparison approach to risk evaluation involves comparisons with the risk levels associated with a range of hazards for which risk statistics are available. It is often assumed that the risk comparison approach to risk evaluation can be extended to include comparisons with 'acceptable risk' levels. Accordingly it is assumed that acceptable (and unacceptable) risk levels can be determined from the risk statistics for existing (accepted) risks. To account for the wide range of observed risk levels for various activities, attempts have been made to relate 'acceptable risks' to the associated 'benefits', and different levels of acceptable risk have been suggested for voluntary and involuntary activities, and for individual and societal risks.

The distinction between acceptable risks for voluntary and involuntary activities was popularised by Starr [1969], who formulated 3 hypotheses which are sometimes referred to as 'laws of acceptable risk': (1) the public is willing to accept voluntary risks roughly 1,000 times greater than involuntarily imposed risks; (2) the statistical death rate from disease appears to be a psychological yardstick for establishing the level of acceptability of other risks; and (3) the acceptability of risk appears to be crudely proportional to the third power of the benefits (real or imagined).

Starr's conclusions have been challenged by other researchers, and in a later paper co–authored by Starr [Starr, Rudman and Whipple, 1976], risk–benefit relationships have been treated with circumspection, and it has been noted that in assuming 'voluntary' risks "the controlling parameter appears to be the individual's perception of his own ability to manage the risk–creating situation...[and]...the individual exposed to an involuntary risk is fearful of the consequences, makes risk aversion his goal, and therefore demands a level for such involuntary risk exposure as much as one thousand times less than would be acceptable on a voluntary basis". This interpretation of the voluntary/involuntary distinction is more consistent with the interpretations of other researchers who have concluded that the apparent aversion to involuntary risk can be better explained by the higher potential for catastrophe and inequity that often accompany that type of risk [Slovic and Fischhoff, 1981]. Furthermore, it is generally agreed that an important determinant of acceptable risk is the acceptability of the process that generated the risk [Fischhoff et al, 1980; Green, 1981]. Hence, a distinction based on the degree of control has been proposed as a generalisation of the voluntary/involuntary distinction [Lathrop, 1982].

Nevertheless, technologists find it convenient to assume that an involuntary risk to an individual (at greatest risk) is negligible (unconditionally acceptable) if it is similar to the risk due to a natural hazard (annual mortality rate approx. 10^{-6}) and it is excessive (unconditionally unacceptable) if it is similar to the risk due to disease (annual mortality rate approx. 10^{-3} for a 30 year old). Accordingly, the acceptability of an intermediate risk (annual mortality rate 10^{-6} to 10^{-3}) depends on the particular circumstances (associated benefits, etc).

To account for the societal impact of multiple deaths caused by a single event, risks involving multiple fatalities have been compared using frequency–consequence curves (from WASH 1400). Furthermore, it has been suggested that levels of societal risk are acceptable if the relevant frequency–consequence curves lie below certain limiting lines (Farmer Curves [Farmer, 1967]). However, it has been shown that frequency–consequence curves are deceptive, and the total risk limits (expected costs) implied by Farmer Curves are critically dependent on the magnitude of the worst possible event [Reid, 1986; Reid, 1987].

It should also be noted that Farmer Curves imply no special aversion to accidents causing large numbers of deaths. This is inconsistent with public attitudes which show an especially strong aversion to risks associated with multiple fatalities and catastrophes. Various researchers have attempted to describe this aversion by equating N lives lost simultaneously to N^m lives lost individually $(m > 1)$.

2.2. Cost–Effectiveness

Whereas the risk comparison approach to risk evaluation is based on the assumption that the acceptability of a risk depends primarily on the estimated level of the risk, the cost–effectiveness approach to risk evaluation is based on the assumption that the acceptability of a risk is related to the cost effectiveness of risk reduction. For life–threatening risks, the cost–effectiveness of risk reduction is related to the marginal cost of saving a life. Comparisons can be made between the marginal costs of saving lives for a variety of risks, assuming the various costs and lives saved are comparable. It should be noted that private and public expenditure on safety is not strongly dependent on the cost–effectiveness of risk reduction (as assessed by the methods of probabilistic risk assessment), and procedures used to save particular lives (e.g., search and rescue procedures) generally have higher marginal costs than procedures used to save statistical lives (e.g., road safety procedures).

2.3. Cost–Risk–Benefit Analyses

For the purposes of cost–risk–benefit analyses, it is assumed that all costs, risks, and benefits (including those based on aesthetic and moral values) can be expressed in terms of monetary values. The evaluation of economic costs and benefits is relatively straightforward (ignoring considerations of justice, equity, social welfare, environmental impact, etc.), but the evaluation of risks is difficult because it is necessary to assign a monetary value to life.

Attempts to assign a value to life are generally based on the evaluation of societal and personal values (revealed preferences, implied preferences, and expressed preferences) or economic values (human capital). It is generally assumed that life can be assigned a value that is independent of the nature of the risk environment, and only the expected costs and benefits are considered.

The analysis of revealed preferences is based on the assumption that preferences (values) can be revealed by the analysis of accepted cost–risk–benefit tradeoffs. Accordingly an apparent value of life can be assessed on the basis of statistical cost, risk and benefit data for any activity involving risk.

The analysis of implied preferences is based on the assumption that legal decisions are based implicitly on the societal value of life. Accordingly the implied value of life may be inferred from legal rulings concerning compensation for the loss of life.

Expressed preferences are determined directly by asking people to express their preferences concerning cost–risk–benefit tradeoffs for hypothetical (quantified)

costs, risks, and benefits. The expressed preferences are analysed to obtain apparent valuations of life.

According to economic models, life is treated as an economic commodity or resource. Economic measures of the societal value of life include: discounted expected future earnings (gross productivity); discounted expected future earnings less consumption (nett productivity); and discounted losses imposed on others due to the death of an individual. A similar measure of the personal value of life is the discounted expected future consumption.

A considerable amount of work has been carried out on procedures for valuing lives (for a review see [Zeckhauser, 1975]). However, the analysis of preferences yields inconsistent results, and economic models are unrealistic and they yield ridiculous results (e.g., discounted nett productivity would indicate that the lives of the young and elderly are worthless). Furthermore, all the procedures are fundamentally flawed from a philosophical point of view [Shrader–Frechette, 1985].

3. Conclusions

3.1. State–of–the–Art of Probabilistic Risk Assessment

Techniques for probabilistic risk assessment can be divided into techniques for risk analysis, risk evaluation and risk management, based on quantitative (probabilistic) measures of hypothetical risks. Some of the techniques for calculating the hypothetical quantities in probabilistic risk assessments are elegant and sophisticated, and they give an impression of authority and precision. However, there are fundamental flaws and gaps in the philosophy of probabilistic risk assessment, and the results are not necessarily realistic, reliable, or even meaningful.

The fundamental problem with quantitative risk assessment is that risks cannot be characterised solely in terms of quantitative parameters. In reality, the nature of risks is extremely complex, and the perception (and acceptability) of risks is affected by many non–quantifiable factors. Clearly, probabilistic risk assessment concerns only a few of the many factors relevant to realistic and reliable risk assessments. Therefore, probabilistic risk assessments are incomplete and not necessarily reliable.

Nevertheless, proponents of probabilistic risk assessment argue that it provides essential information for informed decision making. Furthermore, they claim that it provides uniquely 'rational' results, based on the separation of objective and subjective components. Accordingly, the process of risk assessment (including analysis, evaluation and management) is viewed as an objective process, and the characteristics of objectivity and subjectivity are assigned to the parameters of the probabilistic risk assessment model. It is presumed that the objective parameters include probabilities, direct financial benefits and direct financial costs, and the subjective parameters include personal and societal evaluations of intangible benefits and costs.

However, the claims of objectivity are simplistic and unrealistic. In reality, most apparently objective parameters are subjective to some degree (e.g., the probabilities upon which probabilistic risk assessment is based are actually subjective (Bayesian) probabilities, not objective (frequentist) probabilities). Therefore, experts commonly disagree on the estimation of supposedly objective values. Furthermore, experts tend to underestimate the uncertainty in their own estimates [Fischhoff et al, 1981; Rowe, 1980].

Also, it is unrealistic to presume that the fundamental processes of risk assessment are objective. Probabilistic risk analyses are based on analytical models

which reflect the predilections of the analysts (emphasizing or ignoring various factors so that the analyses give the 'right' answers), and the techniques of probabilistic risk evaluation and risk management are based on a technological paradigm that is fundamentally flawed from a philosophical point of view [Shrader–Frechette, 1985] .

It should be recognised that subjectivity, per se, is not objectionable. However, the disguised subjectivity of probabilistic risk assessments is potentially dangerous and open to abuse, if it is not recognised.

Criticisms of probabilistic risk assessments concern not only the completeness and reliability of the results, but also the effects of probabilistic risk assessments on the decision–making processes of risk management. Critics argue that probabilistic risk assessments focus attention on the factors that can be quantified, thereby diverting attention away from critically important considerations such as the controllability of the risk–producing processes (and associated social changes), and the distributions of costs and benefits (equity).

From a pragmatic point of view, the usefulness of probabilistic risk assessments can be assessed from their impact on the management of technological risks. The record shows that probabilistic risk assessments have been used mainly to confirm and justify predetermined conclusions, and they have failed to resolve differences of opinion concerning acceptable risks [Conrad, 1980; Linnerooth, 1983; Lowrance, 1985; Rowe, 1980]. Thus the numerical results of probabilistic risk assessments are not necessarily useful. However, the process of probabilistic risk assessment can be useful as a catalyst for making explicit assessments of risks (accounting for non–quantifiable factors) within a larger review process. The nature of the larger review process is all important, with regard to the reliability of decisions.

3.2. Probabilistic Risk Assessments and Risk–Based Criteria for Design

It would be convenient if Building Regulations could specify acceptable risk levels for risk–based performance requirements (or a basis for their determination), and building designers could demonstrate compliance on the basis of probabilistic risk assessments. However, acceptable risk levels are difficult to define (they cannot be determined analytically for they depend on value judgements), and probabilistic risk assessments are not sufficiently reliable to allow designs based on their independent use by designers (unlike, for example, the accepted methods of structural analysis, which are reliable if not accurate). Accordingly, probabilistic risk assessments cannot provide convenient solutions to the difficult problems of building risk management. Conclusions concerning the development of risk–based criteria for design, and the appropriate role of probabilistic risk assessments are as follows.

1.   The level of acceptable risk is not a constant. It depends on many factors, including the controllability of the risk, and the distribution of the associated costs and benefits.
2.   Acceptable risk levels cannot be reliably predicted by the methods of probabilistic risk assessment.
3.   The methods of probabilistic risk assessment are suitable for the investigation of hypothetical results for research purposes. Hypothetical results should not be used for design purposes, unless they are fully supported by relevant and reliable empirical evidence.

4.   In most cases, the methods of probabilistic risk assessment are not sufficiently reliable for use in design, because the process of probabilistic risk analysis is based on simplistic models of complex behavior, and the processes of risk evaluation and risk management are based on contrived and unreliable measures of risk.

5.   In particular cases, probabilistic risk analyses might be suitable for Code–writing or design purposes, provided: (1) the nominal risks estimated by the methods of probabilistic risk analysis are not sensitive to reasonable variations of the assumptions upon which the hypothetical risk model is based; and (2) the relationship between nominal risks and real risks can be reliably predicted. (Note that acceptable nominal risks can not be defined independently of the particular methods of risk analysis.)

3.3. Formulation of Risk–Based Design Codes
In view of the unreliability of the methods of probabilistic risk assessment, the following recommendations are given concerning the formulation of risk–based performance–oriented Building Regulations.

1.   Building Regulations should not specify acceptable risk levels, nor allow designs based on probabilistic risk assessments by building designers.
2.   Performance requirements specified in Building Regulations should be risk–based. Associated risks (and acceptable risks) should be considered explicitly by the writers of the Regulations. It should be stated in the Regulations that the performance requirements are based on accepted (unspecified) risk levels.
3.   Acceptable risk levels for particular hazards should be assessed on the basis of the current (accepted) risk levels. The variability of acceptable risk levels should be recognised. Proposals for changes to current risk levels should be assessed on the traditional basis of expert judgement, public comment, and relevant government policies (noting that the political process is the proper process for the management of social conflicts concerning risks imposed on sections of the community).
4.   Decisions concerning acceptable levels of risk should be based on the explicit consideration of all relevant risk characteristics. Decisions should not be based on single–valued measures of risk such as those contrived according to the methods of probabilistic risk evaluation.
5.   Risk levels associated with design methods should be assessed with regard to the nominal risks estimated from probabilistic risk analyses, provided the methods of probabilistic risk analysis have been shown to be reliable (with regard to empirical results).
6.   Building risk levels which cannot be reliably assessed by methods of probabilistic risk analysis should be assessed on the basis of relevant empirical evidence (e.g., for fire safety systems). Risk levels should not be estimated on the basis of extrapolation beyond the limits of empirical data.
7.   If the risk levels associated with a particular design method cannot be reliably assessed from probabilistic risk analysis or empirical evidence (including actual performance records), then that design method should not be used.

3.4. Further Work
Further work should be carried out to assist Code–writers involved in the development of risk–based design Codes. In particular, general guidelines should be developed for the interpretation and use of probabilistic risk assessments. The

guidelines should be simple and practical, and they should include general procedures and checklists concerning: hazard identification; the determination of critical risk parameters; the identification of quantitative modelling assumptions; the treatment of uncertainty and randomness; sensitivity analyses; the interaction of mechanical response models and human response models; the validation of theoretical models with respect to data obtained from the observation of real systems; the assessment of non–quantifiable risk factors; the ranking of alternatives; the assessment of proneness to failure; and procedures for critical review of risk–based decisions. Also, particular performance requirements should be reviewed to determine the relevant risk factors to be considered in the determination of the acceptable risk levels, and to determine appropriate risk parameters.

4. Acknowledgement

This paper is based on work carried out at the University of Sydney for the Building Industry Commission (New Zealand) under Building Research Contract 87/2 (administered by the Building Research Association of New Zealand).

5. References

[1] STARR, C., 'Social Benefit versus Technological Risk', Science, No. 165, 1969, pp. 1232–8

[2] STARR, C., et al, 'Philosophical Basis for Risk Analysis', Annual Review of Energy, Vol. 1, 1976, pp. 629–662

[3] FISCHHOFF, B., et al, Acceptable Risk, Cambridge University Press, 1981

[4] LATHROP, J.W., 'Evaluating Technological Risk: Prescriptive and Descriptive Perspectives', in the Risk Analysis Controversy, Kunreuther and Ley, eds, IIASA, Springer–Verlag, 1982

[5] UNITES STATES ATOMIC ENERGY COMMISSION, 'Reactor Safety Study: An Assessment of Accident Risks in United States Commercial Nuclear Power Plants', Report No. WASH–1400, N.C. Rasmussen, Chairman, National Technical Information Service, U.S. Department of Commerce, Springfield, Va., U.S.A. 1980

[6] FARMER, F.R., 'Siting Criteria – A New Approach', Containment and Siting of Nuclear Power Plants, Symposium Proceedings, IAEA, Vienna, 1967

[7] REID, S.G., 'Frequency–Cost Curves and Derivative Risk Profiles', Research Report No. R514, University of Sydney, School of Civil and Mining Engineering, January 1986, 25 pp.

[8] REID, S.G., 'Frequency–Cost Curves and Derivative Risk Profiles', Risk Analysis, Vol. 7, No. 2, 1987, pp. 261–267

[9] ZECKHAUSER, R., 'Procedures for Valuing Lives', Public Policy, Vol. 23, No. 4, 1975

[10] SHRADER–FRECHETTE, K.S., Risk Analysis and Scientific Method, D. Reidel, Holland, 1985

[11] ROWE, W.D., 'Risk Assessment Approaches and Methods', in Society Technology and Risk Assessment, Conrad ed., Academic Press, London, 1980

[12] LINNEROOTH, J., 'Risk Analysis in the Policy Process' in Risk Analysis and Decision Process' in Risk Analysis and Decision Processes,

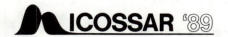 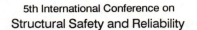
SOME CONSIDERATIONS ON THE ACCEPTABLE PROBABILITY OF
FAILURE

J.K. Vrijling [1,2]

1. Dept. of Civil Engineering, Delft University of
   Technology, Delft, Holland
2. Rijkswaterstaat, Ministry of Public Works
   P.O.Box 20.000, 3502 LA Utrecht, Holland

Abstract

The acceptable failure probability of technical
structures and systems is studied in this paper.
The problem is approached from two points of view:
the personal and the social point of view. The
different view points lead to different criteria.
It is proposed to adopt in a practical case the
most stringent of the criteria as a guide line for
decision making.

Keywords

Acceptable failure probability, risk, optimization.

1. Introduction

This paper is the result of a study carried out by Working Group
10 "Probabilistic method" of the Technical Advisory Committee for
Dikes and Flood defenses. It is intended to be a step on a long
road to establish as objective as possible a set of norms for
levels of risk.
The paper examines the social acceptability of failure of (civil
engineering) structures, installations and activities. Besides,
economic factors play a part in the assessment of acceptability.

Two approaches are adopted for determining a socially acceptable
level of risk, namely:
-    a mathematical-economic method which, with the main emphasis
     on loss expectation, leads to an economic optimum;
-    a method based on accident statistics which, with the main
     emphasis on the expected number of deaths, leads to an
     (apparently) accepted accident probability.

The two approaches lead to sometimes essentially different
results. This study does not, however, establish a particular
standpoint with regard to acceptable levels of risk, for the
determination of such levels should be the result of the political
process. The study should instead be viewed as an instrument to
assist the process of opinion-forming as to a justifiable level of
safety.
The present report has been compiled in the context of an
investigation into the safety of the dikes and flood defenses. The
approach presented is however equally applicable to other cases
involving risk, such as road and rail traffic and transportation,
industrial installations, power stations operating on whatever
source of energy, etc.

2. Personally acceptable level of risk

The smallest component of the socially accepted level of risk is
the personal assessment of risks by the individual. In the
personal sphere the appraisal, i.e., balancing the desired
benefits against the risk associated with them, is often
accomplished quickly and unconsciously.
Also, a correction is quickly made if the appraisal turns out to
be incorrect.
The result of an attempt to establish a model of this appraisal
procedure is represented in fig. 1., presupposing an objective
rational balancing of the benefit - both the direct personal and
the social benefit - against the risk of expected loss
(probability times consequence).

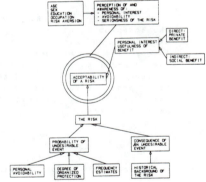

Fig. 1.:
Theoretical model of the assessment of risk in the individual
sphere

The probability component of the risk is estimated on the basis of
the individual's own experience or of the reported experience of
others. Some idea of the possible consequences is also derived
from these sources. That is why forming a personal opinion with
regard to new activities is often difficult due to lack of
historical data. In such cases the information is derived from
pronouncements by "experts" and from the visible degree of
protection.
An important aspect is the degree of voluntariness with which the
risk is endured. In the case of non-voluntariness the individual
can make his appraisal in accordance with his own set of
standards, but any adjustment of the choice in the event of an

unfavorable result is outside his sphere of influence. The two
points compel him to adopt a skeptical attitude towards non-
voluntary risks.
The aspect of non-voluntariness together with the non-
availability of historical data and the lack of clarity as to the
nature of the benefit to be gained may explain the social
resistance to modern sources of energy such as LNG and nuclear
energy.

Psychometric research has so far not attained the operation of the
model presented in fig. 1. A solution could consist in presuming
the appraisal process of each individual to be consistent and in
considering that the result of this process can yield an
indication of his preferences. In a diagram of "the consequence"
consisting in losing one's life the statistic of causes of death
provides a source, which reveals the average result of the
individual appraisals of benefit and risk. An unavoidable risk is
the probability of dying by natural causes. In the Western
countries this probability for a person under 60 years of age is
about 10^{-3} per year. For other activities the personal acceptance
of risk is arrived at by dividing the annual number of casualties
by the number of participants in the activity concerned.

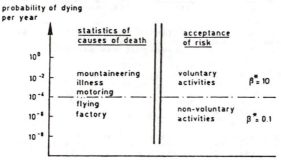

Fig. 2:
Personal risks in Western countries, deduced from the statistics
of causes of death, related to the number of participants per
activity

The personal risk levels for some activities are indicated in
fig. 2. The fact that these levels show stability over the years
and are approximately equal for the Western countries would seem
to indicate a consistent pattern of preference.
The ranking of the risk levels is not surprising either. The
probability of losing one's life in normal daily activities such
as motoring of working in a factory is one or two orders of
magnitude lower than the normal probability of dying. Only a
purely voluntary activity such as mountaineering entails a higher
risk.
In view of the consistency and the stability - apart from a
slightly downward trend due to technical progress - of the death
risks presented, it would appear permissible to deduce therefrom a
guideline for decisions with regard to the personally acceptable
risk.
The permissible probability of an accident associated with
activity i is:

$$P_{f_i} = \frac{\beta^* \cdot 10^{-4}}{P_{\alpha|f_i}} \qquad (1)$$

where $P_{\alpha|f_i}$ denotes the probability of being killed in the event
of an accident. In this expression the policy factor β^* varies
with the degree of voluntariness with which the activity is
undertaken and ranges from 10 in the case of complete freedom of
choice to 0.1 in the case of an imposed risk.

3. Socially acceptable level of risk

What a democratic society accepts in terms of risks is in
principle the aggregate, or sum total, of all individual
appraisals. The aggregated version of the model presented in fig.
1 would have to provide the answer. Although it can be said, at
the social level, for every project in the widest sense the social
benefits are balanced against the social costs (including risk),
this process of appraisal cannot be made explicit. The social
optimization process is accomplished in a tentative way, by trial
and error, in which governing bodies make a choice and the further
course of events shows how wise this choice was.

If a socially acceptable risk level must be determined for a
particular project, a solution can be reached only via
considerable simplification of the problem.

One way to achieve this is to convert the problem to a
mathematical-economic decision problem by expressing all the
consequences in monetary terms. The second approach deduces an
acceptable level of risk from accident statistics.

Standard of appraisal based on mathematical-economic optimization
The problem of the acceptable level of risk can be formulated as
an economic decision problem.

The expenditure I for a safer system is equated with the gain
made by the decreasing present value of the risk.
The optimal level of safety is reached at the minimal cost
solution.

$$\min (C) = \min [I(P_f) + P.V. (P_f) \cdot S] \qquad (2)$$

where:
C   = total cost
P.V. = present value operator
S   = damage

This problem has been solved for many practical situations. One of
the best known examples in the Netherlands is the approximation of
the optimal probability of inundation of Holland by Van Dantzig
[1,2].
The result formed the basis for the political choice of the
return period of 10.000 y for the design flood in the Delta-
project. If despite ethical objections, the value of a human life
is rated at s, the amount of damage is increased to

$$P_\alpha|_{fi} \cdot N_{pi} \cdot s + S \qquad (3)$$

where N_{pi} = number of participants.

This extension makes the optimal failure probability a decreasing
function of the expected number of deaths.
The problem of the valuation of human life is in this paper
solved by choosing the present value of net national product per
inhabitant. The advantage of taking the possible loss of lives
into account in economic terms is that the safety measures are
affordable in the context of the national income.

A limitation of the mathematical-economic approach is that it
presupposes the total loss in the event of a failure to be small
in comparison with the economy as a whole. In fact it is the
confidence in the economy that makes repair a viable proposition.

Standard of appraisal based on accident statistics
The second approach to determine the socially acceptable level of
risk starts from the proposition that the result of a social
process of risk appraisal is reflected into the accident
statistics. It seeks to derive a standard from these.

The standard of appraisal, for socially acceptable risks should be based on a model for the social perception of risk. The model should show that the particularly low probability of fatal accidents is perceptible to members of the community. Secondly, the model should be able to explain the inverse proportionality between the permissible probability of an accident and the number of deaths involved.

As a model hypothesis it is assumed that an individual assesses the social risk level on the basis of the events within his circle of acquaintances. Assuming for the moment that the average circle of fairly close acquaintances can be out at 100 persons, the probability of a death occurring within that circle in consequence of <u>natural causes</u> is equal to:

$$P_n(\text{death}) = 10^{-3}/\text{yr} * 100 = 0.1/\text{yr} \qquad\qquad (4)$$

Similarly, the probability of one death among the acquaintances due to a <u>road accident</u> in the Netherlands, with a population of $14 \cdot 10^6$ and a number of casualties in 2200 in 1980 is:

$$P_{ra}(\text{death}) = \frac{2200/\text{yr}}{14 \cdot 10^6} * 100 = 1,4 \cdot 10^{-2}/\text{yr} \qquad (5)$$

Through the instrument of the circle of acquaintances the particularly low probabilities of a fatal accident, which appear socially acceptable, are perceptible. The recurrence time is within the order of magnitude of a human life span.

In seeking to establish a norm for the acceptable level of risk for civil engineering structures it is more realistic to base oneself on the probability of a death occurring within the circle of acquaintances due to a non-voluntary activity in the factory, on board a ship, at sea, etc. which is approximately equal to:

$$P_{i\alpha}(\text{death}) = \frac{200/\text{yr}}{14 \cdot 10^6} * 100 = 1,4 \cdot 10^{-3}/\text{yr} \qquad (6)$$

If this observation-based frequency is adopted as the norm for assessing the safety of activity i, then with due regard to $\beta^* = 0.1$ for the non-voluntary character:

$$\frac{\Sigma \; N_{p_i} \cdot P_{d|f_i} \cdot P_{f_i} \cdot 100}{14 \cdot 10^6} < \beta^* \cdot 1.4 \cdot 10^{-2} \qquad (7)$$

After re-arranging this expression, and adopting a rather arbitrary distribution over for example 20 categories of activities, the following norm is obtained for an activity i in situations pertinent to the Netherlands:

$$\boxed{P_{f_i} \leq \frac{\beta^* \cdot 100}{N_{p_i} \cdot P_{d|f_i}}} \qquad\qquad (8)$$

This norm should be interpreted in the sense that an activity is permissible as long as it can be expected to claim fewer than β^*. 100 deaths per year.

However the formula looks only to the expected number of death and does not account for the dispersion, which will certainly influence acceptance.

The risk aversion can be represented mathematically by adding a confidence requirement to the norm. For this purpose, the mathematical expectation of the number of deaths, $E(N_{di})$, is increased by the desired multiple of the standard deviation before the situation is tested against the norm:

$$E(N_d) + k \cdot \sigma(N_d) \leq \beta^* \cdot 100 \qquad\qquad (9)$$

where:
k   =   risk aversion index

For a correct determination of the mathematical expectation and
the standard deviation of the number of deaths occurring annually
in the context of activity i, it is necessary also to take into
account in how many independent places N_A the activity under
consideration is carried out.The number of such independent places
is of no influence on the expectation of the number of deaths, but
it does affect the dispersion.

After some rearrangement using the binomial distribution we
obtain the result for the permissible probability of failure.

If a value of k=3 is provisionally adopted for the reliability
requirement a further simplification is possible. For large values
of N_A the formula degenerates into a simple norm in which the
acceptable probability of failure is inversely proportional to the
number of deaths.
For $N_A = 1$ the requirement is more rigorous, but the formula a
retains a simple form (right side). The acceptable failure
probability is inversely proportional with the square of the
number of dead.

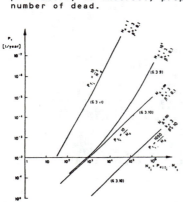

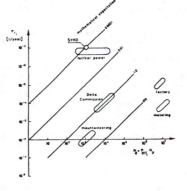

Fig. 3:
The social safety norm for
some values of β^* and N_A;
the probability of failure
is marked on the vertical
axis; the horizontal axis
indicates the number of
deaths in case of failure

Fig. 4:
The risks of some activities
in the Netherlands; the
probability of failure per
system is marked on the
vertical axis, the horizontal
axis indicates the number
of deaths in case of failure

In order to test the model, the acceptable failure probability is
plotted for several values of N_a, β and k = 3 as a function of the
number of dead in case of a failure (see fig. 3). On the same axes
some activities, considered acceptably safe in the Netherlands,
have been plotted (fig. 4).
The agreement between the norm derived in this study for
reasonable values of N_a and 0.1 < β x * < 10 and the risk accepted
in practice seems to support the model.

4. A concept of acceptable risk

Several concepts of determining the acceptable level of risk have
been presented in the foregoing section. One approach, based on
accident statistics, has been given for the personally acceptable
level of risk. The socially acceptable level of risk has been
approached in two ways.
First, the mathematical-economic approach of the material risks
weighed against the cost of safeguarding against them. Second, an
approach based on a model of social perception of risk was
developed.

In assessing the safety of a system the three approaches should
all be investigated:
-    The <u>personally</u> acceptable risk which a member of the
     community is on average prepared to accept. In simplified
     form this risk is represented by eq. (1).
-    The <u>economically</u> optimal level of risk, where the value of a
     human life must be taken into account. An objective
     measure of the value of a human life is the present value of
     the net national product per head. The optimal level is
     attained if the marginal cost of safety measures is just
     equal to the marginal benefit (eq. 2).
-    The <u>socially</u> acceptable level of risk, on the basis of the
     assumed risk aversion model, which leads to the eq. (9)
     giving an evaluation of the acceptable probability of
     failure.

The most rigorous of the three criteria should be adopted as the
governing criterion.
To illustrate the proposal procedure, it has been applied to
Central Holland. The results are represented in fig. 5. Given the
total number of people (N_d = 50000) that will drown in case of an
inundation, the socially acceptable failure probability based on
accident statistics is the most rigorous criterion and equals
$4.5 \cdot 10^{-9}$ per year. The outcome of the political process is plotted
as a circle in fig. 5.
To show that the discussions are not finalized in the
Netherlands, the criteria for LPG systems issued by the Ministry
of Housing and Planning [3] are given in fig. 6.

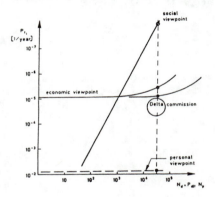

Fig. 5.:
Application of the three safety requirements to the acceptable
probability of inundation of Central Holland; The social point of
view is the most stringent

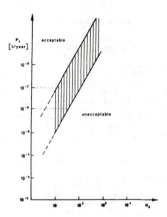

Fig. 6.:
Acceptability of group risks as deduced from the LPG Integral Note
(18233 Nos. 1 and 2). This norm must be complemented with a
maximum personal risk of 10^{-6} - 10^{-7} per year [4.4].

Literature:
1.   DANTZIG, V.D, KRIENS, J. Het economisch beslissingsprobleem
      inzake de beveiliging van Nederland tegen stormvloeden.
      Report of Delta Commission, Part 3, Section II.2, The Hague,
      1960.
2.   DANTZIG V.D, Economic Decision Problems for Flood
      Prevention, Econometrica 24,276.287, New Haven, 1956.
3.   MINISTRY OF HOUSING AND PLANNING, LPG Integral Study (in
      Dutch), The Hague, 1985.

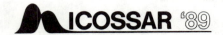
SAFETY PLANS FOR BUILDINGS, STRUCTURES AND TECHNICAL FACILITIES

Dr. sc. techn. Miroslav Matousek

Department of Safety, Quality Assurance and
Environmental Impact, Wenaweser + Wolfensberger AG
Zurich, Switzerland

ABSTRACT

Safety is an important quality feature and depends
on all activities within the building process. The
safety concept must therefore consider all phases
of the building process, from the intention to
build to the demolition. The safety plan has
particular significance, guiding those in position
of responsibility and comprising the following
elements: safety goals, systems analysis, risk
analysis, planning of safety measures, quality
assurance and feedback. Preparation and implementation of safety plans are discussed in detail,
including examples.

KEYWORDS

Structural damage, collapses, safety plan, safety
goals, hazards, risk/cost ratio, feedback.

1. INTRODUCTION

Accidents and collapses indicate the hazards, concerned apartment
houses, parking garages, sports facilities, bridges, industrial
plants, etc. (referred to below simply as structures). A few
recent examples: the collapse of the Kongresshalle in West Berlin,
the collapse of an ice-skating stadium in Ticino (Switzerland),
the collapse of a gallery in a Kansas City hotel, the collapse of
a steel bridge in Connecticut, the collapse of the reinforced
concrete roof of a public swimming pool in Uster (Switzerland),
etc. In addition to these spectacular damages, which attracted
wide publicity in the press, there is a steady stream of failures
and damages, especially during construction. The damages involve
not only **direct material damage** but **personal injury, damage to the
environment and consequential damage** (delay, project interruption,

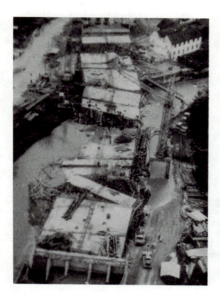

Collapse of bridge after
pouring concrete, caused
by failure of scaffolding

loss of market share, etc.). In
certain cases the environmental
and consequential damage can be
several times greater than the
direct material damage.

Of course, measures have been
and continue to be implemented
to avoid certain hazards.
However, the failures that have
occurred in recent years are
evidence that existing safety
considerations - based mainly on
calculations - have been inadequate.

This unacceptable situation has
prompted research efforts
troughout the world, including
Switzerland (1). The results of
such research indicate that
damage is due mainly to human
error, and that the scope of the
term "safety" as well as the
safety concept itself must be
extended to the building process
as a whole (2)(3)(4). The fruits
of the safety research are already being incorporated into
standards, leading to risk-
oriented solutions to existing
safety problems (5)(6).

2. THE SAFETY PLAN AS A MANAGENEMT TOOL

A systematic **safety-planning** is the presupposition for
guaranteeing safety. The hazards associated with a given structure
must be determined, and optimal measures devised to reduce them.
The determined measures must then be precisely planned,
implemented, and monitored. Safety planning must not be confined
to the utilisation phase alone, but must encompass the
construction process itself: witness the numerous injuries to
persons and property that occur during construction.

The essential information of safety-planning is to be documented
in a **safety plan** (5) to serve as a guide to those in positions of
responsibility for safety. It should enable them to deal systematically with the problem of safety and to select the safety
measures adopted on a risk-oriented basis. The safety plan should
allow the people involved in the building process to answer the
following safety-relevant questions:

* What safety requirements exist? What **safety goals** are established for the structure?

* How is the structure delimited and composed as a **system**?

* What **hazards** are associated with the structure and what hazardous situations (**hazard scenarios**) may occur?

* Which hazards are eliminated by what **safety measures**, and which
  are consciously **accepted as risks**?

* What **errors** regarding established safety measures could arise in
  the course of planning, design, construction, utilization and
  maintenance, and what can be done to prevent them?

* What efforts are being made to evaluate the experience gained
  and to ensure that safety always reflects the state of the art?

These questions, serving as well as a checklist for determining
the safety of existing structures, indicate the purpose of the
safety plan and its preparation.

3. ELEMENTS OF THE SAFETY PLAN

3.1 Definition of safety goals

Safety goals are often established on a basis of safety requirements incorporated into laws, standards, guidelines, recommendations, and the like. If these are insufficient, safety goals
should be established on the basis of **risk acceptance** and the **acceptable costs of safety**. Risk acceptability constitutes a minimal
safety requirement and relates to risk to human beings, risk to
the population, risk of a major accident, and environmental risk.
Safety costs, on the other hand, should reflect optimal
utilization of available resources as well as a balanced ratio
between safety costs and risk reduction (7).

3.2 Systems analysis

The structure is defined as a system, distinguished by the following main components: load-bearing elements, finishes, utilities
(air conditioning, ventilation, etc.), operating equipment and
furnishings, waste disposal, staff, and the environment.

Systems analysis reveals the components of the system, what
quality features they have, and how they work.

3.3 Hazard identification and risk evaluation

The need for systematic hazard identification was developped long
ago in sectors like electrical engineering, space, and the
chemical industry. Consequently, a number of hazard and risk identification methods have been developed: preliminary risk analysis,
failure mode effect analysis, hazard and operability analysis
(HAZOP), fault-tree analysis, etc.

In the case of a given structure under study, primary attention is
generally devoted to hazards already incorporated into laws,
regulations, standards, guidelines, and the like. Their specific

purpose is to determine the hazards already pinpointed by experience. In cases requiring detailed study, hazards are determined using the methods listed above.

Basically, the universe of hazards can be divided into the following hazard groups:

* Hazards caused by the **natural environment** (avalanches,
  hurricanes and tornados, landslides, earthquakes, etc.)

* Hazards posed by **technical environment**, including fire, vehicle
  collision, explosion, overloading, etc.

* Hazards due to failure of **structural components**, such as
  columns, beams, foundations, reinforcements, installations, etc.

* Hazards as a result of **sabotage and criminal activity**

* Hazards caused by materials hazardous to **human health** as well as
  the risk of **accidents in the workplace**.

At any given time, one or more hazards may exist. Similarly, different hazards can arise at different times. To confer a logical
structure on these hazardous situations, they are described by so-
called **hazard scenarios** (5). Each hazard scenario consists of a
leading hazard and an accompanying circumstance. The leading
hazard is always assumed to be extreme in its effect, form, and
magnitude. The accompanying circumstance characterizes any hazards
and factors that occur at the same moment in time and hence accompany the leading hazard. Any hazard scenarios found are evaluated
as risks. The **probability of occurrence** as well as the **damage** are
the two critical evaluation factors.

3.4 Planning of safety measures

The identificated risks are evaluated in terms of the safety goals
and measures for their reduction are investigated. In determining
what measures are to be employed, the existing risks must be
reduced to acceptable levels. If such reduction is not possible,
the result may be a change of utilization and even ban on the
project itself.

If acceptable risk values are attained, a decision is made on the
basis of the balance between risk reduction and safety cost, to
determine which risks can be eliminated by measures (action) and
which are to be consciously accepted (7). The accepted risks are
then documented in greater detail, including specific information
on risk bearers, risk monitoring, steps to be taken in the event
of an accident (emergency plans), and damage control measures.

3.5 Quality assurance

The established safety measures have to be planned in detail, and
then implemented, and maintained. At each step in the building
process, and for each activity of those involved, there is a possibility of error. Systematic quality assurance is therefore required to eliminate these errors. Quality assurance consists of

error prevention and error detection and correction (monitoring and supervision plans, monitoring instructions, checklists, etc.).

3.6 Feedback

The safety plan must always be kept right up to date. New scientific and technical discoveries as well as practical experience must be reflected in the safety plan. In particular, systematic problem evaluation as well as actual instances of damage acquire particular significance in this regard. The section headed "Feedback" therefore provides additional information on how to obtain the required information and make the corresponding improvements.

4. APPLICATION OF SAFETY PLANS - EXAMPLES

Initially it is the builders (later, the owners) who are responsible for safety. It is therefore in their interest to entrust safety plan development to the planning team and to commit the required resources.

Depending on the problem at hand, safety plans can be applied to individual structural components only (load- bearing structures, utilities, operating equipment) or to entire structures including production. They can thereby deal with only one hazard (e.g. impact, earthquake) or the entire universe of hazards. Safety plans can be devised for both planned and existing structural and technical facilities. The broad areas of application of safety plans will now be described in greater detail using three practical examples:

Example 1: Safety plan for a distribution center

The distribution center in question is of enormous importance to individual production facilities. Loss of the distribution center would therefore cause not only personal injury and property damage, but considerable consequential damage as well (suspension of operation and loss of market). It was therefore necessary to investigate the safety problem systematically and establish specific measures. A safety plan was devised, setting safety goals for personal risks (laws and standards), property damage risks (regulations, optimal risk reduction), and consequential damage risks (making arrangements for temporary distribution). The system components encompassed the entire structure, including the operating equipment. Hazard identification included hazards from the natural and technical environments as well as hazards caused by criminal activities, materials hazardous to human health, and accidents. The safety measures were optimally established in the light of the risk reduction/safety cost relationship, taking available insurance into account. The accepted risk was documented and damage control measures proposed (emergency measures, temporary distribution). Then the quality assurance measures as well as feedback were established.

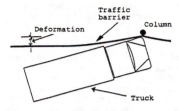

Hazard scenario 1 :
Extreme deformation of traffic
barrier and impact against column

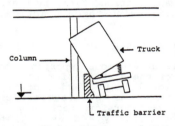

Hazard scenario 2 :
Break trough traffic barrier
and impact against column

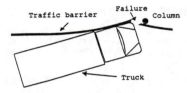

Hazard scenario 3 :
Angular position of truck
and impact against column

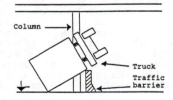

Hazard scenario 4 :
Tipping of truck over barrier
and impact against column

Example 2: Safety plan for impact hazards

The columns of a tunnel entrance structure were endangered by vehicular impact. Deliberate measures were required to counteract the effect of such impacts. Emphasis was on human safety, protecting not only motorists but pedestrians as well. In view of the high frequency of heavy trucks, far more extensive measures were required than those prescribed by existing Swiss SIA Standard 160 and provided in Draft SIA Standard 160. Safety regarding existing impact hazards was systematically studied and a safety plan developed. This plan focused on the following points: The safety goals were established on the basis of acceptable risk (deaths/year) and rescue costs (Swiss francs per individual rescued). The system was delimited and the components (load-bearing structure, traffic, people, environment) described. The impact hazard was determined on the basis of individual features (speed, impact angle, vehicle weight, vehicle frequency, accident rate, impact area, etc.) and the hazardous situations described in greater detail on the basis of eight hazard scenarios (four of which have been included as illustrations). On the basis of the risk reduction/safety costs relationship, measures were established (design measures, traffic barriers, column dimensioning) and the accepted risks, including damage control measures, were documented in detail. Then the quality assurance and feedback procedures were established.

Example 3: Safety plan for rehabilitation of a railroad crossing

At two existing railroad crossings, the question arose as to whether automatic crossing gates should be installed or whether they could be replaced by an underpass. In addition to the efficiency of the

underpass (no waiting time, accessibility to police and fire-
brigade), safety was especially important as a quality feature.
This was studied in depth and a safety plan was developed. As
described above, the latter consisted of the following elements:
safety goals (acceptable risk in deaths/year and rescue costs in
Swiss francs per individual rescued), systems analysis (location,
traffic counts) hazard determination (collision, highway
accidents, degree of severity), measures planning (crossing gates,
underpasses, risk reduction/safety cost relationship), quality
assurance (planning, construction, maintenance) and feedback.

5. CONCLUSIONS

Knowledge gained from safety research is already contributing to
the drafting of standards. Terms such as failure probability,
safety index, design condition, hazard scenarios, human error,
acceptable risks, and safety plan are already part and parcel of
Swiss standards. It was not easy to leave the traditional path
behind and implement the new ways of thinking about safety.

The next step - putting the new safety ideas into practice - still
lies ahead. Incorporation of safety-related data into safety plans
focuses attention on hazards and the measures provided for
reducing them, together with the risks involved. Safety becomes a
function of the hazards eliminated, and the concept of safety
therefore assumes a concrete form. Consequently, safety planning
represents a new task and hence a new challenge for engineers.

6. REFERENCES

(1) International Conferences on Structural Safety and
    Reliability, ICOSSAR, Reports, 1969, 1977, 1981, 1985

(2) MATOUSEK M., SCHNEIDER J.: Untersuchung zur Struktur des
    Sicherheitsproblems bei Bauwerken, Institut für Baustatik und
    Konstruktion ETH Zürich, Bericht Nr. 59, Birkhäuser Verlag,
    Basel und Stuttgart, 1976

(3) MATOUSEK M.: Massnahmen gegen Fehler im Bauprozess, Institut
    für Baustatik und Konstruktion ETH Zürich, Bericht Nr. 124,
    Birkhäuser Verlag, Basel und Stuttgart, 1982
    Von der University of Waterloo in Englisch übersetzt:
    "Measures against Errors in the Building Process", Canada
    Institute for Scientific and Technical Information, Ottawa,
    1983

(4) MATOUSEK M., SCHNEIDER J.: Gewährleistung der Sicherheit von
    Bauwerken - ein alle Bereiche des Bauprozesses erfassendes
    Konzept, Institut für Baustatik und Konstruktion ETH Zürich,
    Bericht Nr. 140, Birkhäuser Verlag, Basel und Stuttgart, 1983

(5) SIA NORMEN, Schweizerischer Ingenieur- und Architekten-Verein, Zürich, insbesondere:
* Weisung SIA 260, Sicherheit und Gebrauchsfähigkeit von Tragwerken, Weisung an seine Kommissionen für die Koordination des Normenwerkes, Entwurf 1982
* Norm SIA 160, Einwirkungen auf Tragwerke, Entwurf 1988
* Norm SIA 162, Betonbauten, Entwurf 1988

(6) General principles on reliablity for structures, International Standard ISO 2394, the International Organization for Standardization, second edition, 1986

(7) SCHNEIDER J.: Zwischen Sicherheit und Risiko, Schweizer Ingenieur und Architekt, Heft 18, 1988

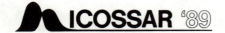
MODELLING OF DAMAGE DUE TO STRUCTURAL FAILURE

S.A. Timashev, V.V. Vlasov

Institute of Machinery, Ural Division of the
USSR Academy of Sciences, 91 Pervomaiskaya St.,
620219 Sverdlovsk GSP-169,   USSR

ABSTRACT

A multistage method of simulating failure situations associated with collapse of structures is
described. Method is applied to evaluate statistical characteristics of the size and cost of
damage and average casualties in a case   study.

KEYWORDS

Damage, Cost, Structural Failure, Modelling, Reliability, Optimization.

In problems of determining the optimal reliability level R_{opt}
of mechanical systems (MS), the accuracy of its evaluation depends largely on the confidence of all the essential
factors that affect the lifetime of a system — from design to
in-service failure.

A vital factor here is allowance for damage due to possible
system failure, a parameter which, according to the kind of
failure, type of system is essentially a random variable (rv).
Virtually all available suggestions concerning the assessment

of the size of damage, C_d, due to the failure of MS boil down
to specific methods of gathering and processing data on the
size of C_d in terms of actual failures. But all of the methods
proposed reveal a number of fundamental drawbacks: nonrepre-
sentativeness and statistical inhomogeneity of the ensemble of
damage realizations; as a rule, incomplete and only overall
allowance for important factors that have effect on the size
of C_d, in which one cannot isolate the fraction of damage per
individual optimizable system components.

Under these conditions, the only way to reliably estimate the
damage C_d due to the failure of a MS and its individual assemblies is by simulating the failures and the damage arising
/1/, /2/. The technique involves multiple, detailed, and comprehensive simulation of the random event — the system's
failure and the body of its major technical and economic consequences constituting the damage C_d — and subsequent averaging over the ensemble of damage realizations.

To determine the damage in the case of system failures that
are not associated with wrecks, collapses, caving-in, etc., it
suffices to know the set D of all possible "unitary" system
impairments d_i and of costs of their elimination $u(d_i)$, which
we will call unitary damages. Then every particular system
failure D_j may be represented as a determinate set (intercep-
tion) of unitary impairments

$$D_j = \bigcap_i k_i^{(j)} d_i \, , \qquad\qquad (1)$$

where $k_i^{(j)}$ stands for the quantitative characteristics of impairments d_i in failure D_j (for example, the length l of
burst welds or fatigue cracks). For specific parts, structures, and known external effects, these characteristics are
found in the general case by algorithms developed in fracture
mechanics and deformable solid bodies mechanics.

In this case the damage C_d due to failure D_k is

$$c_d^{(j)} = \sum_i k_i^{(j)} u(d_i) \, . \qquad\qquad (2)$$

For each real system, the quantities k_i are rv's. Their distribution densities $f(k_i)$ may be determined by statistical
dynamic methods. Then, having tables of d_i and of $u(d_i)$ and
performing a Monte Carlo simulation of realizations of the
quantities k_i, one can obtain from formula (2) any requisite
number of damage C_d realizations and, by invoking statistical
analysis methods, find all the required characteristics of the
damage due to each possible system failure.

The merits and advantages of the method of simulating failure
situations are especially prominent in cases of failures associated with the breakdown of machines, with the collapse of
structures with heavy consequences, including noneconomic ones
(casualties, moral, prestigious, strategic and other losses).
For a well-founded evaluation and prediction of possible damage
due to failures of this kind, one needs to construct a zone of
destruction (ZD) and to determine all the important technical,
economic, and noneconomic consequences of the destruction in
that zone. These tasks constitute two major segments of the
damage simulation method /1/.

The present paper treats applications of this method in the
case of failures associated with the collapse of structures,
since this is the most complicated case of damage evaluation
(for machine components the lines of argument and algorithms
are simpler), and typifies all regularities in damage estimation from failures - wrecks.

The method of simulating wreck situations is realized in several
stages. At the first step one studies the mechanism of falling
of a structure and solves a relevant ballistic problem. From
its solution, one finds zones of destruction (ZD) and equipment
impairments. At the second step one determines, from the value
of the collision energy and from the parameters obtained at the
preceding step, the character, dimensions, and probability of
impairment for the entities located in the ZD. At the third
step one estimates the size of unitary economic damages, at the
fourth step the ballistic problem is multiply reproduced by
simulation with various realizations of the random quantities
involved in it; for each realization one determines the character and dimensions of the impairments of the entities located
in the ZD. At the fifth step one estimates the size of damage
in each realization, and at the sixth step the statistic characteristics of the size of damage are found by averaging over an
ensemble of realizations. At the seventh step "normative"
values of the sizes of damage are obtained by averaging over
the types of structures and buildings.

Allowing for the fact that in the complete-cost-of-system functional /1/ the size of damage C_d is multiplied by a very small
wreck-failure probability, there is no need for very high accuracy, a situation which permits simplification of many complicated algorithms of steps 1 to 4. The necessary number of
simulations at the fourth step for estimating the average of
damage $\bar{U}$ with required accuracy ε and confidence α, is
found by the formula $N = Z_\alpha^2 \sigma_u^2/\varepsilon^2$, where Z_α is the quantile
of normal distribution for confidence α and σ_u^2 is the variance of the size of damage, refined in the course of iteration.

For each of the steps enumerated above, a relevant algorithm
has been developed which is realized in a computer program.
Below we give a brief schematic outline of these algorithms.

Solving the ballistic problem for any structure involves three
main steps. At the time zero, t = 0, a failure occurs: either
the structure undergoes destruction at some point or is torn
off from one of the supports by the action of forces. The destruction probabilities for the structure, $P_s(t)$, and for the
braces that support it, $P_b(t)$, are problems of reliability
theory proper and are assumed to be known. After the moment
t = 0 the structure (or part of it) of length L begins to move
by the action of its own weight Q = mg, under the weight of
the other structures that rest on it, and by the action of the
elastic resistance forces of the braces. The structure con-
tinues to move up to some moment t = t_1, when the last brace
is destroyed and the structure occupies a spatial position
which is characterized in the general case by Euler angles φ,
ψ, θ /3/ and their velocities at the moment t_1. The quantities $\varphi(t_1)$ ..., $\dot{\theta}(t_1)$ (here and henceforth the dot above the
function denotes its differentiation with respect to time) are
evidently random variables. The laws of their distribution
are established in each particular case by a careful analysis
of the layout of the structure and of actual wrecks.

For a specific set of realizations of random quantities $\varphi(t)$,
..., $\dot{\theta}(t_1)$ we explore the second stage of the falling of the
structure under its own weight alone in the time interval from
t_1 to t_2, when one of the ends of the body, A, hits the floor
of the building. During the fall of end A the other end of
the body, B, may: (a) slip down from its support; as a result, a free fall of the body will occur; (b) stay on the
support until the moment t_2, then the process of falling may
be viewed as the motion of a heavy solid body around a fixed
point B. In both cases the motion of the body is described by
Lagrange second-order equations with relevant modifications
/3/. These equations are solved by a computer under the ini-
tial conditions $\varphi_0 = \varphi(t_1)$, ..., $\dot{\theta}_0 = \theta(t_1)$.

On solving the second step at the moment t = t_2, we obtain the
coordinates of the point of impact of end A (ξ_A^2, ζ_A) on the
floor plane, and also the values of the Euler angles $\varphi(t_2)$,
$\psi(t_2)$, $\dot{\theta}(t_2)$ and their velocities, which are used at the
third stage of problem solving — the falling of end B.

The third stage begins at the moment one of the angular points
A_i of end A hits the floor. Assuming that at the moment of
impact the body comes to a halt, i.e., the angular velocity of
the points of the body $\omega(t_2)$ = 0, we arrive at the result
that at this moment, in keeping with D'Alambère's principle
/3/, each element of the body, of mass dm and with radius-
vector r is acted upon by an inertial force equal to
-dmW(r), where W(r) is the total acceleration of this point
of the body.

After end A has hit the floor, the braces keeping end B on the support become destroyed (or the structure itself is destroyed) by the impact action of the inertial forces that act during time τ, and end B falls down. Again, this fall of the structure obeys Lagrange equations of motion of a heavy solid body about a fixed point A_1. At the third stage of the falling of the structure the initial conditions for the Lagrange equations have the form

$$\varphi_o = \varphi(t_2), \; \psi_o = (t_2), \; \theta_o = \theta(t_2), \; \dot{\varphi}_o = \dot{\varphi}(\tau),$$

$$\dot{\psi}_o = \dot{\psi}(\tau), \; \dot{\theta}_o = \dot{\theta}(\tau) , \tag{3}$$

where $\dot{\varphi}(\tau)$, $\dot{\psi}(\tau)$, $\dot{\theta}(\tau)$ are the velocities of the Euler angles at the moment $t_3 = t_2 + \tau$, which are found by solving the system of equations

$$\omega(t_2 + \tau)I = \Omega , \quad \Omega = -v^{-1} m\tau \iiint_V rW(r)dr \tag{4}$$

with I and v being the inertia tensor and the volume of the body, respectively. On solving, for end B, the Lagrange equations with the initial conditions (4) we obtain the coordinates of impact point B (ξ_B, γ_B).

Thus, as a result of solving the ballistic problem we obtain in general case the coordinates of the impact points of ends A and B, their velocities, the energy of the body at the moment it hits the floor and the total falling time $t_t = T$.

These data constitute the initial information necessary to construct the ZD of a fallen structure. All large-scale-production building structures (beams, columns, trusses, plates, etc.) have flat rectangular faces and, after one of the vertices has hit the floor, will certainly fall on one of these faces. If, for the fallen structure, all the three steps of the ballistic problem have been solved and, therefore, the position of the points of impact of both ends, A and B, against the floor is known, then the ZD is a rectangle. One of its sides of length L coincides with the straight line that connects points A and B, and the other side is equal to the height of the structure h. By simulating the direction of fall of the structure relative to the AB line, it is possible to obtain a uniquely determined ZD for each accident realization. All the items of equipment that are inside the determined zone are considered to be striken with allowance for the "force shadow" factor, when individual units of equipment are shielded against an impact by the higher machines.

In a number of cases the distribution densities of random Euler angles and of their velocities at the moment t_1, which are needed for simulating the initial conditions of the ballistic problem, cannot be determined with sufficient accuracy. For such cases an algorithm of constructing the possible ZD has been developed. In it the third step of the ballistic

problem is not solved and all the possible versions of fall of
the AB edge are united to form the zone sought as a circular
sector of radius L with the center at point $A(0, \mathcal{Z}_A)$, central
angle $2\mathcal{X}$ and two strips of length L and width h attached to
the extreme radii. The angle $\mathcal{X}$ is determined from some empirical or statistical considerations.

Adopting the hypothesis that the direction of fall of the AB
edge of the structure is equiprobable, the destruction probability P_i for the i-th unit of equipment that has got into the
ZD is

$$P_i = (\mathcal{X}_i + c\,\mathcal{X}_s)/2\mathcal{X}\,, \quad \mathcal{X}_s = h/L\,, \tag{5}$$

where $\mathcal{X}_i$ is the angle between the extreme positions of the
radius AB which touch that part of the i-th unit of equipment
which has got into the ZD and is not shielded from an impact
by other machines; c = 0.5 provided that $\mathcal{X}_i$ allows for the
"force shadow", otherwise c = 1.

The size of damage in each realization of the wreck situation
should be determined with allowance for the full group of
events — all possible versions of falls of the structure (or
its parts) and the other bodies that rested upon it. Each
event from the full group has its own ZD and manifestation
probability. Unification of the ZD of all full-group events
yields the ZD of the realization of the wreck situation. For
the ZD a formula has also been obtained which specifies the
destruction probability of equipment in ZD with allowance for
possible destruction by several structures that have collapsed.

The size of damage depends on the category of repair (minor
and medium repairs, overhaul and complete replacement) of im-
paired equipment. A table has been compiled which defines
the category of repair with allowance for the type of equipment and for the qualitative assessment of the degree of its
impairment (elastic rebound - concussion, plastic deformations,
penetration - destruction). To permit use of the table, formulas of threshold values of the force and velocity with
which the structure hits against the equipment have been
derived.

To facilitate the evaluation of the size of damage in the
realization of a wreck situation, we introduce the helpful
concept of unitary damage, which implies the cost of reconditioning: (a) for the characteristic unit impairment of a
single structure; (b) of a unit of every type of equipment,
as well as losses due to the downtime of equipment per unit
time. With the impairments being known, these damages are
sufficiently easy to determine from data of price-lists, hand-
books, and special cost accounts.

The size and character of impairments are determined by the
technique outlined above, whereupon further economic computations are made on the basis of the postulate that follows [1].

TABLE: STATISTICAL CHARACTERISTICS OF DAMAGE C_d DUE TO COLLAPSE OF STRUCTURES

Name of structure	Average C_d	Standard S_{C_d}	C_d/iC_s	Contribution of individual components to C_d, %				Average casualties
				C_r	C_e	C_{de}	C_{ds}	
Metallic truss, C_s=1.045								
i = 1	72.58	33.36	69.4	21.5	68.2	6.5	3.7	13.8
Reinforced concrete roof slab, C_s = 0.357 i = 1	12.4	13.35	34.7	7.4	71	10.4	11.2	1.02
i = 8	59.5	47.32	20.7	12.5	80.3	4.9	2.4	
Aluminum panels, C_s=0.410 i = 1	2.78	0.46	6.9	37.4	9.5	2.7	50.4	
i = 8	12.45	1.65	3.8	67.2	16.0	5.6	11.2	
Block of 15 sheets of corrugated steel deck, C_s = 0.705	5.88	1.07	8.1	56.6	14.8	4.8	23.8	
Reinforced concrete column C_s = 0.695, i = 1								
– corner	32.82	–	47.2	56.9	29.4	5.2	8.5	23.5
– extreme-row	55.41	–	79.7	56.1	30.8	8.0	5.1	
– middle-row	92.30	–	123.1	42.5	38.9	15.6	3.0	
Crane girder with crane, C_s = 0.513, i = 1	13.11	3.71	25.6	9.6	65.8	10.9	10.7	6.83

Note: C is the cost of one installed structure, i the number of structures that have collapsed, C_r the cost of structure reconditioning, C the cost of equipment repair, C_{de}, C_{ds} the profit losses due to the downtime of the equipment and of the shop. All cost estimates made in 1977, and shown in thousands of roubles.

<u>Postulate.</u> The working hours and cost of reconditioning struc-
tures and equipment depend on only the size and character of
impairments and do not depend on the causes responsible for the
impairments.

The probability and number of injured persons are determined by
the working places involved in the ZD. The expectation of the
total random size of damage in a wreck situation U is

$$U = \sum_{i=1}^{N} \left\{ \sum_{k=1}^{K_i} \sum_{l=1}^{L_{ik}} U_{lik} + \sum_{j=1}^{M_i} C_{rj} \right\} P_i , \quad (6)$$

where L_{ik} is the number of units of equipment involved in zone
Z_{ik}, C_{rj} and M_i are the cost of reconditioning and the number
of structures and mechanisms that have collapsed in an event
S_i, and P_i is the event S_i manifestation probability to be determined by solving reliability problems.

At the last step one determines the statistical characteristics
of the quantity C_s by averaging over the ensemble of the damage
realizations obtained. As an example, a numerical realization
has been performed of the algorithms and programs developed to
estimate damage for different versions of collapse of some
structures under conditions of an aggregate of repair shops in
one of the plants with medium- and high-accuracy metal working
equipment. Statistical characteristics of the size of damage
for each type of failure have been obtained. We have analyzed
the spread in estimates of damage, the contribution of the
various components to its size, and the relation of damage to
the cost of a collapsed structure. A comparison has been drawn
of damages estimates obtained on the basis of simulating ZD and
solving ballistic problems and of those obtained without simulation, by ordinary crude upper-bound estimations from a small
body of available information about accidents of a given type.
Part of this information is presented in the Table.

REFERENCES

1. TIMASHEV, S.A.: Reliability of large mechanical systems,
   Moscow, Nauka, 1982 (in Russian)
2. TIMASHEV, S.A. and VLASOV, V.V.: Method for estimating the
   size of damage due to different kinds of failures of structures, in book: Investigations in the field of reliability
   of engineering structures, Leningrad, Lenpromstroiproekt,
   1979 (in Russian)
3. APPEL, P.: Theoretical mechanics, Vol.2, Moscow, Fizmatgiz,
   1960 (in Russian)

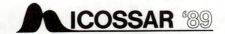
GENERAL EUROPEAN PRINCIPLES OF CODES CONCERNING RELIABILITY

L. Östlund

Department of Structural Engineering
Lund Institute of Technology, Lund, Sweden

ABSTRACT

The requirements for reliability cover safety, service—
ability and durability and consider different causes and
types of failure. They may concern the occurrence of a
failure or the duration or number of times of unfitness for
the intended use of the structure. The measures to ensure
reliability are design, choice of material and protection
with regard to durability and quality assurance measures.
Design includes choice of structural system and material,
dimensioning and detailing. The dimensioning process is
discussed. Probabilistic dimensioning methods and the
method of partial coefficient are mentioned.

KEYWORDS

Reliability, safety, serviceability, durability, failure,
design, dimensioning, stochastic, probabilistic.

1. INTRODUCTION

Within Europe much work has been performed to harmonize the codes concerning load
bearing structures. Some results have been reached and today the trends in the
development of the codes are similar in the different countries. This is
especially the case for the codes concerning the basic principles of reliability.
The work is still going on in order to improve the results and to try to give them
a more general application. The content of the following sections should not be
regarded as a summary of existing codes but as a description of the problems
discussed within different committees concerned with reliability and similar
subjects. The principles which are presented are not new from a scientific point
of view but an attempt has been made to introduce also some fairly new ideas into
the generally accepted system. Of course, some points of view given in the
following reflect the personal opinion of the author.

2. STRUCTURAL PERFORMANCE REQUIREMENTS

2.1   General aspects

A structure shall be designed and constructed with an appropriate degree of reliability so that it will not become unfit for the use for which it is intended during its anticipated life. If this very general requirement is not fulfilled the structure is considered to have failed.

The failure may be caused by
—   an unfavourable combination of several factors such as actions, material properties and geometrical properties
—   a more or less unexpected exceptional action overlooked in the design
—   a gross human error committed during planning, design, construction or use of the structure.

A structure may become unfit for use due to
—   collapse
—   other more or less severe damage
—   malfunction of the structure

The concept of reliability is often divided into safety and serviceability. Then collapse and the more severe types of damage represent failures associated with safety. The less severe types of damage and malfunction represent failures associated with serviceability. There is, however, no sharp distinction between safety and serviceability.

The requirements shall be fulfilled during the anticipated life of the structure. This means that the durability of the structure in its working environment shall be such that deterioration of material will not lead to an unacceptable probability of failure.

2.2   Safety requirements

The safety requirements may be summarized in the following way. The choice of the structural system, the design of the structural parts and the construction of the structure shall be such that any hazardous event will not cause collapse or severe damage of the structural system or part of it.

The requirements shall be fulfilled with an appropriate probability, which should be chosen with regard to the expected consequences of failure in terms of risk to human life of injury, economic losses and the degree of social inconvenience. The amount of expence and effort required to reduce the probability of failure should also be considered.

In a limit state design the requirements for safety are related to the ultimate limit states.

2.3   Serviceability requirements

Two types of serviceability failures can be distinguished:
—   irreversible failures which, once they have occurred, last till the structure has been repaired
—   reversible failures which remains only as long as the causal action is present.

Considering these two types of failure the serviceability requirements may be expressed in one of the following two ways:

—   a structure and its structural parts shall be designed and constructed in such a way that actions and other influences, liable to occur during its construction and use, will not cause permanent damage or permanent malfunction (irreversible failures) of structural or non—structural parts or equipment

—   a structure and its structural parts shall be designed and constructed in such a way that no temporary cracking, deformation or vibrations (reversible failures) will cause discomfort for the users or malfunction of building parts or equipment. This should be valid with exception for an acceptably short duration or for an acceptably small number of occurrences.

In a limit state design the requirements for serviceability are related to the serviceability limit states.

2.4   Consideration of durability

The durability aspects may be considered by one of the two additional requirements.

—   It should be ensured, with an appropriate degree of probability, that the resistance of a structure will not decrease with time due to deterioration of material.

—   A foreseen decrease of the resistance of a structure with time shall be considered at the design.

3.   MEASURES TO ENSURE RELIABILITY

3.1   Different kinds of measures.

The problem of ensuring an acceptable reliability is complex and has many aspects. The design of the structure and its structural parts is only one of these aspects. Another aspect is the choice of material, structural form and protection with regard to durability. Further a certain extent of quality assurance measures is neccessary to ensure that the real structure does not deviate in an unacceptable way from the originally planned structure.

3.2   Design

The design of a structure contains the choice of structural system and material, dimensioning and detailing.

Dimensioning, i.e. the determination of the dimensions of a structure by calculations (or sometimes by testing) is generally regarded as the main measure to avoid failure due to an unfavourable combination of several factors (the first cause mentioned in section 2.1). But the choice of structural system and detailing are also important especially from an economical point of view.

To reduce the probability of failure due to unexpected exceptional actions (the second cause mentioned in section 2.1) it is necessary that the structure is chosen so that it has a certain degree of insensitivity. Therefore, in this case the choice of structural system is most important.

3.3 Durability aspects

In the first case mentioned in section 2.4 the problem of defining a probability of failure of a structure can be simplified so there is
— a probability of failure if the resistance is assumed not to decrease significantly
— a probability that the resistance will decrease significantly.

Approximately the probability of failure should be the sum of the two probabilities.

In the second case of section 2.4 the resistance is assumed to decrease with a certain degree of regularity. This means that the probability of failure, defined, for example, for each year of the service life of the structure, will increase with time. Then it is not clear which value should be chosen to be compared with an acceptable probability of failure. Sometimes the maximum value is chosen which means that the resistance should not be allowed to decrease to a value smaller than that used in the design calculations.

The requirements for durability should be valid during a beforehand chosen period of time, without inspections and maintenance being necessary during this period. It is often necessary or convenient to chose this period of time equal to the whole intended life of the structure. In other cases the period of time may be chosen shorter, but it should not be chosen shorter than the interval between programmed inspections and, if necessary, maintenance. In such cases it is also necessary to chose the structural system and the detailing in such a way that inspection and maintenance are possible without difficulties.

In connection with the design, an acceptable durability may be verified in two ways:
— by calculation with methods similar to those used for strength
— by principles concerning choice of structural system and material and rules for detailing.

Today the latter way is predominant. In the future, however, the use of calculation will probably be more common. Then the calculations concerning strength and durability may perhaps be integrated in the same process.

3.4 Quality assurance measures

The aim of the quality assurance measures is to ensure that the completed structure is consistent, to a reasonable extent, with the requirements and the assumptions made during the design. In this way the quality assurance measures are an important part of codes concerning reliability.

Quality assurance measures may include:
— adapting the tasks to the qualifications of individual persons and organizations
— giving correct and sufficient information about relevant parts of the project to those who participate in the building process
— defining clearly the tasks and the responsibilities of the personnel
— ensuring appropriate working conditions
— quality control.

To a great extent the quality assurance measures aim at reducing the probability of occurence of human errors (the third cause mentioned in section 2.1).

4. DIMENSIONING

4.1   The dimensioning process

For a simple case the dimensioning process can be illustrated as is shown in FIG 1.

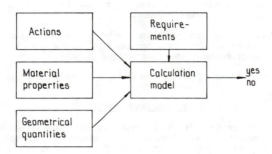

FIG 1. The dimensioning process

The basic variables (input variables) are in most cases actions, material properties (for example strength) and geometrical quantities. They are normally represented by stochastic or deterministic models which deviate more or less from reality and consequently they are affected with model uncertainties. The properties of the actions have generally a more or less pronounced temporal variability. Sometimes also the material properties or the geometrical quantities vary with time (compare section 2.4 and 3.3).

The basic variables are introduced into the calculation model which gives the criteria for no failure of the structure. The reliability requirements are also introduced in terms of, for example, an acceptable probability of failure. The result of the calculation gives the answer (yes or not) to the question whether the reliability is sufficient or not. The calculation model should be regarded partly as a physical or mechanical model and partly as a stochastic model affected with model uncertainties. Sometimes testing is substituted for calculation. In such cases certain problems occur concerning the choice of the input values which should be given to the test specimens.

The process described in this section can be used also for durability calculations (compare section 3.3) if the input variables are adapted to this type of calculations.

4.2   Dimensioning methods

The basic variables, mentioned in the previous section, and also the calculation models have in principle a stochastic character. In some cases, especially concerning serviceability, also the requirements may contain stochastic parameters. Thus the dimensioning problems are to a great extent of a stochastic character and from this point of view probabilistic dimensioning methods ought to be suitable. In some codes probabilistic methods are supposed to be used as one alternative for the calculations.

The probabilistic methods are, however, in general fairly laborious and therefore there is a need for a simplified method. Thus the method of partial coefficients is introduced as the normal dimensioning method for structures. In this method the uncertainties of the basic variables and the calculation models are considered with aid of factors (partial coefficients) which normally are attached to the values of the actions and the material properties. Sometimes additive elements are used instead of factors, especially for the geometrical quantities but also for actions, for example, to consider uncertainties of the water pressure due to a varying water level.

For the treatment of some special problems the method of partial coefficients is sometimes not very suitable. This may, for example, be the case for some dynamic problems and for some problems concerning fatigue.

The problem of combinations of actions is under discussion and has not yet got a good solution.

5. FUTURE DEVELOPMENT

The codes concerning reliability will probably not be subjected to radical changes during the next ten years, but some development, for example, concerning durability is likely to take place. The content of codes can be expected to be introduced into computer programs to an increasing extent which may, in certain respects, prevent the development. Therefore it is important not to make the codes concerning reliability very detailed but to confine their content to the general principles.

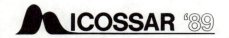
Background studies for the determination of
strength functions in Eurocode 3 - Design rules
for Steel-structures

G. Sedlacek* and J.W.B. Stark**

*   University of Technology Aachen, D
**   University of Eindhoven, NL

ABSTRACT

At present common unified rules for the design of
Civil engineering structures are being prepared on
the initiative of the Commission of the European
Communities. Eurocode 3 which gives design rules
for steel structures contains strength functions
that have been calibrated with tests to achieve an
equal   level   of   reliability. The evaluation
procedure for obtaining characteristic and design
values of strength is presented and some examples
of evaluations are given.

KEYWORDS

Eurocode 3, Calibration of strength functions,
test evaluation, characteristic values, design
values, safety factors.

1.   Introduction

The Commission of the European Communities has
initiated the development of unified technical
rules for the design and erection of Civil
Engineering structures in order to overcome the
obstacles to a common market in this field, caused
by different national technical standards.

The harmonisation program comprises Engineering
Services for which Design Codes were prepared as
well as Products in the construction area, for
which Product-Standards and Quality Standards are

being developed or Agréments are foreseen which
specify the certification procedures for non
standardized products or new production techniques, fig. 1.

The program for the design rules includes a set of
Eurocodes the draft of which has followed general
principles concerning reliability and safety laid
down in Eurocode 1. The project has been coordina-
ted by a Eurocode Coordination group that aimed
at achieving a two-dimensional harmonisation: the
one across the barriers of national design
traditions, the other across different materials
and ways of construction.

Fig. 2 gives a survey on the different Eurocodes,
which all follow the limit state design concept.
Eurocode 3 deals with the design of steel
structures the foreseen contents of which is given
in fig. 3.

So far the draft part 1 of Eurocode 3 - General
rules and rules for buildings - is almost
completed. The list of contents of part 1 of
Eurocode 3 is shown in fig. 4.

The draft of Eurocode 3 is mainly based on the
1978 ECCS-Recommendations, other source documents
were the publications of international technical
organisations as ISO, IIW, CIDECT or national
standards. After a first draft was completed the
member states were consulted and a revision was
undertaken on the basis of the member states'
comments with liaison engineers that explained the
member states' position and helped in getting
agreements and compromises where the positions of
member states were conflicting.

During this stage of harmonisation a powerful tool
for getting agreements proved to be a standard
evaluation procedure to derive design resistances
from tests that was prepared by a CEB-ECCS-Working
Group. This procedure has been used to check and
compare different proposals for strength functions
to facilitate decisions and to achieve a uniform
level of reliability of the strength rules
throughout the Code.

The evaluation studies, that have been performed
on the basis of a large number of test results
for almost all resistances in EC 3 will be
published by the Commission of the European
Communities.

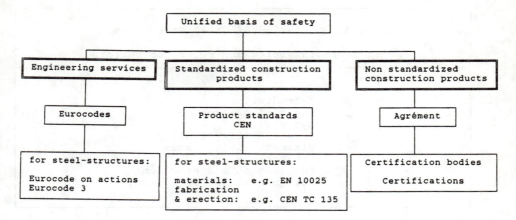

Fig. 1: Outline of the Eurocodes
and Product standards

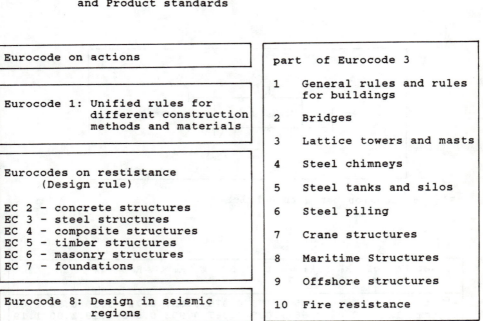

Fig. 2: Outline of all Eurocodes

Fig. 3: Different parts of
Eurocode 3

Eurocode 3 - Design of Steel Structures

Part 1 General rules and rules for buildings

Chapter

1	Introduction
2	Basis of design
3	Material
4	Serviceability limit states
5	Ultimate limit states
6	Connections
7	Fabrication and erection
8	Design assisted by testing
9	Fatigue

Fig. 4: List of contents of part 1 of Eurocode 3

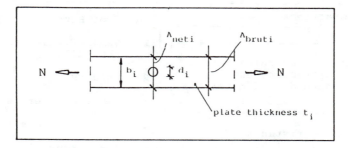

Fig. 5: Tension bar with a hole

subset	No. of tests	k_s	k_d	$\bar{b}$	S_δ	R_k	R_d	γ_M	ΔK	γ_M*
all t	32	1.86	3.35	1.105	0.070	0.837	0.727	1.15	0.94	1.08
t= 6mm	12	2.03	3.66	1.085	0.087	0.903	0.676	1.19	1.00	1.19
t=10mm	14	1.99	3.60	1.096	0.039	0.867	0.771	1.12	0.91	1.03
t=15mm	3	1.86	3.35	1.156	0.107	0.793	0.672	1.19	1.01	1.20
t=20mm	3	1.86	3.35	1.181	0.022	0.883	0.797	1.21	0.83	1.00

Fig. 7: Extract from the evaluation of fracture
resistance tests for the tension members
with holes made of StE 460.

2. Evaluation procedure for test results

The evaluation procedure may be best explained by
the example of a tension bar, fig. 5, where the
net section strength is modeled by

$$R_t = A_{net} \cdot fu \qquad\qquad (1)$$

where A_{net} = net section area
           fu = tensile strength of the material

For all tests available, where $A_{net\,i}$ and fu_i as
well as the experimental strengths $r_{e\,i}$ were
measured, a comparison between $r_{e\,i}$ and $r_{t\,i}$ is
performed, where $r_{t\,i}$ are the results from the
strength model R_t , see fig 6.

The statistical evaluation leads to a corrected
strength model

$$R = \bar{b} \cdot A_{net} \cdot fu \cdot \delta \qquad\qquad (2)$$

where $\bar{b}$ =   mean value correction of the
                      model
           δ = error term with the standard
                 deviation s_δ

The design strength function for the net section
strength can be derived when from prior knowledge
the mean values and the standard deviations for
A_{net}
                $A_{net\,m}$, $\sigma_{A_{net}}$
and for fu
                fu_m, σ_{fu}

are known. In assuming full statistical indepen-
dance of all scattering parameters, the coefficients of variation may be determined conservatively as

$$V_{Rt} = \sqrt{V^2_{A_{net}} + V^2_{fu}} \approx \sigma_{\ln Rt}$$
$$\qquad\qquad\qquad (3)$$
$$V_R = \sqrt{V^2_{Rt} + s^2_\delta} \approx \sigma_{\ln R}$$

The characteristic values of the strength
functions in Eurocode 3 are defined as 5%
fractiles (k_N = 1,64).

With the weighting factors

$$\alpha_R = \frac{V_{Rt}}{V_R} \;;\; \alpha_\delta = \frac{s_\delta}{V_R}$$

the characteristic value of the net section
strength may be expressed by

$$r_K = \bar{b} \cdot A_{net\,m} \cdot fu_m \cdot$$
$$\exp\,(-1,64 \cdot \alpha_{Rt} \cdot V_{Rt} - k_s \cdot \alpha_\delta \cdot s_\delta - 0,5 \cdot V_R^2) \qquad (4)$$

①   Assumed Design model $R_t = A_{net} \cdot f_u$

②   Comparision of the test results r_{ei} with the results r_{ti} of the design model (all A_{neti} and f_{ui} are measured):

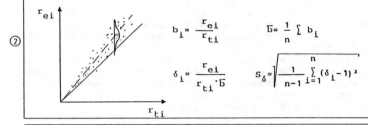

$$b_i = \frac{r_{ei}}{r_{ti}} \qquad \bar{b} = \frac{1}{n} \sum b_i$$

$$\delta_i = \frac{r_{ei}}{r_{ti} \cdot \bar{b}} \qquad s_\delta = \sqrt{\frac{1}{n-1} \sum_{i=1}^{n} (\delta_i - 1)^2}$$

③   Corrected Design model $R = \bar{b} \cdot A_{net} \cdot f_u \cdot \delta$

④   Determination of the variation coefficient for R_t and R:
- Input from prior knowledge: $A_{netm}, \sigma_{A_{net}}$ f_{um}, σ_{f_u}
- Coefficients of variation:

$$v_{R_t} = \sqrt{v_{A_{net}}^2 + v_{fu}^2} \approx \sigma_{\ln R_t}$$

$$v_R = \sqrt{v_{R_t}^2 + s_\delta^2} \approx \sigma_{\ln R_t}$$

- Weighting factors:

$$\alpha_{R_t} = \frac{v_{R_t}}{v_R} \qquad ; \qquad \alpha_\delta = \frac{s_\delta}{v_R}$$

⑤   Determination of the characteristic values r_k and the design values r_d:

$$r_k = \bar{b} \cdot A_{netm} \cdot f_{um} \cdot \exp(-1.64 \cdot \alpha_{R_t} \cdot v_{R_t} - k_s \cdot \alpha_\delta \cdot s_\delta - 0.5 \cdot v_R^2)$$

$$r_d = \bar{b} \cdot A_{netm} \cdot f_{um} \cdot \exp(-3.04 \cdot \alpha_{R_t} \cdot v_{R_t} - k_d \cdot \alpha_\delta \cdot s_\delta - 0.5 \cdot v_R^2)$$

$$\gamma_M = \frac{r_k}{r_d}$$

k_s, k_d = fractile factors for 75% predicting probability
           (for $n \to \infty$ $k_s = 1.64$, $k_d = 3.04$)

⑥   Determination of the correction factor Δk and the safety factor γ_M^* to be applied to the strength function
$R_n = A_{netn} \cdot f_{un}$ fed with nominal values for A_{net} and f_u:

$$\Delta k = \frac{R_n}{r_k} \qquad ; \qquad \gamma_M^* = \frac{R_n}{r_d} = \Delta k \cdot \gamma_M$$

Fig. 6: Steps for the evaulation procedure for the determination of restistances from tests.

where k_s is the fractile factor for a level of convergence of the prediction of 75% (k_s = 1,64 for $n \rightarrow \infty$).

When nominal values $Anet_{nom}$ and fu_{nom} are used for calculating the strength

$$R_{nom} = Anet_{nom} \cdot fu_{nom} \qquad (5)$$

the ratio

$$\Delta k = \frac{R_{nom}}{r_K} \qquad (6)$$

indicates how far these nominal values deviate from the target characteristic values.

The design values of the net section strength are calculated for a safety index β = 3,80 and a separation factor α_R = 0,8 for the resistance side which in connection with α_s = -0,7 for the action side allows to define the design resistance independantly on the action side.

It follows

$$r_d = \overline{b} \cdot Anet_m \cdot fu_m \cdot$$
$$\exp(-3,04 \cdot \alpha_{Rt} \cdot v_{Rt} - k_d \cdot \alpha_\delta \cdot s_\delta - 0,5 \cdot v_R{}^2) \qquad (7)$$

where k_d is the fractile factor for a predicting probability of 75% (k_d = 3,04 for $n \rightarrow \infty$)

The safety factor γ_M to be applied to the characteristic value is

$$\gamma_M = \frac{r_K}{r_d} \qquad (8)$$

If the safety factor is to be applied to the nominal value of the strength function it is indicated as $\gamma_M{}^*$ and can be calculated from

$$\gamma_M{}^* = \frac{R_{nom}}{r_d} = \gamma_M \cdot \Delta k. \qquad (9)$$

This evaluation procedure that looks rather simple only gives reasonable results when it is used with sufficient knowledge of the actual physical limit state phenomenae that are modeled by the simplified expression

$$R_t = Anet \cdot fu.$$

The limit state in the net section of a tension bar may be either controlled by toughness or strength and the geometrical and material parameters that may influence the behaviour in the net section and are not included in the simplified

strength formula should be taken into account by
choosing adequate subsets of the test population,
for which these parameters can be taken as
approximately constant.

3.   Typical example for the application of the evaluation method

For tension members made of StE 460 which were
shaped as demonstrated in fig. 5 tests have been
carried out to determine the fracture loads. An
extract of the evaluations that were conducted
for the resistance function $R_v = 0,9 \ F_u \cdot A_{net}$ is
shown in fig. 7. From all evaluations the partial
safety factor $\gamma_M{}^* = 1,25$ could be derived.

4. References

/1/   Eurocode 3 - Unified rules for the design of
       steel structures Part 1: General rules and
       Rules for buildings
       Commission of the European Communities, Dec.
       1988.

/2/   Background Document for chapter 2 (Basis of
       Design) of Eurocode 3. Commission of the
       European Communities 1989

/3/   M. Kersken-Bradley: ECCS-CEB-Notes on the
       evaluation of test results, Munich 1985.

/4/   Procedure for the evaluation of test
       results, TNO-Delft Annex Z to Eurocode 3.

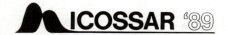
On the CICIND Code

Hermann Bottenbruch 1)

ABSTRACT

The CICIND Model Code for Concrete Chimneys was created by
CICIND, an international technical association which has as one
of its main aims the creation of harmonized internationally
accepted codes for industrial chimneys. The CICIND Model Code is
based firmly on probability theory. It uses the method of partial safety factors. These safety factors are determined in such
a way that all chimney structures designed according to the
Model Code show a uniform failure probability of approximately
1/10000 in 50 years.

KEYWORDS

failure probability, ultimate limit state, industrial
chimneys, extreme value distribution of wind velocity

1. INTRODUCTION

CICIND is an international technical organisation which has as
its main aim the harmonisation of national codes for the design
and construction of chimneys. It was formed as a direct consequence of a collision between two codes which were considered as
a possible basis for the design of a particular chimney: The
French code, which was familiar to the consulting engineer, and
the USA code, which was familiar to the chimney owner. The USA
code resulted in a wall thickness at the chimney bottom which
was by a factor of 2 larger than that required by the French
code. This happened in 1973. The two engineers who were concerned with the resolution of the collision roused the interest
of other experts who were familiar with still other national
codes for chimneys. In the course of those considerations CICIND
was established.
One of the achievements of CICIND is the creation of chimney
codes which are based on the theory of probability of failure.
There exists a CICIND Model Code for Concrete chimneys (completed in 1984) and another CICIND Model Code for Steel Chimneys
(completed in 1988), see Lit. (1 through 4).

1)   Chairman and Major Shareholder of KARRENA GmbH
      Rüdigerstraße 20, D-4000 Düsseldorf 30

To my knowledge they are still the only codes for building
structures which are strictly based on probability theory.
The basis for the determination of the probability distributions
of the variables considered and the failure probability of the
structures which are designed according to these codes are
documented in the Commentaries to the CICIND Model Codes.

I will restrict this paper to the presentation of the safety
concept of the CICIND Model Code for Concrete Chimneys. I do
this not only because I was (and still am) the chairman of the
committee which wrote this Model Code, but also because the
safety philosophy behind the CICIND Model Code for Steel Chimneys is the same as for the CICIND Model Code for Concrete
Chimneys.

I simplify even further: I restrict my report on the CICIND
Model Code to the presentation of the safety concept for the
horizontal cross section and the influence of dead load and
wind, which problem really was the start of CICIND. The complete
material is contained in the Model Code (16 pages DIN A4 format)
and the Commentaries (58 pages DIN A4 format), see Lit. (1,2).

Several research contracts were passed by CICIND to university
professors with a reputation in failure probability and/or
material behaviour in the ultimate limit state in order to get a
firm basis for our undertaking.

2. THE PROBLEM
The problem looked simple at the beginning: We just had to find
a criterion for "sufficient and necessary safety". But in practice, the problem was more complex: Experts from 10 countries,
who were using different codes for their daily design work,
contributed to the Code and had to find an agreement. Experts
from almost 20 different countries had to be convinced that this
was the correct solution for the problem of "sufficient and
necessary safety". This was necessary during the process of
creating the Code and in the final general assembly of the
CICIND association when the votes were cast on the acceptance.

Certain aspects simplified the problem when comparing it to
general building codes. The probability distribution of the
loads was rather simple: The dead load is simply the weight of
the chimney and would not be affected by any variable vertical
load which could possibly be applied to the structure during its
lifetime. The other action applied to the cross section is wind.
The probability distribution of this phenomenon had been investigated in numerous scientific papers.

Another aspect which simplified the investigation was the simplicity of the structures. Although they vary in size from 20 m
height and 1 m diameter to 400 m height and 35 m diameter, they
are always cantilevers with dead load as normal force and horizontal wind force producing moments.

3. BASIC PHILOSOPHY

3.1 Determination of the Acceptable Failure Probability
The basic philosophy to be applied was in dispute for several
years: One approach was to minimize the total cost of the
structure including the cost of a failure multiplied by the
failure probability. This approach was finally abondoned, because this would have meant to attribute a monetary value to
human life. The Model Code is now based on a failure probability
of 1/10000 in 50 years. The practical meaning of this number is
commented in Chapter 6.1.

3.2 Admissible Stresses, Partial Safety Factors or Direct
      Design for a Given Failure Probability
The use of admissible stresses was abandoned because it would
not lead to a uniform failure probability because of the non-
linearity of the system. It became clear very early that we
would use the system of partial safety factors for actions and
material properties in the Model Code. The direct design for a
given failure probability for each structure would be too complicated with the tools available at present.

3.3 Computation Techniques used in the Determination of the
      Safety Factors
We used a "trial and error method" to determine reasonable
values for the safety factors. About 25 sets of safety factors
were used in evaluating the integral which determines failure
probability, Eq. (1). From these sets, the "best fit" was selected by prudent judgement: Since a completely uniform failure
probability could not be achieved, we selected a set which in
the majority of the cases had a failure probability of less than
1/10000. However, we allowed larger values in some instances in
order to avoid substantial oversafety in the majority of the
cases.

The set which was finally selected and the consequences of this
selection are described in Chapter 5.

The following integral was used to determine the failure probability Pf:

$$Pf = \int pc \cdot ps \cdot pt \cdot pmo \cdot pm \cdot pn \cdot cu \cdot dpc \cdot dps \cdot dpt \cdot dpmo \cdot dpm \cdot dpn \quad (1)$$

The integral, Eq.(1), is extended over the space of all possible
combinations of the variables.

The variables in Eq. (1) have the following meaning:

pc   probability distribution of concrete strength
ps   probability distribution of steel strength
pt   probability distribution of wall thickness
pmo   probability distribution of modelling factor
pm   probability distribution of moment from wind
pn   probability distribution of normal force from dead
          load
cu   failure factor; this factor is 1 if a combination
          of the variables leads to failure, otherwise 0.

The integral itself was evaluated for all reasonable combinations of cross-sections, loads and material strengths which are
in ultimate limit state when inserting the characteristic values
of the respective distributions. By using normalized variables,
the amount of computation work and comparison of resulting sets
of failure probabilities could be kept within tolerable limits.
The integral itself was evaluated by a numerical procedure which
we called "the direct integration method". In this method all
functions are represented by step functions with 4 to 15 steps
so that the integration is replaced by a finite sum of about
25000 elements. During the committee meetings we had a terminal
on the conference desk with a link to a powerful computer. Most
questions concerning the effect of certain decisions on the
resulting safety could be answered within a few minutes. One
example of the result of such a computation is Fig. 2, Chapter
6.

The modelling factor is an instrument which was introduced to
model the following "human factor" in safety considerations:
Each computation of a real structure starts with forming a model
for that structure which in itself may be oversafe or undersafe.
Normally, the engineer will conceive the model in such a way
that it is safer than the real structure. However, the engineer
may also underestimate certain effects in the modelling process.

The modelling factor is a multiplier for both normal force and
wind when determinig the failure factor cu. From Table 2, Chapter 5, it can be seen that we assumed an overestimation of the
resulting stresses and strains in the model compared to reality
in 95% of the cases with a standard deviation of 5 %. This
means that underestimations of considerable size and occurrence
are taken into account in our model.

4. THE DEFINITION OF ULTIMATE LIMIT STATE

4.1   The Ultimate Limit State of Reinforcing Steel
The ultimate limit state of reinforcing steel is defined by a
strain of 5/1000.

4.2   The Ultimate Limit State of Concrete
There are two failure mechanisms for concrete which can be
explained by considering two theoretical extreme cases:

Case 1:   In this case, we have no dead load, the only force
          acting on the cross section is wind. Thus, failure
          occurs within a fraction of a second, when the
          structure oscillates under the influence of wind.
          Creep does not occur. The ultimate limit state ist
          defined by point P1 of Fig. 1.

Case 2:   In this case, we have only dead load, wind is of no
          importance. Dead load is a longterm effect and will
          cause creep.   Thus, the ultimate limit state is
          defined by point P2 of Fig. 1.

For the general case of dead load combined with wind load,
CICIND applies the "gliding material law", see combination of
curve C2 and C3 in Fig. 1. The ultimate limit state is defined
by point P5.

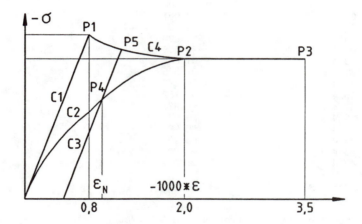

Figure 1: Stress- strain relationship for concrete, "Gliding Material Law"

Curve C1: Short-term load (single wind gust in ultimate limit state)
          without longterm load
Curve C2: Longterm load (from dead load)
Curve C3: Combination of longterm and short-term load.
Curve C4: Curve defining ultimate limit state for the combination of
          longterm and short-term load.
Point P5: Ultimate limit state of concrete

5.   THE PROBILITY DISTRIBUTIONS OF THE VARIABLES

5.1   The Probability Distribution of the Wind Load
CICIND takes as the characteristic wind speed that wind speed, which has a return period of 50 years. This assumption along with the further assumption on the interdependence of wheather situations leads to an extreme value distribution of the Gumbel type. In order to give an idea of the numerical values of that probability, the probability distribution for the German North-Sea coast with a characteristic wind speed of 26.4 m/s is given in Table 1.

Table 1
Cumulative Probability Distribution of 50-Years Extreme Wind for the German North Sea Coast

Wind Speed m/s	Cumulative Probability (%)
19. 4	0. 000
21. 4	0. 001
23. 4	3. 300
25. 4	35. 938
27. 4	73. 565
29. 4	91. 202
31. 4	97. 275
33. 4	99. 175
35. 4	99. 752
37. 4	99. 925
39. 4	99. 978
41. 4	99. 993
43. 4	99. 998
45. 4	99. 999
47. 4	100. 000

5.2   The Probability Distribution of other Variables.
CICIND assumes Gaussian distribution for all variables except wind. Table 2 gives the distributions which were assumed for the variables.

Table 2
Probability Distribution of Variables Used in the
Determination of Failure Probability

Variable	Charateristic Value	Standard Deviation (% of mean value)
Concrete Strength	5 % fractile	12
Steel Strength	5 % fractile	3
Wall Thickness	Mean value	3
Modelling Factor	95 % fractile	5
Dead Load	Mean value	4. 3

6. THE RESULTS

6.1   The Saftety Factors
The set of safety factors in Table 3 turned out to fulfill
sufficiently well and better than all other investigated sets
the requirement of uniform, sufficient and necessary safety.

Table 3
Safety and material factors of the CICIND Model Code
for Concrete Chimneys

Variable	Safety factors
Concrete Strength	1. 5
Steel Strength	1. 35
Wall Thickness	1. 0
Modelling Factor	1. 0
Dead Load	1. 0
Wind	1. 85

The safety factor for wind seems rather high, but it is simply a
consequence of the required failure probability of 1/10000 in 50
years and the assumptions for the probability distributions.   A
failure probability of 1/10000 in 50 years is impossible in
normal human understanding. The probability of a dice showing 14
times the 6 in 14 consecutive throws is of that magnitude if you
allow 10 sec for each throw. You would have to continue that
experiment 10000 times for 50-year periods in order to see the
event coming.

6.2   The Failure Probabilities
The failure probabilities for a typical set of variables which
result from the safety factors in Table 3 are shown in Fig. 2.

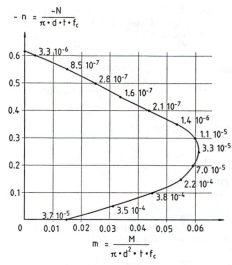

Figure 2: Failure probability for chimney cross sections designed according to CICIND Model Code for Concrete Chimneys.

N = Characteristic value of dead load
M = characteristic value of moment from wind load
fc = characteristic value of concrete strength (30 MPa)
fs = characteristic value of steel strength (400 MPa)
t = characteristic value of wall thickness
d = diameter of cross section
ratio of reinforcement: 0.5%

6.3 Comparison with National Chimney Standards
A comparison, Lit. (5), of the CICIND code with 3 other chimney codes shows that the other codes are in most cases oversafe, in some cases greatly oversafe and in some cases undersafe.

Appendix, Literature

(1)   The CICIND Model Code for Concrete Chimneys
       Part A: The Shell, October 1984
(2)   Commentaries for the CICIND Model Code for Concrete
       Chimneys, April 1987
(3)   The CICIND Model Code for Steel Chimneys, May 1988
(4)   Commentaries for the CICIND Model Code for Steel
       Chimneys, 1989
(5)   Noakowski, P.: Comparison of Chimneys, Constructed according to Various Standards, Proceedings of the 5. International Chimney Congress, 1984.
These documents are available from CICIND- 136 North Street-
Brighton BN 1 1R6 - England.

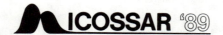

5th International Conference on
Structural Safety and Reliability

RELIABILITY OF STEEL COLUMNS SUBJECTED TO BUCKLING.
COMPARISON OF INTERNATIONAL STANDARD CODES OF DESIGN.

Jean-Pierre Muzeau

Laboratoire de Génie Civil, Université B. Pascal
C.U.S.T., B.P. 206 - 63174 Aubière Cedex - France

ABSTRACT

Standard codes of design represent the phenomenon of
buckling by means of a curve taking into account
random structural imperfections. The author suggests
a probabilistic method to evaluate and compare the
homogeneity of the reliability offered by different
international standard codes of design in the case of
steel columns subjected to compression loads for
different values of slenderness. He shows that the
effects of lateral displacements due to buckling can
be considered in the study of the reliability of
simply compressed columns.

KEYWORDS

Homogeneity of reliability; buckling curves; steel
columns; effect of imperfections; codes of design.

1. INTRODUCTION

Imperfections of structural members and intensity of applied loads are
essential in the phenomenon of buckling. A geometrically non linear
elasto-plastic model has been developed [1]. It is able to calculate
very precisely stresses and displacements in a structure even if snap-
through or plastic hinges appear. The model requires a discreteness
in only a fairly small number of elements and no numerical integration.
It is thus possible to associate it with a Monte-Carlo simulation
technique and to obtain a sufficiently precise evaluation of a
reliability index in a very short time.

The reliability of steel columns depends on the buckling curve
considered. In this paper, different international standard codes of
design are studied. American [2], [3], British [4], Canadian [5],
European [6], French [7] and Polish [8] codes are compared in the case
of structural elements supporting an axial load. Deterministic columns
are obtained from each code in relation with their own buckling curve.

Then, simulations are realized including random defects. The variations of the values of a reliability index, calculated for different values of slenderness, permit to evaluate the homogeneity of the reliability offered by each buckling curve.

2. MODEL OF CALCULATION

The study is based upon the calculation of double-hinged steel columns loaded with an axial compressive force. All the columns are realized with French **IPE** or **HE** hot-rolled sections (figure 1) of **E 24** steel grade (design stress : **fy = 235 MPa**). It is assumed that no buckling occurs about the minor axis.

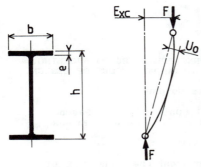

The probabilistic defects or imperfections taken into account are :
- out-of-straightness, **Uo**,
- variation of dimensions of cross-section **Δh**, **Δb** and **Δe**,
- variation of elastic limit **Re** and Young's modulus **E**,
- defect in the position of the applied load **Exc**,
- variation of the intensity of the applied load **F**.

Comparison of the influence of all these defects had been shown in reference [1] or [9].

Figure 1. Notations

2.1 – Random variables

The eight independent random variables are defined in table 1.

Imperfections	Random variable	Distribution	Mean value
Dimension of cross-section	Δh Δb Δe	Uniform	0 mm
Eccentricity of load	Exc	Gaussian	0 mm
Out-of-straightness	Uo	Gaussian	0 mm
Elastic limit	Re	Log-normal	263.2 MPa
Young's modulus	E	Log-normal	210000 MPa
Axial load	F	Log-normal	Variable

Table 1. Random variables

Furthermore, it is assumed that no cross-sections, eccentricities of applied loads and defects of straightness will be found out of the standard tolerance limits. This hypothesis is realistic considering the controls carried out in the ironworks or on sites. If a simulated column does not comply with these conditions, it is discarded and a new one is computed. The coefficients of variation of the elastic limit and Young's modulus are taken equal to 6 %. The coefficient of variation of the load is chosen equal to 10 %. The standard deviations **s** of the

random variables **Uo** and **Exc** are obtained from the relationship :
2s = T, if **T** is the value of the tolerance limit chosen as the
characteristic value. If l is the length of the column, the maximum
value of **Exc** is equal to 0.35 % l. The maximum value of the defect **Uo**
depends on the considered I or H hot-rolled section (table 2).

IPE rolled sections	HE rolled sections (HEA or HEB)	Out-of-straightness tolerance limits
IPE 80 to IPE 160		Uo ≤ 0.30 % l
IPE 180 to IPE 360	HE 100 to HE 360	Uo ≤ 0.15 % l
IPE 400 to IPE 600	HE 400 to HE 600	Uo ≤ 0.10 % l

Table 2. Out-of-straightness tolerance limits

2.2 - Limit states

Safety margins of strength S^S and safety margins of serviceability S^D
are considered. They are linked to the studied cross-sections and they
take into account the positive or negative sign of deflections as
follows :

$$S_1^S = 1 - \frac{N}{Np} - \frac{M}{\emptyset Mp} \quad \text{and} \quad S_2^S = 1 - \frac{N}{Np} + \frac{M}{\emptyset Mp}$$

$$S_3^S = 1 - \frac{M}{Mp} \quad \text{and} \quad S_4^S = 1 + \frac{M}{Mp}$$

$$S_1^D = d_1 - d \quad \text{and} \quad S_2^D = d_1 + d$$

where **N** and **M** are respectively the axial load and the bending moment
produced by buckling and **Np** and **Mp** are the plastic limits (random
variables function of the area of each cross-section). Coefficient ∅
depends on the cross section considered. It is taken equal to 1.22 in
the case of IPE and to 1.11 in the case of HE. **d** and d_1 are the deflection
and the allowable deflection.

2.3 - Index of reliability

The non-linear elasto-plastic mechanical model gives the components
of the acting effect **N**, **M** or **d** relative to each simulated column and
it is compared to the random resistant or limit effect **Np**, **Mp** and d_1
through the relationships stated above. It is to be noted that the
possibility of collapse due to a plastic hinge is controlled. Because
of the possible symmetry of the deflection, the reliability indexes
are evaluated by :

$$\beta^S = \text{Min} \left[m\{\tfrac{1}{2}(S_1^S + S_2^S)\} / s\{\tfrac{1}{2}(S_1^S + S_2^S)\} ; m\{\tfrac{1}{2}(S_3^S + S_4^S)\} / s\{\tfrac{1}{2}(S_3^S + S_4^S)\} \right]$$

$$\beta^D = \left[m\{\tfrac{1}{2}(S_1^D + S_2^D)\} / s\{\tfrac{1}{2}(S_1^D + S_2^D)\} \right]$$

where m{.} and s{.} are respectively the mean value and the standard
deviation of each safety margin. They are computed by means of 1000
simulations.

No probability of failure is deduced from these indexes because it

would be without significance. The method is only used to compare
different mechanical behaviours under identical hypothesis.

3. COMPARISON OF THE BUCKLING CURVES

The comparison is founded on the study of different French rolled-
sections : IPE 80 to IPE 600, HEA 100 to HEA 600 and HEB 100 to HEB
600 for slenderness λ contained between 10 and 240.

The **factored compression load** $Fc = Np/kc$ is deterministically
calculated from the buckling curve described in each standard code.
In order to realize a comparison, a reference is chosen : all codes
are supposed to give the same value of the allowable load relative to
a slenderness of 10 and equal to the deterministic value of **Np** (table
3). Then, the mean value of the **unfactored compression load F** is
obtained from the ratio $F = Fc/1.5$.

λ	LRFD 86 [2]	AISC 78 [3]	BS 5950 [4]	CSA 76 [5]	EUROC 3 [6]	CM 80 [7]	PN–B/03 [8]
10	1.000	1.000	1.000	1.000	1.000	1.000	1.000
20	1.014	1.025	1.003	1.030	1.005	1.005	1.023
30	1.039	1.056	1.025	1.067	1.045	1.043	1.050
40	1.074	1.095	1.051	1.113	1.092	1.100	1.085
60	1.181	1.195	1.129	1.240	1.224	1.225	1.184
80	1.348	1.340	1.282	1.440	1.445	1.450	1.354
100	1.600	1.556	1.577	1.796	1.797	1.810	1.678
120	1.971	1.902	2.031	2.297	2.284	2.310	2.339
140	2.523	2.506	2.615	2.850	2.893	2.940	3.184
160	3.294	3.273	3.307	3.459	3.610	3.700	4.159
180	4.169	4.143	4.100	4.135	4.428	4.560	5.264
200	5.147	5.114	4.989	4.996	5.342	5.510	6.498
220	6.228	6.188	5.971	5.989	6.352	6.580	7.863
240	7.412	7.365	7.046	7.056	7.456	7.690	9.358

Table 3. Calculated buckling coefficients **kc**

The curves obtained for each code computed for the **IPE 300** section as
an example, are shown in figure 2. Ranges of variations of β^S
corresponding to this profile are given in table 4.

Standard Code	Slenderness from 10 to 120	Slenderness from 10 to 240
LRFD Code	1.0	4.8
AISC 1978	1.9	5.0
BS 5950	0.9	3.2
CSA Code	1.4	2.4
French Additif 80	1.5	5.0
Eurocode 3	1.4	4.1
Polish Code	1.7	11.4

Table 4. Ranges of variations of β^S

These results show that the best homogeneity of reliability is obtained
by the Canadian code ranking before the British one for slenderness
between 10 and 240. They also show that for small slenderness (from

10 to 120), British and LRFD codes give the best results. The difference between the new and the former American code is not very important on the considered set of slenderness but the new LRFD code gives better results for small values of λ.

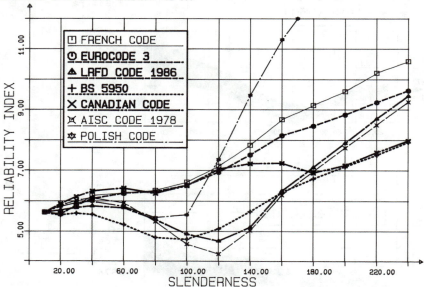

Figure 2 - Comparison of international standard codes for IPE 300

Now, considering all the existing hot-rolled French **IPE** and **HE** profiles, computing gives results covering a zone of a variable area represented in the following figures 3 to 6 obtained for the European, American, British and Canadian Standard Codes.

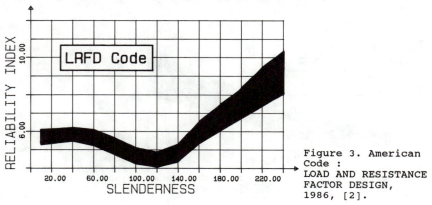

Figure 3. American Code :
LOAD AND RESISTANCE FACTOR DESIGN, 1986, [2].

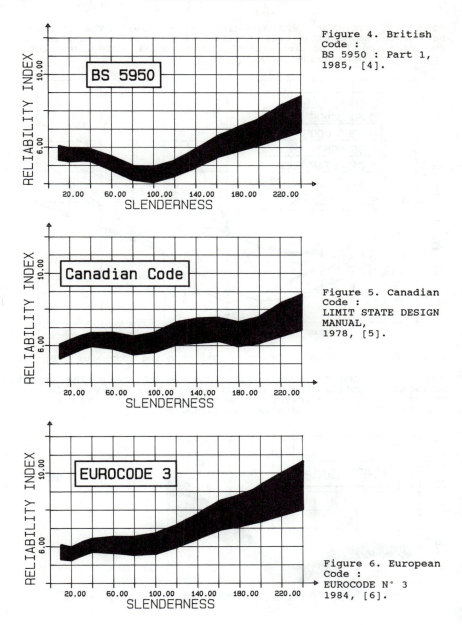

Figure 4. British
Code :
BS 5950 : Part 1,
1985, [4].

Figure 5. Canadian
Code :
LIMIT STATE DESIGN
MANUAL,
1978, [5].

Figure 6. European
Code :
EUROCODE N° 3
1984, [6].

If a code gave a perfect homogeneity of reliability, the results would be a linear horizontal zone. We can then consider that the Canadian Code offers a rather good homogeneity compared to the European one and that the American code ranks last of all for the considered hypothesis, after the British.

The model also includes the reliability linked to a limit state of serviceability β^D in the calculation of a simply compressed column as explained in paragraph 2. The results obtained with the American LRFD code [2] relative to an IPE 300 section are shown in figure 7. In this example, the axial loads are decreased by a ratio of 1.6 to take into account the difference of factor relative to the serviceability limit state in comparison with the ultimate one. The allowable deflection is classically chosen equal to 1/200.

Taking into account this limit state, it is to be noted that the values of its relative index of reliability decrease very fast when slenderness increases (it is nearly constant for λ going from 140 to 240). This explains why builders are very careful for large slenderness. It is still difficult to compare ultimate and serviceability limit states because they are attached to different domains of probability of failure. However, these results emphasize why it could be dangerous to try to get lower values of index β^S in the ranges of larges slenderness because those of β^D would decrease at the same time. Nevertheless, it must be observed that the chosen column is double-hinged and that these boundary conditions are the most unfavourable regarding deflection.

Considering all these results, it is possible to propose a buckling curve leading to a quite homogeneous reliability of axially loaded columns founded on LRFD code for λ bounded by 10 and 40, on Eurocode 3 in the range of average slenderness and on Canadian code in the range of large ones (higher than 120) modified to be continuous (factor 1.017). The results are shown in figure 7.

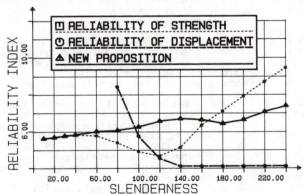

Figure 7 - Comparison of indexes

This curve is obtained with the following relationships :

if : $\lambda r = \lambda (fy/E)^{1/2}/\pi$ and $\lambda o = 1 + 0.34(\lambda r - 0.2) + \lambda r^2$, then :

$0 < \lambda \leqslant 40$	$1/Kc = 0.658^{\lambda r^2}$
$40 < \lambda \leqslant 100$	$1/Kc = 1.017 [\lambda o - (\lambda o^2 - 4\lambda r^2)^{1/2}] / (2\lambda r^2)$
$100 < \lambda \leqslant 180$	$1/Kc = 1.017 [-0.111 + 0.636/\lambda r + 0.087/\lambda r^2]$
$180 < \lambda \leqslant 240$	$1/Kc = 1.017 [0.009 + 0.877/\lambda r^2]$

4. CONCLUSION

This paper proposes a method to calculate the efficiency of standard codes of design from the point of view of reliability. It is founded on the association between a very precise mechanical model taking into account second order effects and a simulation technique. It makes possible to consider many structural or loading imperfections. The method is applied to evaluate the quality of different curves of buckling but it is possible to use it in many other cases. For example, effects of lateral buckling, influence of factored loads or effects of corrosion could be computed in the same way and it makes possible to determine very reliable rules of design. So, this method gives an alternative to increase the knowledge of the homogeneity of the reliability of steel structures. Although it is not usual to consider the effects of deflection to design a column axially loaded, this paper emphasizes the problem of large slenderness and accounts for the care taken by designers in their dimensioning.

5. REFERENCES

[1] **MUZEAU J.P.** : "Modèle de l'influence d'imperfections sur la sécurité des structures métalliques en comportement non-li-néaire. Comparaison de règlements internationaux", Thèse de Doctorat ès Sciences Physiques, Université Blaise Pascal, Clermont-Ferrand, France, 1987.

[2] **A.I.S.C.** : "Load and resistance factor design", American Institute of Steel Construction, 1st edition 1986.

[3] **A.I.S.C.** : "Manual of steel construction", American Institute of Steel Construction, 8th edition, 1978.

[4] **BS 5950** : "Standard use of steelwork in buildings", British Standard, 1985.

[5] **C.S.A.** : "Limit State Design Manual", Canadian Institute of Steel Construction, 1978.

[6] **C.E.E.** : "EUROCODE N°3 : Règles unifiées communes pour les constructions en acier", Rapport EUR 8849 DE, EN, FR, 1984.

[7] **C.T.I.C.M.** : "Règles de calcul des constructions en acier, Additif 80", Revue Construction Métallique, n° 1-1981.

[8] **PN-80/B-03200**: "Steel structures. Design rules", Polish Standard Code, Warszawa, 1984.

[9] **FOGLI M., MUZEAU J.P., LEMAIRE M.** : "Reliability of structures subjected to buckling in a probabilistic context", ICOSSAR'85, 4th International Conference on Structural Safety and Reliability, Vol. 1, pp. 77-85, Kobé, Japan, 1985.

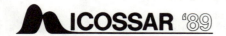
PROBABILISTIC TRAFFIC LOAD MODELS AND
EXTREME LOADS ON A BRIDGE

B. JACOB[*], J.B. MAILLARD[**] and J.F. GORSE[***]

[*] Head of Structural Reliability Section, LCPC,
[**] Research engineer, LCPC, [***] Assistant, LCPC,
Laboratoire Central des Ponts et Chaussées, Paris, France

ABSTRACT

This paper first contains a summary of the intensive traffic measurements collected in France and
Europe, and of the probabilistic distributions
adopted for the main parameters. The evaluation of
the maximum gross weight and axle load over a long
time period is then deduced, using the extreme
value distributions. Two traffic models describing
the total load on a bridge and its maximum values
are presented. Comparisons with the results of
simulations are made using the real traffic measurements. One of these models is then proposed for
calibrating the future codes.

KEYWORDS

Traffic loads, bridge loading, probabilistic load
models, extreme load, code calibration.

1.INTRODUCTION

Recently several countries have revised their bridge load codes
because of traffic and structure evolutions. The new load models
have to be adapted to the modern and automatic bridge calculations
as well as to the advanced notions of structural safety.

Some research was developed on the probabilistic methods in order
to describe the real loads and to calibrate the design models
[3],[5],[6]. Such an approach has been adopted in France from 1980
to evaluate and rewrite the outdated existing code. In the last two
years the work increased widely in the EEC countries, due to the
decision of the European Commission to prepare an Eurocode on
"Traffic loads on bridges".

Statistical data from measured traffics and probabilistic models
are presented to evaluate the extreme vehicle, axle and on bridge
total loads over any time period. The results compared for short
time periods with the simulations yield a code calibration method.

2.TRAFIC MEASUREMENTS AND DISTRIBUTIONS

Before building a traffic load model, a representative set of data
was collected on a large scale. Thanks to the low cost Weigh-in-
Motion system, developed in France by the LCPC from 1980 and using
the piezo-electric cables [7], a large database of 150 traffic
records was filled up. More than two million trucks have been already recorded on about 50 locations, including highways, national and
secondary or urban roads. Similar measurements were made in Italy
and Spain, and a few in West-Germany. Some statistical traffic
results for 8 European locations [2] are given in the next table :

Location	Type	Year	Truck flow v/24hs	Axle loads (kN)		Gross Weights (kN)		
				mean	max/d	1mode	2mode	max/d
BD PERIF.	hwy u.w	83	8076	61	210	120	380	610
AUXERRE v1	hwy	86	2630	83	195	190	410	630
CHAMONIX	NR	87	1204	71	155	140	400	570
EPONE	SR	87	327	56	170	100	280	510
LYON	u.w	87	1232	59	195	120	–	590
BROHLTAL (D)	hwy	84	4793	59	165	160	400	650
FIANO (I)	hwy	87	4000	57	145	200	420	590
GUITIRIZ (E)	NR	87	873	62	190	120	340	430

Statistical data of European traffics

Using the large number of histograms computed for each traffic
record, containing most of the time more than 10,000 trucks,
various probability density functions (PDF) were fitted for each
relevant parameter.

In a flowing traffic the speeds are well fitted by gaussian PDF,
with two modes if cars and trucks are both recorded. The
inter-trucks distances follow a gamma distribution with scattered
mean values and large coefficients of variation. We then have a
relationship between the time intervals t, the distance D and the
speed S, considered as random variables : D = S t. The mean
traffic flow ϕ is then defined by the mean values :

$$\phi = 1 / \overline{t} \simeq \overline{S} / \overline{D} \tag{1}$$

where the approximation is valid if S and D or t are independent,
i.e. in a free traffic.

On the main roads and highways, the gross weight (W) distribution
of the trucks can be fitted by a 2-modes gaussian PDF as shown in
the Henry's diagram (figure 1). The first mode contains the small
trucks and the empty or partially loaded vehicles, while the second
one involves the fully loaded half-trailers and drawbar vehicles.
The mean values and standard deviations of these modes depend on
the traffic as well as the proportions of the modes. The maximum
gross weight over one day is highly concentrated around 600 kN, but
falls down to 450 kN in Spain or on some secondary roads.

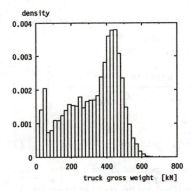
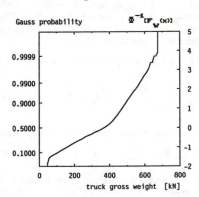

Figure 1 : Gross weights : density and Henry's diagram

The axle load distribution is asymetric, with a mean value around
65 kN and a standard deviation close to 30 kN, but an upper tail
decreasing like a gaussian PDF (figure 2). The maximum axle load
over one day is also weakly scattered around 185 kN in France and
Spain, and 140 kN in Italy. The respective legal limits are 128 kN
and 98 kN.

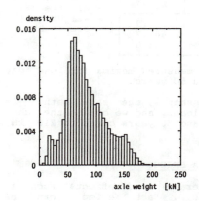
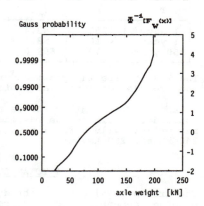

Figure 2 : Axle loads : density and Henry's diagram

3.EXTREME TRUCK AND AXLE LOADS

Ditlevsen [4] proposed an approach to this question considering a finite population of trucks. The simpler method described here is based on the idea that the loading cases are infinite and the truck load PDF decreases like a gaussian one (figure 1).

If the second mode of W (proportion p) governs the maximum gross weight of N independent trucks, this maximum has the distribution of the largest of $n = pN$ independent gaussian variables with parameters $\overline{W}_2$ and $\sigma(W_2)$, denoted Y_n.

The well-known asymptotic distribution of maximum is in our case a Gumbel function [1] :

$$F_{Y_n}(y) \simeq \exp(-\exp(-a_n(y-u_n)))\quad \text{for large n, with :}$$

$$a_n = \frac{\sqrt{2Ln(n)}}{\sigma(W_2)} \quad , \quad u_n = \overline{W}_2 + \sigma(W_2)\left[\sqrt{2Ln(n)} - \frac{Ln(Ln(n)) + Ln(4\Pi)}{\sqrt{2Ln(n)}}\right] \quad (2)$$

and the first moments of Y_n are given by :

$$\overline{Y}_n \simeq u_n + \gamma/a_n \quad , \quad \sigma(Y_n) \simeq \Pi \, / \, a_n\sqrt{6} \quad , \quad \text{with } \gamma = 0.5772 \quad (3)$$

The characteristic truck weight $x(\alpha)$ for a period T is the value having a probability α to be exceeded during T. If N is the total truck occurences during T and $n = pN$, $x(\alpha)$ is the α-fractile of Y_n, or the ε-fractile of W, given by :

$$1 - F_W(x_n(\alpha)) = \varepsilon = 1 - (1-\alpha)^{1/n} \simeq \alpha/n, \text{ for large n.} \quad (4)$$

For $p = 0.5$, $\phi = 2500$, $\alpha = 5\%$, $\overline{W}_2 = 400$ kN and $\sigma(W_2) = 55$ kN the maximum truck weights are :

T	n	$\overline{Y}_n$	$\sigma(Y_n)$	$x_n(\alpha)$
1 day	1250	583.4	18.7	616.9
1 year	4.10^5	658.1	13.9	683.7
50 years	2.10^7	699.0	12.2	721.6

all the weights are in kN.

The results are in agreement with the measured maxima for one day and provide realistic values for 1 and 50 years.

Because of the similar PDF decreasing shapes, the same method is applied to estimate the maximum axle loads, and we obtain the characteristic values for 1 day, 1 year and 50 years close to 210 kN, 240 kN and 260 kN.

4.TOTAL LOAD ON A BRIDGE

The trucks passing on a bridge generate load effects such as bending moments, shear forces, etc... Each effect can be characterized by its influence surface $z(x,y)$ such that:

$$E_t = \sum_{i=1}^{N_t} z(x_{it}, y_{it}) \ W_i \quad \text{at any time t.} \tag{5}$$

N_t is the random variable counting the axles on the bridge at t. The total load Q on a bridge section with a length L and a width l is a particular effect leading to easy calculations in (5), because the influence surface is given by $z(x,y) = 1$ for $0 \leq x \leq L$ and $0 \leq y \leq l$:

$$Q = \sum_{i=1}^{N} W_i \quad \text{and} \quad f_Q(x) = \delta_o + \sum_{n>0} \text{Prob}(N=n) \ f_w^{*n}(x) \tag{7}$$

The traffic is assumed to be stationary and the vehicles independent on each other. The trucks are considered instead of the axles, with their occurences having a Poisson density, in accordance with the gamma density of the distances (parameters close to 1). f_w is the gaussian PDF of the gross weights given in §2. The CDF (cumulative distribution function) of Q can be written as :

$$F_Q(x) = H_o + \sum_{n>0} \text{Prob}(N=n) \sum_{i=0}^{n} C_n^i (1-p)^i p^{n-i} \ \Phi(m_{ni}, \sigma_{ni}, x) \tag{8}$$

with $\quad \text{Prob}(N=n) = e^{-\lambda} \dfrac{\lambda^n}{n!}$, $\quad \lambda = L/\overline{D}$, $\quad \Phi(m,\sigma,x)$ describing the gaussian CDF. Its parameters are :

$$m_{ni} = i \ \overline{W}_1 + (n-i) \ \overline{W}_2 \quad \text{and} \quad \sigma_{ni}^2 = i \ \sigma(W_1)^2 + (n-i) \ \sigma(W_2)^2 \tag{9}$$

The maximum Q of Q during any time T is obtained by splitting [0,T] in K equi-intervals, where $K = [T/\tau] \simeq T/\tau$, and τ is the bridge crossing time ($\tau = L/\overline{S}$); in this case:

$Q = \underset{k=1}{\overset{K}{\text{Max}}}(Q_k)$ with Q_k being K independent variables with the density f_Q. The characteristic value $Q(\alpha)$ of Q during T is :

$$1 - F_Q(Q(\alpha)) = \epsilon = 1 - (1-\alpha)^{1/K} \simeq \alpha/K \tag{10}$$

This equation is easily solved numerically, using (8), by the Newton method [8]. The most probable number of vehicles producing the total load $Q(\alpha)$ is obtained by the way.

For $L \leq 80$ m this number increases far too much because of trucks overlaping. The simple Poisson process neglecting the vehicle lengths can be modified in the following way : if $t_o = D_o/\overline{S}$ represents the minimum time interval between two successive trucks arrivals (D_o = minimum vehicle size plus inter-trucks distance), the density of the time intervals t is :

$$f_t(t) = \mu \ e^{-\mu(t-t_*)} \quad \text{with} \quad \mu = \phi \ / \ (1-\phi t_o) \tag{11}$$

Then the law of the random number of vehicles is :

$$\text{Prob}(N \leq n) = \text{Prob}\left(\sum_{i=1}^{n+1} t_i > \tau\right), \text{ where } \sum_{i=1}^{n+1} \mu(t_i - \tau) \text{ is a Gamma}(n+1,1).$$

The Poisson models are only valid for free traffics. In order to account for the jam situations, we consider the case where N is deterministic : $N = n_{max} = [L/D_o]$.

Equations (8) and (10) are still valid with minor changes ; in (10), 1/K now represents the probability of a jam.

We now introduce the maximum uniformly distributed load (UDL) : $q = Q/(Ll)$. Figure 3 shows the results provided by both Poisson models and the jam model, for a single traffic lane. With the modified Poisson model, the discontinuities for the short spans are due to the constant truck length, and may be avoided by taking the upper envelope. For m independent lanes, (8) and (10) are still valid by replacing Prob(N=n) by Prob(N=n,m) computed sequentially :

$$\text{Prob}(N=n,m) = \sum_{i=o}^{n} \text{Prob}(N=i,m-1) \cdot \text{Prob}(N=n-i,1) \tag{12}$$

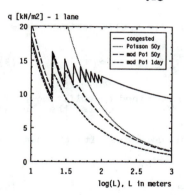
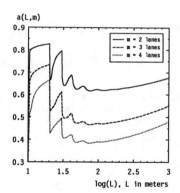

Figure 3 : Maximum uniformly distributed load
and lateral degressive coefficients

With both simple Poisson and jam models, we have for the UDL : $q(L,m) = q(mL,1)$. Most of the existing codes contain coefficients to account for the decrease of $q(L,m)$ with respect to m. Such coefficients may be defined by :

$$a_m(L) = \frac{Q(L,m)}{m\,Q(L,1)} = \frac{q(L,m)}{q(L,1)} \tag{13}$$

and are illustrated in the figure 3 for the modified Poisson model. They are not very sensitive to the loaded length L, but would be higher with the jam model.

5.EXTREME LOAD EFFECTS

The general equation (5) may be treated by numerical simulation for a few days period [7] : the LCPC's computer program TRAFID dispaches the measured traffics on any influence surface and provide the time history and histograms of the load effects. The probability density f_E is obtained by fitting an histogram with an analytical function ; the extreme values or fractiles are evaluated by an extrapolation. But this approach is difficult because f_E is usually far from being gaussian, and the shapes of the influence surfaces are numerous.

There are two ways to generalize the former methods :

- the first one is to split the surface $z(x,y)$ in areas which may be approximated by pieces of planes. A particular case with the simple Poisson model and constant truck loads was treated in [9].

- the second way consists in combining short period simulations and the method of § 4. For any effect e of influence surface z, an equivalent load φ is defined as :

$$\varphi = \frac{e}{S} \text{ , where } S = \int_{0 \leq x \leq L} \int_{0 \leq y \leq l} z(x,y) \, dx \, dy \text{ , (if } S \neq 0) \qquad (14)$$

φ depends on the total load over the part of the bridge where $z(x,y)$ is relatively high. In the following, q and φ are considered as functions of L and T, e being the maximum effect reached during a period T (T=1 is assumed to be the period of the simulation). Our simulations [8] prove for a large set of various real surfaces, that the mean value of $\varphi(L,1)$ remains close to the $q(L,1)$ one. Nevertheless, the sharper the surface, the higher the ratio $\varphi(L,1)/q(L,1)$. Furthermore the extrapolating ratio $q(L,T)/q(L,1)$ remains fairly constant with respect of L (see figure 3). Therefore it is easy to verify that $\varphi(L,T)/\varphi(L,1)$ is nearly independent of L. Finally the maximum load effect is well approximated for any surface by : $e(T) \simeq e(1) \, q(T)/q(1)$, where q(T) is computed with the models of the § 4, and e(1) results from the simulation.

Such an hybrid method seems to be much better than a simple extrapolation because f_E is rather complicated with a slow and chaotic decreasing upper tail. Only if N reaches its maximum n_{max} (in the jam model or for short spans, $L \leq 100$ m), the decreasing f_E becomes nearly gaussian, as the truck loads do. But it is very unlikely to be the same for free traffics and $L \geq 100$ m.

6.CONCLUSION

The extensive traffic measurements made in EEC countries for years provide a fruitful data base for fitting probabilistic load models with random variables. They provide, with the help of the theory of extremes or of extrapolations, very likely estimations of the maximum truck or axle loads, up to 50 or 100 years, if stationariness is assumed. The Poisson process combined with the random distribution of the loads gives an acceptable model to compute the total load on a bridge and its maximum over any time period, for long spans. For short spans a modification is introduced to account the vehicle dimensions, and to avoid unlikely high loads.

Finally an hybrid method combines the simulations with the real
measured traffics to obtain accurate extreme load effects over
short periods (few days), with the previous models which give the
maximum of the total load for any time. The maximum load effects
can then be predicted by a coefficient of extrapolation from the
first results. At the same time degressive coefficients are proposed for the multi-lanes bridges. Once pertinent models and parameters are chosen (free or jam traffic, truck size and weights,
flows...), some curves are established for the calibration of any
design load model. Such curves were already used in France [8] for
pointing out some minor lacks of the old existing code, and are now
the basis to build and evaluate the proposed models of the future
Eurocode 9 part 12 on "Traffic Loads on Bridges".

REFERENCES

[1] ANG, A.H., TANG, W.H., "Probability Concepts in Engineering
    Planning and Design", J. Wiley & Sons, New-York, 1984.

[2] BRULS, A., JACOB, B., SEDLACEK, G., and al.,
    "Traffic Data of EEC Countries", Report of the SG 2,
    Eurocode 9-12 "Traffic Loads on Bridges", 1989.

[3] BUCKLAND, P.G., NAVIN, P.D, ZIDEK, J.V., McBRYDE, J.P,
    "Proposed Vehicle Loading of Long-Span Bridges", Journal of the
    Structural Division, ASCE,106(4), pp.915-932, 1980.

[4] DITLEVSEN, O., "Distribution of Extreme Truck Weight",
    Structural Safety, n°5, Elsevier, 1988.

[5] GHOSN, M., MOSES, F., "Reliability Calibration of Bridge Design
    Code", J. of Structural Engineering, ASCE, 112(4), 745-763,1986.

[6] HARMAN, D.J., DAVENPORT, A.G., "A Statistical Approach to
    Traffic Loading on Highway Bridges", Canadian Journal of Civil
    Engineering, 6(4), pp. 494-513, 1979.

[7] JACOB, B., SIFFERT, M., "An High Performant WIM System and its
    Applications", International Symposium on Heavy Vehicle Weights
    and Dimensions, RTAC, Kelowna, BC Canada, 1986.

[8] JACOB, B., CARRACILLI, J.B., MAILLARD, J.B, GUERRIER, F,
    "Evaluation de la Charge Uniforme 1,2A(1)", Rapport int. LCPC,
    Septembre 1987.

[9] MAILLARD, J.B., GORSE, J.F., "Charge Extrême de Trafic sur un
    Pont", Bull. liais. des labos PC, n° 162, 1989.

[10] TUNG, C.C., "Random Response of Highway Bridges to Vehicles
     Loads", J. of the Engineering Mechanics Division, ASCE,5,1967.

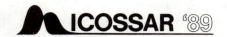 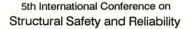

RELIABILITY-BASED LRFD FOR BRIDGES : THEORETICAL BASIS

Masanobu Shinozuka* , Hitoshi Furuta** , Susumu Emi***
and
Masakuni Kubo****

*   Department of Civil Engineering and Operations Research,
      Princeton University, Princeton, NJ, USA
**   Department of Civil Engineering, Kyoto University,
      Sakyo-ku, Kyoto 606, Japan
***   Engineering Division, Hanshin Expressway Public
      Corporation, Chuoh-ku, Osaka 541, Japan
**** Design Division, SOGO Engineering Inc., Higashiyodogawa-ku,
      Osaka 533, Japan

ABSTRACT

   This paper develops a theoretical basis for obtaining
probability-based load combination criteria for structural design.
For choosing the load combinations and for determining the factors
to be considered in a load and resistance factor design (LRFD) format, the notion of limit state probability diagram is introduced to
help establish the interrelationship among the limit states, limit
state probabilities, target limit state probabilities and load combinations. This paper presents a practical procedure of numerical optimization of these factors based on the structural reliability theory in
order to upgrade design codes for highway bridges.

KEYWORDS

Bridge; load combination; load and resistance factor design; reliability; structural safety.

1. INTRODUCTION

   In the practice of structural design, both extreme and abnormal loading conditions
must be considered. This requirement possibly results in a large number of load combinations in the design criteria. Furthermore, the load and resistance factors specified in codes

are usually determined by code committees primarily on the basis of collective judgment and experience. Hence, a more rational procedure is needed to justify the number of load combinations and to determine appropriate load and resistance factors. Following earlier works in this area [1]-[3], this paper further develops the procedure for obtaining probability-based load combination criteria for structural design. While the analysis primarily centers on the LRFD methodology, its implementations extend to a re-examination of the basic issues associated with the concept of structural safety and design.

While current reliability-based LRFD criteria are more rational than those of allowable stress design at least from the probabilistic point of view, there still seems to be a number of issues that require further investigation. In this respect, one of the more important issues is the rather arbitrary way in which particular load combinations are chosen for design purposes. To be on the safe side, one tends to cover all possible combinations of all the conceivable loads. Indeed, this appears to be the case for nuclear power plant design where the grave consequences of failure warrants the utmost care in selecting such load combinations. Probabilistically speaking, one can obviously enumerate all the possible load combinations that are mutually exclusive.

In general, highway bridges are subjected to dead, live, wind and earthquake loads, thermal force, earth pressure, and so on. Recognizing that frequent micro-tremors, constantly present breezes and ordinary temperature do not really constitute earthquake, wind and thermal loads respectively, the structure will be subjected to a set of mutually exclusive load combinations. However, identifying a certain load combination as part of such a mutually exclusive set does not necessarily warrant considering these combinations for structural design. Indeed, for the choice of load combinations to be considered for design, limit state probabilities must be essentially taken into consideration. In this connection, the limit state probability diagram [2] is used to help establish the interrelationship among the limit states, limit state probabilities, target limit state probabilities and load combinations.

The target limit state probabilities are specified taking into consideration the level of limit state probabilities of structures designed by the current method of allowable stress design.

For determining optimum values of load and resistance factors, an objective function is considered so as to measure the difference between the target limit state probabilities and computed limit state probabilities of structures designed by means of LRFD format using an initial set of load and resistance factor values. A new set of load and resistance factor values are determined in the direction of maximum descent with respect to the objective function, and these steps are repeated until a set of load and resistance factor values that minimize the objective function is found.

2. LIMIT STATE PROBABILITY DIAGRAM

Structural limit states represent various states of undesirable structural behavior. For example, the service level stress σ_a usually represents a limit state germane to the yield stress σ_y divided by a (material) safety factor. Similarly, the yield stress σ_y and ultimate stress σ_u represent limit states which, however, have more direct physical significance than the service level stress. These limit states indicate differing degrees of undesirability of structural behavior measured in terms of a physical quantity; stress σ in this case.

Dealing only with the stress-based limit states σ_α ($\alpha = a,y,u$) for simplicity, the limit state probabilities can be written as

$$P_\alpha = \sum_j Prob\,[\,\sigma > \sigma_\alpha \mid F_j\,]\;Prob\,[\,F_j\,] \tag{1}$$

where $Prob\,[\,\sigma > \sigma_\alpha \mid F\,]$ is limit state probability conditional to event F and $Prob[F]$ is probability of event F. F_j is event of a mutually exclusive load combination j ($j=1,2,...,J$). The

target limit state probabilities P_α^* are then introduced associated with σ_α.

The notion of a limit state probability diagram [2] as shown in Fig.-1 is introduced; It plots the common logarithm of the probability $P(x)$ for the response state to exceed x at least once in the structure's lifetime, as a function of x. Curve B_i in Fig.-1 indicates $P(x)$ for structure i. Note that such a curve depends on the structure, thus the super- or subscript i. When x assumes specific limit state values such as $x = \sigma_a$, σ_y or σ_u, $Prob[\sigma > x]$ represents the corresponding limit state probabilities. The target limit state probabilities P_a^*, P_y^* and P_u^* are indicated respectively by points A_a, A_y and A_u in Fig.-1.

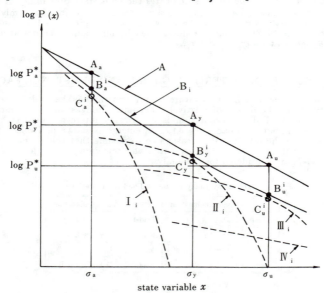

Fig.-1 Limit State Probability Diagram

While it is not a well-recognized notion, Ref.[2] suggests that conceptually the safety of a class of structures for which the design code is intended to be used should be specified by a target limit state probability curve $P^*(x)$, as designated by A in Fig.-1. If the state of structural behavior is to be described by more than one variable, say by x and y, the safety should be specified by a target limit state surface $P^*(x,y)$, where y represents, for example, deflection. Since it is impractical to prescribe the entire curve $P^*(x)$ as a safety requirement, and even if, one could do that, since it is also impractical to verify if curve B_i is below curve A for all values of x, one chooses a few values of x to perform such a check. In the present paper, $x = \sigma_a$, σ_y and σ_u are chosen as an example. Curves I_i, II_i, III_i and IV_i in Fig.-1 represent the limit state probabilities under a mutually exclusive load combination j ($j = I,II,III,IV$) for structure i. The limit state probability P_{ijm} for limit state m corresponds to each circled point designated by C_a^i, C_y^i and C_u^i in Fig.-1, respectively, associated with the limit states σ_a, σ_y and σ_u.

If the curves I_i, II_i, III_i and IV_i indeed take the relative positions as sketched in Fig.-1, then the load combination I controls the limit state probability P_a as designated by B_a in Fig.-1, II the limit state probability P_y and III the limit state probability P_u, as B_y and

B_u in Fig.-1 respectively, where super- or subscript i is omitted for simplicity. Note that, in this case, the load combination IV does not really control any of the limit state probabilities. If all the structures to be designed under the design code exhibit this trend, then the load combination IV does not have to be considered in the design, and the load combination I should be considered only for σ_a, II for σ_y and III for σ_u respectively. In fact, this is the conceptual basis for permitting the allowable stress to be increased when combinations of primary and secondary loads are considered in the classical allowable stress design.

As mentioned above, the curve B_i indicates the limit state probability of structure i. Then, it is expected that a more rational design code provides more uniformly limit state probabilities for all the structures to be designed. In this spirit, the load and resistance factor values in the LRFD format are determined by minimizing the deviations between the curves B_i and A at $x = \sigma_a$, σ_y and σ_u.

3. PROCEDURE FOR DETERMINING LOAD AND RESISTANCE FACTORS

Dead load D, live load L, temperature load T and earthquake load E are considered in this study where the design of relatively short-span bridges under the jurisdiction of Hanshin Expressway Public Corporation is of primary interest. Then, eight mutually exclusive load combinations (load-j ; $j = 0,1,2,...,7$) emerge as shown in Table-1. Load-0 can be eliminated from the set since it is reasonable to assume that bridges can sustain their own dead load without failure. Consequently, the remaining seven load combinations (load-1 to load-7) are considered for the analysis.

Table-1 Mutually Exclusive Load Combinations

	load combination		load combination
load-0	D	load-4	D + L + T
load-1	D + L	load-5	D + L + E
load-2	D + T	load-6	D + T + E
load-3	D + E	load-7	D + L + T + E

At the outset, it is prudent to consider all these seven load combinations for each limit state m ($m=1,2,...$). In fact, Table-2 shows seven design equations corresponding to seven load combinations for limit state m. The coefficients β_{sm} ($s = D,L,T,E$) denote the transformation coefficient from load to load effect, and D_n, L_n, T_n and E_n represent design

Table-2 Example of Design Equations

	design equations (safety formats for m-th limit state)
code-1m	$X_m \geq \beta_{Dm}\, \gamma_{D1m}\, D_n + \beta_{Lm}\, \gamma_{L1m}\, L_n$
code-2m	$X_m \geq \beta_{Dm}\, \gamma_{D2m}\, D_n + \beta_{Tm}\, \gamma_{T2m}\, T_n$
code-3m	$X_m \geq \beta_{Dm}\, \gamma_{D3m}\, D_n + \beta_{Em}\, \gamma_{E3m}\, E_n$
code-4m	$X_m \geq \beta_{Dm}\, \gamma_{D4m}\, D_n + \beta_{Lm}\, \gamma_{L4m}\, L_n + \beta_{Tm}\, \gamma_{T4m}\, T_n$
code-5m	$X_m \geq \beta_{Dm}\, \gamma_{D5m}\, D_n + \beta_{Lm}\, \gamma_{L5m}\, L_n + \beta_{Em}\, \gamma_{E5m}\, E_n$
code-6m	$X_m \geq \beta_{Dm}\, \gamma_{D6m}\, D_n + \beta_{Tm}\, \gamma_{T6m}\, T_n + \beta_{Em}\, \gamma_{E6m}\, E_n$
code-7m	$X_m \geq \beta_{Dm}\, \gamma_{D7m}\, D_n + \beta_{Lm}\, \gamma_{L7m}\, L_n + \beta_{Tm}\, \gamma_{T7m}\, T_n + \beta_{Em}\, \gamma_{E7m}\, E_n$

loads, respectively.

In general, the load factors γ_{sjm} are determined according to procedure summarized below.

1) Consider a set of N representative structures i ($i = 1,2,...,N$).

2) Identify M limit states ($m = 1,2,...,M$) and specify the corresponding target limit state probabilities P_m^*.

3) Select the loads and load combinations to be considered, and specify design loads D_n, L_n, T_n and E_n.

4) Under the assumption that determination of cross-sectional geometry is the sole purpose of design, a specific geometry A_{ijm} is obtained for structure i under design equation code-jm corresponding to load combination j and limit state m. Among all resulting geometries, the most conservative geometry is chosen as the design for structure i. Since the design equations involve the load factors γ_{sjm}, the design (geometry) is a function of these load factors.

5) Evaluate, with the aid of Eq.(1), the limit state probability P_{im} for structure i thus designed under limit state m. An objective function $\Omega = \sum_{m=1}^{M} \Omega_m$ is constructed in which Ω_m measures the total deviation between the target P_m^* and estimated P_{im} for all the representative structures. All optimum values of γ_{sjm} are chosen so as to minimize the objective function. A typical example of Ω, similar to that suggested in Ref.[1], is given below.

$$\Omega = \sum_i \sum_m w_{im} \left(\frac{\log P_{im} - \log P_m^*}{\log P_m^*} \right)^2 \tag{2}$$

where w_{im} is a weighting factor.

4. APPROXIMATE PROCEDURE OF DETERMINATION FOR BRIDGES

Although the above procedure is quite general, it becomes rather impractical when the total number of load factors to be determined increases beyond three (3) or four (4). In the case of Fig.-1 and Table-2 in which M=3 and J=7, the number of load factors γ_{sjm} to be determined are fifty-seven (19 × 3=57). Limiting the current analysis to the design of highway bridges, the past experience tends to dictate that only one load combination is dominant for each limit state in the sense that limit state probability for that limit state is primarily attributable only to that load combination. In fact, the current Japanese Specifications for Highway Bridges specifies four load combinations involving D, L, T and E as shown in Table-3. It is the authors' interpretation that each of these load combinations is associated with either allowable stress (σ_a), yield stress (σ_y), or ultimate stress (σ_u), as also shown in Table-3. Furthermore, it is postulated that the load factors appearing in those load combinations that are dominant in a particular limit state m are determined by minimizing the measure of deviation Ω_m between P_{im} and P_m^* involving all the representative structures.

$$\Omega_m = \sum_i w_{im} \left(\frac{\log P_{im} - \log P_m^*}{\log P_m^*} \right)^2 \tag{3}$$

Obviously, the dominance of a particular combination of loads for a particular limit state does not necessarily materialize in reality, and therefore the above interpretation is most probably too simplistic. Nevertheless, this approximation is used in the companion paper [4].

Table-3 Example of Dominant Load Combinations
to Limit State for highway bridges

	load combination	limit state (in case of stress)
load-1	D + L	σ_a
load-4	D + L + T	σ_y
load-3	D + E	σ_u
load-6	D + T + E	σ_u

5. CONCLUSIONS

The limit state probability diagram suggests the interrelationship among the limit state, limit state probabilities, target limit state probabilities and load combinations. More importantly, the limit state probability diagram as introduced here provides a much more global interpretation of the safety of a structure. Note that the state of structural response may take undesirable values at different structural locations, depending on the load combinations. Therefore the limit state probability diagram is not necessarily constructed with respect to a specific point in the structure.

Taking all these observations into consideration, the present paper develops an approximate procedure for determining load and resistance factors. In this procedure, an approximate objective function which measures the extent of deviation of limit state probability curves from the target limit state probability curve will be minimized with respect to a representative set of structures for which the design code is being developed. The applicability of the procedure developed here will be demonstrated in a companion paper [4]. The question of how to specify the target limit state probabilities still remains elusive, however.

ACKNOWLEDGEMENT

This study was performed under the activities of Hanshin Expressway Public Corporation's HDL-Committee. The authors wish to thank the late Professor Ichiro Konishi, Chairman of the Committee for his support of this study. The authors also wish to thank all other members of the committee, particularly Mr.H.Ishizaki as the program coordinator for their valuable comments. The first author wishes to acknowledge partial support of National Center for Earthquake Engineering Research forwards his participant in this study under contract NCEER-88-1003.

REFERENCES

[1] ELLINGWOOD,B., GALAMBOS,T.V., MACGREGOR,J.G. and CORNELL,C.A.: "Development of a probability based load criterion for American National Standard A58", NBS Special Publication, No.577, US Dept. of Commerce, 1980

[2] SHINOZUKA,M.: "Load Combination and load resistance factor design", Safety and Quality Assurance of Civil Engineering Structures, Proc. of IABSE Symposium, Tokyo, pp.65-69, Sept. 1986

[3] EMI,S. and AKETA,O.: "Probabilistic Load and Resistance Factor Design", Structural Safety and Reliability, Proc. of ICOSSAR'85, Kobe, pp.II309-318, May 1985

[4] Kawatani,M. et al.: "Reliability-Based LRFD for Bridges : Determination of Load Factors", Proc. of ICOSSAR'89, San Francisco, Aug. 1989

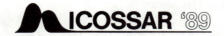
PROBABILISTIC EVALUATION OF LOAD FACTORS FOR STEEL
RIGID—FRAME PIERS ON URBAN EXPRESSWAY NETWORK

W.Shiraki, S.Matsuho and P.N.Takaoka

Dept.of Civil Engineering, Tottori University, Tottori, Japan

ABSTRACT

A procedure is proposed for calculating the optimal load
factors for steel rigid-frame piers using the extended level
2 reliability method. Twelve typical types of pier
structures are selected out of the existing expressway
bridge structures. Four actual load components(dead, live,
temperature, and earthquake load) are considered. By
numerical calculations, the effectiveness of the proposed
procedure is demonstrated.

KEYWORDS
Reliability based design; load factors; rigid-frame piers;
highway bridges; load combinations.

1. INTRODUCTION

Nowadays, it is a world-wide tendency to introduce the LFDM (Load Factor
Design Method) based on the reliability theory into the structural design
standards instead of the conventional ASDM (Allowable Stress Design Method).
In Japan, the practical investigations have been started five or six years ago
for introducing the LFDM into the design standards for highway bridges[1].
With this situation in mind, we are investigating the reliability analyses and
probabilistic design methods of highway bridges [2-8].
    In our early study [5], we pointed out a shortcoming such that the safety
indicies of the pier structures designed by the ASDM differ considerably from
each other, depending on the model type of structure. In this study, having
recourse to the LFDM, a procedure is proposed to determine the optimal values
of load factors for pier structures, for preselected target safety indicies,
for various loading cases (load combinations).
    In analysis, twelve typical types of steel rigid-frame piers are selected
out of the existing actual bridge structures constructed on the Hanshin
(Osaka-Kobe) Expressway Network and are so modeled that they are amenable to
analysis. Four actual load components: dead load (D), live load (L),
temperature load (T), and earthquake load (E), are considered. The three loads
except D are modeled by the Borges-Castanheta (B-C) load processes, based on
observation data on actual loads in Hanshin area.

In numerical calculations, the optimal values of load factors are determined for three preselected target safety indices, and for seven loading cases. Comparing the results of design calculation using the optimal load factors with the results according to the current design code, the effectiveness of the proposed procedure is demonstrated.

2.MODELING OF PIER STRUCTURES

A typical type of highway bridge structural system is selected out of the existing systems on the Hanshin Expressway Network as shown in Fig. 1, where the three-span continuous steel box girder bridges are supported by the steel rigid-frame piers. In this study, the pier structures for transverse direction are considered in analysis. Based on actual data for existing systems, twelve pier structures are so modeled that they are amenable to analysis. In modeling, the combination of three basic parameters, i.e. the span length of superstructure L=40,60,80m, the total height of pier H=10,20m, and the total width of pier W=20,30m, are considered. The principal dimensions and configurations of these twelve models are listed in Table 1 and demonstrated in Figs. 1 through 3.

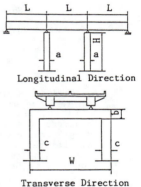

Longitudinal Direction

Transverse Direction

Fig. 1 Typical Type of Highway Bridge Structural System

Table 1 Twelve Models of Rigid-Frame Piers

(unit: m)

Model No.	L	H	W	h	ℓ	a	b	c
1	40.0	10.0	20.0	9.17	18.5	2.00	1.67	1.5
2	"	"	30.0	8.75	28.0	"	2.50	2.0
3	"	20.0	20.0	19.17	18.0	"	1.67	2.0
4	"	"	30.0	18.75	27.5	"	2.50	2.5
5	60.0	10.0	20.0	9.17	18.5	3.00	1.67	1.5
6	"	"	30.0	8.75	28.0	"	2.50	2.0
7	"	20.0	20.0	19.17	18.0	"	1.67	2.0
8	"	"	30.0	18.75	27.5	"	2.50	2.5
9	80.0	10.0	20.0	9.17	18.5	4.00	1.67	1.5
10	"	"	30.0	8.75	28.0	"	2.50	2.0
11	"	20.0	20.0	19.17	18.0	"	1.67	2.0
12	"	"	30.0	18.75	27.5	"	2.50	2.5

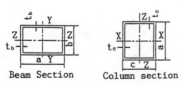

Beam Section   Column section

Fig. 2 Cross-Section of Rigid-Frame Pire

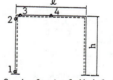

Fig. 3 Analytical Model for Transverse Direction

The wall thickness of beam and column sections, t_b and t_c, of these twelve piers, are determined by the conventional allowable design formats shown in Table 2 [9], and the results of design calculations are summarized in Table 3. In Table 2, D_n, L_n, T_n, E_n= the nominal value of each load component D,L,T and E, respectively; $\alpha_D, \alpha_L, \alpha_T, \alpha_E$ = factors which convert each load

componet into corresponding stress level; ϕ = the augmentation factor of the allowable stress. In the design calculations, four checking points 1 through 4 (see Fig. 3) are considered, and each two different wall thickness t_b, t_b' for beam and t_c, t_c' for column (see Fig. 4), respectively, are considered for more practical analysis. In early study [5], the pier structures with uniform t_c and t_b were analyzed for simplicity.

The characteristics of steel material used for pier structure are as follows; the grade of steel = SM50Y; the allowable stress σ_a = 206 GPa; the yield stress σ_y = 353MPa; the Young's modulus E_o =206 GPa; the linear coefficient of expansion α = 1.2 10^{-5}/$^\circ$C; and the unit weight ρ = 7.69 10^{-2}N/cm^3.

Table 2   Current Design Fomulas

Code	Current Design Formulas	ϕ
1	$\alpha_D \cdot D_a + \alpha_L \cdot L_a \leq \phi \cdot \sigma_a$	1.00
2	$\alpha_D \cdot D_a + \alpha_L \cdot L_a + \alpha_T \cdot T_a \leq \phi \cdot \sigma_a$	1.15
3	$\alpha_D \cdot D_a + \alpha_E \cdot E_a \leq \phi \cdot \sigma_a$	1.50
4	$\alpha_D \cdot D_a + \alpha_T \cdot T_a + \alpha_E \cdot E_a \leq \phi \cdot \sigma_a$	1.70

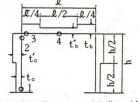

Fig.4 Pier Structure with Nonuniform wall Thickness

Table 3   Results of Design Calculations According to Current Design Code

Model No.	Column Section				Beam Section				Pier Weight
	checking point No.1		checking point No.2		checking point No.3		checking point No.4		
	Code No.	t_c(mm)	Code No.	t_c'(mm)	Code No.	t_b(mm)	Code No.	t_b'(mm)	(kN)
1	2	16.9	1	28.6	1	22.3	1	13.6	488
2	2	26.2	1	38.8	1	28.6	1	21.9	990
3	3	16.4	1	18.8	1	19.0	1	12.5	683
4	3	15.3	1	25.4	1	22.7	1	22.2	1135
5	2	18.5	1	31.0	1	24.2	1	14.8	680
6	2	28.9	1	42.6	1	32.0	1	24.5	1362
7	3	18.7	1	21.1	1	20.7	1	13.4	958
8	3	17.7	1	28.7	1	25.4	1	24.5	1563
9	2	20.0	1	33.3	1	26.0	1	15.9	892
10	2	31.4	1	46.0	1	34.9	1	26.8	1766
11	3	20.7	1	23.0	1	22.5	1	14.3	1263
12	3	19.8	1	31.5	1	27.7	1	26.6	2029
Total									13809

3. MODELING OF ACTUAL LOAD COMPONENTS

Based on the histrical data on earthquakes in Hanshin area, the actual earthquake load E is modeled by the limiting spike type of Borges-Castanheta (B-C) load model. In the same manner, the actual live load L and the actual temperature load T are modeled by the mixed type of B-C load model. The actual dead load is assumed to be deterministic. Based an extensive investigation on actual conditions of various loads acting on urban expressway bridges [1], the characteristics of the B-C process of each load component are determined as follows:

Earthquake Load, E   Actual earthquake load is modeled as $E=S_A/g$, where $S_A=$ linear acceleration response spectrum; and g= acceleration of gravity. The cumulative distribution function (CDF) of S_A is expressed as

for natural frequency of structure =0.5sec

$$F_{SA}(x) = 1 - \exp\left[- \{(x-41.28)/34.24\}^{0.913}\right] \quad (41.28 < x)$$

for natural frequency of structure =0.7sec

$$F_{SA}(x) = 1 - \exp\left[- \{(x-25.88)/26.12\}^{0.879}\right] \quad (25.88 < x)$$

$$\left.\begin{array}{c}\\ \\ \\ \\ \end{array}\right\} \quad (1)$$

for natural frequency of structure =1.0sec

$$F_{SA}(x) = 1 - \exp\left[- \{(x-17.19)/18.05\}^{0.850}\right] \quad (17.91 < x)$$

$$(unit: cm/sec^2)$$

In evaluation of Eq.(1), the occurence of earthquake is assumed to be Poisson process, and its average return period is considered to be greater than 2 years. Furthermore, it is assumed that the magnitude is greater than 5.0, the ground condition is Grade 2 and the damping ratio of structure is 0.05. The uncertainty of attenuation law is not considered.

Dead Load, D   As dead load D, only the own weight of structure is considered and it is assumed to be deterministic. To take its variability into consideration, however, the design value of D is calculated by the formula $D=D'(1+\delta)$, where $D'=$actual weight of the structure calculated on the basis of the unit weight of the material and the volume of the members; and $\delta =0.05$ for the superstructure, 0.10 for pier structure.

Live Load, L   The actual live load is modeled as the support reaction on the piers by using the Monte–Carlo simulation technique. The probability of occurence, p, and the basic time intervals, τ_L, are taken as 0.75 and 6 hours, respectively. Given that the load occurs the CDF of its amplitude, $F_L^*(x)$, is expressed as

$$F_L^*(x) = 1 - \exp\left[- (x/56.49)^{2.342}\right] \quad (x > 0 \ ; \ unit:ton) \quad (2)$$

This CDF is evaluated for two supports on the pier.

Temperature Load, T   Actual temperature load is modeled as the temperature difference such that actual temperature of structure minus 15°C. The parameters p and τ_T are taken as 0.75 and 6 houres, respectively. The CDF of the temperature difference, $F_T^*(x)$, is expressed as

$$F_T^*(x) = 0.5 + 0.5 \ \Phi \ \{ (x-13.2) /4.4\} \quad (x > 0 \ ; \ unit:°C) \quad (3)$$

4. LOAD COMBINATION ANALYSIS AND RELIABILITY ANALYSIS

In load combination analysis and reliability analysis of the model rigid-frame pier structures under combined action of the earthquake, dead, live and temperature load components, the Turkstra's rule in connection with the B–C processes [10] and the extended level 2 reliability method are used. In this study, seven load combination cases of actual load components shown in Table 4 are considered.

(1) **Evaluation of Reliabilty of Pier Structure**   Assuming that the stress σ^* in the ultimate limit state of member is the yieled stress $\sigma_y=353$MPa, the safety index β is evaluated as

$$\beta = (\sigma^* - \sigma_D - \sum_{i=1}^{3} C_{xi} \cdot \mu_{xi}') / \sum_{i=1}^{3} C_{xi} \cdot \sigma_{xi}' \cdot \alpha_i \quad (4)$$

$$\alpha_i = C_{X_i}\, \sigma_{X_i}{}' \Big/ k$$

$$k = \Big(\sum_i C_{X_i}{}^2 \sigma_{X_i}{}'^2 \Big)^{1/2}$$

$$\mu_{X_i}{}' = x_i^* - \Phi^{-1}\{F_{X_i}(x_i^*)\}\cdot\sigma_{X_i}{}'$$

$$\sigma_{X_i}{}' = \phi[\Phi^{-1}\{F_{X_i}(x_i^*)\}]\Big/ f_{X_i}(x_i^*)$$

$$x_i^* = F_{X_i}^{-1}\{\Phi(\beta\,\alpha_i)\}$$

(5)

where X_1, X_2 and X_3= live, temperature and earthquake load, respectively; $x_i^* =$ X_i -coordinate of the design point; F_{X_i} and f_{X_i} = the CDF and the PDF of X_i, respectively; $\Phi(\cdot)$ and $\phi(\cdot)$= the standard normal distribution function and density function, respectively; C_{X_i}= the factor which converts the load X_i into stress level; and σ_D = the deterministic stress for the dead load.

By solving the Eqs.(4) and (5) iteratively for β, the safety index for the four checking points 1, 2, 3 and 4 (see Fig.3), can be determined. In this study , the minimum value among the safety indices for these four checking points is considered to be the safety index of the pier structures.

Table 4 Combination Cases of
Actual Load Component

Case	Actual Load Combinations
1	D + L
2	D + T
3	D + L + T
4	D + E
5	D + L + E
6	D + T + E
7	D + L + T + E

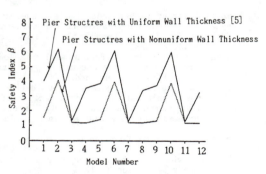

Fig.5 Safty Indices of Twelve Pier Structures
for Load Combination Case 7

(2) <u>Reliability of Pier Structures</u> For the load combination Case 7, the safety index β was calculated for each of the twelve model piers shown in Tables 1 and 2. In numerical calculations, the lifetime of pier structures was assumed to be 50 years. The results obtained are shown by dotted line in Fig. 5. The solid line is the results obtained in our early study [5].

The tendency of variability of β is not so much as the results obtained in [5], but the safety index β considerably differs from each other, depending on the model type of pier structure. For the models Nos. 2,6 and 10, the safety indices are comparatively large, while for other models, the safety indices are very small.

As it was pointed out in our early study [5], the reason for this difference lies in the augmentation factor ϕ =1.50 or 1.70 (see Table 3) of allowable stress which is used in considering the earthquake load. According to the current design code which is based on the ASDM, the allowable stress is uniformly augmented regardless of the type of structure. On the other hand, the model Nos. 2,6 and 10 are not affected significantly by actual load effects due to earthquake compared with other models. Consquently, the

values become larger than those of other models. The current design code does not insure consistent level of safety for difference type of pier structures subjected to earthquake load.

5.PROBABILISTIC EVALUATION OF LOAD FACTORS FOR PIER STRUCTURES

In the previous chapter, the numerical examples pointed out that the conventional allowable design method has a shortcoming. This shortcoming can be greatly reduced by the load factor design method (limit-states design method).

In this chapter, let us propose a procedure for calculating optimal load factors for a target safety index β_T preselected for the ultimate limit state, in such a manner that the preselected value of β_T can be maintained regardless of the model type of structure.

Table 5 lists seven formats of load factor design method, used in this study. γ_D, γ_L, γ_T and γ_E are the load factors for load components D, L, T and E, respectively.

Step 1: A target safety index β_T is preselected.
Step 2: For a load combination case considered, take out the corresponding design format from Table 5, and guess adequately values of the load factors.
Step 3: Using the design format with the load factors guessed in Step 2, cross-sections of the model pier structure i (i= 1,2,.....12; see Table 1) are determined.
Step 4: For the Pier Structure i thus determined, the actual safety index β_i is calculated, under the action of the corresponding loads.
Step 5: Calculate the squared difference Ω between the values of preselected β_T and actual β_i, defined by

$$\Omega = \sum_{i=1}^{m} (\beta_i - \beta_T)^2 \tag{6}$$

in which m= the number of types of model pier structure (in our study m=12, see Table 1).
Step 6: Repeat through Step 2 and 5 until Ω becomes minimum, by re-choosing (variating) a set of load-factor values for each iteration.

According to this procedure, numerical calculations were carried out. In numerical calculations, the target safety indices were taken as β_T=2.5, 3.0 and 3.5.

Table 5 Load Factor Design Formulas

Case	Load Factor Design Formulas
1	$\gamma_D \cdot \alpha_D \cdot D_* + \gamma_L \cdot \alpha_L \cdot L_* \leq \sigma^*$
2	$\gamma_D \cdot \alpha_D \cdot D_* + \gamma_T \cdot \alpha_T \cdot T_* \leq \sigma^*$
3	$\gamma_D \cdot \alpha_D \cdot D_* + \gamma_L \cdot \alpha_L \cdot L_* + \gamma_T \cdot \alpha_T \cdot T_* \leq \sigma^*$
4	$\gamma_D \cdot \alpha_D \cdot D_* + \gamma_E \cdot \alpha_E \cdot E_* \leq \sigma^*$
5	$\gamma_D \cdot \alpha_D \cdot D_* + \gamma_L \cdot \alpha_L \cdot L_* + \gamma_E \cdot \alpha_E \cdot E_* \leq \sigma^*$
6	$\gamma_D \cdot \alpha_D \cdot D_* + \gamma_T \cdot \alpha_T \cdot T_* + \gamma_E \cdot \alpha_E \cdot E_* \leq \sigma^*$
7	$\gamma_D \cdot \alpha_D \cdot D_* + \gamma_L \cdot \alpha_L \cdot L_* + \gamma_T \cdot \alpha_T \cdot T_* + \gamma_E \cdot \alpha_E \cdot E_* \leq \sigma^*$

Table 6 Load Factors for Code 4 (D+E)

β_T	γ_D	γ_E	Ω
2.5	1.05	1.66	0.00009
3.0	1.05	1.96	0.00006
3.5	1.05	2.31	0.00005

Table 7 Load Factors for Code 6 (D+T+E)

β_T	γ_D	γ_T	γ_E	Ω
2.5	1.05	1.06	1.61	0.04802
3.0	1.05	1.10	1.91	0.03128
3.5	1.05	1.14	2.25	0.02169

Table 8 Load Factors for Code 7 (D+L+T+E)

β_T	γ_D	γ_L	γ_T	γ_E	Ω
3.0	1.08	0.54	0.19	1.68	0.02159
	(1.07)	(0.06)	(0.50)	(1.95)	(0.00033)

( ):load factors of piers structures with uniform wall thickness t_b, t_c

Table 9 Influence of the lifetime of Pier Structures on the Load Factors for Code 7, β_T=3.0, and span length L=60m

Lifetime (years)	γ_D	γ_L	γ_T	γ_E	Ω
30	1.07	0.53	0.13	1.61	0.00131
50	1.07	0.50	0.21	1.72	0.00117
70	1.07	0.50	0.20	1.78	0.00076

Table 10 Results of Design Calculations According to Proposed Load Factor Design Formulas for β_T=3.0

Model No.	Column Section				Beam Section				Pier Weight (kN)	Safety Index β
	checking point No.1		checking point No.2		checking point No.3		checking point No.4			
	Code No.	t_c(mm)	Code No.	t_c(mm)	Code No.	t_b(mm)	Code No.	t_b(mm)		
1	7	23.6	7	23.9	7	20.3	7	7.0	446 (91)	2.89
2	7	21.9	7	26.6	7	20.6	7	11.8	683 (69)	3.06
3	7	26.7	7	22.8	7	26.3	7	6.3	887 (130)	3.02
4	7	23.5	7	24.1	7	23.1	7	11.1	1118 (99)	3.08
5	7	25.8	7	26.2	7	22.3	7	7.7	628 (92)	3.04
6	7	24.4	7	29.6	7	23.4	7	13.4	950 (70)	3.05
7	7	29.8	7	25.4	7	28.8	7	6.9	1238 (129)	3.00
8	7	26.5	7	27.2	7	26.0	7	12.5	1545 (99)	3.02
9	7	27.9	7	28.3	7	24.2	7	8.4	829 (93)	3.00
10	7	26.6	7	32.3	7	25.8	7	14.8	1242 (70)	2.99
11	7	32.5	7	27.8	7	33.1	7	7.5	1625 (129)	2.93
12	7	29.2	7	30.0	7	28.6	7	13.7	2010 (99)	2.99
Total									13201 (96)	

( ) : percentage to the pier weight obtained according the current design code

Several calculation results for load factors are shown in Tables 6 through 9. Tables 6 and 7 indicate that the load factor for earthquake γ_E increases with the increase of β_T. In Table 8, the comparison of load factors of pier structures with nonuniform and uniform wall thickness for β_T=3.0 is shown. The influence of the variable wall thickness of member on the load factors is considerably large. Therefore, it should be emphasised that the practical modeling of structures is very important. In calculations above, the lifetime of structures was taken as 50 years. In Table 9, the influence of the lifetime of structures on the load factors is shown for code 7, β_T=3.0 and span length L=60m. This result indicates a slight influence of the lifetime on the load factor of earthquake γ_E.

Table 10 shows the results of design calculation using the proposed optimal load factors. As it is seen from these results that most of the actual safety indices β_i agree well with the target safety index β_T=3.0. The shortcoming inherent in the current ASDM is overcome. The ratio of the total weight of twelve pires which are designed using optimal load factors to the total weight according to the current ASDM is 96%.

6.SUMMARY AND CONCLUSIONS

A probabilistic evaluation of load factors for steel rigid-frame piers supporting three-span continuous box girder bridge is performed using the extended level 2 reliability method. The main results of the analysis are as follows.
  (1) The optimal load factors corresponding to the seven formats were calculated for various lifetimes of pier structures, for various models of actual load components and for various target safety indices.
  (2) The safety indices for model piers which are designed using the optimal load factors obtained above agreed well with the target safety indices.
  (3) The influence of the variable wall thickness of member, and of the lifetime of pier structures on the optimal load factors were demonstrated.
  (4) The results of design calculation using the optimal load factors were compared with results according to the current design code.

ACKNOWLEDGEMENTS

This study was made possible by the use of the observed data offered by the Committee on Design Loads, Hanshin Expressway Public Corporation. The authers are thankful to the Committee for kindness and corporation.

REFERENCES

[1] HDL Committee, Hanshin Expressway Public Corporation: "Report on Investigation of Design Load Systems on Hanshin Expressway Bridges", 1986, (inJapanese).
[2] Matsuho, S., Shiraki, W., Takaoka, N. and Yamamoto, K.: "A Probabilistic Evaluation of Vehicular Loads", Proc. of ICOSSAR'85, pp.I490-494, 1985.
[3] Shiraki, W., Matsuho, S., Takaoka, N. and Yamamoto, K.: "Reliability Analysis of Various Types of Girder Bridges on the Urban Expressway Network Using Theory of Random Processes and Simulation Method", Proc. of ICOSSAR'85, pp.I572-576, 1985.
[4] Takaoka, N., Shiraki, W. and Matsuho, S.: "A Probabilistic Evaluation Method of Load Combination in Structural Design", Reports of the Faculty of Engineering, Tottori University, Vol.19, No.1 pp.55-63, 1985.
[5] Shiraki, W., Matsuho, S. and Takaoka, N.: "Reliability Analysis of Highway Bridges and Modeling of Design Vehicular Load", Jour. of Structural Engineering, Vol.33A, pp.749-760, 1987, (in Japanese).
[6] Shiraki, W., Matsuho, S. and Takaoka, N.: "Load Combination Analysis and Reliability Analysis of Steel Rigid-Frame Piers Supporting Bridges Constructed on Urban Expressway Network", Proc. of ICASP-5, pp.206-213, 1987.
[7] Takaoka, N., Shiraki, W. and Matsuho, S.: "Reliability Analysis of Urban Expressway Bridge Girders and Modeling of Design Vehicular Load", Proc. of 20th Midwestern Mechanics Conf., pp.422-427, 1987.
[8] Takaoka, N., Shiraki, W. and Matsuho, S.: "Safety Evaluation of Highway Bridges Using Observed Data and Simulation Technique", Proc. of 3rd conf. on "Safety of Bridge Structures", pp.381-386, 1987.
[9] Hanshin Expressway Pbulic Corp.: "Design Standards(Ⅱ), 1980 (in Janpanese).
[10] Christensen, P.T. and Baker, M.J.: "Structural Reliability and Its Applications", Springer-Verlag, 1982.

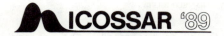
ULTIMATE LIMIT STATE DESIGN
FOR EARTHQUAKE RESISTANT STEEL FRAMES

K. Ohi*

* Inst. of Industrial Science, University of Tokyo, Tokyo, JAPAN

ABSTRACT

This paper describes a basic methodology to assess
the reliability of steel framed structures subjected to ultimate earthquakes. After a discussion on
past damage criteria, a modified energy-based criterion is chosen herein. The important load effect
parameter is the energy input, and it is related to
the Fourier square amplitudes of ground acceleration. Explicit formulas to predict the mean and
the variance of the energy input are presented for
non-stationary random earthquakes. Finally, the
energy-based criterion is modified into a simple
strength checking format for practical purposes.

KEYWORDS

Earthquake Resistant Design; Steel Structure;
Energy Input; Random Vibration; Collapse Mode

1. INTRODUCTION

A widely accepted design philosophy for building structures
subjected to severe earthquakes is to allow them to undergo inelastic deformation but not to reach to a complete collapse. The main
objective of this paper is to describe a basic methodology for such
ultimate earthquake resistant design of steel moment-resisting
frames. The ultimate limit state for a rather ductile steel frame
is reasonably formulated as a reach to an early stage of strength
deterioration after developing a yield hinge mechanism.

Under an earthquake-like repeated loading, several irreversible
plastic deformations are accumulated in the post-buckling behavior
of steel members, and they cause the strength deterioration of the
frame. Then, the complete cumulative damage criterion was used in
the past design method for steel frames[1]. Also for R/C members,
the linear combination of maximum deformation and hysteretic energy
absorption is recently used to indicate structural damage[2]. First
in this paper, the relationship between these criteria are dis-

cussed, and a modified energy criterion is presented. The energy input still plays a significant roll in this criterion, and a simple method to predict its statistical parameters is also presented under random earthquakes. Finally, the energy-based criterion is converted into a strength design format for practical purposes.

2. DAMAGE CRITERIA OF STRUCTURAL ELEMENTS

The safety checking format based on the damage index proposed for R/C structural elements by Park and Ang[2] has the following original form:

$$D_a \geq \frac{\delta_M}{\delta_U} + \frac{\beta}{\delta_U Q_Y} \int dE_P \qquad (1)$$

where D_a : allowable damage index
  δ_M : maximum deformation
  δ_U : ultimate deformation in static loading
  Q_Y : yield strength
  β : a non-negative constant
  $\int dE_P$: hysteretic energy absorption excluding potential energy

The ultimate limit state for a steel structural elements is defined herein as an early stage of strength deterioration. Before this stage the steel member can carry its initial yield strength. Then, the same idea with eq.(1) is also expressed as follows:

$$\delta_{UP} \geq (1-\zeta) \delta_{MP} + 0.5 \zeta \int |d\delta_P| \qquad (2)$$

where $\int |d\delta_P|$: cumulative plastic deformation
  δ_{UP}, δ_{MP} : excluding elastic deformation from δ_U, δ_M
  ζ : a non-negative parameter equal or less than 1.0, which indicate how cumulative the damage model is.

The format given by setting ζ =1 in eq.(2), which corresponds to the infinite values of D_a and β in eq.(1), agrees with the complete cumulative damage criterion for steel structural elements used by Akiyama[1]. The second term coefficient 0.5 is derived from the assumption that the cumulative plastic deformation will occur equally in the both directions before the strength deterioration.

The two response variables in eq.(2), δ_{MP} and $\int |d\delta_P|$, are apparently defined in quite different manners: the former is the extreme value and the latter is integrated from the whole history of the displacement. It is experienced from response analyses, however, there is a considerable positive correlation between them. Fig.1 shows a typical correlation between the two variables observed in the response data of an elastoplastic SDOF system to earthquake-like random excitations. Considering such positive correlation, the following format modified from eq.(2) is adopted instead of dealing with the two explicit response variables independently.

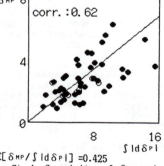

corr. : 0.62

δ_{MP}

$\int |d\delta_P|$

$E[\delta_{MP} / \int |d\delta_P|] = 0.425$

Fig.1 Correlation of δ_{MP} and $\int |d\delta_P|$

$$E_U \geqq E_P \ R_1 \tag{3}$$

where E_U : energy absorption capacity of steel member under monotonic loading (regarded as the resistance)

E_P : hysteretic energy absorption of steel member under seismic loading (regarded as the load effect)

R_1 : reduction factor which is a function of the displacement history profile represented by the term $\delta_{MP}/\int |d\delta_P|$ and a choice of the cumulative damage criterion:

$$R_1 = 0.5 \ \zeta + (1-\zeta) \ \delta_{MP}/\int |d\delta_P| \tag{4}$$

Obviously, (i) R_1 would not depend on ζ if $\delta_{MP}/\int |d\delta_P| = 0.5$, and also (ii) R_1 would not depend on $\delta_{MP}/\int |d\delta_P|$ if $\zeta = 1$. The uncertainty of R_1 parameter would decrease in these situations as well as with the high correlation between δ_{MP} and $\int |d\delta_P|$. Both of these situations would be more plausible for steel members than for R/C members, because $\delta_{MP}/\int |d\delta_P|$ close to 0.5 as shown in Fig.1 is often experienced in seismic responses of a building with medium or long period, and also because an irreversible plastic deformation is believed to appear more significantly in the post-buckling behavior of steel members.

For a single-element structure (more detailed discussion will be given later for a frame system), the format is rewritten as:

$$E_U \geqq E_I \ R_0 \ R_1 \tag{5}$$

where E_I : energy input fed into a structural system by earthquakes

R_0 : reduction factor due to energy dissipation of miscellaneous damping effects

3. ENERGY INPUT MADE BY RANDOM EXCITATION

Consider a SDOF oscillator subjected to a ground acceleration denoted by $y(t)$. Energy input E_I is defined as the work done by the effective excitation term $-m \ y(t)$ after Akiyama[1]:

$$E_I = -m \int v(t) \ y(t) \ dt \tag{6}$$

where m : mass of SDOF oscillator

v : response velocity relative to the ground

$\int$: integral from minus infinite to plus infinite (common expression for all the integrals hereafter.)

According to the frequency-domain expression[3] by use of the power theorem in the Fourier transformation, eq.(6) is rewritten as:

$$E_I = \int W(\omega) \ |Y(\omega)|^2 \ d\omega \tag{7}$$

$$\text{with } W(\omega) = -m \ Real(H(\omega))/2\pi \tag{8}$$

where ω : circular frequency

$Y(\omega)$: Fourier transformation of $y(t)$

$H(\omega)$: complex transfer function to obtain velocity response from the ground acceleration

Eq.(7) implies that an energy input to an oscillator is obtained by integrating the Fourier square amplitudes of the ground acceleration together with the weighting function $W(\omega)$, which is called 'energy admittance[3]' of the oscillator. For a passive oscillator, the area of the energy admittance is a half of the oscillator mass:

$$\int W(\omega)\,d\omega = m/2 \qquad (9)$$

The energy admittance of a viscously damped linear SDOF system (Fig.2) is given as:

$$W(\omega)=\frac{m\,h\,\omega_0\,\omega^2}{\pi\,[(\omega^2-\omega_0^2)^2 + 4\,h^2\,\omega_0^2\,\omega^2]} \qquad (10)$$

where h: fraction of critical damping; ω_0: natural circular frequency

By assuming the oscillator properties are deterministic, the mean and the variance of the energy input can be obtained by the following formulas:

$$E[E_I] = \int W(\omega)\,E[|Y(\omega)|^2]\,d\omega \qquad (11)$$

$$V[E_I] = \int\int W(\omega_1)W(\omega_2)CV_{Y:2}(\omega_1,\omega_2)\,d\omega_1 d\omega_2 \qquad (12)$$

$$CV_{Y:2}(\omega_1,\omega_2)=E[|Y(\omega_1)|^2|Y(\omega_2)|^2] -E[|Y(\omega_1)|^2]E[|Y(\omega_2)|^2] \qquad (13)$$

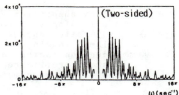

Fourier Square Amplitude Spectrum of Ground Acceleration

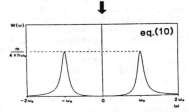

Energy Admittance of Structural System

$$E_I = \int W(\omega)\,|Y(\omega)|^2\,d\omega$$

Energy Input to Structural System

Fig.2 Frequency-domain Analysis of Energy Input

When the mean and the covariance of the Fourier square amplitudes of ground acceleration can be evaluated from an suitable set of earthquake records, the above two formulas provide a direct evaluation of the load effects in the energy-based limit state design. Considering the lack of such information about destructive earthquakes, a classical non-stationary random process is used herein for the earthquake model: a white noise $n(t)$ is first multiplied by a deterministic shape function $a(t)$, and then modulated by a deterministic linear filter. For this type of ground acceleration process, eq.(11) leads to:

$$E[E_I] = 2\,\pi\,S_0\,A_2(0)\int W(\omega)|F(\omega)|^2\,d\omega \qquad (14)$$

where S_0: constant power spectral density of white noise $n(t)$
        $A_2(\omega)$: Fourier transformation of square shape function that is,
        $F(\omega)$: transfer function of deterministic linear filter

Corresponding to a few proposed filter shapes[4][5][6], explicit formulas for the expected energy input fed into a viscously damped linear oscillator can be derived from eq.(14), and the results are shown in Table 1.

Hereafter, the case of no filter will be discussed for the simplicity. The expected energy input in this case is obtained by substituting eq.(9) and $F(\omega)=1$ into eq.(14):

$$E[E_I] = \pi \ m \ S_0 \ A_2(0) \tag{15}$$

By assuming n(t) is Gaussian, the following variance formula is derived from eqs.(12) and (13):

$$V[E_I] = 8 \ \pi^2 \ S_0^2 \ \int \ U(\omega) \ |A_2(\omega)|^2 \ d\omega \tag{16}$$

where $U(\omega) = \int W(u)W(u+\omega) \ du \tag{17}$

$$= \frac{m^2 \ h \ \omega_0}{4\pi(1-h^2)} \cdot \left[\frac{(3-4h^2)\omega^2 - 4\omega_0^2}{(\omega^2-4\omega_0^2)^2 + 16 \ h^2 \ \omega_0^2 \ \omega^2} + \frac{1}{4 \ h^2 \ \omega_0^2+\omega^2} \right]$$

(for a viscously damped linear system)

Corresponding to the two proposed shape function[6][7], explicit COV formulas for the energy input to a viscously damped linear system can be derived from eqs.(16) and (17), and the results are shown in Table 2. These formulas are also applicable to filtered noise if a slowly varying function of ω is used for $|F(\omega)|^2$. (Note: As for more sophisticated excitations including non-Gaussian process, eqs.(11)-(13) shall be directly used.)

Table 1 Expected Energy Input to Viscously Damped Linear System

Ref.	$|F(\omega)|^2$	$E[E_I] \ / \ \pi \ m \ S_0 \ A_2(0)$
[4]	$\dfrac{\omega_g^4}{(\omega_g^2 - \omega^2)^2 + 4h_g^2\omega_g^2\omega^2}$	$\dfrac{\omega_g^3 \ (\omega_0 h + \omega_g h_g)}{h_g \ [(\omega_g^2-\omega_0^2)^2+4\omega_g\omega_0 h_g h(\omega_0^2+\omega_g^2)+4\omega_g^2\omega_0^2(h_g^2+h^2)]}$
[5]	$\dfrac{\omega_g^4 + 4h_g^2\omega_g^2\omega^2}{(\omega_g^2 - \omega^2)^2 + 4h_g^2\omega_g^2\omega^2}$	$\dfrac{\omega_g^2 \ [\ \omega_g(\omega_0 h+\omega_g h_g) + 4\omega_0 h_g^2(\omega_0 h_g+\omega_g h) \ ]}{h_g \ [(\omega_g^2-\omega_0^2)^2+4\omega_g\omega_0 h_g h(\omega_0^2+\omega_g^2)+4\omega_g^2\omega_0^2(h_g^2+h^2)]}$
[6]	$\dfrac{\alpha \ \omega_g \ (\alpha^2 + \omega_g^2 + \omega^2)}{(\alpha^2 + \omega_g^2 - \omega^2)^2 + 4\alpha^2\omega^2}$	$\dfrac{\omega_g \ [\ \alpha(\omega_g^2+\alpha^2+\omega_0^2) + 2\omega_0 h(\alpha^2+\omega_g^2) \ ]}{(\alpha^2+\omega_g^2-\omega_0^2)^2+4\omega_0(\alpha+\omega_0 h)[\alpha\omega_0+h(\alpha^2+\omega_g^2)]}$
	1 (No filter)	1

Table 2 COV of Energy Input to Viscously Damped Linear System

Ref.	a(t)	COV $[E_I]$ ($\sqrt{V[E_I]}$ / $E[E_I]$)	
[7]	$a_1 \cdot t \ e^{-ct}$	Square root of $\dfrac{c}{8(1-h^2)} \cdot \left[\dfrac{8c^2+9c\omega_0 h+3\omega_0^2 h^2}{(c+\omega_0 h)^3} + \right.$	
		$\left. \dfrac{7c^5(1-2h^2)+c^3[c(1-2h^2)+\omega_0 h](c+2\omega_0 h)-\omega_0 h(3\omega_0^2+8c^2+6ch\omega_0)\cdot(2ch+\omega_0)^2}{(c^2+\omega_0^2+2c\omega_0 h)^3} \right]$	
[6]	$a_2 \cdot e^{-ct}$	$\sqrt{\dfrac{c}{1-h^2} \cdot \left[\dfrac{1}{c+\omega_0 h} + \dfrac{c(1-2h^2)-\omega_0 h}{c^2+\omega_0^2+2c\omega_0 h} \right]}$	

As for a MDOF oscillator, the total energy input   can be
defined by the sum of the energy inputs at all the degrees of
freedom in any generalized coordinate system:

$$E_I = {}_i\Sigma\ E_{Ii} = {}_j\Sigma\ E_I{}^{(j)}$$

(18)

where E_{Ii}: i-th energy input described in a usual coordinate system
         relative to the ground.

$$E_{Ii} = - m_i \int v_i(t)\ \ddot{y}(t)\ dt$$

(19)

$E_I{}^{(j)}$: j-th energy input described in a classical normal mode
         system, i.e. the energy input made by the same ground
         acceleration to a SDOF system with j-th effective mass,
         j-th natural frequency, and j-th modal damping factor.

After evaluating the mean and the variance of   by the
SDOF formulas already given, the mean and the variance of the total
energy input   can be obtained by the following formulas:

$$E[E_I] = {}_j\Sigma\ E[E_I{}^{(j)}]$$

(20)

$$V[E_I] \doteqdot {}_j\Sigma\ V[E_I{}^{(j)}] \qquad \text{(independence of each } E_I{}^{(j)} \text{ is assumed)} \quad (21)$$

Considering that the sum of each modal effective mass is equal
to the total mass denoted by M, the following expression of the
mean total energy input can be obtained to a white noise process
modified by a deterministic shape function:

$$E[E_I] = \pi\ M\ S_0\ A_2(0)$$

(22)

As for an oscillator with hysteresis, any equivalent linearization
technique can be applied to the results of the linear system, but it is
often unnecessary to obtain the exact
values of the equivalent linear parameters for the practical purposes:
(1) In case a slowly varying filter is
used, the mean of the energy input is
not sensitive to the equivalent-
linear damping but mainly controlled
by the equivalent-linear period. A
modification for short period range
shown in Fig.3(a) may be conservative
enough. (2) The COV of the energy
input can be rounded up as a con-
stant(0.4-0.5) as shown in Fig. 3(b).
This value depends on the duration.

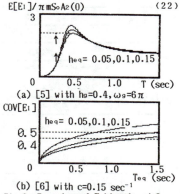

$E[E_I]/\pi m S_0 A_2(0)$

$h_{eq} = 0.05, 0.1, 0.15$

(a) [5] with h_g=0.4, ω_g=6π

$COV[E_I]$

h_{eq}= 0.05, 0.1, 0.15

(b) [6] with c=0.15 sec^{-1}

Fig.3 Examples of Tables 1 and 2

4. ENERGY ABSORPTION CAPACITY WITH RESPECT TO COLLAPSE MODE

Past experimental results on steel structural members enable
at least to evaluate their ductility under monotonic loading.
Recently, an attempt is being made to remake empirical formulas on
such ductility and also to evaluate their prediction errors[8]. For
example, Fig.4 shows the correlation between one of such formulas
and the past experimental data in Japan, where the ductility of H-
shaped beam bounded by the lateral buckling failure is focused[9].

A energy absorption capacity of a steel frame depends not only on the deformability of each members but also on the collapse mode which occurs during earthquakes. Strictly speaking, the energy absorption capacity of the frame should be evaluated for all the possible collapse modes, and the reliability of the frame should be assessed by considering the probabilities of occurrence for each collapse modes. Since there remain much difficulties to do this, it would be a practical strategy so far to check the following format only for the most likely collapse mode:

$$\theta_U \cdot {}_k\Sigma \ M_{Pk} \geqq E_I \ R_0 \ R_1 \tag{23}$$

where ${}_k\Sigma$: summation over all the yield hinge points formed in the most likely collapse mode.
         M_{Pk}: moment resistance of the k-th yield hinge
         θ_U: the minimum plastic rotation capacity among the yield hinge points formed in the most likely collapse mode.

The most likely collapse mode may be searched by one of the proposed methods[10] after proportioning of each members in a preliminary design. A simplified analysis is recommended, however, to search the most likely collapse mode in such preliminary design process. It is well known that an elastic fundamental mode is often very dominant in the dynamic loading path during an earthquake for a low-rise steel building, even if it undergoes a moderate inelastic deformation[11]. This property allows a static analysis under a proportional load to search the most likely collapse mode[12]. In this design process, the energy-based criterion eq.(23) can be rewritten into the following strength design format:

$$Q_B \geqq E_I \ R_0 \ R_1 \diagup \theta_U \ H \ \{p\}^T\{\delta_P\} \tag{24}$$

where Q_B: base shear strength, this strength should be obtained by plastic analysis under a static loading proportional to the vector $\{p\}$.
         H: story height at the lowest story where collapse occurs.
         $\{p\}$: proportional load vector normalized by base shear force.
         $\{\delta_P\}$: plastic displacement vector in the obtained collapse mode, which is normalized by the plastic displacement at the lowest story.

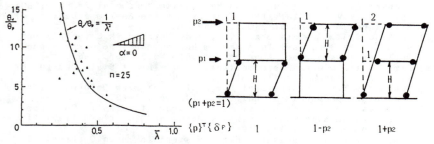

Fig.4 Beam Ductility   (Nakamura[9])   Fig.5 Collapse Modes in 2-story Frame
                                                   and $\{p\}^T\{\delta_P\}$ values

These parameters for two-story frame are illustrated in Fig.5. It is noteworthy that the proportional load vector need not represent the dynamic load process precisely, so long as it can detect a proper collapse mode. The design process, which satisfies eq.(24) under any proportional load vector, will also satisfy the overall energy criterion eq.(23) in the detected collapse mode. The scalar product of $\{p\}$ and $\{\delta_P\}$ is equal to 1.0, if the local collapse occurs at the first story. This product will have a greater value for the overall collapse in case of the weak-beam frame, and this will result in a smaller required strength.

5. CONCLUDING REMARKS

(i) After comparison of Park-Ang's damage criterion and Akiyama's, the former can be interpreted as a reduction of insignificant portions from the total amount of hysteretic energy absorption. A modified energy-based criterion is used herein for steel members.
(ii) Statistical parameters of energy input can be related to those of Fourier square amplitudes of ground acceleration by the concepts of energy admittance. Several formulas are presented to predict the mean and variance of the energy input under non-stationary random excitation. Especially for the white noise modified by a shape function, the mean energy input does not depend on the parameter of the structure except the total mass.
(iii) The energy-based criterion can be converted into a simple strength checking format in a preliminary design process, where only a plastic analysis under static proportional load is required.

REFERENCES

[1] Akiyama,H.: Aseismic Ultimate Design of Building Structures, Tokyo University Press, Tokyo, JAPAN, 1980.
[2] Park,Y.-J. and Ang,A.H.-S.: "Mechanistic Seismic Damage Model for Reinforced Concrete," ASCE, Jour.of Struct.Engrg., Vol.111, No.4, pp.722-739, April 1985.
[3] Ohi,K. and Tanaka,H.:"Frequency-domain Analysis of Energy Input Made by Earthquakes," Proc. of 8WCEE, San Francisco, 1984.
[4] Amin, M. and Ang, A.H.-S.: "Non-stationary Stochastic Model of Earthquake Motions," ASCE, Jour.of Engrg.Mech., No.2, Apr.1968.
[5] Tajimi,H.: "Statistical Method of Determining the Maximum Responses of Building During an Earthquake," Proc.of 2WCEE, 1960.
[6] Bolotin, V.V.: "Statistical Theory of the Aseismic Design of Structures," Proc. of 2WCEE, 1960.
[7] Bogdanoff, J.L., Goldberg, J.E., and Bernard, M.C.: "Response of a Single-Structure to a Random Earthquake-Type Disturbance," Bull. of Seismo. Soc. America, Vol.51, No.2, Apr. 1961.
[8] Limit State Design Sub-committee, AIJ: Limit State Design for Steel Structures (Draft), AIJ, Oct. 1988.
[9] Nakamura, T.: "Statistical Evaluation of Beam Deformability," Commentary prepared for Ref.[8], AIJ, Dec. 1983.
[10] Ditlevsen,O.: "Reliability with respect to Plastic Collapse," Proc.of ASCE Spec.Conf. on Prob. Mech. and Struct. Reliability, pp.57-60, Berkeley, Jan. 1984.
[11] Ohi, K. and Takanashi, K.: "Hysteresis Loops Observed in Earthquake Response Tests on Steel Frame Models," Trans. of AIJ, Jour.of Struct. and Construct.Engrg., No.394, Dec. 1988.
[12] Ohi,K., Gang,Y., Takanashi,K: "Failure Mode Control of Steel Frame Subjected to Random Ground Motions," JCOSSAR, Tokyo, 1987.

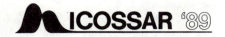

CODE-IMPLIED STRUCTURAL SAFETY FOR EARTHQUAKE LOADING

David Elms*, Tom Paulay* and Sachio Ogawa**

*   Professor of Civil Engineering, University of
     Canterbury, Christchurch, New Zealand

**   Engineer, Shimizu Corporation, Mita 43 Mori Bldg,
     13-16, Mita 3-chome, Minato-ku, Tokyo 108, Japan

ABSTRACT

A method is described for calculating the probability of failure of reinforced concrete frame
structures due to earthquake loading, for use in
code calibration exercises. The failure criterion is
taken to be maximum interstorey displacement. This
is related to basic structural variables through the
medium of the cumulative inelastic energy or damage
energy for each storey, computed by first finding
the total damage energy for a structure through the
use of single degree of freedom inelastic analyses,
and then by determining the fraction of energy in
each storey by an elastic random vibration analysis.
The procedure is checked against the results of full
time history analyses and is found to give
satisfactory results.

KEYWORDS

Building codes, reinforced concrete, earthquakes,
failure, probability, risk

1.   INTRODUCTION

A major problem in the development of structural codes in limit
states or LRFD format is to ensure the code is internally
balanced; that is, that the design equations for different load

for, say, wind load provisions to be more severe than the requirements for gravity or earthquake loads, taken over the total range of structures to which the code applies.

Precisely what is meant by "internally balanced" needs careful and detailed discussion and will not be dealt with here. Its relevance to this paper is that in achieving it, as part of a code calibration process, reliability indices or probabilities of failure must be calculated for a series of different trial structures designed to the load combinations of the proposed code.

For most loadings, such reliability calculations are reasonably straightforward using a first-order second-moment (FOSM) approach (National Bureau of Standards, 1980). However, the FOSM method cannot sensibly be used for earthquake loads, which is a serious problem for countries such as New Zealand where earthquake considerations dominate the design of most structures. The difficulty arises from the fact that the FOSM approach can only be applied easily to the very simplest of structures. Beyond that, matters rapidly become complicated (Ogawa and Elms, 1986). For gravity and perhaps wind loads, the reliability of single members alone is sufficient, so that the FOSM technique can be used. Structural failure due to earthquake, on the other hand, is more a function of the behaviour of the structure as a whole, and not of a single member. The problem is made more acute by the fact that the behaviour is dynamic, not static, and can be understood in detail only by considering complete time histories of inelastic response for a series of different earthquake motions. For these reasons, it is exceedingly difficult to assess structural reliability due to earthquake loading.

Most code development exercises, such as the major effort by the National Bureau of Standards (1980), have merely noted the problem and bypassed it as being too hard. However, this is unsatisfactory as it means that major sections of supposedly rational codes are ill-founded and unreliable. The consistency of each code as a whole then becomes suspect, following the "Principle of consistent crudeness" (Elms, 1985) which essentially says that the quality of any model is primarily governed by that of its crudest part.

There is thus a major need for the development of a relatively simple and speedy technique for the assessment of the reliability of structures subjected to earthquakes, for code calibration purposes. The present paper discusses an attempt to fulfil the need. It is restricted in this instance to reinforced concrete structures, though in principle the approach could be adopted for any material.

2.   DEVELOPMENT OF A PERFORMANCE FUNCTION

2.1   Failure Criterion

Structural failure in an earthquake can happen in many ways, so that a straightforward failure criterion is not immediately

obvious. The situation is simplified by considering only collapse of the structure as a whole and, further, that collapse only occurs due to excessive sway of a single storey. It can be assumed that in New Zealand column shear or crushing failures would not occur because of the common use of the capacity design method, a dominated-mode design approach which would ensure the only failure to occur would be hinging in beam members. In addition, it is assumed that the critical parameter is the residual storey sway deformation after an earthquake rather than an instantaneous maximum occurring during shaking, because instability, which will ultimately lead to collapse, is essentially a static rather than a dynamic problem.

The two factors which will primarily lead to storey sway collapse for a reinforced concrete building are the deterioration of strength due to large imposed deformation and, to cyclic loading, and to the so-called P-δ effect. It is difficult to predict the deterioration quantitatively because the number of cycles for different deformation levels depends on the earthquake, and the deterioration for a given number of cycles at a given deformation level depends on the detailing and materials used and varies between structural members.

A simple assumption based on laboratory tests is that strength would degrade rapidly if the maximum interstorey deflection angle γ_{max} exceeds 3/100. This is taken as the basic failure criterion for reinforced concrete frame buildings.

2.2   Choice of a Performance Function

Unfortunately, from the point of view of calculating a probability of failure, it is not possible to write a performance function using the failure criterion directly, because of the difficulty of expressing γ_{max} in terms of basic load and resistance variables. A new measure of performance must thus be found. Such a measure must fulfil four requirements:

(a)   It must be able to be used in a performance function.

(b)   It must describe the performance of the structure as a whole and be directly relatable to structural damage.

(c)   It must be expressable in terms of the basic load and resistance variables.

(d)   It must be a function of only a relatively few basic variables to allow a ready means of calculation.

The quantity chosen to fulfil these requirements is cumulative plastic strain energy. We shall call it the "damage energy", E_d. It may be defined as the non-recoverable energy due to plastic deformation integrated over the entire time history of an earthquake. It is very similar to the damage measure developed by Akiyama (1985).

However, in practice failure will relate not to the overall damage energy but to the damage energy concentrated in a critical storey. The approach used for obtaining storey damage energy is first to find the damage energy for the whole structure and then to obtain the fraction of that energy in each storey.

2.3 Total Damage Energy

The strategy adopted is to obtain the overall damage energy by carrying out time history analyses of a set of single degree of freedom elasto-plastic models each corresponding to one of the first few modes of elastic vibration of the complete structure.

The total energy input E_T for an _elastic_ multi degree of freedom (MDOF) system can be expressed in terms of the energy inputs $_s E_{Tj}$ for single degree of freedom (SDOF) systems corresponding to the modes j of the MDOF system by the relation

$$E_T = \sum_{j=1}^{n} \left({_s}E_{Tj} \cdot {_{ef.}}M_j \right) \qquad (1)$$

where $_{ef.}M_j$ is the effective mass of the jth mode. We now make two assumptions: that Eq.1 holds for the total energy input of inelastic systems, and that the total energy input for inelastic problems is dominated by damage energy. Using these assumptions, the total damage energy E_{dT} absorbed by a MDOF system becomes

$$E_{dT} = \sum_{j=1}^{n} \left({_s}E_{dj} \cdot {_{ef.}}M_j \right) \qquad (2)$$

where $_s E_{dj}$ is the cumulative plastic strain energy input into a SDOF system with unit mass corresponding to the jth mode of the MDOF system. The damage energy can be standardised by dividing by the structure weight w_T. The ratio E_{dT}/w_T will be called the "standard damage energy".

Equation (2) and its underlying assumptions were checked by comparison with the results of full time history analyses. For this purpose, 3, 7 and 30-storey reinforced concrete buildings were designed in some detail using New Zealand code requirements and the capacity design (i.e. weak beam) approach. The three models were subjected to five earthquake records, and the resulting standard damage energy values were found by Eq. (2) and by full time history analyses. Figure 1 shows the results. The earthquake records used were, from top to bottom, El Centro 1940 N-S, Parkfield 1966 N65E, San Fernando 1971 Pacoima Dam S14W, Bucharest 1977 N-S and El Centro 1979 Imperial County Services Building N-S. It can be seen that the time history results, shown by the chained lines, are very close to the results of Eq. (2), so verifying the model. The diagram also shows that, particularly for the 30-storey model, the effect of higher modes is significant. This contrasts with Akiyama's (1985) conclusion that the first mode alone is important. His analysis, however, did not consider long-period buildings.

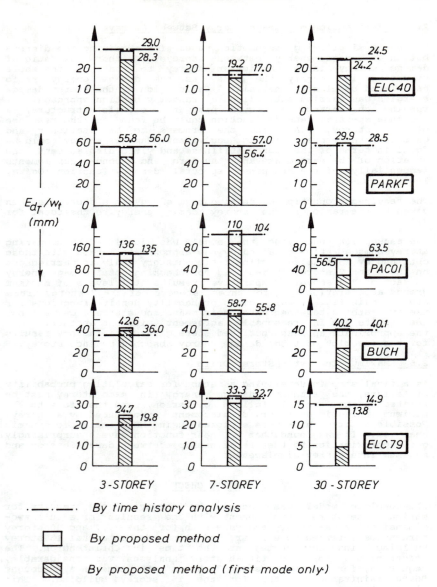

— · — By time history analysis

☐ By proposed method

▨ By proposed method (first mode only)

Figure 1 Comparison of total standard damage energy values

2.4 Distribution of Damage Energy Between Storeys

The basic simplifying assumption used for obtaining the distribution of damage energy between storeys is that the ratio of damage energy in a particular storey to that of the whole structure is roughly the same as the damage energy ratio calculated by elastic analysis. Calculation of the storey damage energies was carried out using a random vibration approach. By running an earthquake record past the elastic structure, a response spectral density function could be found for the response of each storey. Assuming the process to be stationary and Gaussian, the storey elastic equivalent damage energy could be found in terms of the storey stiffness and yield displacement, the duration of the earthquake and the zero- and second-order moments (essentially) of the response spectral density function (Ogawa, 1988).

The required proportion of total damage energy in a storey is then given by the ratio of the storey damage energy to the total for all storeys.

The assumption underlying the model was justified by comparing storey damage ratios obtained by the approximate method with those obtained by a detailed nonlinear time-history analysis carried out on the 3 and 7 storey frames used for checking total damage energy in Section 2.3 above. Comparative results at 3 levels of maximum ground acceleration γ_{max} for three storeys of the 7 storey frame are shown in Fig. 2, which gives probability density functions for the two methods together with their means and standard deviations. The results are in reasonable agreement with one another. Thus the approximate approach outlined here gives satisfactory results for the fraction of total damage energy absorbed by each storey.

2.5 Maximum Inelastic Storey Drift

As a final step in developing a method for calculating probability of failure, the standard damage energy in each storey must be related to the overall failure criterion, and therefore to the maximum inelastic drift or displacement experienced by a storey. Possible relationships were explored using MDOF inelastic dynamic analyses. It was found that a linear function gave a surprisingly good approximation to the relationship between damage energy and maximum interstorey displacement.

3. MODEL CHECK

The complete model was checked by comparing the results for maximum interstorey displacement with the results for a full two-dimensional time-history analysis both for the 3, 7 and 30 storey structures referred to earlier, and also for an actual 15 storey building being constructed at the time in Christchurch. The effort needed for the time-history analyses was considerable, requiring, for instance, some 7 hours CPU time on a Burroughs B6900 mainframe computer for the 15 storey building. This illustrates the need for a simplified method.

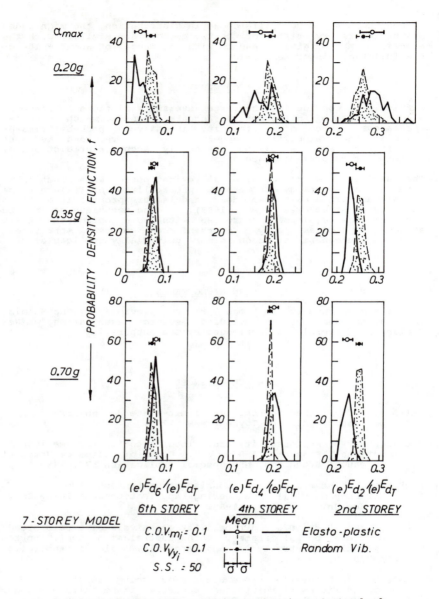

Figure 2   Comparison of storey distribution of standard damage energy

Though there are some differences in distribution, the magnitudes of the interstorey drifts for the two methods are similar, with the simplified method giving results with a range of about ± 10% of the full time-history analysis.

4. CONCLUSIONS

Particularly considering the large overall variations in response due to the use of different earthquake records, which is by far the largest source of variability in analysing probabilities of failure, the simplified model can be recorded as most satisfactory, and as giving reasonably accurate predictions of maximum interstorey displacements.

The next step is to develop a revised performance function in terms of damage energy to enable probabilities of failure to be calculated. This has been done and failure probabilities found for several structures (Ogawa, 1988). The technique can now be used for code development purposes with some confidence. As it stands it applies only to reinforced concrete structures. However, the general approach is more universally applicable.

5. ACKNOWLEDGEMENTS

The generous support for the project provided by the Shimizu Corporation and by the Building Research Association of New Zealand is gratefully acknowledged by the authors.

6. REFERENCES

AKIYAMA, H., "Earthquake-resistant limit-state design for building", Univ. of Toyko Press, 1985.

"Development of a probability based load criterion for American National Standard A58", U.S. Dept of Commerce/National Bureau of Standards, NBS Special Publication 577, 1980.

ELMS, D.G., "The Principle of Consistent Crudeness", Proc. Workshop on Civil Engineering Applications of Fuzzy Sets, Purdue Univ. pp 35-44, 1985.

OGAWA, S. and ELMS, D.G., "Probabilistic analysis of a reinforced concrete portal frame", 10th Australasian Conf. on the Mechanics of structures and Materials, Adelaide, Australia, pp 75-80, 1986.

OGAWA, S., "A simplified procedure for assessing failure probabilities of reinforced concrete frame buildings under earthquake loading", Civil Engineering Research Report 88/11, University of Canterbury, New Zealand, 1988.

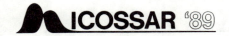
PRESENT AND FUTURE DEVELOPMENTS IN STEEL DESIGN CODES

Theodore V. Galambos, F. ASCE
Civil and Mineral Engineering Department, University of Minnesota,
Minneapolis, Minnesota, U.S.A.

ABSTRACT

Steel structures design codes with load and resistance
factors (or "partial factors") which are based on first-
order probabilistic concepts are now in use in many parts
of the world. This paper will look first at the current
design codes and their level of reliability and economy.
Special emphasis will be on codes used in the USA.
Following this, the current and planned code-development
activity is discussed and the impact on the future methods
of structural design is explored. Finally, some needed
research for a better rationalization of the steel design
codes is presented.

KEYWORDS

steel structures, design codes, reliability, calibration

1. Introduction

Design "codes", also called "standards" or "specifications",
evolved from rules in use by individual designers or contractors in
the early part of this century because it was necessary to establish
uniformity of cost and safety. Such codes continue to evolve as we
change our way of building and as we learn more about the behavior
of the structures and the environment under which they exist.

The past three decades (1960's to 1980's) have seen a
transition in structural design codes from an "allowable stress"
philosophy to a "limit states" philosophy, starting with the
concrete design codes in the USA and continuing on with the various
steel design standards, and, hopefully, ending the decade of the
1980's with new design methods for masonry and timber structures.
This paper will describe the present status of steel design codes
and it will speculate on possible future developments.

2. Present (1989) Status of Codes

In most countries design codes are national standards, while in the
USA they are largely industry sponsored specifications which are
adopted in whole or in part by national or regional building
authorities.
Regardless of the individual difference, such codes are approved
through various levels, starting with a committee of "experts" and
ending up with a legal authority, so that the final document is a
consensus of the opinion of all parties concerned. Because of this,
changes in codes are extremely slow. Exceptions are when an error is
discovered and public safety demands a quick remedy.

One of the most difficult concepts to get through the filter of the
consensus process has been the change from allowable stress design
(ASD) to limit states design (LSD). The former describes the status
of the structure in its everyday condition: loads are credible and
the structure behaves linearly. The latter considers the hopefully
never-to-be-encountered collapse of the structure where loads are
catastrophic and structural behavior is non-linear.

This process has been a particularly time-consuming one for the
specification of the American Institute of Steel Construction (AISC),
which regulates the design of hot-rolled steel building structures.
The AISC issued an entirely new design specification in 1961 which
was a limit states design specification in all but name and format.
This code remained essentially unchanged for a quarter century.
During this time those who wanted to believe in allowable stress
design could do so in confidence because the format was ASD, while
those who liked limit states design could bask in the assurance that
it really was such a code because it was based on LSD philosophy.
While the arrangement maximized the number of satisfied users, it was
a difficult task to teach engineering students the duality of this
design code. In 1986 the AISC issued its new limit states code,
named Load and Resistance Factor Design (LRFD), without, however
withdrawing the previous ASD specification. To make matters more
complex, a new ASD code will be issued in 1989 or 1990 which will be
an upgrading and reorganization of the last (1978) ASD specification.
For the foreseeable future, then, perhaps until the end of the
Twentieth Century, there will be a dual system of steel design
specifications in the USA. This will, no doubt, result in a
considerable amount of confusion. A similar situation will prevail
for the American Iron and Steel Institute (AISI) specification for
cold-formed steel construction. The transition in other countries is
smoother. For example, in Canada there was a set time of years after
which the ASD code was no longer legal. The present situation is
thus that almost everywhere in the world it is possible to design
steel structures by a limit states design standard, and that in many
places this is the only legal method.

The new generation of codes have attempted to accomplish at least three aims: 1) rationalization of code structure, 2) incorporation of new research results and 3) utilization of probability-based multiple load and resistance factors, also called sometimes "partial factors." All the new codes with which this author is acquainted get rather high marks in satisfying all three aims.

The rationalization of code structure permits efficient computerization and integration into knowledge based information systems. Many new results are available from research on the behavior of structures and structural elements so that ultimate strength models can be specified with greater precision than before in these codes. The analytical formulations of these models are rather more complex than in previous codes, but it is assumed that the designer will make heavy use of computers. The trend is thus toward more mechanics and less empiricism in the new codes.

One of the major differences between ASD and the LSD codes is the use of multiple load factors by which the nominal load effects are multiplied, and resistance factors which modify the nominal resistance. In some codes these factors are determined by direct calibration of the new code to reference designs in the old code (e.g., the US Load Factor Design specification for steel highway bridges (AASHTO, 1983), while in others more-or-less use is made of calibrations involving first-order reliability methods (FORM) (e.g., the LRFD Specification of AISC (1986) and the Canadian codes CSA, 1984, and OHBDC, 1983.

In 1989, then, a new generation of steel design codes is in place almost everywhere in the world. These codes are gaining good acceptance. Many classes in engineering schools are teaching only these codes and a number of textbooks have been, or will soon be, published for educational use. More and more designers are using the new methods, and designers appear to like the outcome. Computer programs and "expert systems" based on the new methods are appearing with increasing frequency in commercial advertisements. What to many in the steel design community in the USA seemed like a revolutionary new method in the 1970's will have become the orthodox method before the end of this century.

3 Future Developments and Research Needs

Many of the researchers in the area of reliability - based structural analysis and design can point with pride to the application of their ideas in the new steel design codes. Special recognition goes to those researchers who in the late 1960's and early 1970's developed and refined the first-order second-moment method (Basler, Benjamin, Cornell, Ang, Lind, to mention only a few), which is the basis for most of the modern steel design specifications. This is a good example of academic research going directly into practical application. However, because it takes a long time for the adoption of new structural standards, the present (1989) "new" codes have locked into them the state of knowledge of

structural reliability as it existed in the late 1960's and early
1970's. There has been a vast, almost exponential increase in
research on structural reliability in the past twenty years, and this
expansion is likely to keep going for the foreseeable future. One of
the chief problems is to find ways to introduce these new insights
and procedures on a continuing basis into structural steel design
codes.

Following are some other shortcomings of the "new" LSD Specification
of the AISC (1986):

1)   An early idea (Ravindra, 1978) was that design optimality would
      result if all components of the structure had the same
      reliability of not exceeding the limit state. This is an
      extension of the older ideas of the "one horse shay" or the
      "fully stressed" design methodologies. It was recommended that
      the new steel design code should be designed to have a target
      reliability index β_T = 3.0.   The agrument for this was based
      on calibration to "good" designs from the previous code when
      the loading is gravity loading. It was soon discovered that
      when steel structures under combined wind and gravity loads
      were designed for β_T=3.0, the resulting structures were much
      heavier than structures designed by the previous code.
      Since no convincing argument could be mustered to demonstrate
      that these structures were unacceptable, a new calibration
      exercise was performed (Galambos, 1982, Ellingwood, 1980, 1982)
      and target reliability indices of β_T = 3.0 for gravity loads,
      β_T=2.5 for combined wind and gravity loads and β_T = 2.0 for
      combined gravity and earthquake loads were used as the basis
      for the 1982 ANSI A58.1 Load Code. The philosophical
      underpinning of uniform reliability for all components was then
      abandoned, with no convincing intellectual proof that the new
      scheme was a valid one. This proof is still missing, and it
      may anyway be irrelevant in the light of newer insights into
      systems reliability. From an intuitive standpoint the
      decisions to use the variable target reliability indices make
      sense. It would be nice, however, to have a rigorous
      intellectual justification based on systems reliability
      concepts.

2)   Actual component reliability indices as they were determined by
      FORM methods for the design provisions in the AISC 1986 LRFD
      specifications (see Table 1, Galambos, 1988) are quite
      different again from the target values recommended by
      Ellingwood (1980, 1982). Table 1 gives the β's for a live-to-
      dead load ratio of 1.0 for the key structural steel components.
      In general, connectors and connections tend to have a higher
      reliability than members. This was deliberate to continue the
      past practice of providing higher "safety factors" to joints.
      There is, however, no tendency which distinguishes the varying
      seriousness of the consequences of failure (e.g., columns are
      not more reliable than beams). For beams and columns of
      frequently occurring dimensions the average β is about 2.7,
      which is a deliberate reduction from the target value of 3.0 to

make LRFD, on the average, more economical than the ASD designs.

3) The AISC LRFD (1986) Specification does not explicitly address the concepts of systems reliability. There are a number of intuitive notions which address some systems problems (load combination schemes, frame-stability and second-order analysis requirements, etc.), but on the whole there are many glaring inadequacies. For example, the member in a redundant structure has the same reliability as a critical column or hanger whose failure would result in a catastrophe.

4) Emphasis is placed entirely on "ultimate" limit states, and almost no criteria are given for "serviceability" limit states. Modern steel construction tends to be light and of great strength so that the controlling limit states for many structures are those of deformation and motion.

In the light of the enumerated shortcomings is it true that the enormous effort to change to the new design codes was in vain? Of course not. Considerable progress has been made, and the new codes are far better than the previous ones. The following suggestions are made to improve structural steel design in the future:

1) The present generation of LRFD specifications will surely last into the first decade of the Twenty-first Century, and therefore it is important to remedy the most obvious inadequacies. The following research directions should be greatly expanded and expeditiously implemented in the codes.

   a) To the extent that this is possible, concepts of systems reliability should be incorporated. For example, penalties for critical non-redundant elements and rewards for redundant structures could be easily included by using "systems" factors (Galambos, 1988).

   b) Despite our vast knowledge of member behavior and our enormous computational capabilities we still use very primitive models of structural analysis is the design office (e.g., slabs and walls are ignored in frame stability calculations). Available research knowledge on behavior and limit states of steel structural systems should be transferred to everyday are in the design office. Codes should provide rewards to encourage such applications.

   c) The greatest economic gain could be achieved by placing design against the serviceability limit states on a sound basis. Many steel structures are controlled by such limit states. These problems should be much more easy to solve than the ultimate limit states: the structures are still

linear and much more data on serviceability loads exist.
A major concerted and reasonably funded research effort of
about five years should bring in very high economic
rewards.

2)   It is evident to this author that what we now know about
     structural reliability is far too rich to be confined in the
     traditional LSD design codes. Work should now commence on the
     next generation of structural design standards for
     implementation in the next Century. Such standards should be
     based entirely on requirements of explicit reliability limits.
     Just as today the designer applies structural analysis in
     everyday design, so the future designer should use
     reliability analysis. The following research efforts are
     possible paths toward the realization of the new codes:

     a)   Development of code models to test out format and content.

     b)   Assembly of statistical data for presentation in handbooks
          (material properties, member and joint resistances,
          loads).

     c)   Development of rational models for probability-based
          design against earthquakes and wind.

     d)   Refinement of system resistance models, including damage
          concepts.

Many of the ingredients for the realization of a truly reliability
based code already exist. What is needed foremost is a massive
education program first for the leaders of the structural engineering
profession, then for the engineering students and professionals.

Conclusion

Current (1989) reliability-based LSD steel design specifications are already dated, even though they are a major improvement over the ASD codes. Suggestions have been made for the short-term improvement of the LSD codes, and research is recommended to implement a truly probabilities code for the 21st Century.

TABLE 1 RELIABILITY INDEX BETA FOR A.I.S.C. F.R.F.D. SPECIFICATION
(LOGNORMAL BETA; FOR Ln/Dn=1.00)

TYPE ELEMENT	BETA
tension member, yield limit state	3.0
tension member, fracture limit state	4.1
rolled beam, flexural limit states	2.8
rolled beam, shear limit state	3.4
welded beam, flexural limit states	3.3
welded beam, shear limit state	3.3
welded plate girder, flexural limit state	2.6-2.9
welded plate girder, shear limit states	2.3
columns	2.7-3.6
high strength bolts, tension	5.0-5.1
high strength bolts, shear (bearing-type joints)	6.0-6.0
high strength bolts eccentric joint	4.8
high strength bolts, slip-critical joints	
(standard holes)	1.6-2.0
fillet welds	4.4
eccentrically loaded welded joints	3.9
welded connections, flange bending	2.6
welded connections, local web yielding	4.1
welded connections, web buckling	4.0
welded connection, web crippling	2.9-2.9

References

(1)   AASHTO: "Standard Specifications for Highway Bridges", 13th
        Edition 1983, Washington, D.C.

(2)   AISC: "Load and Resistance Factor Design Specification for
        Structural Steel Buildings", 1st Edition 1986, Chicago, Ill.

(3)   AMERICAN NATIONAL STANDARDS INSTITUTE, "Minimum Design Loads
        for Buildings and Other Structures", ANSI A58.1, 1982.

(4)   CSA "Steel Structures for Buildings-Limit States Design"
        Canadian Standards Association, 1984.

(5)   ELLINGWOOD, B., GALAMBOS, T.V., MACGREGOR, J.G., and CORNELL,
        C.A., "Development of a Probability-Based Load Criteria for
        American National Standard A58" NBS Special Publ. No. 577,
        National Bureau of Standards, Washington, D.C., June 1980.

(6)   ELLINGWOOD, B., MACGREGOR, J.G., GALAMBOS, T.V., and CORNELL,
        C.A., "Probability-Based Load Criteria: Load Factors and Load
        Combinations", Journal of the Structural Division, ASCE, Vol.
        108, No. St5, May 1988, pp 978-979.

(7)   GALAMBOS, T.V., ELLINGWOOD, B., MACGREGOR, J.G., and CORNELL,
        C.A., "Probability Based Load Criteria: Assessment of Current
        Design Practice", Journal of the Structural Division, ASCE,
        Vol. 108, No. St5, May 1982, pp 959-977.

(8)   GALAMBOS, T.V., "Reliability of Structural Steel Systems"
        Structural Engineering Report No. 88-06, Civil and Mineral
        Engineering Department, University of Minnesota, Minneapolis,
        1988.

(9)   OHBDC "Ontario Highway Bridge Design Code", Ontario Ministry of
        Transportation and Communications, 1983.

(10)   RAVINDRA, M.K, and GALAMBOS, T.V.: "Load and Resistance Factor
        Design for Steel", Journal of the Structural Division, ASCE,
        Vol. 104, No. St 9 Sep. 1978, pp 1337-1354.

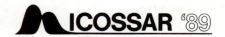
PROBABILISTIC BASIS FOR BRIDGE DESIGN CODES

Andrzej S. Nowak

Department of Civil Engineering, University of Michigan, Ann
Arbor, Michigan, USA

ABSTRACT

There is a growing interest in the development of
rational design criteria for bridges. The
fundamental problem is evaluation of the loads
and load carrying capacity. However, load and
resistance parameters (load components, material
properties, dimensions) are random variables.
Therefore, the reliability is a convenient
measure of structural performance. Traditional
approach was based on design of structural
members rather than systems. In the paper, bridge
load, resistance models, limit states and
calibration procedure are reviewed. Most
important research needs for the development of
the bridge design code are also formulated.

KEYWORDS

bridges; limit states; loads; load factors; LRFD;
reliability; resistance; resistance factors.

1. INTRODUCTION

The current generation of bridge design codes is based on
deterministic models of the structural behavior. The acceptance
criteria are specified for members and they are usually expressed
in terms of the allowable stresses. However, there is a growing
need for rationalization of the design procedures. The major
reasons for change include:

- growth of heavy traffic in terms of weights and frequencies,
  also changes in truck configurations;

- computerized methods of bridge analysis often reduce or remove
  some of the historic safety margins. The degree of the inherent
  redundancy of the bridge should be evaluated;

- new structural materials and technologies;

STRUCTURAL SAFETY AND RELIABILITY

- need for a consistent basis for comparison of competitive bids between materials (e.g. steel, concrete and timber);

- new data on material properties, behavior of structural members and bridge loads (tests, surveys and measurements);

- new developments in the area of theory of reliability and availability of practical procedures for code optimization.

Most of the quantities that enter into engineering calculations involve uncertainties due to randomness of design parameters, imperfect analytical models or human errors. Because of these uncertainties, absolute reliability is not an attainable goal. Selection of the optimum reliability level is an economical problem. Lower reliability results in frequent failures, while higher reliability requires more initial costs (material and labor). Therefore, structural reliability is a convenient acceptability criterion in the development of a design code.

The objective of this paper is to identify the major parameters involved in the development of the bridge design code and to formulate the probability-based procedure for calculation of load and resistance factors. The US and Canadian data base is used as an illustration of the statistical models and methodology.

Behavior of the bridge is determined by realization of load components, members load carrying capacities and members interaction (system behavior). Load components include dead load, live load (static and dynamic), environmental loads and abnormal loads. In bridge analysis, the effects of these loads are considered, such as bending moment, shear force, torsion, overturning moment and so on. On the other hand, the load effects are resisted by structural members depending on the moment carrying capacities of girders, shear capacity, torsional strength, stabilizing moment and so on.

2. LOAD MODELS

The stresses and strains in structural members are usually a result of combined action of several load components. In shorter span bridges, the total effect is dominated by live load and in longer bridges dead load governs. Environmental effects such as wind, earthquake, temperature, ice or water may lead to a structural damage, or even collapse. Simultaneous occurrence of extreme values of these various load components is rather unlikely. Following Turkstra's rule [1], the maximum occurrence of one is simultaneous with average values of others. In some cases there is also a negative correlation, e.g. a very strong wind causes the reduction of traffic, and thus live load is small (if any).

In practice, the major load combination for bridges is dead load and live load. The former is usually well controlled and its

statistical parameters are known. However, the latter involves a considerable degree of uncertainty. Live load is probably the most controversial issue in the current development of bridge design codes.

The dead load can be considered as normally distributed, with the bias factor (mean-to-nominal ratio) varying from 1.03 for factory made members (precast girders, steel girders) to 1.05 for cast-in-place members (slab, cast-in-place girders). The coefficient of variation varies from .04 to .08 [2]. The weight of asphalt depends mostly on the thickness. For example, in Ontario, the mean thickness was found equal to 75mm, with the coefficient of variation about .25. Dead load constitutes about 40-50 percent of the total load for shorter spans (15-20 percent for timber decks), but the percentage may be as high as 80 for long spans.

Live load depends mostly on axle load, truck configuration, traffic pattern within the lane, multi-lane presence, distribution of load to girders and traffic volume. The effect of trucks on the bridge is a time-variant dynamic force. A considerable research effort was directed at modeling truck weights. The data comes from truck surveys, weigh-in-motion (WIM) studies, police files (over-weight citations) and special permits.

The live load model can be based on the truck survey results. In Ontario, the survey covered 10,000 heavy vehicles [3], which corresponds to about two week traffic for class A highway. The data includes axle weights and spacings. The data was processed [4], in particular the lane moments were calculated for all the surveyed trucks. Various spans were considered, ranging from 3 to 60 m. The results are shown in Fig. 1 on a normal probability scale (inverse normal distribution). For easier comparison, the truck moments are divided by the corresponding design moments in Ontario [5]. The extrapolated lane moments are also shown. The number of heavy trucks crossing the bridge in 50 years is estimated at 20 million. This corresponds to the marked probability level.

The analysis showed that the maximum live load per girder (girder moment) is caused by two trucks simultaneously on the bridge. WIM studies indicate [6] that the probability of such an event is about 100 times smaller than for one truck. The corresponding probability level is shown in Fig. 1. Using Turkstra's rule the maximum effect is a combination of a fully loaded vehicle and an average one. Value of the girder moment depends on the curb distance and spacing betwen the trucks. A typical density function of the curb distance for a two lane interstate highway (one direction only) is shown in Fig. 2 [2]. The mean maximum 50 year girder moment can be calculated using the influence lines, as the expected value for various curb distances.

More studies are needed to establish the live load model for various sites, time periods, distribution of truck weight, multiple presence in one lane and in adjacent lanes, and transverse

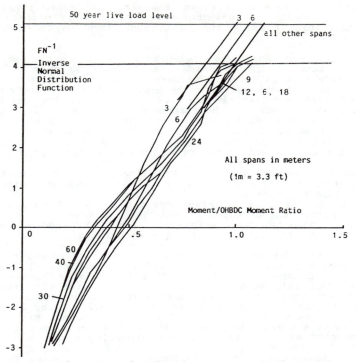

Fig. 1   Lane Live Load Distributions for Various Spans.

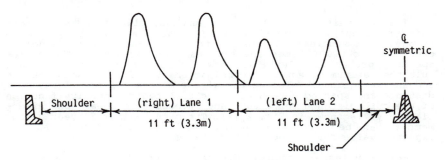

Fig. 2   Typical Density Function for a Curb Distance.

position of truck on the bridge. Additional studies must be carried out in order to estimate the growth in future live loads.

   Another controversial issue is the dynamic load. There are
many definitions of the dynamic load. Traditionally it is considered as an equvalent static force (moment or shear force), added
to other static loads. In AASHTO [7], impact factor is calculated
as a function of the span length and in OHBDC [5] as a function of
the natural frequency of the bridge. Actual realization of stress
or strain history shows a considerable variation. The major reasons are sufrace roughness, bridge dynamics (truck crossing time
is comparable with the bridge natural frequency of vibration) and
vehicle dynamics (suspension system and mass). It is very difficult to predict the percentage effect of these three components.
Bridge tests carried out in Ontario indicate that the average
value of dynamic load is about 15 percent of live load with the
coefficient of variation often exceeding .8.

   Further research is needed to establish the relationship
between the truck weight and dynamic load. It is expected that
dynamic factor is smaller for heavier vehicles. More data is
needed to estimate the effect of dynamic (short-term) forces on
the structural behavior. Most of the available data was obtained
for long-term loads.

   The environmental loads include wind, earthquake, snow, ice,
temperature, water pressure, etc. The basic data has been gathered for building structures, rather than bridges. However, in
most cases the same model can be used. Some special bridge
related problems can occur because of the unique design conditions, such as foundation conditions, extremely long spans, or
wind exposure.

3. RESISTANCE MODELS

   The bridge resistance models were derived from tests, measurements and analysis. The structural behavior is described by load-
deformation relationships, in particular moment-curvature curves.
They involve variation due to uncertainties in material properties
and dimensions. Typical moment-curvature curves are shown in Fig.
3 for a composite steel girder and in Fig. 4 for a prestressed
concrete girder. The solid line corresponds to the mean relationship, and the dashed lines correspond to one standard deviation
above and below of the mean.

4. LIMIT STATES

   A bridge fails when it cannot perform its function any longer.
The function includes not only structural aspects such as to carry
safely traffic loads for expected periods of time and to provide a
comfortable passage (minimize deflections and vibrations), but
also to provide aesthetic values to the users. The structural per-

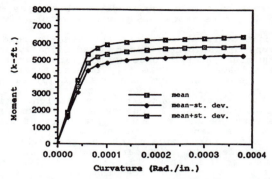

Fig. 3   Moment-Curvature Curve for a Composite Steel Girder.

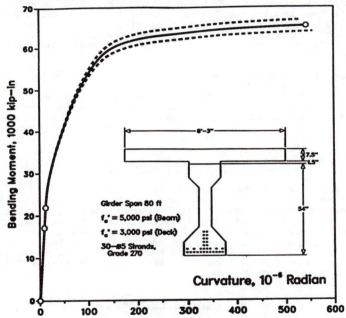

Fig. 4   Moment-Curvature Curve for a Prestressed Concrete Girder.

formance can usually be expressed in terms of mathematical equations, i.e. limit state functions, involving various parameters of material properties, dimensions and geometry.

In practice, formulation ofthe limit state function is very difficult. Both load and resistance are functions of multiple parameters. These parameters vary randomly, some are difficult to quantify (eg. geometric configuration), they are often time-variant (eg. live load or effect of corrosion) and/or correlated. Some bridge limit states require special formulations, for example, fatigue, concrete cracking, corrosion effects, vibrations or deflections. Furthermore, the performance of a structural member is described by a set of limit state functions.

For a typical bridge girder, the limit state functions can be formulated for various conditions, including:

- Bending; resistance can be defined as the moment carrying capacity for the girder, and load as the maximum load resisted by the girder (a joint effect of dead load, live load, environmental effects, and so on).

- Shear.

- Buckling; overall and local.

- Deflection; excessive deformations may affect the user's comfort or cause some structural problems (eg. increased dynamic truck load).

- Vibrations; may affect the user's comfort.

- Accumulated damage conditions, including corrosion (rebars in reinforced concrete beams and slabs, pretressing tendons, structural steel sections), fatigue (mostly affects steel sections, prestressing tendons and rebars), cracking (concrete) and other forms of material deterioration.

Each limit state is associated with a set of limit state functions, which determines the boundaries of the acceptable performance. Load effect is not a single force but a series of forces corresponding to the truck axles passing over the bridge. The capacity requires a special formulation. For a given traffic density ADTT (average daily truck traffic) and cumulative distribution functions of axles weights, resistance can be expressed in terms of the number of years to failure (expected life time).

In general, bridge members are interconnected and they share the loading. Load distribution is not necessarily linear and it may vary depending on load magnitude, point of application or degree of deterioration. Limit states for the whole structure, as for individual members, express the boundry lines between safe (acceptable) realizations and failure. The critical conditions for bridges can be defined in terms of maximum deformations (eg.

deflections), vibrations, or condition of individual members
(girders and slabs). Usually reaching a limit state by one of the
memebrs does not mean the bridge also reached its limit state.
Traditional design and analysis of bridges is based on identification of the governing limit state in each member. Safety provisions are applied to ensure an adquately low probability of occurrence for these limit states. In many bridges this approach is
excessively conservative. Yet, in the evaluation of existing
structures there is a need for accurate estimate of capacity. The
tool for calculation of the reliability for the whole bridge is
provided by the theory of system reliability.

5. CODE CALIBRATION PROCEDURE

The development of a design code includes five basic steps:

(1) definition of the scope,
(2) definition of the code objective,
(3) selection of the target reliability level,
(4) selection of the code format,
(5) calculation of the code parameters.

The procedure is described in [8]. It is convenient to calibrate the design provisions using as a reference the reliability
level of existing structures, designed according to current codes.
The performance of these structures can be evaluated by the
maintenance staff of state and county bridge authorities.

REFERENCES

1. Turkstra, C.J., 1970, "Theory of Structural DEsign Decisions",
   Study No. 2, Solid Mechanics Division, University of Waterloo,
   Ontario, p. 124.
2. Nowak, A.S. et al. 1988. "Risk Analysis for Evaluation of
   Bridges", Report UMCE 88-7, University of Michigan, Ann Arbor.
3. Agarwal, A.C. and Wolkowicz, M., 1976, "Interim Report on 1975
   Commercial Vehicle Survey", Research and Development Division,
   Ministry of Transportation and Communications, Downsview,
   Ontario.
4. Nowak, A.S. and Zhou, J-H., 1985, "Reliability Models for
   Bridge Analysis", Report No. UMCE85-3, University of Michigan.
5. OHBDC, "Ontario Highway Bridge Design Code", Ministry of Transportation and Communications, Downsview, Ontario, 1983.
6. Ghosn, M. and Moses, F., 1984, "Bridge Load Modeling and
   Reliability Analysis", Report No. R 84-1, Department of Civil
   Engineering, Case Western Reserve University.
7. AASHTO, "Standard Specifications for Highway Bridges", Washington, DC, 1983.
8. Nowak, A.S. et al. 1987. "Design Loads for Future Bridges",
   Report FHWA/RD-87/069, University of Michigan, Ann Arbor.

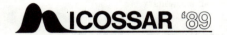
LRFD for Engineered Wood Construction:
ASCE Pre-Standard, Wood Industry Initiative

Joseph F. Murphy, Ph.D., P.E.

Consulting Engineer
Structural Reliability Consultants
Post Office Box 56164
Madison, WI, 53705-9464 USA

Abstract

The structural wood community in the United States
is in the process of developing a reliability-based
consensus Load and Resistance Factor Design (LRFD)
standard for engineered wood construction. This
presentation will provide an historical view of this
process, starting in 1976, through the forming of an
American Society of Civil Engineers (ASCE) task committee
from 1983 to 1987, to the status of a Wood Industry
Initiative started in 1988. Hurdles, obstacles, and
concerns specific to wood and wood construction are
highlighted. Near and long-term goals are presented.

Keywords

wood; construction; reliability; design; code; LRFD;
standard; building; history; timber

1. Introduction

The wood community in the USA is presently developing a Load and Resistance
Factor Design (LRFD) code for engineering with wood and prefabricated wood
components. This paper will review the events leading to the LRFD code as well
as special considerations that the wood community has had to adress.

2. Background

The development of reliability based limit states design for wood in the
USA can be traced back to the Fall of 1976. On September 30, 1976 the
Committee on Wood of the American Society of Civil Engineers (ASCE) formed a
Subcommittee on Reliability Limit States Design. The next year, Purdue
University gave a Seminar on Structural Reliability and Design, March 9-11,

1977, which was well attended by people from the wood community. This was followed by a Wood Design Workshop in Madison, Wisconsin, November 28-30, 1977.

As shown by the flurry of activities, there was immediate interest by the wood community on reliability based limit states design for wood. There was, however, a lot of concern about how to incorporate the duration of load or creep rupture effect (recognized by the wood community) into limit states design as well as giving credit to producers who continually monitor their product verified with destructive sampling versus those producers who only sort by visual grading procedures.

On October 23, 1978, Bruce Ellingwood from the National Bureau of Standards, visited the Forest Products Laboratory (FPL) in Madison, Wisconsin, seeking input from the FPL for the development of reliability based load factors to be specified in ANSI A58.1 'Minimum Design Loads for Building and Other Structures' [1]. Again the problem of handling load duration surfaced. Dr. Ellingwood, using then available test data and the load duration curve with its variability, analyzed the reliability of glued-laminated and other heavy timber structures [2]. The wood community set about researching load duration and damage accumulation models for wood.

In the next five years, the ASCE Subcommittee on Reliability Limit States Design held technical sessions, wrote research needs and general papers, and brought together researchers interested in the subject.

On October 5-6, 1983, there was a 'Structural Wood Research Workshop: State-of-the-Art and Research Needs' meeting held in Milwaukee. Here, the wood community unanimously approved the following recommendation:

'The ASCE Committee on Wood should undertake the development of a reliability-based design (RBD) document, with background commentary, as a pre-standardization activity with recommended model code provisions.'

On October 19, 1983, the ASCE Committee on Wood approved the formation of a Task Committee on Load and Resistance Factor Design for Engineered Wood Construction to write a pre-standard document.

3. ASCE Pre-Standard Report

The ASCE Task Committee began its task on October 1, 1984 and was transformed back to Subcommittee status on October 1, 1987 after successfully completing its task. The 'Load and Resistance Factor Design for Engineered Wood Construction, A Pre-Standard Report' was published by ASCE the Fall of 1988 [3].

A few points regarding this ASCE Task Committee and its report should be noted. When the task committee began its work there was not a method to account for load duration. Also three decisions were made early so that the task could be accomplished in the short three year time frame. The three decisions were:

1. The LRFD pre-standard would only address single-member design.

2. The LRFD pre-standard would use the loads, load factors, and load

statistics as specified in ANSI A58 [1] and NBS SP577 [2].

3. The LRFD pre-standard would use single-member strengths, NOT ultimate stresses (e.g., flexure or moment capacity, NOT modulus of rupture).

The reason for the first decision was that single member reliability was well established [4] and systems reliability could be accommodated later by a 'systems' factor. Two reasons to accept the ANSI A58 loads were that loads SHOULD be independent of material resistance and the task committee should concentrate its efforts on material resistance effects. To dispense with fictitious equivalent section moduli, linearized moduli of rupture, and to accommodate future wood based prefabricated products and to emphasize measurable performance, member strength was used, NOT member ultimate stress.

During the course of ASCE task committee work a method was developed to account for load duration in a relatively straightforward manner. Consider the following example LRFD safety checking equation:

$$\emptyset\, R_n = 1.2\, D_n + 1.6\, L_n$$

where 1.2 and 1.6 are load factors on nominal dead and live loads, D_n and L_n, and $\emptyset$ is a resistance factor on nominal resistance R_n. For specific probability distributions of dead and live loads, D and L, and resistance, R, a reliability index β can be calculated as a function of the resistance factor $\emptyset$ using first-order second-moment level 2 (FOSML2) reliability methodology [4] (see Figure 1). This function by FOSML2 only uses a short-term strength resistance distribution and is the reliability WITHOUT load duration.

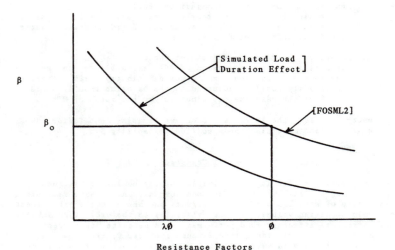

Figure 1 - Reliability Curves - with and without Load Duration

If now the same safety checking equation is used but a numerical simulation is run with a stochastic load process, a specific damage accumulation model, and a short term resistance distribution, probability of survival can be calculated and, with a normal transformation, a reliability index can be computed as a function of the resistance factor. This is the lower curve in Figure 1 and is the reliability WITH load duration. (Note that the statistics for the load process have to be consistent with the load magnitudes used in the FOSML2 analysis.)

For any specific reliability β_o the ratio of $\emptyset$ with load duration to the $\emptyset$ without load duration is defined as λ, the load duration factor. The safety checking equation for wood becomes:

$$\lambda \, \emptyset \, R_n = 1.2 \, D_n + 1.6 \, L_n$$

where λ is a load duration or creep rupture adjustment factor, $\emptyset$ accounts for short term strength variability (using FOSML2) and R_n is a nominal value of the short term strength distribution.

The general concept of using the ratio of $\emptyset$ values (at a specific β_o) can be used in the development of ANY adjustment factor from some original or baseline distribution. Consider the following:

1. Quality Improvement Factor
    A resistance distribution is 'improved' by thinning its low tail by better sorting practices or proofloading. The $\beta-\emptyset$ curve shifts to the right and the quality improvement factor, ρ, is greater than 1.

2. Environment or Treatment Adjustment Factor
    Accounting for environment or treatment shifts the $\beta-\emptyset$ curve. The resulting ratio of $\emptyset$'s can be considered an environment or treatment adjustment factor.

3. Baseline Distribution Adjustment Factor
    If a resistance distribution of 25% variability is chosen as a baseline distribution, any other distribution with different variability would be adjusted to the baseline (and it would use the $\emptyset$ of the baseline distribution) by the ratio of $\emptyset$'s.

The concept of using the shift in $\beta-\emptyset$ curves as adjustment factors is general in nature, and maintains the specified reliability β_o.

4. Nominal Resistance

The final consideration is the definition of the nominal resistance of the short term strength distribution. The wood community has always recognized the variability of wood and wood based products and has used the 5th percentile when determining design values. If one calibrates to the present reliability index of current satisfactory wood performance and uses the 5th percentile as the nominal resistance, one can get resistance factors, $\emptyset$, greater than 1. If a higher percentile is used as the nominal, the resistance factors will be less than if the 5th percentile is used as the nominal.

There are two approaches that the wood community can take to calibrate to present satisfactory wood performance while keeping the resistance factors less than unity.

One approach would be to define the nominal resistance as a higher percentile than the 5th, say the 25th, and separate wood and wood based products according to variability ranges, e.g., 15% or less is low, 15%–25% is medium, and 25% or above is high. This would give three resistance factors depending on low, medium or high variability classification.

Another approach would be to define the nominal resistance as a constant times the 5th percentile times a baseline adjustment factor. As an example, in bending, consider the baseline variability to be 22.3%, the constant is 1.214.

$$R_n = 1.214 \, R_{05}$$

Calibrating to present reliability might yield

$$\emptyset = 0.85$$

Consider a product with a variability of 28.1%. Its nominal resistance is defined by

$$R_n = B \; 1.214 \; R_{05}$$

where the baseline adjustment factor B for a variability of 28.1% might be

$$B = 0.933$$

In general, products with variabilities higher than the baseline would have baseline adjustment factors less than 1 while products with variability lower than the baseline would have baseline adjustment factors greater than 1. This approach allows one resistance factor for all wood and wood based products based on ONE baseline variability. A product with a variability different than the baseline would be adjusted in its nominal resistance value. Typically, for any visual grade of lumber, compressive strength is less variable than bending strength which in turn is less variable than tensile strength. Different baseline variabilities can be chosen for each property reflecting this trend, yielding different resistance factors of, for example,

compression $\emptyset_p = 0.90$,

bending $\emptyset_b = 0.85$, and

tension $\emptyset_t = 0.80$.

The nominal resistance values for each property would be determined thusly:

compression $P_n = B_p \, P_{05}$,

bending $M_n = B_b \, M_{05}$, and

tension $T_n = B_t \, T_{05}$,

where B_p, B_b and B_t would be functions of the variabilities of compression, bending and tension, respectively, each chosen so that the target reliability index is met. These B's have the constant multiplier incorporated in them.

On October 1, 1987, the ASCE Subcommittee on Reliability and Wood Design replaced the ASCE Task Committee on LRFD for Wood Construction. The subcommittee is concentrating on the two areas of serviceability limit states and systems reliability.

Purdue University held a Reliability Engineering Symposium (for Wood) on April 21-23, 1986.

5. Wood Industry Initiative

In Spring 1986, a Select Industry Task Group on Reliability—Based Design was formed under the auspices of the National Forest Products Association. In 1987 they reviewed requested proposals to develop an LRFD manual for the Wood Industry. On July 17, 1987, the Task Group described their intentions:

'Statement of Intent, Reliability—Based Design Project'

...

'The short—term goal is to develop a reliability—based LRFD specification for engineered wood construction by the end of 1989. This specification will be user—tested and designer oriented (practical). It will duplicate the format of the new 1986 AISC Load and Resistance Factor Design Manual for Steel Construction as closely as possible, and will include additional considerations for wood design. The ASCE load and resistance factor pre—standard document for engineered wood construction and developments in the Canadian code will be utilized as primary resources.

'It is intended that the content of numerous current wood design specification documents will be incorporated into the new wood LRFD manual to provide the designer with one convenient design document. It is further intended that each current specification—writing association will continue to maintain its portion of the new wood LRFD manual.

'The new wood LRFD manual will not initially be developed under a formal concensus procedure so that completion of this project can be assured by the end of 1989. It is the intent, however, to submit the new wood LRFD document to an appropriate concensus organization. It is also intended that the final document will be adopted by all building code jurisdictions in the United States.'

...

On June 30, 1988 the National Forest Products Association (NFPA) signed a master contract with Engineering Data Management, Inc., a consulting firm in Fort Collins, CO, to produce a reliability—based Load and Resistance Factor Design manual for wood in three years. The engineer—tested work product will be comparable to the first 220 pages of Chapter 6 of the Steel LRFD Manual [5].

6. ASCE Standards Committee

At its October 22-23, 1988, meeting, the ASCE Board of Direction, approved the proposal submitted by the Committee on Wood to create an ASCE standards activity to develop consensus Standard Design Codes for Engineered Wood Construction According to Limit States Criteria (LRFD). This standards activity will be under the auspices of the Technical Council on Codes and Standards and will utilize the prestandardization document [4] developed by the Committee on Wood and hopefully a draft specification developed by the timber industry as resource documents.

As of January 9, 1989, an ASCE News article and press releases to many trade and professional organization publications had been issued calling for members to serve on the standards committee. Membership on ASCE's standards committees is open to all interested persons, including those that will be affected by the standard. Criteria for committees includes maintaining a balance between consumer, producer, and general interest groups and membership by a number of code officials. ASCE membership is not required.

ASCE is attempting to coordinate this consensus standards activity with the wood industry effort toward the development of an LRFD specification. Since ASCE's membership consists of engineers, designers, researchers and professors most likely to use a LRFD Standard for Engineered Wood Construction in their design work, in the inspection and acceptance of structures, and in their consulting, research and/or teaching activities, a joint Wood Industry-ASCE consensus standard has been proposed. To date the Wood Industry as a group has shown little interest in a jointly-owned consensus standard. (It should be stressed that engineers, researchers and other representatives in the Wood Industry have always actively participated as individuals in ASCE and its Committee on Wood.)

7. Summary

ASCE has provided leadership and organizational support of reliability-based wood design for the last 13 years culminating in 1988 with publication of 'A Pre-Standard Report on Load and Resistance Factor Design for Engineered Wood Construction' and formation of a consensus Standards Committee open to all interested persons.

Building on these 13 years and the ASCE Pre-Standard Report, the Wood Industry (comprised of and controlled by producers) is in the process of developing a proprietary Wood Design Manual under a non-consensus procedure. To meet a self-imposed three year time limit, the Wood Industry will also develop (in a non-consensus manner) their own material characterization methodologies to give resistance values compatible with their Wood Design Manual.

It is the author's opinion that the Wood Community would be best served if the Wood Industry and the ASCE Standards Committee coordinate their unique resources and expertise, and develop a common strategy to advance a consensus Standard on Load and Resistance Factor Design for Engineered Wood Construction.

8. References

[1] American National Standards Institute: 'American National Standard Minimum Design Loads for Buildings and Other Structures (ANSI A58.1 -1982),' American National Standards Institute, New York, 100 p. 1982

[2] Ellingwood, B., T.V. Galambos, J.G. MacGregor, and C.A. Cornell: 'Development of a Probability Based Load Criterion for American Standard A58 Building Code Requirements for Minimum Design Loads in Buildings and Other Structures,' National Bureau of Standards Special Publication 577, Gaithersburg, MD, 222p. 1980

[3] ASCE Task Committee on Load and Resistance Factor Design for Engineered Wood Construction: 'Load and Resistance Factor Design for Engineered Wood Construction, A Pre-Standard Report,' American Society of Civil Engineers, New York, 177p. 1988

[4] Thoft-Christensen, P. and M.J. Baker: 'Structural Reliability and Its Applications,' Springer-Verlag, New York, 267p. 1982

[5] American Institute of Steel Construction: 'Manual of Streel Construction, Load and Resistance Factor Design, First Edition,' American Institute of Steel Construction, Inc., Chicago, 1986

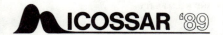
RELIABILITY BASES FOR CODES FOR DESIGN
OF REINFORCED CONCRETE STRUCTURES

Ross B. Corotis
Bruce Ellingwood

Department of Civil Engineering, The Johns Hopkins University, Baltimore, MD
21218

Andrew Scanlon

Department of Civil Engineering, The Pennsylvania State University, University
Park, PA 16802

ABSTRACT

The American Concrete Institute Standard 318, Building Code Requirements for Reinforced Concrete [1,2], has permitted the design of reinforced concrete structures in accordance with limit state principles using load and resistance factors since 1963. A probabilistic assessment of these factors and implied safety levels is made, along with consideration of alternate factor values and formats. A discussion of construction safety issues is included.

KEYWORDS

Buildings (codes); Design (buildings); Limit states; Loads (forces); Probability theory; Safety; Structural engineering.

1. History

Prior to the 1963 edition of ACI Standard 318, reinforced concrete design in the United States was based on linear elastic working stress design principles. Loads were selected from published standards, and stresses were computed on the basis of transformed sections in which the steel was replaced by equivalent concrete utilizing the modular ratio, $n = Es/Ec$. The appropriateness of the linear analysis, as well as an acceptable level of safety, was ensured by keeping stresses below 40 to 45 percent of their ultimate strength values.

Due to the highly nonlinear behavior of concrete near ultimate stress levels, the linear approach could not be extrapolated to give a realistic assessment of true safety levels. ACI introduced the concept of member analysis under conditions of incipient failure in an Appendix to the 1956 edition of ACI 318 [3], and into the standard itself in 1963 [1].

Ultimate Strength Design (currently termed simply Strength Design) accounts for the nonlinear behavior of concrete and an idealized elastic-plastic constitutive model for steel. Safety is ensured by using scaled-up loads, obtained by multiplying standard loads by specified load factors, and scaled-down resistances, found by multiplying nominal resistance values by strength reduction factors.

The developers of ACI 318-63 recognized that uncertainties in loads and strengths had not been treated rationally in working stress design. They noted that larger load factors should be applied to live, wind and earthquake loads, which can be highly variable, than to permanent loads, which are relatively predictable. Similarly, strength reduction factors were introduced independent of the load requirements to account for variability in strength and the nature of the failure mechanism. However, no useable probabilistic methods for code preparation were available in the early 1960's [4]. The load factors were set with the idea that the probability of exceeding the factored gravity loads would be approximately 0.001, while the strength reduction factors were set so that the factored strength was approximately equal to the 0.01 fractile of the actual strength. It should be emphasized that these probabilities were completely judgmental, since there was essentially no substantiating data available at the time the load and strength reduction factors were set for ACI 318-63.

During the late 1960's, probabilistic methods were developed and refined in response to the need to consider uncertainty explicitly and rationally [5,6]. Proposals to introduce explicit second-moment probabilistic information into ACI Standard 318 date back at least 20 years [7,8]. In these procedures the designer would select a desired safety index, β, and then perform the design with a knowledge of the means and standard deviation of both the load and resistance variables. The safety index positioned the mean (or median) resistance a sufficient number of standard

deviations above the mean load to ensure the target reliability desired for design. These approaches never were given serious consideration by ACI Committee 318 for adoption in subsequent editions of the ACI Standard 318.

2. The Current Situation

The load factors currently in use reflect probabilistic considerations only in a heuristic fashion. The basic combination without lateral load is

$$U = 1.4D + 1.7L \tag{1}$$

in which D and L are the dead and live load, respectively. The higher value associated with the live load is due to its greater variability. The basic combination involving wind load, W, is

$$U = 0.75(1.4D + 1.7L + 1.7W) \tag{2}$$

The load combination factor, 0.75, is intended to produce a combined value with a reasonable probability of occurrence, since it is considered extremely unlikely that the design level live load and wind load would occur at the same time. It may be noted, however, that there can be no rational basis for reducing the dead load, which acts independently of other loads on the structure.

Strength reduction factors on the resistance also have been selected heuristically. Their values for each type of failure (e.g., flexure, shear, bearing, compression), reflect the variability of the dominant controlling parameter (e.g., steel yield strength, concrete shear strength, concrete compression strength) and, to a degree, the mode of failure (ductile or brittle).

Although the current ACI 318 format takes into account relative uncertainties among both the loads and the resistances, there are a number of changes that would improve the consistency and the uniformity of reliability. Some of these will be discussed next.

3. Possible Changes to ACI 318-83

3.1 ANSI A58.1-1982 Load Factors

The 1982 edition of the American National Standard, Minimum Design Loads for Buildings and other Structures [9], contains load combinations to be used in checking strength limit states. The basic combinations without lateral load are:

$$1.4D \tag{3}$$

$$1.2D + 1.6L \tag{4}$$

The basic combination involving wind load is

$$1.2D + 0.5L + 1.3W \tag{5}$$

A comparison of Eqs. 1 and 3 and Eqs. 2 and 4 shows that the ANSI load factors are significantly different from the current ones in ACI Standard 318. In order to have one set of load factors for all construction materials, it has been proposed [10] that ACI adopt the ANSI load factors, as the American Institute of Steel Construction has done in its new Specification for Load and Resistance Factor Design [11]. Such a change would necessitate adjustments to the resistance side of the design equation.

3.2 Recalibration of ACI Strength Reduction Factors

With the adoption of the ANSI load factors, but no change in the format of ACI 318, current safety levels can be approximated by adjustment of the strength reduction factors. A comprehensive illustration of a reliability-based calibration of code formats to current acceptable practice has been completed [12,13]. Subsequently, a proposal for converting the ACI strength reduction factors to a set that is consistent with the load requirements in ANSI A58.1-1982 was published [14]. To date, ACI Committee 318 has been reluctant to adapt the ACI resistance criteria to the common load requirements in ANSI A58.1-1982.

3.3 Adoption of Partial Resistance Factors

The resistance of reinforced concrete structures arises from the combination of various effects, including tension, shear, and compression in both the concrete and the steel. Therefore, the single overall resistance factor has the same shortcoming that a single load factor would have, failing to recognize the differing degrees of uncertainty inherent in different variables. Somewhat more flexibility is obtained by assigning a resistance factor to each material characteristic (ϕ_s to steel yield strength, ϕ_c to concrete compressive strength, ϕ_v to concrete shear strength, etc.). Such an approach leads to a more uniform reliability over a variety of load ratios and member designs [15]. In addition, it is possible to introduce a multiplicative overall resistance factor that is dependent on the mode of failure [16]. This overall factor has the advantage of being able to incorporate additional mode-specific information. For instance, the equations used for predicting shear failure are much less accurate than those for flexure in underreinforced beams, and, in addition, the former occurs with fewer warning signs. Using this approach, a general design equation would be of the form

$$1.2D + 1.6L < \phi(R)(\phi_s f_y, \phi_c f_c{}', \phi_v V_{cr}) \tag{6}$$

It is also possible to introduce a factor on dimensions as one of the partial resistance factors. Since the standard deviation of dimensional variables tends to be relatively constant over a wide range of specified dimensions, the dimensional factor would logically be of an additive form, i.e., $d \pm \delta_d$. The partial factor format is widely used in Europe and recently has been adopted for use in the new Canadian code for concrete [17].

3.4 Example

Typical representative beams and columns have been designed in accordance with three different formats [18]. The first format is the current ACI318-83 code, the second is the proposed revision to the code by MacGregor [14], in which ANSI A58-1982 load factors are introduced along with commensurate changes in the overall resistance factors, and the third uses current ANSI load factors and partial resistance factors (without overall factors). Only gravity load cases were considered; consideration of wind forces would be necessary before criteria could be finalized.

A synopsis of the results follows for flexure, shear and compression. For flexural analysis of underreinforced beams ranging in depth from 18 in. (0.46 m) to 34 in. (0.86 m), the ACI method gave reliability indices varying from 3.01 to 3.78, the proposed revision (with an overall factor of $\phi = 0.85$), gave values from 2.88 to 3.32, and the partial factor format (with $\phi_s = 0.9$, $\phi_c = 0.6$, and $\delta_d = 0.5$ in. $= 13$ mm) gave values from 2.99 to 3.37. Moreover, the partial factor results were insensitive to variations in steel reinforcement ratio from 20% to 50% of the balanced steel ratio. A significant possibility of compressive failure existed for steel ratios above 65% of the balanced ratio, ρ_b, and consideration might be given to decreasing the current allowable value of 0.75 ρ_b.

An analysis of a similar range of beams, both singly and doubly reinforced, was conducted also. The ACI provisions for shear including limits on maximum stirrup spacing, gave reliability index values ranging from 3.20 to 5.07. Values for the proposed revision varied from 3.16 to 5.03 ($\phi = 0.75$), while those from the partial factor format varied from 3.39 to 5.02 (using the same partial resistance factors as with the flexural analysis).

Columns analyzed had square cross sections of 14 in. (0.36 m) and 18 in. (0.46 m), eccentricity ratios, e/h, from 0.0 to 1.25 e_b/h, and gross reinforcement ratios of 0.01 and 0.03. The current ACI provisions gave reliability index values from 2.96 to 4.70, the proposed method gave values from 2.75 to 4.47 ($\phi = 0.65$), and the partial factor method gave values from 3.08 to 4.07 (same factors as before).

From this study, it appears that partial factors of $\phi_s = 0.9$, $\phi_c = 0.6$, and $\delta_d = 0.5$ in. (13 mm) make it possible to attain reliability indices in a relatively narrow

range over a wide range of possible member design situations.

4. Safety During Construction

While the scope of ACI 318 includes both design and construction, the safety requirements of Chapter 9 address only those load and strength requirements that pertain to the service life of the structure. No specific guidance is given with regard to the construction phase, often the most critical time in the life of the structure. This is understandable given the complex nature of the construction process and the difficulty of setting out specific rules that can be applied to a wide range of construction procedures. However, since the majority of structural failures occur during construction, there is clearly a need to address the issue of whether to expand the current safety requirements to include the construction phase.

Reliability methods can in principle be applied to the construction phase. However, attention must be paid to the strong dependence on time of the development of load and resistance of a concrete structure as construction proceeds. Safety checks of the structure in service are made on the completed structure. During construction, safety depends on the strength and stability of the partially completed structure and on the adequacy of temporary support systems. In addition, if the strength of the structure is strongly dependent on concrete strength, the time dependence of concrete strength should be taken into account.

Since a complete stochastic model of a structure during construction would be intractable and not suitable for codification, it will be necessary to evaluate safety at discrete times in the construction process. A considerable amount of research is needed to develop the data and methodology to apply reliability concepts and develop code safety requirements for the construction phase.

The special nature of the construction phase suggests that some issues might be treated differently from in-service conditions. Most construction failures can be attributed to human error rather than to inherent variability in load and resistance parameters. Due to the relatively short time span of construction, the role of inspection and monitoring has obvious importance. For example, it may be possible to incorporate active load control over the man-made loads such as stored materials and equipment loads, allowing the use of a deterministic load value or a truncated probability distribution of the load. Reliability analyses on the other hand may be useful in identifying the sensitivity of various parameters to effects of human error.

5. Conclusions

The reinforced concrete industry has been a leader in the United States in design philosophy based on ultimate structural performance. The factored design method has permitted reasonably consistent safety for a complicated, composite material. However, the formal methods of probabilistic risk analysis were not available at the time ACI adopted its current load and resistance factor format. With other materials specifications and the ANSI load specification introducing factor design, this is an opportune time for ACI to update its standard, improve its consistency, and extend its applications to the construction phase.

REFERENCES

1. ACI Committee 318, "Building Code Requirements for Reinforced Concrete," (ACI 318-63), American Concrete Institute, Detroit, 1963, 144 pp.

2. ACI Committee 318, "Building Code Requirements for Reinforced Concrete," (ACI 318-83), American Concrete Institute, Detroit, 1983, 111 pp.

3. ACI Committee 318, "Building Code Requirements for Reinforced Concrete," (ACI 318-56), American Concrete Institute, Detroit, 1956.

4. Winter, G. "Safety and Serviceability Provisions in the ACI Building Code." in Concrete Design, US and European Practice, SP-59, American Concrete Institute, Detroit, MI, 1979, pp. 35-49.

5. ACI Publication SP-31, "Probabilistic Design of Reinforced Concrete Buildings," Detroit, 1972, 260 pp.

6. "A Proposal for an ACI Probabilistic Code Format," from ACI Committee 348, Subcommittee E, October, 1972, 19 pp.

7. Cornell, C.A., "A Proposal for a Probability-Based Code Suitable for Immediate Implementation," Memorandum to ASCE and ACI Committees on Structural Safety, MIT, Cambridge, Mass., August, 1967.

8. Turkstra, C.J., "A Modified Form of Cornell's Code Proposal," Memorandum to ASCE and ACI Committees on Structural Safety, McGill University, Montreal, January, 1968.

9. "Building Code Requirements for Minimum Design Loads for Buildings and Other Structures," (ANSI A58.1-1982), American National Standards Institute, New York, 1982, 103 pp.

10. Statement and Resolution of ACI Committee 348, Adopted September 28, 1983.

11. "Manual of Steel Construction, Load and Resistance Factor Design," First Edition, American Institute of Steel Construction, Chicago, 1986.

12. Ellingwood, B., et al. "Probability Based Load Criteria: Load Factors and Load Combinations." J. Str. Div., ASCE 108(5), 1982, pp. 978-997.

13. Galambos, T.V., et al. "Probability Based Load Criteria: Assessment of Current Design Practice." J. Str. Div., ASCE 108(5), 1982, pp. 959-977.

14. MacGregor, J.G. "Load and resistance factors for concrete design." J. Amer. Concrete Inst., Vol. 80, No. 4, 1983, pp. 279-287.

15. Ellingwood, B. "Safety Checking Formats for Limit States Design." J. Str. Div., ASCE 108(7), 1982, pp. 1481-1493.

16. Corotis, R.B., "Probability-Based Design Codes," Concrete International, Vol 7, No. 4, April, 1985, pp. 42-49.

17. "Design of Concrete Structures for Buildings," A National Standard of Canada, (AN3-A23.3-M84), Canadian Standards Association, Toronto, 1984, 281 pp.

18. Israel, M., Ellingwood, B., and Corotis, R., "Reliability-Based Code Formulations for Reinforced Concrete Buildings," Journal of Structural Engineering, ASCE, Vol. 113, No. 10, 1986, pp. 2235-2252.

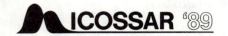

5th International Conference on
Structural Safety and Reliability

LIMIT STATES DESIGN IN MASONRY

CARL. J TURKSTRA

Dept. of Civil and Environmental Engineering,
Polytechnic University, 333 Jay Street,
Brooklyn, New York, 11201

ABSTRACT

This paper reviews some aspects of limit states design for
masonry structures based primarily on experience with a new
Canadian code.   Steps in an advanced second order safety
index analysis are summarized, the calibration process is
reviewed and the process of establishing code values of
material capacity reduction factors is discussed.

KEYWORDS

Limit states design, masonry, reliability analysis, partial
factors

1. INTRODUCTION

The design of masonry structures in North America is currently based on the
principles of working stress design. Since theoretical and experimental
information for masonry structures is rather limited, masonry design is
relatively traditional. Thus, for example, existing design formulas for the
analysis of eccentrically loaded walls are not based on conventional stress-
strain assumptions and engineering mechanics but on approximate relationships
fitted to test data.

In the United States, Ellingwood (1, 2) has examined reliability aspects of
masonry design with a view to development of a limit states design format. In
Canada, a new limit states masonry code based on reliablity concepts has been
proceeding through the approval process for several years.

Development of engineered masonry codes requires a return to first principles.
Theoretical work and test data must be reviewed and new, rational approaches
to behavior for a number of design cases must be developed before a meaningful
reliability analysis can be performed.

Reliability analysis for masonry is itself quite involved. there are many, generally nonlinear, limit states functions. Cases include (1) brick and block walls loaded eccentrically about their weak or strong axis, (2) plain and reinforced sections, (3) capacity in shear combined with variable axial forces, and (4) masonry beams. Even after detailed investigation many questions remain.

The purpose of this paper is to outline the development process for the Canadian code and indicate several highlights that might be of particular interest in reliability analysis. Only ultimate limit states design are considered - although serviceability problems are responsible for most failures in masonry construction, the state of knowledge is very limited and most requirements are simply rules of good practice.

2. PRELIMINARY STUDIES

The first stage in code development is to specify design values of basic material properties and other parameters. In some cases such as strength, these are "characteristic" values corresponding to fractiles of test data. In other cases such as elastic modulus, average values might be given.

For masonry design, the strength of a prism made of several units in stack bond with the mortar to be used is taken as the basic masonry strength for correlation to wall strength. For the new code, 5% fractiles were chosen for characteristic strengths. Simplified relationships between units/mortars/prism strengths and between prism strength/unit types/modulus of elasticity were developed through correlation studies.

To analyze eccecentric load effects, moments of inertia and modulii of elasticity must be provided. Since the stress-strain curves of masonry vary from nearly linear for bricks with certain mortars to more nearly parabolic for concrete blocks with other mortars, a considerable degree of simplification is involved.

It is essential to note that there is a great deal of freedom of choice in code development. Different assumptions will lead to different modeling errors and hence different design strength factors. In other words, similar designs can result from different sets of assumptions and numerical factors.

3. EXAMPLE - PLAIN WALLS

3.1 Behavior and Modeling Errors.

As an example of code development, consider the wall shown in Fig. 1 loaded eccentrically and laterally to produce combined axial stress and minor axis bending. This is a particularly difficult case because of the influence of nonlinear P- Δ effects.

The first step is to establish a short section interaction diagram which, in effect, involves assuming a stress - strain block in compression. It is known that the tensile strength of masonry is zero. The simplest assumption is a linear stress block in which case the equations of the interaction

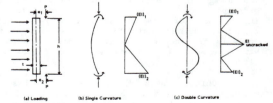

Fig. 1 BASIC LOAD CASE AND STIFFNESS ASSUMPTIONS

diagram can be developed from elementary mechanics. However, the predictions of linear theory become increasingly conservative relative to test data as load eccentricity increases (5, 6, 7).

This is an example of modeling error - i.e. a systematic bias in the relationship between theoretical prediction and experiment coupled with a random variation of test results about predicted values.

The second stage in analysis is to develop a rational method to predict the variation of wall capacity with slenderness. Physically, this is a very complex problem because, as bending moments increase, cracking takes place and spreads throughout portions of the wall. As cracking increases, stiffness decreases, lateral deflections increase and P - Δ effects increase the moments etc.

As a simple approach, a linear analysis of lateral displacements was used to calculate secondary moments (3, 4). An equivalent EI equal to the average over the wall height assuming a possible mechanism at failure was adopted (Fig. 1).

With the preceding assumptions, the theoretical capacity for available wall tests was calculated and the ratio of theoretical capacity to experimental capacity determined for each wall. The ratios were then plotted against slenderness ratio and end eccentricities to expose systematic variations. As expected, there was an average tendency to more conservative predictions as eccentricity effects increased. To correct this bias, theoretical predictions were multiplied by the mean/calculated values shown in Table 1.

TABLE 1: MODELING ERROR - PLAIN WALLS

ECCENTRICITY	MEAN/CALCULATED	C.O.V.	DISTRIBUTION
E/T < 1/6	.95-1.5 E/T	0.14	NORMAL
BRICK E/T > 1/6	.70-.9(E/T-1/6)	.14+.96(E/T-1/6)	NORMAL
E/T < 1/6	1.0-.9 E/T	0.18	NORMAL
BLOCK E/T > 1/6	1.0-.9 E/T	.18+.72(E/T-1/6)	NORMAL

With these average corrections, the histogram of residual modeling error shown in Fig. 2 was constructed. More detailed analysis of the variance of this modeling error lead to the estimates of variance given in Table 1.

3.2 Calibration

With the basic behavior mechanisms established, a limit state or G function can be developed for each case. These functions involve not only the random variables dealing with uncertainty in material properties, modeling errors and loads but must also consider uncertainties in analysis and eccentricity (in the load components) and in workmanship (in the resistance components).

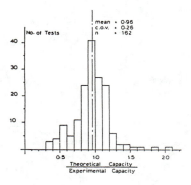

Fig. 2 MODELING ERROR AFTER CORRECTION
Plain Concrete Block Masonry

Shown in Table 2 is the set of factors used for plain masonry walls subjected to dead and live loads. A number of factors including especially allowances for workmanship are based on very limited data and judgment.

TABLE 2. RANDOM VARIABLES FOR CALIBRATION STUDIES

UNCERTAIN VARIABLES	MEAN/SPECIFIED	C.O.V.	DISTRIBUTION
Dead Load	1.00	.07	Normal
Live Load	.70	.30	Extreme I
Load Analysis	1.00	.20	Normal
Inspection			
Special	1.00	.10	Normal
Regular	.80	.15	Normal
None	.70	.20	Normal
Load Eccentricity	1.00	.20	Normal

The situation with respect to inspection is particularly interesting. While there is little data, there is enough to show that workmanship has a pronounced affect on wall strength with as much as 50 % variations in average strength. However, designers are often reluctant to assume responsibility for field supervision and fear inspections will not be reliable. Contractors and industrial representatives admit to only the best quality. In the final draft code, variations due to workmanship were not mentioned - a result that renders most debates about the accuracy of the rest of the design process rather academic.

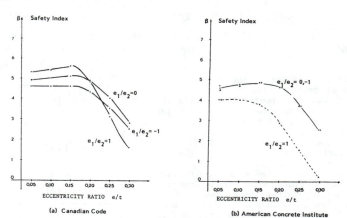

Fig. 3. CALIBRATION STUDIES
Plain Brick Masonry

Safety index analysis was performed using an advanced second order safety index analysis including the effects of distribution functions. Shown in Fig. 3 are typical calibration results for brick walls designed by the old Canadian Code and the ACI masonry code for a height to thickness ratio of 15 and a live to dead load ratio of 1.0. It is seen that safety levels depend on curvature , decreasing rapidly for single curvature. This is especially evident for the ACI code applied to thin walls because the ACI capacity reduction formula for slenderness makes no allowance for curvature. In general, safety levels were high relative to results obtained for working stress steel and concrete codes.

To complete calibration, safety index values were plotted as functions of live to dead load ratio and slenderness ratio. For the existing Canadian code it was found that safety increased with slenderness - a result that was judged desirable.

3.3 Development of New Code

Many alternative code formulations are possible, varying in terms of behavioral assumptions and simplicity. For plain masonry, it was decided to use the linear stress block assumption, zero tensile strength and the P - Δ analysis used in calibration. As mentioned previously, the linear stress block assumption tends to underestimate strength as eccentricity increases. This bias was ignored in the code formulation but was considered when the safety levels implicit in the code was calculated.

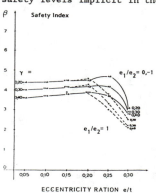

Fig. 4 shows a plot of safety index versus eccentricity for values of the resistance factor γ_m applied to the characteristic masonry prism. To obtain an index level of about 4 in this case, a γ_m value of 0.3 could be used. Fig 4. also suggests an upper level of about 0.25 for admissible eccentricity ratios. Safety levels changed very little the ratios of dead to live load and height to slenderness.

In the process of such studies, it is essential to understand the reasons for variations in safety levels. In Fig. 4, for example, uncertainty is dominated by uncertainties in workmanship, material strength, modeling errors and loads for small eccentricity. As eccentricity increases, the related uncertainty comes into play as the section begins to crack and the effective

Fig. 4. SAFETY INDEX - NEW CODE
Plain Brick, Uninspected

depth of the section becomes a variable. Thus uncertainty in eccentricity comes to dominate for large moments. It is for this reason that an upper bound on eccentricity ratio was retained in the code.

4. REINFORCED WALLS

Development of a procedure for reinforced walls followed a similar path. One major difference was an even greater lack of test data. Because of the number of variable involved, comprehensive safety analysis could not be performed - specific cases had to selected to indicate behavior.

In this case, the use of a rectangular stress block to conform to concrete practice was suggested. Any reasonable stress block can be used if the

reduction factors on resistance are adjusted accordingly. Of course the results would not be identical but they would be reasonably similar.

The difference between the interaction diagrams for linear and rectangular stress blocks depends on the level of reinforcement. The linear case is relatively more conservative but differences depend on the eccentricity ratio.

A major issue for reinforced masonry was the choice between the ACI approach which uses global Φ factors to reduce calculated axial force and moment capacities and a european approach which reduces masonry strength by a factor γ_m and steel strength by a factor γ_s before calculations are performed.

The advantage of separate γ factors can be see from Figs 5 and 6 which show the

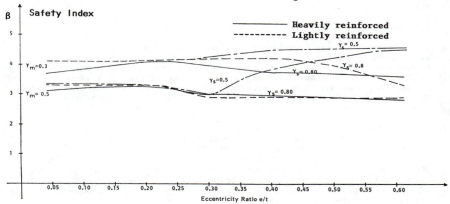

Fig. 5. NEW CODE, REINFORCED CONCRETE BLOCK, UNINSPECTED

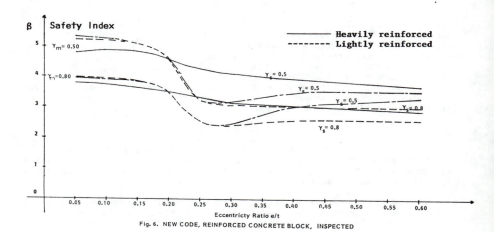

Fig. 6. NEW CODE, REINFORCED CONCRETE BLOCK, INSPECTED

results of a safety index analysis for the proposed code with wall height to
slenderness of 15 and dead to live load ratio of 1.0. In the region of small
eccentricity, safety if governed by the masonry strength and hence can be
controlled by the factor γ_m. For large eccentricities, safety is governed by
the variability of steel strength and hence can be controlled by the factor γ_s.
Results for eccentricity ratios around o.3 are not reliable because a
correction was not made for cases where two limit states (compression and
tension failure) give similar results.

5. SUMMARY

Development of a complete limit states design code based on reliability
concepts is a major undertaking. Without a reasonable understanding of the
behavior mechanisms involved, reliability analysis can be of limited use.

Code development is frustrating for many reasons, First of all, there are so
many special cases that one can not hope to examine more than a few in detail.
For the Canadian code for example, the problem of shear capacity was approached
in a very simple, lower bound way with a rather ad hoc choice of γ factors. A
second frustration is the poor state of knowledge in many cases. Masonry beam
behavior and the strength and stiffness of masonry connectors and anchors are
examples of such situations.

Inevitably there are what might be called organizational frustrations.
Practicing engineers have developed ways of doing things and are very concerned
about liability. Preference for the rectangular stress blocks and reluctance
(i.e. refusal) to admit different classes of inspections involve such factors.

Finally, many questions have no unique answer. Codes can be very complex or
they can be extremely naive. They can cover a vast field of construction or
they can be limited to, for example, short walls with small eccentricities.
Different assumptions and formats can lead to similar results. In all such
cases, it is essential to keep in mind that no code is more than an advisory
document. Designers always have the freedom and the responsibility to adapt
requirements to the needs of a particular situations.

6. REFERENCES

1. Ellingwood, B., "Analysis of Reliability for Masonry Structures," J. St.
    Div., ASCE, Vol.197, No.ST5, pp 757-773, 1980.

2. Ellingwood, B. and Tallin, A.G., "Probability Based Design for Engineered
    Masonry   Construction,"   Proc.   4th   ASCE   Specialty   Conference,
    Probabilistic Mechanics and Structural Reliability, Berkeley, pp 82-86,
    1984.

3. Ojinaga, J., and Turkstra, C.J., "The Design of Plain Masonry Walls," J.
    CSCE, Vol.7, No.2, pp 233-242, 1980.

4. Ojinaga, J., and Turkstra, C.J., "Design of Reinforced Masonry Walls and
    Columns for Gravity Loads," J. CSCE, Vol.9, No.1, pp 84-95, 1982.

5. Yokel, F.Y., "Strength of Load Bearing Masonry Walls," Journal of the
    Structural Division, ASCE, Vol. 94, No. ST5, pp 1593-1609, 1971.

6. Yokel, F.Y., Mathey, R.G., and Dikkers, R.D., "Compressive Strength of
    Slender   Concrete   Masonry   Walls,"   National   Bureau   of   Standards,
    Washington, DC, Building Science Series 33, pps 28, 1970.

7.   Yokel, F.Y., Mathey, R.G., and Dikkers, R.D., "Strength of Masonry Walls
     Under Compressive And Transverse Loads," National Bureau of Standards,
     Washington, DC, Building Science Series 34, pps 68, 1970.

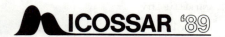
RELIABILITY-BASED ANALYSIS OF SHALLOW FOUNDATIONS

R.M. Bennett*, J.D. Hoskins III*, W.F. Kane*

*Department of Civil Engineering, The University of Tennessee,
Knoxville, TN 37996-2010

ABSTRACT

The effects of different parameters on the
reliability of shallow foundations are examined.
It is found that the footing depth, the footing
length, and the load ratio have almost no effect
on the reliability. The soil type does have a
significant effect of the reliability. Footings
on cohesionless soils have a reliability index
on the order of 0.9; footings on frictionless
soils have a reliability index on the order of
2.0; footings on ϕ-c soils have a reliability
index on the order of 3.4. Thus, the factor of
safety should vary with soil type in order to
achieve uniform reliability.

KEYWORDS

Bearing capacity; shallow foundations; safety; reliability.

1. INTRODUCTION

The ultimate capacity of shallow foundations which fail
through a general shear failure is determined using the bearing
capacity equation [1,2]. Conventional design methods are based
on meeting a minimum factor of safety, usually 3.0, for all
types of soil. This approach can lead to widely varying
probabilities of failure, or conversely, reliability indices.
The reason for these differing levels of safety is a variable
level of uncertainty in foundation design. This uncertainty is
due to footings being subjected to varying load ratios, such as
live load to dead load, diverse soil types, such as cohesionless
or frictionless, and different mechanisms of resistance
contributing to the capacity, such as surcharge or depth. A
more rational approach is the use of load and resistance factor
design. This format can lead to the design of footings with
more consistent levels of safety.

Prior to developing appropriate load and resistance factors
for design, it is necessary to determine the reliability level
inherent in current practice. This process is often referred to

as code calibration. In the study described herein, reliability analyses are conducted on numerous footings under a variety of conditions to examine current levels of safety in shallow foundation design.

2. PROBLEM DESCRIPTION

The capacity of a rigid shallow foundation with a width of 2.0m and varying length and depth is determined using the Terzaghi bearing capacity equation [2]. For a given factor of safety an allowable load is determined. This load is assumed to be composed only of varying ratios of dead and live load.

Three types of soil are considered: a cohesionless soil, a frictionless soil, and a ϕ-c soil. Nominal soil properties used in the analysis are summarized in Table 1. The presence of a water table is not considered.

Soil statistics used are representative of those reported in the literature [e.g., 3,4,5]. The soil cohesion is assumed to follow a beta distribution with a mean value equal to the nominal value, a coefficient of variation of 0.40, a coefficient of skewness of 0.65, and a lower bound of 0.0. The soil friction angle is also assumed to follow a beta distribution with a mean value equal to the nominal value, a coefficient of variation of 0.25, a coefficient of skewness of 0.41, and a lower bound of 0.0. The cohesion and friction angle are assumed to be correlated with a correlation coefficient of -0.3. Statistics and distribution type for the dead load and live load are obtained from [6].

The reliability analyses are conducted using second order methods [7]. The transformation of the correlated beta random variables to independent standard normal random variables is accomplished using the method postulated in [8].

3. RESULTS

Figure 1 shows the results for footings of infinite length, zero depth, live load to dead load ratio of 1.0, and varying factors of safety. The reliability clearly increases with increasing factors of safety. For a given factor of safety, there is a surprisingly different reliability index with different types of soil. Cohesionless soils (c=0) have much lower safety levels than ϕ-c soils, with frictionless soils (ϕ=0) being between the two.

Figure 2 shows the results for footings of infinite length, zero depth, factor of safety of 3.0, and a varying live load to dead load ratio. There is almost no variation with different load ratios, which is different from structural components, where the reliability significantly decreases with increasing live load to dead load ratio [5]. The reason for the invariance with footings is that the variability, as well as the direction cosines, are much greater for the soil than for the load.

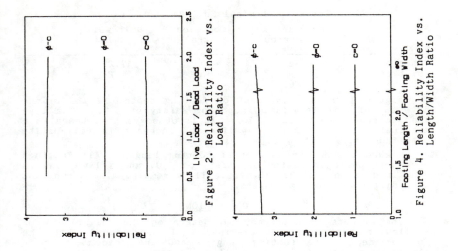

Figure 2. Reliability Index vs. Load Ratio

Figure 4. Reliability Index vs. Length/Width Ratio

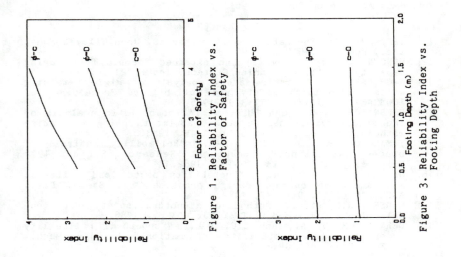

Figure 1. Reliability Index vs. Factor of Safety

Figure 3. Reliability Index vs. Footing Depth

Table 1:   Nominal Soil Properties

Soil Type	Friction Angle (degrees)	Cohesion (kPa)	Unit Weight (kN/m3)
Cohesionless	32	0	17.28
Frictionless	0	80	17.28
φ-c	26	45	17.28

Figure 3 shows the effect of footing depth. All footings have an infinite length, a factor of safety of 3.0, and a live load to dead load ratio of 1.0. It is observed that there is a marginal, but not really significant, increase in reliability with increasing footing depth.

Figure 4 shows the effect of footing length. All footings have zero depth, a factor of safety of 3.0, and a live load to dead load ratio of 1.0. The reliability remains almost constant with different footing lengths.

4.   CONCLUSIONS

The effects of different parameters on the reliability of shallow foundations were examined. It was found that the footing depth, the footing length, and the load ratio have almost no effect on the reliability. The soil type does have a significant effect on the reliability. Footings on cohesionless soils have a reliability index on the order of 0.9; footings on frictionless soils have a reliability index on the order of 2.0; footings on φ-c soils have a reliability index on the order of 3.4. Thus, the factor of safety should vary with soil type in order to achieve uniform reliability.

REFERENCES

[1] TERZAGHI, K.: Theoretical soil mechanics, John Wiley & Sons, Inc., New York, 510 p., 1943.
[2] HANSEN, J.B.: "A revised and extended formula for bearing capacity," Bull. 28, Danish Geotech. Inst., 21 p., 1970.
[3] A-GRIVAS, D.: "Seismic safety analysis of rigid retaining walls," Structural Safety and Reliability, Proceedings of ICOSSAR '85, III:311-III:319, 1985.
[4] BIERNATOWSKI, K., and PULA, W.: "Probabilistic analysis of the stability of massive bridge abutments using simulation methods," Struc. Safety, 5(1), 1-15, 1988.
[5] YAMAMOTO, M., and ANG, A.H-S.: "Reliability analysis of braced excavation," Structural Research Series 497, University of Illinois, 205 p., 1982.
[6] GALAMBOS, T.V. et al.: "Probability based load criteria: assessment of current design practice," J. Struc. Div., ASCE, 108(ST5), 959-977, 1982.
[7] MADSEN, H.O. et al.: Methods of structural safety, Prentice-Hall, Inc., Englewood Cliffs, NJ, 403 p., 1986.
[8] DERKIUREGHIAN, A., and LIU, P-L.: "Structural reliability under incomplete probability information," J. Engr. Mech., ASCE, 112(1), 85-104, 1986.

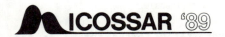
RELIABILITY OF MITER GATES

S. Toussi*, P. F. Mlakar** and A. M. Kao***

*   Bureau of Engineering, Board of Education, New York, USA
** Structural Division, JAYCOR Corporation, Vicksburg, MS, USA
*** US ARMY Construction Engineering Lab., Champaign, IL, USA

ABSTRACT
A new approach for the reliability calculation of
miter gates is presented. The reliability and
hazard functions are related to the surficial wear
of miter gates on their horizontal sill. A gate's
present and future deterioration can then be
assessed when this surficial wear is detected.
Three miter gates, experiencing operational
problems, are examined. The significant
similarities between the predicted and observed
results are outlined.

Keywords
Corrosion; Hazard function; Miter gates; Overload
factor; Reliability function.

1. INTRODUCTION

Miter gates are the most frequently employed type of lock gates
used in the United States. A miter gate consists of two
symmetrical leaves that are supported at the lock walls (see Fig.
1). The operating machinery for miter gates usually consists of
a large gear wheel and a sector arm revolving in a horizontal
plane. The forces to be overcome by the gate-operating machinery
are friction, wind loads, surges, hydraulic drag forces, and head
differentials created by the leaves moving through the water.

Wear, corrosion, and distortion create structural instability
and operational disfunctioning for miter gates [1]. While
corrosion is mostly experienced in the structural members, wear
is a problem of the operational components. Distortion is caused
by impulsive forces or loss of section of the structural members.
Water leakage due to the loss of tight contact between the
sealing blocks is an indication of distortion. The impact of
debris on the blocks or wear of operational components lead to
water leakage.

Damage due to each problem contributes to the increase of water
leakage. The important consequence of leakage is the loss of
contact between the leaves and the lock walls or between the
miter posts of the two leaves. This is an indication of damage
which causes the overstressing of the elements which do maintain
a tight contact. The consequence is an additional deflection of
the miter gate from its normal location at the mitered position.
The extent of the surficial wear of the horizontal sill is a
measure of this deflection. On the other hand, this deflection

represents an increase of the hydrostatic forces acting on the structural elements. The increase of hydrostatic forces is introduced by an overload factor. The yielding of a horizontal girder is considered to be the mode of failure and based upon which, the reliability and hazard functions are developed. Because a miter gate's reliability is directly related to a measurable deflection, its present and future deterioration can be predicted. This formulation is applied to three miter gates experiencing operational difficulties. The similarities between the predicted and observed results are discussed.

2. FORMULATION

Although a number of sophisticated methods are available, a statically determinate equilibrium analysis is used in this formulation [2]. The compressive stress, f_a, and the compressive bending stress, f_b, of a typical horizontal girder is related to water pressure due to differential head, the girder's geometrical parameters and the overload factor. The margin of safety against yielding is then defined as:

$$Ms = f_y - f_a - f_b \qquad (1)$$

Where f_y is the yield stress of the structural steel. The geometry of the girder is presumed to be known with confidence. All the uncertainty in resistance is included by taking the yield stress to be a random variable. The mean value is presumed to equal the nominal grade stress. The coefficient of variation is taken to be 0.01 in accordance with statistical results of AISI [3]. All the uncertainty about the loading will be modeled through considering the overload factor to be a random variable. A coefficient of variation of 0.20 is assumed in order to reflect a large uncertainty about both the overloading and its effect on the girder response. With this consideration and the geometry of

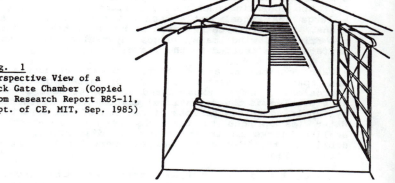

Fig. 1
Perspective View of a
Lock Gate Chamber (Copied
from Research Report R85-11,
Dept. of CE, MIT, Sep. 1985)

a particular gate being available, the reliability and the hazard functions are calculated. The overloading factor is not, however, directly observable. It is thus necessary to relate this factor to some manifestation of the gate condition which can be measured and used to assess the gates deterioration condition. Therein, the overload factor is related to the downstream miter-end deflection of the gate in the mitered position. Applying the method of a unit load, the flexural deflection and the compressive deformations are obtained. Thus, the reliability function is estimated from a measurable deflection.

3. APPLICATION

Wheeler Lock: The General Joe Wheeler Lock and Dam is located in Nashville. During the Fall of 1985, the large lower gate of the main chamber experienced loss of contact between the quion blocks and the gate monolith in the lower portion of the left leaf. The results of the reliability analysis are presented in Fig. 2. The reliability function decreases near an overload factor of 1.80 which corresponds to a net deflection of one inch. In the opinion of the operating staff, the extent of surficial wear was about one inch greater than the previous diver observation during normal operations. This coincidence of observation and prediction suggests that if the variation in mitering had been periodically measured, then some warning of the operational problem might have been possible.

Olwer Lock: The William Bacon Oliver Lock and Dam is located in the Mobile District. During the late Summer of 1985, the left leaf of the lower gate dragged severely on the sill during operation. A significant decrease in the reliability function occurs at a value of 1.60 for the overload factor which corresponds to a 0.6 inch net deflection at the miter end. This is consistent with the level at which experienced Mobile District personnel would become concerned about the condition of this particular gate.

Allegheny and Monongahela Locks: The concentration of this study is on a pair of standard leaves which were renovated and reinstalled during 1986. They are used for a practical illustration of the applicability of the proposed technique to relatively small gates. The reliability function (see Fig. 4) indicates that the condition of the gate requires attention when the overload factor approaches a value of 1.70. At this level of overloading, the net deflection is too small to measure under prototype conditions.

4. SUMMARY

The reliability of a deteriorating miter gate is directly related to the net deflection of its mitered position. This approach can be of practicality for a large number of miter gates. The proposed technique may not be, however, well suited for the gates one of which is illustrated in the third example.

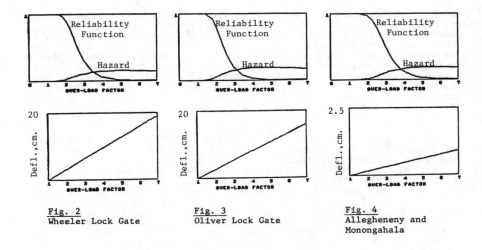

Fig. 2
Wheeler Lock Gate

Fig. 3
Oliver Lock Gate

Fig. 4
Allegheneny and
Monongahala

5. ACKNOWLEDGMENTS

The work upon which this paper is based was supported by the U.S. Army Construction Engineering Research Laboratory.

6. REFERENCES

1. Mlakar, P. F., "A Statistical Study of the Condition of Civil Works Steel Structures", Report No. J650-84-001, JAYCOR, Structural Division, Vicksburg, MS.

2. Department of the Army, "Engineering and Design, Lock Gates and Operating Equipment", Engineering Manual EM 1110-2-2703, Washington Dc.

3. American Iron and steel Institute, 1978, "Proposed Criteria for Load and Resistance Factor Design of Steel Building Structures", AISI Bulletin No. 27.

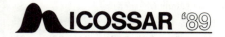
SECURE LANDFILLS: RELIABILITY-BASED ANALYSIS AND DESIGN

Charles L. Vita

Earth Consultants, Inc., 1805-136 NE, Bellevue, WA 98005 USA

ABSTRACT

A framework is outlined for reliability-based
analysis and design of secure landfills.
Reliability, R, is the estimated probability
landfill leakage, Ql, is below an allowable limit,
Qf; i.e., $R = P[Ql<Qf]$. A second moment, Bayesian
approach, with maximum entropy methods, is used.
Because in-service experience with secure landfill
performance is limited, the methodology aims to
complement and extend existing landfill analysis
and design practice to help optimize planning and
design to achieve cost-effective, reliable
landfills. The methodology provides a rational,
systematic framework for (1) estimating and
communicating landfill reliability, and (2) making
effective trade-offs between alternative actions.

KEYWORDS

Flexible Membrane Liner (FML), Hazardous Waste,
Landfill, Liners, Probability, Reliability.

INTRODUCTION

Hazardous waste land disposal requires a secure landfill to
isolate waste materials from the environment, particularly
groundwater. Secure landfills use three isolation/containment
systems: a cover, a primary liner, and a secondary liner. Liners
consist of a leachate collection and removal system (LCRS) underlain by a very low permeability leakage barrier. Barriers commonly
use a FML (flexible membrane liner) or geomembrane. In composite
liners the FML is underlain by a low permeability compacted soil
(typically clay). Covers are similar to liners except they have a
surface water collection and removal system (SWCRS), not a LCRS;
this cover-liner distinction is implicit in this paper.

The reliability methodology aims to optimize proposed or existing facilities and achieve cost-effective, reliable landfills. Reliability, R, is the estimated probability landfill leakage, Ql, is below an allowable limit, Qf: R = P[Ql<Qf]. Facility/site-specific criteria for leakage performance, Qf, and reliability, R-goal (e.g., R-goal = 99% = P[Ql<Qf=100 gallons per day]) can be based on risk management of landfill leakage effects. Then, for a given design, R is compared to R-goal: (1) R<R-goal, means estimated performance is inadequate, and must be increased by appropriate combinations of: (a) improved design, siting, construction, or operation, including retrofits or corrective/remedial action, and (b) improved analysis, i.e., improved information or modeling to reduced uncertainty. (2) R>R-goal, means estimated performance is adequate. Cost-effectiveness may be improved by refining the design to have equal marginal cost-reliability for each variable significantly affecting cost and reliability. This paper outlines ongoing methodology work [1,2].

LANDFILL RELIABILITY (R) & LANDFILL LEAKAGE (Ql)

R depends on Ql. Both are time (t)-dependent. Approximating PDF[Ql] as normal, a reliability estimate, R(t), is calculated from the standardized normal PDF, $f_U(u)$, where $u = (Ql-E[Ql])/V[Ql]^{.5}$, integrated from u(Ql=0) to the reliability index, $B(t) = (Qf-E[Ql(t)])/V[Ql(t)]^{.5}$. $f_U(u)$ can be truncated to eliminate impossible Ql values (e.g., Ql<0; and renormalized, via constant k, to assure $f_U(u)$ properly integrates to one. To a first approximation:

$$R(t) = (k/\sqrt{2\pi}) \cdot \int_{u(Ql=0)}^{B(t)} \exp(-u^2/2) \, du \tag{1}$$

Approximating PDF[Ql] as normal can be justified from (1) the central limit theorem, because liner leakage is from a sum of holes, (2) maximum entropy arguments [3], if only the expected value and variance are known, and (3) practical simplicity. Reasonable PDF[Ql] alternatives include lognormal, beta, and Monte Carlo simulated distributions.

Ql depends on the following. (1) Sources of leachate, in terms of flow $[L^3/t]$ or $[L^3]$, Q: external Q from cover infiltration, Qe; and internal Q from the waste or groundwater, Qi. (2) The fraction of total leachate absorbed by the waste, D. (3) Cover and liner system efficiencies, C, the ratio of flow removed from the system in a controlled manner, Qr, to flow entering the system, Qd (C = Qr/Qd): cover efficiency, Cc; primary liner efficiency, Cp; and secondary liner efficiency, Cs. C can range from zero (total leakage) to one (no leakage), the waste management aim. A second moment approximation for the expected value and variance of Ql is:

$$E[Ql] = Qle \{1 + x + y + z\} + Qli \{1 + z\} \tag{2a}$$

$$V[Ql] = (Qle + Qli)^2 \{CV[1-Cp]^2 + CV[1-Cs]^2 + CV[1-D]^2 + 2z\} \tag{2b}$$
$$+ Qle^2 \{CV[Qe]^2 + CV[1-Cc]^2\} + Qli^2 CV[Qi]^2$$
$$+ 2(Qle + Qli)Qle \{x + y\}$$

where: $x=p_{Cc,Cp}CV[1-Cc]CV[1-Cp]$; $y=p_{Cc,Cs}CV[1-Cc]CV[1-Cs]$;
$z=p_{Cp,Cs}CV[1-Cp]CV[1-Cs]$; $Qli=E[Qi]E[1-D]E[1-Cp]E[1-Cs]$;
$Qle=E[Qe]E[1-Cc]E[1-D]E[1-Cp]E[1-Cs]$.

Note in Eq 2: landfill leakage is also secondary liner leakage; primary liner leakage can be obtained by setting Cs=0; and cover leakage can be obtained by setting Cp=Cs=D=0.

All factors in Eq 2, depend on time, t. Qe is t-averaged over an appropriate analysis period, e.g., yearly. Efficiencies Cc, Cp, Cs, or C(t) generally, decay with time, from operational and post-closure mechanical, chemical, and biological loadings. C(t)s are modeled using an initial, t=0 value, C(0), a long-term (t=1) essentially steady state value, C(1), and a decay term, X(t), describing C(t) from C(0) to C(1): C(t)=C(1)+[C(0)-C(1)]X(t). Similarly for internal leachate flow, Qi(t) transitions, modeled by term Xi(t), from Qi(0) to Qi(1). D time-decays to zero, as the landfill approaches a uniform field capacity moisture content.

LINER EFFICIENCIES (C)

C depends on the configuration and performance of liner components: slopes, toes, floors, sumps, and so on. A component-based C-model is formulated by Vita [1,2]. Components are selected to have reasonably homogeneous liner design and loading conditions. Except for sumps, component efficiency, C_x, is: $C_x=(iK1W)/[iK1W+\Sigma(SFK2)]$; where: S = FML hole size; K1 = LCRS hydraulic conductivity; K2 = FML subgrade hydraulic conductivity; F = flow factor, dependent on S and FML subgrade; i = effective liner slope; W = component flow path width; Σ = Sum of random number (N) of random leaks (SFK2).

$E[\Sigma SFK2]$ and $V[\Sigma SFK2]$ are estimated as follows, using the concept of liner Leak Occurrence Mechanisms (LOMs). LOMs cause liner holes (punctures, bursts, tears, rips, etc.) and permeation or diffusion (formulated as a special case of hole leakage) due to: (1) post-QC/QA uncorrected FML holes and seam-unbonds, or (2) excessive internal or external mechanical or chemical loadings.

Each LOM is assumed to act over a representative loading area, 1A, and produce a load, L. L is resisted by liner resistance, T. T and L are correlated and time-dependent. They includes effects of design details, liner component interactions, liner history, creep, complex chemical or mechanical stress states, and so on. A liner hole occurs where L>T. The mean and variance of loads, E[L] and V[L], and resistance, E[T] and V[T], are estimated using second moment formulations of available FML and geotechnical analysis and design models [1]. A safety margin, SM=T-L, expected value and variance are then calculated from: E[SM]=E[L]-E[T], and $V[SM] = V[L]+V[T]-p_{L,T}(V[L]V[T])^{.5}$. Then, hole probability, P[H], the probability L>T in any give 1A, is: P[H] = P[T<L] = P[SM<0]. PDF[SM] can be approximated as normal, based on maximum entropy, given only E[SM] and V[SM] estimates; a beta distribution can be used if upper and lower bounds on SM are also known.

LOM loading areas are assumed to occur as Poisson events having uncertain average occurrence rate 1Ar, estimated by E[1Ar] and CV[1Ar]. Therefore, liner holes are approximated as Poisson events, occurring at an uncertain average hole rate Hr, with $E[Hr]=E[1Ar]P[H]$ and $CV[Hr]^2=\{CV[1Ar]^2+CV[P[H]]^2+CV[1Ar]CV[P[H]]\}$.

For practical convenience, and not unreasonably, Hr is assumed gamma distributed. Then, hole number, N, for given liner area or seam length, A, follows a negative binomial distribution [4] with:

$$E[N] = E[Hr] \; A \tag{3a}$$

$$V[N] = E[N](1 + E[N]CV[Hr]^2) \tag{3b}$$

Hole size, S, estimates are described by E[S] and V[S]. These can be estimated as E[g(SM)] and V[g(SM)], where g(SM) is a relation between S and SM dependent on design, loading and deformation limiting conditions [1]. If only E[S] is available, a maximum entropy V[S] can be made: $V[S]=E[S]^2$ or CV[S]=1. Where only an upper limit hole size, S^+, and lower limit hole size, S^-, can be estimated, the maximum entropy PDF[S] is uniform and the moment estimates are $E[S]=(S^+ + S^-)/2$, and $V[S]=(S^+ - S^-)^2/12$.

Associated second moment estimates of hydraulic parameters K2 and F are made and used to evaluate E[SFK2] and V[SFK2] for each LOM. From these, the moments of Σ(SFK2), the sum of a random number (N) of random leaks (SFK2), are estimated. To a first approximation, N and SFK2 are uncorrelated; then [5]:

$$E[\Sigma SFK2] = E[SFK2] \; E[N] \tag{4a}$$

$$V[\Sigma SFK2] = E[SFK2]^2 V[N] + E[N] \; V[SFK2] \tag{4b}$$

For multiple LOMs, ΣSFK2 is the sum of holes from each LOM. Assuming J independent LOMs and corresponding sum of holes, the expected value and variance of the combined effect are approximated by Eq 5 (where the sum is from j = 1 to J):

$$E[\Sigma SFK2] = \sum E[\Sigma SFK2]_j \tag{5a}$$

$$V[\Sigma SFK2] = \sum V[\Sigma SFK2]_j \tag{5b}$$

Conditional probability can be used to account for any significant dependence between LOMs or between N and SFK2 for a given LOM [1]. Eq 5 results and second moment estimates of K1 and i are used with second moment formulations of the C_x and C models to estimate E[C] and V[C] for the cover (Cc), primary liner (Cp) and secondary liner (Cs) [1]. Results are input to Eq 2 and Eq 1 to estimate R.

CITED REFERENCES

[1] Vita, C.L. and K.G. Haskell: "Flexible Membrane Liner Reliability Analysis Methodology," 4 vols, Draft Report to U.S. Environmental Protection Agency, Cincinnati, Ohio, Contract No. 68-03-4040, 1987.

[2] Vita, C.L.: "Reliability-Based Design Of Secure Landfills," In Symposium On Reliability-Based Design In Civil Engineering," Lausanne, Switzerland, 1988.

[3] Harr, M.E.: Reliability-Based Design in Civil Engineering," McGraw-Hill, New York, N.Y.

[4] Raiffa, H. and R. Schlaifer: Statistical Decision Theory, Harvard University Press, Cambridge, MA, 1961.

[5] Ross, S.M.: Stochastic Processes, Wiley, New York, N.Y, 1983.

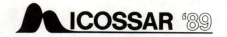
RELIABILITY MODELS OF THE MOHR FAILURE CRITERION FOR MASS CONCRETE

K.V.Bury* and H.Kreuzer**

* Dept.of Mech.Eng.,Univ.of Brit.Col.,Vancouver,B.C.,Can.
** Consultant,1311 McNair Dr.,North Vancouver,B.C.,Canada

Abstract

Failure in biaxially stressed concrete is here
modelled by the Mohr criterion. The reliability of
a stress state is evaluated in terms of 2 load
variables and 1 or 2 strength variables. The relation between the mean safety factor and the reliability index is demonstrated and compared for compression and tension/compression stress states.
Formulation of the reliability model is shown to
critically affect the results.

Keywords
Concrete, Mohr criterion, reliability, safety factor

1.Introduction

This study concerns the safety assessment of mass concrete structures such as dams. A biaxial stress state is assumed for a particular section of the structure. The failure state is modelled by
a Leon parabola [1], see Fig.1, in terms of shear stress τ , normal
stress σ , and tensile (flexural) strength f_t and compressive
strength f_c of concrete:

$$\tau^2 = Q \cdot (f_t + \sigma) \qquad (1)$$

where $Q = f_c + 2 \cdot f_t - 2 \cdot \sqrt{f_t \cdot (f_t + f_c)}$

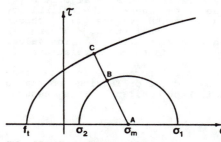

FIG1.: PRINCIPAL STRESS CIRCLE AND
MOHR FAILURE ENVELOPE

This failure envelope is chosen because it accounts for
the presence of shear often
observed at the onset of
cracking in mass concrete
structures [2]. The state of
stress is represented by a
Mohr circle in terms of prin-
cipal stresses σ_1 and σ_2 , see
Fig.1. In this study it is
conservatively assumed that
failure develops with the mean
princ.stress $\sigma_m = (\sigma_1 + \sigma_2)/2$ remaining constant. Thus the
safety of the given stress
state is assessed with respect
to the point C on the failure
envelope closest to σ_m .

Two approaches to safety assessment are compared. One is the traditional safety factor in terms of the mean values of stresses and material strength, see Fig.1:

$$FS = \overline{AC}/\overline{AB} \qquad\qquad (2)$$

The other is the probabilistic approach in terms of the Hasofer-Lind reliability index β which recognizes stresses and material properties as random variables since neither are usually known with certainty. The relations between β and FS are demonstrated, with uncertainty levels of loads and concrete strengths as parameters, in terms of coefficients of variation (C).

2.Analysis

A medium-strength concrete is assumed, with mean strength values $\overline{f_c}$ $=300$ and $\overline{f_t}=37.5\,kg/cm^2$. These values define a specific mean failure envelope (1), so that the safety factor (2) can be computed for a given stress state σ_1 , σ_2. The pure compression state is represented by the principal stress ratio $\sigma_2/\sigma_1 = 0.2$, while tension/compression is given the ratio -0.2. These two ratios encompass frequently encountered ultimate stress states. For the probabilistic analysis the loads (and hence stresses) are assumed to be Normally distributed random variables, while the strength properties are assumed to be Log-Normally distributed. The former choice essentially implies usual loads (eg.dead-weight) on the structure, rather than extreme loads such as earthquakes. The latter choice implies relatively high quality control of the concrete during construction. Three uncertainty levels are assumed for loads: $C_L=0, 0.1, 0.2$ representing a typical range of C-values for usual loads on mass concrete structures. Similarly, three uncertainty levels for concrete strengths are assumed: $C_R=0.05, 0.1, 0.2$; this range of values is typical for concrete materials, and corresponds to a practical range of sample scatter for concrete tests.

Reliability analysis is now a widely known technique [3] for assessing the safety of structures under uncertainty of information; it is assumed that the reader is sufficiently familiar with it to appreciate the following remarks. The failure surface for the reliability problem indicated by Fig.1 is taken to be:

$$M = \overline{AC} - \overline{AB} = 0 \qquad\qquad (3)$$

and is expressed in terms of four random variables σ_1, σ_2, f_t , and f_c :

$$M = Q \cdot \sqrt{f_t + (\sigma_1 + \sigma_2)/2 - Q^2/4} - (\sigma_1 - \sigma_2)/2 = 0 \qquad (4)$$

with $Q = Q(f_t , f_c)$ defined in (1). The Rackwitz-Fiessler algorithm [4] is applied to this 4-variable model to generate the reliability index β , corresponding to: mean strength values $\overline{f_c} , \overline{f_t}$, chosen strength variability C_R , chosen mean principal stress $\overline{\sigma_1}$ giving $\overline{\sigma_2}$ for a chosen σ_2/σ_1 ratio, and chosen load variability C_L . The values of the chosen variables are varied to yield the $FS-\beta$ relations.

In the practice of concrete engineering, information on the tensile strength value is sometimes scarce or lacking, so that a ratio of ultimate strength values is kept constant. This assump-

tion was made in a further analysis, with $f_c/f_t=8$ corresponding to typical experience. This implies that f_c and f_t are perfectly correlated, so that there are now only 3 random variables in the model; hence this model represents a decreased uncertainty level compared to the 4-variable model, since f_c and f_t are not linearly related in (4). Thus in problems where these strength properties are of similar influence on the structure's safety, the 4-var.model is a more conservative one.

3.Results

The general influence of uncertainty in material strength on the $FS-\beta$ relation is shown on Fig.2 for a fixed level of load uncertainty. Fig.3 shows the influence of load uncertainty for a fixed level of strength uncertainty. It is seen that for pure compression $(\sigma_2/\sigma_1 = 0.2)$, β -values are lower for the 4-var.model than for the 3-var.model, at any given FS-value. This result is as expected since in this stress region both strength parameters influence the shape of the failure envelope, and the parameter decoupling of the 4-var.model represents a higher level of overall uncertainty. However, for the tension/comp. state $(\sigma_2/\sigma_1 = -0.2)$ these results are reversed: the 4-var.model indicates a higher reliability at a given FS-value than the 3-var.model. This phenomenon is due to the fact that, when the stress circle is close to the apex of the failure envelope, f_t dominates the reliability analysis, f_c having little influence on the shape of the envelope in that region. Thus the strength-decoupling of the 4-var.model yields the higher assessed reliability measure β , whereas the strength-correlation in the 3-var. model gives undue influence to f_c , resulting in the lower reliability assessment. This result is significant in that it in-

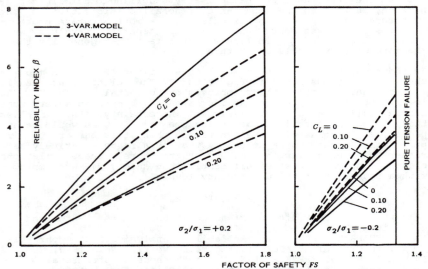

FIG.3: THE INFLUENCE OF LOAD UNCERTAINTY ON THE FS-β RELATION FOR THE THREE-
AND FOUR-VARIABLE MODELS, AT CONSTANT RESISTANCE UNCERTAINTY C_R=0.15.

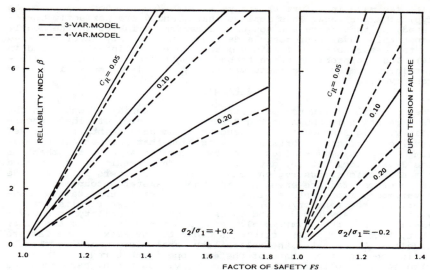

FIG.2: THE INFLUENCE OF RESISTANCE UNCERTAINTY ON THE FS-β RELATION FOR THE
THREE- AND FOUR-VARIABLE MODELS, AT CONSTANT LOAD UNCERTAINTY C_L=0.05.

dicates the importance of realistically modelling the failure phenomenon, as well as the importance of detailed information on the
structure. It implies, for example, that a more accurate probabilistic safety assessment would result if separate C-values were
assigned to f_c and f_t, in line with the fact that generally there
are more data available on f_c than on f_t.

In addition, the following general observations [2] may be made on
the FS–β relations shown in Figs.2 and 3. The relations are nearly
linear; that is, reliability improves proportionally with FS. At
lower uncertainty levels an increase in FS improves safety more
substantially than at high levels of uncertainty; this implies
that β is a more refined safety measure than FS, as would be expected, and accords with the experience that it is easier to improve the safety of a structure that is already relatively safe
rather than one of marginal safety. Furthermore, these relations
clearly indicate how significantly the safety measure improves
with a reduction of the uncertainty levels.

4.References

[1]   LEON,A.: "Uber die Scherfestigkeit des Betons", Beton und
      Eisen, Vol.34, No.8, pp.130-135, 1935.
[2]   KREUZER,H.,BURY,K.V.: "Reliability Analysis of the Mohr Fai-
      lure Criterion",ASCE,Jrl.Eng.Mech.,Vol.111,No.3,1989.
[3]   MADSEN,H.O.,KRENK,S.,LIND,N.C.: "Methods of Structural
      Safety", Prentice-Hall, Englewood Cliffs, N.J., 1986.
[4]   RACKWITZ,R.,FIESSLER,B.: "Structural Reliability Under Com-
      bined Load Sequences", Comp.and Struct.,Vol.9,pp.489-94,1978.

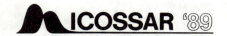
RELIABILITY ANALYSIS OF REINFORCED CONCRETE STRUCTURES
FOR SERVICEABILITY LIMIT STATES

Zhao Guofan[*]   and   Li Yungui[**]

* Prof. , Dept. of Civil Engineering, Dalian University
   of Technology, Dalian, 116024, China
** Doctoral Candidate, Dept. of Civil Engineering, Dalian
   University of Technology, Dalian, 116024, China

ABSTRACT

   Some   problems of reliability analysis of
reinforced concrete structures in serviceability
limit states are discussed, and two reliability
analysis methods for serviceability limit states
are presented. The applications of the presented
methods are illustrated by taking maxmium crack
width limit state of reinforced concrete structure
in flexure as examples. The calculated examples
have shown that the presented methods might be
more rational.

KEYWORDS

Serviceability; limit state; reinforced concrete;
reliability; fuzzy set; crack.

1. INTRODUCTION

   In the design of structures, the limit states of a structure
are usually classified as ultimate strength limit state and
serviceability limit state. The ultimate strength limit state
indicates that the whole structure or its members has come to the
state where no more loads can be born. The serviceability limit
state is the state in which crack width or deformation of the
structure or its members has reached the allowable value in
serviceability or in durability.

   Changes in analysis, design and construction practice in recent
years have led to well integrated structural systems built with
high strength, lightweight construction materials. Such systems are
relatively   flexible   and lightly damped.   In comparison with
traditional construction they may be prone to objectionable static

or dynamic structural movement. Design experience shows that in some case the serviceability limit states are more important . However, most research efforts have been directed towards the reliability analysis of structures in ultimate strength limit states. Our discussion in this paper will centre on the reliability analysis of reinforced concrete structures for serviceability limit states (crack width or deformation).

The calculations for serviceability limit states require that the maximum crack width or deformation of a reinforced concrete member should not exceed the allowable value, i.e. $X<[X]$, where X is the maximum crack width or deformation of the reinforced concrete member under service loads, and $[X]$ is the allowable crack width or deformation. It is usually stipulated that $[X]$ is a constant in current codes. However, in practical engineering, it is known that $[X]$, the maximum crack width or deformation which does not lead to failure of a member in service, is not a constant. In order to carry out the reliability analysis of a structure in serviceability limit states, two methods are presented in this paper based on approximate probability theory and fuzzy sets theory, respectively.

2. APPROXIMATE PROBABILITY ANALYSIS FOR THE RELIABILITY OF REINFORCED CONCRETE STRUCTURES FOR SERVICEABILITY LIMIT STATES

Similar to ultimate strength limit states, the reliability margin M in serviceability limit states can also be expressed as $M=[X]-X$, where $[X]$ and X are random variables. The mean value μ_M and the standard deviation σ_M of M can be derived from the mean value and standard deviation of $[X]$ and X. Thus, failure is associated with $M<0$, and survival is associated with $M>0$. Since the allowable failure probability in a serviceability limit state is usually bigger than 10^{-3}, the probability distributions of $[X]$ and X have less effect on the failure probability, so it is usually assumed that $[X]$ and X have normal distributions. The reliability index β can be calculated by using first order and second moment reliability method.

$$\beta = \mu_M / \sigma_M \tag{1}$$

[Example 1] The expression of maxium crack width of reinforced concrete members adopted in the code for design of reinforced concrete structures in harbour engineering in China is as follows

$$X = W_{max} = Q \frac{C_1 C_2 C_3 (30 + \phi)}{A_s E_s d(0.28 + 10 \rho)} \tag{2}$$

where C_1, C_2, C_3 are constants, Q is the moment at the calculated section, ϕ is the diameter of steel bar, A_s and E_s are the area and modulus of elasticity of steel bar, d is effective depth of the section and ρ is the content of steel bar. According to reference [3], Eq.(2) can be rewritten as

$$X = K_z K_p K_Q K_a X_k \tag{3}$$

where K_z is the uncertainty between maximum crack width of a real

member and that of the member in laboratory. The mean value μ_{kz} is taken as 1.0 and the coefficient of variation δ_{kz} is taken as 0.07.

K_p is the uncertainty of the calculation formula, $\mu_{kp}=0.95$, $\delta_{kp}=0.34$.

K_Q is the uncertainty of load effects, $\mu_{kQ}=0.8--0.95$, $\delta_{kQ}=0.1--0.25$.

K_a is the uncertainty of materials and geometry, μ_{ka} and δ_{ka} can be calculated according to the last term of Eq.(2).

X_k is the characteristic value of X.

The mean value and coefficient of variation of X can be given as follows

$$\mu_x = \mu_{kz}\mu_{kp}\mu_{kQ}\mu_{ka}X_k \tag{4}$$

$$\delta_x = \sqrt{\delta_{kz}^2 + \delta_{kp}^2 + \delta_{kQ}^2 + \delta_{ka}^2} \tag{5}$$

The statistic parameters of allowable maximum crack width [X] which make the structure unserviceable, should be determined using sufficient statistical data. According to the above parameters, the reliability analysis can be carried out by using Eq.(1).

3. FUZZY PROBABILITY ANALYSIS FOR THE RELIABILITY OF REINFORCED CONCRETE MEMBERS FOR SERVICEABILITY LIMIT STATES

In the above section, an approximate probability analysis method for serviceability limit states has been given by considering the randomness which affects the service of a structure or its member. This consideration is in agreement with that of the current code. However, not only randomness but also fuzziness affects the service of a structure. For example, for the crack width limit state of a reinforced concrete structure, the load effects are random variables, but the failure criterion has a certain fuzziness. It is reasonable to regard the failure of a reinforced concrete structure in its serviceability limit state as a fuzzy random event. So fuzzy sets theory may be used to evaluate the failure probability of a structure for serviceability limit states.

When using fuzzy sets theory to evaluate the failure probability of a structure in a serviceability limit state , it is important to determine the relationship function s(z) of the failure criterion and the probability density function f(z) of the load effects. Because the failure probability of a structure in a serviceability limit state is usually greater than 10^{-3}, it is reasonable to assume that the probability density function f(z) is a normal density function with a mean value μ_w and a standard deviation σ_w. Determination of the relationship function s(z) is a difficult problem. It is always done by means of fuzzy statistic test. However, in the case of lack of data, it may be done on the basis of experience. For exmaple, it can be given as follows [4]

$$s(z) = \begin{cases} 0 & z \leqslant a \\ (z-a)/(b-a) & a < z \leqslant b \\ 1 & z > b \end{cases} \tag{6}$$

Where, a and b are constants to be determined on the basis of experience and some design codes. According to the fuzzy sets theory, the failure probability P_f is given as follows

$$P_f = \int_{-\infty}^{+\infty} f(z)S(z)dz \tag{7}$$

[Example 2] For the maximum crack width limit state of a reinforced concrete beam, assuming that the maximum crack width which is produced by loads, has a normal probability distribution, where $\mu_w = 0.1291mm$, $\sigma_w = 0.0475mm$, and assuming further that the relationship function s(z) is given as Eq.(6), where a=0.10mm, b=0.35mm. We have $P_f = 1.486 \times 10^{-2}$.

4. CONCLUSIONS

(1) The reliability analysis of reinforced concrete structures for serviceability limit states has been discussed in this paper, and two reliability analysis methods for serviceability limit states, an approximate probability analysis method and a fuzzy probability analysis method, have been presented.

(2) The consideration of randomness in serviceability limit states is in agreement with that of the current code as well as recent research. The consideration of fuzziness in serviceability limit states might be a new approach. However, the fuzzy probability analysis method presented in this paper should be developed and improved with further research.

(3) There is little data on serviceability limit states. It is necessary to investigate further to improve the reliability analysis of reinforced concrete structures for serviceability limit states.

REFERENCES

[1] Naaman, A. E. and siriaksorn, A., Reliability of Partially Prestressed Beams at Serviceability limit states, PCI Journal, Vol.27, No. 6, 1982, pp. 66-85.
[2] Ellingwood, B. ,Probability-based Criteria for Serviceability Limit States, Proceedings of ICOSSAR'85, Kobe, Japan, 1985, pp.137-146.
[3] Guofan Zhao, Causes and Reliability Analysis of Cracking in Concrete Structures, Journal of Dalian University of Technology, No.3, 1984.
[4] Tieyu Tang, Fuzzy Probability Analysis for the Reliability of the Crack Control in Reinforced Concrete Members, Journal of Dalian University of Technology, No. 2, 1986.
[5] G.F. Zhao et al, Structural Reliability Analysis, Publishing House for the Hydraulic and Electric-power Industry, Beijing, 1984.

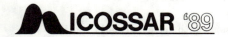

RELIABILITY ANALYSIS OF AXIALLY COMPRESSIVE MEMBERS

Ji-Hua Li', Zheng-Zhong Xia'

* Prof. of Civil Engineering, Chongqing Institute
of Architecture and Engineering, Chongqing,
People's Republic of China

ABSTRACT

In the steel structure design code (TJ17-74) in China,
a single curve was adopted for designing axially
compressive members and the reliability of the
column analyzed previously was based on this curve.
However the present international tendency is to use
multiple column curves and it is more resonable in
design than using a single curve. We had computed 93
curves by use of maximum strength theory. This paper
analyzed the reliability of column based on these
curves.

KEYWORDS

Steel axially loaded columns; single column curve;
residual streess; out-of-straightness;multiple column
curves; maximum strength; reliability; safety index.

1. INTRODUCTION

In deriving the code TJ17-74 column curve, it is assumed that the
Euler hyperbola is valid at the stress up to the proportional limit
of the column stress-strain relation, and above the proportional

limit,   the plastic strength is assumed to be a parabola which is
determined by tests. This column-strength had then be reduced by
a constant factor of safety to obtained allowable-stress formula.
This method can't reflects the actual working condition of   various
columns. The current method is using multiple column curves. We   had
computed 93 column curves by use of maximum strength theory.

2. MULTIPLE COLUMN CURVES

Recent studies have demonstrated the desirability of including of
effect of initial imperfections in the theory instead of   in   the
factor of safety. 93 maximum strength curves are computed for a wide
range of column shapes and types. Each curve is based on an   actual
measured residual stress distribution and an assumed initial out-of-
straightness at mid-height of δ_c=0.001L, where L is the   length of
column. Results are presented herein as column strength curves a, b
and c, the arithmetic-mean curves that present three subgroups of 93
curves.

These curves a, b and c compared with the curves commended by
ECCS and Lehigh University and TJ17-74 are shown in Fig.1.

By the mathematical statistics method, we obtained three set of
equations through which the more exactly approximate value of curves
a, b and c can be computed.

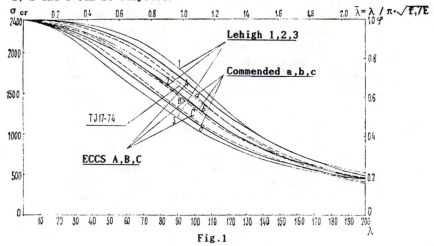

Fig.1

3. RELIBILITY INDEX

The limit state equation for dead load(D) + live load(L) is

$$0.92R_k = 1.2C_rG_k + 1.4C_LL_k$$

in which R_k ······ nominal resistance
                 C_r ······ coefficient of dead load effect
                 C_L ······ coefficient of live load effect
                 G_k ······ nominal dead load
                 L_k ······ nominal live load

According to the analysis of statistical parameters of steel column (here omitted) we can calculate reliability index.

Reliability index are computed by the method of first order and second moment as following Table:

curve	$\mu\lambda$	20	40	60	80	100	120	140	160	180	200	$\overline{\beta}$
	0.25	3.19	3.23	3.37	3.34	3.16	3.08	2.97	2.88	2.93	2.93	3.11
	0.50	3.30	3.34	3.36	3.38	3.24	3.18	3.10	2.98	3.06	3.07	3.20
a	1.0	3.20	3.21	3.23	3.24	3.11	3.08	3.02	2.94	2.99	3.00	3.10
	2.0	3.07	3.07	3.08	3.06	2.98	2.96	2.92	2.85	2.88	2.89	2.98
	0.25	3.18	3.18	3.18	3.05	2.88	2.96	3.01	2.99	2.99	3.05	3.06
	0.5	3.30	3.30	3.30	3.20	3.06	3.11	3.14	3.11	3.11	3.20	3.18
b	1.0	3.20	3.19	3.18	3.11	3.01	3.03	3.04	3.03	3.02	3.07	3.09
	2.0	3.06	3.05	3.05	2.99	2.91	2.92	2.93	2.92	2.91	2.95	2.97
	0.25	3.14	3.27	3.30	3.29	3.23	3.31	3.30	3.24	3.24	3.18	3.25
	0.5	3.33	3.36	3.36	3.34	3.28	3.32	3.32	3.27	3.26	3.25	3.31
c	1.0	3.20	3.20	3.20	3.21	3.13	3.16	3.16	3.13	3.12	3.11	3.16
	2.0	3.07	3.07	3.06	3.04	2.99	3.01	3.01	2.98	2.98	2.98	3.02

$$\rho = \frac{\text{live load effect}}{\text{dead load effect}}$$

Curve a $$\beta = \frac{3.11+3.20+3.10+2.98}{4} = 3.10$$

Curve b $\qquad \beta = \dfrac{3.06+3.18+3.09+2.97}{4} = 3.08$

Curve c $\qquad \beta = \dfrac{3.25+3.31+3.16+3.02}{4} = 3.19$

4. CONCLUTIONS

This paper analyzed the reliability of column based on multiple curves and found that the safety index β of middle slenderness ratio column designed by the code TJ17-74 is rather low, which is conformed to the general regularity. The present reliability analysis is abviously better than the previous one.

The further work is studying the properties of heavy plates for the column of tall buildings.

REFERENCES

[1] A.H-S.Ang and C.A.Cornell: Reliability Basis of Structure Safety and Design. Journal of the Structural Division, ASCE, 100, Sept., 1974, PP.1755-1769.

[2] T.V.Galambos: The Basis for Load and Resistance Factor Design Criteria for steel Building Structures. Canadian Journal of Civil Engineering, Volume 4, 1977.

[3] D.J.L. Kennedy and M.Gad Aly: Limit State Design of Steel Structures Performance Factors. Canadian Journal of civil Engineering, Volume 7, 1980.

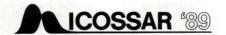
LIFETIME BEHAVIOR OF WOOD STRUCTURAL SYSTEMS

W. M. Bulleit and P. J. Vacca, Jr.

Associate Professor and Former Graduate Student
Department of Civil and Environmental Engineering
Michigan Technological University
Houghton MI   49931

Abstract

The lifetime behavior of wood floors and flat roofs was examined
using Monte Carlo simulation.   The number of members which
failed over the systemlifetime was examined.

Keywords

Systems, structural systems, wood, statistical analysis.

1. Introduction

Reliability based design of wood structural systems requires knowledge
about the lifetime behavior of the system.   Wood systems present some unique
difficulties in obtaining this information: 1) Wood systems require a complex
analysis which includes partial composite action and two-way behavior. 2) The
load duration effect under random load sequences must be considered.   3) Wood
systems are highly redundant so first member failure behavior is necessary but
not sufficient information.

The objective of the research discussed in this paper was to gain
knowledge about wood floor and flat roof systems over a lifetime of 50 years.
Design stress levels [1] and overload conditions were considered.   Load
duration, random member properties and random load histories were included.

2. Background

It was realized some time ago that efficient use of wood in wood systems
required analysis procedures which included partial composite and two-way
action.   The system analysis procedure chosen for this research was FEAFLO [2]
because it has been well studied and was available in the literature.

The load duration effect was a necessary inclusion. Due to the nature of the simulations to be performed, a cumulative damage model had to be chosen which would minimize computing time. It has recently been shown that the so-called exponential damage rate model (EDRM) adequately models load duration under random load sequences [3]. EDRM was chosen with parameters taken from the literature [3]. Thus the damage accumulation rate is:

$$\frac{d\alpha}{dt} = \exp[-40.0 + 49.75S]$$

where S is the stress ratio $\sigma(t)/MOR$, with $\sigma(t)$ being the stress history and MOR is the short term ultimate bending stress. The damage parameter, α, ranges from 0.0 to 1.0. When $\alpha \geq 1.0$, failure occurs.

The member property simulations were performed using lumber property information from Pellicane [4]. The Johnson SB distribution was used for both MOE and MOR since that distribution best fits those properties in many cases. The correlation between MOE and MOR can now be dealt with in the simulation [5]. A correlation coefficient of 0.7 was used in the simulations.

The load history data was taken from the literature. The maximum annual ground snow load data was obtained from Ellingwood and Redfield [6]. The conversion of ground snow loads to roof snow loads was performed following O'Rourke and Stiefel [7], assuming a heated and moderately sheltered structure. Wood roofs for two locations in Michigan, Flint and Marquette, were examined. The in time variation of the annual snow load was assumed to be triangular for Flint and parabolic for Marquette. The floor load data were taken from Chalk and Corotis [8]. The floor load model consisted of sustained and extraordinary live load events assumed to be rectangular pulses. Nominal dead load was determined for each system design and assumed to be the mean dead load. The coefficient of variation of dead load was 0.10. The dead load was assumed constant over the life of the structure.

3. Results

All systems consisted of 50.8 x 203.2 mm, No. 2 or better Douglas fir members spaced 406.4 mm on centers. The roofs had 12.7 mm thick plywood sheathing and the floors had 19.05 mm thick plywood sheathing, each attached with 8d nails. Each of the three systems, a residential floor, a roof in Flint and a roof in Marquette, were designed in accordance with the NDS [1]. The nominal design loads were, respectively, 1.92 kPa, 1.01 kPa and 2.01 kPa where the last two loads are 0.7 times the respective nominal ground snow load [9]. A load duration factor of 1.0 was used in the floor designs and value of 1.15 was used for the roofs [1].

The span of the members was then controlled by the amount of overstress desired for a particular simulation. Four simulations were performed for the three systems. Two were designed with no overstress, i.e. design stress levels, where one had material properties from in-grade tests on lumber from a wide geographical area and one from the so-called Inland region [4]. In general, the smaller the area from which lumber is harvested, the less the variation in the strength distributions. Because of this, the systems built from Inland Douglas fir were simulated for 100 years rather than 50 years. The other two simulations were for 20 and 40 percent overstress using the member property data for lumber from the wide geographical area.

Each of the above sets of simulations consisted of 100 systems, with one exception where 200 systems were simulated. During the system lifetime, when a member failed the failed member had its stiffness reduced to 10 percent of its original value and the system stiffness matrix was regenerated. This is the same reduction used by Criswell [10]. The reduction of stiffness in actual wood system tests is highly variable and is also not a single event [11].

Table 1 shows the number of systems which sustained one or more failed members during their lifetime. The column headed system failure shows the number of systems which sustained adjacent member failures during their lifetime. The occurrence of adjacent member failures was found to be a good indicator of system failure by Criswell [16]. From the present study it is difficult to make generalizations; but the indications are that once adjacent members have failed, the system is very likely to fail completely under the next large load event. The other columns show the number of members which failed over the system lifetime without producing system failure. The actual number of failures seems high. This is likely due to the approach used to model the annual snow load history.

Table 1. Fifty Year Member Failures for an 11 Member Wood System.

Simulation set (100 systems each)	Number of systems where n members failed without producing system failure				Number of systems where system failure occurred*
	n=1	n=2	n=3	n=4	
Marquette					
100%	27	4	0	0	4
120%	28	4	1	0	23
140%	20	11	5	0	34
100% In.**	20	1	3	0	4
Flint					
100%	13	1	0	0	5
120%	14	6	1	0	11
140%	17	8	1	0	19
100% In.**	8	5	0	0	5
Floors					
100%	12	2	0	0	9
120%	19	4	0	0	7
140%	19	5	1	1	24
100% In.***	29	1	0	0	3

* System failure defined as rupture of two adjacent members.
** Systems simulated for 100 years.
*** 200 floors simulated for 100 years.

4. Conclusion

Simulation of the lifetime behavior of representative wood floors and roofs have been performed. Design stresses and overstresses were considered. The systems simulated behaved similarly at design stress levels, but overload behavior was less similar.

5. Acknowledgements

This material is based upon work supported by the Cooperative State Research Service, U.S. Dept. of Agriculture under Grant No. 87-FSTY-9-0420. Any opinions, findings, conclusions, or recommendations expressed in this publication are those of the author and do not necessarily reflect the view of the U.S. Dept. of Agriculture.

References

[1] National Forest Products Association. National Design Specification for Wood Construction. NFPA, Washington, D.C., 1986.

[2] Thompson, E. G., Vanderbilt, M. D., and Goodman, J. R., "FEAFLO: A Program for the Analysis of Layered Wood Systems," Computer and Structures, Vol. 7, 1977, pp. 237-248.248.

[3] Hendrickson, E. M., Ellingwood, B., and Murphy, J., "Limit State Probabilities for Wood Structural Members," Journal of Structural Engineering, ASCE, Vol. 113, No. 1, Jan., 1987, pp. 88-106.

[4] Pellicane, P. J., "Goodness-of-Fit Analysis for Lumber Data," Wood Science and Technology, Vol. 19, 1985, pp. 117-129.

[5] Pellicane, P. J., "Application of the SB Distribution to the Simulation of Correlated Lumber Properties Data," Wood Science and Technology, Vol. 18, 1984, pp. 147-156.

[6] Ellingwood, B., and Redfield, R., "Ground Snow Loads for Structural Design," Journal of Structural Engineering, ASCE, Vol. 109, No. 4, Apr., 1983, pp. 950-964.

[7] O'Rourke, M. J., and Stiefel, U., "Roof Snow Loads for Structural Design," Journal of Structural Engineering, ASCE, Vol. 109, No. 7, July, 1983, pp. 1527-1537.

[8] Chalk, P. L., and Corotis, R. B., "Probability Model for Design Live Loads," Journal of the Structural Division, ASCE, Vol. 106, No. ST10, Oct., 1980, pp. 2017-2033.

[9] American National Standards Institute, American National Standard Minimum Design Loads for Buildings and Other Structures, ANSI A58.1-1982, New York, NY, 1982.

[10] Criswell, M. E., "Selection of Limit States for Wood Floor Design," Probabilistic Mechanics and Structural Reliability, ASCE, 1979, pp. 161-165.

[11] Wheat, D. L., Gromala, D. S., and Moody, R. C., "Static Behavior of Wood-Joist Floors at Various Limit States," Journal of Structural Engineering, ASCE, Vol. 112, No. 7, July, 1986, pp. 1677-1691.

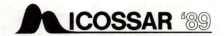
RELIABILITY SYNTHESIS OF CFRP LAMINATED PLATE

Shuichi TANI *   and   Shigeru NAKAGIRI **

*   Central Research Laboratory, Mitsubishi Electric Corporation,
    8-1-1, Tsukaguchi-honmachi, Amagasaki, Hyogo, 661, Japan
** Institute of Industrial Science, University of Tokyo,
    7-22-1, Roppongi, Minato-ku, Tokyo, 106, Japan

ABSTRACT

The paper presents a formulation of shift synthesis to improve the reliability of fiber reinforced
plastic (FRP) laminated plates with uncertain
elastic constants and lamina strength. The synthesis is carried out by taking the fiber orientations as design parameters so as to enhance "the
augmented reliability index", which is defined
negative when the expectation point lies in the
failure domain. The failure of each lamina is assumed to be subjected to the Tsai-Hill's law. The
validity of the presented method is verified by
the numerical example of three-layered CFRP plate
under the biaxial and bending load condition.

KEYWORDS

Composite materials; reliability; shift synthesis;
CFRP; reliability index; fiber orientations;
uncertainty.

1. INTRODUCTION

The tailoring design enables FRP plates to control their
mechanical behavior so that the optimum design have been investigated so far in a deterministic sense. In order to use FRP laminated plates as reliable structural members, it is necessary to
find the optimal fiber orientations in presence of the uncertainties of the material properties and stacking parameters.

In this paper, we deal with the reliability synthesis to
improve the reliability of the FRP plates whose elastic constants
and strength values are uncertain. The fiber orientation of each
lamina, as design parameter, is changed so as to increase "the
augmented reliability index β'" to the aimed value. The index β'

is defined negative extensively in the case that the expectation
point lies in the failure domain and is equal to the conventional
reliability index of the Advanced First-Order Second Moment
(AFOSM) method in the safe domain. Employing the index β' thus
defined, we can treat the structures which are estimated
unsafe at the initial design.

The numerical example is concerned with the CFRP laminated
plate under the biaxial and bending load condition. It is
verified that the reliability synthesis can be achieved so that
the fiber ofientations result in the enhancement of the approximate index.

2. FORMULATION OF RELIABILITY SYNTHESIS FOR FRP LAMINATED PLATE

The constitutive equation of the laminated plate based on the
classical lamination theory [2] is employed in the formulation.
The relationship between the resultant forces and mid-plane
strains and the one between the moments and curvatures are
governed by the four independent elastic constants, layer
thickness and fiber orientations. The in-ply stresses can be
evaluated from the orthotropic stress-strain relationship.

In this paper, the failure of each lamina is assumed to be
governed by the Tsai-Hill's law [4]. The performance function G
is defined by eq.(1).

$$G = 1 - \left\{ \left(\frac{\sigma_L}{\sigma_L^C} \right)^2 - \frac{\sigma_T^C}{\sigma_L^C} \cdot \frac{\sigma_L}{\sigma_L^C} \frac{\sigma_T}{\sigma_T^C} + \left(\frac{\sigma_T}{\sigma_T^C} \right)^2 + \left(\frac{\tau_{LT}}{\tau_{LT}^C} \right)^2 \right\} \qquad (1)$$

In the above, σ_L, σ_T and τ_{LT} are in-ply stresses in the fiber
orientation, transverse direction and shear, respectively, and
superscript "c" means the critical values of lamina strength.
The uncertainty of any elastic constant or lamina strength X can
be expressed in the following form by the small independent probabilistic variable ξ with zero mean. The upper bar means the
deterministic term.

$$X = \overline{X} (1 + \xi) \qquad (2)$$

The probabilistic variables ξ are to be converted into the standardized probabilistic variables η with zero mean and unit
variance in order to evaluate the reliability index according to
the AFOSM method [1]. The variation of performance function due
to the fluctuation of the stresses and strength values is given
by eq.(3) in the form of the first-order approximation.

$$G = \overline{G} + \sum_{i=1}^{N} \frac{\partial G}{\partial \eta_i} \eta_i \qquad (3)$$

where N indicates the total number of probabilistic variables
taken.

In many cases, we have to treat unsafe structures at the initial design. In the reliability synthesis, it is therefore
necessary to employ such an index which is defined negative
extensively in the unsafety domain. Based on the linear function
with respect to the probabilistic variables such as eq.(3), the
augmented reliability index β' is defined as follows.

$$\beta' = \overline{G} \cdot \left\{ \sum_{i=1}^{N} \left(\frac{\partial G}{\partial \eta_i} \right)^2 \right\}^{-\frac{1}{2}} \qquad \begin{array}{l} > 0 \text{ (safe)} \\ < 0 \text{ (unsafe)} \end{array} \qquad (4)$$

The index β' thus defined can be used as the approximate one of the AFOSM method in the safe domain and can be differentiated by the design parameters. By taking the fiber orientations θ_j as the design parameters, the variation of β' is given by eq.(5) in the same form as eq.(3).

$$\beta' = \beta_i' + \sum_{j=1}^{M} \frac{\partial \beta'}{\partial \theta_j} \triangle \theta_j \qquad (5)$$

In the above, M and $\triangle \theta_j$ are the total number of design parameters and the change of fiber orientations, respectively, and β_i' means β' at the initial design. The design parameters can be determined by the minimization of the functional π defined by eq.(6).

$$\pi = \sum_{j=1}^{M} (\triangle \theta_j)^2 + \sum_{k=1}^{L} \mu_k (\beta_a' - \beta_k') \qquad (6)$$

where L is the total number of the indices β_k' to be enhanced and μ_k is the Lagrange multiplier corresponding to β_k'. The determination should be iterated by means of renewing the initial design to overcome the deficiency of the first-order approximation included. The similar scheme is employed in the search of the design point and reliability index of the AFOSM method [3].

3. NUMERICAL EXAMPLE

The numerical example deals with three-layered CFRP plate under the biaxial and bending load condition. The expectations of the elastic constants and the strength values are listed in tables 1 and 2, respectively. The fiber orientations and each layer thickness of 0.1 mm are taken deterministic. Both elastic constants and strength values are taken probabilistic, and all their coefficients of variation are assumed as 10%. Figures 1 and 2 show the changes of the fiber orientations and the values of β', respectively. The aimed index values β_a' are set equal to 5. It is seen that the augmented reliability index β' in each layer is improved from a negative value to the positive value larger than 3. Figure 3 indicates the change of the index β' of each failure mode such as σ_L, σ_T and τ_{LT} criterion. It is seen in Fig. 3 that the index β' of each failure mode is enhanced as desired, while the synthesis is conducted by taking solely the damage of the lamina expected by the Tsai-Hills' law into account.

4. CONCLUSION

The shift synthesis of fiber orientation aiming at the enhancement of β' can be achieved by rather small number of iterations. When the shift is completed, the value of β' is confirmed in good approximation of a result of Monte Carlo simulation.

Table 1. Elastic constant expectations

E_L (GPa)	222
E_T (GPa)	5.7
G_{LT} (GPa)	4.2
ν_{LT}	0.26

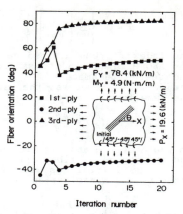

Table 2. Lamina strength expectations

	σ_L^C (MPa)	σ_T^C (MPa)	τ_{LT}^C (MPa)
Tensile Strenth	1000	20	40
Compressive Strenth	700	90	

Figure 1. Change of fiber orientations

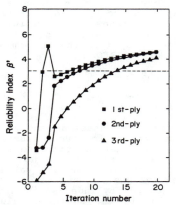

Figure 2. Convergence of reliability indices β'

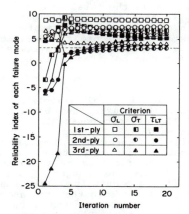

Figure 3. Change of β' on each failure mode

REFERENCES

[1] Hasofer, A.M. and Lind, N.C. : "Exact and Invariant Second-Moment Code Format," Proc. ASCE, Vol.100, No.EM1, pp.111-121, 1974.

[2] Jones, R.M. : "Mechanics of Composite Materials", McGraw-Hill Kogakusha, pp.147-236, 1975.

[3] Shinozuka, M. : "Basic Analysis of Structual Safety," Proc. ASCE, Vol.109, No.ST3, pp.721-740, 1983.

[4] Tsai, S.W. : "Strength Characteristics of Composite Materials," NASA, CR-224, pp.5-17, 1965.

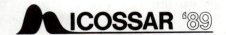
LOAD RESISTANCE FACTOR DESIGN
OF ALUMINUM STRUCTURES

Andrew J. Hinkle and Maurice L. Sharp

Alcoa Laboratories, Aluminum Company of America,
Alcoa Center, PA 15069

ABSTRACT

Load and Resistance Factor Design (LRFD) is a
reliability-based approach to design with
uncertainty.   It involves the explicit
consideration of limit states, multiple load and
resistance factors and implicit probabilistic
determination of reliability.   Most aluminum
designs, however, are based on the an allowable
stress design code. This paper provides progress
of work underway to define a strategy for
calibrating a LRFD code for aluminum structures
that will not penalize existing markets. End uses
for aluminum structures and corresponding dead-to-
live ratios are provided. Based on this and
previous work, a factor is proposed to be applied
to the structures resistance. The factor is
dependent upon the importance of the structure
being designed.

KEY WORDS

Specifications, Structures, Aluminum, Design,
Reliability, LRFD code.

1. INTRODUCTION

The present U. S. code for aluminum structures [1] is an
allowable stress design code covering many types of structures.
In effect, it covers the same domain as several steel
specifications, AISI and AISC specifications for building type
structures with light gauge and heavy construction respectively,
and AASHTO specification for highway type construction.
Successful design by this code for the past 20 years has proven
that the factors of safety for the aluminum specifications are
satisfactory for most applications in aluminum.

For aluminum building structures, Galambos [2] determined a set
of resistance factors based on material property, geometry and
response of the various elements considered in the existing
aluminum structural code. Test data were used when available,
otherwise experience with similar types of steel structures was
employed. The resistance factors were determined by calibrating
against the current allowable stress code to obtain a more
uniform safety index. The LRFD provisions, as calibrated, were
not adopted because some of the applications would need to be
more conservatively designed than required by the allowable
stress code. The manufacturers felt this was unacceptable
because their product was satisfactory as designed.

The work of Galambos [2] covers many of the factors needed to
calibrate an LRFD aluminum code. This paper provides some
additional points to be considered in the calibration. We will
begin by discussing representative uses of aluminum and typical
design considerations for aluminum structures. With this
background, some new issues on calibration will be discussed and
recommendations will be made. The intent of this work is to help
establish procedures for calibration so that the final LRFD code
will not unintentionally penalize an application by requiring a
more conservative (or a less conservative) design.

2. REPRESENTATIVE USES OF ALUMINUM

The uses of aluminum in various markets reflect the advantageous
attributes of aluminum compared to the competing materials of
steel, organic composites and plastic. Examples of some of these
attributes are: 1) light weight; density is about 2.7×10^3
Kg/m^3, 2) corrosion resistance, which allows very thin aluminum
parts to be used with minimal maintenance, and 3) ease of
applying long-term finishing techniques for architectural
variety. Structural codes and specifications are used primarily
in the building and construction market, and the transportation
market. However, structural design is also becoming increasingly
important in the containers and packaging market as new and
improved containers are designed for buckling and other
structural limit states.

Historically, the building and construction market, and the
transportation market have approximately 25% of the total
aluminum market. Some of the representative uses in these
markets are highway sign supports, curtain wall mullions,
spaceframes for roofs, roofing and siding, bridge decks, coal
cars, and automotive parts. Some of these uses are as primary
structural members where the cost of failure may be high, while
other uses are as secondary members where the cost of failure may
be lower. Because of this variation in use of aluminum, it is
practically impossible to make a simple uniform calibration for
LRFD code that does not penalize some market.

3. CALIBRATION ISSUES

Because one of the advantages of designing with aluminum is its
low self weight, typically aluminum structures have lower dead-
to-live load ratios than comparable steel structures. A study
determining the dead-to-live load ratio of typical aluminum

structures was conducted using the types of structures described above. The variety of applications, ranging from primary structural members to sheathing, are believed to be representative of the uses of aluminum. Close examination of the types of structures and their dead-to-live load ratios indicates that the magnitudes of the dead-to-live load ratio are related to the type of structure, i.e., primary or secondary. The lower values, 0.01 - 0.07, are for sheathing type applications (roofing, siding, automotive trunk lid) (considered here to be secondary) whereas structures with higher dead/live load ratios, 0.17 - 0.30, are those providing the main framing of the structure (coal car frame, long span roof frame, etc.). Because there are two different classes of structures, it is reasonable to calibrate the LRFD code at two different load ratios or to incorporate a factor for the importance of the structure.

Another calibration issue is to define the load equations and load factors that are appropriate for design. The ANSI A58.1 recommendations [3] are specifically for building type structures, while the specification [1] has usefulness for a wider variety of structures. (This issue will not be addressed in this paper.)

4. DETERMINATION OF φ FACTORS

Cases given in [4] were used to explore calibration possibilities. First the value of φ found by Galambos [2] to best fit all types of structures was investigated for different dead-to-live load ratios. Generally, with this calibration the LRFD provisions require more material for structures with low D/L (say <0.4) ratio and less material for structures with higher D/L ratios (say >0.4), compared to the existing allowable stress requirements. In some cases the differences can be relatively large (20% on load capacity), and in some cases small (same load capacity). The initial calibration results in a LRFD provision giving more conservative designs for low D/L ratios (generally secondary members).

To our knowledge, the product designs made with the present allowable stress code are adequate. Thus, we should try to minimize differences between the allowable stress code and new LRFD code that would result in either less conservative (not as safe) or more conservative (less competitive) designs. A few percent change in weight could make the product less or not competitive with that of another material. After consideration of several similar ways of calibration, we propose to accomplish a better fit between LRFD and the allowable stress code by incorporating a structural type factor, ψ. Therefore the design equation becomes

$$\sum \gamma_i S_i \le \psi \sum \phi_n R_n \qquad (1)$$

The advantage of this method is that more than one structural type can be easily accommodated.

The study of representative design examples indicates that dead-to-live load ratios up to about 0.3 are of interest for aluminum structures and that an average primary structure has a ratio of approximately 0.22. We propose, therefore, that a ϕ factor for primary structures be determined in the dead-to-live load range of 0.2-0.25. This ϕ-factor would become the basis of calculation for all load ratios. In addition, a structural type factor, ψ, is proposed to account for structural types and varying load ratios. This preliminary study indicates that when the above recommendation is followed, ψ=1.0 for primary structures and ϕ=1.05 for secondary structures, reasonable comparisons to the current design practice can be made. A systematic study of all provisions needs to be done before the final calibration is complete.

5. CONCLUSIONS

1. The present Aluminum Association Specifications applies to all types of aluminum structures, light gauge through heavy construction. Thus, it is not possible to use one calibration for a LRFD code for all cases without introducing more or less conservation into some product design.

2. A review of current applications shows that aluminum structures have a low dead-to-live load ratio, 0.30 or less, and that the dead-to-live load ratios for secondary members (sheathing, etc.) are much less than those for primary framing (generally less than 0.10).

3. The introduction of a structural type factor is proposed, the value of the factor dependent on dead-to-live load ratio. This provides a means to develop a more realistic safety index consistent with present design practice.

4. The load factor and load equations will be examined, particularly the wind loading cases. These studies are underway.

6. REFERENCES

[1] Aluminum Association: "Specifications for Aluminum Structures", The Aluminum Association, Inc., 900 19th Street, N.W., Washington, DC, 20006, Fifth edition, December, 1986.

[2] Galambos, T. V.: "Load and Resistance Factor Design for Aluminum Structures", Research Report No. 54, Structural Division, Civil Engineering Department, Washington University, St. Louis, MO, May, 1979.

[3] Ellingwood B., Galambos, T. V., MacGrefor, J. G. and Cornell, C. A.: "Development of a Probability Based Load Criterion for American National Standard A58", National Bureau of Standards Special Publication 577, National Bureau of Standards, Dept. of Commerce, Washington, DC, 20234, June, 1980.

[4] Aluminum Association: "Illustrative Examples of Design", The Aluminum Association, Inc., 900 19th Street, N.W., Washington, DC, 20006, April, 1978.

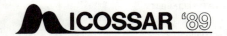
STUDY ON RELIABILITY OF ACTUAL STEEL STRUCTURE USING MEASURED DATA

Tetsuro ONO[*], Hideki IDOTA[*], Toshiki OHYA[**] and Tohru TAKEUCHI[**]

* Dept. of Architecture, Nagoya Institute of Tech., Nagoya, Japan
** Building Construction & Urban Development Division
Nippon Steel Corporation, Tokyo, Japan

ABSTRACT

The measurement of uncertainties was taken on an actually
constructed steel structure. The stress and strain of each
part of the structure are calculated based on the conditions
assumed in the design stage by the numerical analysis. The
differences between the measured data and calculated values
are statistically analyzed. The reliability level of an
actual steel structure is discussed using the measured data.

KEYWORDS

Actual Structure, Measurement of Strain, Human Error,
Construction Error.

1. INTRODUCTION

To assure the safety of structures, it is necessary to consider various
uncertainties that apply to the structure. For the rational design of
structures, it is important to quantitatively determine these uncertainties and
incorporate them in the design method of structures. Among the uncertainties
that affect the safety of actual structures are those that have been rarely
handled in the past, such as human error in the design stage, construction
error, and load estimation error. It can be understood from past accidents that
such uncertainties have a serious impact on the safety of structures.

In this study, the axial strain developed in the columns of an actual steel
structure was measured in the construction stages and after completion. The
differences between the theoretical values based on the assumptions made in the
design stage and the measured values are analyzed. The reliability level of the
structure is investigated using the measured data.

2. METHODS OF MEASUREMENT

The building measured in the present work is an office building with eighteen floors above the ground in Nagoya, Japan. The building is of steel skeleton construction above the ground. The steels used are JIS G 3106 SM 50 and SM 53. The axial strain in the columns was measured. Welding strain gauges were used in view of gauge maintenance and durability. The general data of the building and the strain measuring positions is shown in Fig. 1. The axial strain in the columns was determined by averaging the strains measured with two gauges. A thermocouple was attached near each strain gauge to measure the temperature during the strain measurement and to compensate the measured strain data with the temperature. During the construction of the building, the dead load and live load were measured in detail at each strain measured in detail at each strain measurement and were used as load data for stress analysis. The measurements were made 55 times from March 12, 1986 to November 15, 1988. The building was completed when the 30th measurement was made, and the tenants moved into the building when the 40th measurement was made.

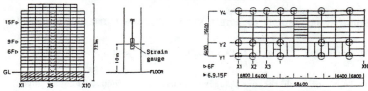

Fig. 1 The Structure of Investigation and Strain Measuring Position

3. STRESS CALCULATION

Stress calculation was performed using the structural model employed in the design stage and the loads recorded during the strain measurement. The axial strain in the columns was computed, deterministically and stochastically. The Three Point Estimate method[1] that produces accurate results in a short computing time was used as a stochastic analysis method. Nippon Sekkei Design Office's NASCA was used as the structural analysis program for the stress calculation. The stochastic stress analysis handles the sectional area A and Young's modulus E of steel members, dead load S_{DL} and live load S_{LL} as random variables. The means and coefficients of variations in statistics A and E were set at $\mu_A = 0.986a_n$ and $\mu_E = 2.094$ and $\delta_A = 0.05$ and $\delta_E = 0.084$, respectively, where a_n is the nominal sectional area of the steel member. The coefficients of variations in S_{DL} and S_{LL} were set at $\delta_{SD} = 0.10$ and $\delta_{SL} = 0.40$, respectively.

4. RESULTS OF MEASUREMENT AND THEIR STATISTICS

Fig. 2 shows the axial strain on the 6th and 9th floors. The dates of measurement are plotted along the horizontal axis and the axial strain of each column($\times 10^{-6}$) is plotted along the vertical axis. The solid line indicates the measured axial strain em and the three broken lines indicate the axial strain calculated by the stress analysis. The upper, middle, and lower broken lines correspond to $\mu_{cd} + \sigma_{cd}$, μ_{cd}, and $\mu_{cd} - \sigma_{cd}$, respectively. As more measurements are made, the randomness of axial strain more closely falls within the range of $\pm\sigma_{cd}$. The proportions between the randomness of uncertainties associated with design and construction and that of mechanical properties, sectional area, and load can be determined accordingly.

The ratio $\varepsilon_m/\varepsilon_c$ of the measured axial strain ε_m to the deterministically

calculated axial strain ε_c was statistically processed to clarify the variability of difference between the measured and calculated axial strains. The mean and standard deviation of the $\varepsilon_m/\varepsilon_c$ ratio were computed for each floor. The statistical results of the measured data are given in Table 1. The mean varies with the measuring section and tends to be large on the 6th floor and small on the 9th floor. The standard deviation is about 0.3 to 0.4 during the construction of the building and is about 0.1 to 0.2 after the completion of the building.

Fig. 3 shows histograms of the $\varepsilon_m/\varepsilon_c$ ratio for each floor during the construction of the building. The mean varies among the three floors in question but the standard deviation is nearly the same for each floor.

Table 1 Statistical Results of Measured Data

	All (6th–55th)			During Construction (6th–29th)			After Completion (30th–55th)		
	Mean	S.D.	n	Mean	S.D.	n	Mean	S.D.	n
6,9,15F	1.090	0.368	1609	1.191	0.535	666	1.018	0.174	943
6F	1.268	0.346	701	1.438	0.438	324	1.123	0.109	377
9F	0.896	0.279	480	0.889	0.409	194	0.901	0.127	286
15F	1.016	0.304	428	1.049	0.444	148	0.998	0.200	280

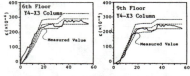

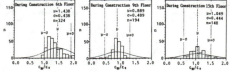

Fig. 2 Axial Strain   Fig. 3 Histograms of $\varepsilon_m/\varepsilon_c$ ratio

5. MODELING OF MEASURED DATA AND RELIABILITY OF ACTUAL STRUCTURE

The statistics of strain ε_e in each part of an actual structure can be modeled as shown below when classified by the nature of the population that causes the randomness:

$$\varepsilon_e = \varepsilon_c + M + \Delta_b + \Delta_f + \Delta_c \tag{1}$$

where ε_e is a random variable that indicates the axial strain of each column determined from the strain measurement in a sufficient number of buildings; ε_c is a random variable that indicates the strain in each part of the building obtained from deterministic stress analysis; M is a constant that indicates the difference between the mean of ε_e and ε_c; Δ_b is a random variable that indicates the randomness of strain between different buildings; Δ_f is a random variable that indicates the randomness of strain between the floors of a building; and Δ_c is a random variable that indicates the randomness of strain on a floor. M is a quantity that varies with the construction grade of the building and quality control at the construction site, among other factors. Generally, Δ_b, Δ_f, and Δ_c are random variables with zero mean and are statistically independent. To apply the results of statistical processing in the foregoing section, Eq. (1) is transformed as follows:

$$\varepsilon_0 = \varepsilon_e/\varepsilon_c = 1 + m + \delta_b + \delta_f + \delta_c \tag{2}$$

where $m = M/\varepsilon_c$, $\delta_b = \Delta_b/\varepsilon_c$, $\delta_f = \Delta_f/\varepsilon_c$, and $\delta_c = \Delta_c/\varepsilon_c$. The statistical data ε_m

obtained from the measurements cover a single building, so that ε_m corresponds to $\Delta_f + \Delta_c$ in Eq. (1). Thus, Eq (2) is rewritten as

$$\varepsilon_0 = 1 + m + \delta_b + r_m \qquad (3)$$

where $r_m = \varepsilon_m / \varepsilon_c$. Based on the results of this measurement, the standard deviation of r_m is $\sigma_r = 0.488$ during the construction and is $\sigma_r = 0.197$ after completion. Since the measured data is obtained only from a single building here, m = 0 is assumed.

When design and construction errors are taken into account, the reliability index β_T, a target to impart the reliability index β_S to a given structure, is related to the reliability index β_S that does not take the design and construction errors into account as follows:

$$\beta_T = \frac{\beta_S \sqrt{\sigma_d^2 + \sigma_g^2}}{\sigma_g} \qquad (4)$$

where $\sigma_g^2 = \sigma_A^2 + \sigma_E^2 + \sigma_{DL}^2 + \sigma_{LL}^2$ and $\sigma_d^2 = \sigma_e^2 - \sigma_g^2$. When the measured data under discussion is applied, $\sigma_d^2 = \sigma_b^2 + 0.488^2 - \sigma_g^2$ during the construction, $\sigma_d^2 = \sigma_b^2 + 0.197^2 - \sigma_g^2$ after the completion, and $\sigma_g = 0.418$ during the construction and after the completion. Fig. 6 shows the relationship between β_S and β_T with σ_b as a parameter. During the construction, β_T is considerably higher than β_S and after completion, β_T can be considerably reduced. When $\sigma_s = 0.4$, for instance, $\beta_T = \beta_S$. If σ_b is less than 0.4, the randomness associated with the design and construction of the building can be covered by the safety factor that allows for the randomness of a sectional area, mechanical properties, and load.

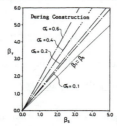 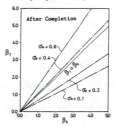

Fig. 4   Relationship between Reliability Index β_T and β_S

6. CONCLUSIONS

The axial strain developed in the columns of an actual building was measured. A simple stochastic model was established for the measured data and was used to study the reliability of structures. At the current time, the measured data is only available for a single building. When similar measurements are continuously taken on more buildings, it will become feasible to quantify the uncertainties and evaluate the safety of actual structures more clearly.

ACKNOWLEDGEMENT
    We would like to thank Mr.T.Nakano, Nippon Steel Corporation; Mr.H.Matsuzaki, Mitsui Construction Co.Ltd.; and Mr.S.Makishi, Nihon Sekkei Design Office, for their support of the measurement.

REFERENCE
[1] ROSENBLUETH, E. : "Point Estimates for Probability Moments", Proc. of National Academy of Science of U.S.A., Vol.72, 1975.

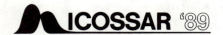
STATISTICAL PROPERTIES OF PRESTRESSED CONCRETE BEAMS IN SHEAR

William M. Bulleit and James E. Rintala

Associate Professor and Former Graduate Student
Department of Civil and Environmental Engineering
Michigan Technological University
Houghton, MI 49931

Abstract

Monte Carlo simulations were performed on prestressed concrete
rectangular and tee-section beams in shear produced by a single
concentrated load at mid span. The truss analogy model was
used to determine the shear capacity of the beams. The data
from the simulations was used to obtain the first four
statistical moments of the shear capacity.

Keywords

Prestressed concrete, beams, shear strength, statistical
analysis.

1. Introduction

The advent of reliability based design codes has necessitated the
determination of statistical properties for many structural members and their
respective failure modes. The available information on the statistics of
shear strength for reinforced concrete is limited, particularly for
prestressed concrete [1,2].

The objective of this research was to examine the statistical properties
of the shear strength of prestressed concrete beams in shear. The truss
analogy method [3,4,5] was chosen as the procedure for determining the shear
strength. Along with the statistical properties, a model error was determined
by comparing model results to test results available in the literature [5,6].

2. Background

2.1 Shear Strength Determination

The truss analogy method [3,4,5] is a mechanically based model founded on
the assumption that a concrete beam at ultimate capacity will behave like a

truss with its diagonals in compression. The compression chord of the analogous truss consists of the compression zone of the concrete and any compression steel. The tension chord of the truss is composed of prestressed and nonprestressed tension reinforcement. The stirrups in the beam act as the vertical web members in tension. The truss diagonals are the concrete bounded by and parallel to the inclined tension cracks present at ultimate capacity. The analogous truss can fail in three possible modes. Tension yielding of the stirrups or crushing of the concrete diagonals, i.e. a shear failure. The last two modes correspond to a flexural failure: exceedence of the strength of the tension chord, strand rupture in a prestressed member, or crushing of the concrete in the compression chord. A full description of the method is beyond the scope of this paper but can be found in the references [3,4,5].

In the United States, the shear capacity equations in ACI-318 [7] are commonly used. The ACI equations for shear strength of prestressed concrete members are based on a truss analogy model with the inclination of the diagonal tension cracks limited to 45 degrees [4].

2.2 Model Error

The mathematical models used to predict the response of a structural member are based on certain assumptions. Due to these assumptions, some error exists between the predicted and actual response, even if all the input variables are known. Thus a model error factor can be estimated as the ratio of the actual response to the predicted response based on the model [8]. This model error can then be incorporated into reliability analyses as a random variable.

3. Results

3.1 Model Error

The truss analogy analysis routine developed for this research was able to analyze sections loaded at mid span by a single concentrated load. Prestressing tendons could be either straight or harped at mid span. The beams from Ref. [6] were loaded at third points. These beams were modeled as a beam with a concentrated load at mid span which was two-thirds as long as the original beam. With this approximation, the twenty-nine beams in Refs. [5,6] were each analyzed using the truss analogy method. The resulting model error had a mean value of 1.12 and a coefficient of variation of 0.17. Using mathematical probability plotting, the lognormal distribution fit the model error data best when compared to the normal or Weibull distributions. Best fit was determined by comparison of the coefficient of determination for the linear regressions performed in the probability plotting. The value of the coefficient of determination for the lognormal, normal and Weibull distributions, respectively, were: 0.986, 0.940 and 0.885.

3.2 Shear Strength Simulations

Four sections from the PCI Design Handbook [9] were simulated: two rectangular sections, a 12RB36 with a span of 7.32 m and a 16RB40 with a span of 9.14 m; and two single tee-sections, a 168-D1, 3.05 m wide by 0.914 m deep section, on a span of 18.9 m, and a 208-D1, 3.05 m wide by 1.22 m deep section, on a span of 24.4 m. In all cases the nominal concrete strength was 34.48 MPa. The stirrups consisted of Gr. 40, #3 bars with a yield strength of

275.8 MPa. The selection of a lower grade stirrup was necessary to prevent minimum stirrup area from controlling the design [7]. When minimum stirrup area controlled, the simulations showed no shear failures, ultimate flexural capacity controlled. Of the above four sections, the two tee-sections also showed no shear failures. Thus, a shortened version of the 3.05 m x 1.22 m tee was simulated. It had a span of 15.24 m.

The simulations were performed using data compiled by Siriaksorn and Naaman [10]. Three ratios of average/nominal concrete cylinder strength were used, 0.67, 0.92 and 1.17, and two coefficients of variation at each level, 0.1 and 0.25. These values bracket the statistics used in reference [2]. Each of the three beams was simulated 3 times with a 1000 replications each for the 6 combinations of average/nominal cylinder strength and coefficient of variation.

The simulations performed using high average/nominal cylinder strength produced either very few or no shear failures. The most shear failures occurred at low average/nominal shear strength and high coefficient of variation. The nominal shear strength of each of the three sections was determined per ACI [7]. The range of average/nominal shear strength was 1.5 to 1.65 for the rectangular sections and 1.65 -> 1.8 for the tee-sections. The high end of each range was obtained from the few shear failures for high average/nominal cylinder strength. The coefficient of variation ranged from 0.03 to 0.08. The high range came from simulations using low average/nominal cylinder strength and high coefficient of variation.

The third moments showed distribution shapes which were only very slightly skewed. Both left and right skewnesses occurred with left skewness predominating. The coefficient of kurtosis ranged from 2.0 to 3.0 with values near 3.0 most common.

Considering the relatively small coefficients of variation and the values of the third and fourth moments, a normal or lognormal distribution seems reasonable. It could also be argued that a Weibull distribution might be appropriate due to the brittle nature of shear failures. The assumption of a Weibull distribution is not precluded by the statistics calculated in this study. From a practical reliability standpoint, it is unlikely to make much difference which distribution is chosen.

From this study, the ratio of average shear strength to ACI nominal shear strength for realistic prestressed concrete beams can be taken as 1.5 with a coefficient of variation of 0.05.

References

[1] Mirza, S. A., MacGregor, J. G., "Statistical Study of Shear Strength of Reinforced Concrete Slender Beams," Journal of the American Concrete Institute, ACI, Vol. 76, No. 11, November 1979, pp. 1159-1178.

[2] Ellingwood, B., Galambos, T. V., MacGregor, J. G., Cornell, C. A., Development of a Probability Based Load Criterion for American National Standard A58, NBS Special Publication 577, U.S. Department of Commerce, Washington, D.C., June 1980.

[3] Collins, M. P., Mitchell, D., "Shear and Torsion Design of Prestressed and Non-Prestressed Concrete Beams," PCI Journal, Prestressed Concrete Institute, Vol. 25, No. 5, September/October 1980, pp. 32-100.

[4] Ramirez, J. A., Breen, J. E., Review of Design Procedures for Shear and Torsion in Reinforced and Prestressed Concrete, Research Report 248-2, Center for Transportation Research, University of Texas at Austin, March 1984.

[5] Rezai-Jorabi, H., Regan, P.E., "Shear Resistance of Prestressed Concrete Beams With Inclined Tendons," The Structural Engineer, Vol. 64B, No. 3, September 1986, pp. 63-74.

[6] Elzinaty, A. H., Nilson, A. H., Slate, F. O., "Shear Capacity of Prestressed Concrete Beams Using High-Strength Concrete," Journal of the American Concrete Institute, ACI, Vol. 83, No. 3, May/June 1986, pp. 359-368.

[7] ACI Committee 318, Building Code Requirements for Reinforced Concrete, ACI 318-83, American Concrete Institute, Detroit, 1983.

[8] Thoft-Christensen, P., Baker, M. J., Structural Reliability Theory and Its Applications, Springer-Verlag, New York, 1982.

[9] Prestressed Concrete Institute, PCI Design Handbook, Third Edition, Prestressed Concrete Institute, Chicago, Illinois.

[10] Siriaksorn, A., Naaman, A. E., Reliability of Partially Prestressed Beams at Serviceability Limit States, Thesis presented to the University of Illinois at Chicago Circle in partial fulfillment of the requirements of the degree of Doctor of Philosophy, Chicago, IL, 1980.

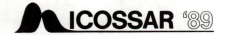
Reliability-oriented Materials Design of Composite Materials

Hidetoshi Nakayasu[1], Zen'ichiro Maekawa[2] and Rüdiger Rackwitz[3]

Abstract

A basic formulation for evaluating safety index of unidirectional fiber reinforced laminate is presented. The proposed formulation is based on in-plane failure criteria of composite laminate and FORM/SORM. It enables one to evaluate the stochastic behavior of composite laminate with any lamination angle under multi-axial stress or strain condition.

Introduction

There have been numerous research works of reliability analysis whose objectives are concerned with the evaluation of variability of characteristic features of composite materials. Most of these are can not construct the method of materials design under the complicated multi-axial load conditions probabilistically.

One of the major objectives of this paper is to offer a practical tool for stochastic materials design of unidirectional composite laminate with material characterization under the complicated static load states. Quadratic polynomial criteria and FORM/SORM(first/second order reliability method)[2] are applied to construct the evaluation model of probabilistic safety of composite laminates.

Formulation for Reliability Analysis

(i) In-plane stress-strain relation

When the coordinate system of composite laminate is defined as Fig.1, the stress-strain retationships are expressed by the equations

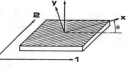

$$\boldsymbol{\sigma}_x = \mathbf{Q}_x \boldsymbol{\varepsilon}_x, \quad \boldsymbol{\sigma}_1 = \mathbf{Q}_1 \boldsymbol{\varepsilon}_1,$$
$$\boldsymbol{\varepsilon}_x = \mathbf{S}_x \boldsymbol{\sigma}_x, \quad \boldsymbol{\varepsilon}_1 = \mathbf{S}_1 \boldsymbol{\sigma}_1 \qquad (1)$$
$$\boldsymbol{\sigma}_x = \{\sigma_x, \sigma_y, \sigma_s\}^T, \quad \boldsymbol{\sigma}_1 = \{\sigma_1, \sigma_2, \sigma_6\}^T,$$
$$\boldsymbol{\varepsilon}_x = \{\varepsilon_x, \varepsilon_y, \varepsilon_s\}^T, \quad \boldsymbol{\varepsilon}_1 = \{\varepsilon_1, \varepsilon_2, \varepsilon_6\}^T \qquad (2)$$

Fig.1 Coordinate system

---

[1] Kanazawa Institute of Technology, Kanazawa 921, Japan
[2] Kyoto Institute of Technology, Kyoto 606, Japan
[3] Technical University of Munich, Munich 8000, West Germany

where $\mathbf{Q}_x$ is in-plane stiffness matrix on x-y-s coordinate system and $\mathbf{S}_x$ is compliance matrix which is inverse of

$$\mathbf{Q}_x = \begin{bmatrix} Qxx & Qxy & 0 \\ & Qyy & 0 \\ sym. & & Qss \end{bmatrix}, \qquad \begin{array}{ll} Qxx = \eta\, Ex, & Qyy = \eta\, Ey, \\ Qxy = \eta\, \nu_x\, Ey, & Qss = Es, \\ \eta = (1 - \nu_x^2 Ey/Ex)^{-1} \end{array} \qquad (3)$$

The coordinate transformation matrices

$$\mathbf{T}_1 = \begin{bmatrix} m^2 & n^2 & 2mn \\ n^2 & m^2 & -2mn \\ -mn & mn & m^2-n^2 \end{bmatrix}, \qquad \mathbf{T}_2 = \begin{bmatrix} m^2 & n^2 & mn \\ n^2 & m^2 & -mn \\ -2mn & 2mn & m^2-n^2 \end{bmatrix} \qquad (4)$$

enables one to transfer stress/strain components from 1-2-6 system to x-y-s system as

$$\boldsymbol{\sigma}_x = \mathbf{T}_1\, \boldsymbol{\sigma}_1, \qquad \boldsymbol{\varepsilon}_x = \mathbf{T}_2\, \boldsymbol{\varepsilon}_1 \qquad (5)$$

where $m=\cos\theta$, $n=\sin\theta$ and θ is lamination angle of composite laminate. On the other hand, the transformation of stiffness/compliance matrices can be easily shown using the orthogonal relationship $\mathbf{T}_1^{-1} = \mathbf{T}_2^{\mathsf{T}}$, $\mathbf{T}_2^{-1} = \mathbf{T}_1^{\mathsf{T}}$ such as

$$\mathbf{Q}_1 = \mathbf{T}_2^{\mathsf{T}}\mathbf{Q}_x\mathbf{T}_2, \qquad \mathbf{S}_1 = \mathbf{T}_1^{\mathsf{T}}\mathbf{S}_x\mathbf{T}_1 \qquad (6)$$

(ii) Failure criteria and limit state function

     Quadratic polynomial failure criteria is most suitable to reliability-oriented materials design of composite laminate, since the limit state is expressed by only one state function including various failure modes. Threfore, the failure probability of unidirectional composite laminate is formulated as

$$Pf = P(g(\mathbf{X}) \leq 0) \qquad (7)$$

     There are four famous failure criteria on maximum work, which are proposed by Hill[1], Chamis[4], Hoffman[3], Tsai-Wu[5]. Limit state functions corresponding to each criterion are described below respectively on stress and strain space.

(a)Stress space

        Hill and Chamis: $g(\mathbf{X}) = \boldsymbol{\sigma}_x^{\mathsf{T}} \mathbf{F}_{A,x} \boldsymbol{\sigma}_x - 1$ (8)

        Hoffman and Tsai-Wu: $g(\mathbf{X}) = \boldsymbol{\sigma}_x^{\mathsf{T}} \mathbf{F}_{A,x} \boldsymbol{\sigma}_x + \mathbf{F}_{B,x}^{\mathsf{T}} \boldsymbol{\sigma}_x - 1$ (9)

(b)Strain space

        Hill and Chamis: $g(\mathbf{X}) = \boldsymbol{\varepsilon}_x^{\mathsf{T}} \mathbf{G}_{A,x} \boldsymbol{\varepsilon}_x - 1$ (10)

        Hoffman and Tsai-Wu: $g(\mathbf{X}) = \boldsymbol{\varepsilon}_x^{\mathsf{T}} \mathbf{G}_{A,x} \boldsymbol{\varepsilon}_x + \mathbf{G}_{B,x}^{\mathsf{T}} \boldsymbol{\varepsilon}_x - 1$ (11)

where $\mathbf{F}_{A,x}$ and $\mathbf{F}_{B,x}$ are

$$\mathbf{F}_{A,x} = \begin{bmatrix} Fxx & Fxy & 0 \\ & Fyy & 0 \\ sym. & & Fss \end{bmatrix}, \qquad \mathbf{F}_{B,x} = \{\, Fx\;\; Fy\;\; 0\, \}^{\mathsf{T}} \qquad (12)$$

and $\mathbf{G}_{A,x} = \mathbf{Q}_x^{\mathsf{T}} \mathbf{F}_{A,x} \mathbf{Q}_x$, $\mathbf{G}_{B,x}^{\mathsf{T}} = \mathbf{F}_{B,x}^{\mathsf{T}} \mathbf{Q}_x$ (13)

(iii) Reliability formulation of failure probability integrals

     FORM/SORM[2] can easily evaluate the multi-dimensional probability integral

$$P(g(\mathbf{X}) \leq 0) = \int_{Dg} dF_X(\mathbf{x}) \qquad (14)$$

using Rosenblatt transformation

$$\mathbf{X} \rightarrow \mathbf{U}: \quad \mathbf{X} = T\,(\mathbf{U}) \quad \text{or} \quad \mathbf{U} = T^{-1}\,(\mathbf{X}) \qquad (15)$$

Each safety index β_i corresponding to approximated failure surface ∂Dgu and ∂Dqu is obtained respectively as follows.

(1) FORM: $\partial Dgu \approx \{u : \alpha^T(u-u^*)=0\} = \{u : 1(u)=\alpha^T u + \beta =0\}$,

$\beta_1 = \| u^* \| = \min(\| u \|) = -\alpha^T u^*$ for $\{u : g_u(U) \leq 0\}$,

$Pf_1 \approx \Phi(-\beta_1)$, $\alpha = \mathrm{grad}\{g_u(u^*)\} / \| \mathrm{grad}\{g_u(u^*)\} \|$ (16)

(2) SORM: $\partial Dqu \approx \{u : q(u)=1/2(u-u^*)^T H(u-u^*)+\alpha^T(u-u^*)=0\}$,

$\beta_2 = -\Phi^{-1}(Pf_2)$,

$$Pf_2 \approx \Phi(-\beta_1) \prod_{j=1}^{n-1} (1-\Psi(-\beta_1)\kappa_j)^{-1/2}, \Psi(\cdot)=\phi(\cdot)/\Phi(\cdot)$$ (17)

Nnmerical Analyses

When the load condition of unidirectional composite laminate is shown by Fig.2

$\sigma_1=P_1^*/(B \cdot H)$, $\sigma_2=P_2^*/(L \cdot H)$,

$\sigma_6=P_6^*/(B \cdot H)$ (18)

where SF_1, SF_2, SF_3 are safety factors

$P_1^*=P_1/SF_1$, $P_2^*=P_2/SF_2$, $P_6^*=P_6/SF_3$ (19)

Table 2 shows the values of stochastic features of parameters and design variables which is of T300/5208.

Table 3 shows the calculation results of reliability indecies for the case of multi-axial loading condition with shear load.   From this table

1) There are few significant differences among β_1, β_2 and β_3.   This results suggests that β_1 by FORM is the sufficient measure for reliability-oriented materials design of composite laminate, since the limit state function of composite laminate has stable behavior for searching beta point.

2) The results of reliability analysis based on stress and strain space are equivalent each other, though each case has different number of basic random variables.

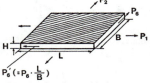

$P_6'(=P_6 \cdot \frac{L}{B})$

Fig.2 Load conditions

Table 2 Stochastic features
of random variables

Parameter	Expectation		Standard deviation	CV
$\sigma_{L.t}$	1500	(MPa)	150	0.10
$\sigma_{L.c}$	1500	(MPa)	180	0.12
$\sigma_{T.t}$	40	(MPa)	4.40	0.11
$\sigma_{T.c}$	246	(MPa)	19.68	0.08
σ_{LT}	68	(MPa)	4.08	0.06
k_{12}	Value calculated by Table 1		-	0.01
E_x	181000	(MPa)	9050	0.05
E_y	10300	(MPa)	515	0.05
E_s	7170	(MPa)	358.5	0.05
ν_x	0.28	(MPa)	0.0028	0.01
θ	0	(deg)	0.1	-
B	30	(mm)	1.5	0.05
L	150	(mm)	7.5	0.05
H	2	(mm)	0.1	0.05
P_1	90000	(MPa)	4500	0.05
P_2	120000	(MPa)	600	0.05
P_6	4080	(MPa)	204	0.05

Fig.3 describes the behavior of β_1 curve corresponding to various load condition and lamination angle.   Comparisons of failure criteria can be also evaluated by this figures.   Typical differences are shown under shear load, and multi-axial load conditions.   As a total rough conclusion from various calculations, Tsai-Wu or Hoffman criteria are superior among many criteria along with reliability-based materials design.

As an application to stochastic materials design of composite materials, stochastic failure envelopes which are β contour diagram on in-plane strain space are drawn in Fig.4 for the new design diagram.

Table 3 Reliability levels for various failure criteria
($SF_1=SF_2=SF_3=2.5$; $\sigma_1=604.3$, $\sigma_2=16.1$, $\sigma_6=27.4$ MPa)

Criteria	β_1	β_2	β_3	Pf_1	Pf_2	Pf_3
TW(S)	3.0692	3.0473	3.0367	1.07×10^{-3}	1.15×10^{-3}	1.20×10^{-3}
HO(S)	2.4690	2.4555	2.4456	6.77×10^{-3}	7.03×10^{-3}	7.23×10^{-3}
HI(S)	4.0188	3.9388	3.9330	2.93×10^{-5}	4.10×10^{-5}	4.20×10^{-5}
CH(S)	6.8868	6.8049	6.8214	2.87×10^{-12}	5.08×10^{-12}	4.54×10^{-12}
MS	6.5339	6.5363	6.5239	3.22×10^{-11}	3.17×10^{-11}	3.44×10^{-11}
TW(N)	3.0692	3.0473	3.0367	1.07×10^{-3}	1.15×10^{-3}	1.20×10^{-3}
HO(N)	2.4690	2.4555	2.4456	6.77×10^{-3}	7.03×10^{-3}	7.23×10^{-3}
HI(N)	4.0188	3.9388	3.9330	2.93×10^{-5}	4.10×10^{-5}	4.20×10^{-5}
CH(N)	6.8868	6.8049	6.8214	2.87×10^{-12}	5.08×10^{-12}	4.54×10^{-12}
MN	6.9650	6.9660	6.9653	1.65×10^{-12}	1.64×10^{-12}	1.65×10^{-12}

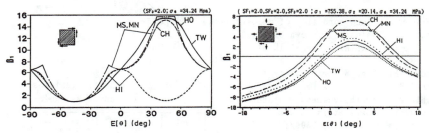

Fig. 3 Safety index and lamination angle under verious criteria

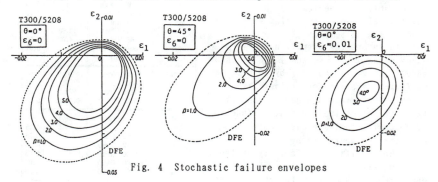

Fig. 4 Stochastic failure envelopes

Acknowledgment

 This work was partly supported by the Science Research Fund of the Ministry of Education, Science and Culture of Japan.

References
[1] Hill, R.: "The Mathematical Theory of Plasticity", Clarendon, Oxford, pp. 317-331, 1950.
[2] Hochenbichler, M. et. al.: "New Light on First- and Second-Order Reliability Methods", Structural Safety, Vol. 4, pp. 267-284, 1987.
[3] Hoffman, O.: "The Brittle Strength of Orthotropic Materials", J. Compos. Mater., Vol. 1, pp. 200-206, 1967.
[4] Sinclair, J. H. and Chamis, C. C.: "Fracture Models in Off-axis Fiber Composites", Proc. 34th SPI, Section 22-A, pp. 1-11, 1979.
[5] Tsai, S. W. and Wu, E. M.: "A General Theory of Strength for Anisotropic Materials", J. Compos. Mater., Vol. 5, pp. 58-80, 1971.

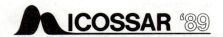
OPTIMAL INSPECTION PLANNING FOR FATIGUE
DAMAGE OF OFFSHORE STRUCTURES

H.O. Madsen [*], J.D. Sorensen [**] and R. Olesen [***]

[*]   Danish Engineering Academy, Lyngby, Denmark

[**] Aalborg University Centre, Aalborg, Denmark

[***] A.S Veritas Research, Hovik, Norway

ABSTRACT.

A formulation of optimal design, inspection and maintenance against damage caused by fatigue crack growth is formulated. A stochastic model for fatigue crack growth based on linear elastic fracture mechanics is applied. Failure is defined by crack growth beyond a critical crack size. The failure probability and associated sensitivity factors are computed by first-order reliability methods. Inspection reliability is included through a pod (probability of detection) curve. Optimization variables are structural design parameters, inspection times and qualities. The total expected cost of design, inspection, repair and failure is minimized with a constraint on the life time reliability.

KEY WORDS

Fatigue; crack growth; optimization; reliability; inspection; repair.

1. INTRODUCTION

Probabilistic methods for fatigue crack growth have been developed and applied in probability-based design and for reliability updating based on inspection results, see e.g. [1-5]. Reliability against fatigue crack growth is achieved through efforts in design, inspection and repair or replacement. These efforts all introduce cost and the minimum total expected cost solution is of interest. An optimization problem can be formulated which can further include a constraint on the smallest allowable reliability, e.g. as specified by a regulatory body. Contributions to the formulation and solution of this optimization problem can be found in e.g. [6,7].

This paper presents a consistent formulation of cost optimal fatigue design, inspection and repair. The analysis is based on a description of fatigue crack growth utilizing linear elastic fracture mechanics. Reliability and sensitivity calculations are performed by a first-order reliability method and the optimization is carried out by a general non-linear optimization algorithm.

2. SAFETY AND EVENT MARGINS

A one-dimensional description of the crack size a is employed. Crack growth is described by Paris' equation with the stress intensity factor K calculated by linear elastic fracture mechanics

$$\frac{da}{dN} = C(\Delta K)^m, \quad \Delta K = Y(a)\sqrt{\pi a}\, S > 0, \quad a(N=0) = a_0 \tag{1}$$

a_0 is the crack size after fabrication inspection, N is the number of stress cycles, C and m are material constants, and $Y(a)$ is the geometry function. A Weibull distribution with random distribution parameters is used for the distribution of the stress ranges S. This is a relevant choice for the long term distribution for offshore jacket structures.

$$F_S(s) = 1 - \exp(-(s/A)^B), \quad s > 0 \tag{2}$$

The stress ranges depend on the structural optimization parameters introduced in a later section. The number of stress cycles per unit time is ν, and the safety margin M for failure - defined as crack growth to a critical size a_c - before time t is, [2]

$$M(t) = \int_{a_0}^{a_c} \frac{dx}{Y(x)^m (\pi x)^{m/2}} - C\nu t A^m \Gamma(1+\frac{m}{B}) \tag{3}$$

The first inspection at time T_1 leads to a crack detection or no crack detection. An event margin H is defined as

$$H = \int_{a_0}^{a_{d1}} \frac{dx}{Y(x)^m (\pi x)^{m/2}} - C\nu T_1 A^m \Gamma(1+\frac{m}{B}) \tag{4}$$

The event margin is negative when a crack is detected and is otherwise positive. Values for the smallest detectable crack size a_d for different inspections are mutually independent. When a crack is detected and repaired at time T_1, the safety margin after repair is

$$M^1(t) = \int_{a_R}^{a_c} \frac{dx}{Y(x)^m (\pi x)^{m/2}} - C\nu(t-T_1) A^m \Gamma(1+\frac{m}{B}), \quad t > T_1 \tag{5}$$

The geometry function is identical before and after repair. The material parameter C is either fully dependent or completely independent before and after repair. Crack sizes after repair a_R are mutually independent. A notation is introduced to describe the sequence of repair/no repair events. With repair at times T_1 and T_2 and no repair at T_3, the safety margin for $T_3 < t \le T_4$ is as an example

$$M^{110}(t) = \int_{a_R}^{a_c} \frac{dx}{Y(x)^m (\pi x)^{m/2}} - C \nu (t-T_2) A^m \Gamma(1+\frac{m}{B}), \qquad T_3 < t \le T_4 \qquad (6)$$

and the event margin for crack detection at time T_4 is similarly

$$H^{110} = \int_{a_R}^{a_{d4}} \frac{dx}{Y(x)^m (\pi x)^{m/2}} - C \nu (T_4 - T_2) A^m \Gamma(1+\frac{m}{B}) \qquad (7)$$

The crack growth formulation has been extended to include a possible positive threshold value for the stress intensity factor range in (1), a constant corrosion rate, and a crack initiation period, [8].

3. REPAIR STRATEGY; EVENT TREE

The following strategy for repair is selected: All detected cracks are repaired. In [8] also other repair and replacement strategies are included.

With n inspections performed at times $T_1, \ldots, T_n$, the total number of different repair courses is 2^n, see fig. 1.

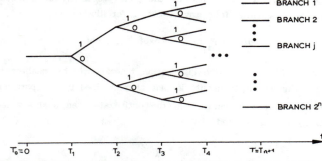

Figure 1. Repair realizations. 0 denotes no repair, while 1 denotes repair.

4. FAILURE AND REPAIR PROBABILITIES

The failure probability before time t is $P_F(t)$. The corresponding reliability index is

$$\beta(t) = -\Phi^{-1}(P_F(t)) \tag{8}$$

In terms of the safety and event margins, the failure probability before time t is as an example

$$P_F(t) = P(M(T_1) \leq 0) + P(M(T_1) > 0 \bigcap H > 0 \bigcap M^0(t) \leq 0) \tag{9}$$

$$+ P(M(T_1) > 0 \bigcap H \leq 0 \bigcap M^1(t) \leq 0), \quad T_1 < t \leq T_2$$

and similarly for other inspection time intervals and the life time T. With n inspections between 0 and T, $2^{n+1}-1$ parallel systems are analysed to compute the failure probabilities.

The expected number of repairs $E[R_1]$ at time T_1 is identical to the probability of repair at time T_1. It is at time T_2

$$E[R_2] = P(M(T_1) > 0 \bigcap H > 0 \bigcap M^0(T_2) > 0 \bigcap H^0 \leq 0) \tag{10}$$

$$+ P(M(T_1) > 0 \bigcap H \leq 0 \bigcap M^1(T_2) > 0 \bigcap H^1 \leq 0)$$

and similarly for other inspection times. With n inspections between 0 and T, 2^n-1 parallel systems are analysed to compute the repair probabilities.

5. DEFINITION OF INSPECTION QUALITY

The inspection quality is defined by the *pod* (probability of detection) curve $p(a)$ for which an exponential form may be chosen for illustration.

$$p(a) = F_{a_d}(a) = 1 - \exp(-\frac{a}{\lambda}), \quad a > 0 \tag{11}$$

The pod curve is identical to the distribution function of the smallest detectable crack size a_d. The inspection quality is thus characterized by the parameter λ. λ can take values between 0 and ∞. In the optimization an auxiliary measure of inspection quality q is introduced.

$$q = \frac{1}{\lambda} \tag{12}$$

q can take values in the interval $[0;\infty[$. $q=0$ corresponds to no inspection, while $q=\infty$ corresponds to a perfect inspection where infinitely small cracks are found.

6. FORMULATION OF OPTIMIZATION PROBLEM

The number of inspections n during the life time T is selected beforehand. This is done to avoid an optimization problem with a mixture of integer and real valued optimization variables. The analysis is repeated for several values of n and the resulting optimal costs are compared. The n-value with the smallest total expected cost is the optimal value. Inspection times and qualities are optimization variables together with a structural design parameter vector z.

The following cost items are included: initial cost, $C_I = C_I(z)$, inspection cost, $C_{IN} = C_{IN}(q)$, cost of repair, C_R, and cost of failure, $C_F = C_F(t)$. Inspection and repair costs are assumed to increase with the rate of inflation. The difference between the desired rate of return and the rate of inflation is assumed constant r. The cost of failure may be a function of time.

The optimization is formulated as a minimization of the total expected cost, with a constraint on the reliability index for the life time and simple constraints on the optimization parameters:

$$\min_{t,q,z} \quad C_I + \sum_{i=1}^{n}(C_{IN}(q_i)\ (1-P_F(T_i)) + C_R E[R_i])\frac{1}{(1+r)^{T_i}} \tag{13}$$

$$+ \sum_{i=1}^{n+1} C_F(T_i)\ (P_F(T_i)-P_F(T_{i-1}))\ \frac{1}{(1+r)^{T_i}}$$

s.t $\quad \beta(T) \geq \beta^{min}$

$\quad t^{min} \leq t_i = T_i - T_{i-1} \leq t^{max}, \quad i=1,2,\ \ldots,n$

$\quad 0 \leq T - \sum_{i=1}^{n} t_i \leq t^{max}$

$\quad q^{min} \leq q_i \leq q^{max}, \quad i=1,2,\ \cdots,n$

$\quad z^{min} \leq z_i \leq z^{max}, \quad i=1,2,\ \cdots,k$

The possibility of predetermining one or more of the inspection times and qualities as well as elements in z is available.

The optimization problem is solved for each n using the NLPQL algorithm [9]. The value of the objective function and of the constraints are computed in a separate routine. This routine calls upon the reliability analysis program PROBAN [10] for analysis of $2^{n+1}-1$ parallel systems for calculation of failure probabilities and 2^n-1 parallel systems for calculation of expected repair cost. PROBAN provides a reliability index calculated by a first-order reliability method [11] for each parallel

system together with exact partial derivatives of the reliability index with respect to λ_i, T_i and z_j. From these partial derivatives, the partial derivatives with respect to q_i, t_i and z_j are easily derived. Possibilities are included to perform the optimization with only the most important branches in the event tree in fig. 1 included.

7. OPTIMIZATION AFTER INSPECTION

A more general formulation than presented in fig. 1 has different inspection times and qualities in different branches. The number of optimization variables is thereby increased drastically. To overcome this problem, a procedure is here chosen where the inspection plan is first optimized at the design stage. When the result of the first inspection is known, a new optimal inspection plan is determined applying this information in addition to the information available at the design stage. The various failure probabilities and probabilities of repair are then conditional probabilities, conditioned upon the result of the first inspection. Actual crack measurement results can be considered at this stage. For each inspection result being available, the tree of possibilities in fig. 1 is reduced to one half of its size as the actual branch at the inspection time is known. With inspection results available at times $T_1, \cdots, T_{j-1}$, the optimization problem is formulated as

$$\min_{t,q} \quad \sum_{i=j}^{n} (C_{IN}(q_i) \ (1-P_F(T_i)) + C_R E[R_i]) \frac{1}{(1+r)^{T_i}} \tag{14}$$

$$+ \sum_{i=j}^{n+1} C_F(T_i) \ (P_F(T_i)-P_F(T_{i-1})) \ \frac{1}{(1+r)^{T_i}}$$

s.t $\quad \beta(T) \geq \beta^{min}$

$$t^{min} \leq t_i = T_i - T_{i-1} \leq t^{max}, \quad 0 \leq T - \sum_{i=1}^{n} t_i \leq t^{max}, \quad q^{min} \leq q_i \leq q^{max}, \quad i=j,...,n$$

where failure and repair probabilities are computed conditioned upon the results of the first j-1 inspections, [4].

8. AN EXAMPLE

The results of an example are presented. The selected input data are described in [8] and are to some extent representative for a joint in an offshore steel jacket structure. Figure 2 shows the different cost items as a function of the number of inspections, and it shows the change in reliability index with time for the case of n=4 inspections. If the inspection at time T_1 does not result in a crack detection, a new optimization is performed as described in the previous section. Figure 3 shows

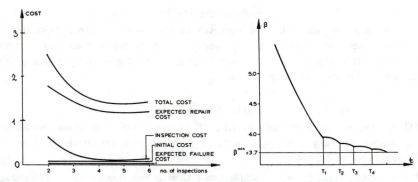

Figure 2. Cost functions at optimal solution; reliability index with time.

the change in reliability index with time for the new optimal solution and for the optimal solution determined with no crack detection in the first two inspections.

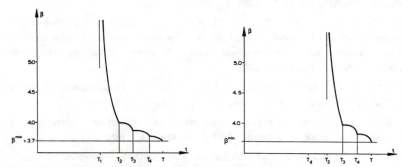

Figure 3. Updated function of reliability index with time.

9. CONCLUSIONS

A procedure for optimal design, inspection and repair of a fatigue sensitive element has been presented. Fatigue crack growth has been described by Paris' equation and failure been defined as growth to a critical size. Reliability calculations and associated sensitivity calculations have been performed by a first-order reliability method. Inspection times and qualities as well as structural design parameters are the optimization variables. A standard non-linear optimization routine is used. The optimization is first carried out at the design stage and later updated each time new inspection information becomes available. The repair strategy as presented is that all detected cracks are repaired, but other strategies have also been implemented.

10. ACKNOWLEDGEMENT

The paper is based on work supported by the research program "Reliability of Marine Structures" sponsored by A.S Veritas Research, SAGA Petroleum, STATOIL and CONOCO Norway, and by the research program "Marine Structures" sponsored by the Danish Technical Research Council.

11. REFERENCES

[1] KOZIN, F. and BOGDANOFF, J.L., "A Critical Analysis of Some Probabilistic Models of Fatigue Crack Growth," *Engineering Fracture Mechanics*, Vol. 14, No.1, pp. 59-89, 1981.

[2] MADSEN, H.O., "Random Fatigue Crack Growth and Inspection," in *Proceedings*, ICOSSAR'85, IASSAR, Vol.I, pp. 475-484, 1985.

[3] PALIOU, C., SHINOZUKA, M. and CHEN, Y.-N., "Reliability and Durability of Marine Structures," in *Proceedings*, Marine Structural Reliability Symposium, Arlington, SNAME, pp. 77-90, 1987.

[4] MADSEN, H.O., SKJONG, R., TALLIN, A.G., and KIRKEMO, F., "Probabilistic Fatigue Crack Growth Analysis of Offshore Structures with Reliability Updating Through Inspection," in *Proceedings*, Marine Structural Reliability Symposium, Arlington, SNAME, pp. 45-56, 1987.

[5] DITLEVSEN, O., Random Fatigue Crack Growth - A First Passage Problem," *Engineering Fracture Mechanics*, Vol. 23, No.2, pp. 467-477, 1986.

[6] SKJONG, R., "Reliability-Based Optimization of Inspection Strategies," in *Proceedings* ICOSSAR'85, IASSAR, Vol.III, pp. 614-618, 1985.

[7] THOFT-CHRISTENSEN, P. and SORENSEN, J.D., "Optimal Strategies for Inspection and Repair of Structural Systems," *Civil Engineering Systems*, Vol. 4, pp. 94-100, 1987.

[8] MADSEN, H.O., "Theoretical Manual PRODIM - PRObability-based Design, Inspection and Maintenance," A.S Veritas Research Report No. 88-2019, Hovik, Norway, 1988.

[9] SCHITTKOWSKI, K., "NLPQL: A FORTRAN Subroutine Solving Constrained Non-Linear Programming Problems," *Annals of Operations Research*, 1986.

[10] TVEDT, L., "User's Manual PROBAN - PRObabilistic ANalysis," A.S Veritas Research Report No. 86-2037, Hovik, Norway, 1986.

[11] MADSEN, H.O., KRENK, S. and LIND, N.C., *Methods of Structural Safety*, Prentice-Hall Inc., Englewood Cliffs, New Jersey, 1986.

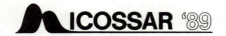

5th International Conference on
Structural Safety and Reliability

OPTIMAL STRATEGIES FOR DESIGN, INSPECTION, AND REPAIR OF FATIGUE-

SENSITIVE STRUCTURAL SYSTEMS USING RISK-BASED ECONOMICS

Paul H. Wirsching[*] and Yi Torng[**]

[*]Professor, [**]Graduate Research Assistant
Department of Aerospace and Mechanical Engineering
The University of Arizona, Tucson, Arizona 85721

ABSTRACT

The fatigue reliability and maintainability process for a marine structure
(caisson platform) is studied. Reliability analysis, performed by Monte Carlo,
quantifies the improvement in reliability which can be realized by a program
of periodic inspection and repair. Yet, when economic value analysis is
performed for this structure, the optimal strategy is to design using a large
safety margin and avoid an expensive maintenance program.

KEYWORDS

Fatigue, reliability, inspection, economics, crack growth.

1. INTRODUCTION

Metallic structures subjected to dynamic loads are vulnerable to fatigue. Consider a "series" structure in
which there are J fatigue-sensitive points and, therefore, J failure modes. In general, the nominal
stress at each point, produced by a random loading process Q(t), will be different. Fatigue is assumed
to be the only failure mode. Moreover, it is assumed that fatigue is crack propagation only and that
crack growth is described by the fracture mechanics fatigue model.

The general problem of ensuring integrity of structural systems relative to fatigue can be summarized
by the following: (1) Fatigue (and fracture) is the principal mode of failure in many structural and
mechanical systems. (2) Fatigue design factors are subject to considerable uncertainty and therefore
reliability methods are appropriate as a tool for making design decisions. (3) System reliability can be
improved by a maintenance program of periodic inspection and repair or replacement. (4) However, a
maintenance program can be expensive. (5) Ultimately, the goal of the design process is to specify a
design, inspection, and repair strategy to minimize life-cycle costs. Several papers have been written
on this topic, e.g., [1-5].

The goal of the study described herein is to develop a method for performing an analysis of the fatigue reliability and maintainability process for a series structural system. This analysis provides supporting information for an economic value analysis in which a design and inspection strategy can be constructed to minimize life-cycle costs.

2. THE PHYSICAL PROBLEM AND PROCESS MODEL

A model of the fatigue process for a single element in shown in Fig. 1. The fatigue stress is assumed to be a narrow-band process. Stress range S is a random variable having a Weibull distribution over the service life. The material resistance of strength is assumed to be described by a fracture mechanics fatigue model. In this analysis, the Paris law with a threshold level is employed.

The inspection process is defined by (1) the probability of detection (POD) curve associated with a given inspection type, (2) the number and spacing of the inspections, and (3) the quality of the member after the repair. For this exercise, the following assumptions are made for the inspection-repair process: (1) Any number of inspections can be specified. (2) The POD curve is specified for each inspection type. (3) Pre-installation inspection can be performed. (4) The pre-installation POD curve will be different from the in-service POD curve. (5) Repairs will be performed on any detected crack. (6) After repairs, the crack size distribution will be the same as the initial crack size distribution.

3. METHOD OF SOLUTION

An analytical-numerical approach to the fatigue reliability inspection process using FORM (first-order reliability methods) has been developed by Madsen et al. [1, 2]. However, Monte Carlo simulation looks attractive relative to the analytical approach because of the complexity of the process. Specific advantages of the Monte Carlo approach are: (a) complex crack growth and fracture processes can be accommodated; (b) all issues of dependency, e.g., construction of distribution of post-inspection crack sizes, are automatically avoided; (c) the many possibilities and options with respect to inspection/repair strategies can easily be treated; (d) other failure modes, e.g., fracture from an overload, can easily be included; (e) it is flexible in that almost any probability question can be answered quickly with a minimum of programming; and (f) it can be employed effectively to get a physical feel for the fatigue-inspection process. Furthermore, the process becomes even more complicated when a redundant structure is considered, making the Monte Carlo approach look even more attractive.

4. PROBABILISTIC CRACK GROWTH LAW

In the simulation process, it is necessary to compute, for a given joint, crack length a for cycle life N (i.e., inspection), given a specific set of random design factors. A technical issue is the method of computing a given N. Employing the fatigue strength model of Fig. 1, cycles N to develop crack length a can be expressed as [6]

$$N = \frac{1}{CB^m \overline{S^m}} \int_{a_0}^{a} \frac{dx}{G(x)Y^m(x)(\pi x)^{m/2}} \tag{1}$$

where

$$G(a) = \overline{S_0^m}(a)/\overline{S^m} , \tag{2}$$

$$\overline{S_0^m} = \int_{s_0(a)}^{\infty} s^m f_S(s)ds , \tag{3}$$

$$s_0(a) = \Delta K_{th}/Y(a)\sqrt{\pi a} , \tag{4}$$

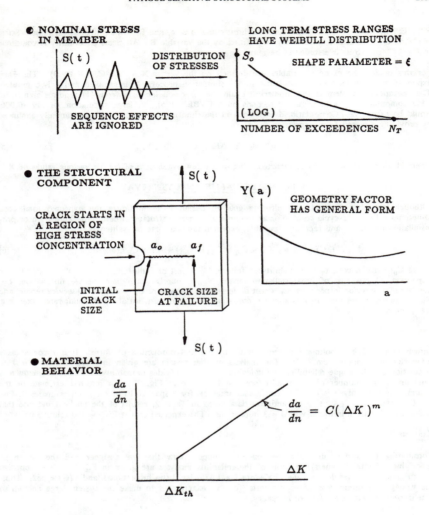

Fig. 1. A summary of the basic model used for the life prediction program.

$\overline{S^m} \equiv E(S^m)$, which is Eq. 3 with the lower limit equal to zero, and f_S = the pdf of stress range. Stress modeling error (bias and uncertainty) is defined by the variable B. All stress uncertainty is contained in B [7]. Mean stress is assumed to be zero.

For this model, the random variables are defined by the vector $X = (a_0, C, \Delta K_{th}, B)$. The Paris exponent m is assumed to be constant. In the simulation process, X is sampled and a is computed. The operational problem is that numerical evaluation of crack length a from (1) requires significant CPU computer time (e.g., about 10 seconds on a CYBER 175). A large simulation, of say 10,000, would be expensive. Computational efficiency in the simulation process can be substantially enhanced by providing a closed-form expression for a:

$$a = H(N, a_0, Y, \Delta K_{th}, C, B, m, \overline{S^m}) \,, \tag{5}$$

where H is an empirical form constructed using only a few evaluations of (1) for various values of X.

5. ECONOMIC VALUE ANALYSIS (EVA)

Although not always stated, the ultimate goal in design is to minimize the expected total cost, sometimes called "life-cycle cost." For an example of a marine structure where post-fabrication or pre-installation inspection and repair are possible, expected total cost can be written as

$$E(C) = C_0 + p_f C_F + C_{I0} + C_{R0} E(N_{R0}) + IC_I + C_R E(N_R) \tag{6}$$

where C_0 = initial cost, p_f = probability of failure, C_F = cost of failure, C_{I0} = cost of pre-installation inspection, C_{R0} = cost of pre-installation repair, N_{R0} = number of repairs before installation, I = number of in-service inspections, C_I = cost of in-service inspection, C_R = cost of in-service repair, and N_R = number of repairs. In this analysis, considerations of interest, inflation, discount rate, taxes, etc. are ignored.

Example 1

Shown in Fig. 2 is a bottom-founded caisson drilling and production platform. It is modeled as a three-element series system. Data for analysis of that system are given in Table 1 (also t = 1.0). Simulation of the fatigue reliability-maintainability process provides probability of failure estimates as a function of the number of inspections (see t = 1.0 curve in Fig. 3). An intuitive estimate of the accuracy of the analysis is that 90% confidence intervals for β span ±0.10 of the point estimates of Fig. 3. Employing an EVA using (6) and cost data given in Table 2, results of the analysis indicate that life-cycle costs are minimized with I = 3 inspections. The expected cost is E(C) = 460 relative units.

Example 2

Considering the system in the above example, suppose now that the designer had the option of specifying t (plate thickness). Results of the reliability analyses are given in Fig. 3. The optimal design from the results of the EVA are t = 1.475, I = 0 (no inspections), and E(C) = 362. Thus, the strategy for minimizing life-cycle costs, for this example, is to make the system large enough so that inspection and repair are not necessary.

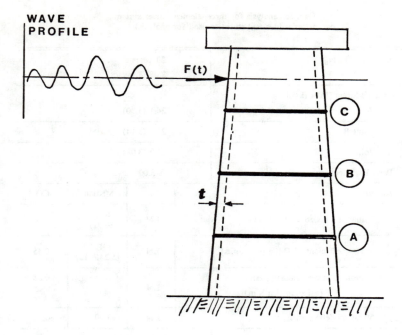

Fig. 2. Caisson platform (simplified for example) having three fatigue-sensitive points.

Table 1. Data for analysis for three-element series system
(parameter values and statistics used for analysis).

· Service Life, N_S	20 years 2.03E8 cycles			
· Miner's Stress [MPa (ksi)]				
Joint A	24.1 (3.50)			
Joint B	21.7 (3.14)			
Joint C	21.2 (3.08)			
· Paris Exponent, m	3.00			
· Random Variables		Dist	Median	COV
Initial crack size, a_0 [mm (in)]		LN	0.25 (0.01)	1.00
Paris coefficient, C [MPa (ksi) units]		LN	10E-12 (5.24E-10)	0.45
Threshold stress intensity, ΔK_{th} [MPa$\sqrt{m}$ (ksi $\sqrt{in}$)]		LN	4.4 (4.0)	0.08
Stress modeling error, B		LN	1.00	0.25
· Geometry Factor	Y(a) = 1.12			

· POD Curve: Pre-installation [a in mm]
    POD = $\Phi\left[\dfrac{a - 0.04}{0.0132}\right]$
	· POD Curve: In-Service [a in mm]
	    POD = 1 - exp(-a/0.051)			

Table 2. Estimated costs for examples.

	Cost (Relative Units)
C_0	300 + 80 (t - 1)
C_F	10,000
C_{I0}	2.5
C_{R0}	2.0
C_I	5.0
C_R	50.0

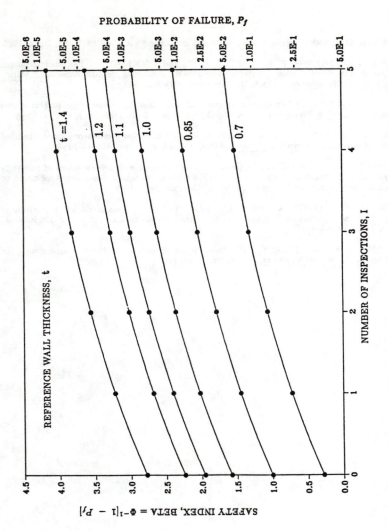

Fig. 3. Safety index as a function of number of inspections for various values of wall thickness. The three-element series system.

6. REFERENCES

[1]   MADSEN, H. O.: "Random fatigue crack growth and inspection," ICOSSAR'85, Kobe, Japan, 1985.

[2]   MADSEN, H. O., SKJONG, R. K., TALLIN, A. G., and KIRKEMO, F.: "Probabilistic fatigue crack growth analysis of offshore structures with reliability updating through inspection," Marine Structural Reliability Symposium (SNAME), Arlington, Virginia, 1987.

[3]   YANG, J.N. and CHEN, S.: "An exploratory study of retirement for cause for gas turbine engine components," Journal of Propulsion, AIAA, Vol. 2, No. 1, pp. 38-49, 1986.

[4]   YANG, J. N. and CHEN, S.: "Fatigue reliability of gas turbine engine components under scheduled inspection maintenance," Journal of Aircraft, AIAA, Vol. 22, No. 5, pp. 415-422, 1985.

[5]   WIRSCHING, P. H. and ORTIZ, K.: "Optimal economic strategies with considerations of reliability of fatigue-sensitive structural systems," Proc. NSF Workshop "Research Needs for Application of System Reliability Concepts," Civil Engr., Univ. of Colorado, Sept. 1988.

[6]   WIRSCHING, P. H., ORTIZ, K., and CHEN, Y. N.: "Fracture mechanics model in a reliability format," Proc. 6th International Symposium on Offshore Mechanics and Arctic Engineering, Houston, Texas, March 1987.

[7]   WIRSCHING, P. H. and CHEN, Y. N.: "Considerations of probability based fatigue design for marine structures," Marine Structures, Vol. 1, pp. 23-45, 1988.

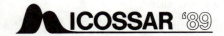
STRATEGIES FOR BRIDGE INSPECTION
USING PROBABILISTIC MODELS

Jamshid Mohammadi[*] and G. John Yazbeck[**]

* Illinois Institute of Technology, Chicago, Illinois, USA
** American University of Beirut, Beirut, Lebanon

Abstract

This paper describes the application of probability
analyses in planning an inspection program for highway bridges. When damage in the form of fracture
and cracking is the major concern, the inspection
time intervals may be formulated in terms of the
load cycles, parameters describing the damage growth
and a pre-defined critical damage level at which an
inspection becomes necessary. This paper discusses
the use of probabilistic modeling for estimating the
critical time intervals between inspections. The
probabilistic analysis is used mainly because the
damage process is a random phenomenon. A method is
developed and applied to the problem of cracking in
reinforced concrete slabs in highway bridges.

Key Words

Bridges, Concrete, Fracture, Inspection, Probability

1. Introduction

Highway bridges are periodically inspected so that any damage,
deterioration, loss of effective strengths in members, missing
fasteners, fracture, cracking, etc. can be detected and repaired
on time. Serviceability of a bridge to a great extend depends on
the frequency and the quality of the inspection. An effective
inspection program for a bridge requires a careful planning based
on potential modes of failure in the structural elements, history
of major structural repairs done for the bridge and of course the
frequency and the magnitude of the applied load (i.e., truck
traffic). The inspection planning also depends on the available

funds.

Major failures in structural components in a bridge often happen as a result of a timely application of the load alone, or combined with a hostile environment. As a result, it is often difficult to assess the extend of damage occurred on a component at any given time. This is especially true when the inspection is done purely based on visual examination of the component and the experience of the inspection personnel. Even with a highly experienced inspector, it is difficult to ascertain when a potential failure can occur. This is because of uncertainties associated with behavior of the structural component and its resistance to the distress conditions influential on the failure, and human errors and biases inherent in the inspection process. An effective planning for the inspection process should properly consider not only the potential modes of failure but also the time-dependency of the failure modes, the level of damage that is considered critical (and thus requires an immediate repair), different levels of uncertainties as described earlier and of course the available funds for the inspection. A combined probabilistic-optimization modeling is therefore necessary for this purpose. This paper explains such a model. Specifically, the application of the model to the problem of fracture in deck slabs of highway bridges is described. A mathematical scheme is then developed that can be used in planning an inspection program for a given bridge and predicting the time intervals for inspection based on the growth of cracks with time and the potential for damage to the deck slab as a result of repeated application of the load and the hostile environment.

The fracture of the reinforced concrete members are affected by the applied load, existing cracks, reinforcement, size of the member and time. As in other materials, the extend of fracture due to the repeated application of the load is highly random. In the case of fracture in a reinforced concrete slab, the cracked section becomes highly vulnerable to moisture, ice, salt, etc., which cause corrosion of the rebars or deterioration of concrete, when the cracks grow to a certain critical size. At this stage, not only the section suffers from accumulation of damage due to the repeated load, but also it may lose its effective strength because of the hostile environment. The critical crack size can be used as a limiting value in developing a model for the inspection of reinforced concrete slabs in highway bridges. The model presented in this paper uses a specific probability level as the accepted probability that cracks exceed a pre-determined critical level. Then using a strain damage accumulation relation for the slab, the time at which the cracks size reaches this critical level is evaluated and used as a basis for the required inspection time intervals.

The extension of the model to account for the optimization of the inspection cost within the available inspection funds is described in the paper. The application of the method for a comprehensive inspection planning for the entire bridge considering potential failures in different components, however, becomes complicated. The formulations and numerical illustrations

presented in the paper, at this time, are limited to the fracture
mode of failure of the deck slab only.

2. Inspection Intervals, Formulation and Limitations

A common program of inspection for highway bridges comprises
of a periodic inspection and condition assessment conducted usually
once every two years and additional inspections in cases where an
unusual load occurs on the bridges. During the inspection, the
conditions of components are assessed and the results are then used
in a comprehensive structural analysis of the bridge to develop
bridge rating in accordance with the AASHTO specification for
rating [1].   A regular program of inspection even though
satisfactory for some bridges, may not be adequate for many others.
Depending on the general condition of a bridge, shorter inspection
intervals may become necessary. In fact, the inspection intervals
for a given bridge must be selected such that the bridge can
continue providing an un-interrupted reliable service to the
public. On this basis a pre-determined limit state may be used as
a target value at which the condition of the component becomes
structurally critical so that an inspection may become necessary.
A formulation may then be developed to determine an average
inspection interval for the component based on this critical
condition. Determination of this inspection interval also depends
on the potential modes of failure of the component and the effect
of repeated load applications and the hostile environment on the
component. Internal stresses developed as a result of truck loads
in bridge components are, however, relatively small [2].
Therefore, except for rare cases of failure because of unusual
loads (such as earthquakes and bridge impact with ship and barges),
most failures occur due to repeated action of the load. In this
mode of failure, a component develops damage in the form of cracks
which grow with the repeated load application. As a result, the
damage reduces the serviceability of the bridge component to a
level where a fracture failure occurs or the damage is so severe
that the hostile environment may rapidly deteriorate the component
and its load bearing capabilities. For example, a crack in the
reinforced concrete deck slab may become an ideal situation for
further damage to rebars because of attacks by salt and other
deicing chemicals. In the case of a steel girder, damage because
of fracture may grow much faster in a severely cold weather
environment.

In cases where the nature of damage and its growth with time
(i.e., with the frequency of load application) are known, an
inspection interval may then be selected based on the extend of
damage to the component and the state at which the component
serviceability has reduced to a critical level as described
earlier. Theoretically, the inspection interval depends on, and
can be formulated in terms of, the following factors, among others.

o   The initial damage state of the component (its condition at
     the start of the inspection program),

o   A limit state at which the damage is considered to be
     critical,

o   Damage model and damage rate of growth,

o   Environmental factors that may expedite the damage growth,

o   The general structural condition of the component, its useful resistance capacity and reduction in the capacity because of corrosion, deterioration, local failures, etc.

At the practical level, however, development of a straightforward relationship between the inspection intervals and the above factors is difficult; and a trial and error approach becomes necessary. Furthermore, the solution process more appropriately requires a probabilistic formulation. This is because the damage growth is a random phenomenon which is influenced by the distribution of the applied load and the component's geometry and material properties which are also random variables. As a result, the damage growth to a specified limit state can only be defined with a probability value. This probability value will enter the formulation for the inspection intervals as will be described later.

Theoretically, the inspection time intervals can be formulated in terms of the stress distribution, load cycles and parameters describing the damage growth. In cases where the relation between the damage, its rate of growth and the load cycles is relatively simple, a closed form solution may be obtained for the inspection intervals. This is especially true for the case of steel plate girders for which the damage model in the fatigue and fracture mode of failure is relatively well established. In such cases, recent statistical models developed for the prediction of the remaining useful life of bridge components (e.g., Ref. [2]) can be used as a basis for establishing the inspection time intervals.

In more complicated cases, such as the case of a reinforced concrete deck slab, the damage due to fracture not only depends on the load cycles but also many other factors such as the internal strains in concrete and steel, material stiffness, size of steel rebars, etc. A factor such as stiffness also changes as cracks grow and the component deteriorates [3,4]. This would cause even more complexity in the problem. The inspection time interval, T, in this case may be written in the general form of Eq. 1.

$$T = f(p_F, W_L, N, S, \epsilon_s, \epsilon_c, E, a_1, a_2, \ldots, D_0) < T_0 \qquad (1)$$

In this equation, p_F=a probability level, W_L=limit state in terms of crack width, ϵ_s, ϵ_c=strain at steel and concrete, respectively, E=concrete modulus, a_i=constants describing damage growth and D_0=initial damage in terms of the crack width. The strain and modulus values change with N. As it is explained in the next sections, a closed-form solution to Eq. 1 is difficult to obtain. Thus a trial and error approach for solving Eq. 1 becomes unavoidable.

3. Inspection Planning for R.C. Deck Slab

3.1 Damage Formulation

For the reinforced concrete deck slab, serviceability is measured with the crack width. Reference [3] uses the following relation for the crack width, $W_{max,N}$, in terms of the number of stress cycles N.

$$W_{max,N} = c_2 (\epsilon_{s,N} - \epsilon_{r,N}) (h_2/h_1)_N \qquad (2)$$

in which $\epsilon_{s,N}$=steel strain at the crack, $\epsilon_{r,N}$=average concrete strain between the cracks after N cycles of the load, and c_2=constant. The ratio h_2/h_1 depends on the geometry of the crack section and is written as:

$$h_2/h_1 = (h - C)(d - C) \qquad (3)$$

where h=depth of the slab, C=depth of the neutral axis and d=effective depth of the section. Following Ref. [3], in terms of the static crack width W_{max}, the load cycle-dependent crack width can be written as:

$$W_{max,N}/W_{max}=[c_2(\epsilon_{s,N}-\epsilon_{r,N}) (h_2/h_1)_N]/[c_1(\epsilon_s-\epsilon_r)h_2/h_1]+W_0/W_{max} \qquad (4)$$

in which c_1=constant, ϵ_s and ϵ_r=strain values at the static load condition, and W_0=crack width at the start of the inspection program or at the previous inspection when a series of consecutive inspections are considered. The last term in Eq. 4 was added to account for the initial crack width in the inspection program. Values of the ratio c_2/c_1 are reported in Ref. [5] based on experiments conducted on wire-mesh reinforced mortar specimens.

The relation between strain values after N cycles of the load and at the static conditions is also given in Ref. [3] in the following form.

$$(\epsilon_{s,N}-\epsilon_{r,N}) = (\epsilon_s-\epsilon_r) (\epsilon_{s,N}/\epsilon_s) [1+1.22 (\log_{10}N)^2/47] \qquad (5)$$

Combining Eqs. 2-5, the following equation can be written.

$$W_{max,N}/W_{max}=1.2 (h_2/h_1)_N (\epsilon_{s,N}/\epsilon_s) [1+1.22 (\log_{10}N)^2/47]/(h_2/h_1)+ W_0/W_{max} \qquad (6)$$

Another alternative to this relation is the one suggested by the European Committee of Concrete. This relation is further discussed in Ref. [4].

At every cycle of the load, an internal analysis of the cross section of the slab needs to be done to evaluate the strain values. To do this, the "cyclic-dependent secant modulus", E_N, must be used. The modulus E_N depends on the stress range, the number of load cycles and the average concrete strain. A relation for E_N is given in Ref. [3].

3.2 Estimation of Inspection Time Intervals

Development of a comprehensive model for the estimation of inspection time intervals requires an optimization process, for which the objective function is described in terms of the total annual cost for inspection. A series of constraints must then be included in the process. Constraints such as a serviceability requirement for cracking, limitation on stresses and applied loads, structural geometry and integrity requirement, deck slab thickness requirement, local code provisions, etc. [4]. In this discussion, however, such detail optimization process is avoided. The inspection intervals are obtained based on (i) a target value for the crack width (W_L) which describes the limit state, and (ii) a probability level (p_F) describing the accepted probability that the crack exceeds W_L, i.e.

$$P(W_{max,N} \leq W_L) = 1 - p_F \qquad (7)$$

The limiting value, W_L, can be selected based on the performance criteria of the bridge deck slab, as described by the design code or the bridge engineer and the potentials for slab exposure to the hostile environment. The probability level $1-p_F$ is merely a measure for confidence and reflects the effect of uncertainties associated with the damage process, material, geometry, etc. as described earlier. The process of finding the inspection time intervals is as follows.

o   Select p_F and W_L

o   Use a trial and error approach solving Eqs. 6 and 7 simultaneously to find the mean number of cycles N which causes $W_{max,N}$ to exceed W_L with p_F probability. In doing this process, parameters involved in Eq. 6 are treated as random variables.

o   Find the average time for inspection by using the statistics of average daily truck traffic and the average number of cycles N.

4. Numerical Illustrations

The damage formulation and process described in Section 3 was applied to a bridge deck slab with the following dimensions and other characteristics (Ref. [4] presents a detailed description of the bridge and its components).

Depth (h) = 203 mm, concrete compressive strength=24.1 N/mm^2, rebars=#5 (16 mm diameter) @ 102 mm intervals with yield capacity of 289 N/mm^2, concrete weight = 2323 kg/M^3, and W_L = 0.3 mm. Three different values for the average hourly truck traffic are considered. These are respectively 500, 1000 and 1500 vehicles/hour. Also, p_F = 0.01 was used in the analysis. The internal stress and the slab analysis for the static load was conducted using the AASHTO [6] recommended design loads. The trial and error procedure described in Section 3 was then used for this slab; Eqs. 5-7 were then solved. The results indicated that for

p_f=0.01 and 500 vehicles/hour truck traffic, the required inspection interval is 4 years. This is based on the assumption that the truck traffic is uniform in any given hour and there is no other requirement for inspection. For average hourly traffic of 1000 vehicles, the required inspection time interval is 3.4 years. With an initial crack equal to 20% of the limiting crack width (W_L), the inspection interval time must be reduced substantially. These results are shown in Fig. 1.

5. Conclusions

This paper describes a probabilistic model for estimation of inspection time intervals for bridge components. The formulation is based on a desired serviceability level which is described in terms of a limit state and a damage model. The method was applied to the case of reinforced concrete deck slabs. In this formulation, the effect of environmental factors that may expedite the damage growth or may introduce other modes of failure are not considered. The inspection intervals obtained according to this formulation thus can be used as a base line for planning an inspection program for the deck slab. It is, however, emphasized that a more elaborate formulation should properly consider the optimization of the cost of inspection, limitations on the number of inspections and priorities for inspection as prescribed by the code. The sate of initial damage before an inspection program starts is also important and needs to be included in the optimization process. In a case where a regular program of inspection is considered for a bridge, the outcome of an inspection (in terms of damage detected or undetected) will definitely influence the damage formulation and the minimum time until the next inspection.

The formulation presented was applied to a deck slab in a highway bridge. The purpose of the numerical illustration presented was only to demonstrate the applicability of the method. The values derived for the inspection intervals are, of course, subjected to limitations and adjustments so that the effect of the aforementioned factors can properly be considered in the process.

References

[1]   American Association of State Highway and Transportation Officials, "Manual for Maintenance Inspection of Bridges," Third Edition, 1979.

[2]   MOSES, F., SCHILLING, C. G., and RAJU, S. K., "Fatigue Evaluation Procedures for Steel Bridges," NCHRP 299, Transportation Research Board, Washington, D.C., 1988.

[3]   SHAH, S. P. (Editor), Fatigue of Concrete Structures, Publication SP-75, American Concrete Institute, Detroit, Michigan, pp 133-175, 1982.

[4]   YAZBECK, G. J., "Damage Assessment of Reinforced Concrete Deck Slabs in Highway Bridges and Strategies for Inspection," M.S. Thesis, Department of Civil Engineering, Illinois Institute

of Technology, Chicago, Illinois, 1987.

[5]   BALAGURU, P., NAAMAN, A. E., and SHAH, S. P., "Ferrocement in
      Bending," Part II: Fatigue Analysis, Report No. 77-1,
      Department of Materials Engineering, University of Illinois,
      Chicago, Illinois, Oct., 1977.

[6]   American Association of State Highway and Transportation
      Officials, "Standard Specification for Highway Bridges," 1977.

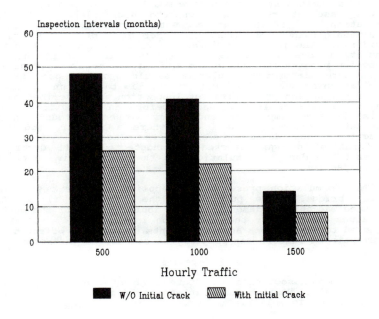

Fig. 1 Estimated inspection time intervals

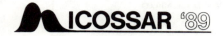
DAMAGE ASSESSMENT BASED ON REPAIRABILITY CRITERION

By Fabian C. Hadipriono M. ASCE
Department of Civil Engineering
The Ohio State University, Columbus, Ohio 43210

ABSTRACT

This paper discusses the assessment of damage of
protective structures based on the repairability
criterion. The variables of the repairability are
the amount of repair, repair cost, repair time, and
resource availability. Their values were determined
based on the asessment of the experts. These values
are incorporated in the production rules and
contained in the knowledge bases of a Fuzzy
Reasoning Expert System (FRES). The modus ponens
deduction technique is employed for partially
matching the user's input to the production rules.

KEYWORDS

Expert System, Fuzzy sets, Damage, Repairability.

INTRODUCTION

An approach using expert system for determining the damage level
of protective structures was developed at The Ohio State University
[3]. In this paper, the use of this approach based on the
repairability criterion is demonstrated. A shorter version of this
paper is presented in Reference [5]. Repairability is defined as the
extent of repair required for a damaged structure. In this study the
structure is assumed damaged by the impact of an explosive load, and
subsequently, repair is required to restore the structure to the
original or an acceptable condition. The variables of repairability
is usually assessed by experts using qualitative subjective
judgment.

VARIABLES AND VALUES OF REPAIRABILITY

In an earlier study [2,6], the variables of repairability are
determined as: amount of repair, repair time, repair cost, and
resource availability. The values of these variables are obtained
from experts. The expertise level of the experts and the damage
level of the structures were also obtained. An example of an expert

assessement is shown as follows:

IF   amount of repair is <u>major</u> AND
      repair time is <u>long</u> AND
      repair cost is <u>moderate</u> AND
      resource availability is <u>scarce</u>
THEN damage level of the structure is <u>severe</u>

   This example is called a production rule. The underlined values
of the variables are linguistic, and therefore, constitute fuzzy
sets. Each expert provides ten production rules associated with ten
structures observed. These production rules are contained in the
knowledge base of the expert system.

FUZZY PRODUCTION RULES

   When generating the production rules, for the same structure,
expert asessments may produce different values. Hence, a consensus
for the assessment of a particular structure is obtained through
the use of the fuzzy average operation [3].

   The fuzzy consensus is obtained from averaging the fuzzy sets in
relation to the weight of each set. The weight could represent the
expertise of the expert or the importance of damage criteria. The
fuzzy average can be found from: $A = \sum (A_i \times W_i)/\sum W_i$; where: A is
the consensus of assessments; A_i is the individual assessment; W_i
is the importance of damage criteria or expertise level of the
expert; and $i=1,2,3,...,n$. Since A_i and W_i are fuzzy sets, the
arithmetic manipulations should follow a special process based upon
Zadeh's extension principle [7]. This principle extends the
ordinary algebraic operations to fuzzy algebraic operations.

   In general, a fuzzy set may have a negative, positive, or
neutral characteristic. Examples are <u>poor,</u> <u>good,</u> or <u>fair,</u>
respectively. However, if the models representing the original
values are of different shapes, then the characteristic of the
average value may not be easily determined. The following ranking
index can be employed: $I = A_r$ [3]; where A_r is the area enclosed to
the right of the membership function. In this paper, the author
uses Blokley's rotational models to represent the values of the
amount of repair, repair time, repair cost, and resource
availability [3]. Information provided by a user will be partially
matched with the fuzzy production rules.

PARTIAL MATCHING USING FUZZY LOGIC

   Consider the following production rule: IF the amount of repair
(AR) is <u>very</u> <u>severe</u> (VSE), THEN the damage level (DL) is <u>severe</u>
(SEV). When a fact shows that AR is VSE, the consequent (THEN
statement) is then realized. However, when observation shows that
"AR is SEV," then partial matching is in order. This can be
performed by the Modus Ponens Deduction (MPD) technique which
consists of: (1) Truth Functional Modification (TFM), (2) Inverse
Truth Functional Modification (ITFM), and (3) Luckasiewics
Implication Relation (LIR). All of these are elaborated on in

earlier papers [4]. A brief discussion is presented here.

TFM is a logic operation that can be used to modify the membership function of a linguistic value in a certain proposition with a known truth value. Suppose that damage level (DL) is "negligible" (NNE) and is believed to be "false," (FA). This proposition can be expressed as:

P:(DL is NNE) is FA; NNE$\subset$DL, FA$\subset$T

where DL is a variable (universe of discourse), T is the truth space, and NNE and FA are the values of DL and T, respectively. The symbol $\subset$ denotes "a subset of." Modification of this proposition yields

P':(DL is DL1); DL1$\subset$DL

where DL1 is a value of DL and whose membership function, is solved graphically as shown in Figure 1a where the fuzzy set NNE and FA are represented by Blockley's models [1].

ITFM is a logic operation that can be used to obtain the truth values of a conditional proposition. Suppose a proposition, P, is expressed as: "damage level is negligible given damage level is severe"; then the proposition can be rewritten as:

P: (DL is NNE)$\mid$(DL is SEV); NNE,SEV$\subset$DL

The ITFM reassesses the truth of (DL is NNE) by modifying this proposition to yield:

P': (DL is NNE) is T1; T1$\subset$T

where T1 is the new truth value for (DL is NNE). The membership function of T1 is obtained through the graphical process shown in Figure 1b.

LIR is a logic operation whose task is to obtain the truth value of a consequent provided that the information about the antecedent and its truth value are available. Consider again the above example:

P: (AR is VSE)$\supset$(DL is SEV)
P': (AR is SEV)

The ITFM operation leads to:

P": (AR is VSE) is T1 $\supset$ (DL is SEV)

which can be simplified by LIR operation into:

P": (DL is SEV) is T2

and solved by the TFM operation:

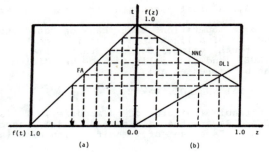

Figure 1a TFM Graphical Solution

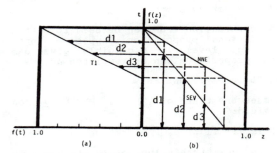

Figure 1b ITFM Graphical Solution

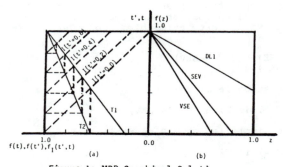

Figure 1c MPD Graphical Solution

P": (DL is DL1)

where T1 is the truth value of (AR is VSE), T2 is the truth value of (DL is SEV), and DL1 is the new value of damage level, DL. The graphical solution for this example is presented in Figure 1c. Further discussion of this can be found in Reference [3].

A similar process is repeated to partially match the input values of the variables repair time, repair cost, and resource availability with the related rules. The damage level in relation to repairability is obtained by taking the average of the four damage level values (related to the four repairability variables). The importance of the variables becomes the weight factor. Here, the importance of all variables are assumed equal. The whole procedure is repeated when the same input is partially matched with the other repairability-related rules. Hence, ten damage values (related to ten structures) will be obtained. Their average value becomes the final result. The characteristic of the final value is determined from the ranking index discussed earlier.

THE PROGRAM

Fuzzy Reasoning Expert System (FRES) is written in TURBO PASCAL for use in microcomputers. It consists of a main program and five supporting programs, namely, DAMAGE ASSESSMENT, BUILD MODEL, ASSIGN WEIGHT, BUILD RULE, and UTILITY. The architecture of FRES is shown in Figure 2.

The program DAMAGE ASSESSMENT performs the partial matching using the MPD operations. Here, the fuzzy average operation, fuzzy arithmetic, and fuzzy characteristic rating procedures are employed to obtain the damage level based on functionality. Note that in order to perform these operations, DAMAGE ASSESSMENT will have to obtain the data from MODEL file, WEIGHT file, UTILITY file, and the KNOWLEDGE BASE. BUILD MODEL allows the knowledge engineer to create his/her own models following the pattern of Blockley's models (Model 1) or this author's models (Model 2). This program also allows the knowledge engineer to modify the models by rotating the lines that represent the linguistic value (for Model 1) or by shifting the breaking points along the diagonals (for Model 2). This procedure is performed in a MODEL file.

ASSIGN WEIGHT allows the knowledge engineer to provide the weight for use in the fuzzy arithmetic operations performed in DAMAGE ASSESSMENT program. Here, the weights represent the importance of the damage criteria. This is used for obtaining the overall damage level based on all criteria. The process of assigning the weights is performed in a WEIGHT file. The program BUILD RULE is used to construct the production rule in the knowledge bases. The program limits the rule to the input from eight experts for ten protective structures. In the KNOWLEDGE BASES, ten rules are allocated for repairability. The expertise level of the expert is identified in the production rule. The UTILITY program defaults the fuzzy set models that have been created in BUILD MODEL and the weights that have been assigned in

ASSIGN WEIGHT. A UTILITY file is created when using this program.

FRES is written in a user's friendly style with menu windows and HELP files. lists the computer code of FRES. The knowledge bases contain the experts' knowledge compiled from the questionnaires [2]. The package is also furnished with graphical display to demonstrate the linguistic values of the variables of repairability and damage level. The use of FRES is demonstrated in the examples below.

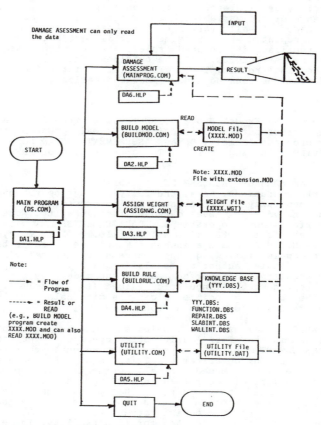

FIGURE 2   The Architecture of FRES

<u>EXAMPLE</u>

This example demonstrates the use of FRES for assessing the damage level of a protective structure based on repairability criterion. For this purpose, Blockley's fuzzy set models were employed. The user needs only to assess the values of the damage variables based on his/her observations on a protective structure. For example, a user assesses the amount of repair as being 100% <u>major</u> (MAJ). Then the user assesses both repair time and repair cost as being 100% <u>moderate</u> (MOD). Furthermore, the user determines the resource availability as being 100% <u>abundant</u> (ABD). This information leads to a result as shown in Figure 3. The ranking index is computed as 4.85, which is close to <u>moderate.</u>

OVERALL DAMAGE LEVEL BASED ON REPAIRABILITY DATA

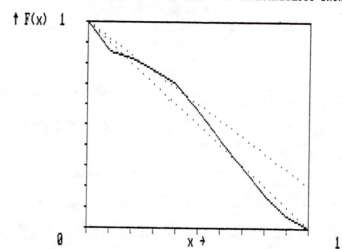

Overall Ranking Index : 4.85 between :
Moderate (5.00) and Light (4.00)
Push any key to continue or "F2" to print

Figure 3. Result of Damage Level for Repairability

<u>SUMMARY AND CONCLUSIONS</u>

Ten protective structures were investigated to create the production rules in repairability. Their assessments were acquired to construct the production rules, which are contained in the knowledge bases of FRES. Due to the variability in the assessments of the experts, a consensus of opinion is needed. This consensus can be reached by averaging these assessments in relation to the expertise level of the experts. The averaging process, however,

should be treated using fuzzy algebra. The modus ponens deduction technique is used as the inference mechanism for partially matching the information provided by the user to the rules contained in the knowledge bases.

The partial matching and averaging processes eliminate the potential combination explosions of the production rules. Also, the partial matching process suggests that the system does not need a large number of rules. An adequate number of rules from experts with high level of expertise will produce sufficiently accurate results. The system displays the result graphically, easily understood by a user with little knowledge of the fuzzy reasoning concept. It also provides the knowledge engineer the capability for constructing and modifying the models; adding, modifying, and deleting the production rules; and assigning the weights of the variables.

ACKNOWLEDGMENT

The author of this paper wishes to thank the Universal Energy System and the United States Air Force Office of Scientific Research for sponsorship of this study.

REFERENCES

[1] Blockley, D.I., The Nature of Structural Design and Safety, 1st Edition, John Wiley and Sons, New York, 1980, 365 pp.

[2] Hadipriono, F.C., Development of A Rule-Based Expert Sysytem for Damage Assessment of Air Force Base Structures," A Report Submitted to the Universal Energy System, Contract No. F49620-85-C-0013, August, 1986, 20 pp.

[3] Hadipriono, F.C., "Fuzzy Reasoning Expert System (FRES) for Assessing Damage of Protective Structures," A Research Report Submitted to Universal Energy System, Contract No. F49620-85-C-0013/SB5851-0360, March, 1988, 337 pp.

[4] Hadipriono, F.C. and Ross, T.J., "Towards A Rule-Based Expert System for Damage Assessment of Protective Structures," Second International Fuzzy Set Association Congress, Tokyo, Japan, July, 1987, pp 156-159.

[5] Hadipriono, F.C. and Ross, T.J., "Expert System for Repairability of Protective Structures," Proceedings of the North American Fuzzy Information Processing Society, NAFIPS '88, San Francisco, California, June, 1988.

[6] Ross, T.J., Wong F.S., Savage, S.J., and Sorensen, H.C., "DAPS: An Expert System for Damage Assessment of Protective Structures, " in Expert System in Civil Engineering, by Kostem, C.N. and Maher, M.L., Seattle, Washington, 1986.

[7] Zadeh, L.A., "Fuzzy Logic and Approximate Reasoning," Synthese, Vol 30, 1975.

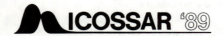

5th International Conference on
Structural Safety and Reliability

THE EFFECT OF REPEATED INSPECTIONS ON THE RELIABILITY
DEGRADATION DUE TO FATIGUE CRACK GROWTH

Akira Tsurui*, Akira Sako**,
Taisei Isobe* and Takeyuki Tanaka*

*   Dept. of Applied Mathematics and Physics, Kyoto University, Kyoto 606 Japan
**   Mitsubishi Heavy Industry Ltd., Hiroshima Works, Hiroshima 733 Japan

ABSTRACT

The effect of repeated in-service inspections on the reliability degradation due to fatigue crack growth is investigated based upon a stochastic
fracture mechanics. A method to evaluate the failure probability under repeated inspections is first discussed on the assumption that the
component is exchanged when cracks are detected. The results are then
applied to investigate how the reliability of structures behaves under
repeated ultrasonic inspections.

KEYWORDS

Reliability; fatigue crack growth; random loading; in-service inspections.

1 INTRODUCTION

It is a matter of course that such an important structure as aircraft should generally
be maintained in a highly reliable state. For this purpose, it will be a standard practice
to detect fatigue cracks by in-service inspections before their size reaches a certain critical
value, and to exchange the components if cracks are detected. It should be noted, however,
that the inspections are far from perfect and that too many inspections bring about the
exceeding decrease of availability and much economical load. Standing on this point of view,
we previously studied the effect of periodic inspections on the reliability degradation due to
fatigue crack growth under stationary random loading with the aid of the stochastic fracture
mechanics[1].

Practically, however, we often come across the case in which the inspections are not necessarily performed periodically. Therefore, under the assumption that an aperiodic inspection
program is prespecified, in this study a method to evaluate the effect of in-service repeated
inspections on the reliability degradation is developped and discussion is made on how the
failure probability behaves for the case of ultrasonic inspection method. It has been found
that the reliability is significantly affected by the inspection policy.

2 STOCHASTIC CRACK GROWTH MODEL

In order to take various statistical properties of the crack growth process into the reliability analysis under repeated inspections, it is inevitable to give a stochastic model of fatigue crack length distribution. Consider a crack growth model

$$\frac{dX}{dn} = \epsilon Z_n^\nu X^{\nu/2} \tag{1}$$

where X denotes the nondimensional crack length, Z_n the nondimensional random stress amplitude after n cycles of loading, ϵ a small positive parameter and ν a material constant usually assuming a value not less than 2.0. It should be paid attention to that this equation results from the well-known Paris-Erdogan's crack growth law.

After some mathematical manipulation[2][3], from this equation we can get the probability $W(x, n \mid x_0)$ that a crack of initial size x_0 is in the range $(0, x)$ after n cycles of stationary random stressing as follows:

$$W(x,t \mid x_0) = \int_0^x w(\xi,t \mid x_0)d\xi = \Phi\left[\frac{x_0^{-\lambda} - x^{-\lambda} - \lambda t}{\lambda\sqrt{\mu t}}\right], \tag{2}$$

where,

$$\lambda = \frac{\nu}{2} - 1, \quad t = \epsilon \mathbf{E}[Z_n^\nu]n, \quad \mu = \frac{2\epsilon}{\mathbf{E}[Z_n^\nu]}\int_{-\infty}^0 \{\mathbf{E}[Z_n^\nu Z_{n+n'}^\nu] - \mathbf{E}[Z_n^\nu]\mathbf{E}[Z_{n+n'}^\nu]\}dn'. \tag{3}$$

Here $\Phi[\cdot]$ means the standardized normal distribution function.

Now, assuming that the time t_0 when the crack size reach a certain specific length $\tilde{x}$ obeys a two-parameter Weibull distribution and that the initial crack length x is related to t_0 through $t_0 = c(\tilde{x}^{-\lambda} - x^{-\lambda})$, we will here fix a probability density function $g(x)$ of the initial crack length[4]. The solid curve in Fig.1 shows an example of this density function $g(x)$, which will be made use of in what follows.

3 RELIABILITY EVALUATION METHOD

In this section, we will develop a method to evaluate the reliability of a component under repeated inspections. Let t_j be the time interval between $(j-1)$-th and j-th inspections. $\mathbf{D}_j$ is introduced to express the event that at least one crack is detected at the j-th inspection, $\mathbf{U}_j$ the event that no cracks are detected at the j-th inspection, $\mathbf{S}_j$ the event that the failure did not occur from the start of operation up to the j-th inspection time and $\mathbf{F}_j$ the event that the failure occured by the j-th inspection time. If we pay attention to the j-th inspection time, we can easily see that $\mathbf{F}_j$, $\mathbf{S}_j \cap \mathbf{D}_j$, $\mathbf{S}_j \cap \mathbf{U}_j$ constitute a disjoint and exhaustive system of events. As the result, the relationship

$$P_F^{(j)} + P_D^{(j)} + P_U^{(j)} = 1 \tag{4}$$

holds for arbitrary j, where

$$P_F^{(j)} = \mathbf{Pr}[\mathbf{F}_j], \quad P_D^{(j)} = \mathbf{Pr}[\mathbf{S}_j \cap \mathbf{D}_j], \quad P_U^{(j)} = \mathbf{Pr}[\mathbf{S}_j \cap \mathbf{U}_j]. \tag{5}$$

Further, suppose that $\tilde{u}_j(x)$ and $u_j(x)$ mean crack length density functions just before and just after the j-th inspection, respectively, and both satisfy the same normalization condition

$$\int_0^{x_c} \tilde{u}_j(x)dx = \int_0^{x_c} u_j(x)dx = 1 - P_F^{(j)}, \tag{6}$$

where x_c stands for a prespecified maximum crack length beyond which the structural component fails. If the component is immediately exchanged for new one at the j-th inspection time when the event $\mathbf{S}_j \cap \mathbf{D}_j$ takes place, then the relationship

$$u_j(x) = \begin{cases} g(x) & \text{for } j = 0; \\ P_{\mathrm{D}}^{(j)} g(x) + P_{\mathrm{U}}^{(j)} \tilde{u}_j(x \mid \mathbf{S}_j \cap \mathbf{U}_j) & \text{for } j > 0, \end{cases} \tag{7}$$

holds, where $\tilde{u}_j(x \mid \mathbf{S}_j \cap \mathbf{U}_j)$ denotes a conditional crack length density function under the condition that the component did not fail up to the j-th inspection time and no cracks are detected at the j-th inspection. If $D(x)$ means the crack detection probability for a crack of size x, Bayes's formula leads to

$$P_{\mathrm{U}}^{(j)} \tilde{u}_j(x \mid \mathbf{S}_j \cap \mathbf{U}_j) = \{1 - D(x)\} \tilde{u}_j(x). \tag{8}$$

Therefore, making use of the relation

$$P_{\mathrm{D}}^{(j)} = \int_0^{x_c} D(x) \tilde{u}_j(x) dx, \tag{9}$$

we can interpret eq.(7) as a bridge between $u_j(x)$ and $\tilde{u}_j(x)$.

On the other hand, since the function $\tilde{u}_j(x)$ can be easily connencted with the function $u_{j-1}(x)$ through eq.(2) as

$$\tilde{u}_j(x) = \int_0^{x_c} w(x, t_j \mid x_0) u_{j-1}(x_0) dx_0 , \quad (j > 0) \tag{10}$$

the series of functions $u_j(x)$ $(j = 0, 1, 2...)$ can be recursively calculated.

Now, with the aid of these functions, we will calculate the failure probability $H(t)$ of the component up to time t. For the convenience sake, we will introduce the time interval s_k from the start of operation to the k-th inspection time as $s_k = \sum_{j=1}^{k} t_j$ $(s_0 = 0)$. For the interval $s_k < t \leq s_{k+1}$, as the difference between $H(t)$ and $P_{\mathrm{F}}^{(k)}$ is nothing but the probability that the component fails in the time interval (s_k, t),

$$H(t) = P_{\mathrm{F}}^{(k)} + \int_0^{x_c} \{1 - W(x_c, t - s_k \mid x_0)\} u_k(x_0) dx_0. \tag{11}$$

4 NUMERICAL EXAMPLE

Here, we will give an example to show the feasibility of the method proposed in the preceding section. So, according to Harris[5], we will introduce a crack detection probability curve for ultrasonic inspection technique as

$$D(x) = 1 - \frac{1}{2} \mathrm{erfc}\left\{\alpha \ln\left(\frac{x}{x^*}\right)\right\}, \tag{12}$$

where α and x^* are adequate parameters. The dotted curve in Fig.1 corresponds to a typical one $(\alpha = 2.1, x^* = 1.0)$, which will be made use of in the numerical calculation.

Figure 2 shows the temporal variation of the failure probability $H(t)$ of the component up to the time t. The curve (1) corresponds to a case in which we perform no inspections. We can see from this curve that the component is supposed to fail certainly by $t = 4.0$, unless inspections are performed. The curve (2) gives the failure probability under periodic inspections $(t_1 = t_2 = t_3 = t_4 = 1.25)$, but the curves (3) $(t_1 = 1.667, t_2 = t_3 = 1.167, t_4 = 1.0)$

and (4) $(t_1 = 2.0, t_2 = t_3 = t_4 = 1.0)$ are corresponding to aperiodic inspections. Small circles in the Figure mean the inspection time.

These curves show that the inspections have significant effect on the reliability degradation and that the inspection policy considerably affects the final reliability level. In the present stage, however, we can not directly obtain the optimum inspection policy except for trial-and-error method.

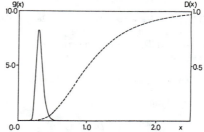

Fig.1 Probability density function of the initial crack length and crack detection probability
($\alpha = 2.1$, $x^* = 1.0$)

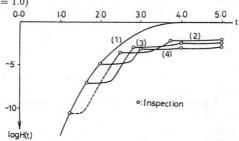

Fig.2 Reliability degradation under repeated inspections ($\lambda = 0.5$, $\mu = 0.02$, $x_c = 231$)

REFERENCES

[1] Tsurui,A. and Sako,A., "Reliability Analysis of Fatigue Crack Growth Processes under Stationary Random Loading", Recent Studies on Structural Safety and Reliability (CJMR, Vol.5), (1989) Elsevier, pp. 153-165.

[2] Tsurui,A. and Ishikawa,H., "Application of the Fokker-Planck Equation to a Stochastic Fatigue Crack Growth Model", Structural Safety, Vol.4 (1986), pp.15-29.

[3] Tsurui,A., Ishikawa,H. and Utsumi,A., "Theoretical Study on the Distribution of Fatigue Crack Propagation Life under Stationary Random Loading", Proc. of Fatigue 84, Vol.2 (1984), pp.949-957.

[4] Shinozuka,M., "Development of Reliability-Based Aircraft Safety Criteria: An Impact Analysis", Technical Report AFFDL-TR-76-36, Vol.1 (1976).

[5] Harris,D.O. and Lim,E.Y., "Application of a Probabilistic Fracture Mechanics Model to the Influence of In-Service Inspection on Structural Reliability", ASTM STP 798 (1983) pp.19-41.

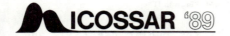
ANALYSIS OF BRIDGE MAINTENANCE AND RATING PROCEDURES

Youssef Hachem,[1] Student Member, ASCE
Konstantinos Zografos,[2] Associate Member, ASCE
Merhdad Soltani,[2] Member, ASCE

Abstract

In the United States, bridges are routinely rated to evaluate their existing conditions. The formula used to evaluate bridges is called the Sufficiency Rating Formula. The formula uses nineteen variables based on the bridge inspection record. The nineteen parameters in this formula are examined and an improved formula is developed using regression and history of bridges in the State of Florida.

Introduction

Bridge management consists of inspection and maintenance. Only the inspection process is discussed in this paper. The inspection process provides information about the existing condition of a bridge. The data provided ranges from the location to the rating of all structural elements of the bridge as the super and substructure.

The rating of a bridge is obtained from the United States Department of Transportation/Federal Highway Administration (USDOT/FHWA) Highway Bridge Replacement and Rehabilitation Program, and is called the Sufficiency Rating Formula (USDOT/FHWA 1979). The Formula consists of four major factors, structural adequacy and safety (S1), serviceability and functional

----
[1]Hazen and Sawyer, P.C., 5950 Washington St., Hollywood, FL 33023
[2]Assistant Professor, Department of Civil and Architectural Engineering, University of Miami, FL 33124

obsolescense (S2), essentiality for public use (S3), and a special reduction factor (S4). Each factor of the Sufficiency Rating Formula carries a different weight and is the result of a group of data collected at the site of the bridge or calculated from the plans of the bridge. The data collected from the bridge site is summarized in the form of a Structure Inventory and Appraisal Sheet (SI&A). Factors S1, S2, and S3 carry maximum percentages of 55%, 30%, and 15% respectively. The special reduction factor (S4) carries a maximum percentage of 13%. The sufficiency rating is therefore computed as

$$0 \leq S1 + S2 + S3 - S4 \leq 100$$

A rating of 100% means completely sufficient bridge and a rating of 0% means completely insufficient bridge.

Sufficiency Rating Prediction Equation

An attempt was made to form an equation to predict the sufficiency rating of bridges using polynomial regression. The data used was from the inspection records of the State of Florida bridges. The bridge data set was randomly divided into two data sets containing 8064 (set A) and 2818 (set B) bridges. General linear models method (GLM) was performed on data set A. GLM is a special case of analysis of variance method (ANOVA) that nullify the effect of unbalanced data (Betha, Duran and Boullion 1985, Draper and Smith 1981). The dependent or response variable was the sufficiency rating and the independent variables were:

- Age (years)
- Age^2
- Age^3
- ADT
- ADT^2
- ADT^3
- Approach Road Width
- Skew (degrees)
- Number of spans, main unit
- Number of lanes
- Number of approach spans
- length of maximum span (feet)
- Total horizontal clearance (feet)
- Structure length (feet)
- Bridge width (feet)
- Deck width (feet)

The age of a bridge is defined as the time period elapsed from its last major reconstruction (not from its initial construction), therefore, eliminating the effect of rehabilitation.

The developed model was significant at 1% level (Alpha = 0.01). The model obtained was tested on data set B, resulting in a Pearson Product-Moment Correlation Coefficient (r) of 0.67, i.e., this model accounts for 67% of the Sufficiency Rating of bridges.

Taking the model a step further, thirteen blocking effects were used to reduce the error term in the model. They were:

-Type of service	-Structure type, main unit
-Highway system	-Structure type, approach spans
-Custodian	-Administrative jurisdiction
-County	-Functional classification
-Owner	-Structure open/closed
-City/town	-Wearing surface
-State highway district	

The following fourteen objective independent variables were used:

-ADT	-Bridge roadway width
-Age	-Number of approach spans
-Age^2	-Length of maximum span
-Age^3	-Structure length
-Detour length	-Deck width
-Horizontal clearance	-Approach roadway width
-Minimum vertical clearance over bridge	
-Minimum vertical clearance under bridge	

The general linear models method (GLM) was performed on data set A. The model was significant at 1% level. The model was used to predict the sufficiency rating on data set B producing $r = 0.84$, i.e., this model is highly correlated and it accounts for 84% of the sufficiency rating of bridges. Therefore, r has improved due to the blocking effect. All the independent variables were significant at the 5% level. The predicted versus actual sufficiency rating values are plotted in Figure 1. Note: for a perfect correlation the points should form a line that passes through the origin with a slope = 1. For further validation, the residual values versus predicted values are shown in Figure 2. The points are scattered randomly around the Zero axis which approves of the model's adequacy.

Conclusion

The Sufficiency Rating Formula was examined using a history of bridges in Florida. All the variables used in the model are objective thus eliminating any subjective judgments. This model can be best used as a screening tool to identify bridges that need inspection in the future. This model can be used as a part of the planning process rather than a decision making tool.

Future work includes the development of an algorithm

to determine optimal inspection intervals, and life cycle cost/risk assessment to determine an effective and economical bridge design.

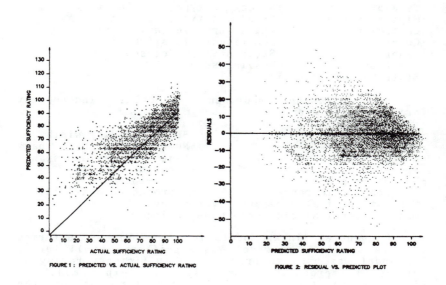

FIGURE 1 : PREDICTED VS. ACTUAL SUFFICIENCY RATING

FIGURE 2: RESIDUAL VS. PREDICTED PLOT

References

Betha,R.M.,Duran,B.S. and Boullion,T.L.(1985). Statistical Methods For Engineers and Scientists, Second edition,Marcel Dekker, New York.

Draper,N. and Smith,H. (1981). Applied Regression Analysis, Second edition, John Wiley and Sons,New York.

USDOT/FHWA (1979). Recording and Coding Guide for the Sturcture Inventory and Appraisal of the Nation's Bridges, Washington, D.C.

Acknowledgement

The authors would like to acknowledge Mr. John Ahlskog of FHWA and Mr. Larry Sessions of Florida DOT for their input in this research.

The opinions and judgements mentioned in this paper are the authors solely, without any liability to the persons acknowledged above.

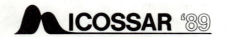
PARAMETER IDENTIFICATION AND STATE ESTIMATION
OF STRUCTURE MODELS IN SHAKING TABLE TEST

Zhi-Wen BAO [*]   Wen-Yue SHI [**]

* Associate Professor, Dept. of Civil Engineering
** Teaching Assistant, Dept. of Automation,
   Tsinghua University, Beijing, China

ABSTRACT

In this paper physical parameters and inner states
of structure can be estimated from the input and
responses in shaking table test by means of
Extended Kalman Filter(SIEKA).Comparing with other
methods this algorithm of structure identification
by (SIEKA) does not need accurate initial
estimates and is good at identifying physical
parameters and inner states directly and
dynamically. Not only every stiffness, and every
damping of the structural model, but also every
storey displacement can be estimated. According to
this method, the results of SIEKA's application to
two chained-multi-degrees-of-freedom structure
models in shaking table test are given.

KEYWORDS

PARAMETER IDENTIFICATION; STATE ESTIMATION;
SHAKING TABLE TEST; EXTENDED KALMAN FILTER;
STATE VECTOR; STATE EQUATION; STOREY DISPLACEMENT

INTRODUCTION

In recent years, some techniques for structural identification
have been developed and applied to the damage evaluation of
structures. However, most of techniques for parameter
identification are those of modal parameter identification and
other techniques for physical parameter identification which need
rather accurate initial estimates and a lot of computation time.
Comparing with the modal frequency , the stiffness is more
sensitive and direct to show the state and damage of the
structure. In this paper , the algorithm of structural
identification by means of Extended Kalman Filter (SIEKA) is
proposed.

THE PRINCIPLE OF THE ALGORITHM OF STRUCTURAL IDENTIFICATION
BY MEANS OF EXTENDED KALMAN FILTER (SIEKA)

In a control system the movement of a system is often expressed
in the form of state-variable. This method can also describe the

movement of structures. The states and unknown parameters can be estimated by means of this method. [1] The movement of a time-invariant vibration system can be expressed by a set of first-order differential equations (i.e. state equations) as follows:

$$\dot{s} = \mathcal{P}(s, \theta, u) \tag{1}$$

where, s —— state-vector, θ —— parameter-vector
u —— input of system

When the parameter-vector is unknown, both θ and s can be considered as an extended state vector, i.e. :

$$x = \begin{bmatrix} s \\ \theta \end{bmatrix} \tag{2}$$

Then, the state equation is as follows:

$$\dot{x} = f(x, u) = \begin{bmatrix} \mathcal{P}(x, u) \\ 0 \end{bmatrix} \tag{3}$$

The masses, stiffnesses, dampings are considered as a state vector in the algorithm of structural identication by means of Extended Kalman Filter. If the structure is time-invariant , the parameter state vector will be a constant one. Both the parameter state vector and the real state vector are expressed as an extended state vector. Therefore the extended state vector will be estimated optimally or suboptimally by means of Extended Kalman Filter Algorithm. [1][2]

Since the algorithm is realized by using computers, the state equation and the observation equation are expressed in discrete-time form as follows:

$$x(t_{k+1}) = f(x(t_k), u(t_k), t_k) \cdot \tau + x(t_k) + w(t_k) \tag{4}$$
$$z(t_k) = h(x(t_k), u(t_k), t_k) + v(t_k) \tag{5}$$

The current state $x(t_k|t_k)$ will be estimated recursively by means of a sequence of signals measured before the current state in the SIEKA.

Where, $w(t_k)$ —— white noise sequence
$v(t_k)$ —— white noise sequence
$Q(t_k)$ —— covariance matrix of $w(t_k)$
$R(t_k)$ —— covariance matrix of $v(t_k)$

The formulas of the Extended Kalman Filter Algorithm are as follows:

state predictor:

$$\hat{x}(t_{k+1}|t_k) = \hat{x}(t_k|t_k) + f(\hat{x}(t_k|t_k), u(t_k), t_k) \cdot \tau \tag{6}$$
$$P(t_{k+1}|t_k) = \Phi(t_{k+1}, t_k; \hat{x}(t_k|t_k)) \cdot P(t_k|t_k) \cdot \Phi^T(t_{k+1}, t_k; \hat{x}(t_k|t_k)) + Q(t_k) \tag{7}$$

gain matrix calculation:

$$G(t_{k+1}, \hat{x}(t_{k+1}|t_k)) = P(t_{k+1}|t_k) \cdot H^T \cdot (t_{k+1}, \hat{x}(t_{k+1}|t_k)) \cdot$$
$$(H(t_{k+1}, \hat{x}(t_{k+1}|t_k)) \cdot P(t_{k+1}|t_k) \ H^T(t_{k+1}, \hat{x}(t_{k+1}|t_k)) + R(t_{k+1}))^{-1} \tag{8}$$

state filter:

$$\hat{x}(t_{k+1}|t_{k+1}) = \hat{x}(t_{k+1}|t_k) + G(t_{k+1}, \hat{x}(t_{k+1}|t_k)) \cdot$$
$$(z(t_{k+1}) - h(\hat{x}(t_{k+1}|t_k), u(t_{k+1}), t_{k+1})) \tag{9}$$
$$P(t_{k+1}|t_{k+1}) = (I - G(t_{k+1}, \hat{x}(t_{k+1}|t_k)) \cdot H(t_{k+1}, \hat{x}(t_{k+1}|t_k))) \cdot P(t_{k+1}|t_k) \tag{10}$$

where, τ —— sampling time interval ; Φ —— state transition matrix

$$\Phi(t_{k+1}, t_k; \hat{x}(t_k|t_k)) = I + \tau \cdot F \Big|_{\substack{x=\hat{x}(t_k|t_k) \\ u=u(t_k)}}$$

$$H(t_{k+1}, \hat{x}(t_{k+1}|t_k)) = H \Big|_{\substack{x=\hat{x}(t_{k+1}|t_k) \\ u=u(t_{k+1})}}$$

$$F = \left(\frac{\partial f_i}{\partial x_j} \right) \ ; \quad H = \left(\frac{\partial h_i}{\partial x_j} \right) \ ; \quad \tau = t_{k+1} - t_k$$

According to equation (6)—(10), if the initial estimates of the system parameters and states are given, the recursive algorithm

can be performed on the basis of a sequence of signals measured.
In practice, this algorithm does not need accurate initial
estimates which are only determined approximately instead.

PARAMETER IDENTIFICATION OF STRUCTURAL MODELS IN SHAKING TABLE TEST

The application of SIEKA to two structural models in the shaking
table tests is introduced in this paragraph. Both models are made
of reinforced concrete. The first one is with five-degrees-of-
freedom . The model was excited several times by the earthquake
waves of different amplitudes. The maximal values of each
earthquake wave are 0.28g, 0.79g and 1.26g respectively. A set of
signals measured are absolute acceleration in each degree-of-
freedom of the model and the acceleration of the shaking table.
Each mass is given. A set of parameters are as follows:

$$\left[\frac{k}{m} , \frac{c}{m} ; \quad i = 1,2,3,4,5 \right]$$

The results of physical parameters which are estimated by means of
this algorithm are listed in the Table 1. Each stiffness change
(relative) is shown in Fig. 1. According to the estimated physical
parameters the model parameters could be calculated. And their
results are listed in Table 2. After earthquake waves the model
was excited by a white noise, the modal frequencies obtained from
FFT in this case are listed in Table 3. The second model on
shaking table is the three-degree-of-freedom structural system.
The estimated storey displacement-time curves of the second model
are shown in Fig.2. Other results are ommitted here.

CONCLUSION

On the basis of the Structural Identification Extended Kalman
Filter Algorithm (SIEKA) the physical parameters and inner states
of structures can be estimated from a sequence of signals
measured.   It is shown that SIEKA is an effective algorithm for
parameter estimation and damage evaluation of structures.
The estimated stiffness parameters possess corresponding accuracy.
The accuracy of estimated damping parameters will be discussed in
future.

REFERENCES

1. Zhi-Wen BAO, Wen-Yue SHI, Parameter Identification Of
Structural Models By Means Of Extended Kalman Filter Algorithm,
Proc. of 7-th International Modal Analysis Conference,1989,U.S.A.
2.   Wen-yue SHI, Graduate Thesis Of Tsinghua University,
Structural   Identification Based On Extended Kalman   Filter
Algorithm And Application, 1986

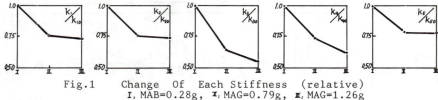

Fig.1   Change Of Each Stiffness (relative)
          I, MAB=0.28g, I, MAG=0.79g, I, MAG=1.26g

Table 1-(a) Each stiffness estimated by SIEKA when the model 1 was excited by earthquak waves of different amplitude

	maximum acceleration value of earthquake wave		
	0.28g	0.79g	1.26g
k1 (kg/cm)	1658	1277	1215
k2 (kg/cm)	1865	1415	1341
k3 (kg/cm)	2353	1502	1286
k4 (kg/cm)	2248	1689	1412
k5 (kg/cm)	5014	3926	3950

Table 2-(a) The calculated modal frequencies based on the estimated stiffnesses of the model 1.

	maximum acceleration value of earthquake wave		
	0.28g	0.79g	1.26g
f1 (Hz)	7.969	6.788	6.393
f2 (Hz)	21.717	18.877	18.243
f3 (Hz)	33.682	28.647	27.483
f4 (Hz)	41.253	35.543	34.337
f5 (Hz)	48.245	41.661	39.796

Table 1-(b) Each damping estimated by SIEKA when the model 1 was excited by earthquake waves of different amplitude.

	maximum acceleration value of earthquake wave		
	0.28g	0.79g	1.26g
c1(kg/cm/s)	0.31	1.00	1.59
c2(kg/cm/s)	2.63	0.86	1.34
c3(kg/cm/s)	0.75	0.76	0.76
c4(kg/cm/s)	0.00	1.88	2.25
c5(kg/cm/s)	6.08	4.83	6.44

Table 3. After suffering earthquake waves the model 1 was excited by a white noise, the modal frequencies were obtained from FFT analysis of responses to the white noise.

	maximum acceleration value of earthquake wave		
	after 0.16g	after 0.40g	after 0.79g
f1 (Hz)	8.2	7.4	7.4
f2 (Hz)	20.7	19.7	19.4
f3 (Hz)	35.8	33.1	31.8

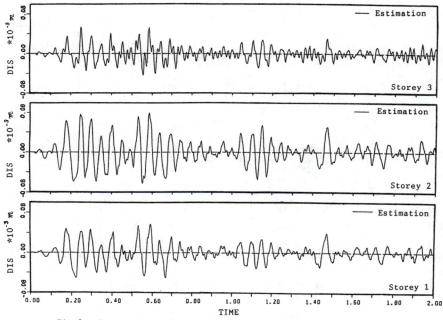

Fig.2   Estimated Storey Displacement Time Curves of the Three-Degrees-of-Freedom Structure Model (0.11g)

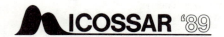

5th International Conference on
Structural Safety and Reliability

Bayesian Reliability Analysis of Structures
with Multiple Components

Yukio Fujimoto[1], Hiroshi Itagaki[2], Seiichi Itoh[3],
Hiroo Asada[4] and Masanobu Shinozuka[5]

[1]Associate Prof., Dept. of Naval Architecture and Ocean
Engrg., Hiroshima University, Higashi-Hiroshima 724, Japan.
[2]Prof., Dept. of Naval Architecture and Ocean Engrg.,
Yokohama National University, Yokohama 240, Japan.
[3]Researcher, Full-Scale Test Sec., Airframe Division,
National Aerospace Laboratory, Mitaka, Tokyo 181, Japan.
[4]Head, Full-Scale Test Sec., Airframe Division,
National Aerospace Laboratory, Mitaka, Tokyo 181, Japan.
[5]Sollenberger Prof., Dept. of Civil Engrg. and Operations
Research, Princeton University, Princeton, NJ 08544, USA.

Abstract

This   study   develops   a   method   of   Bayesian
reliability   analysis   to   establish   an   optimal
inspection   schedule of structures with multiple
components using the data collected during in-
service   inspections.   Expressions   for   the
probabilities of various events that take place at
the time of an inspection are derived. On the
basis of these probabilities, the optimum time of
which the next inspection should be performed is
computed. A numerical example is presented in
order to demonstrate the effectiveness of the
proposed method.

Introduction

     The estimation of the structural reliability is performed by
considering that the structural failure, fatigue crack initiation
and propagation and crack-detection capability are characterized
by a number of random variables. The probabilistic characteristics
of these random variables are obtained from service experience of
similar structures and from various kinds of tests.   It should be
mentioned, however, that the previously mentioned data are usually
scarce. The Bayesian reliability analysis methodology (Itagaki
and Yamamoto 1985; Shinozuka et al. 1981) considered an effective
method to overcome this scarcity of data since a part of the
method consists of estimating the joint probability density

function of the random variables involved in the problem.

Analytical Model

The following assumptions (Paliou et al. 1987) are made for
the present model of a structure consisting of multiple
components(members):

1. All members are classified into several groups SL-J(J=1,N)
   according to their stress level,
2. Each member has one fatigue-critical location,
3. No stress redistribution is considered in the structure after
   the occurrence of member failure,
4. Failed members and members with detected cracks are replaced
   or repaired back to their initial state,
5. A crack of length a_0 is initiated at time t_c. The probability
   density and cumulative distribution functions(PDF and CDF) of
   t_c are denoted by $f_c(t_c|\beta)$ and $F_c(t_c|\beta)$,
6. A fatigue crack of length a at time $t(>t_c)$ is computed by,

$$a(t-t_c|c) = a_0 \cdot \exp\{c(t-t_c)\} \tag{1}$$

7. The probability of detecting a crack of length a is given by,

$$D(a) = 1-\exp\{-d(a-a_0)\} \tag{2}$$

8. Two kinds of static failures are considered. Their respective
   failure rates(FR) and reliability functions(RF) at time t
   $(>t_c)$ are given below. $T_\ell (<t_c)$ is the time when a member is
   replaced or repaired.
   1. Static failure before fatigue crack initiation

   $$\text{FR : } h_0 = e^r \quad (3), \qquad \text{RF : } U(t-T_\ell) = \exp\{-e^r(t-T_\ell)\} \quad (4)$$

   2. Static failure after fatigue crack initiation

   $$\text{FR : } h(t) = e^{qt+r} \quad (5), \qquad \text{RF : } V(t-t_c) = \exp\{-[e^{q(t-t_c)+r} - e^r]/q\} \quad (6)$$

Reliability after the j-th Inspection and Bayesian Analysis

At the time of the j-th inspection, T_j, of a certain member,
one of the events listed in Table 1 may occur. It is assumed that
a member is put into service at time T_ℓ, a fatigue crack is not
detected in it during inspections at $T_{\ell+1}$, $T_{\ell+2}$,...., T_{j-2} and T_{j-1}
and the member does not fail until $T_{j-1}(>T_\ell)$.

The reliabilities at time $t^*(>T_j)$ of a member which is
repaired at T_j and of a member not repaired at T_j are given
respectively by:

$$R(t^* : \text{Repair}) = \{1-F_c(t^*-T_j|\beta)\}\cdot U(t^*-T_j) + \int_{T_j}^{t^*} f_c(t-T_j|\beta)\cdot U(t-T_j)\cdot V(t^*-t)dt \tag{14}$$

$$R(t^* : \text{No repair}) = P[\text{Survival}]/P[B_2 : j] \tag{15}$$

$$P[\text{Survival}] = \{1-F_c(t^*-T_\ell|\beta)\}\cdot U(t^*-T_\ell) + \int_{T_j}^{t^*} f_c(t-T_\ell|\beta)\cdot U(t-T_\ell)\cdot V(t^*-t)dt$$
$$+ \sum_{i=\ell}^{j-1} [\int_{T_i}^{T_{i+1}} f_c(t-T_\ell|\beta)\cdot U(t-T_\ell)\cdot V(t^*-t)[\prod_{k=i+1}^{j} \{1-D[a(T_k-t|c)]\}]dt] \tag{16}$$

Table 1. Event and Probability of Occurrence

Event occurring at the j-th inspection performed at T_j	Probability of occurrence
1. Failure Event[A:j] : Union of $E_{1,j}$ and $E_{2,j}$ $E_{1,j}$: The member has failed within $[T_{j-1}, T_j]$ before crack initiation. $E_{2,j}$: The member has failed within $[T_{j-1}, T_j]$ after crack initiation.	$P[A:j] = P_{1,j} + P_{2,j}$ $\qquad$ (7) $P_{1,j} = \int_{T_{j-1}}^{T_j} f_c(t-T_\ell \mid \beta)\{U(T_{j-1}-T_\ell)-U(t-T_\ell)\}dt$ $\quad$ (8) $\qquad + \{1-F_c(T_j-T_\ell \mid \beta)\}\{U(T_{j-1}-T_\ell)-U(T_j-T_\ell)\}$ $P_{2,j} = \sum_{i=\ell}^{j-2} [\int_{Ti}^{T_{i+1}} f_c(t-T_\ell \mid \beta)U(t-T_\ell)\{V(T_{j-1}-t)-V(T_j-t)\}$ $\qquad \times [\prod_{k=i+1}^{j-1} \{1-D(a(T_k-t \mid c) \mid d)\}]dt]$ $\qquad + \int_{T_{j-1}}^{T_j} f_c(t-T_\ell \mid \beta)U(t-T_\ell)\{1-V(T_j-t)\}dt$ $\quad$ (9)
2. Detection Event[$B_1(a_j):j$] : $E_{3,j}(a_j)$ $E_{3,j}(a_j)$: A fatigue crack of length $[a_j, a_j+\Delta a]$ is detected.	$P[B_1(a_j):j] = P_{3,j}(a_j)$ $\quad = f_c(t_c-T_\ell \mid \beta)U(t_c-T_\ell)V(T_j-t_c) \mid dt/da_j \mid$ $\qquad \times [\prod_{k=\ell+1}^{j-1} \{1-D(a(T_k-t_c \mid c) \mid d)\}]D(a_j \mid d)\Delta a$ $\quad$ (10) $t_c = T_j-1/C \cdot \ln(a_j/a_0), \quad a(T_k-t_c \mid c) = a_0 \cdot \exp\{c(T_k-t_c)\}$ $\mid dt/da_j \mid = 1/C\, a_j$
3. Nondetection Event[$B_2:j$] : Union of $E_{4,j}$ and $E_{5,j}$ $E_{4,j}$: A fatigue crack has not been initiated within $[T_\ell, T_j]$. $E_{5,j}$: A fatigue crack can not be detected.	$P[B_2:j] = P_{4,j} + P_{5,j}$ $\qquad$ (11) $P_{4,j} = \{1-F_c(T_j-T_\ell \mid \beta)\}U(T_j-T_\ell)$ $\qquad$ (12) $P_{5,j} = \sum_{i=\ell}^{j-1} [\int_{Ti}^{T_{i+1}} f_c(t-T_\ell \mid \beta)U(t-T_\ell)V(T_j-t)$ $\qquad \times [\prod_{k=i+1}^{j} \{1-D(a(T_k-t \mid c) \mid d)\}]dt]$ $\quad$ (13)

As an example, the scale parameter β in the two-parameter Weibull density $f_c(t_c \mid \beta)$ and parameter c in eq.(1) are considered to be uncertain and will be estimated by the proposed Bayesian method. The initial prior joint PDF $f_J^0(\beta_J, c_J)$ for each SL-J group is assumed to be uniform

$$f_J^0(\beta_J, c_J) = 1/(\beta_{2J}-\beta_{1J})(c_{2J}-c_{1J}) \qquad \beta_{2J} \geq \beta_J \geq \beta_{1J}, \; c_{2J} \geq c_J \geq c_{1J} \qquad (17)$$

When the j-th inspection has been completed, the posterior joint PDF is calculated as follows:

$$f_J^j(\beta_J, c_J) = LF_J^j \cdot f_J^0(\beta_J, c_J)/\int\int(\text{Numerator})d\beta_J dc_J \qquad (18)$$

$$LF_J^j = \{P_J[A:j]\}^{m_{AJ}} \cdot \{P_J[B_1(a_j):j]\}^{m_{B1J}} \cdot \{P_J[B_2:j]\}^{m_{B2J}}$$
$$\times \prod_{g=1}^{n_{AJ}} \{P_J[A:k_g]\} \cdot \prod_{g=1}^{n_{B1J}} \{P_J[B_1(a_{k_g}):k_g]\} \qquad (19)$$

$$M_J = m_{AJ} + m_{B1J} + m_{B2J}$$

where M_J is the number of members in the SL-J group, n_{AJ} is the number of members which have failed within $[T_\ell, T_{j-1}]$, and n_{B1J} is the number of members with detected cracks at any inspection within $[T_\ell, T_{j-1}]$.

The expected reliability of the structure at t* can be derived from eqs.(14), (15) and (18) as follows:

$$\bar{R}_M(t*) = \prod_{J=1}^{N} \bar{R}_{M,J}(t*) \tag{20}$$

$$\bar{R}_{M,J}(t*) = \int\int \{R_J(t* : \text{Repair})\}^{m_{AJ}+m_{B1J}} \{R_J(t* : \text{No repair})\}^{m_{B2J}}$$

$$\times f_J^j (\beta_J, c_J) \, d\beta_J dc_J \tag{21}$$

The (j+1)-th inspection is performed at t* according to the following equations, where R_{design} is the specified minimum reliability of the entire structure:

$$\bar{R}_M(t*) \geq R_{\text{design}} \quad (22), \quad T_{j+1} = t* = \bar{R}_M^{-1}(R_{\text{design}}) \quad (23)$$

Numerical Example

In the numerical example considered, the number of stress-level groups is N=3. The stress-level of the SL-2 group is 90% of the stress-level of SL-1 while that of SL-3 is 80%. The total number of structural members is M_T=100 and groups SL-1, SL-2 and SL-3 have 20, 30 and 50 members, respectively. A typical inspection schedule of a structure is shown in Fig.1. The solid line corresponds to an inspection schedule calculated with the estimated values of the uncertain parameters β and c. The dotted line corresponds to an inspection schedule computed with the true values of β and c. Finally, as for the posterior joint PDFs of SL-1 to SL-3 after the sixth inspection are concerned, their modal values concentrate around the corresponding true values.

Conclusions

This report describes the effectiveness of the Bayesian reliability analysis to assess the optimal inspection schedule of a structure with multiple components by evaluating data collected during in-service inspections.

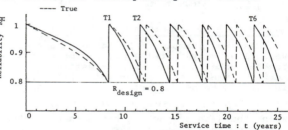

Figure 1.   Inspection Schedule and Reliability of a Structure
[Uncertain parameters : β and c]

Further studies will be focused on developing computer programs to be employed to actual in-service structures and to respond to several requirements of the inspection schedule.

References

Itagaki, H., and Yamamoto, N. (1985). "Bayesian analysis of inspection on ship structural members." Proc. ICOSSAR'85, 3, 533-541.

Paliou, C., Shinozuka, M., and Chen, Y. N. (1987). "Reliability and durability of marine structures." J. Struct. Engrg., ASCE, 113(6), 1297-1314.

Shinozuka, M., Itagaki, H., and Asada, H. (1981). "Reliability assessment of structures with latent cracks." Proc. US-Japan Cooperative Seminar on Fracture Tolerance Evaluation, 237-247.

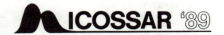
BRIDGE LIVE LOAD MODELS

Robert J. Heywood[*] and Andrzej S. Nowak[**]

[*] Senior Lecturer in Structures, School of Civil Engineering,
Queensland University of Technology, Brisbane Q 4001, Australia
[**]Associate Professor of Civil Engineering, Department of Civil
Engineering, G.G Brown Bldg, University of Michigan, Ann Arbor MI
48109, U.S.A.

ABSTRACT

This paper describes the analysis of weigh-in-
motion data and the conclusions drawn.
Statistical distributions of the effects induced
in typical bridges due to single vehicle events
are presented. Extrapolated serviceability and
ultimate limit state effects are compared with
National Association of Australian State Road
Authorities bridge design loading and the
inconsistencies discussed. The variations around
the world in the definitions of the design life
and return periods for serviceability conditions
are discussed in relation to the results of the
analysis.

KEYWORDS

bridges; live loads; limit states; bending; shear;
design; calibration; statistics; design life.

1. INTRODUCTION

Australia like North America and Europe are developing or have
developed new bridge design codes using Load and Resistance Factor
Design (LRFD) or limit state concepts. Bridge live load models
form a cornerstone of these codes. The development of weigh-in-
motion (WIM) systems has made available a huge data base of truck
configurations and loads. The actual truck effects on bridges can
now be analyzed by modern statistical methods thus providing an
opportunity to check bridge design live loads.

Various WIM techniques based on bridges, culverts, piezo electric
materials and other load sensitive devices are available. The
Australian Road Research Board has developed 'CULWAY'[10,2] which
uses reinforced concrete box culverts as an equivalent static
scale. Trucks moving at highway speed across the culvert trigger

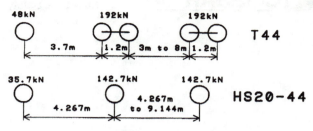

Figure 1: T44 and HS20-44 design loadings

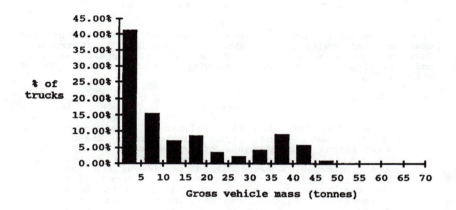

Figure 2: Histogram for gross vehicle mass(t)

tape switches and a data logger records the time of arrival, axle spacing, speed, and classification of the vehicle. Strains recorded when each axle is over the center of the culvert are converted to axle loads via a correlation with strains obtained from the passage of a series of calibration trucks.

2. WEIGH-IN-MOTION DATA

Design bending moments and shears are fundamental. For this reason the analysis of CULWAY data has concentrated on the

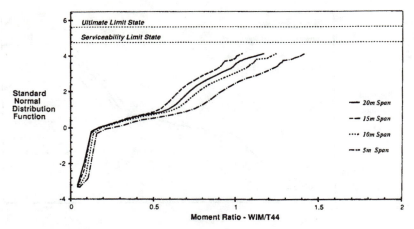

Figure 3: Normal distribution of central moment for simply supported spans

statistical distribution of effects as opposed to the multitude of
parameters required to define a truck.   The WIM data analyzed in
this paper was collected on the Cobb Highway in rural New South
Wales, Australia.   In 1986 and 1987 56,195 truck were sampled
during a total recording time of 350 days.

Figure 2 summarizes the gross vehicle mass in histogram form and
provides a useful comparison for other sites. Figures 3, 4, 5 and
6 present sample distributions of effects in single lane simply
supported beams with spans varying between 5m and 40m. Moments
and shears have been normalized by dividing by the effects induced
by the NAASRA [6] T44 design loading (refer Figure 1).   The
distributions are conditional on a truck being on the bridge in
such a position as to induce a maximum effect.   The T44 loading is
based on the AASHTO [1] HS20-44 loading increased by 35% and the
single drive and trailer axles replaced with tandem axles.

3. EXTRAPOLATION OF LIMIT STATE MOMENTS AND SHEAR FORCES

The draft NAASRA Bridge Design Code [6] has defined the
probability of occurrence of the SLS and the ULS based on a 100
year minimum useful life expectancy of a properly designed and
maintained bridge [4].   These are summarized in Table I.   These
limits have been superimposed on the distributions shown in
Figures 3, 4, 5 and 6 and the distributions extrapolated linearly

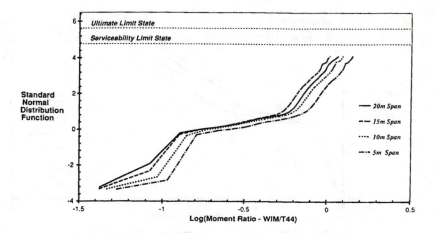

Figure 4: Log-normal distribution of central moment for simply supported spans

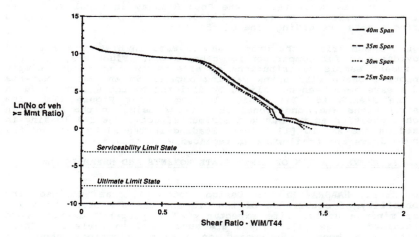

Figure 5: Exponential distribution of maximum shear for simply supported spans

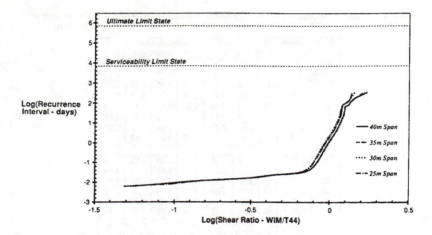

Figure 6: Recurrence interval distribution of maximum shear for simply supported spans

TABLE I: Definition of loads in the NAASRA Bridge Design Code

Type of load	Serviceability	Ultimate
Probability of occurrence in 100 years	1.0	0.05
Probability of occurrence in 1 years	0.05	0.0005
Mean recurrence interval	20 yrs	2000yrs

to give limit state moments and shears for various spans up to 30m (refer Table II). The predictions were not made for spans greater than 30m because of the possible influence of multiple presence and truck and trailer combinations.

The predicted results for the SLS and ULS midspan moments vary considerably. The plot on normal probability paper is concave downwards indicating a log-normal tendency which is confirmed by log-normal plot. The exponential distribution has been used in the calibration [3] of the OHBDC [8] and provides a reasonable fit to the tail of the data.

The ratio of the ULS to the SLS moments is approximately constant for each distribution, but varies from 1.1 to 1.4 over all the distributions considered. The ratio of the largest extrapolation

Table II: Extrapolated limit state bending moments and shears divided by the effect due to the NAASRA T44 design loading.

Span	Serviceability Limit State Distribution				Ultimate Limit State Distribution			
(m)	I	II	III	IV	I	II	III	IV
Moments								
5	1.55	1.63	1.65	1.79	1.75	1.95	1.99	2.42
10	1.37	1.44	1.45	1.58	1.55	1.75	1.76	2.17
15	1.15	1.21	1.22	1.33	1.31	1.47	1.48	1.82
20	1.25	1.32	1.33	1.45	1.42	1.59	1.61	1.98
25	1.34	1.40	1.42	1.53	1.51	1.68	1.71	2.07
30	1.42	1.49	1.51	1.61	1.63	1.80	1.83	2.18
Shears								
5	1.53	1.58	1.61	1.59	1.74	1.86	1.92	2.13
10	1.15	1.17	1.20	1.21	1.31	1.41	1.45	1.65
15	1.29	1.35	1.40	1.38	1.46	1.60	1.69	1.89
20	1.42	1.47	1.52	1.49	1.62	1.76	1.81	2.02
25	1.53	1.53	1.61	1.65	1.76	1.83	1.94	2.31
30	1.53	1.53	1.61	1.65	1.76	1.83	1.94	2.31
	I – *Normal Distribution*		II – *Lognormal Distrib'n*					
	III – *Exponential Distrib'n*		IV – *Recurrence Interval*					

to the ULS (recurrence interval) to the smallest prediction of the SLS (normal distribution) moments is plotted in Figure 7 and is essentially constant at a value of 1.6. For shear the ratio varies between 1.4 and 1.5. These ratios are far smaller than the ultimate load factor of 2.0 specified in the limit state draft NAASRA Bridge Design Code. This will result in the SLS almost always controlling the design with large reserves of strength at the ULS.

A shorter mean recurrence interval for truck live loads may be reasonable. The OHBDC serviceability limit states are based on a small probability of exceedence during recurrence intervals of 1 week for cracking of concrete beams and 1 month for yielding in extreme fibers or slip in bolted connections of steel structures [3]. Switzerland's serviceability loads are based on a daily occurrence interval. These contrast dramatically with NAASRA's 20 years, which has been selected to be consistent with wind and stream flow forces.

Figure 7 presents, graphically, the ratio of the predicted ULS and SLS moments to the effects due to the T44 design loading versus the span of a simply supported girder. This illustrates, along with Table II, that the safety of bridges designed to the NAASRA T44 loading varies with span. The T44 loading significantly underestimates the limit state effects for many circumstances. Therefore it is recommended that the T44 design loading should be revised or the legal limits restructured to permit a more uniform prediction of the effects of traffic loads on a variety of bridge spans.

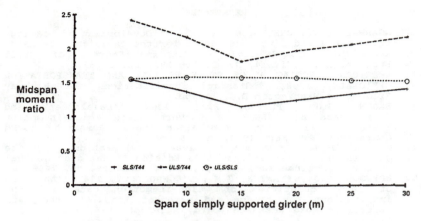

Figure 7: Prediction of SLS and ULS moments for simply supported spans.

4. CONCLUSIONS

Serviceability and ultimate limit state bending moments and shears have been predicted based on the distributions generated from 56,195 truck records collected on the Cobb Highway in Australia. Exponential and log-normal distributions demonstrate reasonable correlation for the tails of the distributions but not over the entire range.

The predicted ULS and SLS effects are not consistent with those generated by the NAASRA T44 design loading. The safety of bridges designed to the T44 loading varies considerably with span. It is therefore recommended that a new bridge design load be developed.

The wide variety of recurrence intervals have been have been adopted by authorities around the world for serviceability limit states. These vary between one day in Switzerland and a proposed 20 years in Australia. The 20 year interval results in the serviceability conditions dominating the design of bridges thus leading to large reserves of strength at ultimate. A one week average recurrence interval is likely to be more appropriate subject to satisfactory fatigue performance being demonstrated.

REFERENCES

1. AGARWAL, A.C., CHEUNG, M.S., 1987 ``Development of Loading-truck Model and Live-load Factor for the Canadian Standards Association CSA-S6 Code,'' Canadian Journal of Civil Engineering, Vol. 14, No. 1, Feb 1987 pp. 58-67.
2. AMERICAN ASSOCIATION OF STATE HIGHWAY AND TRANSPORTATION OFFICIALS, 1983, ``Standard Specifications for Highway Bridges,'' Washington D.C., 394pp.
3. BROWN, J.J., PETERS, R.J., 1988 ``Development and Performance of CULWAY: A Culvert Based Weigh-in-motion System,'' Proc 14th Australian Road Research Board Conference, Aug-Sept 1988, part 6, pp. 86-105.
4. GROUNI, H.N., NOWAK, A.S., 1984 ``Calibration of the Ontario Bridge Design Code 1983 Edition,'' Canadian Journal of Civil Engineering, Vol. 11, No. 4, Dec 1984 pp. 760-770.
5. HEYWOOD, R.J., FENWICK, J.M., O'CONNOR, C., 1987 ``Analysis of large quantities of weigh-in-motion data and the implications for bridge and pavement loads,'' Proc 14th Australian Road Research Board, Aug-Sept 1988, part 6, pp. 181-191.
6. HEYWOOD, R.J., 1988 ``Bridge live load models,'' Civil Engineering Research Report 1988/1, Queensland Institute of Technology, Brisbane,Australia, May, 85 p.
7. NATIONAL ASSOCIATION OF AUSTRALIAN STATE ROAD AUTHORITIES (NAASRA), 1987, ``Draft NAASRA bridge design code'', Australia.
8. NOWAK, A.S., CZERNECKI, J., ZHOU, J. and KAYSER, J.R. 1987. ``Design Loads for Future Bridges'', Report No FHWA/RD-87/069, U.S. Department of Transport, Federal Highway Administration, 1987, p 121
9. ONTARIO MINISTRY OF TRANSPORTATION AND COMMUNICATIONS, 1983, ``Ontario Highway Bridge Design Code,'' Downsview, 357pp.
10. PETERS, R.J., 1986 ``CULWAY - An unmanned and undetectable highway speed vehicle weighing system,'' Proc. 13th ARRB/5th REAAA Conference, 13(6), pp.70-83.
11. SCOTT, G., 1987 ``Weigh-in-motion Technology - Status of CULWAY in Australia,' Australian Road Research Board Seminar on Road Traffic Data Collection using Weigh-in-motion, Melbourne, Australia.

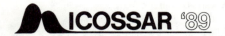
RELIABILITY-BASED BRIDGE LIFE ASSESSMENT

Fred Moses,[1] M. ASCE
Charles G. Schilling,[2] M. ASCE
Surya K. Raju,[3]

[1]Professor of Civil Engineering, Case Western Reserve University, Cleveland, Ohio 44106.
[2]Structural Engineer, 3535 Mayer Drive, Murrysville, PA.15668.
[3]Graduate Student, Case Western Reserve University, Cleveland, Ohio 44106.

ABSTRACT

Fatigue considerations have usually entered design codes as a secondary consideration and often under the guise of serviceability limits. Most engineers felt if the maximum stress ranges remained below the endurance limit, then fatigue would not be a problem. Two main issues have arisen where codes, such as highway bridge codes, need to further consider fatigue. These are: 1) the random stress spectra, which means there is a significant likelihood that some stress ranges will exceed the endurance limit and, 2) the need to respond to the issue of evaluation of remaining safe life of the structure. In both cases, realistic assessments of the true stress ranges must be used in both fatigue life design and evaluation. This paper concentrates on the evaluation or assessment of existing structures and the definition and determination of safe remaining life. The application is to steel highway bridges but other structures may be evaluated in a similar manner.

KEY WORDS

fatigue, evaluation, reliability, damage, stress range, stress cycles, truck volume, distribution factors, impact factors, design life.

INTRODUCTION

Current highway bridge fatigue provisions [1] are not suitable for assessing the remaining safe life of an existing bridge. The codes utilize an artificially low number of stress cycles combined with an inflated stress value to obtain the design section. For example, the code recommends 2 million as a typical number of design cycles with corresponding stress ranges of 8 to 24 ksi (55 to 165 MPA). Actual field measurements show much lower effective stress ranges in the 1-4 ksi range (7 to 27 MPA) but on

STRUCTURAL SAFETY AND RELIABILITY

heavily traveled routes the number of stress cycles may exceed 100 million and even approach 300 million cycles. Through the use of high stresses and low number of cycles, the codes are now calibrated to produce safe design lives. However, they may not be helpful for assessing existing structures because they do not reflect actual site conditions. New fatigue methods described herein were developed with the following aims-

(1) Reflect the actual fatigue conditions observed in highway bridges.
(2) Give an accurate estimate of remaining fatigue life and permit this estimate to be updated in the future to reflect changes in traffic conditions.
(3) Provide consistent levels of reliability.
(4) Permit different levels of effort by the engineer using site-specific data to reduce uncertainties and improve predictions of remaining life.

PROBABILISTIC DAMAGE MODEL

For the fatigue life model, the failure function is expressed as

$$Z = Y_F - Y_S \tag{1}$$

where Y_F is the true fatigue life and Y_S is the specified fatigue life for the detail. Using a linear Miner's rule damage summation the uncertain fatigue life, Y_F, may be derived as a function of several other random variables, leading to the following failure function [2].

$$Z = \frac{X \; N_T}{365 \; (ADTT) \; C} \; [\frac{\gamma \; Z_X \; S}{W \; G \; I \; M \; H}]^3 - Y_S \tag{2}$$

Eq. 2 assumes each truck passage causes a single cycle of stress range which contributes damage according to Miner's rule. The cumulative damage depends on: 1) the number of cycles expressed by the term, ADTT (or average daily truck traffic, since only trucks cause significant stresses) and C, the equivalent number of stress cycles per truck which depend on vehicle and span lengths and 2) the effective stress range which depends on the effective weight (W), dynamic amplitude (I) and the member stress distribution which depends on vehicle configuration (M), amplification due to closely spaced trucks (H), distribution of load to member (G) and section modulus (Z_X). Eq. 2 also contains the fatigue life variable (S) and a variable to represent uncertainty in linear damage model (X). Equation (2) is very similar to the failure function utilized by Moses and Nyman [3] with a few modifications. These are

(1) Introduction of a safety factor γ. This makes the failure function adaptive for both design and evaluation.
(2) A random variable C to account for occurrence of more than one stress cycle per truck passage. This variable is needed for very short spans of less than 30 feet [2].
(3) A random variable, Z_X to permit possible increase in resistance effects not normally considered in design [2].

The exponent in the failure function (eq. 2) is a result of the assumed linear log S-N curve with a slope of 3 to represent the fatigue strength of welded details [4]. In the failure function in Eq. (2), X, ADTT, C, Z_X, S, W, G, I, M and H are all random variables. N_T is the nominal number of expected stress cycles on the bridge in its lifetime. is a safety factor to calibrate an acceptable safety index.

The varied truck population introduces a variable-amplitude stress spectra. Random variable M, called the moment ratio, reflects the effect of axle spacing and axle weight distribution on the stress spectra. The moment ratio is defined as the ratio of the average influence factor due to actual truck spectrum to the influence factor of a representative fatigue design vehicle. The design vehicle was selected by Schilling [5] such that the moment ratio is close to 1.0 for varying spans and bridge geometries. Truck traffic data measured at 12 sites [6] were used to evaluate the average moment ratios and coefficients of variation (COV). See Table 1 for data used.

The variable W reflects the uncertainty in the gross weight of the fatigue design truck calculated from

$$W_{eq} = [\Sigma \; f_i \; W_i{}^3]^{1/3} \qquad\qquad (3)$$

where f_i is the percentage of trucks within the weight interval W_i. Eq. 3 is based on Miner's rule and the assumed exponent of 3 mentioned above. It provides that the fatigue damage caused by a given number of passages of this equivalent truck is the same as the fatigue damage caused by the same number of different weight trucks in the actual traffic stream. The effective fatigue truck weight was calculated from recent weigh-in-motion (WIM) studies [6] that included 30 sites nationwide.

Random variable H, (Headway factor), reflects the effect of multiple presence of trucks on the bridge. It was found from a simulation program that for most traffic conditions the major fatigue damage comes from only single vehicles on the span. The random variable I, called impact ratio, reflects the uncertainty in the estimation of impact factor for a given site. Random variable G, called distribution ratio, reflects the analysis uncertainty in the structural load response of the component being checked based on the presence of a single truck on the span.

Random variable Z_X reflects the variability in the effective section modulus compared to the actual value. The proposed procedures [2] give factors for composite and noncomposite members, which account for beneficial effects not normally considered in design, such as unintended composite action, contributions from nonstructural elements and direct transfer of load through the slab to the supports. Random variable C represents the equivalent number of stress cycles due to a single truck passage on the bridge. Estimates of C are given for different spans and span types. Random variable ADTT represents the average daily truck traffic in shoulder lane at the site.

STRUCTURAL SAFETY AND RELIABILITY

Random variable S reflects the uncertainty in the estimation of fatigue strength curves. For convenience it is taken as the ratio of the true fatigue strength at 2 million cycles to the design value obtained from the design S-N curve. The statistical properties of S are calculated from test results reported at Lehigh University [4].

TARGET RELIABILITY AND SAFETY FACTORS

A safe remaining life is expressed herein through the Safety Index, Beta, (β); defined as:

$$\beta = \frac{\bar{Z}}{\sigma_Z} \qquad (4)$$

A target beta is defined as the value of beta (or the corresponding probability of failure) that is acceptable for evaluation. Ideally, the selection of target beta should be based on economic considerations involving cost of construction, interest rates, and cost of repair, inspection or failure. An alternative adopted in the study is to compute betas for existing bridges that are performing satisfactorily in service, then use these computed betas as the target for future evaluations. The factors in the code, however, should allow for site conditions, changes in traffic weights and volume over time, and consequences of failure such as presence of redundancy. Furthermore, the changes should adjust parameters to provide consistent reliabilities among different detail categories, span geometries and traffic conditions.

Average betas in the current AASHTO design were evaluated using 13 typical design cases with different volumes, detail categories, spans, impact factors, girder distributions and span types. Betas for redundant members were found to vary between 0.7 to 3.6. For example, the beta value of 0.7 corresponded to a site with high average impact factor of 1.20 due to a rough deck surface and a high girder distribution factor of 0.50 for a short span of 30 feet (9.1m). The beta of 3.5 corresponded to a site condition with a low average impact factor of 1.10 and a low girder distribution factor of 0.30 for a 60' (18.3m) continuous span.

These examples demonstrate quite strongly the advantages of the proposed methods where consistent betas are obtained over the entire range of application. This is especially important in bridge evaluation where site or traffic conditions may be accurately determined if warranted by the economic alternatives of load restriction, repair or replacement often considered in bridge management studies. From the study, the average beta is determined to be 2.0 (probability of failure = 2.3%). This is taken as the target beta for redundant members. A target beta of 2.0 is close to 95th percentile values used traditionally in fatigue design codes. A similar analysis was done for nonredundant members where those betas are different due to different allowable stresses in the code [1]. Betas are found to vary between 1.50 to 5.35. A target safety index is fixed as 3.0 (probability of failure = 0.1%) for nonredundant members.

Safety factors were derived corresponding to these target betas for evaluation and design [2]. Adjustments in the safety factors are introduced to make the evaluation procedures flexible enough to allow the engineer to incorporate site-specific data. For example, the engineer has an option to use the measured truck weight spectrum at the site so there is a lesser uncertainty regarding the effective stress range at the detail [2]. Similarly, several other options regarding estimation of effective stress and estimation of the girder distribution factor are given.

ILLUSTRATION

The evaluation of the safe remaining life (see Fig. 1) involves finding the nominal stress range for the detail under consideration [2], 1) using the fatigue design truck, 2) best estimates of the impact factor based on deck or abutment conditions and 3) the appropriate analysis or girder distribution factor, which can be found in various different ways (table, graph, finite elements, etc). The safe fatigue life can then be evaluated corresponding to the nominal stress range using the appropriate detail constant developed from the current AASHTO detail categories and a safety factor. [Figure 1 shows that there is an option in the evaluation to check whether the bridge may be assumed to have essentially infinite life as estimated by the safety index associated with the likelihood of stresses exceeding the endurance limit [2]].

To illustrate, the evaluation method was applied to predict the safe life of a bridge example provided by A.G Lichtenstein and Associates. The main girder developed a fatigue crack across the entire tension flange. The girder has 3 spans, which are 150', 180' and 150' (45.7m, 54.9m and 45.7m). A cover plate detail at 60 (18.3m) feet from the intermediate support in the middle span is being checked. Best estimates of impact factor, girder distribution factor from the proposed methods [2] and the truck traffic are used. Safe life (Y_f) is calculated using the S-N fatigue life curve.

$$Y_f = \frac{K \times 10^6}{T_a \, C \, (R_S \, S_r)^3} = \frac{2.9 \times 10^6}{840(10)(1.75 \times 3.8)^3} = 12 \text{ years} \qquad (5)$$

where K is the detail constant for the fatigue category under consideration, T_a is the truck volume in trucks/day in the shoulder lane, C is the number of cycles per truck passage, R_S is the reliability or safety factor and S_r is the nominal stress range for the critical location. Equation (5) show that the detail has a safe life of only 12 years using the data shown in Table (2). Note that "safe" here corresponds to a beta of 3.0 or a probability of only 0.1% (nonredundant member) that the fatigue life will be less than 12 years. Since there were several locations on the bridge with other similar details, it is not surprising that cracks were detected in the structure at the present age of 25 years. The proposed evaluation method was used for several other examples, which are discussed in reference 2.

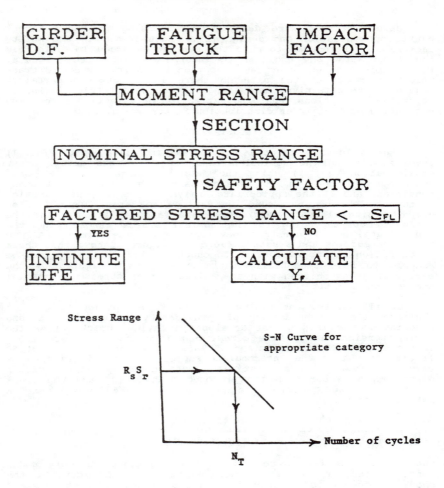

Figure 1 - Fatigue Life Evaluation Flow Chart
$$Y_f = N_T/\text{ADTT} (365) \, C$$

TABLE 1 DATA BASE

Variable	Mean	COV%	Ref.
M-Moment ratio			
Span-30' (9.1m)	.925	5.2	3
-60' (18.3m)	.965	5.5	
-90' (27.4m)	.9675	3.3	
-120'(36.6m)	.977	2.2	
W-Effective weight	54 kips (240 KN)	10	6
H-Headway	1.03	.6	3
I-Impact	typically 1.1-1.3	10	3
G-Girder distribution	Ref. 7	13	2
Z_x-Section Modulus	Ref. 2	10	2
C-Cycles per vehicle	Ref. 2	5	2
ADTT-Truck Volume	Ref. 2	10	2
S-Fatigue Scatter	AASHTO(nominal)	varies	4
X-Miner Modelling	1.0	15	2

TABLE 2 EVALUATION PARAMETERS FOR ILLUSTRATION

Moment range due to fatigue truck	= 1320 kft (1790 KNm)
Girder distribution factor	= 0.72
Impact factor (smooth surface)	= 1.10
Moment range at the detail evaluated	= 1045 kft (1417 KNm)
Section modulus (nominal)	= 3345 in^3 (.055 m^3)
Nominal stress range on girder	= 3.8 ksi (26.2 MPa)
Reliability factor (nonredundant)	= 1.75
Truck traffic per day (shoulder lane)	= 840
Cycles per truck passage	= 1.0
Detail constant (K) for category E	= 2.9
Present age	= 25 years

CONCLUSIONS

The new evaluation methods provide the following advantages.

1. They realistically reflect actual fatigue conditions in highway bridges.
2. They use analysis procedures similar to present AASHTO specifications.
3. They provide consistent levels of reliability.
4. They are based on extensive recent research.
5. The evaluation method gives a reasonable estimate of safe remaining life of an existing bridge.
6. The evaluation method permits different levels of effort to reduce uncertainties and improve the prediction of remaining life.

REFERENCES

[1] AASHTO, American Association of State and Transportation Officials (1984). "Standard Specifications for Highway Bridges", Washington D.C.
[2] Moses F., Schilling C.G. and Raju S.K. (1988). "Fatigue Evaluation Procedures for Steel Bridges", NCHRP 299 Transportation Research Board, Washington D.C.
[3] Nyman W.E and Moses F. (1985). "Calibration of Bridge Fatigue Design Model", Journal of Structural Engineering, ASCE, VOL. 11, No. 6.
[4] Keating P.B. and Fisher J.W. (1985). "Review of Fatigue Tests and Design Criteria on Welded Details", Fritz Engineering Laboratory report 488-1(85) Lehigh University, Bethlehem, Pa.
[5] Schilling C.G. (1984). "Stress Cycles for Fatigue Design of Steel Bridges", Journal of Structural Engineering ASCE, Vol. 110, No. 6.
[6] Snyder R.E., Likins G.E. and Moses F. (1985). "Loading Spectrum Experienced by Bridge Structures in the United States", Report FHWA/RD-85/012, Bridge Weighing Systems Inc., Warrensville, OH.
[7] Schilling C.G. (1982). "Lateral-Distribution Factors for Fatigue Design", Journal of Structural Division, ASCE, VOL. 108, No. ST9.

ACKNOWLEDGEMENTS

The research reported here was sponsored by the Transportation Research Board as part of the NCHRP program.

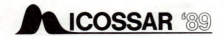
DEVELOPMENT OF A SAFETY ASSESSMENT SYSTEM
FOR STEEL BRIDGE ERECTION WORK

Shigeo Hanayasu* and Yoshitada Mori**

*   Senior Research Officer, Research Institute
    of Industrial Safety, Ministry of Labor,
    5-35-1, Shiba, Minato-ku, Tokyo 108 JAPAN
** President, Scaffolding and Construction
    Equipment Association, Kenchiku-Kaikan 5F,
    5-26-20, Shiba, Minato-ku, Tokyo 108 JAPAN

ABSTRACT

In order to secure safe execution of steel bridge
erection work, development of a comprehensive
checking system to set up safety countermeasures
at the stage of construction planning prior to the
initiation of the work was achieved. The developed
system   for evaluating   construction   planning
includes four steps: 1) Collection of various data
for assessment ; 2) Ranking of the potential risks
of the steel bridge erection work ; 3) Check the
basic safety measures for execution of work ; and
4) Check the safety measures against inherent
accidents in steel bridge erection work.

KEYWORDS

Steel Bridge Erection, Safety Assessment, Project
Planning, Occupational Accidents, Risk Management

1. INTRODUCTION

   In bridge construction work, accidents involving many workers
injured such as collapse of frame structures, falling, struck by
objects take place occasionally.
   According to the analysis of causes of these accidents, there
exists some cases where safety precautions at the stage of work
planning seem to have been inadequate. In other words, if appropriate safety measures were established during the execution plan-

ning stage and carried out in accordance with them, many accidents could have been avoided.

Therefore, to prevent future accidents during construction, it is important to set up precautionary measures during the work planning stage prior to the initiation of the work.

In this connection, Article 88 in the Japanese Industrial Safety and Health Law prescribes that the contractor who is going to execute a specified large scale construction work should notify the construction plan to the Labor Minister or the Chief of the Labor Standard Inspection Office prior to the commencement of the said work [1].

So far as the bridge construction concerned, work requiring notification of the construction plan to the Labor Minister is the one for bridges with a maximum span of 500 m or more (1000 m or more in case of suspension bridges), while work needed to notify the plan to the Chief of the Labor Standard Inspection Office are ones for bridges with a maximum span of 50 m or more.

In addition to the notification of a plan, Article 88 also prescribes that the contractor shall use fully qualified persons in making a plan of above-mentioned bridge work. After receiving the work plan, the Labor Minister or the Chief of the Labor Standard Inspection Office may make necessary recommendations or requests on matters concerning the prevention of occupational injuries to the contractor who submitted the notification.

The aim of this system is to promote the advance evaluation of the work plan from the view point of safety for attainment of the basic safety level in construction site.

With this thinking in mind, development of a comprehensive checking system ensuring safety measures at the work planning stage, which we call here " safety assessment for construction work " procedure was carried out.

This paper presents a brief outline of the developed safety assessment system for steel bridge erection work [2].

2. PROCEDURE OF THE SAFETY ASSESSMENT

The developed safety assessment system is a sequential process to check the necessary safety measures against all anticipated accidents by predicting the potential risk that are inherent to steel bridge erection work. The procedure of the safety assessment system is described using the following four steps :

2.1 Collection of basic data

At this step, basic materials and data necessary to perform the safety assessment for steel bridge erection work are collected and compiled.

Typical of these are :
1) Documents of design, various drawings, schedule, work plan
2) Topographical, geological and meteological survey
3) Survey results of existing structures or buildings nearby, access roads to the site, environment, substructures, temporary facilities, etc.
4) Work records of similar-type bridge erections
5) Safety and health laws and other related safety guides and practice, accidents information

2.2   Ranking of the potential risk of the steel bridge erection
       work and check the safety measures against evaluated risks

     At this step, quantitative evaluation of the potential risks
inherent to a specific bridge erection work are achieved by taking
the construction site's characteristics into consideration. Then,
check whether the safety measures to ensure safe execution of the
work are established according to the risk classification analyzed
by a numerical evaluation.
     In bridge construction work, generally, many site character-
istics including the erection method, type of structure, etc. have
already been fixed by the order when a contract was awarded. Hence
are vitally important to assess the given conditions at the
initial stage of the safety assessment.
     In order to evaluate the potential risk of a specific steel

Table 1. Risk Evaluation Table for Steel Bridge Erection Work

Element	Condition	Score
Type of structure	a. Other than I girder or box girder b. I girder or box girder	2 0
Maximum length of span	a. Exceed 100 m b. Between 50 m and 100 m c. Less than 50 m	2 1 0
Plane shape of structure	a. Curved bridge with large curvature b. Curved bridge with moderate curvature c. Straight bridge	2 1 0
Clearance below girder	a. Exceed 20 m b. Between 10 m and 20 m c. Less than 10 m	2 1 0
Bridge erection method	a. Cable-supported erection method b. Supported method other than bent (Cantilever,launching,erection girder) c. Bent supported erection method d. Large-scale block erection method	4 2 1 0
Available crane	a. Cable crane b. Other than cable crane	1 0
Topographic features	a. Different foundation level for support b. Same foundation level for support	1 0
Conditions of foundation	a. Pile foundation for temporary support b. Concrete base for temporary support c. Timber or iron-plate base for support	2 1 0
Season and regional area of erection	a. Erection time between June & October 1 Region of Okinawa, Kyushu, Shikoku 2 Region of Chugoku, Kinki, Chubu 3 Region of Hokuriku, Kanto, Touhoku, and Hokkaido b. Erection time in November to next May	 3 2 1 0
Period of erection work	a. Exceed 6 months b. Between 3 to 6 months c. Less than 3 months	2 1 0

(Risk point is obtained by summing up identified element score)

Table 2.   Rank Classification of Risks

Assessment point	Risk rank
15 or more Between 10 and 14 Between 5 and 9 4 or less	Rank I (Very high) Rank II (High) Rank III (Probable) Rank IV (Low)

bridge erection work numerically, 10 primary elements of construction site's conditions were selected out. To analyze the
conditions of the elements in conjunction with the potential risks
of the construction work, descriptive criteria on the conditions
and assigned evaluation raw score were prepared for each element.

Table 1 shows the 10 selected site elements and their detailed
description of the element conditions and the corresponding raw
scores for each site element.

The numerical potential risk assessment point can be obtained
by summing up all identified raw scores. The evaluated risk scores
is then set into a four-rank classification from I (very high) to
IV(low). Table 2 exhibits the relation between evaluated raw score
and rank classification of potential risks.

After accomplishment of the potential risk evaluation of the
bridge erection work, safety measures in accordance with the risk
classification will be confirmed. The safety measures considered
in this step include the placement of a qualified manager as well

Table 3.   Safety Measures for each Risk Classification

Rank \ Measure	Rank I	Rank II	Rank III	Rank IV
Designate experienced project planner and manager	More than 10 years of planning or execution of steel bridge erection	More than 7 years of planning or execution of steel bridge erection	More than 5 years of planning or execution of steel bridge erection	Experienced in planning or execution of steel bridge erection
Select appropriate subcontrac-tor with experience	More than 10 years of execution of same type steel bridge erection	More than 7 years of execution of same type steel bridge erection	More than 5 years of execution of similar type steel bridge erection	Experienced in execution of similar type steel bridge erection
Placement of an experienced foreman	Experienced in execution of same type steel bridge erection as a foreman	Experienced in execution of same type steel bridge erection	Experienced in execution of a similar type steel bridge erection work	
Supervision system of contractor	Carry out intensive and regular safety inspections and patrols by experts	Carry out regular safety patrols and inspections of construction site		

as a foreman responsible for the safe execution of the work, the
selection of appropriate subcontractors, and an introduction of a
suitable superintendence system at the site.

Table 3 illustrates a detailed explanation of safety measures
corresponding to each rank classification. Before starting the
steel bridge erection work, the awarded contractor should fulfill
the prescribed safety measures.

2.3   Check the basic safety measures

At this step, examine whether the appropriate safety measures
are already in force or are going to be taken with respect to the
basic matters that are indispensable for the safe execution of
steel bridge erection work.

For this purpose, comprehensive checklists were prepared,
including more than 95 major items (further broken down into 278
items) as necessary requirements to ensure the safe execution of
the steel bridge erection work.

The basic items included in the tables in checklists cover a
wide range of steel bridge erection work. Major groups of items
are : 1) Management; 2) Circumstances of site; 3) Steel bridge
structures; 4) Erection method, machines and equipment ; 5)
Temporary facilities; 6) Slab deck construction; 7) Paint-work;
and 8) Stock yards, access roads, etc.

Each item has a detailed guideline to help assess whether or
not the corresponding safety measures in the work plan is
adequate. A study should also be made as to when and how the
proposed safety measures are to be implemented. Table 4 depicts a
portion of the items of basic matters with a detailed explanation
for assessment, in which work control organization requirements
are presented. Many of these items listed as the basic safety
matters, were selected from the articles related to bridge
construction work in the Industrial Safety and Health Law
established in the year of 1972 and other authorized engineering
practices and specifications.

After conducting the checking the basic matters, corrective

Table 4.   Safety Measures for Basic Matters

Basic matters related to control system	Details of assessment
1. Chief engineer or supervisory engineer	1. Assign a chief engineer or a qualified supervisory engineer with sufficient experience and knowledge in steel bridge erection work
2. Control system regulations	1. Prepare control system regulations specifically prescribing duties for all members including subcontractors
3. Selection of subcontractors	1. When selecting subcontractors, consider with his achievement of experienced work and safety results
4. Work schedule plan	1. Prepare work schedule plan including preparation period, after finishing period and holidays

actions, including modification of the execution plan, should be
taken if the necessary safety measures are not yet established in
the construction plan.
   These safety measure assessment tables for the basic matters
prepared here are not only useful at the work planning stage for
safety evaluation, but also can be effectively used as the checklists during the execution of bridge construction stage as well.

2.4   Check the safety measures against inherent accidents

   After checking the basic safety requirements for steel bridge
erection work, as the final stage of safety assessment, check to
see whether proper safety measures against anticipated inherent
accidents associated with the steel bridge erection work are
already undertaken in compliance with the potential risk grade.
   From the accidents investigation in recent steel bridge
construction work, major accidents that resulted in many workers
being injured or large amount damage to work performance were
identified as collapse of structures, accidents due to falling and
struck by objects during handling of materials.
   For accidents due to collapse, check to see whether suitable
analyses of structural safety of the steel bridge during its
erection stage in terms of stress, safety factor, and stability of
the structures have already been performed.
   Table 5 displays the required items to be examined and necessary load combinations for the analyses of the structural safety
related to bent. These necessary items to be examined total 97,
which are classified into three classes : 1) items related to main
bridge structures; 2) items related to type of erection method;
and 3) equipment and facilities for erection.

Table 5. Necessary Items and Load Combinations for Structural
          Safety Analysis of Steel Bridge Erection Work

Items to be considered related to bent	Load combination to be considered							Result		
	P	p	U	I	H	W	O	Actual value	Allowable value	Safety Index
1. Check bearing stress of bent	◯	◯	◯		◯			A	B	B/A
2. Check stability of bent	◯	◯	◯		◯			A	B	A/B
3. Check the bent collapse due to eccentric or horizontal load	◯	◯	◯		◯			A	B	B/A
4. Check bearing stress of foundation	◯	◯	◯		◯			A	B	B/A

(Loads indicated by circle should be considered for the analyses)
where P : Load of main bridge structure under erection
       p : Load of machines and instrument for erection
       U : Unbalanced load during erection, I : Impact load
       H : Horizontal load for stability checking
       W : Wind load, O : Others

Table 6. Evaluation Table on Structural Safety Analysis
for Accidents due to Collapse

Consequence of examination	Grade	Measures
1 Without having consideration of items 2 Lack of loads needed to examine 3 Safety indices on tensile stress and bending stress are less than 100 4 Safety index on safety factor is less than 100 5 Safety index on buckling stress is less than 105 6 Safety index on stability is less than 100	Bad	Modification of the execution plan
7 Safety indices on tensile stress and bending stress are between 100 and 124 8 Safety index on safety factor is between 100 and 124 9 Safety index on buckling stress is between 105 and 132 10 Safety index on stability of structures on/in ground is greater equal 100	Fair	Carry out extensive safety controls on structural safety during erection
11 Safety indices on tensile stress and bending stress is greater equal 125 12 Safety index of safety factor is greater equal 125 13 Safety index of buckling stress is greater equal 133 14 Safety index of stability of structure supported by another structure is greater equal 100	Good	Not particular

where :

$$\text{Safety Index on stress} = \frac{\text{allowable stress}}{\text{actual stress}} \times 100$$

$$\text{Safety Index on safety factor} = \frac{\text{actual safety factor}}{\text{allowable safety factor}} \times 100$$

$$\text{Safety Index on stability} = \frac{\text{actual stability value}}{\text{allowable stability value}} \times 100$$

Table 7. Safety Measures for Accidents due to Falls

Item	Details of assessment
Structure of scaffoldings	1 Prepare the assembly drawings of the scaffolds
	2 Install the working floor with a 40 cm or more width without any gaps
	3 Install a fence of 75 cm height or more at the edge of working floor or around the openings
	4 Install stairs or other relevant ascending and descending facilities with handrails
	5 Set up a schedule for assembly and disassembly of the scaffolds

By making use of these tables, officers, who are responsible
for making or checking the execution plan of work, can find the
necessary items to be examined as well as the load combinations in
the analyses of the structural safety during the erection stage.

After the completion of the structural safety analyses, the
consequences of the examination should be evaluated according to
the evaluation table as given in Table 6. If the results of the
examination were graded into bad or fair, additional safety
measures to improve the execution plans should be carried out.

For accidents due to falls, checklists including 37 items with
the detailed guidelines have been prepared. Table 7 gives a portion of the list of necessary checking items related to scaffolds
with a detailed explanation. After a thorough checking is carried
out, additional safety measures should be taken if the adopted
safety measures in the execution plan are judged inadequate.

For accidents during handling of heavy materials, another
checklists including 27 items as necessary requirements with the
detailed explanation have also been prepared. Similar to accidents
due to falls, preventive measures for handling accident should be
carefully checked whether they are going to be implemented in
compliance with the detailed explanation.

3. CONCLUDING REMARK

After accomplishment of the development of safety assessment
system for bridge erection work, the Ministry of Labor of Japan
has approved this assessment system and released for notification
in 1986. The Ministry of Labor recommends to contractors, who
have full responsibility for the safety of workers, to employ this
system to evaluate the project plans before they start the bridge
erection work [3].

Since then, two large steel bridge erection work plans were
submitted to the Labor Minister and reviewed by this system. Also,
considerable numbers of small bridge plans have been submitted to
the Labor Standard Inspection Offices and reviewed. However, at
this time, it is rather difficult to evaluate the effects of this
system because of the short period since its publication.

ACKNOWLEDGMENTS

This assessment system was developed by the committee on the
development of safety assessment system for steel bridge erection
work under the supervision of the Ministry of Labor. Special
appreciation goes to Messrs. S. Takaoka, H. Nose, R. Matsuoka and
S. Sugie for their outstanding contribution to the development of
this system.

REFERENCES

[1] Industrial Safety and Health Law and Related Legislation of
     Japan: Japan Industrial Safety and Health Association, 1983.
[2] Guideline on the Safety Assessment for Steel Bridge Erection
     Work, (in Japanese): Japan Construction Safety and Health
     Association, 1986.
[3] Notification on Safety Assessment for Steel Bridge Erection
     Work, (in Japanese): Ministry of Labor, 1986.

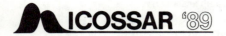
REDUNDANCY EVALUATION OF STEEL
GIRDER BRIDGES

Dan M. Frangopol * and Rachid Nakib **

* Dept. of Civil Eng., University of Colorado, Boulder, CO 80309-0428, USA.
** Dept. of Civil Eng., University of Colorado, Boulder, CO 80309-0428, USA.

ABSTRACT

The focus of this paper is on the use of redundancy measures in the evaluation of bridge systems. The emphasis is on the definition of system redundancy measures suitable for use by bridge evaluators and designers. The deterministic and probabilistic redundancy factors of an existing steel girder bridge, Colorado State Bridge E-15-AF, under corrosion damage scenarios are calculated using nonlinear finite element analysis. Effects of different damage states on both bridge redundancy and bridge reliability are estimated.

KEYWORDS

Bridge design; bridge evaluation; corrosion; damage; finite elements; girders; nonlinear analysis; system redundancy; structural reliability.

1. INTRODUCTION

An important problem with existing highway bridges in the United States is the need to evaluate their load-carrying capacity at any given time. This is a complex problem that involves the ability of the bridge to withstand damage by redistributing loads. Although this ability is the result of a redundancy system built into the bridge at the design stage, it has been acknowledged that the current AASHTO design specifications [1] do not address in a systematic way the question of bridge redundancy. Furthermore, these specifications do not offer an objective measure for redundancy. The need for such a measure, however, is obvious. In this context, the present paper reviews several deterministic and probabilistic definitions of structural system redundancy and quantifies both the deterministic and the probabilistic redundancy factors of an existing steel girder bridge by considering corrosion damage scenarios.

2. STRUCTURAL SYSTEM REDUNDANCY

2.1 General

Redundancy in a structure has been generally defined as the absence of critical components whose failure would cause collapse of the structure. This implies that the problem of structural redundancy should be discussed in conjunction with "fail-safe" structures. However, there are considerable differences of opinion about the definition of structural redundancy. A review of the definitions available in the literature [2-15] is given in the following subsections.

2.2 Deterministic System Redundancy Measures

Several deterministic measures of system redundancy have been suggested in [4,5,9], among others. These include:

(a) *Degree of indeterminacy* (i.e., classical definition of indeterminacy used in structural analysis), I, defined as

$$I = F - E \tag{1}$$

in which F and E are the number of unknown reactive forces and of independent equilibrium equations, respectively. Unfortunately, this definition of redundancy has no applicability in assessing the overall strength capacity of a structural system.

(b) *Reserve strength factor*, $R_{reserve}$, defined as

$$R_{reserve} = Q_{intact}/Q_{nominal} \tag{2}$$

in which Q_{intact} and $Q_{nominal}$ are the ultimate strength of the undamaged structural system and the nominal applied load on this system, respectively.

(c) *Residual strength factor*, $R_{residual}$, defined as

$$R_{residual} = Q_{damaged}/Q_{intact} \tag{3}$$

in which $Q_{damaged}$ is the ultimate strength of the damaged structural system.

A statically determinate structure would have no residual strength, $R_{residual} = 0$, after failure of any single component. On the other hand, failure of one component of a statically indeterminate structure will not necessarily constitute a complete loss of the load-carrying capacity for the structure (i.e., $R_{residual} > 0$).

(d) *Redundancy factor*, R, defined as

$$R = Q_{intact}/(Q_{intact} - Q_{damaged}) = 1/(1 - R_{residual}) \tag{4}$$

The redundancy factor R depends on: the loading; the damaged members; the amount of damage in each member; and the material behavior of the damaged as well as of

the intact members. As shown in Fig. 1 (b) the redundancy factor R of the bridge in Fig. 1 (a) correlates directly with the overall bridge strength in a damaged condition [4]. In general, the redundancy measure (4) ranges from a value of 1, when the structure has completely lost its strength (i.e., collapse), to ∞, when structural damage has no effect on the residual strength of the structure. It is interesting to note that in some cases of brittle behavior a damaged structure could have a higher ultimate strength than that of the intact structure (i.e., $Q_{damaged} > Q_{intact}$). In these particular cases of unexpected favorable behavior, when a structure increases its strength if it is damaged, the redundancy factor R is negative. Examples of such cases are given in [12].

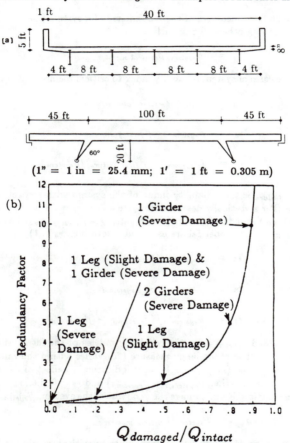

$$(1" = 1 \text{ in } = 25.4 \text{ mm}; \ 1' = 1 \text{ ft } = 0.305 \text{ m})$$

Fig. 1: (a) Bridge example and (b) Associated redundancy factor.

2.3 Probabilistic System Redundancy Measures

The computational experience reported recently by Moses [10] indicates that the effect of redundancy on system reliability depends on system geometry (e.g., series, parallel), material behavior (e.g., brittle, ductile), load and strength uncertainties (e.g., coefficients of variation), and correlations. For example, the existence of redundant members may not give benefit to the structure safety if all members are designed against their behavior limit and if load uncertainty overhelms resistance uncertainty [10].

Several probabilistic measures of the redundancy of a system have been suggested in [3,4,7,8,9,13], among others. These include:

$$(a) \qquad r_1 = \beta_{system}/\beta_{critical\ member} \qquad (5)$$

in which r_1 is the ratio of system reliability index to critical member reliability index.

$$(b) \qquad r_2 = \beta_{system}/\beta_{first\ failure} \qquad (6a)$$

$$r_3 = \beta_{system} - \beta_{first\ failure} \qquad (6b)$$

$$r_4 = (\beta_{system} - \beta_{first\ failure})/\beta_{system} \qquad (6c)$$

in which $\beta_{first\ failure}$ is the reliability index of the system with respect to any first member failure. For a statically determinate system $r_3 = r_4 = 0$ (i.e., $r_2 = 1$). On the other hand, a high redundant system (i.e., $\beta_{system} >> \beta_{first\ failure}$) has almost the same reliability given a member failure as it had before (i.e., $r_4 \simeq 1$).

$$(c) \qquad r_5 = \beta_{system,\ i}/\beta_{system,d} \qquad (7a)$$

$$r_6 = \beta_{system,\ i} - \beta_{system,d} \qquad (7b)$$

$$r_7 = \beta_{system,\ i}/(\beta_{system,i} - \beta_{system,d}) \qquad (7c)$$

in which $\beta_{system,i}$ and $\beta_{system,d}$ are the reliability indices of the intact and damaged system, respectively. The redundancy measure (7c) varies within the range between zero and ∞ with $r_7 = \infty$ (i.e., $r_5 = 1, r_6 = 0$) indicating no effect of damage on the reliability of the structure (i.e., $\beta_{system,d} = \beta_{system,i}$) and $r_7 = 0$ (i.e., $r_5 = 0, r_6 = \infty$) indicating catastrophic effect of damage (i.e., collapse of the system: $\beta_{system,d} = -\infty$).

$$(d) \qquad r_8 = P_{system}/P_{critical\ member} \qquad (8)$$

in which r_8 is the ratio of system failure probability to critical member failure probability.

$$(e) \qquad r_9 = P_{system}/P_{first\ failure} \qquad (9)$$

in which $P_{first\ failure}$ is the first member failure probability.

$$(f) \qquad\qquad r_{10} = (P_{damage} - P_{system})/P_{system} \qquad\qquad (10)$$

in which P_{damage} is the probability of damage occurrence to the system (e.g., failure of any member). According to the redundancy definition (10) a system is considered nonredundant ($r_{10} = 0$) if $P_{damage} = P_{system}$. It is expected in this case that no damage will occur before the system failure. On the other hand $P_{damage} > P_{system}$ indicates that the system possesses residual reliability measured in probabilistic redundancy ($r_{10} > 0$).

3. REDUNDANCY AND RELIABILITY OF AN EXISTING STEEL GIRDER BRIDGE

The example bridge used for this study is the Colorado State Bridge E-15-AF. This bridge has a simple span of 90 ft (27.43m) and a width of 36 ft (10.97m). The four steel girders are spaced at 9 ft (2.74m). The concrete deck is modeled using 143 quadrilateral shell elements with a thickness of 8.5 in (21.6cm), while each of the four girders is modeled using 39 two-noded beam elements. The total number of degrees of freedom within the finite element bridge model equals 1794. The interaction between girders and deck is accounted for through the 3-D finite element modeling. The detailed description of the geometrical and mechanical properties of the bridge, the inelastic stress-strain relationships for both steel and concrete, the finite element modeling, the statistics of loads and strengths, and the side-by-side position of the two HS-20 trucks which induce maximum stresses in the bridge are given in [12]. The live and impact loads are incremented progressively by the factor λ until reaching collapse.

Bridge redundancy and reliability were evaluated in the presence of corrosion. The damage factor for corrosion is given as $D.F. = (A_i - A_d)/A_d$, where A_d and A_i are the cross sectional areas of the damaged (i.e., corroded) and intact girders, respectively [6]. Fig. 2 shows the effect of corrosion damage factor on the maximum stress in bridge girders under dead load for three different damage scenarios as follows: case (1), all girders corroded uniformly; case (2), internal girders corroded uniformly; and case (3), external girders corroded uniformly. As expected, case (1) has the most effect on the stress level. When only the internal girders are corroded (i.e., case (2)), the stress level in the internal girders is always higher than that of the external girders. On the other hand, when only the external girders are corroded (i.e., case (3)) internal girders have higher stresses for damage factors less than about 0.50.

Redundancy and reliability of the Colorado State Bridge E-15-AF were evaluated based on the ultimate strength capacity of this bridge in the presence of uniform corrosion of all girders. Inelastic bridge load-deflection response curves (under dead (D), live (L) and impact (I) loads) for various corrosion damage factors (i.e., 0.00; 0.25; 0.50, and 0.75) were obtained and results are shown in Fig. 3. Based on these curves, deterministic redundancy factors R (see (4)) were calculated and the results are presented in Table 1, where λ_{intact} and $\lambda_{damaged}$ are the ultimate load increment

factors for the intact and damaged bridge, respectively, and

$$R = \lambda_{intact}/(\lambda_{intact} - \lambda_{damaged}) \tag{11}$$

In addition, the bridge reliability index β versus the load increment factor λ is shown in Fig. 4. The effect of damage on bridge reliability varies with the rate of section loss. As shown in Fig. 4, a 25% corrosion under dead, live and impact loads (i.e., $\lambda = 1$), results in β shifting from 4.75 to 3.40, whereas 50% corrosion results in the deterioriation of β to a value as low as 1.35. For these two cases, the values of the associated probabilistic redundancy factor (see (7c)) are $4.75/(4.75 - 3.40) = 3.52$ and $4.75/(4.75 - 1.35) = 1.40$, respectively.

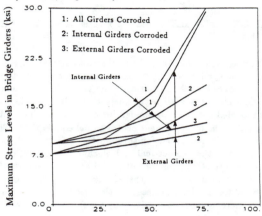

Fig. 2:   Effect of corrosion on maximum girder stress for three damage
          scenarios (1 ksi = 6.89 MPa).

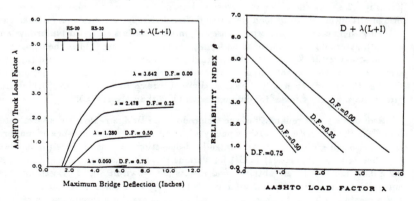

Fig. 3:   Inelastic load-deflection bridge response   Fig. 4:   Bridge reliability index

Table 1:   Redundancy factors for E 15-AF 4 girder bridge, under dead, live and impact loads

Corrosion damage factor, D.F.	Load increment factors		Redundancy factor, R
	λ_{intact}	$\lambda_{damaged}$	
0.00		3.642	∞
0.25	3.642	2.478	3.13
0.50		1.280	1.54
0.75		0.060	1.02

4. CLOSURE

An attempt has been made to evaluate existing bridges by using system redundancy measures. It should contribute to a wider acceptance of these measures by bridge evaluators and designers.

5. ACKNOWLEDGMENT

Financial support of this study by the National Science Foundation under grant ECE-8609894 is gratefully acknowledged.

References

[1] American Association of State Highway and Transportation Officials (AASHTO), "Standard specifications for highway bridges", Washington, D.C., 1984.

[2] BENNETT, R.M., ANG, A.H-S. and GOODPASTURE, D.W.: "Probabilistic safety assessment of redundant bridges", in *Structural Safety and Reliability* (Eds. I. Konishi, A.H-S. Ang and M. Shinozuka), IASSAR, Vol. III, Kyoto, Japan, 1985, pp. 205-211.

[3] FENG, Y.S. and MOSES, F.: "Optimum design, redundancy and reliability of structural systems", *Computers and Structures*, Vol. 24, No.2, 1986, pp. 239-251.

[4] FRANGOPOL, D.M. and CURLEY, J.P.: "Effects of damage and redundancy on structural reliability", *J. Struct. Eng.*, Vol. 113, No. 7, ASCE, NY, 1987, pp. 1533-1549.

[5] FRANGOPOL, D.M. and GOBLE, G.G.: "Development of a redundancy measure for existing bridges", in *Bridge Evaluation, Repair and Rehabilitation* (Eds. A.S. Nowak and E. Absi), University of Michigan, Ann Arbor, MI, 1987, pp. 365-375.

STRUCTURAL SAFETY AND RELIABILITY

[6] FRANGOPOL, D.M., NAKIB, R. and FU, G.: "Bridge reliability evaluation using 3-D analysis and damage scenarios", in *Probabilistic Methods in Civil Engineering* (Ed. P.D. Spanos), ASCE, NY, 1988, pp. 217-220.

[7] FU, G.: " Modeling of lifetime structural system reliability", Report No. 87-9, Dept. Civ. Engrg., Case Western Reserve Univ., Cleveland, OH, 1987.

[8] FU, G. and MOSES, F.: "Lifetime system reliability models with application to highway bridges", in *Reliability and Risk Analysis in Civil Engineering* (Ed. N.C. Lind), Vol. 1, 1987, pp. 71-78.

[9] FURUTA, H., SHINOZUKA, M. and YAO, J.T.P.: "Probabilistic and fuzzy interpretation of redundancy in structural systems", Presented at First Intern. Fuzzy Systems Assoc. Congress, Palma de Malorca, Spain, 1985.

[10] MOSES, F.: " New directions and research needs in system reliability research" in *New Directions in Structural System Reliability* (Ed. D.M. Frangopol), University of Colorado, Boulder, CO, 1989, pp. 6-16.

[11] MOSES, F. and YAO, J.T.P.: "Safety evaluation of buildings and bridges", in *The Role of Design, Inspection, and Redundancy in Marine Structural Reliability* (Eds. D. Faulkner, M. Shinozuka, R.R. Fiebrandt and I.C. Frank), National Research Council, Washington, D.C., 1984, pp. 349-385.

[12] NAKIB, R.: "Reliability analysis and optimization of multistate structural systems", Thesis presented to the University of Colorado at Boulder, in partial fulfillment of the requirements for the degree of Doctor in Philosophy, Boulder, CO, 1988.

[13] NORDAL, H., CORNELL, C.A. and KARAMCHANDANI, A.: "A structural system reliability case study of an eight-leg steel jacket offshore production platform", Proceedings Marine Structural Reliability Symposium, Arlington, VA, 1987, 24 pp.

[14] NOWAK, A.S.: "Risk analysis for evaluation of bridges", Res. Rep. UMCE 88-7, Dept. Civ. Eng., University of Michigan, Ann Arbor, MI, 1988.

[15] PARMELEE, R.A. and SANDBERG, H.R.: " If it's redundant, prove it", *Civ. Eng.*, ASCE, Oct. 1987, pp. 57-58.

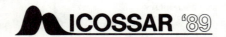
DAMAGE ASSESSMENT OF BRIDGE STRUCTURES
BY SYSTEM IDENTIFICATION

Chung-Bang Yun*, Won Jong Kim** and Alfredo H-S. Ang**

*   Dept. of Civil Engineering, Korea Advanced Institute of Science and
Technology, P.O. Box 150, Cheongryang, Seoul, Korea
** Dept. of Civil Engineering, University of California at Irvine, Irvine,
CA 92717.

ABSTRACT

A method for the damage estimation of bridge structures under
seismic ground excitation is developed based on system identification
approach. The dynamic behavior of a damaged structure is represented
by a nonlinear hysteresis model. Estimation of the nonlinear structural
parameters is carried out by the extended Kalman filter.

KEYWORDS

bridge structure, damage estimation, extended Kalman filter, nonlinear
hysteresis model, state vector, system identification

1. INTRODUCTION

The identification of the degree of deterioration or damage of structures becomes increasingly important, particularly in connection with the maintenance, repair and rehabilitation of
existing civil engineering structures. Decisions as to the degree of rehabilitation of a structure
are difficult and may involve high cost. The proper bases for such decisions require a method
for determining the existing conditions of the structure.

In this study, a method is developed for the assessment of seismic damage of bridge
structures using a system identification approach. In the assessment of structural damage, two
major tasks are involved. One is the development of damage model for describing the degree
of damage. The other is the identification of structural parameters in the damage model. Many
studies have been reported about damage model[1,2,3,4]. Under a severe seismic loading which
can cause damage to a structure, the load-deformation relationship of the structure is nonlinear.
Energy is also dissipated through hysteresis under repeated cyclic loadings. Also, the damage is
very likely to be accompanied by the degradation of the strength as well as the stiffness. In

order to incorporate the nonlinear behavioral characteristics into the damage model, the analytical nonlinear restoring force model developed by Bouc[5] and Baber and Wen[6] is used in this study. Then the structural parameters associated with the nonlinear behavior under ground excitation is evaluated by using the extended Kalman filtering algorithm[7,8,9]. Example studies are performed for an idealized structural model with discrete masses on the basis of artificially generated time histories of structural responses.

2. MODELLING OF NONLINEAR BEHAVIOR OF BRIDGE

Idealized Model of Bridge

An idealized structural model of a bridge as shown in Fig.1 is used in this study. By using a discrete mass model, the bending moment at node j, M_j, is approximated as

$$M_j = E \, \overline{I}_j \rho_j \tag{1}$$

where

$$\rho_j = u_j{''} = \frac{2}{l_j + l_{j+1}} \left\{ \frac{1}{l_j} u_{j-1} - (\frac{1}{l_j} + \frac{1}{l_{j+1}}) u_j + \frac{1}{l_{j+1}} u_{j+1} \right\} \tag{2}$$

$$\overline{I}_j = \frac{I_j l_j + I_{j+1} l_{j+1}}{l_j + l_{j+1}} \tag{3}$$

and u_j is the vertical displacement at node j; ρ_j and $u_j{''}$ are the curvature at node j; l_j is the length of the element (j); E is the Young's modulus; and $\overline{I}_j$ is the average value of the area moment of inertia of the two elements (j) and ($j+1$).

From force equilibrium at node j (in Fig.1b), the relationship between the external force F_j and the shear forces can be obtained as

$$F_j = V_j - V_{j+1} \tag{4}$$

with

$$V_j = \frac{1}{l_j}(M_{j-1} - M_j) \tag{5}$$

where V_j is the shear force at both ends of the element (j).

Then, using Eq. 5, Eq. 4 can be rewritten as

$$F_j = \frac{1}{l_j}M_{j-1} - (\frac{1}{l_j} + \frac{1}{l_{j+1}})M_j + \frac{1}{l_{j+1}}M_{j+1} \tag{6}$$

Non-linear Restoring Moment

Using the hysteretic restoring force model of Baber and Wen[4], the nonlinear restoring bending moment is represented as

$$M_{NL_j} = E \, \overline{I}_j(\alpha_j \rho_j + (1.0 - \alpha_j) z_j) \tag{7}$$

where the nonlinear hysteretic function z_j is modelled as

$$\dot{z}_j = \frac{A_j(t)\dot{\rho}_j - v_j(t)\{\beta_j|\dot{\rho}_j||z_j|^{n_j-1}z_j + \gamma_j\dot{\rho}_j|z_j|^{n_j}\}}{\eta_j(t)} \tag{8}$$

$$
\begin{aligned}
A_j(t) &= 1.0 - \delta_{A_j}\varepsilon_j(t) \\
v_j(t) &= 1.0 + \delta_{v_j}\varepsilon_j(t) \\
\eta_j(t) &= 1.0 + \delta_{\eta_j}\varepsilon_j(t) \\
\dot{\varepsilon}_j(t) &= E\,\bar{I}_j(1.0 - \alpha_j)\dot{\rho}_j z_j
\end{aligned}
\tag{9}
$$

and α_j, β_j, γ_j and n_j are constants related to the hysteretic restoring moment characteristics at node j; δ_{A_j}, δ_{v_j} and δ_{η_j} are constants related to the strength degradation; and $\varepsilon_j(t)$ represents the dissipated hysteretic energy. In this study, n_j is assumed to be 1.0.

The nonlinear restoring moment in Eq. 7 can be conveniently rewritten as

$$M_{NL_j} = E\,\bar{I}_j\rho_j + E\,\bar{I}_j(1.0 - \alpha_j)(z_j - \rho_j) \tag{10}$$

Utilizing Eqs. 6 and 10, the nonlinear restoring vertical force vector, $\{F_{NL}\}$, can be obtained as

$$\{F_{NL}\} = [K_0]\{u\} + \sum_{i=1} \{B_j\}E\,\bar{I}_j(1.0 - \alpha_j)(z_j - \rho_j) \tag{11}$$

where $[K_0]$ is the stiffness matrix of the undamaged linear structure and $\{B_j\}$ is the coefficient vector for the nonlinear restoring force at node j which is determined by Eq. 6.

Equation of Motion for Earthquake Loading

The equation of motion of the bridge model under a vertical earthquake excitation can be obtained as

$$[M_0]\{\ddot{u}\} + [C_0]\{\dot{u}\} + [K_0]\{u\} + \sum_{i=1}^{n_y} \{B_j\}E\,\bar{I}_j(1.0 - \alpha_j)(z_j - \rho_j) = -[M_0]\{1\}\ddot{u}_g \tag{12}$$

where $[M_0]$, $[C_0]$ and $[K_0]$ are the mass, damping and stiffness matrices of the structure before yielding occurs; $\{1\}$ is a column vector with elements being equal to 1.0; $\ddot{u}_g$ is the vertical ground acceleration; and n_y is the number of nodes expected to yield.

In this study, for computational efficiency, the dynamic analysis is carried out using the superposition of the natural modes of the undamaged linear structure; namely,

$$\{u\} = \sum_{i=1}^{n_f} \{\phi\}_i q_i \tag{13}$$

where $\{\phi\}_i$ is the i-th mode which is normalized so that $\{\phi\}_i^T[M_0]\{\phi\}_i = 1.0$; q_i is the generalized modal coordinate for the i-th mode; and n_f is the number of modes included in the analysis.

Using the orthogonality conditions of the modes, Eq. 12 can be rewritten in terms of the

modal coordinates $\{q\}$,

$$\ddot{q}_i + 2\xi_i\omega_i\dot{q}_i + \omega_i^2 q_i + \sum_{j=1} \{\phi\}_i^T \{B_j\} E \overline{I_j}(1.0 - \alpha_j)(z_j - \rho_j) = \Gamma_i \ddot{u}_g \tag{14}$$

$$\Gamma_i = -\{\phi\}_i^T[M_0]\{1\} \tag{15}$$

where ω_i and ξ_i are the natural frequency and damping ratio for the i-th mode; and Γ_i is the modal participation factor for the earthquake load. Eq. 14 is coupled with respect to the modal coordinates $\{q\}$, since z_j and ρ_j are in terms of the structural displacement and velocity $\{u\}$ and $\{\dot{u}\}$ as shown in Eqs. 2 and 8.

3. PARAMETER ESTIMATION USING EXTENDED KALMAN FILTER

In this study, it is assumed that the structural parameters and the modal quantities of the undamaged linear structure are known. It is also assumed that structural damage induced by strong ground motion is limited to two locations; i.e. at the mid-span and near the quarter span (nodes 5 and 8 in Fig.1a).

Based on the formulation for dynamic analysis using the first three modes, the equation of motion, Eq. 14, can be converted into a nonlinear state equation as

$$\{\dot{X}\} = f(X,\ddot{u}_g;t) \tag{16}$$

where $\{X\}$ is the state vector consisting of the modal coordinates q_j and $\dot{q}_j$, the hysteretic functions and dissipated energy at the two damaged locations, and the unknown parameters as augmented state variables,

$$\{X\} = \{x_1, x_2, x_3, x_4, x_5, x_6, x_7, x_8, x_9, x_{10}, x_{11}, x_{12}, x_{13}, x_{14}, x_{15}, x_{16},$$
$$x_{17}, x_{18}, x_{19}, x_{20}, x_{21}, x_{22}\}$$

$$= \{q_1, q_2, q_3, \dot{q}_1, \dot{q}_2, \dot{q}_3, z_5, \varepsilon_5, z_8, \varepsilon_8, \alpha_5, \beta_5, \gamma_5, \delta_{A_5}, \delta_{v_5}, \delta_{\eta_5},$$
$$\alpha_8, \beta_8, \gamma_8, \delta_{A_8}, \delta_{v_8}, \delta_{\eta_8}\} \tag{17}$$

and

$$f_i = 0. \qquad \text{for } i = 11, 12, \cdots, 22 \tag{18}$$

It is assumed that the time histories of the input ground acceleration and the vertical displacements and velocities at nodes 5 and 8 are available as the observational data. Then the response observation equation can be written as

$$\{Y(k)\} = [C_1, 0]\{X(k)\} + \eta(k) \tag{19}$$

where $\{Y(k)\}$ is the observational vector at $t = k\Delta t$; $\{X(k)\}$ is the state vector at $t = k\Delta t$; $[C_1]$ is the (4×6) matrix related to the mode shape coefficients for nodes 5 and 8; $[0]$ is a null matrix of size (4×16) and $\eta(k)$ is the observational noise vector with covariance ∇_η.

The basic algorithm of the extended Kalman filter[7,8] is a recursive process for estimating the conditional expectation of the state vector(so called the filtered state) based on the

observational data for the excitations and responses. Applying the extended Kalman filter to the above state and observation equations(Eqs. 16, 17 and 19), the estimated values of the unknown parameters can be obtained as components of the filtered state, if the initial value of the state and its error covariance matrix are given. However, they are unknown, particularly because the state vector contains the unknown parameters as in Eq. 17. Hence, improper initial values for the system parameters may cause divergence of the filtering algorithm. Several remedial schemes were proposed[10,11,12]. Among them, the weighted global iteration procedure is employed in this study.

4. NUMERICAL EXAMPLES

Identification of the nonlinear parameters associated with the hysteretic restoring moments are carried out for an idealized bridge structure subjected to the San Fernando earthquake(at Pacoima Dam site, 1971, S16E; Fig.2). The simplified bridge structure was analyzed during the first 10 seconds with the time step of 0.02 second. The structural properties and input parameters used are summarized in Table 1. The exact time histories of vertical displacements calculated at Nodes 5 and 8 are shown in Fig.3a. The corresponding hystereses of the bending moments at the two nodes are shown in Fig.3b and 3c. Figures 3a and 3b indicate that the structure underwent a very severe yielding at the mid-span(Node 5) during the earthquake excitation. On the other hand, at Node 8 which is near the quarter span, it experienced a moderate nonlinear behavior.

Identification of the nonlinear parameters were carried out based on the time histories of the ground motion and structural response in Fig.3. In the present example, it is assumed that the response observations are noise free. Table 2 compares the exact and estimated values for the nonlinear parameters. Fig.4 shows the displacements and nonlinear hysteresis curves corresponding to the case with the estimated parameters. Compared with the exact values, the estimated parameters particularly for those for Node 8 does not appear to be satisfactory. However, the estimated nodal displacements are found to be virtually same to the exact ones. The estimated hystereses also show good agreements with the exact ones. Parametric studies reveal that the structural responses are not affected sensitively by the changes of the parameters associated with Node 8 in the vicinities of the current values. Therefore, refinement of the algorithm is desired to obtain better estimates for the insensitive parameters such as those for Node 8. Further investigations are also suggested for the cases with more realistic structural models and with observational noise.

5. CONCLUSION

A method for the damage estimation of bridge structure is developed based on the system identification approach. The behavior of a damaged structure is represented by an analytical nonlinear restoring force model. Parameter identification is carried out by the extended Kalman filter. The example analysis indicates that the present method predicts the nonlinear behavior of the structure very well even for the case with severe nonlinearity. However, the estimated parameters are not quite satisfactory, particularly for those which are insensitive to the structural response. Refinement of the algorithm is desired.

ACKNOWLEDGEMENTS

This research was jointly supported by the Korea Science and Engineering Foundation and the National Science Foundation of the United States. Those supports are gratefully acknowledged.

Table 1. Structural Parameters

Structural Properties	
$l_j\ (m)$	3.0
$E\,\overline{I_j}\ (KN \cdot m^2)$	1.143×10^7
Nodal Mass $(KN \cdot sec^2/m)$	27.63
Damping Ratio	0.1
Natural Period(sec)	
T_1	0.513
T_2	0.132
T_3	0.061
Nonlinear Parameters	
$\alpha_5 = \alpha_8$	0.0
$\beta_5 = \beta_8$	1500
$\gamma_5 = \gamma_8$	1500
$\delta_{A_5} = \delta_{A_8}$	0.01
$\delta_{v_5} = \delta_{v_8}$	0.02
$\delta_{\eta_5} = \delta_{\eta_8}$	0.02

Table 2. Estimated Parameters

	Exact Values	Initial Guesses	Estimated Values
α_5	0.0	0.0	0.0
β_5	1500	0.0	1462
γ_5	1500	0.0	1236
δ_{A_5}	0.01	0.0	0.0105
δ_{v_5}	0.02	0.0	0.0175
δ_{η_5}	0.02	0.0	0.0066
α_8	0.0	0.0	0.0
β_8	1500	0.0	1297
γ_8	1500	0.0	1122
δ_{A_8}	0.01	0.0	0.003
δ_{v_8}	0.02	0.0	0.004
δ_{η_8}	0.02	0.0	0.001

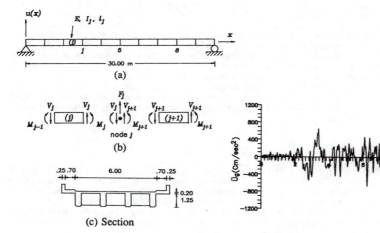

(a)

(b)

(c) Section

Fig.1. Idealized Bridge Model

Fig.2. Earthquake Excitation(San Fernando Earthquake at Pacoima Dam, 1971)

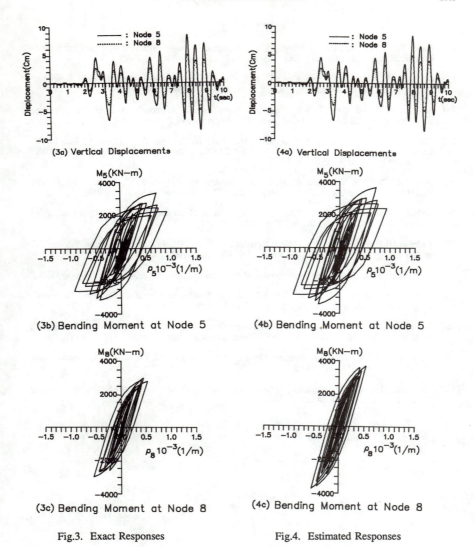

(3a) Vertical Displacements

(4a) Vertical Displacements

(3b) Bending Moment at Node 5

(4b) Bending Moment at Node 5

(3c) Bending Moment at Node 8

(4c) Bending Moment at Node 8

Fig.3. Exact Responses

Fig.4. Estimated Responses

REFERENCES

[1] Park, Y-J., Ang, A. H-S. and Wen, Y.K., "Seismic Damage Analysis of Reinforced Concrete Buildings", *J. of Struct. Eng., ASCE*, Vol. 111, pp.740-757, 1985.

[2] Sues, R.H., Wen, Y.K. and Ang, A. H-S., "Stochastic Evaluation of Seismic Structural Performance", *J. of Struct. Eng., ASCE*, Vol. 111, pp.1204-1218, 1985.

[3] Natke, H.G. and Yao, J. T-P., "System Identification Approach in Structural Damage Evaluation", *ASCE Structural Congress '86*, Preprint 17-1., 1986.

[4] Chung, Y.S., Meyer, C. and Shinozuka, M., "Seismic Damage Assessment of Reinforced Concrete Members", Report No. NCEER-87-0022, National Center for Earthquake Engineering Research, Buffalo, 1987.

[5] Bouc, R., "Forced Vibration of Mechanical System with Hysteresis", *Proc. of the Fourth Conference on Nonlinear Oscillation*(Abstract), Prague, Czechoslovakia, 1967.

[6] Baber, T.T. and Wen, Y.K., "Random Vibration of Hysteretic Degrading Systems", *J. of Eng. Mech., ASCE*, Vol. 107, pp.1069-1088, 1981.

[7] Kalman, R.E and Bucy, R.S., "New Results in Linear Filtering and Prediction Theory", *J. of Basic Engineering, ASME*, Series D, Vol. 83, pp.95-108, 1961.

[8] Jazwinski, A.H., *Stochastic Processes and Filtering Theory*, NY., Academic Press, 1970.

[9] Yun, C.-B. and Shinozuka, M, "Identification of Nonlinear Structural Dynamic Systems", *J. of Structural Mechanics*, Vol. 8, pp.187-203, 1980.

[10] Schmidt, S.F., "Compensation for Modeling Errors in Orbit Determination Problems", Analytical Mechanics Associates, Inc., Westbury, N.Y., Iterim Report 67-16, 1967.

[11] Jazwinski, A.H., "Limited Memory Optimal Filtering", *IEEE Trans. on Automatic Control*, Vol. 13, pp.558-563, 1968.

[12] Hoshiya, M. and Maruyama, O., "Identification of Nonlinear Structural Systems", *Proc. of ICASP-5*, Vancouver, Canada, 1987.

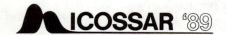
TOWARDS REDUCTION OF THE VULNERABILITY
OF MULTI-TANK STORAGE FACILITIES

K. van Breugel[*] and J.P.G. Ramler[**]

[*] Department of Civil Engineering
     Research, Delft University of Technology, Delft,
     The Netherlands
[**] Research and Development
     Hollandsche Beton Groep NV, Rijswijk,
     The Netherlands

ABSTRACT

Accident scenarios in large scale industrial plants
show that conventionally designed multi-tank storage
facilities are susceptible to domino effects. With
the aim to reduce the probability of domino effects
the potentialities of Prestressed Concrete Pressure
Vessels (PCPV's) have been investigated. For a
fictitious LPG-storage facility, built up of either
conventional Steel Pressure Vessels (SPV's) or
PCPV's, 24 accident scenarios were analysed. The
calculated probability of a domino effect in a SPV-
storage facility was found to be in close agreement
with existing records. In the case of PCPV-
facilities this probability would be zero. Design
and cost aspects of PCPV's are dealt with.

KEYWORDS

pressure vessels, prestressed concrete, domino
effects, safety, design, costs.

1. INTRODUCTION

Increasing scale and complexity of modern industrial facilities, e.g.
large scale storage facilities for hazardous products, compel to

reconsider current safety concepts (a.o. [1]). Recent disasters like
the Mexico-LPG and Bhopal disaster, both in 1984, and a number of
almost and smaller accidents constitute another reason for a re-
evaluation of these concepts. And this not in the least since, as
Baldewicz has once stated very impressively, the majority of past
accidents should be considered merely as precursor events and the
worst event is yet to come! ([2], pp 81-90).
In a re-evaluation of safety concepts the active, passive and
inherent safety concepts should be considered. Broadly speaking
active safety refers to measures and provisions with the aim to
reduce the probability of occurrence of an accident. Passive safety
measures mainly focus on the control of consequences of an accident.
Most challengeing are studies on the possibility of inherently safe
plants [1]. Plant lay-out, fuel consumption and complexity of
installations are some of the main topics dealt with in these
studies. With regard to this "inherently safe" concept Wicks [5]
remarks that for pressurized storage 100% inherent safety is an
unreachable aim. Inherent safety of pressurized storage being
unreachable, passive safety would be the next-best option to be
strived at. Pressure vessels with built-in passive safety could be
built in prestressed concrete. This paper deals with the
potentialities of PCPV's for an anti-domino effect design of a multi-
tank storage facility. Design and cost aspects of PCPV's are briefly
dealt with.

2. INDUSTRIAL ACCIDENTS

2.1 Accident statistics

A valuable record of major industrial accidents has been compiled by
Manuele [3]. He briefly describes one hundred accidents with total
direct property losses exceeding $ 10,000,000 per accident in the
period from 1954 to 1984. Fig. 1a shows a quite remarkable increase
in the number of accidents per decade. This increase can largely be
attributed to the increasing number of industrial plants. Fig. 1b
indicates a tendency of property losses per accident to increase,
which can be carried back to scale effects and increased complexity
of modern industrial facilities.

Fig. 1. Accident statistics
1a. Number of major industrial accidents from 1954
to 1983;
1b. Development of direct
losses per accident.
(data obtained from [3]).

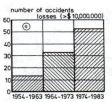

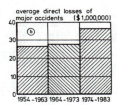

2.2 Vessel failures

The probability of a vessel failure is said to range from 10^{-4} to
10^{-7} [8,10]. From data Bohla [9] took form the TNO-databank "FACTS'
[7] a failure probability of even $3.6 \cdot 10^{-3}$ could be deduced. The wide
scatter in these figures can partly be carried back to insufficiency

of the input data. An other point that should be born in mind when interpreting literature data on failure frequencies refers to the fact that the majority of vessel failures took place in "domino accidents" in multi-tank storage facilities [3,13]. From an evaluation of past multi-tank accidents it is evident that insufficient resistance of conventional steel storage vessels against impact, fire and blast loads constitute a major reason for an escalation of a plant accident, e.g. domino effects. Higher resistance under these extreme load conditions can be obtained with structural components of prestressed concrete. The potentialities of prestressed concrete for minimizing the vulnerability of storage systems for cryogenic use has been recognized since long [18]. To which extent prestressed concrete, in the form of PCPV's, might be of interest for an anti-domino effect design of multi-tank storage facilities will be discussed in more detail in the following sections.

3. DOMINO EFFECT ANALYSIS

3.1 Multi-tank LPG-storage facility - site description

In order to illustrate the potentialities of PCPV's for an anti-domino effect design the probability of a domino effect in a fictitious multi-tank storage terminal for LPG will be investigated. The terminal consists of 12 spheres, 3,000 m^3 each, and one double walled (steel/concrete) cryogenic tank of 36,000 m^3. Distance between spheres 20 m. Total site area is 22,000 m^2. LPG supply by sea ship: 11 landings per year. Transport land-in by tank car (60%), by railway car (20%) and by ship (20%). The plant lay-out is sketched in Fig. 2. For the determination of the probability of a domino effect in this tank park both a probabilistic and a phenomenological approach will be followed. Two alternatives for the spheres are considered. In one alternative the spheres are assumed to be traditionally built SPV's with a wall thickness of 37 mm. In the second alternative sphere-shaped PCPV's are considered with a wall thickness of 600 mm and forseen with a steel liner at the inside. The probabilistic analysis mainly refer to the behaviour of the vessels under hazard loads.

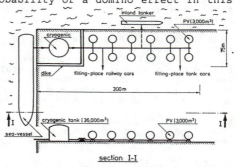

Fig. 2 Lay-out LPG-storage terminal [11]

3.2 Probabilistic approach

3.2.1 Initiating events
In total 24 initiating events and associated consequences have been

analysed. The selected events were assumed to constitute a cluster of events that represent 80% to 90% of the most frequent accident causes. The considered initiating events and event trees are described in detail in [11]. They include the spontaneous failure of an SPV, rupture of piping, BLEVE's (P{BLEVE} = 0.7-1.28 10^{-5}/yr), tornado generated missiles and aircraft crashes (Phantom II; 2.25 10^{-7}/yr). Event probabilities were deduced from literature [8, 10, 11, 12, 14, 15].

3.3.2 Event trees

With respect to the event trees the following assumptions were made:
a. The probability of ignition of a large spill (>100 tons) has been estimated at 70%, of a smaller spill 50%.
b. The probability of an unconfined vapour cloud explosion (UVCE) has been estimated at 70% for large clouds, 50% for medium size clouds (10-100 tons) and 10% for clouds < 10 tons (deduced from [4]).
c. Primary fragments (10-20 pieces) in case of a BLEVE of a steel vessel were calculated according to procedures proposed by Baker et al.[22]. For secondary fragments Baker [22] and Westine [16] have been followed.
d. Blast overpressures have been calculated according to TNO-guide lines [19] (UVCE's) and Held et al. [6] (BLEVE's).
e. An aircraft crash has been recommended to be specified for storage systems for hazardous products by [21]. For a particular plant a consequence analysis must indicate whether such a loading shall be specified or not.

3.2.3 Resistance

Impact: The wide variety in missiles generated by either a gas cloud explosion, a BLEVE or a tornado, compels to the adoption of different impact formulae. The limits of applicability of applied formulae were taken from an extensive literature survey [17]. A PCPV was assumed to fail when the scabbing thickness of the concrete wall is reached. Any favourable effect of a liner at the inside of the wall was not taken into account. The adopted failure criterion must be considered as very conservative (also [20]).

Fire: Well documented accident scenarios justify the assumption that a traditionally built SPV is generally not able to withstand a fire triggered by a BLEVE. On the other hand: the high insulating capacity of a thick walled concrete structure warrants the assumption that, depending on the duration of the fire, a PCPV will survive in such a fire (also [20]).

Blast: With respect to blast resistance it was assumed that both steel and concrete vessels, including their foundation, had been so designed that they were able to withstand blast loads.

3.3 Results probabilistic approach / Steel Pressure Vessels

The probability of a catastrophic failure of one single SPV in the adopted scenarios was calculated at 2.9 10^{-5}/yr. The probability of a **major** domino effect in which all vessels were involved would be 3.5 10^{-4}/yr. Reasons why the probability of twelve vessels to fail in one accident is higher than the probability that only one single vessel out of twelve would fail are:

1. The more vessels are present in a tank park the higher the probability that one of these vessels will fail due to one cause or another.
2. The **conditional** probability of failure, e.g. the probability of failure of a second, a third etc. vessel, given the failure of a first vessel, is much higher than the probability of failure of one single vessel. For SPV's the conditional probability of failure would range from 0.35 to 0.9/yr [8,13].

3.4 Results phenomenological approach / Steel Pressure Vessel

Based on accident statistics an expression has been derived with which, for conventionally built multi-tank storage facilities, the probability of failure of m vessels out of n vessels present in a tank park can be estimated [13]:

$$P\{F_{m \text{ tanks}}\} = n * P\{F_i\} * P\{F_{m>1}\}^{m-1}$$

where $P\{F_i\}$ = initiating event

$$P\{F_{m>1}\} = 0.95 * (1-e^{-0.35*m^{0.5}}) \qquad (= P\{F_{\text{conditional}}\})$$

According to this emperical expression the probability of a major domino effect at the fictitious plant would be $4.1 \ 10^{-6}$/yr. When comparing this figure with the corresponding figure obtained in the probabilistic approach ($3,5 \ 10^{-4}$) one should bear in mind that the probability of failure of twelve vessels in one single accident calculated on the basis of individual behaviour of the vessels would be about 10^{-60} (!). In the light of this figure one may conclude that the results obtained with the probabilistic and phenomenological approach are in fairly good agreement. Further conclusions are that the assumptions made in probabilistic analysis are not far from reality and that for an indication of the susceptibility of a tank park to domino effects the analysis of 24 scenarios already give a quite reliable picture of the actual situation.

3.5 Results probabilistic approach / Prestressed Concrete Pressure Vessels

For a 3,000 m^3 PCPV, with a wall thickness of 600 mm, the probability of failure was calculated at $4.0 \ 10^{-7}$/yr. In this calculation the probability of spontaneous failure of a PCPV in the operational stage due top material defects was assumed to be zero. This assumption recieves justification from a tendon failure analysis. Such an analysis, in which many conservative assumptions were made and dynamic effects were accounted for, resulted in a probability of failure of a prestressing tendon somewhere in the shell of 10^{-31}. The probability of failure of two adjacent tendons would be 10^{-36}. In an other study Bomhard [20] found the probability of failure of one single tendon to be less than 10^{-35}. Parallel coupling of the load bearing element explains why these values are so low. Precondition in above tendon failure analysis is a good concrete quality in order to assure permanent protection of the prestressing steel against chemical attack, e.g. corrosion. Many decades experience with

prestressed concrete have shown, however, that this precondition can
easily be met.
Because of the BLEVE-proof character of a PCPV [20], a major domino
effect did not occur in any of the analysed scenario's. Failure of
more than one vessel in one single accident was imaginable only in
cases of an aircraft crash and a BLEVE of either a tanker
compartment, a tank car or a railway car. In view of the limited
number of primary fragments in case of a BLEVE, however, no realistic
scenario could be composed in which all PCPV's would have been
submitted to a catastrophic impact load in one single accident.

4. DESIGN ASPECTS

A major design problem of PCPV's refers to the anchoring of the
prestressing tendons. Depending on the vessel size and design
overpressure the concept of anchoring cables at span ribs, common
practice for cylindrical vessels, soon reaches beyond the limits of
sound engineering practice. Bomhard [20], therefore, has proposed a
concept where the pressure vessel is built up of several sections
which in the first stage of construction can move relative to each
other (Fig. 3). In this stage the prestressing tendons are installed
in a stress-free state. Inside the concrete vessel a liner is
forseen. By blowing up the "liner-vessel" the joints between the
concrete sections grow wider. When the strains exerted in the
prestressing steel have reached their design value the joints are
fixed, e.g. filled with concrete. After hardening of this concrete
the overpressure is relieved and the concrete is put in a stressed
state.

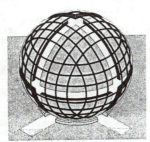

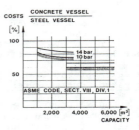

Fig. 3 PCPV: Dywidag-system [20]   Fig. 4 Costs of PCPV's related
                                    to costs of conventional SPV's
                                    (from [20])

5. COST ASPECTS

Several independent studies have revealed PCPV's to be cost competive
with SPV's (e.g. Fig. 4). It is emphasized that extra costs for
enhanced resistance of PCPV's against hazard loads - if required -
are relatively small. This can easily be seen when seperating the

cost factors of PCPV: the major cost factor refers to the prestressing steel, whereas the concrete itself is relatively cheap. Since an increase of impact and fire resistance is obtained mainly by increasing the thickness of the concrete, leaving the amount of prestress constant, the low extra costs for enhanced safety are obvious [8].

6. CONCLUSIONS AND CLOSURE

It was told that at the beginning of this century the streets of New York were so polluted by dung etc. of the many horses in the town, that petrol consuming and smoke producing cars were wellcomed with open arms. The introduction of the car was considered to be the definite solution for the traffic imposed environmental problems of those days. Meanwhile millions of cars, representing tens of millions of horse powers, are responsible for another kind of pollution, certainly not less severe than the former kind. This example is representative for a huge file of problems, including safety problems, inherent to our industrialized world. It teaches us that solutions and concepts, wellcomed in the past, have to be modified or even rejected and replaced by appropriate new ones. For large scale storage facilities for hazardous products this means that in the future due attention must be given to built-in passive safety in order to minimize the vulnerability of these facilities. The results of the probabilistic analysis reported in this paper confirm the expected advantageous features of PCPV's. **PCPV's, because of their high passive safety potential, offer remarkable prospects for control of consequences of industrial accidents. Less vulnerability, hence lower capitalized risks and lower insurance premiums** at even lower **initial costs of construction** constitute a cluster of reasons why adopting for PCPV's for future storage facilities seems to be reasonable. **Function seperation** of constituents, e.d. **prestressing steel for carrying membrane forces** in the operational stage, **a liner to assure tightness and concrete for insulating and shielding** in both the operational and calamity stage, offer the possibility to carry through the **"fitness for purpose"** principle to a high degree and explains why PCPV's are so cost economic.

Acknowledgements: Professor A.C.W.M. Vrouwenvelder of the TU Delft is greatly acknowledged for his kind assistance regarding the probability studies.

7. REFERENCES

[1] KLETZ, T.A.: "Inherently safer plant - the concept, its scope and benefits", Loss Prevention Bulletin, No. 051, pp. 1-8.
[2] WALLER, R.A., et al.: "Low-Probability/High Consequence Risk Analysis". Plemum Press, New York, 1984, 571 p.
[3] MANUELE, F.A.: "One hundred largest losses - A thirty year review of property damage losses in the hydrocarbon-chemical industries", Loss Prevention Bulletin, No. 058, 1984, pp. 1-12.

[4]   WIEKEMA, B.J.: "Vapour cloud explosions – an analysis based on
      accidents", Journal of Hazardous Materials, No. 8, pp. 295-311,
      1984 (Part I); pp. 313-329 (1984) (Part II).
[5]   WICKS, K.M.: "Inherent safety of pressure vessels and systems",
      Loss Prevention Bulletin, No. 053, pp. 7-14.
[6]   HELD, M. , JAGER, E.H., STOLZL, D.: "TNT-Blast equivalence for
      Bursting of pressurized Gas Conventional Vessels", 6th SMIRT
      Conf., Vol. J 10/6, 4 p.
[7]   "FACTS'. TNO – Data Bank, Apeldoorn, The Netherlands.
[8]   BREUGEL,   K.   VAN: "Concrete   and   the   Economy   of   Hazard
      Protection".   Proc.   1st.   Int.   Conf.   on   Concrete   for   Hazard
      Protection, pp. 3-14, Edinburgh, 1987.
[9]   BHOLA,   S.M.:   "The   role   of   inspection   and   certification   of
      chemical plant equipment in the control of accident hazards",
      Proc.   Int.   Symp.   The   Chemical   Industry   after   Bhopal,   IBC
      Technical Service Publication, 1985, pp. 255-269.
[10] LEES, F.A.: "Loss Prevention in the Process Industry", Vol. I,
      II, Butterworth, London, 1980, 1316 p.
[11] RAMLER, J.P.G., TAFFIJN, E.: "Domino-effect voor LPG-opslag-
      tanks" (Domino effect for LPG-storage tanks), TU Delft, Faculty
      of Civil Eng., 1988.
[12] LPG-Integral Study, Ministerie van Volkshuisvesting, Ruimtelijke
      Ordening en Milieubeheer, Voorburg, The Netherlands, 1983.
[13] BREUGEL,   K.   VAN: "Domino   Effects   in   Large   Scale   Storage
      Facilities", IFHP-IULA Symposium Prevention and Containment of
      Large-Scale Industrial Accidents, Rotterdam, 1987, pp. 135-139.
[14] ANSI/ANS-2.12-1978,   "Guidelines   for   combining   natural   and
      external   man   made   hazards   at   power   reactor   sites",   American
      Nuclear Society.
      ANSI/ANS-2.19-1981,   "Guidelines   for   establishing   site-related
      parameters for site selection and design of an independent spent
      fuel storage installation (water pool type), American Nuclear
      Society.
[15] PHILIPS, D.W.: "UK-aircraft crash statistics – 1981 rivision,
      SRD-report R-198, 1981.
[16] WESTINE, P.S.: "Constrained Secondary Fragment Modelling", SWRI
      Final Report for US Army Ballistic Research Laboratory, 1977.
[17] RAMLER, J.P.G.: "Loads and Effects caused by Impact (in Dutch)",
      Delft University of Technology, Delft, 1988, 220 p.
[18] BREUGEL, K. VAN: "Potentialities of Concrete in Storage Systems
      for Liquefied Gases and Hazardous Products", Tenth Int. Congr.
      of the FIP, New Delhi,. Proc. Vol. 4, 1986, pp. 80-98.
[19] TNO-RVO: "Methods for calculation of the physical effects of the
      incidental   release   of   hazardous   materials   (in   Dutch)",
      Directoraat-Generaal van de Arbeid, Voorburg, The Netherlands,
      1979.
[20] BOMHARD, H.: "Concrete Pressure Vessels – The preventive answer
      to the Mexico City LPG disaster", GASTECH '86, Hamburg, pp. 415-
      421.
[21] BUREAU   VERITAS: "Liquefied   gas   storage   installations   under
      atmospheric pressure", Guidance Note, NI 002 CMI, October 1984.
[22] BAKER, W.E. et al.: "Workbook for predicting pressure wave and
      fragment effects of exploding propellant tanks and gas storage
      vessels", NASA CR-1134906, NASA Lewis Center, Nov. 1975.

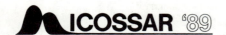
SEISMIC RELIABILITY ANALYSIS
FOR
ELEVATED PRESSURE VESSELS

Albert T. Y. Tung[1]   and   Anne S. Kiremidjian[2]

1. Acting Assistant Professor, The John A. Blume Earthquake Engineering Center,
Department of Civil Engineering, Stanford University, Stanford, California 94305
2. Associate Professor, Department of Civil Engineering, Stanford University,
Stanford, California 94305

ABSTRACT

This paper presents two approaches for assessing the seismic reliability of elevated pressure vessels. A discretized mass mechanical system is used to model the dynamic effects of liquid sloshing. The first approach computes the annual failure probabilities of the structural components at intact state. The second approach obtains the most likely progressive failure paths and the overall system reliability utilizing the system reliability analysis program FAILSF developed at Stanford. The two methods are applied to analyze the seismic performance of an elevated spherical tank of 20m diameter located in the San Francisco Bay Area. The results are compared with the results obtained from a random vibration analysis employing the stochastic equivalent linearization technique.

KEYWORDS

Reliability analysis; system analysis; component analysis; pressure vessels.

1. INTRODUCTION

Elevated pressure vessels are commonly used in refineries and chemical plants to store liquid or gas. Since many of them are located in seismic regions and contain extremely flammable and/or toxic material, their safe performance under seismic load is of critical importance. This paper presents methods for evaluating the seismic reliability of elevated pressure vessels. The methods are applicable to pressure vessels of various shapes although elevated pressure vessels of spherical shapes are the most common and are used for discussion and illustration purposes in this paper.

2. MECHANICAL MODEL

When the content of the vessel is liquid and is not completely full (Figure 1), the effects of liquid sloshing must be accounted for in the dynamic analysis. A two degree-of-freedom mechanical system (Figure 2) can be used to model the forces and moments exerted by the fluid on the tank. The validity of the model's parameters have been substantiated by theoretical and experimental investigations [1]. Neglecting sloshing effects of higher modes and assuming rigid tank, the main parameters are (1) m_c, the convective portion of the total fluid mass m_f; (2) m_i, the inertia mass which is the sum of the structural mass m_s and the stationary portion of the fluid mass; (3) λ_c, the frequency parameter of the convective mass m_c ($\lambda_c^2 = rk_c/gm_c$ where r is the radius of the tank, k_c is the stiffness of m_c and g is the gravitational acceleration). The parameters m_c and λ_c are shown

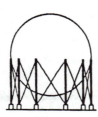

Figure 1. Elevated Spherical
Tank.

Figure 2. Equivalent Mechanical System.

as functions of the ratio h (liquid fill height) over r in Figures 3 and 4 respectively.

Using the seismic hazard curve of the region and the dynamic amplification factor spectrum representative of the local soil condition, the base shear S_B can be computed by means of modal superposition and as a function of the liquid fill height h. The effective mass m_e of the elevated pressure vessel system at a give liquid fill height h can be defined as $m_e = S_B/S_{ai}$ where S_{ai} is the spectral acceleration associated with m_i.

3. COMPONENT RELIABILITY

The resistance R of a structural component is taken as a time invariant independent variable. The induced load L in a structural component is a function of two random variables, the spectral acceleration S_{ai} and the liquid fill height h, such that

$$L = S_{ai}\, s(h) + b(h) \qquad (1)$$

where $s(h)$ is obtained by first performing deterministic structural analysis of the tank system at different liquid fill heights (Figure 5) and then expressing the slopes as a function of h (Figure 6). $b(h)$ is obtained similarly except that it is the intercept rather than the slope of Figure 5.

Assuming lognormal distribution for resistance R [5], the conditional component failure probability at a given pair of S_{ai} and h values is given by

$$
\begin{aligned}
P_c(\text{failure} \mid s_{ai}, h) &= P(R < l \mid s_{ai}, h) \\
&= P((u\,\sigma_{\ln R} + \ln \tilde{m}_R) < \ln l \mid s_{ai}, h) \\
&= P(u < \frac{\ln(s_{ai}\, s(h) + b(h)) - \ln \tilde{m}_R}{\sigma_{\ln R}}) \\
&= \Phi(\frac{\ln(s_{ai}\, s(h) + b(h)) - \ln \tilde{m}_R}{\sigma_{\ln R}})
\end{aligned}
\tag{2}
$$

where $\Phi(\cdot)$ is the cumulative distribution function (CDF) of the standard normal variable; $\sigma_{\ln R} = [\ln((\sigma_R/\mu_R)^2 + 1)]^{1/2}$, $\tilde{m}_R = \mu_R \exp(-\sigma_{\ln R}{}^2/2)$ and μ_R and σ_R are respectively the mean and standard deviation of the component resistance.

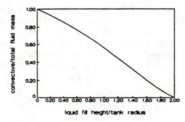

Figure 3. m_c/m_f at Different h/r Ratios.

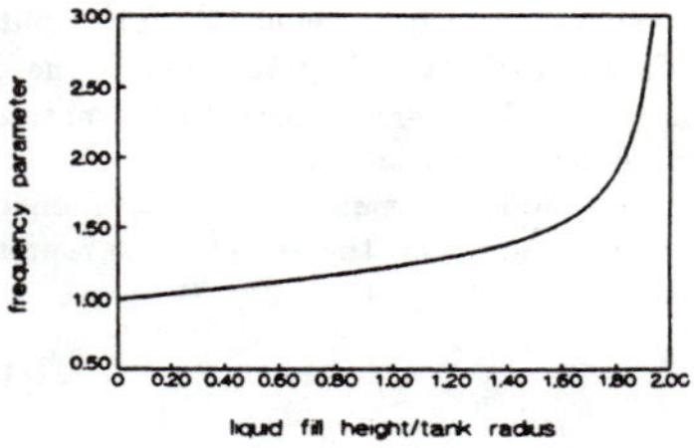

Figure 4. Frequency Parameter, λ_c, as Function of h/r.

With equation 2, the annual probability of failure for an arbitrary component at intact state can be computed as

$$
P_c(\text{failure}) = \int_h \int_{S_{ai}} P_c(\text{failure} \mid S_{ai}, h) f_{S_{ai}|h}(s_{ai} \mid h) f_H(h)\, ds_{ai}\, dh
\tag{3}
$$

where $f_{S_{ai}|h}(s_{ai} \mid h)$ is the conditional probability density function (PDF) of S_{ai} and $f_H(h)$ is the PDF of liquid fill height which may be obtained from past operating records of the vessel.

The conditional PDF of S_{ai} is derived from its complimentary CDF written as

$$
P(S_{ai} > s_a \mid h) = \int_{C_u} \int_{\xi} P(S_{ai} > s_a \mid \xi, h, c_u) f(\xi) f_{C_u}(c_u)\, d\xi\, dc_u
\tag{4}
$$

where $f_\xi(\xi)$ is the PDF of percent critical damping and $f_{C_u}(c_u)$ is the PDF of the uncertainty factor representing the inherent and mathematical modeling errors of S_{ai}.

Expressing the spectral acceleration as the product of peak ground acceleration A_p, the dynamic amplification factor Q_a and including the influence of damping and uncertainty on the value of spectral accelerations, S_{ai} is written as

$$
S_{ai} = Q_a(h)\, (\xi_r/\xi)^{1/2}\, C_u\, A_p
\tag{5}
$$

The conditional probability in equation 4 can then be obtained from the hazard curve, i.e.,

$$P(S_{ai} > s_a \mid \xi, h, c_u) = P(A_p > \frac{s_a}{q_a(h)\,(\xi_r/\xi)^{1/2}\,c_u})$$ (6)

where ξ_r is the average percent critical damping of the supporting frame.

4. SYSTEM RELIABILITY

The system reliability analysis is accomplished by using the reliability analysis program FAILSF developed by the authors. FAILSF is based on FAILUR [6] and SHASYS [7] developed previously at Stanford but with added features and capabilities for considering failure modes such as plastic hinges for structural members and axial, shear and moment failure modes for foundations. The interactive system reliability analysis program FAILSF is applicable for the analysis of general space frame structural systems which satisfy the following requirements: (1) the structure (and the applied loads) can be accurately modeled and analyzed by an finite element program; (2) system failure can be described as combinations of component failures, i.e., the use of minimal cutset representation to define system failure; (3) the failure equation for each component is expressible as a function of independent random variables with known distribution functions. FAILSF uses an element replacement technique to model progressive failure events and the branch and bound algorithm to generate the most likely failure sequences [6,12]. It computes the reliability of the components by a second order approximation of the failure surfaces at the design points while a first order approximation of the the multinormal integral is used to estimate the overall system reliability. FAILSF allows perfectly plastic, perfectly brittle or semi-brittle modes of failure for components. In the analysis, the overall structure is modeled as elastic with components having non-linear post-failure behaviors.

The loading on the structure is represented by the dead load which acts vertically and the live load which acts horizontally. Both are random and are given respectively as

$$D = W\,\{d\}$$ (7)
$$L = S_{ai}\,S'_{ai}\,e(W)\,W\,\{l\}$$ (8)

where W and S_{ai} are envelope PDF's with unit mean for the total weight $(m_c + m_i)g$ and the spectral acceleration respectively; $S'_{ai}(W)$ is the mean spectral acceleration as a function of W; $e(W)$ is the

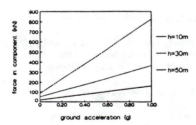

Figure 5. Example of Induced Axial Load in a Structural Component.

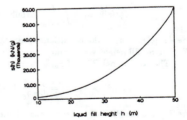

Figure 6. Variation of the Slopes, s, in Figure 5 as a Function of h.

ratio of the effective weight over the total weight of the structure, also a function of W; $\{d\}$ and $\{l\}$ are load patterns for mean vertical and horizontal loads.

At a particular damage state, e.g., after p members have failed, the limit state equation g corresponding to two dimensional failure (e.g., foundation shear) of a member in the next damage state $p+1$ is

$$g(p+1) = R_i$$
$$- \{ [\sum_{k=1}^{n} a_x(n+1,k) R(k) \eta(k) + b_x(p+1,\{l\})] S'_a S''_a(W)$$
$$E(W) W + c_x(p+1,\{d\}) W]^2 + [\sum_{k=1}^{n} a_y(p+1,k) R(k) \eta(k)$$
$$+ b_y(p+1,\{l\}) S'_a S''_a(W) E(W) W$$
$$+ c_y(p+1,\{d\}) W]^2\}^{1/2} \tag{9}$$

For one dimensional failure (e.g., axial load), $a_y(i,j)$, $b_y(i,j)$, and $c_y(i,j)$ in equation 9 above are all zeros. In the same equation, $a_x(i,j)$ is the stress induced in the member at damage state i in the direction x of its member coordinate system by a unit force applied at the node(s) of a failed member or support constraint in damage state j; $b_x(i,\{l\})$ is the stress induced in the member of damage state i in the x direction of its coordinate system by the mean horizontal load defined by load pattern $\{l\}$; $c_x(i,\{d\})$ is similar to $b_x(i,\{l\})$ except that it is induced by the mean vertical load defined by $\{d\}$; $R(i)$ is the resistance capacity of the member of damage state i; $\eta(j)$ is the reduction factor for resistance capacity of the member failed in damage state j ($\eta = 0$ and $\eta = 1$ imply perfectly brittle and perfectly plastic failures respectively).

The above limit state equation is evaluated repeatedly in FAILSF with different coefficients a_i, b_i, and c_i which are computed by the structural analysis module of FAILSF for different components and damage states. Note that the loadings induced in various components are functions of the two random variables S_{ai} and W and if $f_{S_{ai}}(S_{ai} \mid h) = f_{S_{ai}}(S_{ai})$ is assumed, $S'_{ai}(W)$ becomes a constant.

5. RANDOM VIBRATION

To verify the results of the component and system reliability analyses, a random vibration analysis using the recently developed stochastic equivalent linearization technique is performed. The seismic excitation is modeled as a nonstationary (modulated white noise) process [11]. A closed form linearization technique without the use of the Krylov-Bogoliubov assumption, proposed by Wen [13], is employed. Wen's hysteretic restoring force model for nondeteriorating and multi-degree-of-freedom system [3] is applied to the mechanical model shown in Figure 2 (with only m_c having a hysteretic restoring force). A set of recommended hysteretic restoring force model parameter values for steel structures is used [11]. In addition, the initial tangent stiffness and the yield strength of the elevated pressure vessel supporting frame are determined from a static inelastic analysis of the supporting frame. The maximum displacement statistics are derived based on the assumption that the displacement peaks in the time interval (T_1, T_2) can be represented by the Weibull distribution and that the peaks occur indepently and the total number of peaks is large [11]. The threshold exceedence probability is computed from the convolution integral

$$P(D > d) = \int_{h_{min}}^{h_{max}} \int_0^{\infty} P(D > d \mid h, a_p) f_{A_p}(a_p) f_H(h) \, da_p \, dh$$
$$= \int_{h_{min}}^{h_{max}} \int_0^{\infty} (1 - F_{D\mid h, a_p}(d \mid h, a_p)) f_{A_p}(a_p) f_H(h) \, da_p \, dh \tag{10}$$

where $F_{D|h,a_p}(d \mid h, a_p)$ is the maximum peak displacement CDF (Extreme Type I) of the supporting frame and $f_{A_p}(a_p)$ is the PDF of the peak ground acceleration. To incorporate the uncertainties due to the model parameters used in the random vibration analysis, a standard first order approximation is used, i.e.,

$$\text{Var}[\hat{x}] = \sum_i \sum_j \left(\frac{\partial \hat{x}}{\partial p_j} \right)_{\underline{p}^*} \left(\frac{\partial \hat{x}}{\partial p_i} \right)_{\underline{p}^*} \rho_{ij} \, \sigma_{p_i} \, \sigma_{p_j} \tag{11}$$

where p_i is a particular model parameter; ρ_{ij}, is the correlation coefficient between the parameters p_i and p_j; σ_{p_i} and σ_{p_j} are the standard derivations of parameters p_i and p_j respectively; $\underline{p}^*$ is the vector of the mean model parameters; and $\text{Var}[\hat{x}]$ is the variance of the displacement threshold exceedence probability (equation 10).

6. EXAMPLE

A typical liquid containing elevated spherical tank (Figure 1) located in the San Francisco bay area and meeting the current seismic design criteria of the petroleum industry in the U. S. is analyzed for its seismic safety. The dimensions of the tank are given in Table 1. The dynamic amplification factor spectrum for the firm soil of the site is shown in Figures 7.

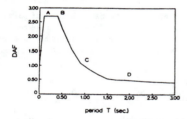

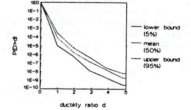

Figure 7. Dynamic Amplifica-   Figure 8.   Exceedence Prob-
tion Factor Spectrum for Firm   ability Versus Ductility Ratio
Soil. u.

For the component analysis, the following failure modes are analyzed: (1) brace buckling, (2) brace yielding, (3) concrete pedestal shear, (4) anchor bolt yielding. The damping ratio ξ is assumed to be normally distributed with mean of 0.05 and COV of 0.12. The uncertainty factor C_u is assumed to be lognormally distributed with unit mean and COV of 0.27. The distribution of liquid fill height is based on the assumption that the volume of the liquid content is uniformly distributed. Since the period (as a function of h) associated with the inertia mass of the tank is confined in a relatively narrow range of 0.09 seconds to 0.38 seconds, the distribution of S_{ai} is taken to be independent of h (see Figure 7) and this simplification is accounted for in the uncertainty factor. $f_{S_{ai}}(s_{ai})$ is obtained by fitting the CDF of S_{ai} with a lognormal distribution which results in a mean of 0.146g and COV of 0.510. Failure modes range from perfectly plastic to perfectly brittle as indicated by the η values in Table 2. The results of the component level analysis are also shown in Table 2.

On the system level, consistent with the component failure analysis, all member resistances are assumed to be lognormally distributed. Also, W is taken to be uniformly distributed and the PDF of S_{ai} is the same as the component analysis but with a mean of unity. The first eight most likely failure paths are used to approximate the overall failure probability. The majority of the most likely failure paths are initiated by brace buckling (7 out of 8). For each failure path, the

occurence probability after 3 to 5 sequential component failures approaches a plateau at which state, the structure has lost approximately 55% of its original (horizontal) stiffness. The overall system reliability index for the tank is 4.322 (P(failure) $= .7737(10)^{-5}$).

Using the Kanai-Tajimi power spectrum density function and a time domain intensity modulation function, the random vibration analysis is performed for four different levels of ground excitation, i.e., $0.25g$, $0.50g$, $0.75g$, and $1.00g$, and at 3m intervals over the liquid fill height range. The main random variables used in the analysis and their statistical properties are given in Table 3. After fitting the CDF (Extreme Type I) to the maximum displacement statistics, the convolution integral (equation 10). is evaluated at five targeted fill heights. A first order approximation (equation 11) is used to evaluate influence of the uncertainties on the exceedence probability $P(D > d)$. Assuming lognormal distribution for $P(D > d)$, the 5% and 95% bounds of $P(D > d)$ are found using the variance of $P(D > d)$ from equation 11. The results are shown in the graph of $P(D > d)$ versus the ductility ratio in Figure 8. Note that the component and system reliability analyses compute the failure probabilities of the components and the overall structure at the first yield of the frame, i.e., at ductility ratio equal to one.

Table 1: Example Elevated Spherical Tank Dimensions.

sphere diameter / min. sphere thickness	20m / 25mm
no. of columns / o. d. / thickness	12 / .915m /12.5mm
bracing dimensions	2L6x6x7/8
liquid fill height, minimum / maximum	3m / 15.25m
anchor bolts, no. per column / diameter	24 / 38mm
concrete pedestal, l x w x h	1.5m x 1.5m x .915m

Table 2: Component Reliability Indices for Spherical Tank in Example.

	brace buckling	brace yielding	pedestal	anchor bolt
η	.3	1.	0.	1.
COV of η	.1	0.	0.	0.
β	4.176	5.183	4.676	$\geq 8.$
P(failure)	$1.486(10)^{-5}$	$1.094(10)^{-7}$	$1.464(10)^{-6}$	$\leq 4.(10)^{-23}$

Note: η is the resistance reduction factor. β is the reliability index.

Table 3: Main Random Variables and Their Statistical Properties of the Random Vibration Analysis.

Parameter	Mean	COV
frame stiffness, k	742520kN/m	0.10
frame strength, f_y	19647kN	0.20
frame damping, ξ	0.05	0.10
ground frequency, ω_g	16.5 rad/sec	0.425
ground damping, ξ_g	0.80	0.426
strong motion duration, t_d	6.9 sec	0.42

7. CONCLUSION

Typically, the pressure vessels and their supporting frames are highly indeterminate structural systems. However, indeterminacy does not imply effective redundancy in the reliability sense. Due to load redistribution, brittle and progressive failure of the components, and the inevitable inclusion

of the weakest components in the most likely failure paths of the structural system, the failure probability of the system's weakest components (e.g., braces) play an influential role in determining the overall reliability of the system. Based on this seismic reliability study of a typical elevated spherical tank, it can be concluded that: (1) added redundancy leads to greater absolute safety but often it does not effectively increase the overall reliability of the system; (2) identifying and strenghtening the weakest component (such as braces) can effectively increase the overall reliability of the system; (3) the failure probability of the weakest component computed using the more cost effective component level reliability analysis provides a reasonable lower bound of the overall system reliability; (4) the component and system reliability analysis results show high degree of consistency with the results of the random vibration analysis. (5) the current seismic design codes for critical industrial structures such as elevated spherical tanks seem to provide an adequate margin of safety.

References

[1] H. N. Abramson. *The Dynamic Behavior of Liquids in Moving Containers*. NASA Special Report 106, NASA, 1966.

[2] Giuliano Augusti, Alessandro Baratta, and Fabio Casciati. *Probabilistic Methods in Structural Engineering*. Chapman and Hall, New York, 1984.

[3] T. T. Baber and Y. K. Wen. Random vibration of hysteretic degrading systems. *Journal of the Engineering Mechanics Division, ASCE*, 107(EM6), 1981.

[4] Stephen H. Crandall, editor. *Random Vibration*. Volume 1, The MIT Press, Cambridge, Massachusetts, 1958.

[5] Bruce Ellingwood and Theodore V. Galambos. Probability-based criteria for structural design. *Structural Safety*, 1, 1982.

[6] Y. Guenard. *Application of System Reliability Analysis to Offshore Structures*. The John A. Blume Earthquake Engineering Center Report 71, Stanford University, 1984.

[7] A. Karamchandani, Y. Guenard, and K. Ortiz. *SHASYS A Software Package for Component and System Reliability Analysis*. The John A. Blume Earthquake Engineering Center Report 78, Stanford University, 1986.

[8] Y. K. Lin. *Probabilistic Theory of Structural Dynamics*. McGraw-Hill, New York, 1967.

[9] H. O. Madsen, S. Krenk, and N. C. Lind. *Methods of Structural Safety*. Prentice-Hall, Englewood Cliffs, New Jersey, 1986.

[10] D. E. Newland. *An Introduction to Random Vibrations and Spectral Analysis*. Longman, New York, 1975.

[11] R. H. Sues, S. T. Mau, and Y. K. Wen. A method for estimation of hysteretic force parameters. *Recent Advances in Engineering Mechanics and Their Impact on Civil Engineering Practice, Engineering Mechanics Division, ASCE*, 1, 1983.

[12] Palle Thoft-Christensen and Yoshisada Morotsu. *Application of Structural System Reliability Theory*. Springer-Verlag, New York, 1986.

[13] Y. K. Wen. Equivalent linearization for hysteretic systems under random excitation. *Journal of Applied Mechanics, ASME*, 47, 1980.

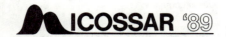
On Reliability of Mechanical Logics and Its Application

SHIBATA, Heki, Professor-Dr.

Institute of Industrial Science,
University of Tokyo
22-1, Roppongi 7, Minato-ku, Tokyo 106 JAPAN

ABSTRACT

Recently, most of logical circuits are understood to consist of electrical components and wiring, however, we are still using various types of mechanical logics at the end of the safety-related system. It is very famous that the Three Mile Island accident occurred and were developed by several causes including the relief valve which was stuck-open.
Such a functional failure of mechanical components is mainly coming from a "self-lock" phenomenon of links in general. This paper deals with the fundamental problems to evaluate the failure or stuck probability of a mechanical link using for safety-related components in NPPs.

KEYWORDS

Mechanical Logics, Valve Sticking, Diesel Engine Failure, Self-lock, Equipment Functional Capacity, TMI Accident, Pilot-operated Relief Valve.

§1   INTRODUCTION

It is very famous that a pilot type relief valve (in Fig.1) didn't close when it was required in Three Mile Island accident. For the safety system, many logical elements are using, and we consider that most of them are electric logical elements. But it should be noticed that various mechanical logical elements are still using even in nuclear power plants. The TMI case is one of

the examples which was enlarged by the functional failure of
mechanical logical components. As the author described in one[1]
of the previous papers, under the strong earthquake occurred in
Tokyo Bay Area in December 1987, many petrochemical processing
plants failed to start their emergency diesel generators for the
demand[2].   A diesel generator consists of many components, both
mechanical and electrical.   Even though malfunction of electrical
components has a responsibility for failing its start as the author
will discuss later, the role of mechanical components including
various kinds of valves is significant for that.
     Recently, ORNL group reviewed the operating experience of
relief valves and block valves[3]. They mentioned that design
details of individual components brought much difference of their
reliability. On the other hand, one[4] of the documents mentioned
only on chattering of electrical relays and malfunction of mercury
switches are significant to plant reliability in their review of
Equipment Functional Capacity. These phenomena are also mechanical
failures in a sense, and the author should mention other types of
mechanical malfunctions caused by responses of electrical
components to earthquake motions[5].
     Many papers[6] refer to the importance of reviewing the plant
specific safety feature. As mentioned before, this "plant
specific" is more critical for mechanical logical components than
electrical components.   The details of electrical components are
more standardized as a mechanism than mechanical components.   He
likes to emphasize the importance of the study on the reliability
of mechanics for a nuclear power plant safety, and very few data on
a specific mechanical component based on its design details.

§2   LOGICAL MECHANISM

     Various kinds of logical mechanisms have been used for the
purpose of interlocking. Electric relays had taken the place of
mechanical logical components between late 1930's and 40's in many
fields.   And electronic devices are now most popular devices to
compose logical relations. However, mechanical devices are still
using in some fields. A kind of relief valves is a good example.
In non-nuclear fields, the logical systems in traffic signal and
brake systems for the railway are famous in this view point. For
brake systems, electric logical systems became popular in late
1950's and electronic one followed to them by ten or fifteen year
later, but most of logics still employ valve-logics in general. A
cam, link, slider, piston and cylinder are mainly using for this
purpose. Hydraulic oil and pressurized air are used for
transmitting force and displacement , and for amplifying them.
Spring and, dash-pot are also significant elements.   But, usually
the effect of inertia force is not used, and a rotating system is
seldom to be used except a cam mechanism.
     A sliding piece is a fundamental form of the logical system.
A set of piston and cylinder is also included in this category.
However, a sliding piece does not transmit a force like a piston.
Its position shows a logical value like a binary code "1" or "0".
A set of pieces expresses a binary code as we know a cylindrical
key mechanism. A sliding mechanism in Fig.2 is one of the simplest
logics by sliding pieces, and used in the railway signal system.
Rocking piece, slider, may be used for the same purpose. More
complicated logics are available by using such mechanisms. If it
is combined with a hydraulic system, more complicated logical
relation can be expressed as the brake system of a train (in

Fig.3).

These examples are static expression of logical value by their positions, but a locking mechanism of a hydraulic snubber is operated by the velocity of its piston. A pilot valve system provides a function of force amplification or breaking the force balance on both sides of sliding piece like a valve disc. Those functions are not so simple to design as recent electronic logical circuits. The mechanism is unique for an each type of items, and it is a representation of the history of its design improvements, or the designers' knowledge.

§3 FAILURE MODES AND MECHANISM OF MECHANICAL LOGICAL COMPONENT

"Sticking" is the main failure mode of these mechamisms. The other mode is "breaking". "Sticking" brings "fixing" a logical variable in a state. "Breaking" brings "free moving" to a piece, that is, a variable becomes free, or uncertain. Sticking is deeply related to the friction coefficient which might be changed by wearing, lacking of lubricant, mixing unintentional powder in lubricant, and so on. It may occur by increasing gaps on sliding parts by wearing like Fig.4. Breaking of a rod may bring also not only free state, but also stuck state. Those states may be induced by external events like an earthquake event, a LOCA event and so on too, as well as their poor design, poor maintenance and aging.

The simplest locking mechanism is explained as follow: Figure 5 indicates the force relation of sliding piece logical mechanism, which was shown in Fig.2. If the resistant force of an inter-connecting bar "i" is f_i, the limit friction coefficient μ_l becomes

$$\mu_l = 1-2\gamma_c \qquad (3.1)$$

where $\gamma_c = f_i/F_a$ is force converting ratio on the sliding mechanism of inter-connecting bar pushed by the bar "A" by the force F_a. The angle of sliding surface is $\pi/4$, or $45°$ is the easiest angle to transmit the force from bar "A" to "i". For this mechanism, the friction coefficient μ_l of sliding surface should be smaller than $(1-2\gamma_c)$ which is decided from the resistant force of bar "i". If bars "A" and "B" are connected to transmit the force from bar "A" to "B" as shown in Fig.2, then F_b is as

$$F_b = \frac{F_a}{4}(1-\mu)^2 \; . \qquad (3.2)$$

Even if $\mu=0$, F_b is reduced to one force of F_a, therefore such system has some limitation without force amplification mechanism. The mechamism in Fig.2 is only a logical mechanism to express $A \cap B = 0$. If so, the failure by sticking may occur in the condition

$$\mu_l \geq 1-2\gamma_c \qquad (3.1a)$$

where $\theta=\pi/4$. Here, we assume that the resisting force of bar "i" comes from the side force by bar "A", then $f_i = \mu_i F_a$, therefore

$$\gamma_c = \mu_i \qquad (3.3)$$

then $\qquad\qquad \mu_l \geq 1-2\mu_i \qquad\qquad\qquad (3.4)$

or $\qquad\qquad \mu_l + 2\mu_i > 1 \qquad\qquad\qquad (3.4a)$

is the condition for sticking.

For the functional failure of such a sliding piece logic system, we may consider several cases:

(i) Even friction coefficients between pieces, and piece and guide, that is, μ_l and μ_i, are small as ordinary operation-basis, the value described in Eg.(3.4a) is larger than unity.

(ii) Any of friction coefficients is catastrophically large as a sliding mechanism.

(iii) Other types of failure like breakage of a bar.

For the stochastic failure evaluation, the procedures are different case by case. For category (i), we can assume the distribution of friction coefficients, but for category (ii), it is the evaluation of probability of occurring such mechanical failure as well as category (iii).

Another example is a slider lock as shown in Fig.3. We consider such a subject for control rod modeling. This case is free-falling of a block in a cylinder. If this is a slider with gap 2δ, then the stuck limit for a friction coefficient

$$\mu_l \geq 2\delta/l \tag{3.5}$$

It seems to be very simple. But our feeling is slightly different from this result for a lubricated slider in slot, that is, the smaller gap is smoother and no trouble. It comes from the effect of the lubricating oil film.

One of the important subject in this field is "self-lock". Three joint rod mechanism as shown in Fig.6, can resist to the buckling-type instability, if there is some amount of friction at their joints. A classical single cylinder reciprocal engine was faced to this problem when the geometrical relation of elements became a three joint straight. Such a relation is called as "self-lock", and often used for mechanical lock systems as Fig.6. Without rotating as " " at A in Fig.6, no way to move down the point B. The friction coefficient of these three hinges is related to its stability. For practical use of this mechanism, an adequate amount of friction is necessary, otherwise it shall be unstable. But if it exceeds some limit it is very slow to start to move by the axial force F , even if the link $\overline{AC}$ rotates over the limit angle at the hinge A. Under a simple assumption,

$$\mu \geq \delta/r \tag{3.6}$$

is given for the critical friction coefficient of the bearing of at each hinge, where r is the radius of journal of the bearing and δ is the off-set at the hinge C. The range of "self-locking" is very narrow for the ordinary condition of the journals. But we need larger off-set to start the link, if the friction coefficient is larger. This relation between gap and friction coefficient caused by aging effect is the similar to a strength-load relation in the reliability analysis. Self-lock type mechanisms are often used for a closure like air-lock, valve and so on.

§4   VALVE FAILURE

Already mentioned, in the case of Three Mile Island the relief valve didn't closed. That valve is made by "Dresser Industries", and its structure is as shown in Fig.1. When it closes by inactivating solenoid, this state is kept by two springs "a" and "b" in the figure. If the solenoid is activated, then lower

spindle of the pilot valve move to open. The steam flow causes the
pressure difference at the both sides of the main valve disc , and
the disc moves to open. If the solenoid will be inactivated, the
both spindles will become free and set back by the spring "b", and
the main disc will be pushed back by the main spring "a". This
means all closing actions are activated by both spring forces. As
known well, "direct repeater" of the action or state of the main
disc of this valve is not provided, and only the position of the
solenoid (plunger) is detected by a switch. As far as shown in the
figure, which was submitted by Dresser Industries, the main spring
"a" seems to be not so strong to the size of the main disc. It may
be probable that the disc will not be able to set back to the
closed position, if there is additional friction force. "Open"
state was continued by sticking of either the main disc or pilot-
spindle (valve). A balance between friction force and restoring
force by spring brings two states "open" or "close", that is,
"false" or "true" for "close" signal in logical sense. Therefore,
the design of restoring force by a spring should cover the increase
of friction coefficient caused by mixing with any powder to
lubricant where may occur often or by surface degrading.

The report[3] from ONL discussed on such subjects based on
their survey. Pilot-operated relief valves are less reliable
compare to air-operated, spring loaded valves. A pilot-operated
valve has two logical operations in series, and there is no reason
to decrease the chance of failure as a whole. And, even the types
are same, their failure rate is different manufacture by
manufacture as the reflection of its detailed mechanisms. The
table quoted in this paper as Table 1 is shown this fact. Another
table (Table 2) shows "failure" is easier to occur in control parts
than mechanical parts. In this case "control failure" includes
controller malfunctions and air/nitrogen actuating pressure loss.

§5   EMERGENCY DIESEL GENERATOR

As the author reported[1], the failure rate for starting by
the demand was very high at the earthquake-1987 in petrochemical
processing plants in Tokyo-bay Area. Also, Nuclear Power Safety
Information Research Center in Japan reported on the failure of
emergency diesel generators on demand and during operation[7]. In
this document, rates of failures and fatal failures of EDG in Japan
are shown. Frequency of failures of starting on demand is
0.45/1000 demands in Japan against 2.9/1000 demands in the U.S..
All fatal failures in Japan (Table 3), 9 events per 432.2 EDG-year,
came from the malfunctions of electronic or electric parts.
However, their failures were related to their mechanical troubles
mainly.

A manufacturer of large marine diesel engines pointed out that
there are some possibilities of failing to start by malfunctions
caused by sticking of pistons used for mechanical interlocks or by
operating of starting air distribution valves incorrect-orderly.
One of nine events above-mentioned included the malfunction of the
main solenoid valve for starting air is a typical mechanical
failure. We should consider that there is a possibility of
mechanical failure or malfunction of the valve mechamism.

§6   EARTHQUAKE AND LOGICAL FAILURE

Various items, which are employing such a logical mechanism,
were tested on a shaking table under some seismic conditions. As a
typical example, the simulation of control rod behaviors for PWR

plants showed that the increase of frictional force at their guide tube by inertia force caused the delay of its insertion by seismic accelerations. And some test results showed such a tendency, but almost no possibility of occurring their sticking by earthquake motions. Also there is a possibility of delaying the action of valves by acceleration either in the axial direction or transversal direction. We are going to test a diesel engine-generator system on the 1000 ton Tadotsu shaking table, NUPEC, next year. This test will include the starting test under seismic conditions. One of the subjects is that "failing-the-start" caused by the disorder of the sequence air-control valve opening. Therefore, some additional tests on a control valve alone may be necessary. Because, failing-the-start of ship diesel engines are not seldom by this reason.

§7   CONCLUDING REMARKS

We are facing the trouble of sticking some mechanisms as daily incidents. This type of trouble in the mechanical engineering field is similar to trouble caused by disconnection of electric circuits in the electric and electronic engineering field. And the probability of occurrence is governed by the details of their design.

Also the functional failure of mechanical logical components is related to the conditions of maintenance, aging and so on. We treat the failure probability of such items as a global one, but the author considers that we must analyze on such failures in the mechanical engineering view point.

After completing to write this paper, the author was received a book "The Reliability of Mechanical Systems"[8]. Most of discussions in the book is PRA type approach. However, the pilot operated valve is analyzed parts by parts. This practice is connecting to the author's view-point.

§8   REFERENCES

1)   SHIBATA, H.: A Basic Research of Plant Operability during Seismic Condition and PRA Study-Seismic Trigger System, *Proc. of 3rd International Topical Meeting on Nuclear Power Plant Thermal-Hydraulics & Operation*, Seoul, Vol.II (Nov. 1988).
2)   Ministry of International Trade and Industry: Report on Effect of Off-Bohso Peninsula (Chiba-ken Toho-oki), Earthquake-1987 to Chemical Eng'g. Plants, KHK (Dec.-1988) 30pp.+Appendix (in Japanese)
3)   MURPHY, G.A. and CLETCHER, J.W. Jr.: Operating Experience Review of Failures of Power Operated Relief Valves and Block Valves in Nuclear Power Plant, NUREG/CR-4692, ORNL/NOAC-233 (Oct. 1987) 75pp..
4)   ANDERSON, N.R. and CHANG, T.Y.: Regulatory Analysis for Resolution of Unresolved Safety Issue A-46, Seismic Qualification of Equipment in Operating Plants, USNRC, NUREG-1211 (Feb. 1987) p.11.
5)   SHIBATA, H. and others: Malfunction of Edi-Current Type Relay under Some Earthquake Conditions, *Bull. of ERS*, Inst. of Ind. Sci., Univ. of Tokyo, No. 15 (March 1982) p.97.
6)   AMICO, P.J.: An Approach to the Qualification of Seismic Margins in Nuclear Power Plants; The Importance of BWR Plant Systems and Functions to Seismic Margins, NUREG/CR-5076, UCRL-15985 (May 1988) p.4 and others.
7)   NII, K.: Failure of Emergency Diesel Generator - on Demand and

in Operation, Nuclear Power Safety Information Research Center, *Proc. for National Symposium* on PSA, IAE-R8807 (Dec. 1988) p.160. 8)   DAVIDSON, J. *ed.*: The Reliability of Mechanical systems, IMechE (1988) Mechanical Eng'g. Pub. Ltd., p.73 and p.113.

Table 1 Pilot-operated Relief Valve; Mechanical
Failures (excludes design events) (Ref.[2])

Manufacturer	Number of Events		
	Failure	Degraded	Total
Crosby (p)	2	5	7
Dresser (p)	8	25	33
Garrett (p)	0	5	5
Copes-Vulcan (a)	3	25	28
Masoneilan (a)	3	7	10
Control components (a)	0	2	2
Unknown	8	8	16
Total	24	77	101

(p):pilot-operated
(a):air-operated (spring close)

Table 2   Number of Events; Pilot-operated
Relief Valve and Block Valve (Ref.[2])

	Failures	Degraded	Total
PORV mechanical	24	77	101
control	61	30	91
design	0	6	6
PORV events total	85	113	198
BV events	15	17	32
Total	100	130	230

Table 3   Failed Components; Critical Failure of
Emergency Diesel Generator (Ref.[6])

Item	Number
Revolution Detector	2
D/D Converter	1
Circuit Breaker	1
Generator Protector	1
Inverter	1
Fuse for Instrument Transformer	1
Trip Mechanism	1
Main Stater Solenoid Valve	1

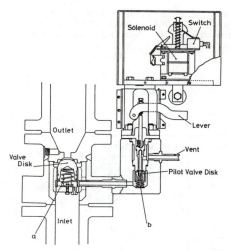

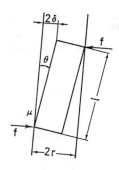

Fig. 1 Schematic Drawing of an Example
of Pilot-operated Relief Valve
(Modified from the figure prepared by
Dresser Industries for Ref.[3])

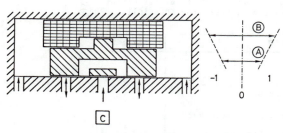

Fig. 4 Sticking of Piece in Guide

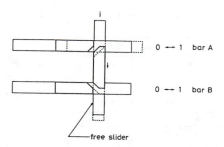

Fig. 2 An Example of Mechanical Logic

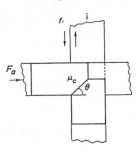

Fig. 5 Mechanics of Sliding
Piece-type

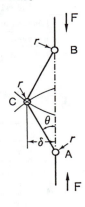

Fig. 3 Hydraulic-type Mechanical Logic

Fig. 6 Mechanics of Self-
lock Mechanism

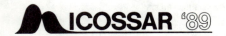
RECENT SEISMIC RISK STUDIES OF
NUCLEAR POWER PLANTS: AN OVERVIEW

Mayasandra K. Ravindra,
EQE Engineering, Inc.
Costa Mesa, California

Abstract

Seismic probabilistic risk assessments (PRA) are
conducted to estimate the frequencies of
occurrence of severe core damage, serious
radiological releases, and consequences in terms
of early fatalities, long term adverse health
effects and property damage, and to identify
significant contributors to plant risks so that
review and upgrading may focus on them.   This
paper provides an overview of the seismic PRA
methodology, describes some of the recent PRA
applications and discusses the role of structural
and geotechnical engineers in the nuclear power
plant PRA.   It discusses the seismic safety issues
that a PRA can address, the available data/models
and the limitations.   It also highlights the
methodological differences in the system
reliability analysis conducted by the seismic PRA
analysts and the structural system reliability
analysts.

Keywords

Nuclear Power Plant, Seismic Risk, Fragility,
Hazard, Probabilistic Risk Assessment, System
Reliability

1. Introduction

The objectives of seismic PRA of a nuclear power plant are to
estimate the frequencies of occurrence of severe core damage,
serious radiological releases, and consequences in terms of early
fatalities, long term adverse health effects and property damage,
and to identify significant contributors to plant risks. The key
elements of a seismic PRA are: seismic hazard analysis, seismic
fragility evaluation, systems/accident sequence analysis and
evaluation of core damage frequency and public risk.

The procedures for conducting the seismic PRA are described in the
PRA Procedures Guide [1]. Since the role of the structural and
geotechnical engineers is mainly in the seismic fragility
evaluation of structures and equipment, the methods used to
estimate the seismic fragilities are discussed below.

2. Seismic Fragility Evaluation

The methodology for evaluating seismic fragilities of structures
and equipment is documented in [2,3]. Seismic fragility of a
structure or equipment item is defined as the conditional
probability of its failure at a given value of the seismic input
or response parameter (e.g., ground acceleration, stress, moment,
or spectral acceleration). Seismic fragilities are needed in a
PRA to estimate the conditional probabilities of occurrence of
initiating events (i.e., large LOCA, small LOCA, RPV rupture) and
the conditional failure probabilities of different mitigating
systems (e.g., safety injection system, residual heat removal
system, and containment spray system). The objective of fragility
evaluation is to
estimate the ground
acceleration capa-
city of a given
component. This
capacity is gene-
rally defined as
the peak ground
motion acceleration
value at which the
seismic response of
a given component
located at a speci-
fied point in the
structure exceeds
the component's
resistance capa-
city, resulting in
its failure. The
ground acceleration
capacity of the
component is estim-

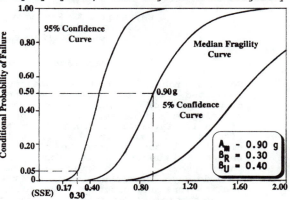

Figure 1: Example Fragility Curves

ated using information on plant design bases, responses calculated
at the design analysis stage, as-built dimensions, and material
properties. Because there are many variables in the estimation of
this ground acceleration capacity, component fragility is

described by a family of fragility curves; a probability value is assigned to each curve to reflect the uncertainty in the fragility estimation (Figure 1).

In the seismic PRAs performed so far, a simple fragility model has been used.   In this model, the family of fragility curves for a structure or equipment corresponding to a particular failure mode is derived using three parameters: A_m, the best estimate of the median ground acceleration capacity, β_R and β_U, the logarithmic standard deviations representing randomness in capacity and uncertainty in median capacity, respectively.   These fragility parameters are estimated for the credible failure modes of the structure or equipment.   Failure modes of structures may include excessive inelastic deformation, partial collapse, complete collapse, bearing failure and base slab uplift.   Table 1 lists the fragility parameters for structures and equipment in a typical nuclear power plant [4].   Failure modes of equipment may include structural failure modes (e.g., bending, buckling of supports and anchor bolt pullout), functional failures (e.g., binding of valve stem and excessive deflection), and electric relay trip and chatter.   Soil failure modes also include slope instability and liquefaction.

3. Recent Applications

Seismic PRAs have been conducted for over 25 nuclear power plants in the US, Europe and Taiwan.   The methodology has evolved in the last ten years.   Recent emphasis in the seismic PRAs has been on detailed review and walkdown of the plant structures and equipment, use of insights obtained from past PRAs, development of simplified methods, and detailed systems modeling.   Plant walkdown is performed to confirm the design details shown on plant and equipment drawings, to verify the adequacy of anchorage and seismic supports for equipment, to identify potential for nonseismic designed objects falling in an earthquake on critical items of equipment, and to obtain measurements on equipment and structures not readily seen on drawings.

The insights obtained from seismic PRAs are documented in [5,6]. Table 2 shows the results of seismic PRAs that have been published.   It is seen that seismic contribution to core damage frequency is not insignificant; however, the seismic contribution to frequencies of early fatalities and late fatalities dominate in some plants.   These studies have shown that only plant-specific analysis will be able to identify weak link elements and plant-unique features which have significant impact on seismic safety of a plant.   A compilation of significant contributors to seismic risk as observed in past seismic PRAs has been done so that future PRAs can focus on them [6].   The plant walkdown procedures are developed based on these insights.

Recent applications include the seismic risk analysis of spent fuel pools [7], development and applications of simplified seismic risk analysis procedures [8] and development of seismic margin review methodologies [5, 9].

Table 1
Fragilities of Key Structures and Equipment - Millstone - 3

Structure/Equipment	A_m (g)	β_R	β_U
1. Loss of Offsite Power	0.20	0.20	0.25
2. Emergency Gen. Encl. Bldg.	0.88	0.20	0.46
3. Refueling Water Storage Tank	0.88	0.30	0.36
4. Emergency Diesel Generator	0.91	0.24	0.43
5. Control Building Collapse	1.00	0.24	0.33
6. Service Water Pumphouse Failure	1.30	0.24	0.49
7. Service Water System Pumps	2.40	0.31	0.53

Table 2
Mean Annual Frequencies of Accidents

Plant	Severe Core Damage		Early Fatalities		Late Fatalities	
	Overall	Seismic	Overall	Seismic	Overall	Seismic
Limerick	2.4E-05	5.9E-06	6.9E-07	5.3E-07	4.2E-05	3.2E-06
Indian Point 2	1.4E-04	2.0E-05	3.7E-07	7.7E-09	5.2E-05	7.7E-06
Indian Point 3	1.9E-04	5.7E-06	5.8E-07	2.6E-09	1.1E-05	6.7E-06
Millstone 3	5.9E-05	8.9E-06	2.1E-06	1.1E-07	8.7E-06	8.1E-06
Oconee 3	2.5E-04	6.2E-05	2.2E-06	1.2E-06	1.7E-04	6.3E-05
Seabrook	2.3E-04	2.3E-05	2.4E-06	5.7E-07	1.7E-04	2.4E-05
Zion	5.2E-05	5.7E-06	1.0E-06	--	5.8E-06	5.6E-06

In the analysis of spent fuel pools [7], it was found that the dominant contributor to seismic risk is a gross structural failure of the spent fuel pool resulting in the spent fuel becoming uncovered with a potential for fire and release of significant amounts of radioactivity. However, the study showed this risk is negligibly small for representative plants.

Based on the experience gained in conducting several seismic PRAs, a simplified seismic PRA methodology [8] has been developed which captures the salient features of the full scope analysis in a cost-effective manner. This methodology has been applied in the NRC-sponsored Accident Sequence Evaluation Program Reactor Risk Reference Document reevaluations often referred to as the NUREG-1150 program. It has been applied in the seismic PRA of Surry and Peach Bottom, a Pressurized Water Reactor and a Boiling Water Reactor respectively.

Using the results and insights from published seismic PRAs, the data from seismic analyses and qualification tests, and the experience on performance of structures and equipment in major earthquakes around the world, procedures for assessing the realistic seismic margins beyond the design levels of existing plants have been developed in the last five years [5,9] and applied on a trial basis to the Maine Yankee Atomic Power Station [10] and Catawba Nuclear Power Plant [11].

4. Seismic Issues

Current emphasis in the nuclear industry is on reevaluation of operating nuclear power plants. The issues that have been identified in the seismic area are:

Unresolved Safety Issues such as the Seismic Qualification of Equipment in Operating Plants (A-46), Spatial Seismic Interaction (A-17), and Seismic Adequacy of Decay Heat Removal System (A-45);

Individual Plant Examination for External Events (IPEEE) to search for plant vulnerabilities;

Seismic margins beyond design basis of existing plants;

Resolution of the Charleston Earthquake and Eastern U.S. Seismicity Issue.

Seismic PRAs could be used to some extent in obtaining cost effective solutions to these issues. Since a detailed walkdown will be performed under A-46 resolution on most plants, it is beneficial to combine the seismic PRA and A-46 walkdown resulting in substantial cost savings.

5. Data For Fragility Evaluation

Three sources of data have become available in the last few years to supplement analytical methods used in the fragility evaluation. Fragility tests where in the equipment is subjected to earthquake excitation levels leading to failure of the equipment have been done for selected equipment items [12]; in many instances, equipment vendors have tested the equipment beyond the qualification levels to actual failure levels [13]. There has also been a concerted effort to collect seismic qualification test data on various types of equipment which has resulted in establishing lower bound seismic capacities of equipment [14]. Another major source of data has been the performance of equipment in industrial and power plant facilities in actual earthquakes around the world [15]. These earthquakes are in many instances larger than the design bases of the plants in the eastern U.S. The equipment are similar and in many instances identical to those existing in nuclear power plants. The experience database could also be used to establish lower bounds on seismic capacities of equipment.

6. Structural System Reliability Considerations

Some major differences exist between the nuclear plant seismic PRAs and the structural system reliability analysis. The quantification of seismic risk is somewhat unique to the seismic PRA which recognizes the capability of an earthquake to affect a large number of components in the plant and to compromise the redundancy of systems and components essential for the safety of the plant. The system considerations are also included in the quantification by describing the accident sequences in terms of the component failures wherein the redundancies in the plant systems are explicitly accounted for. Another feature of the quantification is the uncertainty propagation at all levels of analysis - seismic hazard uncertainty is expressed in terms of a number of hazard curves and the family of fragility curves display the uncertainty in the fragility. This aspect of uncertainty analysis has not received much attention in the reliability analysis of conventional structures. Other differences exist in the definition of systems, component reliability models, failure modes or limit states, system reliability models and probability interpretation as discussed in [16].

7. Problem Areas

The uncertainties in seismic risk estimates are seen to be two to three orders of magnitude; the major source contributing to this uncertainty is the uncertainty in the seismic hazard. Through extensive research currently undertaken by the NRC and EPRI, site-specific seismic hazard curves are becoming available for all nuclear power plant sites in the United States. Admittedly, these hazard curves display wide uncertainties. They do however represent the collective opinions of the best experts in the field.

Other areas requiring attention in future include consideration of gross design and construction errors and operator response in the stressful condition following an earthquake and the validation of seismic PRAs. Some research has been performed in these areas [17, 18, 19].

8. Conclusions

This paper has provided an overview of seismic PRA studies of nuclear power plants. Seismic risk analysis has matured over the years. Recent developments that have contributed to this are the seismic hazard characterization, collection of fragility, experience and qualification test data, recognition of the importance of detailed walkdown, and greater attention paid to systems modeling. It is expected that seismic risk analysis would continued to be used in resolving seismic safety issues in nuclear power plants.

References

[1]   PRA Procedures Guide. NUREG/CR-2300,   Vol 2, 1983.

[2]   Ravindra, M. K., and R. P. Kennedy.   "Lessons Learned from Seismic PRA Studies."   Paper M6/4.   In Proceedings of the Seventh Conference on Structural Mechanics in Reactor Technology, Chicago, Illinois, August 1983.

[3]   Kennedy R.P. and M.K.Ravindra, "Seismic Fragilities for Nuclear Power Plant Risk Studies", Nuclear Engineering and Design, Vol. 79, No. 1 pp 47-68 May 1984.

[4]   Ravindra, M.K., Sues, R.H., Kennedy, R.P., and Wesley, D.A., "A Program to Determine the Capability of the Millstone 3 Nuclear Power Plant to withstand Seismic Excitation above the Deign Basis SSE", Prepared for Northeast Utilities, NTS/SMA 20601.01-R2 November 1984.

[5]   Budnitz R.J., et al "An Approach to the Quantification of Seismic Margins in Nuclear Power Plants", NUREG/CR-4334, UCID-20444, August 1985.

[6]   Ravindra, M.K., "Seismic Probabilistic Risk Assessment and Its Impact on Margin Studies", Nuclear Engineering and Design, 107(1988) 51-59.

[7]   Prassinos P.G., et al. "Seismic Failure and Cask Drop Analyses of the Spent Fuel Pools at Two Representative Nuclear Power Plants", NUREG/CR -5176, UCID-21425, Lawrence Livermore National Laboratory, Livermore, California, June 1988.

[8]   Bohn, M.P. and Lambright, J.A., "Recommended Procedures for the Simplified External Event Risk Analyses for NUREG 1150", Draft Report prepared for USNRC by Sandia National Laboratories, Albuquerque, New Mexico, January 1988.

[9]   NTS Engineering, et al. "Evaluation of Nuclear Power Plant Seismic Margin", NTS Engineering, RPK Structural Mechanics Consulting, Pickard Lowe and Garrick and Woodward Clyde Consultants for EPRI, August 1987.

[10] Prassinos, P.G., R.C. Murray, and G.E. Cummings, "Seismic Margin Review of the Maine Yankee Atomic Power Station - Summary Report", NUREG/CR-4826, UCID-20948, Vol.1 March 1987.

[11] NTS Engineering, et al. "Seismic Margin Assessment of the Catawba Nuclear Station", Volumes 1 and 2, Prepared for the Electric Power Research Institute, Palo Alto, California, August 1988.

[12] Holman, G.S. et al. "Component Fragility Research Program - Phase I Demonstration Tests - Volume 1: Summary Report", Lawrence Livermore National Laboratory Report, Livermore, California, May 1986.

[13] Bandyopadhyay, K. et al. "Seismic Fragility of Nuclear Power Plant Components", NUREG/CR-4659, 2 Volumes, December 1987.

[14] ANCO Engineers, "Generic Seismic Ruggedness of Nuclear Plant Equipment", EPRI NP-5223, Electric Power Research Institute, Palo Alto, California, May 1987.

[15] EQE Inc., "Summary of the Seismic Adequacy of Twenty Classes of Equipment Required for Safe Shutdown of Nuclear Power Plants", Prepared for the Seismic Qualification Utilities Group, February, 1987.

[16] Ravindra M.K., "System Reliability Considerations in Nuclear Power Plant PRAs", Proceedings of NSF Workshop on Research Needs for Applications of System Reliability Concepts and Techniques in Structural Analysis, Design and Optimization, Boulder, Colorado, September 1988.

[17] Ravindra M.K. "Treatment of Gross Errors in Nuclear Plant Risk Studies- A Suggested Approach", in Structural Safety and Reliability - Proceedings of ICOSSAR'85. Edited by I. Konishi, A.H.-S. Ang, and M.Shinozuka, Kobe, Japan, May 1985.

[18] Ravindra M.K., "Sensitivity of Seismic Risk Estimates to Design and Construction Errors", Prepared for the U.S. Department of Commerce, National Bureau of Standards, Gaithersburg, Maryland, September 1987.

[19] Ellingwood, B., "Validation Studies of Seismic PRAs", Proceedings of the Second Symposium on Current Issues related to Nuclear Power Plant Structures, Equipment and Piping, Electric Power Research Institute, Orlando, Florida, December 1988.

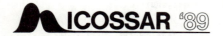 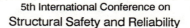
COMPARISON OF A PRACTICAL FRAGILITY ANALYSIS OF A REACTOR BUILDING
WITH NON-LINEAR MONTE CARLO SIMULATION

Toshiyuki Komura*, Masatoshi Takeda*, Tadashi Ando** and Atsushi Onouchi**

* Ohsaki Research Institute, Inc. c/o Nuclear Power Division
  Shimizu Corporation 13-16, Mita 3 Chome Minato-ku, Tokyo 108, Japan
**Chubu Electric Power Co., Inc., Civil & Architectural Engineering
  Department,1, Toshin-cho, Higashi-ku, Nagoya 461-91, Japan

ABSTRACT

There are several conventional methods to assess seismic safety of
nuclear power plants practically. In these methods, non-linear effect
of the structural elements is considered with an inelastic energy
absorption factor $F\mu$. However, since the effect of more realistic non-
linearity of a multi-degree-of-freedom system, such as the phenomenon
of damage concentration, is not considered in the conventionally
estimated factor $F\mu$, it is insufficient to apply the factor to such a
complex system. In this respect, one of the authors has proposed a
method extending the factor to include the realistic non-linear effect
for simple systems.

In this paper, in order to examine the applicability of the
proposed method for complex systems, the fragility analysis of a
typical BWR-type reactor building is performed by using the proposed
method and the Monte Carlo simulation. Through the comparison of these
results, it is shown that the proposed method is sufficiently
applicable, in a practical sense, even to such complex systems.

KEYWORDS

nuclear reactor building; fragility analysis; proposed method; Monte
Carlo simulation; complex system; damage concentration

1. INTRODUCTION

Aseismic design is very important for structural design of nuclear power
plants in Japan. Earthquake response analyses, which are performed in design
stage, use analytical models considering many different effects in detail
according to the seismic condition of construction sites, such as the soil-
structure interactions(SSI). Since many variabilities (e.g. soil material
properties) are involved in the structural response, aseismic design is
conducted conservatively. As a result it may lead to an overdesign against
earthquakes. In this sense, the probabilistic approach is very effective not
only for assessing seismic safety, but for avoiding excessive conservatism.

There are several conventional methods to estimate seismic safety of nuclear power plants[1-6] probabilistically. A safety factor method, the so-called "Zion Method", is well-known as one of the conventional methods. In this method, conservatism included in the design procedures is estimated as a "safety factor" which is a random variable defined as the ratio of actual seismic capacity to actual response. In calculating the failure probabilities of structural elements, such as shear walls, the effects of non-linear response can be considered with an inelastic energy absorption factor Fμ by assuming the response to be linear to various acceleration levels. As shown in Fig. 1, the factor Fμ expresses the effect of non-linearity by enlarging the ultimate strength of structural elements. This treatment is effective to calculate the fragility practically. However, it is insufficient to evaluate multi-degree-of-freedom (MDOF) systems, since the effect of more realistic non-linearity is not considered in the factor Fμ in complex systems, such as shear force redistribution or concentration of non-linear deformation, generally known as the phenomenon of damage concentration.

In addition, it is needed to consider the effects of SSI in evaluating Fμ based on the ground condition of the construction site. For the reasons mentioned above, we found it difficult to apply the conventional method directly. One of the authors has proposed a method extending the factor Fμ considering the effect of MDOF for simple systems[7]. However, its applicability to complex systems with SSI effects has not yet be confirmed. Thus in this paper, we proposed a new approach which refines the conventional method to evaluate the effects of SSI as well as non-linearity of MDOF systems.

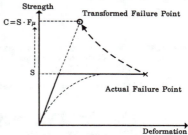

Fig. 1 Conception of inelastic energy absorption factor

The fragility analysis of a typical BWR-type reactor building was performed to examine the applicability of the proposed method to a complex system. The reactor building is modeled as a multi-stick-lumped-mass model with SSI effect. At the same time the Monte Carlo simulation(MCS) of non-linear dynamic response analysis was performed to compare its result with that by the proposed method.

2.FRAGILITY ANALYSIS BY PROPOSED METHOD

2.1 Proposed Method

The proposed method has two features. One is the way to evaluate the seismic response realistically considering only SSI effects in detail. The other is to evaluate the effects of both non-linear response in MDOF systems and SSI on the factor Fμ, unlike the conventional method. The proposed method assumes that all of the random variables are statistically independent and follow the log-normal distribution.

Seismic response of a nuclear reactor building varies due to the variability associated with (a) spectral shape of ground motion, (b) soil material properties, such as shear wave velocity and damping factor and (c) structural material

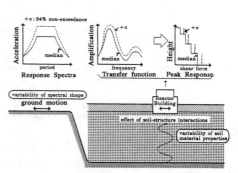

Fig. 2 Evaluation for realistic response

properties, such as the compressive strength of concrete and the yield strength of steel. Therefore, in order to evaluate the realistic response of a structural element (i.e. shear wall) such a response analysis which is illustrated in Fig. 2 is performed considering the above-mentioned variability.

In detail, the structural response is calculated in the following way. First, using the log-normal distribution assumption, the probabilistic characteristics of response in a frequency domain, which are expressed in terms of the median and the logarithmic standard deviation of the transfer function, are calculated considering the variability of (b) and (c). In this analysis, a realistic SSI model is used in accordance with the seismic condition of the model plant. Next, a realistic response R' is calculated by applying the probabilistic method proposed by Romo-Organista assuming the stationarity of the random process [8].

Here, an actual response R may be formatted as follows:

$$R = R' \cdot F_{SP} \cdot F_{ME} \cdot F_{PM} \tag{1}$$

where, F_{SP}, F_{ME} and F_{PM} are assumed to be log-normal random variables, which represent the variability of spectral shape of ground motions, uncertainty associated with the analytical model and analytical method used for the estimation of R', respectively. F_{ME} may be estimated by studying the difference between observed and calculated response values. The factor F_{PM} is needed especially when a deterministic analytical method is used instead of the Monte Carlo simulation. On the contrary, if the best estimation method, such as the best MCS, is used, this factor would not be necessary. In this study, we assume the median value of R' obtained from the probabilistic method to be equal approximately to that from the best MCS. For this reason, the factor F_{PM} is assumed to have a unit median.

Seismic capacity of a shear wall changes depending on the variability of both (d) the ultimate strength of the shear walls and (e) the ultimate deformation. In a case of the reinforced concrete structure, like an ordinary reactor building, the variability is evaluated mainly from the variability in the compressive strength of concrete and the yield strength of steel. In addition to the inherent variability, the evaluated strength contains systematic uncertainty due to the analytical assumption or estimation error. Therefore, in order to evaluate the actual capacity, the variability is considered based on the available experimental data.

Here, the actual seismic capacity C of a shear wall is calculated as follows:

$$C = S \cdot F\mu \tag{2}$$

Like in the conventional method, non-linear effect is expressed as an inelastic energy absorption factor $F\mu$ as follows:

$$F\mu = \sqrt{2\mu_u - 1} \cdot \varepsilon \tag{3}$$

where, μ_u is the ductility ratio corresponding to the ultimate deformation, and the factor ε is a log-normal random variable which denotes the uncertainty associated with the use of this equation.

In this method, the equivalent ductility ratio μ_e is introduced to replace μ_u, which is defined on the perfect elasto-plastic skelton curve, as illustrated in Fig. 3. The reason for the use of μ_e is that the equation (3) is based on the dynamic performance of the elasto-plastic system. From a result of evaluating ultimate ductility based on the available experimental data, median values and logarithmic standard deviations of μ_u are obtained,

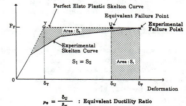

$$\mu_e = \frac{\delta_U}{\delta_Y} \quad : \text{Equivalent Ductility Ratio}$$

Fig. 3   Estimation of ultimate deformation by equivalent ratio

such as shown in Table 1.

The factor ε, which is called a "modification factor" in the proposed method, represents the effect of both non-linearity of MDOF system and SSI on Fμ in such a way as shown in Fig. 4. In this proposed approach, at least one non-linear response analysis is required to estimate the effect of non-linearity in MDOF system, such as the concentration of non-linear deformation, i.e., so-called the phenomenon of damage concentration. At the same time, a linear response analysis is required to calculate the modification factor ε by the following equation:

Table 1   Evaluated results of ultimate deformation based on experimental data

Failure Mode	Equivalent Ductility Ratio (μu)	
	Median	Logarithmic Standard Deviation
Shear Failure	4.0	0.24
Flexure Failure	10.0	0.10

$$\varepsilon = \frac{R_L}{R_{NL}\sqrt{2\mu_e - 1}} \qquad (4)$$

where, R_L and R_{NL} denote linear and non-linear response values, respectively. The response ductility ratio μ_e is calculated by using the force-deformation relationship(i.e. skeleton curve) obtained from the non-linear response analysis. In order to evaluate the modification factor ε, a deterministic SSI model is used. It has the median values of random parameters associated with (a), (b), (c) and (d) described earlier.

The modification factor ε is calculated to each shear wall and the calculated value from the deterministic analysis is assumed to be the median value of ε. Thus, it is necessary to carry out the non-linear response analysis in the level of peak ground acceleration (PGA) in which the response deformation is assumed to reach the median value of ultimate deformation, for instance, μ_e of approximately 4.0. In case of this model reactor building, the required PGA level is about 1500 gals. In this deterministic approach, the median value of Fμ is associated with some amount of systematic uncertainty.

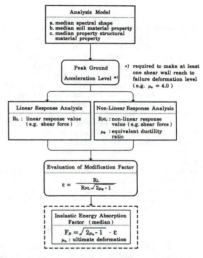

Fig. 4   Evaluation of median value of Fμ

As stated later, the systematic uncertainty has been excluded in comparing the result of the proposed method with that from the Monte Carlo simulation.

The fragility of shear walls is calculated as the conditional probability that the actual response R exceeds the actual seismic capacity C to various levels of PGA by the same way done in the conventional method.

2.2 Analytical Model

To examine the applicability of the proposed method to complex systems, a typical BWR-type reactor model was taken. As shown in Fig. 5, the superstructure is modeled as a multi-stick-lumped-mass system and the effect of SSI is idealized by using rocking and swaying springs attached to the base mat. This is one of the typical analysis models for reactor buildings adopted in Japan, which is generally called the "Rocking and Sway (R & S) model". Figure 6 shows the spectral shape of the ground motion which was evaluated based on the observed data at an actual site in Japan.

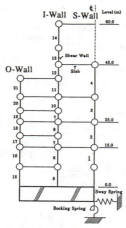

Fig. 5 Analytical model
for BWR type reactor building

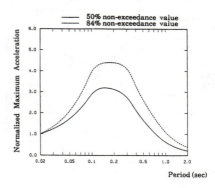

Fig. 6 Distribution of spectral shape

3 MONTE CARLO SIMULATION

3.1 Sampling

Monte Carlo simulation was also performed to obtain more accurate probability of failure. In this simulation, time histories of ground motions were generated from the distribution of (a) which was assumed to have the log-normal distribution, as shown Fig. 6. And non-linear dynamic analysis models were generated whose parameters have the log-normal distribution corresponding to the variability, (b), (c) and (d).

All of these were sampled independently. Whereas, in the sampling of response spectra, a cross correlation characteristic between the spectral values at the different natural periods obtained from the earthquake records was considered, as shown in Fig. 7. Details with regard to the analytical method is mentioned in the reference [9]. From those sample spectra, one hundred time histories of ground motion were generated.

As for the soil material properties, shear wave velocity and damping factor of each layer were sampled independently. To simulate the variability in force-deformation relationship, fifty tri-linear shaped skeleton curves were generated to each shear wall, and one of the generated samples is shown in Fig. 8.

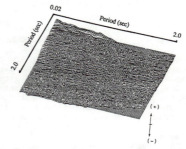

Fig. 7 Cross correlation of spectral values
at different natural periods

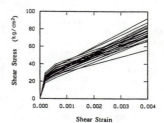

Fig. 8 Sample skeleton curves

3.2 <u>Calculation of Probability of Failure</u>

Monte Carlo simulation of non-linear time history response analysis was performed fifty times. In this simulation, the generated parameters were randomly combined at the two levels of PGA : 1500 gals and 2000 gals.

Probability of failure from the simulation cannot be less than 1/50, since the sample size N is fifty. Therefore, these levels of PGA were determined to give high failure probability referring to the fragility curve obtained from the proposed method. The probability of failure was calculated in such a way as illustrated in Fig. 9.

All response deformation values obtained from the simulation were evaluated as an equivalent ductility ratio μ_e. This ductility ratio μ_e is assumed to reflect the influence of force redistribution and the phenomenon of damage concentration more realistically. Then, the probability of failure was calculated by the ratio of the number of cases N_f, in which the response ductility μ_e exceeded the ultimate ductility μ_u.

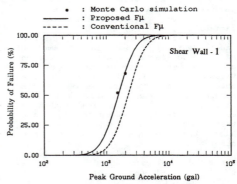

Fig. 9 Calculation of probability of failure

Here, μ_u was generated independently according to its probability distribution, as shown in Table 1. Thus, the probability of failure Pf was calculated by:

$$Pf = \text{Prob} \left[\mu_e \geq \mu_u \right] = \frac{N_f}{N\,(=50)} \qquad (5)$$

4. COMPARISON OF RESULTS

The shear wall-1 resulted in the highest probability of failure both in the proposed method and in the Monte Carlo simulation. This is due to the damage concentration which is mentioned later.

The shear wall has a significant meaning from a viewpoint of aseismic design, since seismic safety of a building can be estimated based on the fragility of such critical element.

The comparison of the fragility obtained by the proposed method with that from the Monte Carlo simulation at two PGA levels is shown in Fig. 10. It can be seen that the median fragility curve obtained by the proposed method coincides closely to the result from the Monte Carlo simulation. Whereas, another fragility curve calculated by using the conventionally evaluated factor Fμ is also shown as a broken line in the same figure. In this curve, the factor Fμ dose not consider the effects of both non-linearity of

Fig.10 Comparison of failure probability

MDOF systems and SSI. The difference shown between these two fragility curves represents these effects.

In these analyses the same variability of parameters were adopted, however the systematic uncertainty, which is not considered in the Monte Carlo simulation, was excluded. For this reason, the factor F_{ME} was excluded in calculating the fragility curve, since the same analytical model, i.e., the "R & S model", was used both in the proposed method and in the simulation.

From the above-mentioned comparison, it is found that the proposed method is quite applicable, in particular, to the shear wall on which non-linear deformation concentrates, like the shear wall-1 which is shown in Fig. 11(b).

To examine this more clearly, the median values of $F\mu$ are compared in the following way. Figure 11 shows the comparison of the median value of $F\mu$ of each shear wall obtained from the proposed method and the Monte Carlo simulation. In this figure, conventionally evaluated median $F\mu$ is also plotted, which has a constant value since the effect of non-linearity of MDOF systems is not considered. The effect of non-linearity in the MDOF system as well as SSI is reflected accurately on the median $F\mu$ from the Monte Carlo simulation.

From this comparison, it is shown that the shear wall-1 and wall-14 have the comparatively small median value of $F\mu$. This is due to the concentration of non-linear deformation. Furthermore, it is also indicated clearly that the proposed approach, as shown in Fig. 4, is very effective to evaluate the median $F\mu$ of such shear walls. Consequently, little difference exists between the failure probability by the proposed method and that by the Monte Carlo simulation.

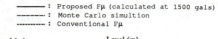

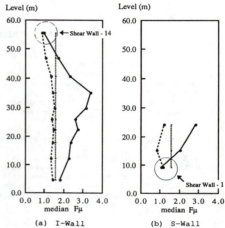

Fig.11 Comparison of median value of $F\mu$

Another significant result obtained from the non-linear Monte Carlo simulation is that the quantitative randomness of $F\mu$, due to the variability of (a) spectral shape of ground motion, (b) soil material properties, (c) structural material properties, (d) non-linear characteristics of each shear wall, had a logarithmic standard deviation of 0.3 approximately. This may be used for the data base in the practical fragility analysis of a reactor building.

Through the comparison of these results, it is found that the proposed method is sufficiently applicable, in a practical sense, to evaluate the seismic fragility of complex systems like nuclear reactor buildings.

5. CONCLUSIONS

In this study, in order to examine the applicability of the proposed method for complex systems, a fragility analysis of a typical BWR-type reactor building was carried out. At the same time, the non-linear Monte Carlo simulation was also performed to compare its result with that of the proposed method. The effects of both non-linearity in MDOF systems and SSI are assumed to be reflected more realistically in the result of the simulation. Through the comparison of both results, the following conclusions are obtained:

(1) Due to the phenomenon of damage concentration, one of the shear walls resulted in the highest failure probability in the proposed method as well as in the Monte Carlo simulation, and both results coincided closely. Taking into consideration that such critical shear wall has a significant meaning from a viewpoint of aseismic design, it has been found that the proposed method is sufficiently applicable, in a practical sense, to evaluate the seismic fragility of multi-degree-of-freedom systems like nuclear reactor buildings.

(2) The quantitative randomness of $F\mu$, due to the variability existing in the spectral shape of ground motions, soil and structure material properties and non-linear characteristics (i.e. skeleton curves) of each shear wall, had a logarithmic standard deviation of 0.3 approximately. This will be available for the data base in the practical fragility analysis of nuclear reactor buildings.

REFERENCES

[1] Smith, P.D., et al.,"Seismic Safety Margins Research Program(SSMRP) Phase I Report Overview", NUREG/CR2015, Vol.1, 1981
[2] ANS/IEEE, "PRA Procedures Guide", NUREG/CR-2300, 1983
[3] Commonwealth Edison Company, "Zion Probabilistic Safety Study", Chicago, Illinois, 1981
[4] Kennedy, R.P., et al., "Probabilistic Seismic Safety Study of an Existing Nuclear Power Plant", Nuclear Engineering and Design Vol.59,No.2, pp.315-338, 1980
[5] Bohn, M.P., et al., "Application of the SSMRP Methodology to the Seismic Risk at the Zion Nuclear Power Plant", NUREG/CR-3428, UCRL-53483, 1984
[6] Budnitz, R.J., et al., "An Approach to the Quantification of Seismic Margins in Nuclear Power Power Plants", NUREG/CR-4334, UCID-20444, 1985
[7] M.Takeda, et al., "On Failure Probability of Multi-Degree-of-Freedom System Considering the Influence of Failure of Other Stories",proceedings of JCOSSAR'87 pp. 345-350, 1987 (in Japanese)
[8] Miguel P.Romo-Organista, "PLUSH A Computer Program for Probabilistic Finite Element Analysis of Seismic Soil-Structure Interaction", EERC-77/01, 1980
[9] T.Okumura, et al., "Earthquake Ground Motions Consistent with Probabilistic Seismic Hazard Analysis", appear in the proceedings of ICOSSAR'89

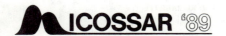
EFFECT OF THE UNCERTAINTY ABOUT THE FATIGUE LIMIT
ON HIGHWAY BRIDGE RELIABILITY

P. B. Keating[*] and L. D. Lutes[*]

[*] Civil Engineering Department
Texas A&M University
College Station, Texas   77843

ABSTRACT

Current fatigue design specifications for welded
steel highway bridges subjected to high traffic
volumes require the design to be capable of
resisting high-cycle, long life conditions without
extensive crack growth. Results from laboratory
fatigue tests of large-scale welded attachments
under random variable loading have indicated that
the existence of a fatigue limit below which no
fatigue cracks develop is assured only if none of
the stress cycles exceed the constant amplitude
fatigue limit. However, the value of the constant
amplitude fatigue limit is not precisely known and
in some cases will differ from the assumed value
due to the limited fatigue test database. This
paper examines the influence of the variability in
the fatigue limit on the fatigue design of bridges.

KEYWORDS

Bridges, Steel, Welded, Fatigue, Extreme life

1. INTRODUCTION

Repeated passages of heavy vehicles may cause fatigue damage in
bridges. While the fatigue behavior of steel bridge components has
been thoroughly investigated resulting in a comprehensive set of
design provisions for new bridges, accurate prediction of fatigue
behavior remains difficult. A primary source of difficulty arises
from the variability that can exist in the value for the fatigue
limit or crack growth threshold for a given detail. This value

determines the difference between the growth (finite life) or the no-growth (infinite life) condition. Many steel bridge components on high-volume routes are designed for infinite life. That is, the anticipated loading in not expected to contribute to fatigue crack growth. The fatigue test database for large-scale test specimens in the high-cycle, long-life region remains quite sparse and is an inadequate basis for accurate fatigue assessment.

2. BACKGROUND

The fatigue design provisions for welded steel structures in current design specifications, such as AISC, AASHTO, AREA, and AWS, were developed from a common fatigue test database generated from a research program sponsored by the National Cooperative Highway Research Program [1,2]. This research program involved constant amplitude fatigue testing of large-scale beams incorporating commonly used welded bridge details. The large number of test specimens allowed for a statistically controlled experiment that revealed that the stress range was the significant stress parameter influencing the fatigue behavior of a given detail type or category. This led to the development of S-N curves to define the relation between the constant amplitude stress range and the number of cycles to failure. In addition, a limited number of tests, performed at lower stress ranges, indicated the existence of a threshold for crack growth or a Constant Amplitude Fatigue Limit (CAFL). A recent review of fatigue test data generated since the original research program has led to a revision of the S-N fatigue design curves [3]. The set of revised curves currently in use is shown in Fig.1.

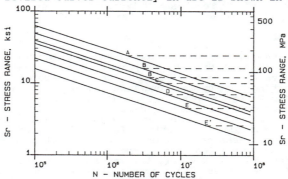

Figure 1:   AASHTO Fatigue Design Curves

The S-N fatigue design curves are based on test data resulting from constant amplitude loading, whereas the loading experienced by bridges is variable in nature. Additional fatigue testing under variable amplitude loading showed that the rate of crack growth for a variable amplitude stress spectrum could be conveniently represented in terms of a single constant amplitude effective stress range (Sr_e) parameter [4]. This allows the use of the constant amplitude S-N curves for both the fatigue design and assessment of highway bridges.

However, it is only recently that the database of large-scale test results has been expanded into the region of fatigue life most representative of the stress conditions experienced by almost all in-service bridges, namely, the long-life region [5,6]. Most bridges are subjected to a large number of stress cycles (100 million or more) of relatively low magnitudes.

The variable amplitude fatigue test results from large-scale specimens have indicated that if any of the stress cycles resulting from a load spectrum exceed the CAFL, fatigue crack propagation is likely to occur and fatigue life estimates should be based on the assumption that all stress cycles contribute to fatigue damage. Stress cycle exceedance rates of the CAFL as low as 0.1% of the total spectrum have resulted in fatigue crack growth consistent with the assumption that all cycles contribute to growth. Therefore, the fatigue analysis of welded details subjected to variable amplitude loading requires that two stress parameters resulting from the load spectrum be considered: the effective stress range, Sr_e, and the maximum stress range, Sr_{max}.

Depending on the values of the two stress parameters and the value of the constant amplitude fatigue limit, three different cases can be encountered. When both Sr_e and Sr_{max} are above the CAFL, fatigue crack propagation occurs and is defined by the S-N curve. If Sr_e is below the CAFL but Sr_{max} is above, crack propagation is initiated by the larger stress cycles in the spectrum and fatigue life can be estimated by assuming all cycles in the spectrum contribute to growth and is similar to the first case. Results from large scale test results have shown that for this case the fatigue resistance is defined as the straight line extension of the S-N curve with a slope constant of 3.0. Only when the entire stress range spectrum is below the CAFL will crack propagation not occur.

3. FATIGUE DESIGN

Bridges designed for primary highway systems will be subjected to moderate to heavy traffic volumes. This requires the design to be capable of resisting high cycle, long life conditions without premature failure and/or extensive crack growth. As the design life approaches the extreme life region, the fatigue design procedure becomes more sensitive to the values of Sr_{max} and the CAFL and the influence of the effective stress range, Sr_e is minimized. The difference between crack growth (finite life) and no growth (infinite life) will depend on the relative values of the first two stress parameters. However, due to the variability and uncertainty that exists for the CAFL and, to a lesser extent, the maximum stress cycle, reasonable estimations for fatigue life are difficult.

Current fatigue design provisions for high-volume highway bridges are based on a maximum design stress range (Sr_{max}) equal to the CAFL. This design procedure minimizes the possibility that bridge details will experience stress cycles that exceed the fatigue limit and initiate crack growth. If it can be reasonably assured that the peak stress range remains below the CAFL, fatigue crack propagation will not occur and infinite life can be expected. Designing for crack growth requires better definitions of the parameters that influence fatigue strength in the extreme life region.

Examination of the AASHTO Fatigue Design Curves, as shown in Fig.
1, reveals that the CAFL occurs at an increasing number of cycles
as the detail severity increases. The CAFL for Category B begins
at approximately 3 million cycles, whereas for the more severe
categories, such as E and E', the CAFL is reached at 12 and 22
million cycles, respectively. Until recently, fatigue design
procedures have led to a design that would provide life in the
extreme life region. This, in part, is due to the use of low
strength details which force the design life into the extreme life
region and conservative estimates of the loading which result in
high values for the design stress range.

However, recent changes in the way bridges are designed have led to
a shift in the region of actual fatigue life. This includes the
elimination of severe or low strength details from a bridge. The
higher fatigue strengths of the details that are used will now
require a fewer number of cycles prior to failure if cracking does
occur. Any changes or variations in the assumptions used in design
(e.g., loading, detail type) will lead to failure in a fewer number
of cycles or shorter time period. In addition, the increased use
of refined structural analysis methods has led to a better
estimation of stresses. This has led to design stresses being
closer in value to actual stresses. This, in a sense, has removed
a portion of the safety factor or reserve inherent in the structure.

4. CONCLUSIONS

In the development of fatigue design procedures, consideration must
be given to the limitations of the database on which the fatigue
strength of welded bridge details is based. Fatigue design for
extreme life or infinite life should not rely on an accurate
estimate of the CAFL. In order to achieve infinite fatigue life in
the face of uncertainty of the CAFL, it is important to design with
an adequate measure of safety such as the current method used in the
AASHTO fatigue design provisions.

REFERENCES

[1]   FISHER, J.W., K.H.FRANK, M.A.HIRT, and B.M.MCNAMEE, "Effect
      of Weldments on the Fatigue Strength of Steel Beams", NCHRP,
      Report 102, 1970.
[2]   FISHER, J.W., P.A.ALBRECHT, B.T.YEN, D.J.KLINGERMAN, and B.M.
      MCNAMEE, "Fatigue Strength of Steel Beams, With Transverse
      Stiffeners and Attachments", NCHRP, Report 147, 1974.
[3]   KEATING, P.B. and J.W.FISHER, "Evaluation of Fatigue Test Data
      and Design Criteria", NCHRP, Report 286, 1986.
[4]   SCHILLING, C.G., K.H.KLIPPSTEIN, J.M.BARSOM, and G.T.BLAKE:
      "Fatigue of Welded Steel Bridge Members Under Variable-
      Amplitude Loadings", NCHRP, Report 188, 1978.
[5]   FISHER, J.W., D.R.MERTZ, and A.ZHONG: "Steel Bridge Members
      Under Variable Amplitude Long Life Fatigue Loading", NCHRP,
      Report 267, 1983.
[6]   KEATING, P.B. and J.W.FISHER: "Fatigue Behavior of Variable
      Loaded Bridge Details near the Fatigue Limit," Bridge Needs,
      Design, and Performance, TTR 1118, pp. 56-64, 1987.

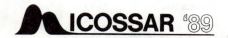
PROBABILITY BASED BRIDGE SAFETY ANALYSIS

J. P. Mohsen* and T. A. Weigel**

* Assistant Professor, Civil Engineering Department, University of Louisville, Louisville, KY, USA
** Associate Professor, Civil Engineering Department, University of Louisville, Louisville, KY, USA

ABSTRACT

Traditional methods of quantifying safety of highway bridges have been associated with deterministic analysis. However, ascertaining the true measure of safety, or reliability, should not follow a deterministic approach but rather a probabilistic one. A procedure is described to develop a probability based reliability analysis to determine the true measure of safety for deteriorating steel truss bridges. Markov series is used to predict the rate of deterioration and future safety index of the bridge.

KEYWORDS

Reliability, Stochastic Modeling, Bridge Evaluation, Bridge Reliability, Safety Index, System Reliability, Bridge Deterioration

1. INTRODUCTION

In the traditional design of highway bridges, where safety and serviceability are the primary considerations, several implicit safety factors are used to account for uncertainties in materials, loads, fabrication details and possible errors in construction. While the design procedures result in bridge structures that are usually safe, these same procedures, when used to evaluate the structural strength of an existing structure, predict capacities and strengths which are far less than their respective true values. True measures of the capacity and strength of an existing bridge are required to evaluate its safety.

The ability of a bridge superstructure to redistribute loads after the failure of one of its components is the reason why several bridges have survived such component failures that theoretically would have caused total collapse. Multiple load path structures exhibit this type of behavior. This redistribution of the loads after the failure of one or

more components needs to also be accounted for in the evaluation procedure.

2. Technical Program

This paper will include a methodology to enable the user to predict the rate of deterioration of various members of a steel truss bridge using a stochastic model. The primary objective is accomplished by carrying out the following tasks:

1) Define the exciting loads as random variables and identify their corresponding probability distribution function;

2) Define resistance parameters of the structure as random variables and identify the corresponding probability distribution function;

3) Quantify the response uncertainties due touncertainty in the properties and loads of the system;

4) Define performance parameters for the structure which adequately measures its condition state, influence its future behavior and reflect its need for maintenance; and

5) Select and quantify a stochastic model for the prediction of the structural behavior in terms of its load resistance parameters under the applied static and traffic load.

The coefficient of variation for the live load is presented by Grouni and Nowak (1). Statistical parameter values that are reported in the literature by Moses(2) and Nowak (3) will be modified using experimental data for the bridge being tested. It will be necessary to modify the load related statistical values provided by other researchers to make it more appropriate and more representative for the structure at hand. Local traffic loads will be surveyed to develop a load spectrum specific to the bridge under investigation. A weigh-in-motion system will be utilized to record the vehicular weights passing through the bridge. This will result in a histogram representing the distribution of actual live load to which the bridge is subjected during the period of survey. The mean for the live load of the specific bridge and its associated coefficient of variance can then be identified for the duration of the survey.

The statistical parameters of resistance will be taken from the available literature. Kennedy and Gad Aly (4) developed ratios of mean to nominal values and their associated coefficient of variation for steel columns as a function of their slenderness ratio. These recommended values will be used to evaluate resistance parameters of the compression members. The statistical characteristics of the tension members will be determined by modifying the values that are provided by Ellingwood (5). The resistance of steel members is treated by Nowak and Tharmabola (6) as lognormally distributed. This assumption will also be made in this research study. Load and resistance distribution parameters will be used to produce reliability indices for the structural elements and also the entire structure as a unit. The reliability index β for each member is evaluated using the following formula:

$$\beta = \frac{m_r - m_s}{\sqrt{S_r^2 + S_s^2}}$$

where m_r and m_s are mean values for resistance and load respectively and S_r and S_s are standard deviations for resistance and load of each member. Reliability analysis will be performed for all of the structural components of the bridge. The relationship between the performance of the elements and that of the entire structure will be investigated using

the system reliability theory. Reliability of the entire bridge structure will be determined by analyzing the interactions and the correlations between the members. In developing the reliability model of the system, the bridge will be considered as a parallel system with two elements (trusses). Each truss is considered as a system in series and therefore it is assumed that failure of any truss causes the overall collapse of the entire truss. The lower chord members, which share the load with stringers and the deck, are exceptions to this assumption.

The most valuable and innovative contribution of this research project is the introduction (application) of Markov chains in order to predict the rate of deterioration of components and future safety index of a bridge. This is accomplished by using the procedure developed by Jiang and Saito (7). The Markov chain, as applied to bridge performance prediction, requires defining states in terms of bridge condition ratings and obtaining the probabilities of bridge condition passing from one state to another. The transition probability matrix of the Markov chain represents these probabilities.

Performance parameters for the structure are defined so that the condition state of the bridge structure can be evaluated. The condition state of the bridge is a measure of the current characteristics that may influence the current or future maintenance needs of the system. In the current research project, the actual cross- sectional area of members, and degree of corrosion of members are used to evaluate the current condition of the bridge. Ten bridge condition ratings are defined as ten states with each condition rating corresponding to one of these ten states. The ratings are compatible with the FHWA bridge rating system.

REFERENCES

1. Grouni, H. N., Nowak, A. S., "Calibration of the Ontario Bridge Design Code 1983 edition," Canadian Journal of Civil Engineering, 11, 1984, pp. 760-770.

2. Moses, F., "Reliability of Structural Systems," Journal of the Structural Division, September, 1974, p. 1813.

3. Nowak, A. S., Zhou, J. H., "Reliability Models for Bridge Analysis," Report UMCE 85-3, Department of Civil Engineering, University of Michigan, Ann Arbor, 1985.

4. Kennedy, D. J. L., Gad Aly, M. M. A., "Limit State Design of Steel Structures-Performance Factors," Canadian Journal of Civil Engineering, 1980, pp. 45-77.

5. Ellingwood, B., et al., "Development of a Probability Based Load Criterion for American National Standards A58," National Bureau of Standards, NBS Special Publication 577, Washington, DC.

6. Nowak, A. S., Tharmabala, T., "Experimental and Analytical Evaluation of Bridges," Proceedings of a Session Sponsored by the Engineering Mechanics Division of the American Society of Civil Engineers, Boston, Massachusetts, October, 1986, p. 17.

7. Jiang, Y., Saito, S., Sinha, K. C., "Bridge Performance Prediction Model Using the Markov Chain," Transportation Research Board, 67th Annual Meeting, January 11-14, 1988, Washington, DC.

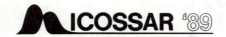
RELIABILITY ASSESSMENT AND RELIABILITY-BASED
RATING OF EXISTING ROAD BRIDGES

H-N. Cho* and A. H-S. Ang**

*   Department of Civil Engineering,
        Hanyang University
        Seoul 133-791, Korea
**   Department of Civil Engineering,
        University of California
        Irvine, CA 92717 USA

ABSTRACT

A practical reliability model for the safety
assessment and rating of existing R.C. bridges are
developed by explicitly incorporating the degree
of deterioration and damages based on available
inspection or test data. The stable configuration
approach is used for an approximate but conservative estimation of the system reliability of existing bridges. It is suggested that the system
reliability index be used as a rating criterion
for bridge rehabilitation. Also, a reliability-
based LRFR (Load and Resistance Factor Rating) is
proposed.

KEYWORDS

Reinforced Concrete Bridges;   Safety;   Rating;
Structural Reliability; Rehabilitation.

1. INTRODUCTION

Some of the road bridges in many countries are severely deteriorated and/or damaged caused by heavy freight vehicles, lack
of proper maintenance, natural hazards or combinations there-of.
Many of these bridges are in need of rehabilitation or replacement. Recently, this has led to an increasing attention to the
problems of the evaluation of reserved capacity and rating of deteriorated or damaged bridge structures.

As a part of an on-going cooperative research between Korea and the U.S.A., reliability models and methods for the safety assessment and rating of existing bridges are being developed. The degree of deterioration or damage estimated on the basis of available field inspection, measurement and test data is explicitly included. The stable configuration approach[1] is used as a basic tool for the system reliability assessment of existing bridges. Also, a practical rating criterion is proposed to provide a guide for the maintenance and rehabilitation of existing bridges[2].

2. LIMIT STATE MODEL FOR EXISTING BRIDGES

The performance function for a structure may be stated as

$$g(\cdot) = R - \Sigma S_i \tag{1}$$

where R is the structural resistance and S_i is the ith load effect.

A realistic safety assessment or rating of existing bridges requires a determination of the degree of deterioration or damage. The evaluation of the resistance, therefore, must reflect such deterioration/damage and the underlying uncertainties. Various approaches, such as Fuzzy set theory[3] or probabilistic measure of structural redundancy[4] have been suggested. A simple quantitative approach is proposed herein for practical application. Normally, the resistances of the deteriorated or damaged members are estimated on the basis of visual inspection and/or in-situ test results supplemented with the engineer's judgement. The true resistance, R, may be estimated as,

$$R = R_n \cdot N_R \tag{2}$$

where R_n = the estimated nominal resistance of a deteriorated or damaged member; N_R = the correction factor for adjusting any bias and incorporating the uncertainties involved in the assessment of R_n.

The mean-value of N_R represents the bias correction, and its C.O.V. represents the uncertainties in the material strength, modelling, and effect of deterioration or damage.

However, if an elaborate technique, such as system identification based on field test data, is available to evaluate the statistical strength and stiffness parameters of a deteriorated or damaged bridge, the resistance R should be evaluated on the basis of the structural parameters obtained through system identification.

Since the dead load and truck loads are the primary loadings on short-span bridges, the load effects may be expressed as:

$$S = C_D D_n N_D + C_L L_n K N_L \tag{3}$$

where C_D, C_L = the influence coefficients for dead and live load effects, respectively; D_n, L_n = the nominal dead and truck loads, respectively; $K = K_s(1+I)$ in which K_s is the ratio of the measured stress to the calculated stress and I is the impact factor, either measured or calculated; N_D, N_L = the respective correction factors for adjusting the biases and uncertainties in the estimated D_n and L_n.

In Eq. 3, the uncertainties in N_L should include those associated with the traffic load model and load test.

3. SYSTEM RELIABILITY OF EXISTING BRIDGES

An existing bridge structure may be rated on the basis of a specified target reliability. This requires the assessment of the reliability of a bridge as a structural system. For this purpose, the stable configuration approach(SCA) will be used to provide a practical and approximate method for estimating the system reliability against ultimate collapse of R.C. bridges, which may be considered to be a brittle mode of failure for which the SCA is particularly effective. By selecting and including only the dominant stable configurations[2] in the calculation, the SCA consistently yields conservative estimates of the system reliability. For instance, the SCA analysis for the R.C. T-beam bridge with five girders may be carried out based on only such dominant stable configurations as those with no or single beam failure. Then, the probability of failure can be obtained from

$$P_F = P[C_0 \ C_1 \ C_2 \ C_3 \ C_4 \ C_5] \qquad (4)$$

where C_i is the damage to the ith configuration with C_0 representing the damage to the original structure.

The system reliability index $\beta_s = -\Phi^{-1}(P_F)$ corresponding to the failure probability obtained from Eq. 4. Then, using the reliability index as a requirement for structural safety, the values of β_s are suggested as a guide for developing rating criteria of deteriorated bridges.

4. RELIABILITY-BASED RATING

A load and resistance factor rating(LRFR) criterion may be developed corresponding to a specified system reliability index β_s. Based on Eqs. 2 and 3, the following rating criterion expressed in terms of the rating factor, RF, which is the ratio of the nominal load carrying capacity, P_n, to the standard design or rating load, P_L, specified in the code may be obtained:

$$RF = \frac{P_n}{P_L} = \frac{\phi'R_n - \gamma_D'C_D D_n}{\gamma_L'C_L K P_L} \qquad (5)$$

where ϕ', γ_D', γ_L' = the nominal resistance, dead load and live

load factors, respectively.

For the rating of an existing deteriorated bridge, the following two load levels of capacity rating of Eq. 5 have to be provided. At the service or lower level, the capacity rating may be referred to as the Service Load Rating(SLR) which corresponds to the allowable safe load level for the normal operation of the bridge. At the over-load or higher load level, the capacity rating may be referred to as the Maximum Over-load Rating(MOR) which corresponds to the absolute permissible load level for special operation (over-load permit) of the bridge.

The nominal safety parameters, ϕ', γ' of the LRFR criterion of Eq. 5 for each load level, SLR or MOR, corresponding to a specified target reliability (for instance, β_{so} = 3.5 for SLR ; β_{so} = 2.5 for MOR) may be calibrated by using the well established procedure for code calibration [2, 5].

5. CONCLUSIONS

The proposed models and methods are practical and relatively simple to apply for the safety assessment and rating of existing bridges. Extensive applications of the proposed model to existing bridges is being undertaken with various numerical investigations. The ongoing research with further numerical investigations and extensions of this study include the incorporation of system identification techniques.

ACKNOWLEDGEMENT

The research reported herein is a part of the U.S.-Korea cooperative research project supported by the KOSEF and NSF. The authors gratefully acknowledge these supports.

REFERENCES

[1] Quek, S.T. and Ang, A.H-S., "Structural System Reliability by the Method of Stable Configuration", SR Series No. 529, Dept. of Civil Eng., Univ. of Ill., Urbana-Champaign, Ill, Nov. 1986.
[2] Shin, J-C., Cho, H-N., and Chang, D-I., "A Practical Reliability-Based Capacity Rating of Existing Road Bridges", J. of Structural Eng./Earthquake Eng., JSCE, Vol. 5, No. 2, Oct. 1988, pp. 245-254.
[3] Yao, J.T.P., "An Unified Approach to Safety Evaluation of Existing Structures", Proc. of a Session, Structural Div., ASCE Convention, Seattle, Washington, Apr. 1986, pp. 22-29.
[4] Frangopol, D.M. and Curley, J.P., "Effects of Redundancy Deterioration on the Reliability of Truss Systems and Bridges", Proc. of a Session, Structural Div. ASCE Convention, Seattle, Washington, Apr. 1986, pp. 30-45.
[5] Ravindra, M.K. and Lind, N.C., "Trends in Safety Factor Optimization", Beams and Beam Columns ed. by R. Narayanan, Applied Science Publishers, Barking, Essex, UK, 1983, pp. 207-236.

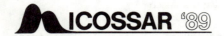
RELIABILITY AND REDUNDANCY ASSESSMENT
OF PRESTRESSED TRUSS BRIDGES

Bilal M. Ayyub[*] and Ahmed Ibrahim[**]

*   Associate Professor, Department of Civil Eng.,
    University of Maryland, College Park, MD 20742, USA.
** Research Assistant, Department of Civil Eng.,
    University of Maryland, College Park, MD 20742, USA.

ABSTRACT

Methods for the assessment of the reliability and
redundancy of post-tensioned trusses are suggested.
The effect of tendon layout on the reliability and
redundancy of post-tensioned trusses is
investigated. Post-tensioning enlarges the elastic
range, increases fatigue resistance, redundancy,
reliability, and reduces deflection and member
stresses.

KEYWORDS

Bridges, Trusses, Post-tensioning, Reliability,
Redundancy, Tendons, Safety

1. INTRODUCTION

Truss bridges are structural systems that are composed of several
elements. The reliability of a truss bridge is a function among
others of the structural strength of the elements, loading and the
level of redundancy. Post-tensioning truss bridges is a means of
creating redundancy, i.e., alternate load paths, in the structural
system. Consequently, structural strength, redundancy and
reliability can be increased. The purpose of this study is to
develop a general method for the reliability and redundancy
analysis of post-tensioned plane trusses with three tendon
layouts. They are a straight tendon, one-drape tendon and two-
drape tendon.

2. SYSTEM RELIABILITY OF POST-TENSIONED TRUSSES

The objective of reliability analysis is to insure the event (R >
L) throughout the useful life of an engineering system where, the
strength or capacity of a structural component R and load effect L
are random variables and are assumed to be statistically
independent. This assurance is possible only in terms of the
probability $P(R > L)$.

In this study, a normally distributed R and L, a linear performance function G of R and L and two potential failure modes are considered: yielding or buckling of a truss member. The failure event E_i can be expressed as

$$E_i = \{ \text{R and L such that } G_i < 0 \} \tag{1}$$

The exact probability of failure or reliability of the system depends on the correlation levels among the failure modes and generally is difficult to evaluate. Therefore, approximations are necessary and the probability of failure or reliability is evaluated in terms of lower and upper bounds. The probability of failure of a nonredundant system can be bounded for small P(E) as follows :

$$\max_{i=1}^{n} P(E_i) \le P(E) \le \sum_{i=1}^{n} P(E_i) \tag{2}$$

The effect of post-tensioning on the reliability of the statically determinate truss system is evaluated, for the purpose of illustration, using three different post-tensioning tendon layouts for the truss of Fig. 1, case (a) where one drape tendon coincides with members (11 and 12), case (b) where two drape tendon coincides with members (7, 2, 3 and 9) in addition to case (c) with straight tendon as shown in Fig. 1.

The addition of the post-tensioning tendon changes the condition of the truss to a statically indeterminate one. The event tree modeling technique is used to evaluate the reliability of the truss system after post-tensioning. The event tree model for case (c) is shown in Fig. 2 with an initiating event, E, and a number of possible failure paths. For a given path to occur, a sequence of subsequent events in the event tree must occur. These subsequent events are mutually exclusive. The probability associated with the occurrence of a specific path is the product of the sequential conditional probabilities of all the events on that path as follows:

$$P(\text{path } j) = \prod_{i=1}^{k} P(E_{ji}) \tag{3}$$

where $P(E_{ji})$ is the conditional probability of occurrence of $i\underline{th}$ event in path j given that all the events $E_{j1}, \ldots, E_{j(i-1)}$ have occurred. This means that the occurrence event of path j is given by the intersection of all the events of the path.

$$\text{Path } j = \bigcap_{i=1}^{k} E_{ji} \tag{4}$$

The lower bound of the probability of failure of a redundant system is the probability of the most likely failure path, i.e., the largest probability of occurrence of all paths. The upper bound is the probability of the union of all the failure paths of

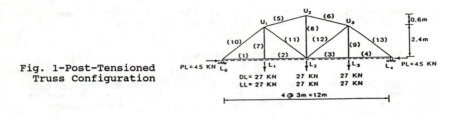

Fig. 1-Post-Tensioned
Truss Configuration

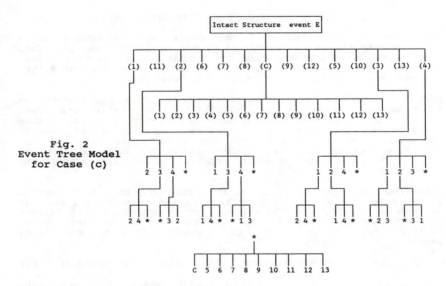

Fig. 2
Event Tree Model
for Case (c)

Table 1-Safety and Redundancy
of Post-Tensioned Truss

Truss System	Probability of Failure, $P_{Fs} \times 10^{-3}$ (Safety Index, β)			Redundancy Factor ψ
(1)	Upper Bound (2)	Lower Bound (3)	Average Bound (4)	(5)
original truss	7.390 (2.438)	1.350 (3.000)	4.370 (2.719)	1.0000
case (a) brittle	7.370 (2.440)	1.350 (3.000)	4.360 (2.720)	1.0023
case (b) brittle	3.820 (2.670)	1.350 (3.000)	2.585 (2.835)	1.6905
case (c) brittle	1.990 (2.880)	0.434 (3.330)	1.212 (3.105)	3.6056

all members that define the intact structure. Therefore, the probability of system failure P_{Fs} can be bounded as

$$\max_{j=1}^{n} P(\text{path } j) \leq P_{Fs} \leq 1 - \prod_{j=1}^{n}(1-P(\text{path } j)) \qquad (5)$$

where the lower bound corresponds to $\rho = 1$ and the upper bound corresponds to $\rho = 0$. The results of the reliability analysis are summarized in Table 1., where the truss components are assumed to exhibit brittle failure modes.

3. SYSTEM REDUNDANCY OF POST-TENSIONED TRUSSES

The classical definition of the degree of redundancy in a structural system is the number of stress resultants and reactions that cannot be determined from the conditions of statics alone. This definition does not constitute an adequate measure of system redundancy as demonstrated by the event tree shown in Fig. 2.

Post-Tensioning results in a truss that is indeterminate to the first degree. Based on the reliability analysis of the truss system, it is evident that redundancy depends on which member fails first and whether an alternative load path exists, and if the other members in that path can sustain the load. It is, therefore, particularly significant that a new viewpoint for the structural redundancy based on system reliability concept be considered.

The effect of post-tensioning on the redundancy of a truss system is evaluated using a redundancy factor. The redundancy factor is defined as the ratio of the average system probability of failure before post-tensioning to the average system probability of failure after post-tensioning. In an equation form, the redundancy factor, ψ, is given by

$$\psi = P_{Fs} \text{ (before post-tensioning)} / P_{Fs} \text{ (after post-tensioning)} \quad (6)$$

For an effective post-tensioning configuration, the redundancy factor ψ must be larger than one. The redundancy factors ψ for the post-tensioned truss using these tendon layouts are shown in the fifth column of Table 1.

CONCLUSIONS

It is evident that a great improvement in the system reliability due to post-tensioning is achieved. For the truss discussed herein, the bottom chord members are the critical ones. It is clearly demonstrated that the approach of evaluating the redundancy of a post-tensioned truss system based on the reliability of the truss system, provides realistic estimates. Although post-tensioning the truss with different tendon layouts makes it redundant to the first degree according to the classic definition of redundancy, the results show that every tendon layout result in different load paths which reflects different levels of structural redundancy.

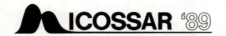
DESIGN METHOD SATISFYING SAFETY REQUIREMENTS FOR VARIOUS LIMIT-STATES BASED ON INFORMATION INTEGRATION METHOD

Shigeyuki MATSUHO*, Wataru SHIRAKI* and Nobuyoshi TAKAOKA*

*Dept. of Civil Eng., Tottori Univ., Koyama, Tottori,(680)JAPAN

ABSTRACT
This paper discusses the applicability of the information integration method to optimum design of structures and the advantages of the above optimum structural design method over other existing methods. By this method, all the constraint requirements can be easily incorporated into consideration in the optimization process. As a simple example, the optimum design of a cantilever beam is shown, and the efficiency of this method is demonstrated.

KEYWORDS
Information integration method; optimum design of structures; cantilever beam; reliability theory.

1. INTRODUCTION

The structural design is essentially aimed at choosing the best geometric and mechanical properties of structures among possible solutions which satisfy both safety and economic requirements. Thus, the structural design constitutes an optimization problem. Such an optimization has so far been performed by minimizing (or maximizing) the objective function under the constraint conditions. However, in such a method, optimization generally becomes difficult with increment of the number of constraint requirements.

On the other hand, recently, an alternative approach of the optimum design was proposed using the concept of integrated information measure[1,2]. The method is called IIM (Information Integration Method). By this method, multiple constraint qualifications can be easily incorporated into consideration in the optimization process.

In this paper, the optimum reliability design of beam is performed using the IIM. Three safety requirements for bending stress, shear stress and deflection and one economic requirement for weight are considered. In a numerical example, the applicability of the IIM to optimum design of structures as well as the advantages of this optimum design method over other existing methods are shown.

2. Information Integration Method

In the information integration method, first a measure of information I_0 is calculated by

$$I_o = \ln p(A) \tag{1}$$

in which p(A) is a probability that a certain event A takes place. The measure of information indicates information, material and energy necessary for causing event A. Second the measures of information for all evaluation items of the structural system are integrated. Finally, the optimum design is performed by minimizing this integrated information measure I.

A parameter which characterizes the state of system for a certain evaluation item is defined as a system parameter for its evaluation item. In this method, the system parameters with fuzziness as well as randomness can be treated. For simplicity of illustration, first a system parameter with uniform distribution as shown in Fig.1 is considered. In this figure, the system range Rs is a possible range which the system parameter can take and the design range Rd is a desirable range demanded for the design. Furthermore, Rc is a common range of both Rs and Rd. Then, the measure of information I_o in Eq.(1) is expressed by

$$I_o = \ln(\ Rs\ /\ Rc\ ). \tag{2}$$

According to this definition, the design for I= ∞ should be rejected and the design for I=0 is the best design.

In the case that the system parameter is assumed to be a variable with fuzziness, a function of satisfaction must be introduced. The value of the function of satisfaction takes 1 for the design with perfect satisfaction, and 0 for the design to be rejected. Fig.2 shows an example of the function of satisfaction for the failure probability. In this figure, the ordinate y is a degree of satisfaction and the abscissa x is a system parameter for failure probability. The degree of satisfaction is considered to be a variable with fuzziness. The upper and lower functions of satisfaction are introduced correspondingly to the upper and lower bounds of the degree of satisfaction (see Fig.2). Using the Rs, Rc in Fig.2, the information measure I_o can be calculated by Eq.(2).

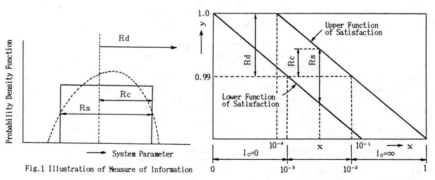

Fig.1 Illustration of Measure of Information

Fig.2 Function of Satisfaction for Failure Probability

The advantages of the IIM are as follows:
1)The optimizing problem is converted to the problem which has

only one objective function without constraint conditions, then all the constraint requirements can be easily incorporated into consideration.
2) System parameters for all the evaluation items can be optimized by minimizing (or maximizing) the information measure integrated with all the evaluation items.
3) The different types of system parameters can be easily evaluated by using the logarithmic scaling for information measure.
4) Fuzziness of the constraint conditions demanded for the design can be incorporated by function of satisfaction.
Moreover, this method may be applied to the fields of expert-system (including a design, a choice of material , an execution of work...), design of entire system (not components) and decision making, etc. In this study, a simple example for the design of beam is considered in the next section.

3. APPLICATION TO OPTIMUM RELIABILITY DESIGN OF BEAM

As a numerical example, the design of steel cantilever beam subjected to a randomly distributed load $\tilde{q}$ is considered. The span length of the beam is L=7(m). The optimum cross-section of cantilever beam is chosen from twenty types of steel cross-sections of JIS (Japan Industrial Standard) shown in Table 1. The properties of this steel is assumed to be as follows: the allowable bending stress σ_a=235.4(MPa), the allowable shear stress τ_a=137.3(MPa) and the Young's modulus E=206(GPa). And the allowable deflection is assumed to be f_a=L/1000=0.007(m). The probabilistic characteristics of the randomly distributed load

Table 1 Elements of JIS I-Beam Sections

Steel No.	standard dimensions (mm)					area of section (cm²)	weight per unit length (N/m)	moment of inertia (cm⁴)	
	H × B	t₁	t₂	r₁	r₂			I_x	I_y
No. 1	100 × 75	5	8	7	3.5	16.43	126.42	283	48.3
No. 2	125 × 75	5.5	9.5	9	4.5	20.45	157.78	540	59.0
No. 3	150 × 75	5.5	9.5	9	4.5	21.83	167.58	820	59.1
No. 4	150 × 125	8.5	14	13	6.5	46.15	354.76	1780	395
No. 5	180 × 100	6	10	10	5	30.06	231.28	1670	141
No. 6	200 × 100	7	10	10	5	33.06	254.8	2180	142
No. 7	200 × 150	9	16	15	7.5	64.16	493.92	4490	771
No. 8	250 × 125	7.5	12.5	12	6	48.79	375.34	5190	345
No. 9	250 × 125	10	19	21	10.5	70.73	543.9	7340	560
No.10	300 × 150	8	13	12	6	61.58	473.34	9500	600
No.11	300 × 150	10	18.5	19	9.5	83.47	641.9	12700	886
No.12	300 × 150	11.5	22	23	11.5	97.88	752.64	14700	1120
No.13	350 × 150	9	15	13	6.5	74.58	573.4	15200	715
No.14	350 × 150	12	24	25	12.5	111.1	854.56	22500	1230
No.15	400 × 150	10	18	17	8.5	91.73	705.6	24000	887
No.16	400 × 150	12.5	25	27	13.5	122.1	938.84	31700	1290
No.17	450 × 175	11	20	19	9.5	116.8	898.66	39200	1550
No.18	450 × 175	13	26	27	13.5	146.1	1127	48800	2100
No.19	600 × 190	13	25	25	12.5	169.4	1303.4	98200	2540
No.20	600 × 190	16	35	38	19	224.5	1724.8	130000	3700

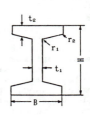

are assumed to be as follows:
the expected value $\bar{q}=3.647(kN/m)$, the variance $D_q=22.75(kN/m)^2$
the autocorrelation function $K_q(\tau)=22.75\exp(-0.3193|\tau|)(kN/m)^2$
in which $\tau=x_2-x_1$ is a difference of abscissas x_1 and x_2 along
the beam.

The safety checking in design is performed based on the
reliability analysis for deflection, bending stress and shear
stress. Four system parameters are considered for the weight of
beam and three failure
probabilities Q_y, Q_M and Q_Q
which correspond to excursions
of allowable levels by maximum
deflection, bending stress and
shear stress owing to above-
mentioned load, respectively.
Thus design is performed by
optimizing all the failure
probabilities and the weight
of beam at the same time.

For simplicity, the
functions of satisfaction for
Q_y, Q_M and Q_Q are assumed to
be identical ones which are
shown in Fig.2, and the
function of satisfaction for
weight of beam is shown in
Fig.3. These functions of
satisfaction mean to keep each
failure probability under 0.01
at least and to keep weight
of steel beam under 844.1(N/m)
per unit length at least.

The calculation results
are shown in Table 2 in which
the columns for "Bending",
"Shear", "Deflection" and
"Weight" indicate the measures
of information for Q_M, Q_Q, Q_f
and the weight of beam,
respectively. And the
integrated information
measures are also shown in the
column "I.I.M.". It is shown
from this table that the steel
cross-section No.11 is the
most optimum one because of
its smallest integrated
information measure. On the
other hand, by the existing
optimum design method , the
most economical section No.7
becomes the optimum solution.

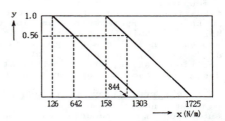

Fig.3 Function of Satisfaction for Steel Weight

Table 2 Integrated Information Measure

Steel No.	Bending	Shear	Deflection	Weight	I.I.M.
No. 1	∞	∞	∞	0	∞
No. 2	∞	∞	∞	0	∞
No. 3	∞	∞	∞	0	∞
No. 4	0	0	∞	0	∞
No. 5	∞	∞	∞	0	∞
No. 6	∞	∞	∞	0	∞
No. 7	0	0	0.5341	0	0.5341
No. 8	∞	∞	∞	0	∞
No. 9	1.01650	0	∞	0	∞
No.10	5.1552	∞	3.4202	0	∞
No.11	0	0	0	0	0
No.12	0	0	0	0.9272	0.92720
No.13	3.3647	∞	1.19310	0	∞
No.14	0	0	0	∞	∞
No.15	1.2430	∞	0	0.4551	∞
No.16	0	2.0210	0	∞	∞
No.17	0	0	0	∞	∞
No.18	0	0	0	∞	∞
No.19	0	0	0	∞	∞
No.20	0	0	0	∞	∞

REFERENCES
[1] NAKAZAWA,H.: "Information Integration Method", Corona
    Publishing Co., Ltd., 1987 (in Japanese).
[2] SUH,N.P., et al.: "On an Axiomatic Approach to Manufacturing
    Systems", J.Eng.Indus.Trans.ASME, Vol.100, pp.127-130, 1978.

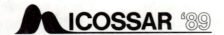
ON THE DYNAMIC RESPONSE OF BRIDGES
BY OSCILLATING VEHICLES ON ROUGH SURFACES

Gross, P. and Rackwitz, R.

Technical University of Munich, Arcisstr. 21, 8000 Munich 2, F.R.G.

ABSTRACT

The spectral theory is applied to the response of bridges under a Poissonian stream of vehicles moving with constant speed over the bridge. The vehicles are modeled as oscillators with random masses, the bridge surface as a Gaussian process with given spectral density. It is found that the surface roughness can produce non—negligible additional loads but, for normal road traffic, no dynamic interaction between vehicles and bridge need to be considered.

KEYWORDS

Random vibration, Traffic loads, Poissonian pulse processes, Road surface roughness, Dynamic interaction

1. INTRODUCTION

A complete model for the stochastic load—load—effect relationships in bridges is still inexistent and would be extremely complex even under the simplifying mechanical assumptions of a linear elastic, viscously damped Euler—beam. However, it is generally agreed that the extreme load effects in the main components of bridges occur either in congested traffic states or when special heavy vehicles are passing the bridge at relatively low speeds. Thus, any dynamic effect is small and can be neglected. On the contrary, dynamic effects need to be considered when determining the extreme load—effects for directly loaded bridge components and especially for the load—effect ranges relevant for fatigue for both directly loaded bridge members and the main girders. A realistic model for this case is interesting per se but especially for the purpose of correctly interpreting load and load—effect measurements on bridges. The dynamics of bridge response have been studied by a large number of authors with emphasis on various aspects (see, for example [1,2,3,4,5]. In this note, the spectral version of the correlation theory of random processes is applied to independent Poissonian trains of damped oscillators with random weights moving with

temporarily constant speed on a random surface of a dynamically reacting, linear elastic bridge.

2. BASIC ASSUMPTIONS AND RESULTS

Assume that the spatial distribution of the (static) vehicle loads (or even axle loads) can be modeled by a filtered Poisson process where A_k is the random amplitude of a pulse:

$$p(\xi) = \sum_{k=1}^{N(\xi)} A_k \, \delta(\xi - \xi_k) \tag{1}$$

If this stream of pulses is moving with constant speed v and the arrival time at $\xi = 0$ is $t = t_k$, there is $p_k(x,t) = A_k \, \delta[x - v(t - t_k)]$ with $x = \xi + vt$. The n–th modal equation of motion of a standard Euler–beam with deterministic, constant stiffness and mass distribution is [7]

$$\ddot{Y}_n(t - t_k) + 2 D_n \, \omega_n \, \dot{Y}_n(t - t_k) + \omega_n^2 \, Y_n(t - t_k) = \frac{1}{M_n} A_k \, u[v(t - t_k)] \tag{2}$$

Y is associated with the k–th pulse. Further, there is $M_n = \int_0^L u_n^2(x) \, \mu \, dx$, $\omega_n = (n\pi/L)^2 \, (EJ/\mu)^{1/2}$, $D_n = c/(2\omega_n)$, c the damping coefficient and $u_n(x)$ the modal shape function. Then, for the n–th modal coordinate we have

$$y_n(t) = \sum_{k=1}^{N(t)} Y_n(t - t_k) \tag{3}$$

Taking the Fouriertransform according to $F_y(\omega) = \int_{-\infty}^{+\infty} y(t) \exp[-j\omega t] \, dt$ with $j = \sqrt{-1}$, the cross spectrum for the steady state with $h_n(\omega)$ denoting the unit impulse response function follows as:

$$S_{y_n, y_m}(\omega) = \lambda \, E[A^2] \, F_{v_n}^*(\omega) \, F_{v_m}(\omega) \text{ with } F_{v_n}(\omega) = \frac{1}{M_n v} \int_{-\infty}^{\infty} h_n(s) \, e^{-j\omega s} ds \int_0^L u_n(x) \, e^{-j\omega x/v} dx \tag{4}$$

The bridge deflection is calculated by

$$w(x,t) = \sum_{n=1}^{\infty} u_n(x) \, y_n(t) \tag{5}$$

from which other response quantities can be calculated. Taking spectral moments the yields the covariance functions of the responses. If, however, the vehicles are modeled as oscillators remaining in contact with the road as in figure 1 the equation of motion needs to be augmented as follows

$$\ddot{Y}_n(t - t_k) + 2 D_n \, \omega_n \, \dot{Y}_n(t - t_k) + \omega_n^2 \, Y_n(t - t_k) = \frac{1}{M_n} [A_k - \ddot{z}(t - t_k) \, m_F] \, u[v(t - t_k)] \tag{6a}$$

where (see figure 1 for notations):

$$m_F \ddot{z}(t) + d_F[\dot{z}(t) - \dot{r}_x(t)] + c_F[z(t) - r_x(t)] = 0 \tag{6b}$$

Of course, the mean of the additional loads and, hence, of the bridge motions vanishes, i.e. $E[\Delta y(t)] = 0$. Therefore, it is sufficient to consider only the additional load:

$$\Delta \ddot{y}_n(t - t_n) + 2 D_n \omega_n \Delta \dot{y}_n(t - t_n) + \omega_n^2 \Delta y_n(t - t_n) = -\frac{1}{M_n} \Delta q_n(t) \, u_n[(v(t - t_n)] \tag{7}$$

with $\Delta q_n(t) = -\ddot{z}(t) \, m_F$. It is then possible to derive the evolutionary spectral density as

$$S_{\Delta y_n, \Delta y_m}(\omega_1, \omega_2, t_1, t_2) = \frac{\lambda}{M_n \, M_m} E[F_{\Delta q}^*(\omega_1) \, F_{\Delta q}(\omega_2)] \times$$
$$\times \int_0^{\min(t_1, t_2)} G_n^*(\omega_1, t_1 - t^*) \, G_m(\omega_2, t_2 - t^*) \exp[j(\omega_1 - \omega_2)t^*] \, dt^* \tag{8}$$

where

$$G_m(\omega_2, t_2 - t^*) = \int_0^{t_2 - t^*} h_m(s_2)\, u_m[v(t_2 - t^* - s_2)]\, \exp[-j\omega_2 s_2]\, ds_2$$

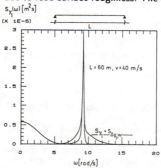

Figure 1: Vehicle—bridge system

and $E[F_{\Delta q}^*(\omega_1)\, F_{\Delta q}(\omega_2)]$ the spectrum of the additional load. As shown in figure ?? the displacement $r_x(t)$ of the vehicle wheel has two components, the surface roughness and the bridge displacement assuming contact between wheel and bridge, i.e.: $r_x(t) = \Delta(vt) + w(x,t)$. $\Delta(x)$ can be assumed to be a homogeneous zero mean Gaussian process with given spectral density, for example, by $S_\Delta(\varpi) = S(\varpi_0)\,[\varpi/\varpi_0]^{-n}$ where $\varpi = \omega/v$ and both the reference value $S(\varpi_0)$ and the exponent n depend on the type and state of the pavement on the bridge. Some calculation then yields for $w(x,t) = 0$, i.e. no dynamic interaction of vehicles and bridge

$$S_{\Delta q}(\omega_1, \omega_2) = E[m_F^2]\, H_F^*(\omega_1)\, H_F(\omega_2)\, v^2\, (\omega_1/v)^2\, (\omega_2/v)^2\, S_\Delta(\omega_1/v,\, \omega_2/v)$$

$$= 2\pi\, E[m_F^2]\, H_F^*(\omega_1)\, H_F(\omega_2)\, v^2\, (\omega_1/v)^2\, (\omega_2/v)^2\, S_\Delta(\omega_2/v)\, \delta[\tfrac{1}{v}(\omega_1 - \omega_2)] \qquad (9)$$

with $H_F(\omega) = (\omega_F^2 + 2jD_F\omega_F\omega)/(\omega_F^2 - \omega^2 + 2jD_F\omega_F\omega)$ the transfer function of the vehicle. The stationary spectral density of the additional load due to surface roughness is then obtained by integrating over ω_1 and dividing by 2π.

Figure 2 shows the spectrum of the first normal coordinate without and with the additional load due to road surface roughness. The left—hand peak corresponds to the static part. The right—hand

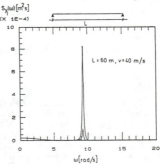

Figure 2: Spectrum of first modal coordinate including effect of surface roughness($E[A] = 48.2$ [kN])

Figure 3: Spectrum of first modal cooridante including effect of mass coupling ($E[A] = 300$ [kN])

peak signifies the dynamic amplification of the beam. It is seen that the surface roughness makes the spectrum significantly broader.

Eq. (9) also gives the result for bridge–vehicle–interaction. One only needs to replace $S_\Delta(\omega_1/v,\omega_2/v)$ by $S_w(\omega_1,\omega_2,x_1,x_2)$ but now with $\Delta(x) = 0$. One finds after some algebra:

$$S_{\Delta y_n, \Delta y_m}(\omega_1,\omega_2,t_1,t_2) = \frac{\lambda}{M_n\,M_m}\, E[m_F^2]\, H_F^*(\omega_1)\, H_F(\omega_2)\, \omega_1^2\, \omega_2^2$$

$$\int_0^{\min(t_1,t_2)} G_n^*(\omega_1,t_1{-}t^*)\, G_m(\omega_2,t_2{-}t^*)\, S_w[\omega_1,\omega_2,v(t_1{-}t^*),v(t_2{-}t^*)]\, \exp[j(\omega_1{-}\omega_2)t^*]\, dt^* \quad (10)$$

On the left–hand side we have the spectrum of individual modes but the right–hand side contains the spectrum of the beam deflections. Therefore, each modal spectrum depends on all other modal spectra implying a fully coupled system of equations. Only if the infinite series for w(x.t) can be replaced by some $w(x,t) \approx u_n(x)\, y_n(t)$, which is admissible if the n–th eigenform dominates, one can derive an explicit lengthy formula which cannot be given here. Superposition of the three types of load effects finally gives the overall spectrum. The stationary limit is obtained by setting $t_2 = t_1 + \tau = t + \tau$ and letting $t \to \infty$. In figure 3 the spectrum of the first modal coordinate including the effect of mass coupling but not of the surface roughness under the same conditions as for figure 2 except that the mean vehicle weight is roughly six times higher than in normal traffic. The effect of vehicle mass and speed is highly non–linear so that dynamic interaction between bridge and vehicles can be neglected for normal road traffic. One now can determine quantities like the ratio of the standard deviations of dynamic and static bridge responses as given in figure 4 where it is demonstrated that the surface roughness produces non–negligible additional loads. They must be included in all fatigue–related considerations.

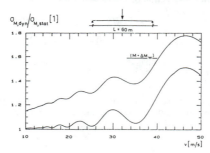

Figure 4: Ratio of static and dynamic standard deviation of mid span bending moment versus speed

References

[1]   Frýba, L.: Non – Stationary Response of a Beam to a Moving Random Force, Journal of Sound and Vibration, 46(3), 1976, pp. 323–338

[2]   Iwankiewicz, R., Sniady, P.: Vibration of a Beam under a Random Stream of Moving Forces, Journal of Structural Mechanics, 12(1), 1984, pp. 13–26

[3]   Hikosaka, H., Yoshimura, T., Uchitani, T.: Non – Stationary Random Response of Bridges to a Single Load Moving on Uneven Road Surfaces, Proc. 26th Japan Nat. Congress for Applied Mechanics, Vol. 26, 1976

[4]   Honda, H., Kobori, T., Yamada, Y.: Dynamic Factor of Highway Steel Girder Bridges, IABSE Proceedings P–98/86, IABSE Periodica 2/1986, pp. 57–75

[5]   Geidner, T., Zur Anwendung der Spektralmethode auf Lasten und Beanspruchungen bei Strassen– und Eisenbahnbrücken, Ber. z. Zuverl. d. Bauw., 37, TU München, 1978

[6]   Lin, Y.K.: Probabilistic Theory of Structural Dynamics, McGraw–Hill, New York, 1967

[7]   Gross, P.: Zur Berechnung der dynamischen Lastwirkung von Balkenbrücken unter bewegten, stochastischen Fahrzeuglasten, Dissertation, TU München, 1989

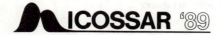
APPROXIMATE RELIABILITY ANALYSIS OF HIGHWAY
BRIDGES UNDER THE ACTION OF RANDOM VEHICLES—IN
RELATION TO AN INTERCHANGE BRIDGE OF CHINA

Ying—Jun Chen* and Naisong Wang**

*   Professor, Civil Engineering Department, Northern Jiaotong
     University, Beijing, 100044, China
* * Graduate Student, Civil Engineering Department, The Universi-
     ty of Calgary, Calgary, Alberta, T2N 1N4, Canada

ABSTRACT

In relation to an interchange curved bridge of China, a simple method
by FSE for analysis of stochastic vibration has been proposed. The random vibration characteristics of bridge structures under the action of
random traffic flow and roughness of road surface can be analysed by
the method. The maximum deflection of bridge under these random
loads which has Gumbel distribution has been obtained. Finally the
problem of first passage time and threshold is solved in a mean sense.

KEYWORDS

Curved Bridge, Random Vibration, Reliability, Response Analysis,
Stochastic Process.

1. INTRODUCTION

In this paper, based on Chinese traffic reality, a method for
the dynamic response analysis for stiffened curved plate bridge,
Fig.1, which has been widely used in interchange bridges in
China, under the action of random traffic flow and rough-
ness of road surface, has been proposed. Generalized filter
Poisson process has been analysed, and the distribution of
maximum deflection response of bridge has been obtained.
This paper is an extension of [1].

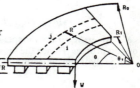

Fig.1 Sectoral Plate
with Stiffened Ribs

2. THE DYNAMIC RESPONSE ANALYSIS OF VEHICLE AND BRIDGE SYSTEM

2.1 Basic Assumpitons

(1)   Only vertical vibration of vehicles is considered, the analytical model is as shown in Fig.2

(2)   Spectral density function of road surface roughness r(vt) is given by eq.(1). [2] [3]

$$S(\Omega) = a / (\Omega^2 + \beta^2) \quad \text{or} \quad S(\omega) = 2\pi Va / (\omega^2 + \alpha^2) \tag{1}$$

in which Ω is the number of waves, a is parameter resrepresenting road surface evenness, β is a small constant, V, ω are velocity and cycle frequency respectively, and $\alpha = 2\pi V\beta$. The autocorrelation function is given by eq.(2).

$$R_r(\tau) = \int_{-\infty}^{+\infty} S(\omega) exp(i\omega\tau)d\omega = \pi a\beta^{-1} exp(-\alpha|\tau|) \tag{2}$$

(3)   Dynamic characteristics of bridge are determined by some lower vibration modes.

2.2 Basic Equations and Solving Method

Let L, T = L / V denote span of bridge and passing time respectively. Refer to Fig.3.

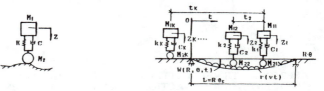

Fig.2   Vehicle Model   Fig.3   Vehicle Flow

According to FSE method, the deflection of bridge is [4]

$$W(R,\theta,t) = [C(R)][\overline{Z}]\{p(t)\} Y(\theta) \tag{3}$$

in which $[\overline{Z}]$ is mode matrix, and for simply supported bridge,

$$Y(\theta) = \sin(\pi\theta / \theta_t), \qquad \theta = V(t - t_l) / R \tag{4}$$

The control equations of vehicle−bridge system are given by eqs.(5−7)

$$\ddot{Z}_l(t-t_l) + 2\xi_l\omega_l\dot{Z}_l(t-t_l) + \omega_l^2 Z_l(t-t_l)$$

$$= \omega_l^2\{U(t,t_l) \{[C(R_a)] [\overline{\Phi}] \{p(t)\} Y(V(t-t_l))\} + \gamma(V(t-t_l))\} \tag{5}$$

$$\dot{Z}_l(0) = \dot{z}_{0,l}; \quad Z_l(0) = z_{0,l}, \quad l = 1,2,...n(t)$$

$$[I][\ddot{p}(t)] + [\overline{C}][\dot{p}(t)] + [\Lambda]\{p(t)\}$$

$$= \sum_{l=1}^{n(t)} [\Phi]^T [C(R_a)]^T U(t,t_l) M_l g[\eta_l - g^{-1}\ddot{Z}_l(t-t_l)] Y(V(t-t_l)) \tag{6}$$

$$\{\dot{p}(0)\} = \{0\}, \qquad \{p(0)\} = \{0\}$$

$$U(t,t_l) = \{\begin{matrix} 1 & 0 \leqslant t - t_l \leqslant T \\ 0 & T < t - t_l < 0 \end{matrix} \tag{7}$$

In which n(t) is a Poisson process with parameter λ, M_l is a random variable with logarithmic normal distribution, while r(vt) is a stationary Gaussian process. [4]

Let deflection of bridge is $\qquad W = W_s + W_d \tag{8}$

in which W_s, W_d represent static and dynamic deflection respectively.

$$W_s = [C][\overline{Z}][\Lambda]^{-1} Y(\theta) \sum_{l=1}^{n(t)} M_l' g[\varphi]^T [C]^T Y(V(t-t_l))$$

$$= [a] \sum_{l=1}^{n(t)} M_l'[b] Y(V(t-t_l)) \tag{9}$$

as M_1', M_2',$\cdots M_b'\cdots$ are independent variables with identic distribution, W_s is a generalized filter Poisson process. Hence the mean, mean square value, and varience of W_s can be obtained.

The dynamic deflection W_d is given by

$$W_d = [C][\overline{Z}]Y(\theta)\{q(t)\} \tag{10}$$

in which
$$\{q(t)\} = \{q_1(t), q_2(t), \cdots q_m(t)\}^T \tag{11}$$

$$q_i(t) = -a_i\omega_i^{-1} \sum_{l=1}^{n(t)} M_l U(t,t_l)$$

$$\cdot \int_0^{t-t_l} \ddot{Z}(\tau)Y(V\tau)exp\ \{-h_i\omega_i(t-t_l-\tau)\}\ sin(\omega_i'(t-t_l-\tau)d\tau$$

$$= \sum_{l=1}^{n(t)} M_l \overline{q}_i(t-t_l) \tag{12}$$

hence
$$W_d = [\overline{a}\ ]\sum_{l=1}^{n(t)} M_l\{\overline{q}(t-t_l)\} \tag{10}'$$

For approximate analysis, taking $N = \lambda$, the mean value of traffic flow process, so that

$$E[W_d^2] = [\overline{a}\ ]^T[\overline{a}\ ]\sum_{l=1}^{N}\sum_{j=1}^{N} E[M_l M_j]E[\{\overline{q}(t-t_l)\}\{\overline{q}(t-t_j)\}^T] \tag{13}$$

as
$$E[M_l, M_j] = \begin{cases} E[M_l^2] & l = j \\ 0 & l \neq j \end{cases} \tag{14}$$

hence
$$E[W_d^2] = [\overline{d}]\sum_{l=1}^{N} E[M_l^2]E[\{\overline{q}(t-t_l)\}\{\overline{q}(t-t_l)\}^T] \tag{13}'$$

while
$$\sigma_{Wd}^2 = E[W_d^2] - E[W_d]^2 = E[W_d^2]$$

from eq.(8)
$$E[W] = E[W_s];\quad \sigma_w^2 = [\sigma_{Ws}^2] + E[W_d^2] \tag{15}$$

$E[\{\overline{q}(t-t_l)\}\{\overline{q}(t-t_l)\}^T]$ can be obtained by using the method suggested in [1].

As W is a stochastic process in time range $[0, T_0]$, We define $W_n = E[W_s] + \gamma\sigma_w$ as nominal deflection, and
$$W_{ne} = MaxW_n \tag{16}$$
in $[0,T_0]$ as the extreme value of nominal deflection. which is a random variable.

If consider the difference of traffic flow in different areas, and also the difference for traffic loads and road condition, λ, $E[M_l]$ and parameter a in eq.(1) must be treated as random variables. For simplicity, assume λ, a, and $E[M_l]$ have uniform distribution. $E[W]$, and σ_w shown in Fig.4—5 are only a member in stochastic process sets.

Within $[0, T_0]$ we can get a series of extremes $\overline{W}_{ne} = \{W_{ne1}, W_{ne2}, \cdots, W_{nem} \cdots\}$. Obviously, $\overline{W}_{ne}$ is a random series. As its numbers are the extreme values of W_n, one may assume it obeys Gumbel distribution. The validity of the assumption will be proved in next section.

According to the assumption, the CDF and PDF of W_{ne} are
$$F_{Wne}(w) = exp\{-exp[-c(w-u)]\};\ f_{Wne}(w) = cexp\{-c(w-u)-exp[-c(w-u)]\} \tag{17}$$
If the data used in obtaining W_{ne} are based on yearly statistics, F_{Wne} will represent yearly extreme deflection. Based on the fact of independent and identic distribution of yearly extremes, CDF of W_{ne} for life time of bridge can be obtained.
$$F_t(w) = [F_{Wne}(w)]^n = exp\ \{-nexp[-c(w-u)]\} \tag{18}$$
The probability of no deflection W_{ne} exceeding x_0 is
$$P\{W_{ne} < x_0\} = F_{Wne}(x_0),\ \text{and the mean first passage time } T_n(x_0) \text{ is given by eq.(19)}$$
$$T_n = [1-F_{Wne}(x_0)]^{-1} = n \cdot [-ln(F_t(x_0))]^{-1} \tag{19}$$
For given T_n and F_{Wne} or F_t, the threshold x_0 can be obtained.

3. NUMERICAL EXAMPLE

Fig.4—5 show the mean and standard deviation of a bridge, the main data for bridge and vehicle are given in Table 1 and Table 2. [4]

Fig.6 shows the outer envelope of deflection based on eq.(16) by taking $\gamma = 2.0$.

Changing parameters λ, $E[M_l]$ and a randomly to obtain $W_{n1}(t)$, $W_{n2}(t)$, $\cdots$ $W_{nl}(t)\cdots$, and those W_{nej} from each $W'_{ni}(t)$ in the range$[0, T_{0i}]$, a number of samples for W_{ne} have been ob-

tained. The histogram of these samples is plotted and shown in Fig.7, Fig.8 shows its CDF.The two parameters to determine Gumbel distribution are $c = 4.206 \; 1 \,/\, cm; u = 2.179 \; cm$ (20)

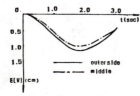

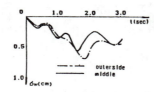

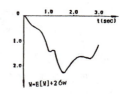

| Fig.4 Mean of Deviation | Fig.5 Standard De-viation of Deflection | Fig.6 Nominal Deflection |

χ^2 Goodness fit of test has also been performed to test the hypothesis. As application of eq.(19), threshold has been calculated for the bridge. Taking $T_n = 50$(years), by inversing eq.(17)

Table 1 Data For Vehicle

Upper Mean Mass(t)	Lower Mean Mass(t)	$E[M^2]$ (t^2)	Damping Coeff. (N / m / s)	Regidity (N / cm)
13.73	2.42	166.90	8168.35	2×8052.63

Table 2   Data for Bridge

θ_τ (rad)	R_0 (m)	R_1 (m)	f(Hz)		
			1	2	3
0.58	40.0	27.08	2.74	7.86	21.6

$x_0 = -[\ln(\ln F_{Wne}^{-1})] \cdot c^{-1} + u$, $F_{Wne} = 1 - 1\,/\,T_n$ (21)
subisituting eq. (20) for c and u in eq. (21) , we get $x_0 = 3.10$ cm.

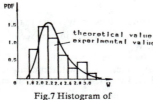

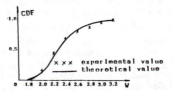

| Fig.7 Histogram of Extreme Deflection | Fig.8 CDF of Extreme Deflection |

REFERENCES

[1]   CHEN, Y.-J. and WANG, N. :"A Simple Method for the Coupled Vibration Problem in the Stochastic Vibration Analysis of Highway Bridge Structures and a Discussion about First-Passage Problem",Symposium JCOSSAR'87, pp 369-374 , 1987

[2]   HONDA, H. , KIDO, T. , KAJIKAWA, Y. , and KOBORI, T. :"Spectral Characteristics of Roadway Roughness on Bridges", Proceedings JSCE, No315, pp.149-155, 1981

[3]   ZHAO, J. H. :" Spectral Analysis of Roadway Roughness", Research Report of Changchun Motor Vehicle Research Institute,China ,1983

[4]   WANG, N. and CHEN,Y.-J. :" Dynamic Analysis of Orthotropic Sectorial Plate and its Stochastic Response", Research Report of Northern Jiaotong University, 1985

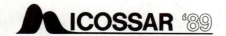
RELIABILITY OF A SUSPENSION BRIDGE
UNDER TRAFFIC FLOW

D. Bryja*   and   P. Śniady*

* Institute of Civil Engineering, Technical University
   of Wrocław, Wrocław, Poland

ABSTRACT

The problem of reliability of a single-span spatial
suspension bridge under random highway traffic is
considered. The displacements of the bridge are described by coupled, non-linear, integro-differential
equations. The stochastic properties of the static
loading are idealized by the filtering Poisson process. The probability density functions of the vertical and horizontal deflections, as well as the reliability estimation are determined for both the
linear and non-linear cases.

KEYWORDS

Suspension bridge; random highway traffic; static
load; reliability; filtering Poisson process

1. INTRODUCTION

   The paper deals with the problem of reliability of a suspension
bridge under the static load caused by highway traffic. One of the
main difficulties connected with that problem is the adequate
idealizing of the traffic jam loading with the random quantities
taken into consideration, such as vehicles headways, weights and
lenghts of cars as well as different types of vehicles involved.
In this paper, the stochastic properties of the live load have
been modelled by means of the filtering Poisson process. Similar
approach has been presented in reference [3]. The response of the
suspension bridge in the space domain is described in the paper by

the set of coupled,non-linear,integro-differential equations as it
has been done in ref.[1]. The probability density function of the
response as well as the reliability estimation are determined by
applying the theory of the filtering Poisson process for both the
linear and non-linear cases. In the latter case the iterative and
linearization method have been used to solve the problem (see [2]).

2. BASIC ASSUMPTIONS AND SOLUTIONS

The analysed single-span bridge consists of a prismatic, open
cross-section thin-walled stiffening girder, which is simply supported and underslung by means of vertical hangers to two whipped
cables (see Fig.1,2). The dead-load curve of the cable forms a parabola, its equation being $z(x) = 4x(1-x)f/1^2$.

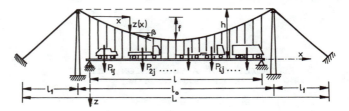

Figure 1.Scheme of an analysed suspension bridge.

The displacements of the bridge under the live load are described by equations which have been derrived on the basis of the combined stress theory of thin-walled rods and flexible cables theory
with the non-linear effects being taken into account. The equations
are as follows ([1]):

$$EJ_y w^{IV} - 2H_o w^{II} + \frac{16kf}{1^2} \int_0^1 wdx - 2kw^{II} \int_0^1 wdx - 2e^2 k\varphi^{II} \int_0^1 \varphi dx = p(x) ,$$

$$EJ_z v^{IV} + \frac{m_b g}{h} v - \frac{m_b gc}{h} \varphi = 0 ,$$

$$EJ_\omega \varphi^{IV} - GJ_s \varphi^{II} - 2H_o e^2 \varphi^{II} + m_b g(b-c)\varphi + \frac{16kfe^2}{1^2} \int_0^1 \varphi dx - \frac{m_b gc}{h} v -$$

$$-2e^2 k\varphi^{II} \int_0^1 wdx - 2e^2 kw^{II} \int_0^1 \varphi dx = m(x),$$

(1)

where $w(x)$, $v(x)$ denote vertical and lateral horizontal displacements respectively, $\varphi(x)$ is the rotation round the shear center A
(Fig.2), $(\cdot)^I = d/dx, EJ_y, EJ_z, EJ_\omega, GJ_s$ are the flexural and torsional
rigidities of the girder and $E_c A_c$ is the longitudinal rigidity of
the cable. Loadings $p(x)$ and $m(x)$ are reffered to the shear centre.
The parameter of the cable is expressed as

$$k = (E_c A_c / L_c)(8f/1^2), \qquad L_c = \int_{(L)} \cos^{-3}\beta dx .$$

(2)

The dead-load cable tension H_o is given by equation $2H_o z^{II} + (m_g + 2m_c)g = 0$, where $m_g g, m_c g$ are the weight of the girder and the weight of the cable with hangers, respectively. To simplify the problem it is assumed that the parameters of the bridge are deterministic.

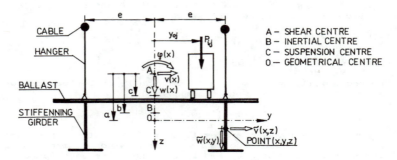

Figure 2. Cross-section of the bridge.

A - SHEAR CENTRE
B - INERTIAL CENTRE
C - SUSPENSION CENTRE
0 - GEOMETRICAL CENTRE

The bridge is loaded by the random highway traffic jam. This traffic is idealized by the trains of concentrated forces (vehicle weights) with random values P_{ij} (Fig.1,2). Each of the trains is located on the separate traffic lane. The distances between the forces are assumed to be random and the number of forces constitutes the stationary Poisson process $N_j(x_j)$ with the parameter λ_j, where $j=1,2,\dots,n_P$ is the number of the traffic lane. It is assumed that $E[P_{ij}^k] = E[P_j^k] = constant$, where $E[\cdot]$ means the expected value of the quantity in brackets and k is the exponent. The functions $p(x)$, $m(x)$ in equations (1) for the above loading process have forms

$$p(x) = \sum_{j=1}^{n_P} \sum_{i=1}^{N_j(x_o)} P_{ij}\, \delta(x-x_{oij}) = \sum_{j=1}^{n_P} \int_0^l P_j(x_o)\delta(x-x_o)dN_j(x_o) \ ,$$

$$m(x) = \sum_{j=1}^{n_P} \sum_{i=1}^{N_j(x_o)} P_{ij} y_{oj}\, \delta(x-x_{oij}) = \sum_{j=1}^{n_P} \int_0^l P_j(x_o) y_{oj}\, \delta(x-x_o)dN_j(x_o),$$

(3)

where δ is the Dirac delta function and y_{oj} is the deterministic coordinate of the traffic lane's location.

In the first step it is assumed that the equations (1) are linear (non-linear parts being neglected). Let $H_v(x,x_o,y_{oj})$, $H_v(x,x_o, y_{oj})$, $H_\phi(x,x_o,y_{oj})$ be the influence functions satisfying the linear equations (1) with $p(x)=1\delta(x-x_o)$, $m(x)=1 y_{oj}\delta(x-x_o)$. Functions H_v, H_v, H_ϕ can be obtained by applying Galerkin method with the sine approximation of displacements (see ref.[2]). The displacements of the cross-section point y,z (Fig.2) have forms

$$\tilde{w}(x,y) = w + y\phi = \sum_{j=1}^{n_P} \int_0^l P_j(x_o)\left[H_v(x,x_o,y_{oj}) + yH_\phi(x,x_o,y_{oj})\right]dN_j(x_o),$$

$$\tilde{v}(x,z) = v - (a+z)\phi = \sum_{j=1}^{n_P} \int_0^l P_j(x_o)\left[H_v(x,x_o,y_{oj}) - (a+z)H_\phi(x,x_o,y_{oj})\right]dN_j(x_o).$$

(4)

The cumulants can be described by the following relationships

$$\varkappa_{\tilde{v}k} = \sum_{j=1}^{n} \lambda_j E[P_j^k] \int_0^l (H_v + yH_\phi)^k dx_o, \quad \varkappa_{\tilde{v}k} = \sum_{j=1}^{n} \lambda_j E[P_j^k] \int_0^l [H_v - (a+z)H_\phi]^k dx_o. \qquad (5)$$

where k is integer. The probability density function ([4]) for th e vertical deflection can be expressed as

$$f_{\tilde{v}}(\tilde{w}, x, y) = \Phi_o(r)\sigma_{\tilde{v}}^{-1} + \sum_{k=3}^{\infty} \frac{(-1)^k}{k!} \varkappa_{\tilde{v}k} \sigma_{\tilde{v}}^{-(k+1)} \Phi_k(r), \quad r = \frac{\tilde{w} - E[\tilde{w}]}{\sigma_{\tilde{v}}},$$

$$\Phi_k(r) = \frac{1}{\sqrt{2\pi}} \frac{d^k}{dr^k} \left[\exp(-\frac{r^2}{2}) \right], \quad E[\tilde{w}] = \varkappa_{\tilde{v}1}, \quad \sigma_{\tilde{v}}^2 = \varkappa_{\tilde{v}2}. \qquad (6)$$

The probability of the event when the limit deflection wd is exceeded can be obtained from the expression

$$P(x, y) = \int_{wd}^{\infty} f_{\tilde{v}}(\tilde{w}, x, y) d\tilde{w}. \qquad (7)$$

After replacing $\tilde{w}$ by $\tilde{v}$ in expressions (6,7) solutions for the ho- rizontal deflection can also be found. For the non-linear equations (1) the solution can be obtained by applying the Gaussian distri- bution $f_{\tilde{v}}(\tilde{w}, x, y) = \Phi_o(r)\sigma_{\tilde{v}}^{-1}$. The expected value and variance should be determined by applying both iterative and linearization methods in the same way as presented in ref. [2].

3. CONCLUSIONS

While idealizing the loading of the bridge under consideration, it has been assumed that the number of vehicles in a traffic lane can be approached by the application of the Poisson process. This assumption enabled to determine the probability density function of the response. However, if traffic is jammed a renewal process with a gamma distribution or a general point stochastic process with the correlations between different headways being taken into account, should rather be used. The acceptance of the filtering Poisson process as a model of vehicles' traffic can have an im- portant practical meaning since it enables to evaluate the lower bound of reliability as it follows from reference [4].

REFERENCES

[1] BRYJA, D.: „Spatial vibration of a suspension bridge under iner- tial moving load", Archiwum Inżynierii Lądowej, Vol. 30, No. 4, pp. 607-627, 1984 (in Polish)
[2] BRYJA, D., ŚNIADY, P.: „Random vibration of a suspension bridge due to highway traffic", Journal of Sound and Vibration, Vol. 125, No. 2, pp. 379-387, 1988
[3] SHINOUZUKA, M., MATSUMURA, S., KUBO, M.: „Analysis of highway bridge response to stochastic live loads", Proc. of J. SCE, No. 344, I-1, pp. 367-376, 1984 (in Japanese)
[4] ŚNIADY, P.: „Dynamic response of linear structures to random stream pulses", Journal of Sound and Vibration, Vol. 129, No. 3, 1989 (in print)

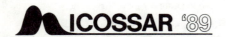
RESEARCH ON STRUCTURAL SYSTEM RELIABILITY OF CABLE-STAYED
P.C. RAILWAY BRIDGE ACROSS HONG-SHUI RIVER

Ning Sun and Mingchu Yao*

* Institute of Railway Engineering, China Academy of
   Railway Sciences, Beijing, P.R. of China 100081.

ABSTRACT

This paper analyses the structural system reliability of cable-stayed bridge by using the modified
all failure modes. The concept of the structural subsystem is used to help the selection of major failure
modes and to facilitate the correlate analysis. A
special structural analysis algorithm considering nonlinear effect is used to caculate the loading response
and a multiple integral method is utilized to estimate
the P_f of each failure mode. According to the design
and test data of the bridge, the structural system
reliability of the first-order bounds and the second-
order bounds for ultimate limit-states are obtained.

KEYWORDS

System reliability; Cable-stayed bridge; Subsystem.

1. INTRODUCTION

Since the structural system reliability theory is fundanmental to
the evaluation of structural system in the sense of safety and economic, and to improve the design of structure, more researchers and
engineers have increased their concern on structural system reliability. As a practical application, the structural system reliability
of a cable-stayed bridge is studied. Hongshui river bridge which
completed in 1980 is the first cable-stayed P.C. railway bridge in

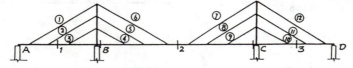

Fig. 1 Schematic drawing of the Hongshui river bridge

China. The bridge comprises three parts: the main girder (48+96+48m), the pylons and the cables. The pylons of the bridge are cast monolithicly with the main girder and are separated from the piers.

2. ESTIMATING METHOD FOR THE SYSTEM RELIABILITY OF THE CABLE-STAYED BRIDGE

The cable-stayed system is separated into three subsystem: the main girder, the pylons and the cables. The relationship between the system failure and different subsystem failure is shown in Fig 2.

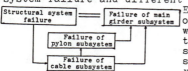

Fig. 2 Relations between System Failure and Subsystem Failure

Each subsystem failure is caused by the occurrence of the basic failure events which includes the shear failures and the bending failures of the critical sections (Fig1) in the main girder subsystem, the eccentric compression failures and the stability failures in the pylon subsystem and the fracture-corrosion failures or the prestress deterioration in the cable subsystem. According to the estimation of the P_f for the basic failure events,[1] the bending failure in the main girder subsystem and the cable failure in the cable subsystem are chosen as the initialing events of the structural failure modes. So, the contents of estimation of the system reliability of cable-stayed bridge include:

1) Estimation of the structural system reliability (P_F) of the main girder subsystem in the ultimate limit-states.

2) Estimation of the failure probability of the cable in the cable subsystem according to the following procedure:

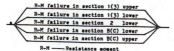

Fig. 3 The failure modes of the Main girder Subsystem

a) caculation of the P_f of pylons followed by the redistribution of the loading response in the pylons.

b) caculation of the system reliability of main girder subsystem which is concerned as a dynamic subsystem for each state of the cable failures.

The method discussed above is based upon the modified all failure mode method (combination of the all failure mode method with the subsystem analysis). The all failure modes of the main girder subsystem is illustrated in Fig. 3. It is found that in the case of the single cable failure, the occurrence probability of the pylon failures can be ignored. The all failure modes of the cable-stayed bridge system are shown in Fig. 4. Fig. 5 shows the analysis procedures of the method.

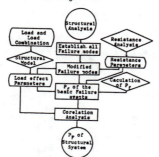

Fig. 5 Flowchart of the Caculation

3. ESTIMATION OF SYSTEM RELIABILITY FOR THE HONGSHUI RIVER BRIDGE

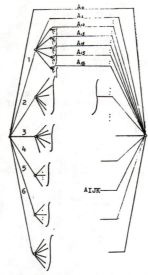

A_I —— The dynamic failure modes of main girder subsystem after the failure of NO. I cable.

Fig. 4 All Failure modes of Cable-stayed Bridge

A speical structural analysis alogrithm considering non-linear effect and auto-caculating the maximum loading response is used to caculate the load effect. As the caculation shows that P_f in each critical sections are so small that the JCSS method would introduced much error, the multiple intergral method[2] is utilized to caculate the P_f in the critical sections.

3.1 Correlation Analysis

The limit-state eouations for bending in the main girder subsystem have the form as below:

$$\left.\begin{array}{l} Z_i = m_{R_i} R - m_{1_{d_i}} S_d - m_{1_{1_i}} S_1 \\ \underline{Z}_i = \underline{m}_{R_i} R - m_{1_{d_i}} S_d - m_{1_{1_i}} S_1 \end{array}\right\} \quad (1)$$

Assumption:
1) Bending strength and dead loads in dif-ferent modes are strongly correlated.
2) Live loads are strongly correlated as the random running trains have the same positions and combibctions for the dif-ferent failure modes when the maximum loading response appeared.
3) R, S_d and S_l are incorrelated.

Z_i and Z_j are correlated. According to these assumptions, $Z_{P1upper}$ and $Z_{P2lower}$, $Z_{P3upper}$ and $Z_{P2lower}$ are correlated.

Caculation shows that the correlation coefficients are all greater than 0.9. In this way, the occurrence probability of the two failure modes is:

$$P_F(F_iF_j) = \min(P_F(F_i), \ P_F(F_j)) \qquad (2)$$

3.2 Caculation of the System Reliability of Main girder Subsystem

According to the above method, the $P(F_i)$ and $P(F_iF_j)$ of different critical sections in the main girder subsystem can be caculated. The second-order bounds of the $P(F)$ for the series main girder subsystem are expressed as:

$$P(F) \geqslant P(F_1) + \sum_{i=2}^{n} \max\{(P(F_i) - \sum_{j=1}^{i-1} P(F_iF_j), \ 0\} \qquad (3)$$

$$P(F) \leqslant \sum_{i=1}^{n} P(F_i) - \sum_{\substack{i=2 \\ j<i}}^{n} \max P(F_iF_j) \qquad (4)$$

For comparison, the first-order bounds of the P(F) is also given by:

$$P(F_i) \leqslant P(F) \leqslant \sum_{i=1}^{n} P(F_i) \qquad (5)$$

The caculation of P_f bounds of the main girder subsystem are summarized in Table 1.

Table 1. Reliability bounds of the main girder subsystem

First-order bounds		Second-order bounds	
$A_1 \leqslant P_f \leqslant B_1$	$C_1 \leqslant \beta \leqslant D_1$	$A_2 \leqslant P_f \leqslant B_2$	$C_2 \leqslant \beta \leqslant D_2$
0.197E-5 0.682E-5	4.350 4.615	0.288E-5 0.669E-5	4.757 4.535

3.3 Cable parallel Subsystem Analyses

The failure of the cables may be caused by the fracture, corrosion failure and the prestress deterioration, which leads to reduction of the reliability in the main girder subsystem. As the P_{f_i} can not be easily caculated, P_{f_i} of each cable is estimated conservatively as:

$$P_{f_i} = 10^{-5} \qquad (6)$$

The estimations of system reliability of the dynamic main girder subsystem dut to the failure of different single cable are obtained.

3.4 Estimation of the system Reliability for Hongshui river bridge

Analysis show that the failure of main girder subsystem and the failure of each dynamic main girder subsystem can be assumed to be incorrelated, therefore, the P_F of the Hongshui bridge is expressed as:

$$P_F = \sum_{i=1}^{k} P_{F_i} \qquad (7)$$

In which P_{F_i} is the P_f bounds of the dynamic main girder subsystem in case of No. i cable failure, i=0 means the P_f bounds of the main girder subsystem. The caculating results of the system reliability for Hongshui river bridge are summarized in Table 2.

Table 2. Reliability bounds of the Hongshui river bridge

First-order bounds		Second-order bounds	
$A_1 \leqslant P_F \leqslant B_1$	$C_1 \leqslant \beta \leqslant D_1$	$A_2 \leqslant P_F \leqslant B_2$	$C_2 \leqslant \beta \leqslant D_2$
0.2157E-5 0.7233E-5	4.338 4.598	0.3251E-5 0.7077E-5	4.344 4.508

4. CONCLUSIONS

(1) The method proposed in this paper, which is called subsystem analysis method, can be used effectively for the estimation of the system reliability of cable-stayed bridge as well as other similar structural system.
(2) Estimating results show that the Hongshui river bridge has high safety and reliability.
(3) The cable-stayed bridge is an advanced type of bridge compared with the other type of bridge in the view of system reliability.
(4) The optimization of structural design based upon system reliability may have better benefit than the optimization based upon the reliability of individual element.

REFERENCES

[1] N. Sun:"Research on the System Reliability of the cable-stayed Bridge", M.Sc thesis, CARS, Aug. 1987, (in chinese)
[2] H.F. Lei and M.C. Yao:"Modified multiple integral method for Caculating Pf", CCES, B & S conf. , Oct. 1987, (in chinese)
[3] G.I. Schueller:"Current Trends in System Reliability", ICOSSAR'85, Vol 1, PP 157—169, 1985

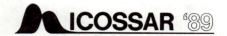
VIBRATION TESTS AND DAMAGE DETECTION OF P/C BRIDGES

J.P.Tang and K.M.Leu

Professor and Graduate Student, Civil
Engineering Dept., National Central
University, Chung-Li, Taiwan, R.O.C.

ABSTRACT

A three-span countryside P/C bridge
has been found to have excessive deflections in two of the three spans. It is
reported that the prestresses in tendons
of two girders of different spans are
lower than design values after investigations. The objectives of the present
studies are (a)to prove that the change
of dynamic characteristics of a bridge
can be used as index for damage detection
and (b)to study the possibility of application of system identification in defining damage quantitatively. It is found
that the change of mode shapes can be
used as an index for damage detection.
In addition, the used of technique of system identification can define the damaged
member in the P/C bridge successfully and
quantitatively.

KEYWORDS

bridge, vibration, system identification,
damage detection, safety assessment.

1. INTRODUCTION

Dynamic characteristics of a damaged structure will
be different to a certain extent from those of the intact
structure. Natke and Yao[1] pointed out that if dynamic
in-situ tests are carried out at certain intervals,change

of the parameters could be detected and interpreted through dynamic analysis. In addition, for purpose of safety
inspection, it is necessary to detect change of modal
frequency inthe order of 0.01Hz for bridges. Tang and Leu
[2] studied a bridge model with continous superstructure
and found that the changes of mode shapes are more sensitive than the changes of natural frequencies as damages
in girders change in values and/or with positions.

Douglas and Reid[3] have presented a method of system
identification. Satisfactory results have been obtained
in treating dynamic response data of a bridge.

The Kim-Yueh bridge, referring to Fig.1, was found to
have excessive deflections in span 1 and span 2. It has
been reported[4] that the prestresses in tendons of girders G II and G IV, referring to Fig.1, are lower than design values after performing compression test of concrete
core samples, measurement of girder deflections and corresponding analysis.

2. VIBRATION TEST AND DATA ANALYSIS

Six seismometers with signal conditioners were used
to collect vertical vibration data of the bridge excited
by two students jumping away from the seat attached with
a rope to the middle point of the rail of downstream side.
Natural frequencies and mode shapes have been estimated
and part of the results are presented in Fig.2. It is noted that the 1st and 2nd modes can be obtained.

3. SYSTEM IDENTIFICATION AND RESULTS

Using a finite element model with 85 nodal points and
assuming different damaged conditions in upstream girder
by assigning lower flexural rigidities, the natural frequencies and mode shapes, using SAP IV program, are computed. It is found that the lower the flexural rigidity, the
higher the magnitude at upstream side and the lower the
magnitude at downstream side for the 1st mode. Opposite
trends have been obtained for the 2nd mode. From measured
data as shown in Fig.2, the downstream side girder of span
II, i.e. G IV is the weakest one among all girders. Since
damaged conditions in girders are not so simple as those
assumed in above analysis, technique of system identification must be used to analyze data such that damages in
girders can be defined.

Using the method proposed by Douglas et al.[3] and the
finite element model mentioned above, the flexural rigidities of girders are chosen as variables to be identified.
Results obtained in the present study and from Ref.4 are
listed in Table 1. It can be seen that girder G IV is the
weakest one and girder G II the second. These coincide
with conclusions in Ref.4. Natural frequencies and mode

shapes have been computed on the basis of identified results and part of them are also presented in Fig.2.
    Applying HS20-44 load on finiteelement model of span
II, the maximum deflection and maximum bending moment
have been computed. It is found that this bridge must be
posted for limited load.

4. CONCLUSIONS

    The change of dynamic characteristics can be used as
an indicator for damage detection of bridges. In addition, the use of technique of system identification can
define the damaged member in the P/C bridge successfully
and quantitatively.

5. REFERENCES

1. Natke,H.G. and Yao,J.T.P.,"System Identification Approach in Structural Damage Evaluation", T.R. CE-STR-
   86-21, CRI-F-2/86, School of C.E., Purdue University,
   1986.
2. Tang,J.P. and Leu,K.M.,"Vibration Measurement and Safety Assessment of Bridges", US-Taiwan Joint Seminar
   on Rehabilitation of Public Works, Taipei, Taiwan,
   R.O.C., Jan. 1988, pp.55-69.
3. Douglas,B.M. and Reid,W.H.,"Dynamic Tests and System
   Identification of Bridges", J. of the St. Div., ASCE,
   Vol.108, No.ST10, Oct. 1982, pp.2295-2315.
4. Chern,C.C. and Chen,C.C.,"Structural Safety Assessment of the Kim-Yueh Bridge", Research Report TS-772-
   01, Taiwan Construction Technology Research Center,
   Taiwan, R.O.C., Oct. 1987.

6. ACKNOWLEDGEMENT

    The present study has been supported by National
Science Council, R.O.C. Through Grant No.NSC-78-0414-P-
008-06B.

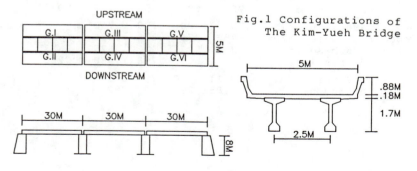

Fig.1 Configurations of
The Kim-Yueh Bridge

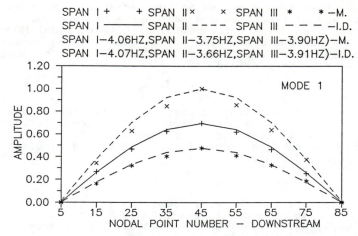

Fig.2 Mode Shapes and Natural Frequencies,
      Results of Experiments and System
      Identification

Table 1. Identified Results in the Present Study
         and Prestresses in Tendons from Ref.4

		$I1, m^4$	$I2, m^4$	$I3, m^4$	$I4, m^4$	Error after Id.
span I	base value	.700	.700	.600	.600	
	1.25 base value	.875	.875	.750	.750	
	0.75 base value	.525	.525	.450	.450	
	Id. results	.696	.657	.486	.700	.0528
span II	base value	.724	.563	.400	.400	
	1.25 base value	.906	.704	.500	.500	
	0.75 base value	.543	.423	.300	.300	
	Id. results	.752	.622	.498	.439	.0939
span III	base value	.700	.700	.700	.700	
	1.25 base value	.875	.875	.875	.875	
	0.75 base value	.525	.525	.525	.525	
	Id. results	.724	.563	.768	.588	.0290
prestresses in tendons		G II 342.88T	G IV 296.96T	G IV 382T		

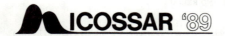
AN OVERVIEW OF THE NASA (LeRC) - SwRI PROBABILISTIC STRUCTURAL ANALYSIS (PSAM) PROGRAM

T. A. Cruse,[*] C. C. Chamis,[**] and H. R. Millwater[*]

[*]Department of Engineering Mechanics, Southwest Research Institute, San Antonio, Texas 78284, USA

[**]National Aeronautics and Space Administration, Lewis Research Center, Cleveland, Ohio 44135, USA

ABSTRACT

The development of a new methodology for probabilistic structural mechanics is reported in several papers within the current **Proceedings**. The methodology combines a fast probability integration algorithm with advanced and general purpose structural analysis programs based on finite element and boundary element algorithms. The resulting programs allow static and dynamic analysis of structures subject to random loading, material properties, geometries (e.g., tolerances), and boundary conditions. The stochastic output of the analyses consists of a user defined region for the probabilistic distribution function (or cumulative distribution function), together with confidence bands. The combined algorithms are sufficiently efficient for the probabilistic analysis of significant engineering structural problems. The individual and combined algorithms are presented in overview form in this paper. Additional, referenced papers support this overview with details and applications.

KEYWORDS

Finite Element, Reliability, Random Variables, Probability Density Function, Cumulative Distribution Function, Fast Probability Integration, Expert System, Random Fields

1. INTRODUCTION

The use of advanced tools for structural analysis is the heart of the design system for aerospace systems. Design criteria, however, still rely on experience-based safety factors which do not relate directly to structural reliability. The current study on probabilistic structural analysis methods (PSAM) has resulted in a new class of tools the designer can use to obtain direct information on the uncertainty of structural performance. The uncertainty can then be easily related back to known or estimated uncertainties in design data.

The paper presents an overview of this NASA contract effort (Contract NAS3-24389), now in its fifth year. Supporting papers which outline the specific elements of the PSAM-developed algorithms accompany this paper in the current Proceedings.

2. APPROACH

2.1 Goals

The Probabilistic Structural Analysis Methods (PSAM) project funded by NASA has the central goal of developing a comprehensive structural analysis system capable of considering general forms of uncertainty in loading, geometry, material behavior, and boundary conditions. The purpose of this analysis system is to support the design of advanced space propulsion hardware capable of operating in severe environments with adequate reliability. The PSAM methods are sufficiently general, however, that they can be applied to a very wide range of structural reliability concerns.

The PSAM project requirements include the development of probabilistic structural analysis methods using finite element, boundary element, and approximate methods. In the finite element approach, the goal is to include several advances in FEM formulation for stiffness and hybrid element formulations, appropriate for modeling thick shells and solids. The formulation is required to be in terms of nodal representations of all variables (loading, geometry, and solution - stress and displacement).

The PSAM project requires that the analysis results predict the entire distribution for a given solution variable. That is, the cumulative distribution function (CDF) must be computed with accuracy into either or both tails of the distribution. The probabilistic calculations required are to be based on two, different probabilistic methods of calculation or simulation.

Three levels of uncertainty modeling for variables such as loading, geometry, or material properties are to be analyzed. Figure 1 portrays these three levels for the example of material stiffness or modulus. Level 1 considers the modulus of each part to be a single sample from a distribution of moduli. Level 2 is to consider the uncertainty description for modulus to be inhomogeneous; i. e., to have different statistics depending on the forging location for the samples. Level 3 considers stochastic relations between such fundamental variables as stress and strain. As shown, the strain may be a random variable for very small strain samples, with the mean value tending towards the displacement-strain (δ/l).

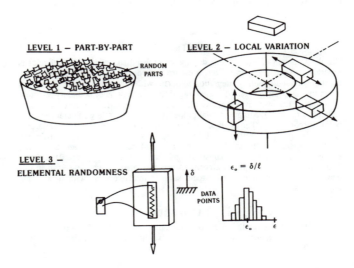

Figure 1. Three Levels of Modeling for Random Variables

2.2 Analysis Modules

Figure 2 highlights the major modules used by the NESSUS (nonlinear evaluation of stochastic structures under stress) software system, illustrated for the finite element method (FEM). These modules include an expert system which facilitates the user interface [1], a preprocessor for random data fields (NESSUS/PRE), a structural modeling module (e.g., NESSUS/FEM) [2], a database for intermediate analysis results, and a module for making fast probability calculations (NESSUS/FPI) [3].

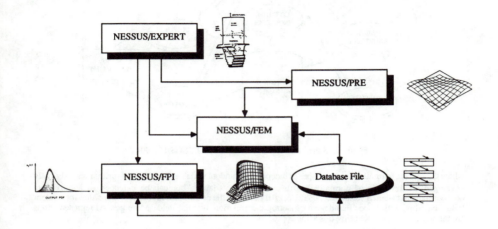

Figure 2. Overview of the NESSUS Code for PSAM

Figure 3 is an overview of the operation of the expert system for NESSUS. The purpose of NESSUS/EXPERT is to manage the use of various knowledge bases associated with specialized information that supports the application of NESSUS. For example, past knowledge of the user-input variables related to execution of the analysis that might lead to non-convergence of solution perturbations, or knowledge of important random variable effects and their associated statistics, form input knowledge bases. Interpretation of random data results based on past experience to highlight accuracy concerns can be included in the output knowledge base.

Probabilistic analysis requires that each random variable must be an independent random variable. For random fields, some correlation of the variables is expected from point to point. Random fields include descriptions of the probabilistic pressure or temperature loads, probabilistic thickness variations, or probabilistic material properties within the structure. As an example, consider aerodynamic loading. Aerodynamic pressure loading is likely to have a high degree of statistical correlation from one region to another, owing to large scale order in variables such as velocity and pressure. On the other hand, turbulent boundary layers generally have statistical aspects that may only be locally correlated, related to the size and distribution of the turbulent eddies.

The decomposition of random fields with partial spatial correlations into independent random variable fields is presented in [4]. The approach is to represent each field in terms of a modal decomposition

$$[C_{ij}] = \sum_{i=1}^{N} \{\phi_i\} \lambda_i \{\phi_i\}^T \qquad (1)$$

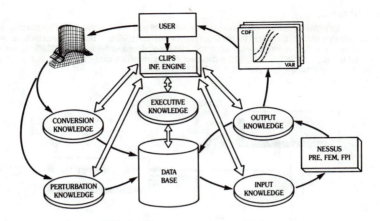

Figure 3. Schematic of the Operation of NESSUS/EXPERT

where the $\{\phi_i\}$ are normalized mode shapes of independent random variable fields. The algorithm was originally developed on the basis of normally distributed random variables, but can be extended to approximate non-normally distributed variables. NESSUS/PRE performs the normalized modal decomposition for random fields, based on a user-specified degree of isotropic correlation over the field. More general representations are theoretically allowable using this same approach.

Figure 4 is a geometric interpretation of the mathematical treatment of random fields for Level 2 analysis. For a field with some spatial correlation, the shape functions for a square region are indicated. A fully correlated random field would have one (rigid) mode shape. The form of the modes for relatively uncorrelated fields might look like the region within the last square, showing a rough pattern indicating very localized correlations.

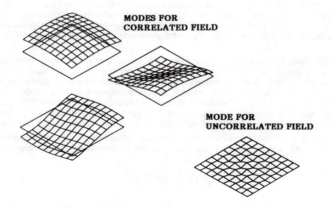

Figure 4. Modal Decomposition of Random Fields

Figure 5 represents the key element of the PSAM algorithms. The representation is given in terms of a simple example of a vibration problem with frequency dependence on material modulus. The random variable is described empirically or in terms of standard cumulative distribution functions. The sensitivity of the solution (frequency) is estimated by sensitivity modeling (response model). Finally, the cumulative distribution function for the frequency is given by a special fast probability integration (FPI) algorithm.

The FPI algorithm was developed [5] for basic reliability calculations for closed-form system models. In the PSAM project, FPI has been adapted very effectively to numerical analysis of complex structures [6].

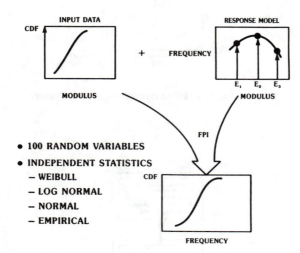

Figure 5. Fast Probability Integration (FPI) Algorithm Uses Sensitivity Data

2.3 NESSUS Structural Analysis Framework

The basis of the NESSUS algorithm, discussed above, is the use of sensitivity analysis of the response model. In general, the response model is a numerical analysis model (e.g., FEM) of the structures subject to the usual deterministic loading and boundary conditions. For linear analysis, we are generally concerned with the maximum stress or the natural frequencies of fundamental modes. Sensitivity analysis seeks to estimate how much the stress or natural frequency depends on each of the design variables (loads, geometry, material properties, boundary conditions). By changing the design variables a small amount, as in Figure 6, the solution changes some amount, and a hyperplane or hyper-(quadratic)surface can be fitted to the solution data at the design point and a suitable number of perturbed-solution points.

NESSUS stores the perturbed solutions in an extended, special purpose database for later retrieval and FPI analysis, as shown in Figure 7. The static and dynamic analyses are handled separately, in order to obtain the most efficient results.

The NESSUS code has specially tailored algorithms for obtaining the perturbed solutions for static and dynamic problems, as described in the previously cited reference by Dias and Nagtegaal. In general terms, the structural analysis routines maintain the full element description of the unperturbed problem, and place the effects of perturbations on the right-hand side of the system equations as pseudo-forces, for the FEM case. Iteration of the system equations is based on the use of the reduced unperturbed solution stiffness equation, as an iterative preconditioner matrix. This approach results in a significant time savings for static problems over the full resolution at the perturbed condition.

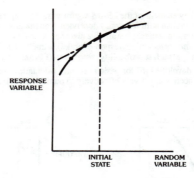

Figure 6. Perturbations Used to Establish Sensitivity Models

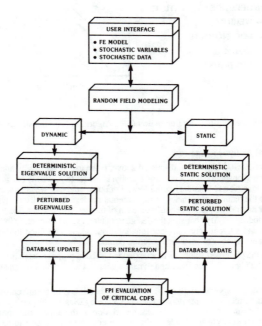

Figure 7. NESSUS Code Structure Overview

An algorithm for the dynamic problem does not yet yield the same relative efficiency, due to the number of eigenmodes required for the iterative solution algorithm. A "smart" subspace iteration scheme where the eigenvectors for the unperturbed system are used as trial vectors for the perturbed systems is currently being used.

The finite element solution package available in NESSUS is summarized in Figure 8. As discussed by Dias and Nagtegaal, the FEM code is based on a mixed variational principle that results in a full nodal solution capability for all solution variables (displacements and stresses). The user interface makes extensive use of keywords for defining data operations. The solution capability has recently been extended to include transient, nonlinear analysis for classical von Mises plasticity, thermoviscoplasticity, and large deformation/displacement conditions. The element formulations emphasize surface node element definitions using strain- and stress-based hybrid elements. The strain-based, enhanced solid element formulation has been successfully used for modeling a range of thick shell test problems.

NODAL SOLUTION STRATEGY
- ALL INPUT DATA
- EQUILIBRIUM ITERATION
- ALL OUTPUT DATA

GENERAL SOLUTION CAPABILITY
- LINEAR, NONLINEAR
- STATIC, DYNAMIC
- IN-CORE SOLUTION
- MULTIPLE ELEMENT TYPES
- RANDOM LOADS
- RANDOM VARIABLES

USER INTERFACE STRATEGY
- KEYWORD DATA STRUCTURE
- INTERFACE TO NESSUS/PRE
- DATABASE TRANSLATORS
- EXPERT SYSTEM

ADVANCED ELEMENT FORMULATIONS
- SURFACE NODE BASED
- HYBRID SHELL/PLATE
- ENHANCED SOLID
- SPECIAL THERMAL LOADS

Figure 8. NESSUS/FEM Code Developed for Probabilistic Modeling

Stochastic thermoviscoplastic material behavior is seen to be an important area of development for PSAM. A report on this is being given in this Conference [7]. The importance of this work is in being able to describe the uncertainty of material stress-strain-failure behavior, based on the statistics of the microstructure of the material.

3. CONCLUSIONS

The development of a new level of probabilistic structural analysis methods has been achieved by the synthesis of standard structural analysis methods with advanced reliability methods. The application of these methods to significant structural systems is under way [8]. The results to date clearly demonstrate the applicability of the PSAM approach to significant design issues related to structural reliability. The uncertainty of structural response realistically includes the effects of all real design variables, in a manner that provides new insights to the design engineer. The importance of each random variable is easily highlighted, based both on the mechanical importance of the variable, and the statistical uncertainty of the variable.

Application of the PSAM analysis methodology to new designs is now envisioned. Design criteria must, however, be reconsidered in probabilistic terms. The design community must now step up to this challenge by careful consideration of past, safety margin based designs. Only in this way can higher performance structural designs be achieved in a manner that maintains sufficient reliability, while reducing weight or increasing stress levels.

4. ACKNOWLEDGMENT

The work presented in this paper was supported by NASA-Lewis Research Center under Contract No. NAS3-24389.

5. APPENDIX

[1]   MILLWATER, H., PALMER K., and FINK, P., "NESSUS/EXPERT - An Expert System for Probabilistic Structural Analysis Methods," 29th AIAA/ASME/ASCE/AHS/ASC Structures, Structural Dynamics, and Materials Conference, Williamsburg, Virginia, April 18-20, 1988 (Paper No. 88-2372).

[2]   MILLWATER, H. R., DIAS, J. B., and RAVEENDRA, S. T., "The NESSUS Code for Probabilistic Structural Analysis," Proceedings of ICOSSAR '89, this volume.

[3]   WU, Y.-T., and WIRSCHING, P. H., "Advanced Reliability Methods for Probabilistic Structural Analysis," Proceedings of ICOSSAR '89, this volume.

[4]   DIAS, J. B., and NAGTEGAAL, J. C., "Efficient Algorithms for Use in Probabilistic Finite Element Analysis," in *Advances in Aerospace Structural Analysis* (eds. O. H. Burnside and C. H. Parr), ASME AD-09, pp. 37-50, 1985.

[5]   WU, Y.-T., "Demonstrations of a New, Fast Probability Integration Method for Reliability Analysis," *Journal of Engineering for Industry*, ASME, Vol. 109, pp. 24-28, February 1987.

[6]   CRUSE, T. A., BURNSIDE, O. H., WU, Y.-T., POLCH, E. Z., and DIAS, J. B., "Probabilistic Structural Analysis Methods for Select Space Propulsion System Structural Components (PSAM)," *Computers & Structures*, Vol. 29, No. 5, pp. 891-901, 1988.

[7]   HARREN, S. V., "A Mechanistically Based Framework for the Construction of Stochastic Constitutive Laws for Nonlinear Materials," Proceedings of ICOSSAR '89, this volume.

[8]   RAJAGOPAL, K. R., "Verification of the NESSUS Code on Turbine Engine Components," Proceedings of ICOSSAR '89, this volume.

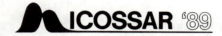
ADVANCED RELIABILITY METHODS FOR PROBABILISTIC

STRUCTURAL ANALYSIS

Y.-T. Wu
Department of Engineering Mechanics
Southwest Research Institute
6220 Culebra Road
San Antonio, Texas 78284

P. H. Wirsching
Department of Aerospace and Mechanical Engineering
The University of Arizona
Tucson, Arizona 85721

ABSTRACT

The probabilistic methods employed in the NASA/SWRI probabilistic
structural analysis program (PSAM) are summarized. An advanced reliability
method capable of treating correlated non-normal variates is employed. This
FPI (fast probability integration) scheme is employed with the most probable
point locus (MPPL) algorithm to construct CDF's of implicit functions of
random variables.

KEYWORDS

Reliability, Probabilistic Mechanics.

1. INTRODUCTION

NASA is currently funding a 5-year program called PSAM (probabilistic structural analysis methods)
aimed at developing probabilistic methods and structural analysis codes for the components of current
and future reusable space propulsion systems. A probabilistic finite element computer program,
NESSUS (numerical evaluation of stochastic structures under stress), is being developed as part of the
PSAM effort. The NESSUS code uses specially developed probabilistic analysis methods coupled with
efficient FEM perturbation algorithms. A summary of the probabilistic methods employed is provided
in this paper.

2. FAST PROBABILITY INTEGRATION (FPI)

In structural reliability analysis, a "performance function" or "limit state function" $g(\mathbf{X})$ is often formulated in terms of a vector of basic design factors, $\mathbf{X} = (X_1, X_2, ..., X_n)$, in which X_i's are independent random variables. The limit state which separates the design space into "failure" and "safe" regions is $g(\mathbf{X}) = 0$. By convention, the probability of failure, p_f, is denoted as

$$p_f = P[g < 0] . \tag{1}$$

An exact solution of p_f requires the integration of a multiple integral denoted as

$$p_f = \int_\Omega f_{\mathbf{X}}(\mathbf{x})d\mathbf{x} , \tag{2}$$

where $f_{\mathbf{X}}(\mathbf{x})$ is the joint pdf of $\mathbf{X}$ and Ω is the failure region. The solution of this multiple integral is, in general, extremely complicated.

Current approaches, based on analytical approximation methods, include first- and second-order reliability methods (FORM ad SORM) [1-8]. Another algorithm was developed for fast and accurate point probability estimates of explicit functions of independent random variables [9-11]. Features of this fast probability integration (FPI) scheme are: (1) It approximates the limit state by a quadratic at the design point. (2) It transforms the quadratic form into a linear one. (3) It employs an optimization routine to approximate non-normal variates as equivalent normals, thereby approximating the limit state as linear in normally distributed design factors. The FPI scheme is described in detail in Ref. [11], so only a brief summary is present here.

Assume that the g-function is <u>linear</u>:

$$g = a_0 + \sum_{i=1}^{n} a_i X_i . \tag{3}$$

If, in addition, all X_i are <u>normally distributed</u> with mean and standard deviation of μ_i and σ_i, respectively, then $p_f = \Phi(-\beta)$, where

$$\beta = \frac{\mu_g}{\sigma_g} = \frac{a_0 + \sum\limits_{i=1}^{n} a_i \mu_i}{\sqrt{\sum\limits_{i=1}^{n} a_i^2 \sigma_i^2}} . \tag{4}$$

For all non-normal variates, we wish to establish "high-quality" equivalent normals. The approach is to determine the equivalent normal of X using curve-fitting, i.e., the original cdf of X, $F_X(x)$, will be fitted by an approximating function H(x). The choice for H(x) is a three-parameter normal [2, 10],

$$F_X(x) \simeq H(x) = A\Phi(u) , \tag{5}$$

where $u = (x - \mu_N)/\sigma_N$ and A, μ_N, and σ_N are the three parameters to be determined. These are computed by solving an optimization problem in which the integral of the square of the difference between H(x) and F(x), multiplied by a suitable weighting function, is minimized [11]. Using these parameters for each non-normal variate, p_f, is estimated as

$$p_f = \Phi(-\beta') \prod_{i=1}^{n} A_i , \tag{6}$$

where β' is computed using (4) and μ_N and σ_N.

For a linear g-function, (6) is sufficient. But, in general, the g-function is non-linear and is linearized as follows: The first step is to approximate the limit state by a second-order polynomial using Taylor's series expansion about the design point, $\mathbf{X}^*$:

$$g(\mathbf{X}) \simeq \sum_{i=1}^{n} a_i(x_i - x_i^*) + b_i(x_i - x_i^*)^2 , \tag{7}$$

where the partial derivatives, a_i and b_i, can be evaluated using numerical methods.

The next step is to linearize (7) as

$$g(\mathbf{Y}) = c_0 + \sum_{i=1}^{n} b_i Y_i \tag{8}$$

in which

$$c_0 = -\frac{1}{4} \sum_{i=1}^{n} \frac{a_i^2}{b_i} \tag{9}$$

$$Y_i = \left[X_i - \left(x_i^* - \frac{a_i}{2b_i} \right) \right]^2 . \tag{10}$$

The later defines the transformation from X_i to Y_i. The log transformations for the X_i's and the g-function can also be included in a selection process to define the best linearization model [10]. Note that the Y_i's are independent, because each Y_i is a function of X_i alone, and that the distribution of Y_i may be computed easily from X_i. The transformation would become too complicated if the second-order cross-product terms were included in (8).

The FPI algorithm as described above has now been employed by several investigators on a wide variety of problems. It has been found to be fast, accurate, and robust. Within the NESSUS system, the FPI has been used mainly for processing explicit response functions which are created at the approximation points determined by an advanced mean-value method.

3. CORRELATED RANDOM VARIABLES

The FPI method as described above provides reliability estimates for the case where $\mathbf{X}$ is a vector of independent variates. If some, or all, of the variables are correlated, a transformation can be made to uncorrelated standard normals prior to the FPI analysis. The method is summarized as follows: Let $\mathbf{X} = (\mathbf{X}_1, \mathbf{W})$, where $\mathbf{X}_1$ is a vector of variates, independent of all others, and $\mathbf{W}$ is a vector of correlated variables. Consider any two correlated variates, W_i and W_j. In general, both are non-normal and they have marginal CDF's, F_i and F_j, respectively. They have correlation coefficient $\rho_{W_i W_j}$. Transform W_i and W_j to correlated standard normal variates u_1 and u_2:

$$W_i = F_i^{-1}[\Phi_i(u_1)] = T_i(u_i) \ . \tag{11}$$

To find $\rho_{u_i u_j} \equiv \rho$ (for simplicity), consider

$$\rho_{W_i W_j} = \frac{E(W_i W_j) - \mu_{W_i} \mu_{W_j}}{\sigma_{W_i} \sigma_{W_j}} \ . \tag{12}$$

Define $H(u_i, u_j) \equiv W_i W_j = T_i(u_i) \cdot T_j(u_j)$. Taking a Taylor's series expansion of H about $(u_1, u_2) = (0, 0)$ and then the expected value, it can be shown that the numerator of (12) becomes

$$\rho_{W_i W_j} \sigma_{W_i} \sigma_{W_j} = \rho\left[H_{11} + \frac{1}{2}(H_{13}+H_{31}) + \frac{1}{8}(H_{15}+2H_{33}+H_{51}) + \frac{1}{48}(H_{17}+3H_{35}+3H_{53}+H_{71})\right]$$
$$+ \frac{\rho^2}{2}\left[H_{22} + \frac{1}{2}(H_{24}+H_{42}) + \frac{1}{8}(H_{26}+2H_{44}+H_{62})\right] \tag{13}$$
$$+ \frac{\rho^3}{6}\left[H_{33} + \frac{1}{2}(H_{53}+H_{35})\right] + \frac{\rho^4}{24}[H_{44}] + \text{H.O.T.}$$

where

$$H_{k\ell} = \frac{\partial^k T_i}{\partial u_i^k} \cdot \frac{\partial^\ell T_j}{\partial u_j^\ell}\bigg|_{u=0} . \tag{14}$$

Upon evaluation of $H_{k\ell}$, (13) is of the form

$$\rho_{W_i W_j} \sigma_{W_i} \sigma_{W_j} = A_1\rho + A_2\rho^2 + A_3\rho^3 + A_4\rho^4 \ . \tag{15}$$

This equation can be solved for ρ. A similar operation can be performed for all pairs of dependent vectors in **W**. Thus, we have transformed **W** into **U**, a vector of standard normals with correlation coefficient matrix C_U.

At this point, an orthogonal transformation is made to obtain uncorrelated standard normals **V** [12]. The original g-function through substitution of **W** = W(U) and **U** = U(V) is now $g'(X_1, V)$. FPI now operates directly on this.

4. ADVANCED MEAN-VALUE FIRST-ORDER METHOD TO OBTAIN CDF'S OF IMPLICIT FUNCTIONS

The probability distribution of a response function (e.g., stresses, displacements) can be calculated efficiently employing the advanced mean-value first-order method (AMVFO), an abbreviated form of the most probable point locus method (MPPL) [13]. The scheme has wide application in probabilistic mechanics and design.

Consider a response function

$$g = g(X) \ ,$$

where **X** is a vector of n random design factors and g can be either explicit or implicit. Determine the CDF of g, denoted as F_g. If g is an explicit function of **X**, then the construction of F_g is straightforward using Monte Carlo or fast probability integration. All numerical reliability methods require many (100 to 10,000) function evaluations; a very fast operation with a digital computer if g is an explicit function of **X**.

However, when g is an implicit function (e.g., g(X) defined only through a finite element code), a single function evaluation may be costly. So, the fundamental question is to construct a high-quality CDF of g with a very minimum of function evaluations.

The MPPL may be close to the optimum in requiring a minimum number of function evaluations relative to the accuracy of F_g. There is no formal proof of this, but intuition suggests that it would be difficult to produce a reasonable estimate of F_g with fewer function evaluations.

A summary description of MPPL, an iterative process, follows:

1. Approximate g as a linear function of X. This requires solutions at the mean value and at small perturbations about the mean to evaluate the parameters of the linear function.

2. Now that an explicit function (albeit approximate) is available, reliability methods (e.g., [11]) can be used to approximate probabilities in selected points in the sample space of g. This first approximation to the F_g, called the mean-value first-order method (MVFO) is, in general, not likely to be accurate.

3. To improve the estimate of F_g, the function g is evaluated at each design point. These are "improved" g's at each probability level. This "first move" in MPPL is called AMVFO. Experience has shown that AMVFO provides remarkably accurate estimates of F_g in most cases.

4. A linear approximation to g can be obtained at each of the design points. This requires, again, perturbed solutions to g. And, again, a fast probability integration method can be employed for point probability estimates in the "second move" to construct F_g. An improved F_g is obtained at each g.

5. For the "third move," the function is evaluated at the design points computed in step 4.

While steps 4 and 5 can be repeated to improve the estimate of g, it has been found that AMVFO generally provides an accurate estimate of F_g.

Shown in Fig. 1 is an example of a construction of the CDF of a performance function Δ. Here, Δ is an explicit function, but it is useful for demonstration purposes. Using the linear approximation and FPI at selected points (mean plus three points on either side), we get a first approximation to F_Δ. As expected, this approximation is not very good.

In the second iteration (AMVFO), we improve F_Δ at each probability level by performing a function evaluation at each design point. Figure 1 shows that the AMVFO approximation is quite good. To construct this CDF, 14 function evaluations were required. This still could be very expensive, but we feel this is about the minimum possible for the high-quality CDF obtained.

5. MODIFICATION OF THE AMVFO SOLUTION FOR NON-MONOTONIC FUNCTIONS

The algorithm described above does not work in cases of a non-monotonic response function with multiple competing most probable points for each response value. However, in such cases, the AMVFO-based CDF would also be non-monotonic, suggesting that a modification is necessary [13]. The modification can be done as follows: (1) For a concave g-function, the CDF plot would identify a minimum value of g, g_{min}. For $g > g_{min}$, two CDF's, F_1 and F_2, can be identified for each g. Let $F_2 > F_1$. Then the modified CDF is simply $(F_2 - F_1)$. (2) For a convex g-function, a g_{max} can be identified. For $g < g_{max}$, the modified CDF is $[1 - (F_2 - F_1)]$.

Shown in Fig. 2 is an example applying the modified procedure. The problem considers a simulated turbine blade model made of a single crystal material. The material orientation is characterized by the three random angles

$$\theta_1 \sim \text{Normal}(-5°, 3.87°) \quad \theta_2 \sim \text{Normal}(-2°, 3.87°) \quad \theta_3 \sim \text{Normal}(3°, 3.87°)$$

in which the two parameters are the mean and the standard deviation. The response function is the first bending natural frequency which is a convex function within the significant probabilistic range.

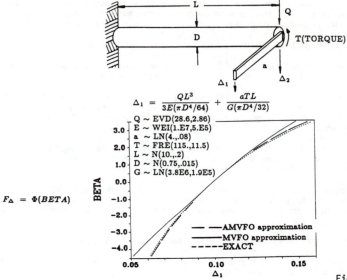

$$\Delta_1 = \frac{QL^3}{3E(\pi D^4/64)} + \frac{aTL}{G(\pi D^4/32)}$$

$$F_\Delta = \Phi(BETA)$$

Figure 1

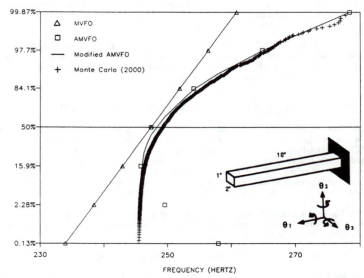

Figure 2

Fig. 2 shows that the original AMVFO solution has a parabolic shape. The modified solution $(F_2 - F_1)$ agrees well with the Monte Carlo solution which was produced by running the NESSUS finite element program 2000 times. An important result is that the frequency distribution is left-truncated and has a strong right tail. The fact that the CDF appears to be accurate at the tail regions and that the error is greater around 0.5 is not surprising because the function is strongly non-monotonic in the 0.5 region. This suggests that following the modification, a mean-value second-order analysis may be needed to improve the accuracy in this region. Further discussions can be found in Ref. [14].

6. REFERENCES

[1]   RACKWITZ, R. and FIESSLER, B.: "Structural reliability under combined random load sequences," Journal of Computers and Structures, Vol. 9, pp. 489-494, 1978.

[2]   CHEN, X. and LIND, N. C.: "Fast probability integration by three parameter normal tail approximation," Structural Safety, Vol. 1, pp. 169-176, 1983.

[3]   DITLEVSEN, O.: "Principle of normal tail approximation," Journal of Engineering Mechanics Division, ASCE, Vol. 107, pp. 1191-1208, 1981.

[4]   FIESSLER, B., NEUMANN, H. J., and RACKWITZ, R.: "Quadratic limit states in structural reliability," Journal of the Engineering Mechanics Division, ASCE, Vol. 105, pp. 661-676, 1979.

[5]   BREITUNG, K.: "An asymptotic formula for the failure probaiblity," DIALOG 6-82, Euromech 155, Danmarks Ingeniorakadem., Lyngby, Denmark, 1982.

[6]   TVEDT, L.: "Two second order approximations to the failure probability," Det Norske Veritas (Norway), RDIV/20-004-83, 1983.

[7]   HOHENBICHLER, M. and RACKWITZ, R.: "Non-Normal Dependent Vectors in Structural Safety," Journal of the Engineering Mechanics Division, ASCE, Vol. 100, pp. 1227-1238, 1981.

[8]   MADSEN, H. O., KRENK, S., and LIND, N. C.: Methods of Structural Safety, Prentice-Hall, 1986.

[9]   WU, Y.-T.: "Efficient methods for mechanical and structural reliability analysis and design," Ph.D. dissertation, Univeristy of Arizona, 1984.

[10] WU, Y.-T. and WIRSCHING, P. H.: "A new algorithm for structural reliability estimation," Journal of the Engineering Mechanics Division, ASCE, Vol. 113, pp. 1319-1336, (1987).

[11] WU, Y.-T.: "Demonstration of a fast, new probability integration method for reliability analysis," Journal of Engineering for Industry, ASME, Vol. 109, pp. 24-28 (1987).

[12] HASOFER, A. M. and LIND, N. C.: "Exact and invariant second moment code format," Journal of the Engineering Mechanics Division, ASCE, Vol. 100, pp. 111-121, 1974.

[13] WU, Y.-T., BURNSIDE, O. H., and CRUSE, T. A.: "Probabilistic methods for structural response analysis," Computational Mechanics of Reliability Analysis (W. K. Liu and T. Belytschko, Eds.), Elmepress International (to appear).

[14] Wu, Y.-T, MILLWATER, H.R., and CRUSE, T.A.: "An advanced probabilistic structural analysis method for implicit performance functions," Proceedings of the 30th AIAA/ASTM/AHS/ASC Structures, Structural Dynamics and Materials Conference, pp. 1852-1859, 1989.

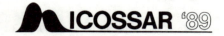
THE NESSUS SOFTWARE SYSTEM FOR PROBABILISTIC STRUCTURAL ANALYSIS

H.R. Millwater*, Y.-T. Wu*, J.B. Dias, R.C. McClung*,
S.T. Raveendra*, and B.H. Thacker***

> *   **Department of Engineering Mechanics**
> **Southwest Research Institute**
> **6220 Culebra Road**
> **San Antonio, TX 78284**

> **   **Division of Applied Mechanics**
> **Stanford University**
> **Stanford, CA 94305**

ABSTRACT

The Probabilistic Structural Analysis Methods (PSAM) project developed at SwRI integrates state-of-the-art structural analysis techniques with probability theory for the design and analysis of complex large-scale engineering structures. An advanced efficient software system (NESSUS) capable of performing complex probabilistic analysis has been developed. A number of software components are contained in the NESSUS system and include: an expert system, a probabilistic finite element code, a probabilistic boundary element code and a fast probability integrator. This paper discusses the NESSUS software system and its ability to carry out the goals of the PSAM project.

KEYWORDS

Probability, reliability, finite element, boundary element, expert system, fast probability integrator.

1. INTRODUCTION

The probabilistic structural analysis methods (PSAM) effort currently under development at Southwest Research Institute provides engineers with a new means for more effective design of structures. PSAM was developed to determine the effect of uncertainties or randomness in the design variables have on the structure. The goal is to develop efficient and robust probabilistic structural analysis software and a PSAM

knowledge base. The development effort is supported by NASA under contract NAS3-24389, titled "Probabilistic Structural Analysis Methods for Select Space Propulsion System Structural Components [1-4]." The immediate application is to assist in the design of the next generation Space Shuttle main engine; however, the technology in PSAM can be applied to nearly all structural analysis.

2. OVERVIEW

An overview of the PSAM program is shown in Fig. 1. The PSAM software system is entitled "NESSUS" for Numerical Evaluation of Stochastic Structures Under Stress. NESSUS consists of a number of analysis modules integrated through a common database. A summary of the components are:

NESSUS/EXPERT - expert system for the PSAM package

NESSUS/FEM - general purpose probabilistic finite element program

NESSUS/PBEM - general purpose probabilistic boundary element program

NESSUS/FPI - fast probability integrator

NESSUS/PRE - preprocessor to decompose correlated random fields

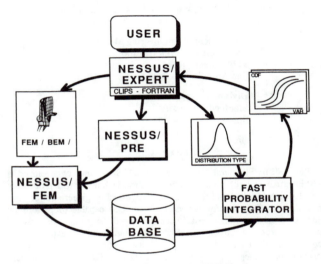

Fig. 1. Overview of the NESSUS System

The structural evaluation tools are a probabilistic finite element code (FEM) and a probabilistic boundary element code (PBEM). These codes obtain the structural response i.e., stress, displacement, natural frequency, etc., and the structural sensitivities of the response with respect to random variables such as geometry, material properties, loads, etc.

The probabilistic analysis is performed using a fast probability integrator program (FPI). FPI obtains the desired response data and statistical data from the database and computes the cumulative distribution function CDF and probability density function PDF.

Other utilities are included in NESSUS such as NESSUS/PRE which uncorrelates partially correlated random fields into a set of uncorrelated random vectors.

An expert system is included to capture and utilize PSAM knowledge and experience. NESSUS/EXPERT is an interactive menu-driven expert system that provides information to assist in the use of the probabilistic finite element code NESSUS/FEM and the fast probability integrator (FPI).

The various programs and their operation will be discussed in some detail to demonstrate the technology of the NESSUS system.

3. EXPERT SYSTEM

Probabilistic analysis of structures will be new to most engineers and a significant amount of new technology has been developed and incorporated within PSAM. The NESSUS finite element, boundary element and fast probability integrator codes contain many options and algorithms unfamiliar to many engineers. As a result, the engineer will not have existing knowledge on which to base decisions. Also, appropriate probabilistic input for random variables is often not known. Therefore, an expert system has been developed and incorporated into NESSUS in order to guide and assist the engineer in the effective and efficient use of probabilistic structural analysis technology.

The majority of the development to date has concentrated on providing an intelligent guide for data input for the finite element code. Several capabilities have been implemented including: an intelligent menu selection procedure to guide the user in appropriate input choices, error checking at various levels to insure functional input data, help screens to explain input formats and provide detailed explanations of keywords, advice and default values of probabilistic input to assist the user in selecting data from which the user is unfamiliar or where a knowledge base is not established.

The expert system serves as a place to compile and exploit PSAM knowledge as it is acquired. Future enhancements will involve links to other analysis packages and the development of solution analysis algorithms to determine the accuracy of the probabilistic solution. Further details on EXPERT can be found in [5].

4. PROBABILISTIC FINITE ELEMENT ANALYSIS

The PSAM project is designed to perform probabilistic structural analysis on realistic engineering structures under complex loading environments. The NESSUS system contains a state-of-the-art nonlinear probabilistic finite element code, NESSUS/FEM, to determine the structural response and sensitivities for such structures. Special algorithms have been developed and implemented to perform sensitivity analysis efficiently [13].

NESSUS/FEM contains state-of-the-art finite element technology. The finite element kernel is based on a mixed method finite element variational statement. The mixed method contains stresses and strains as well as displacements as nodal degrees of freedom; however, only the displacement stiffness matrix is factorized. The solutions for stresses and strains are determined through an iterative technique. This method has many advantages over a purely displacement based approach; the primary advantages being improved stress results in a continuous stress field, see [10,14].

A broad range of analysis capabilities and an extensive element library is present. A summary of the current analysis capabilities include static, eigenvalue, harmonic, random vibration, transient and buckling analyses. A variety of random variables are available such as geometry, loads, material properties, temperatures, boundary conditions, harmonic excitation, power spectral density input, etc. As an example of a geometric random variable, one might have the tilt, lean or twist of a turbine blade, the radius of a fillet, the thickness of a valve flange, etc.

A summary of the current code capabilities is presented below in Fig. 2.

ANALYSIS TYPES	ELEMENT LIBRARY	RANDOM VARIABLES
STATIC	BEAM	GEOMETRY
NATURAL FREQUENCY	PLATE	LOADS
		FORCES
BUCKLING	PLANE STRESS	PRESSURES
		TEMPERATURES
HARMONIC EXCITATION	PLANE STRAIN	
		MATERIAL PROPERTIES
RANDOM VIBRATION	AXISYMMETRIC	ELASTIC MODULUS
		POISSON'S RATIO
	3D SOLID	SHEAR MODULUS
		ORIENTATION ANGLE
NONLINEAR		
MATERIAL		
GEOMETRY		

Fig. 2. NESSUS/FEM Capabilities

Probabilistic structural analysis requires the sensitivity of the structural response to the random variables. These sensitivity calculations can be very expensive; therefore, efficient algorithms are essential for probabilistic analysis of engineering structures. NESSUS/FEM contains special "perturbation" algorithms to obtain the response to a perturbed set of random variables efficiently. Perturbation algorithms provide efficient ways of computing the sensitivity of the structural response to small fluctuations of the random variables about a deterministic value. For static analysis and transient dynamic analysis, the previously factored and stored "unperturbed or deterministic" stiffness matrix is used as an approximation to the "perturbed" stiffness matrix. The solution to the perturbed problem is then computed using an approach similar to a modified-Newton solution algorithm. Only the deterministic stiffness matrix must be factorized. The perturbed solution is determined through back substitution. For eigenvalue analysis, the perturbed eigenvectors are computed as a linear expansion of the unperturbed eigenvectors.

These perturbation algorithms have several advantages over other sensitivity computation techniques. All computations are performed at the element level and the partial derivatives of the element stiffness and load vector do not need to be stored. The perturbation algorithms are independent of element formulation and, therefore, can be introduced into an existing code with a minimum of effort [10,13].

NESSUS has recently been extended into nonlinear material and geometric areas. The extension of the perturbation and solution algorithms to nonlinear problems is straightforward. Random variables include nonlinear parameters such as yield stress and hardening parameters. Further details can be found in [10].

NESSUS/FEM has been validated on a number of test problems and compared with exact solutions. Validation problems have consisted of: static, natural frequency, buckling, harmonic excitation and random vibration. In all cases, excellent results have been obtained. Future validation problems will be performed on nonlinear material and geometry problems. Further details can be found in [6].

5. FAST PROBABILITY INTEGRATION

The probabilistic model implemented in NESSUS is based on an extension of the Rackwitz-Fiessler (R-F) and Chen-Lind (C-L) fast probability integrator (FPI) methods. In addition, a highly efficient algorithm

Advanced Mean Value First Order method (AMVFO) couples the finite element and FPI codes to efficiently obtain CDF's of large scale engineering structures. The AMVFO method has been demonstrated to have significant advantages in efficiency and robustness over traditional first and second order reliability methods (FORM and SORM).

The FPI method implemented in NESSUS is based on an improved version of the Rackwitz-Fiessler (R-F) and Chen-Lind (C-L) techniques. NESSUS/FPI features (a) an improved scheme for constructing the equivalent normal distributions and (b) a quadratic approximation of the original limit state with log transformation options. This approach has been demonstrated to produce improved results over traditional methods at a minimal cost increase [7].

A number of options and features are available in NESSUS/FPI and are listed in Table 1.

Table 1. Summary of NESSUS/FPI Options

ADVANCED MEAN VALUE FIRST ORDER METHOD
LINEAR AND QUADRATIC G FUNCTION APPROXIMATION
CONFIDENCE LEVELS
SENSITIVITY FACTORS
EMPIRICAL DISTRIBUTION FUNCTIONS
LIBRARY OF STANDARD DISTRIBUTION FUNCTIONS
HARBITZ METHOD MONTE CARLO

One very important by-product of FPI analysis is the random variable sensitivity factors.

The sensitivity factors provide the engineer with an indication of the relative significance of the random variables in the analysis. This information can be used to determine the most significant random variables to control and the most cost-effective way to optimize the structure.

Advanced Mean Value First Order Method (AMVFO)

An entirely new solution procedure entitled Advanced Mean Value First Order method (AMVFO) couples the finite element and FPI codes [6]. The AMVFO method is an extremely efficient and robust procedure for performing probabilistic analyses of large-scale complex engineering structures whose solution times are computationally expensive. Although developed specifically for PSAM and probabilistic finite element analysis, AMVFO can be applied to any limit state whether implicit or explicit.

Figure 6 demonstrates the Advanced Mean Value procedure. A Mean Value First Order (MVFO) solution is obtained given the mean value and sensitivities of the response function. The MVFO solution is updated to the advanced mean value first order solution by evaluating the response at the "design point" (or most likely combination of random variables) at each probability level.

6. EXAMPLE PROBLEM

An example problem demonstrating the results of the application of the Nessus software is presented below. A number of other validation problems have successfully been executed with Nessus [6]. In addition, several large scale verification analyses can be found in [15]. More complete details of the AMVFO and other probabilistic methods developed for the PSAM project are documented in [6,7,9].

Example - Cantilever Plate

This problem is a cantilever plate subjected to correlated random static loadings as shown in Fig. 3. The other random variables are elastic modulus, thickness, width and a base spring stiffness as defined in Table 2. The response functions considered are the bending stress, S, at the base and the tip displacement, δ.

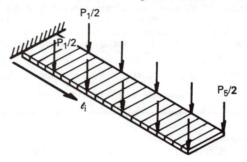

Fig. 3. Cantilever Plate

Table 2. Variables for Problem 1

Variables	Distri.	Median	COV
Modulus, E	Lognormal	10E6 psi	0.03
Length, L	Lognormal	20.0 in	0.05
Thickness, t	Lognormal	0.1 in	0.05
Width, w	Lognormal	1.0 in	0.05
Base Spring, K	Lognormal	1E5 lb-in/rad	0.05
Loads*,			
P_1 to P_5*	Normal	0.1 lb (mean)	0.1

* Note: Correlated with correlation coefficients = exp
　　　　　(-distance between loads / 20)

The finite element model consisted of 20 shell elements with 42 nodes as shown in Fig. 2. The NESSUS "mean" solutions were 0.7648 inches for the displacement and 3657 psi for the stress. These values agreed with theory - 0.7692 inches and 3600 psi, respectively. The differences are 0.5% for the displacement and 1.6% for the stress.

For either the displacement or the stress, the probabilistic solutions were checked by selecting two points in the right tail of the distribution (i.e., cumulative probability > 50%).

The MVFO, the AMVFO, and the first-iteration solutions for the displacement and the stress, respectively, are shown in Figs. 3 and 4. The exact solution shown in the figures was generated by applying Monte Carlo simulation (sample size = 100,000) to the theoretical solutions.

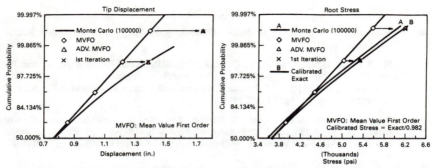

Fig. 4. CDF of the Plate Tip Displacement   **Fig. 5. CDF of the Cantilever Plate Root Stress**

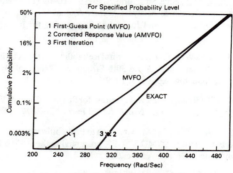

Fig. 6. Iteration Algorithm

By comparing the NESSUS solutions with the adjusted solutions, it can be concluded that the AMVFO and the first-iteration solutions provide excellent probability estimates.

7. CONCLUSIONS

A number of sophisticated computer programs have been developed and integrated to perform probabilistic analysis of structures. State-of-the-art finite element and boundary element codes are available for structural analysis. A fast probability integration program performs probabilistic analysis on the structural response. The codes are coupled through a very efficient "Advanced Mean Value First Order (AMVFO) algorithm. The focus has been on analyzing realistic complex engineering structures; therefore, great emphasis has been placed on developing algorithms for "efficient" analysis. PSAM technology has been successfully applied to a number of validation problems and complex structures.

2290   STRUCTURAL SAFETY AND RELIABILITY

REFERENCES

[1-4]   "Probabilistic Structural Analysis Methods (PSAM)," Annual Reports 1st through 4th, NASA
        Contract NAS3-24389.

[5]   MILLWATER, H.R., PALMER, K., FINK, P., "NESSUS/EXPERT - An Expert System for
        Probabilistic   Structural   Analysis   Methods,"   presented   at   the   29th
        AIAA/ASME/ASCE/AHS/ASC Structures, Structural Dynamics and Materials Conference,
        Williamsburg, Virginia, April 18-20, 1988.

[6]   WU, Y.-T., and BURNSIDE, O.H., "Validation of the Nessus Probabilistic Finite Element
        Analysis Computer Program," 29th AIAA/ASME/ASCE/AHS/ASC Structures, Structural
        Dynamics, and Materials Conference, Williamsburg, Virginia, (Paper No. 88-2372), April 18-20,
        1988.

[7]   WU, Y.-T., "Demonstration of a New, Fast Probability Integration Method for Reliability
        Analysis, Advances in Aerospace Structural Analysis, AD-09 (Proceedings of Symposium on
        Probabilistic Structural Design & Analysis, Winter Annual Meeting of ASME, Miami Beach,
        Florida, November 17-22, 1985); and Journal of Engineering for Industry, ASME New York,
        February 1987.

[8]   MILLWATER, H.R., and WU, Y.-T., "Structural Reliability Analysis Using a Probabilistic Finite
        Element Program, to be presented at the 30th Structures, Structural Dynamics, and Materials
        Conference, Mobile, Alabama, April 3-5, 1989.

[9]   WU, Y.-T., and WIRSCHING, P., "New Algorithm for Structural Reliability Estimation," Journal
        of Engineering Mechanics, Vol. 113, No. 9 (ASCE Paper No. 21770), September 1987.

[10]   DIAS, J.B., and NAKAZAWA, S., "An Approach to Probabilistic Finite Element Analysis Using
        a Mixed-iterative Formulation," AMD Vol. 93, presented at the ASME/SES Summer Meeting,
        University of California at Berkeley, June 20-22, 1988.

[11]   DIAS, J.B., and NAGTEGAAL, J.C., "Efficient Algorithms for Use in Probabilistic Finite
        Element Analysis," Advances in Aerospace Structural Analysis, O.H. Burnside, and C.H. Parr
        (Eds.), AD-09, ASME, pp. 37-50, 1985.

[12]   CRUSE, T.A., and WU, Y.-T., DIAS, J.B., and RAJAGOPAL, K.R., "Probabilistic Structural
        Analysis and Applications," Computers and Structures, Vol. 30, No. 1/2, pp. 163-170, 1988.
        (Presented at the Symposium on Advances and Trends in Computational Structural Mechanics
        and Fluid Dynamics, Washington, DC, October 17-19, 1988).

[13]   DIAS, J.B., NAGTEGAAL, J.C., and NAKAZAWA, S., "Iterative Perturbation Algorithms in
        Probabilistic Finite Element Analysis," to appear in Computational Mechanics of Probabilistic
        and Reliability Analysis, Liu and Belytechko (Eds.), ELME Press.

[14]   NAKAZAWA, S., NAGTEGAAL, J.C., and ZIENKIEWICZ, O.C., "Iterative Methods for
        Mixed Finite Element Equations," AMD Vol. 74, ASME, pp. 57-67, 1985.

[15]   RAJAGOPAL, K.R., DEBCHAUDHURY, A., and NEWELL, J.F., "Verification of Nessus Code
        on Space Propulsion Components," to be presented at the International Conference on Structural
        Safety and Reliability, San Francisco, California, August, 1989.

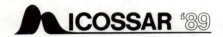

PROBABILISTIC ANALYSIS OF STRUCTURES COMPOSED OF
PATH DEPENDENT MATERIALS

Stephen V. Harren

Senior Research Engineer, Southwest Research Institute, San Antonio, TX, USA

ABSTRACT

A procedure is outlined for calculating the time-evolving CDF's (cumulative probability distribution functions) of structural response variables for cases in which the structure's deformation is history dependent. A structure exhibits history dependent response when it is subject to extreme loadings that cause plasticity, damage, etc., to its constituent materials. The solution procedure presented preserves the history dependence of the structure so that the response variable CDF's obtained are meaningful. Additionally, the procedure is based on the FPI (fast probability integration) method of Wu, which is capable of providing accurate probability estimates in the tails of the CDF's. Finally, the procedure is demonstrated by means of some simple example problems.

KEYWORDS

Probabilistic structural analysis; structural reliability; path dependence; history dependence; plasticity; damage; failure.

1. Introduction

The reliability and integrity of high performance structures and systems subject to extreme conditions is of primary concern to the aerospace and other industries. Additionally, as the operation envelope and performance of such systems are being continually increased, demands on material performance are also increasing. Obviously, the integrity of the system as a whole depends on the reliability of the structure's materials when subject to extreme operating environments. Thus, it is becoming increasingly more important that advanced design procedures and reliability analyses take into account variations in structural behavior resulting from uncertainties in material response.

A consequence of subjecting a material to extreme environments is that damage is induced, e.g., plasticity, void growth, microcracking, etc. Physically, the damage accumulation in a material under load is an evolutionary, history dependent process. The current state of a material, which includes its current state of damage, depends upon the loading history to which it has been subject. Additionally, a material's current response characteristics, which include its current rate of damage accumulation, depend on its current state and its current loading. Mathematically, such a process is termed *path dependent*. Hence, the focus of this paper is to demonstrate how to describe statistically the uncertain response of a structure when it is composed of material undergoing a path dependent process.

2. Scope of Problem

The main tenet taken here is that an element of real material, knowing its history dependent current state and its current loading, knows precisely how to deform and whether or not to fail. In other words, an element of real material does not deform or fail by means of a stochastic process, although of course, the loading history to which it is subject *may be* a stochastic process. If, in fact, the deformation and damage processes occurring in the material were stochastic, then this would be equivalent to having the state of the material element change under the influence of no thermodynamical force, which obviously, is physically unacceptable.

In structural analyses, an element of material is idealized as an homogeneous continuum which obeys some *phenomenological,* mathematical model of deformation and failure. The word "phenomenological" is used because the parameters of such a model usually are not directly related to readily observable, physical properties of the real material (such as grain size, dislocation density, etc.), but merely serve to describe the overall, gross characteristics of the material. Obviously, a phenomenological model is not an exact description of reality, but to maintain physical sensibility in the structural analysis, the continuum, phenomenological idealization of the material, like the real material, should exhibit path dependence, and should not deform or fail by means of a stochastic process.

The parameters appearing in phenomenological models for path dependent materials may be divided into three groups. The first is a set of history insensitive material constants, which theoretically are obtainable in some direct manner from the material's initial state, and whose values are unchanged by any subsequent deformation and/or damage. The second is a set of history dependent "internal" variables, which describe the extent of irreversible (or path dependent) processes that the material has undergone, such as plasticity. In order to model the material's path dependence, the values of these variables (at some time t, say) are determined from evolution equations that must be integrated from time 0 (the initial state) to time t. The third group is a set of state variables, such as stress and temperature. A concise description of the mathematical structure of path dependent phenomenological material models may be found in Section 2 of the paper by Cordts and Kollmann [1].

When performing a deterministic analysis of a structure composed of path dependent material, one applies some prescribed loading history to the structure, whose initial geometry and initial material state are specified. The initial material state is described by the history insensitive material constants and the initial values of the internal and state variables. The solution to the deterministic analysis problem gives the geometry of the structure and the state of the material, i.e., the values of the internal and state variables, as functions of time for $t>0$. Subsequently, variables which are obtained as a solution to a structural analysis problem will be referred to as *response* variables. As alluded to above, to obey the path dependence of the material, the structural analysis solution must be obtained in an incremental, evolutionary fashion by integrating from time 0 to some time t (see, e.g., Section 5 of reference [1]).

When performing reliability calculations for a structure, one needs to find the *statistical* description of some critical response variable. Obviously, finding these statistics requires that quantities which were prescribed in the deterministic analysis (i.e., loading history, initial structural geometry, and initial material state) now be prescribed as *random fields* with specified statistics (a "random field" is an uncertain function of space and/or time). Thus, the loading history is prescribed to be a random function of time, or equivalently, a random field in one dimension (i.e., time). Accordingly, the uncertainties in the structure's initial geometry are specified with a random field in two dimensions (i.e., the structure's surface coordinates). Finally, the material's initial state is described by random fields in three dimensions (i.e., space). In practice though, space and time are discretized, e.g., by finite elements and/or by finite differencing, which reduces these random fields to a set of discrete, correlated random variables (see, e.g., Liu, Belytschko and Mani [2]). Denote these (N, say) random variables by the vector $\mathbf{k}=(k_1,k_2,...,k_N)$. Thus, in the practical case, the uncertainties in the loading history, the structure's initial geometry, and the material's initial state are described entirely by $\mathbf{k}$, whose joint probability density function (PDF) is specified.

3. Solution Procedure

Denote the critical response variable of interest by r, where r depends on the random variables and time, i.e., $r=r(\mathbf{k},t)$. For the statistics obtained for r to be meaningful, they must be based upon *realizations* of r vs. t. By a "realization" is meant a physically possible occurrence, or within the context of the statistical structural/material/loading model, an occurrence that is admissible under the model's governing equations and the prescribed joint PDF of $\mathbf{k}$. Thus, a realization of r vs. t is obtained by subjecting a realizable initial structure, i.e., one with admissible initial geometry and admissible initial material state, to a realizable loading history. In short, a realization of r vs. t is a deterministic solution based on a fixed, admissible value of $\mathbf{k}$.

Now, in order to calculate the statistics of r at some time of interest t_0, the function $r=r(\mathbf{k},t_0)$ needs to be determined. Unfortunately, the path dependence of the problem makes the calculation of this function quite difficult, or at least very time consuming. Conceptually, $r=r(\mathbf{k},t_0)$ is constructed by calculating a realization of r vs. t for each and every fixed, admissible value of $\mathbf{k}$ in turn. This, at least conceptually, requires that infinitely many integrations from $t=0$ to $t=t_0$ be performed. Of course, in practice $r=r(\mathbf{k},t_0)$ is constructed using a finite set of fixed $\mathbf{k}$'s, which is chosen to span adequately $\mathbf{k}$'s range of admissible values. Call these (M, say) fixed $\mathbf{k}$-values $\mathbf{k}_i$, i=1,2,...,M. For each $\mathbf{k}_i$ in turn then, a realization of r vs. t is calculated incrementally from time 0 to the latest time of interest t_L. Obviously, if many realizations have to be calculated (i.e., if M is large), or if incremental finite element analysis has to be used to calculate each realization, then this task is potentially very time consuming. But, if r was not calculated in this manner, then its physical meaning would be lost (as explained in the first two paragraphs of Section 2), and consequently, the results obtained for r's statistics would also be meaningless. In any case, calling the M resulting r vs. t realizations $r_i=r(\mathbf{k}_i,t)$, i=1,2,...,M, at any time of interest t_0 where $0<t_0<t_L$, the M $(\mathbf{k}_i,r_i=r(\mathbf{k}_i,t_0))$ points can be fit to a polynomial of appropriate order to obtain an approximate analytical expression for $r=r(\mathbf{k},t_0)$. Knowing this, the statistics of r at time t_0 can be determined.

Specifically, one can input the function $r=r(\mathbf{k},t_0)$ and the joint PDF of $\mathbf{k}$ into the fast probability integration (FPI) method of Wu [3] to obtain r's cumulative probability distribution function (CDF) at time t_0, which will be denoted as $c=c(r,t_0)$. Finally, performing other polynomial fits to the M $(\mathbf{k}_i,r_i)$ points at various other times allows one to construct the evolution of $c=c(r,t)$ with time for $0<t<t_L$. This procedure will now be demonstrated via some simple examples.

4. Examples

Example 1. Here, the very simple case of one random variable is considered, i.e., N=1 so that $\mathbf{k}=(k_1)$. Now, for convenience, k_1 will be denoted by k, and the PDF of k is taken as

$$p(k) = -6k(k-1) \quad , \tag{1}$$

where the admissible range of k is 0<k<1. To span this admissible range, the k-values k=0.005, 0.250, 0.500, 0.750, and 0.995 are chosen, i.e., M=5. Thus, for each of these fixed k-values, say that a path dependent structural analysis was performed by calculating incrementally from t=0 to t=4 (i.e., t_L=4) to yield the five realizations of r vs. t depicted in Fig. 1. (In actuality, the r vs. t traces shown in the figure are plots of the function

$$r = t^k \quad . \tag{2}$$

Of course, in practice the analytical form of r vs. t would not be known, and eq. (2) is used here only for the sake of demonstration.) Now, say that t_0=2 and t_0=4 are two times of interest. Pursuantly, as depicted in Fig. 1, the functions $r=r(k,2)$ and $r=r(k,4)$ are constructed by fitting a polynomial of the form

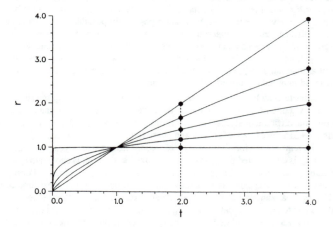

FIG. 1. REALIZATIONS OF r VS. t. Taken from Ref. [4].

$$r = a_1 + a_2 k + a_3 k^2 + a_4 k^3 + a_5 k^4 \tag{3}$$

to the five $(k_i, r_i=r(k_i,2))$ points and the five $(k_i, r_i=r(k_i,4))$ points, respectively. Using these two resulting polynomials and the PDF of k, i.e., eq. (1), in the numerical FPI method of Wu [3] yields the CDF's $c=c(r,2)$ and $c=c(r,4)$ shown in Fig. 2. The continuous curves in the figure are the exact analytical results obtained from eqs. (1) and (2), and the discrete, plotted points are the numerical, i.e., the FPI, results. As can be seen, this procedure works quite well. Just as a final point of interest, note that the analytical solution for the expected value of response is

$$\mu[r(t)] = \frac{6}{(\log t)^3} \{(1+t)\log t + 2 - 2t\} \quad , \tag{4}$$

which illustrates that the expected value of r vs. t does not correspond, most generally, to any realization of r vs. t.

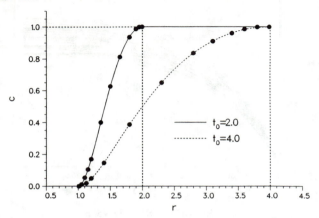

FIG. 2. EXACT AND CALCULATED CDF'S AT TWO DIFFERENT TIMES.
Taken from Ref. [4].

Example 2. This is a somewhat more realistic example in that the analytical form of r vs. t is not known. Here, the problem analyzed is a high temperature uniaxial tension test carried out on a polycrystalline material at a constant strain rate. The material model used is based on the physical mechanism of grain boundary sliding with diffusional accommodation, but the fine details of the model are not important for our present purposes (see reference [4] for more details). The response variable is taken to be the uniaxial stress, and the random inputs are Young's modulus $E=k_1$ and the grain boundary sliding activation energy $Q_b=k_2$, i.e., N=2 and $\mathbf{k}=(k_1,k_2)$. The variable k_1 is taken to have mean 150 GPa and standard deviation 10 GPa, i.e., $\mu(k_1)=150$ GPa and $\sigma(k_1)=10$ GPa. The variable k_2 has $\mu(k_2)=72.5$ kJ/mole and $\sigma(k_2)=0.5$ kJ/mole. It is assumed that k_1 and k_2 are statistically independent, and each is described by a truncated normal distribution with cut-off points at plus and minus three standard deviations from the mean. Twenty five $\mathbf{k}$-values (i.e., M=25) are chosen to span the admissible range of $\mathbf{k}$. These values correspond to all possible combinations of k_1=120.5, 135.0, 150.0, 165.0, and 179.5 GPa, and k_2=71.05, 71.75, 72.50, 73.25, and 73.95 kJ/mole. The 25 realizations based on these fixed $\mathbf{k}$-values are shown calculated out to t_L=150.0 sec in Fig. 3. Thus, $r=r(\mathbf{k},t_0)$ may be constructed by fitting the 25 obtained $(\mathbf{k}_i,r_i)$ points at any time of interest t_0 to the polynomial

$$r = a_1 + a_2 k_1 + a_3 k_1^2 + a_4 k_1^3 + a_5 k_1^4 \tag{5}$$

$$+ a_6 k_2 + a_7 k_1 k_2 + a_8 k_1^2 k_2 + a_9 k_1^3 k_2 + a_{10} k_1^4 k_2$$

$$+ a_{11} k_2^2 + a_{12} k_1 k_2^2 + a_{13} k_1^2 k_2^2 + a_{14} k_1^3 k_2^2 + a_{15} k_1^4 k_2^2$$

$$+ a_{16} k_2^3 + a_{17} k_1 k_2^3 + a_{18} k_1^2 k_2^3 + a_{19} k_1^3 k_2^3 + a_{20} k_1^4 k_2^3$$

$$+ a_{21} k_2^4 + a_{22} k_1 k_2^4 + a_{23} k_1^2 k_2^4 + a_{24} k_1^3 k_2^4 + a_{25} k_1^4 k_2^4 \quad .$$

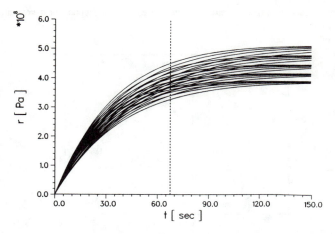

FIG. 3. REALIZATIONS OF r VS. t.

For example, making the fit at t_0=67.5 sec, and using the resulting polynomial and the joint PDF of k in the FPI method yields $c=c(r,67.5\ sec)$, which is shown in Fig. 4. Note that in the figure, a smooth curve is passed through the calculated CDF points, and the results are plotted in terms of the standard normal unit u, i.e.,

$$c = \frac{1}{2} + \frac{1}{2} \operatorname{erf}\left(\frac{u}{\sqrt{2}}\right) \qquad \text{with} \qquad u = \frac{r - \mu(r)}{\sigma(r)} \quad , \qquad (6)$$

where "erf" is the error function. Of course, the CDF at other times $0.0\ sec < t < 150.0\ sec$ can be constructed in the same manner.

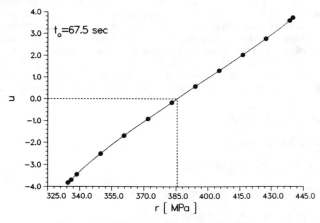

FIG. 4. CALCULATED CDF AT t=67.5 SEC.

Example 3. This example illustrates the procedure for finding the statistics of r when the material model of the analysis contains a failure criterion. As in Example 1, for simplicity the case of one random variable k is considered, whose PDF is

$$p(k) = -\frac{3}{4}(k-2)(k-4) \quad , \tag{7}$$

and whose admissible range is $2<k<4$. Now, suppose that path dependent analyses were performed out to failure for the fixed k-values $k=2$, 3, and 4, which gave the realizations shown by the dashed curves in Fig. 5. (The plotted points at the end of each dashed curve are used to denote failure. Also, the meanings of the solid curves will be explained later.)

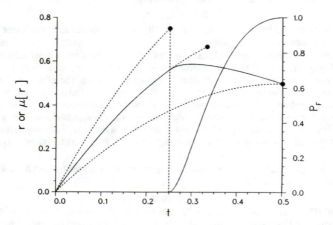

FIG. 5. THE DASHED CURVES TERMINATING AT THE PLOTTED POINTS ARE REALIZATIONS OF r VS. t. THE SOLID CURVE TERMINATING AT THE PLOTTED POINT IS $\mu(r)$ VS. t. THE OTHER SOLID CURVE IS P_F VS. t.

As in Example 1, in actuality the dashed traces in the figure are plots of the function

$$r = kt(1-t) \quad , \tag{8}$$

where the time of failure t_F is taken as $t_F=1/k$. Once again, these analytical forms would not in general be known, and are used here only for the sake of demonstration. In any case, the main point of this example is that, when finding the statistics of r in the interval $(1/4)<t<(1/2)$, not all realizations come into play, i.e., some have "failed out of the problem." This is taken into account by dividing the PDF of k, i.e., eq. (7), by the current probability of survival P_S, and by adjusting the admissible range of k accordingly. The function $P_S=P_S(t)$ is constructed by writing $P_S=1-P_F$, where $P_F=P_F(t)$ is the probability of failure, and by noticing that $P_F=P_F(t)$ corresponds to the CDF of $t_F=t_F(k)$, i.e., to $c=c(t_F)$. Thus, using eq. (7) and $t_F=1/k$, analytically one finds

$$P_F(t) = \frac{1}{4t^3}(1-t)(1-4t)^2 \quad , \qquad \frac{1}{4}<t<\frac{1}{2} \quad , \tag{9}$$

so that the PDF of k is

$$p(k,t) = -\frac{3t^3(k-2)(k-4)}{(2t-1)^2(5t-1)} \quad , \qquad \frac{1}{4} < t < \frac{1}{2} \quad , \tag{10}$$

where the admissible range of k is now $2<k<(1/t)$. In practice $P_F=P_F(t)$ can be constructed by fitting a polynomial to the realized failure points to obtain $t_F=t_F(k)$, and then by inputting the resulting polynomial and eq. (7) into the FPI method to obtain $c=c(t_F)$ (which would provide enough information to construct k's new PDF and new admissible range). Then, as in the previous examples, polynomial fits to the realizations of r vs. t and the appropriate PDF's of k can be used in the FPI method to obtain $c=c(r,t)$. Finally, just to illustrate the behavior of r's statistics in this example, the analytical results for the expected value of r vs. t, i.e., $\mu(r)$ vs. t, and the probability of failure vs. time, i.e., eq. (9), are depicted by the solid curves in Fig. 5.

5. Closing Remarks

A solution method has been described for quantifying the uncertainties in a structure's response, due to uncertainties in loading, geometry, and material, when the path dependent nature of the structure's material cannot be ignored. It is expected that this method will prove useful in assessing the reliability of structures subject to extreme operating environments, since these are usually the conditions under which a structure's material exhibits path dependence. The solution procedure obeys this aspect of the material behavior, and at the same time, keeps the number of incremental solutions required fairly reasonable, since the analyst can decide on the minimum number needed to represent adequately the response's dependence on the random variables. Additionally, this FPI-based procedure is capable of obtaining high degrees of accuracy in the tails of the response distribution (see Cruse et al. [5]), which is a distinct advantage over the more traditional method of moment analysis.

Acknowledgment. The work presented in this paper was supported by NASA-Lewis Research Center under contract NAS3-24389.

6. References

[1]   CORDTS, D. and KOLLMANN, F.G.: "An Implicit Time Integration Scheme for Inelastic Constitutive Equations with Internal State Variables", International Journal for Numerical Methods in Engineering, Vol. 23, No. 4, pp. 533-554, 1986.

[2]   LIU, W.K., BELYTSCHKO, T. and MANI, A.: "Random Field Finite Elements", International Journal for Numerical Methods in Engineering, Vol. 23, No. 10, pp. 1831-1845, 1986.

[3]   WU, Y.-T.: "Demonstration of a New, Fast Probability Integration Method for Reliability Analysis", ASME Journal of Engineering for Industry, Vol. 109, No. 1, pp. 24-28, 1987.

[4]   CRUSE, T.A., UNRUH, J.F., WU, Y.-T. and HARREN, S.V.: "Probabilistic Structural Analysis for Advanced Space Propulsion Systems", To be presented at the 24th ASME International Gas Turbine Conference.

[5]   CRUSE, T.A., BURNSIDE, O.H., WU, Y.-T., POLCH, E.Z. and DIAS, J.B.: "Probabilistic Structural Analysis Methods for Select Space Propulsion System Structural Components (PSAM)", Computers and Structures, Vol. 29, No. 5, pp. 891-901, 1988.

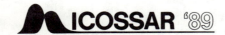
VERIFICATION OF NESSUS CODE ON SPACE PROPULSION COMPONENTS

K. R. Rajagopal*, A. DebChaudhury**, and J. F. Newell***

*   Principal Engineer, Computational Structural Mechanics
**   Specialist, Structural Dynamics
*** Manager, Methods and Procedures Group
    ROCKETDYNE DIVISION, ROCKWELL INTERNATIONAL CORPORATION
    6633 Canoga Avenue, Canoga Park, California 91304

ABSTRACT

Application of probabilistic finite element method to the
structural analysis of typical space propulsion components
such as turbine blades and high pressure ducts is
reported. The study uses Probabilistic Loads obtained
through engine system models and uses NESSUS family of
computer codes to obtain the cumulative distribution
function of structural response variables. The turbine
blade study considered pressure, temperature, speed,
geometry, and material properties as random. The duct
analysis treated damping, random dynamic and periodic
dynamic loads as uncertain. The output responses obtained
include stresses, stress resultants and displacements.
The practical aspects of the application of probabilistic
analysis methods using comparatively large finite element
models are discussed.

KEYWORDS

Finite element, probabilistic methods, stress analysis,
fundamental modes, random vibration, harmonic analysis

1. INTRODUCTION

The purpose of the verification efforts is to apply and verify the
Probabilistic Structural Analysis Methods (PSAM) to the analysis of actual
space propulsion system components. Four components typical of the hardware
found in rocket propulsion engine systems have been chosen for this
application. These components are subject to environments with many random
variables.

A wide range of probabilistic structural analysis tools have been implemented in NESSUS Finite Element and Fast Probability integrator codes. The theoretical background for these computer codes are described in [1, 2, 3]. The verification efforts have been tailored such that different areas of structural mechanics are emphasized in each of the components. This has been done consistent with the primary design requirement for each component. The areas of structural mechanics include 1) linear static analysis, 2) dynamic eigen value analysis, 3) harmonic and random vibration analysis, 4) static material nonlinear analysis and 5) static material and geometric nonlinear analysis. This paper discusses examples of applying the probabilistic analysis methods in the first three areas.

2. TURBINE BLADE STATIC ANALYSIS

A high performance high pressure fuel turbopump turbine blade was considered for this study (Figure 1). The blade is made out of a nickel based super alloy with anisotropic and temperature dependent material properties. The material properties are direction dependent with reference to material axis. The random variables considered in this analysis were system as well as load random variables. A total of eighteen random variables were considered in the analysis and is listed in Table 1, and all distributions were assumed to be normal for this example.

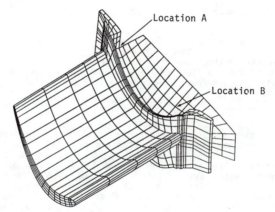

Figure 1. Turbine Blade Finite Element Model

Statistical data for material axis orientations were obtained from a set of one hundred blades. The variations in elastic constants were primarily obtained from expert opinion due to lack adequate data needed over a wide range of temperatures. The nature of the geometric variations in turbine blade is a function of manufacturing methods. For blades such as the one analyzed in this study, majority of geometric differences from blade to blade occur as a rigid body shift about the stacking axis. This shift results in change of primary centrifugal stresses in the blade. The geometric perturbations were introduced as rigid body shifts of lean, tilt and twist angles.

TABLE 1
LIST OF RANDOM VARIABLES

| RANDOM VARIABLE | | | | STANDARD |
NO DESCRIPTION	TYPE	FEM QUANTITIES AFFECTED	MEAN	DEVIATION
1 MATERIAL AXIS ABOUT Z		MATERIAL	-0.087266 RADIANS	0.067544 RADIANS
2 MATERIAL AXIS ABOUT Y	MATERIAL AXIS	ORIENTATION ANGLES	-0.034907 RADIANS	0.067544 RADIANS
3 MATERIAL AXIS ABOUT X	VARIATIONS		+0.052360 RADIANS	0.067544 RADIANS
4 ELASTIC MODULUS	ELASTIC		18.38E6 PSI (126.22E9 Pa)	0.4595E6 PSI (3.168E9 Pa)
5 POISSON'S RATIO	PROPERTY	ELASTIC CONSTANTS	0.386	0.00965
6 SHEAR MODULUS	VARIATIONS		18.63E6 PSI (128.45E9 Pa)	0.46575E6 PSI (3.223E9 Pa)
7 MASS DENSITY	MASS VARIATIONS	MASS	0.805E-3	0.493E-5
8 GEOMETRIC LEAN ANGLE ABOUT X			0.0	0.14 DEGREES
9 GEOMETRIC TILT ANGLE ABOUT Y	GEOMETRY VARIATIONS	NODAL COORDINATES	0.0	0.14 DEGREES
10 GEOMETRIC TWIST ANGLE ABOUT Z			0.0	0.30 DEGREES
11 MIXTURE RATIO LIQUID HYDROGEN/ LIQUID OXYGEN	INDEPENDENT LOAD		6.00	0.02
12 FUEL INLET PRESSURE	VARIATIONS	PRESSURE	30.00 PSI (2.068E5 Pa)	5.00 PSI (.344E5 Pa)
13 OXIDIZER INLET PRESSURE	DEPENDENT	TEMPERATURE	100.00 PSI (6.894E5 Pa)	26.00 PSI (1.793E5 Pa)
14 FUEL INLET TEMPERATURE	LOADS ARE TURBINE BLADE	CENTRIFUGAL	38.5° R (21.39° K)	0.5° R (0.278° K)
15 OXIDIZER INLET TEMPERATURE	PRESSURE TEMPERATURE	LOAD	167.0° R (92.78° K)	1.33° R (0.739° K)
16 HPFP EFFICIENCY	AND SPEED		1.00	0.008
17 HPFP HEAD COEFFICIENT			1.0237	0.008
18 COOLANT SEAL LEAKAGE FACTOR	LOCAL GEOMETRY	TEMPERATURE	1.00	0.1
19 HOT GAS SEAL LEAKAGE FACTOR	FACTORS		1.0	0.5

The engine load variables listed in Table 1 is a representative subset of more than sixty independent random variables that affect engine performance and individual loads on components. The statistics used for these random variables cover normal variations from engine to engine, ground tests, flight to flight and during flight conditions. They do not cover anomaly situations. The probabilistic engine loads model development is the subject of separate independent contract study [4]. These multi-level engine models develop component level loads and their dependence to independent engine loads. Thus, for the case of turbine blades, pressure, temperature and speed variations due to basic random variables and their correlations are captured.

With the perturbation database concept implemented in NESSUS, even large finite element models, such as the one reported here, (Figure 1) approximately 6000 Degrees of Freedom can be processed easily for mean value first order method. For selected critical nodes, improved estimates of the cumulative distribution function can be obtained using Advanced Mean Value First Order Method (ADMVFO). This is illustrated in Figure 2 and Figure 3 for two nodes. The importance factors listed in the Figures can be used in a variety of ways to improve the reliability of the component.

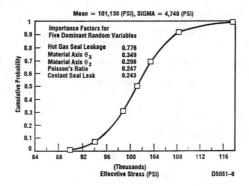

Figure 2. CDF For Effective Stress at Location A

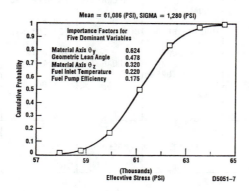

Figure 3. CDF For Effective Stress at Location B

3. TURBINE BLADE FREQUENCY ANALYSIS

The variations in the frequencies of the first few modes of the turbine blade must be controlled to avoid resonance conditions. This can occur due to the periodic excitation forces generated by disturbances to the gas path generated by upstream nozzles and struts. Knowing the distribution of frequencies, it is possible to obtain quantitative probability estimates of interference with engine orders.

A subset of random variables and their statistical parameters defined in Table 1 were used for probabilistic frequency analysis, i.e. the first ten random variables listed in Table 1 were used. Since the stress stiffening effects were included in the analysis, each modal analysis was preceeded by static analysis to obtain correct initial stresses.

The mean value first order and the advanced mean value first order method results are also displayed in Figure 4 for Mode 1. It points out the unsymmetrical nature of the distribution. The finite element model that was used in the analysis had about 2100 degrees of freedom and the eigen values were extracted using subspace iteration technique. Since eigen value extraction procedure is computationally intensive for problems of this size, it is of interest to note that to obtain a 7 point cumulative distribution function with nine random variables required 27 static analyses to obtain correct initial stress followed by 27 eigen value extractions for three modes.

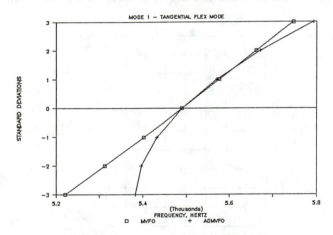

Figure 4.   CDF For Mode 1 Frequency Using MVFO and ADMVFO

4.   VIBRATION ANALYSIS OF DUCTS

Space propulsion system duct components are attached to engine or vehicle structure at multiple support points and are subjected to vibration from more than one source at these attachment points. For the particular duct analyzed here the sources of vibration are the two turbopumps and the three combustors. The combustors are the source of random energy and the pumps produce both random and sinusoidal energy. The variation in vibration loads

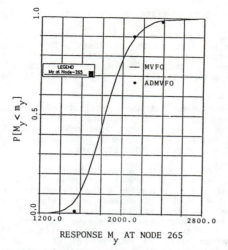

FIGURE 5.   CDF FOR A TYPICAL RESPONSE FUNCTION M_y AT A TYPICAL NODE

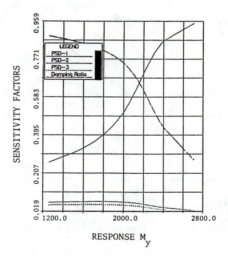

FIGURE 6.   SENSITIVITY FACTORS OF THE RANDOM VARIABLES FOR RESPONSE FUNCTION M_y

are attributed to engine system duty cycle operation, engine system hardware variations, and local component variations within the turbopump or injectors.

The sinusoidal excitations are correlated to the first synchronous speed of the pump 1N. The 1N, 2N, 3N responses are due to rotordynamic imbalance and 4N excitation is caused by 4 primary blades of the pump. The statistics used in this study included data from at least two engines and six tests. The power variations for random vibrations (area of the PSD diagram, units of g^2) has been observed to have a coefficient of variation of 0.2 to 0.8. The variations in the sinusoidal amplitudes are significantly larger with coefficient of variation from 0.5 to 1.0. The high variations in the sinusoids are roughly equally caused by component to component variation and test to test differences.

The finite element model used in this study incorporates two noded Timoshenko beams, elbow flexibility factors with internal pressure and three point excitation, random and harmonic loads applied simultaneously on all the three translational degrees of freedom simultaneously. The dynamic analysis was conducted in the frequency domain using modal superposition technique covering modes and excitation up to 2000 HZ. The results reported here considered the variations in PSD power levels, harmonic excitation frequencies and variations in damping. Table 2 summarizes the statistics of the input random variables. The probability distribution obtained for a typical component (My) at a typical node is shown in Figure 5 and the corresponding sensitivity factors are shown in Figure 6.

TABLE 2
BASIC INDEPENDENT RANDOM VARIABLES (INPUT)

R. V. #	MEAN	CV	DIST. TYPE
PSD - 1	222.0 (g^2)	.73	LN
PSD - 2	73.5 (g^2)	.808	LN
PSD - 3	73.5 (g^2)	.808	LN
PSD - 4	22.5 (g^2)	.20	LN
PSD - 5	54.0 (g^2)	.20	LN
PSD - 6	69.5 (g^2)	.20	LN
HPOT Speed	2940.53 Radians/Sec.	.014	LN
HPFT Speed	3707.08 Radians/Sec.	.01	LN
Damping	.033	.15	LN

TABLE 3
RESPONSE STATISTICS (STRESS RESULTANTS AT A TYPICAL NODE)

COMPONENT	MEAN	CV
1	221.33 (984.9N)	.1412
2	206.16 (917.4N)	.1943
3	373.13 (1660.4N)	.1180
4	1859.7 (210.12NM)	.1135
5	1969.4 (222.51NM)	.1242
6	1026.1 (115.93NM)	.1116

5. OBSERVATIONS ON PROBABILISTIC FINITE ELEMENT ANALYSIS

When compared to conventional deterministic analysis, probabilistic structural analysis and design procedures require many function evaluations. The quality of the probabilistic response prediction also improve with increasing number of function evaluations. For production oriented practical sized finite element models, this results in large demand of computational and data storage resources. Thus, implementation of the most efficient finite element procedures and probabilistic response prediction procedures are a necessity, not just a desirable feature. This includes development of tools for efficient coupling of finite element and probabilistic codes as well as post processing of probabilistic data, without which probabilistic analysis and design tools can put inordinate burden on the practicing structural design engineer. An accurate estimate of the cumulative distribution function of response variables combined with failure models and material strength data can be used to great advantage in improving the reliability of existing designs or new designs of the aerospace components.

6. ACKNOWLEDGEMENT

Probabilistic load models were developed at Rocketdyne and Battelle Columbus Laboratory under the NASA Lewis Research Center contract. The Probabilistic Structural Analysis Methods were developed at Southwest Research Institute under NASA Lewis Research Center contract. The authors appreciate and acknowledge the support given by Dr. Chris Chamis, NASA Program Manager.

REFERENCES

[1] WU, Y. T.: "Efficient Probabilistic Structural Analysis Using Advanced Mean Value Method", 1988 ASCE Specialty Conference, Virginia Polytechnic Institute and State University, Blacksburg, Virginia, May 25-27, 1988.

[2] DIAS, J. B. and NAGTEGAAL, J. C.: "Efficient Algorithms for use in Probabilistic Finite Element Analysis", Advances in Aerospace Structural Analysis, Burnside, O. H. and Parr, C. H., editors, AD-09, ASME, pp 37-50.

[3] CRUSE, T. A., DIAS, J. B., and RAJAGOPAL, K. R.: "Probabilistic Structural Analysis Methods and Applications", Symposium on Advances and Trends in Computational Structural Mechanics and Fluid Dynamics, Washington, DC, October 17-19, 1988.

[4] NEWELL, J. F., KURTH, R. E., and HO, H.: "Composite Load Spectra for Select Space Propulsion System Components", NASA Contractors Report, NASA-CR-179496, March, 1986.

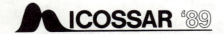
APPLICATION OF PROBABILISTIC RELIABILITY ANALYSIS METHODS TO ARIANE 5 UNEQUIPPED BOOSTER CASE

M. Grimmelt, D. Nendl, E. Sperlich

MAN Technologie AG, München, Dachauerstr. 667, F.R.Germany

ABSTRACT

The paper demonstrates the capabilities of probabilistic structure analysis methods as a tool to support the design and manufacture of the Ariane 5 Unequipped Booster Case within a RAMS
(reliability, availability, maintainability, safety) integrated
development approach. Their applicability to the failure modes
buckling and mountability is shown, which are safety and availability related. In a first approach to the buckling mode the
formulation well-introduced in [4,5] for dimensioning purposes,
is utilized; it seems to yield too large failure probabilities
necessitating further studies. Also, load distribution among parallel members having slip is addressed, for which probabilistic
analysis of the geometric tolerance combination problem is performed. Manufacture tolerances are thus evaluated and improvements, i.e. narrower tolerance bands, specified.

KEYWORDS

probabilistic structural reliability, space structures, buckling,
probabilistic combination of geometric tolerances

1. INTRODUCTION

The Unequipped Booster Case (CPN) is a subsystem of the ARIANE 5
solid propellant booster; two of them form the so-called Solid
Propellant Stage (EAP) of the future ARIANE 5 launcher.
"Design-to-RAMS" (Reliability, Availability, Maintainability,
Safety) is considered an important contribution within the gene-
ral design and development process of this motor case [1].

Its objectives are a RAMS optimized design and the achievement
of a quantitative RAMS target allocated to the motor case. Among
the tools and analyses applied during the RAMS achievement process - which is detailed in a RAMS program plan - are failure
modes, effects, and criticality analyses (FMECA), critical points

logs, reliability analyses. There are tools for RAMS management, design-to-RAMS ("identification, evaluation, elimination
of risks") and RAMS assurance.

This paper concentrates on methods of structural reliability
analysis in order to show their capabilities with respect to
identification of significant failure modes and determination of
most important design parameters; the   results show clearly
which failure modes may be eliminated, or the probability of
which modes has to be reduced; also the methods indicate the
parameters the statistical uncertainty of which should be reduced, which design parameters should be changed if the reliability has to be improved, and which design parameters should be
controlled during manufacture in order to avoid a decrease of
reliability of series products.

After a short presentation of the main features of the motor
case and its failure modes, two of them are selected for detailed consideration. The topics shown should be considered to
be a state of discussion.

2.   MOTOR CASE STRUCTURAL SYSTEM AND FAILURE MODES

The motor case consists of two domes and several cylinders made
from high-strength steel and connected to each other by load
transferring and sealing joints. Some of the connections are
assembled in Guyana. Further attachments connect the case to the
booster skirts and the A5 central stage, see Fig. 2/1. During
pre-launch phases, the booster cases have to carry the total
launcher thus subjected to compression forces and bending moments; during flight, it acts as a pressure vessel and as a
structural component of the launcher.

All possible failure modes are collected in the FMECA; the respective failure rates or probabilities are - in view of the
special type of product - not readily available. Many, however,
are amenable to a mechanical formulation ("limit state
function":g) after establishment of failure criteria. This enables determination of mode failure probabilities by available
methods of probabilistic structural reliability analysis. A selection of the failure modes and the corresponding probabilities
in terms of the reliability indices

$$\beta_i = - \Phi^{-1} (p_{f,i}) \qquad\qquad (2.1)$$

with $p_{f,i} = P (g_i \leq 0)$ (2.2)

and $\underline{X}$:= vector of basic uncertainty variables, are presented in
Tab. 2/1; the index i denotes the mode number. The term "mode"
is used for the elementary level of consideration; the term
"subsystem" stands for a collection of components or a union/
intersection of modes. The important modes are obvious - it is,
however, to be emphasized that the reliability indices in Tab.
2/1 should be considered as relative statements indicating the
potential problem areas; they are preliminary estimates and do

not represent absolute failure probabilities.

Some of the modes were discussed earlier [2] and are therefore
not repeated here; new interesting aspects are presented instead.
These address buckling and geometric tolerance problems.

Failure Mode		Reliab. Index β	
		Mode	Sub-System
2	flange 1 bearing	>10	
3	flange 2 bearing	8	
4	flange 3 bearing	>10	
6	flange 1 tension	>10	
8a	flange 2 tension	>10	
8b	flange 3 tension	>10	
10	pin failure	9.9	
	mechanical connection		7.6
	aft skirt attachment		>10
	front skirt attachment		>10
	buckling	4.4	
11	seal ring: leak during shift/ignition	4÷2	
12	seal ring: no pressure activation	5.4	
	leak test shift	3.7	
1	wall yielding		6.1

TABLE 2/1: Selection of Modes' and
Subsystems' Reliability Indices

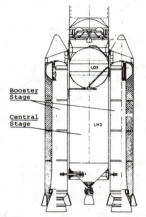

FIGURE 2/1: Launcher Sys-
tem Ariane 5

3. BUCKLING

3.1 Introduction and Mechanical Model

During standby, the boosters are subjected to normal forces and
bending moments so that buckling is a failure mode to be considered. For the first approach, as mechanical model the relation
for the buckling stength σ_K of medium length cylinders, as derived in e.g. [3] and utilized in [4], is chosen

$$\sigma_K = \frac{A}{\sqrt{1+0.01 \ r/t}} \ \frac{E}{\sqrt{3 \ (1- \nu^2)}} \ \frac{t}{r} \qquad (3.1)$$

$$= \varepsilon \ \cdot \ \sigma_{ki} \qquad (3.2)$$

with the uncertain parameters E:= Young's Modulus, r:= radius,
t:= wall thickness, A:= a variable or a function f(r,t) as detailed below, and the deterministic Poisson's ratio ν = 0.3.
This relation comprises the buckling strength of ideal cylinders
and the well introduced reduction factor which accounts for deviations from the ideal configuration. The applied stress is

$$\sigma = \frac{1.05 \ N}{2\pi \ r \ t} + \frac{1.2 \ MF}{\pi r^2 t} \qquad (3.3)$$

with N:= normal force, MF:= bending moment. For the purpose of this paper, no distinction is made for the increase of buckling strength in case of pure bending or also for N-MF-interaction. The "project factors" in Eq. 3.3 are not detailed here. Failure is assumed if

$$g = \sigma_K - \sigma \leq 0 \qquad (3.4)$$

and the corresponding buckling probability is

$$P_{f,b} = P (g \leq 0) \qquad (3.5)$$

with g:= state function which is well introduced.

The formulation presented in [5] is - if compared graphically - by and large identical with Eq. 3.1 except for the magnitude of A. It should be mentioned that - for the numerical values chosen - the elastic range is applicable so that Eq. 3.1 holds.

A was determined from test results [6]; in order to consider regression,

$$A := a + b \cdot r/t + Y \qquad (3.6)$$

with a,b:= regression coefficients, Y:= new random variable with zero mean, was also used.

3.2   Results and Discussions

Since the motor case's wall thickness is dimensioned for internal pressure, the buckling failure probability was in an acceptable range, as a preliminary calculation showed. Since, however, the first stochastic model for the constant A had been a rough estimate and since the sensitivity factors indicated major importance of the variable A (Tab. 3/1), more efforts were spent on its characteristics. The guidelines [4] represent a deterministic approach with A = 0.52; the test information contained in [6] - obviously the basis for the guidelines [4] - was evaluated and Eq. 3.6 established together with numerical input from regression and statistical analysis. Fig. 3/1 shows that the resistable load is very sensitive with respect to changes of (a + b r/t) and the variance of Y. To get more insight, the test results published in [7] - obviously the basis for the guidelines [5] - were also evaluated. It was expected that the two sets of test results mentioned above had resulted from different populations since the guidelines propose different reduction factors (Eq.3.2) in the deterministic approach (A ≅ 0.8 in [5]). Surprisingly, the two sets were rather similar. Due to the limited time available, unfortunately only graphical displays of collected test results were available for evaluation. Also, the applicability of the test results - which indeed are the basis for the general dimensioning procedure in [4,5] - has to be analysed

when used for a structure like the motor case [8].

A statistical evaluation of the two test sets led to the interpretation of the deterministic reduction factors A as 90%
(guidelines [5]) and 99.9% (guidelines [4]) fractiles, respectively. Obviously, different conclusions with respect to a dimensioning rule had been drawn by the authors of [4] and [5], probably in view of different applicable safety factors. But, in ca-
se of the example structure, a safety factor ratio $j_{DASt}/j_{NASA} = 1.3/1.5$ would not cover the difference of the reduction factors.
The deterministic approach, there-fore, remains unsatisfactory.
Moreover, when deterministically dimensioned cylinders are evaluated, too large failure probabilities are encountered as e.g.
Tab. 3/2 shows.

E	r	t	A	N	MF
0.018	≈ 0	0.027	0.98	-0.15	-0.094

TABLE 3/1: Sensitivity Factors for Buckling Mode

j_R	DASt [4]		NASA [5]	
	t	P_f	t	P_f
1.5	6.53	$2.9 \cdot 10^{-5}$	5.51	$1.4 \cdot 10^{-4}$
1.25	6.06	$5.3 \cdot 10^{-5}$	5.12	$3.3 \cdot 10^{-4}$

TABLE 3/2:   Wall Thickness and Failure Probabilities

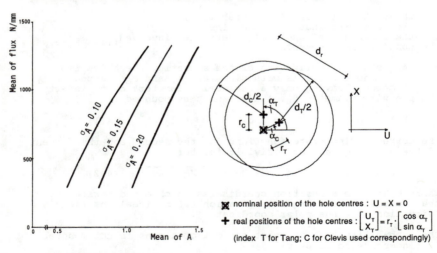

nominal position of the hole centres : $U = X = 0$

real positions of the hole centres : $\begin{bmatrix} U_T \\ X_T \end{bmatrix} = r_T \cdot \begin{bmatrix} \cos \alpha_T \\ \sin \alpha_T \end{bmatrix}$

(index T for Tang; C for Clevis used correspondingly)

Figure 3/1: Allowable applied load as function of mean and variance of A

Figure 4/1: Geometry of interfering Tang and Clevis holes

The conclusions of this short study are as follows: If the safety factor approach is really chosen for dimensioning purposes,
the safety factors should be reconsidered. For certain assumptions with respect to distribution laws, the safety factor may
be as large as j_R = 2.0. Preferable is a probabilistic dimensioning procedure. An elaborate study should be performed including
vast analysis of test results in view of the distribution of A
(or a + b r/t+Y) and the applicability of the results to the
structure considered.

Any procedures mentioned in [9] were not considered because of
the recent arrival of [9].

4. GEOMETRIC TOLERANCES

4.1 Introduction and Mathematical Formulations

The Tang-Clevis (T/C) type connection between the CPN elements
transfers the longitudinal forces acting as a pinned double shear connection. Fabrication tolerances cause play and the fact
that the more than 100 pins never will share in the total force
equally; an optimum is aimed at by reducing the tolerances.
The intention to have little play and the compromise of tolerances leads to the probability that not all pins can be mounted
during the motor case assembly. Furthermore, a relative rotation
of two cylinders due to assembly also reduces pin mountability;
for simplicity, this is not considered here.

The objectives of this analysis are to determine

- the probability of pin mountability,
- the load distribution among pins,
- the influence of the design variables involved,
- possible improvements.

Fig. 4/1 shows a realization of the situation where the two
holes of Tang and Clevis with diameters d_T and d_C respectively,
do not match, leaving only a residual "diameter" d_r to insert
the pin with diameter d_p. A pin cannot be mounted if

NO FIT:= $d_r < d_p$ U $d_C < d_p$ U $d_T < d_p$ (4.1)

The corresponding state functions for the second and third event
of the union Eq. 4.1 are trivial; further

$$d_r = 0.5(d_C + d_T) - \sqrt{(U_C - U_T)^2 + (X_C - X_T)^2} \quad (4.2)$$

where U,X:= hole position coordinates which are described by
the basic variables "radial deviation r from ideal position" and
"direction α" (thus polar coordinates).

The probability that at least one out of n pins is not mountable
is (independence is provided)

Pf,T/C = 1- (1 - P (NO FIT))^n (4.3)

The "goodness" of load distribution among the pins depends on the clearance between residual diameter d_r and pin diameter d_p and on the relative situation of Tang and Clevis holes compared to load direction. The play of the pin (in booster longitudinal direction), i.e. the necessary relative displacement of two cylinders before pin i picks up some load is

$$t'_{xi} = x_{Ti} - x_{Ci} \pm \sqrt{(0.5 \, (d_{Ci} + d_{Ti}) - d_{pi})^2 - (U_{Ci} - U_{Ti})^2} \qquad (4.4)$$

and the difference of the ith pin to the first active pin is

$$t_{xi} = t'_{xi} - t'_{x1} \qquad (i \neq 1) \qquad (4.5)$$

The distribution of t_{xi} provides the law for load distribution among the pins.

4.2 Stochastic Model and Results

Tab. 4/1 contains a numerical example chosen for demonstration. Each geometric variable is subject to fabrication tolerances which can be interpreted as multiples of a standard deviation; different assumptions were made as shown.
The results for mountability are given in Fig. 4/2 for a range of hole position tolerances and two interpretations of the tolerances which were found to be the most important design variables. The smaller the position tolerances are, the better is the mountability. Fig. 4/3 shows the distribution of t_{xi} (Eq. 4.4), the mean of which is representative for pins carrying the nominal load share, and the extremes represent the pins less or harder loaded. The extremely loaded pins are thus interesting; their load share should be describable by extreme value distributions. It can be shown that a reduction of tolerances causes both less total slip as well as smaller extremes, which should be achieved for a more uniform load distribution among the pins. The extreme value distribution is utilized for dimensioning an individual pin.

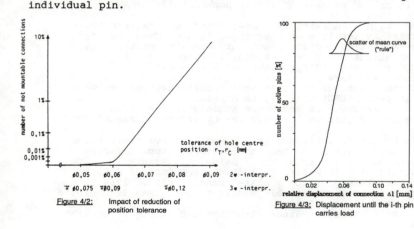

Figure 4/2: Impact of reduction of position tolerance

Figure 4/3: Displacement until the i-th pin carries load

5. CONCLUSIONS

This contribution is an "application paper". It is shown how the methods of structure reliability analysis can be utilized as a development tool. Mode failure probabilities indicate significant modes which have to be studied further, e.g. leak of connection, buckling, wall yielding, pin mountability. Proposals for design criteria and decisions can be made, e.g. reduction of certain tolerances, reduction of imperfections. Specific results are a model for load distribution among parallel members having slip and the proposal for a probabilistic buckling analysis.

The results also provide importance factors which indicate the relative influence of variables on probabilities. They serve two purposes. First, the importance factors indicate which variables have to be changed if mode probabilities have to be reduced, e.g. hole position tolerances. Second, the importance factors help to decide where additional efforts can be made with respect to improvement of statistical uncertainty, e.g. the variable A of the buckling mode. For practical applications, the determination of stochastic models may involve considerable efforts.

	mean	standard deviation	type of distribution
d_C, d_T	24.0105	0.00525	Normal
d_p	23.905	0.0025	Normal
r_T, r_C	0	0.0225	Normal
a_T, a_C	$[0, \pi[$	–	Uniform

Tab. 4/1:   Stochastic Model for Tolerance Combination

6. ACKNOWLEDGEMENT

RAMS work is teamwork. Contributions of and discussions with many colleagues are gratefully acknowledged.

REFERENCES

1. Centre National d'Etudes Spatiaux (CNES): "Management Specification A5-SM-0-50-1 (2): Dependability (RAMS)", Paris, 1988
2. Grimmelt, M.; Rackwitz, R.; Gollwitzer, S.: "Ariane 5 Booster: Reliability of Unequipped Booster Case and of Intersegment Connection", 6th Int. Conf. on Reliability and Maintainability, Strasbourg, 1988, pp.140-145
3. Pflüger: "Stabilitätsprobleme der Elastostatik", Springer V., 1984
4. Deutscher Ausschuß für Stahlbau: "Beulsicherheitsnachweise für Schalen "(DASt-Ri 013), Köln 1980
5. National Aeronautics and Space Administration (NASA): "Buckling of Thin-walled Circular Cylinders" (NASA SP-8007) 1968
6. Bornscheuer: "Plastisches Beulen von Kreiszylinderschalen unter Axialbelastung"; Der Stahlbau 9/1981 pp. 257-262
7. Weingarten V.I.; Morgan, E.J.; Seide, P;: Elastic Stability of Thin-walled Cylindrical and Conical Shells under Axial Compression", AIAA-Journal, Vol. 3, 1964
8. Agatonovic, P.: "Allowable Buckling Load"; MAN Technologie AG - Internal Report/1988
9. DIN 18800E, Part 4: "Steel Structures, Buckling of Shells", 1988

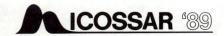
Space Station Freedom
Requirements for Structural Safety,
Reliability and Verification

Alden C. Mackey* and Curtis E. Larsen**

*   Chief, Loads and Structural Dynamics Branch,
    National Aeronautics and Space Administration,
    Lyndon B. Johnson Space Center, Houston, Texas

** Loads and Structural Dynamics Branch, National
    Aeronautics and Space Administration, Lyndon B.
    Johnson Space Center, Houston, Texas

ABSTRACT

The Space Station Freedom (SSF) is a large, manned
vehicle that will be designed to remain in Space
for 30 years. The SSF Program has established
extensive safety, reliability and verification
requirements to ensure the integrity of the SSF
structure during its lifetime. These requirements
are detailed in SSF specification documentation and
are imposed on all elements of the SSF structure.
This paper presents a review of the requirements,
analyses, and testing procedures which must be
implemented in the design and development process.

KEYWORDS

Design Requirements, Safety, Reliability, Verification, Space Station Freedom (SSF)

1.   INTRODUCTION

A permanently manned orbital vehicle has been a major goal of
America's space program since NASA's inception. The Space Station
Freedom (SSF) will acheive that goal when it begins operation in the
mid-1990's. As directed by President Reagan in his 1984 State of
the Union Address, NASA has planned the SSF Program to support
scientific and commercial applications in space, to stimulate the

development of new technologies, to enhance America's space-based operational capabilities, and to maintain U.S. leadership in space exploration into the next century.

As an orbital laboratory, the SSF will provide capability for long-term observation of the Earth's continents and oceans for scientific, commercial, and defense purposes. It will also provide for observation of the Sun, planets, and stars unobscured by the Earth's atmosphere. In addition, the SSF will provide new capability for material processing research and production in the microgravity environment of space.

As an operations base, the SSF will be used to maintain and service unmanned satellites and as a future depot for permanently space-based Orbital Transfer Vehicles (OTV). In the future it may serve as a launch platform for missions to the Moon, planets, asteroids and comets, and as a construction base for the building of space vehicles too large to be launched from the Earth's surface.

The SSF will be constructed in stages over a three-year period, starting in 1995, using 20 flights of the Space Shuttle. Upon completion, this baseline configuration will consist of a main truss boom, 155 meters (508 feet) long; four manned modules containing 878 cubic meters (31,000 cubic feet) of pressurized volume; and eight photovoltaic arrays generating 75 kilowatts of electrical power. See Figure 1. The four pressurized modules will provide living quarters for up to eight crew-persons, and three laboratory facilities built by the U.S., Japan, and the European Space Agency (ESA). Resource nodes will link the modules to provide passage-ways and additional equipment space.

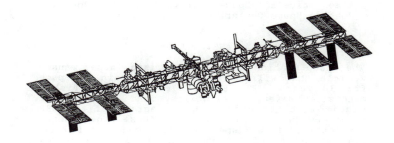

Figure 1.   Space Station Freedom

The main truss boom will incorporate two rotary joints which enable the photovoltaic arrays to track the Sun during orbit while the remainder of the SSF maintains a local-vertical, local-horizontal flight attitude. The truss itself will be constructed in space by astronauts using graphite epoxy tubes or struts connecting to aluminum spheres or nodes to form cubic truss bays 5 meters on each side.

Because it is intended to be operational for several decades, the SSF is being designed to allow for future expansion upon the current baseline configuration. Current plans call for the eventual addition of two truss towers or keels, each 105 meters (345 feet) long, joined by upper and lower booms, each 45 meters (148 feet) long, which would provide much greater space for accommodation of attached payloads and external experiments. The main truss boom will also be lengthened to add two solar dynamic power units for an additional 50 kilowatts of electrical power, greatly enhancing research capabilities. Further expansion plans may include boosting total available power to 300 kilowatts, addition of a satellite maintenance and repair facility, a hangar for space-based OMV's, and a fuel storage depot.

2.   STRUCTURAL SAFETY REQUIREMENTS

All SSF systems are to be designed to meet a fail-safe requirement. This requires that each system and the Space Station as a whole be able to safely tolerate a single failure in any component, including the complete loss of any single pressurized module. The specific requirement for structure calls for the design to be able to tolerate the failure of any single member (which includes truss struts, module external support attachment points, and selected truss nodes or joints) without catastrophic failure, without prevention of Orbiter docking or berthing, and without a loss of strength or stiffness that would jeopardize the crew or the mission. The fail-safe requirement on structure applies to all types of member failures, whether due to overload, space environment effects, or impact damage from micrometeoroids or space debris.

Additional specific requirements are imposed on the design of pressure vessels. Potentially explosive containers must be located outside of habitable areas and must be isolated or protected to prevent their failure from propagating to nearby pressure vessels. All pressure vessels must be designed to leak before rupture and must be able to sustain a wall puncture due to an accident, collision, or impact from micrometeoroids or space debris, without rupture. A conservative factor of safety is to be used in those cases in which a single-point failure mode of operation cannot be eliminated. Systems or components within a pressure vessel must be designed to withstand decompression, repressurization, and differential pressure without creating a hazard. Within the pressurized modules, access must be assured to all walls, bulkheads, hatches, and seals required to maintain pressurization.

In addition to the fail-safe requirement, all primary load-carrying structure must be designed for safe-life such that failure will not occur due to undetected flaws or damage during the life of the vehicle. Specifically, primary structure must be capable of surviving without failure a total number of mission cycles equal to or greater than four times the expected number of service mission cycles. This requirement is allowed to be met by either analysis or test using a rationally derived cyclic loading and temperature spectrum. Design of structure to safe-life criteria is not to preclude the fail-safe design requirement.

3. STRUCTURAL RELIABILITY REQUIREMENTS

Structural reliability requirements for the SSF will establish the overall dependability of the structure during its lifetime.
In general, the reliability of systems such as electronic units is
established by testing the system to determine its life under the
operational environment.   When that is established the unit will
have a reliability number of 1 or less. The structure has the overall reliability requirement that it must not fail during its defined
lifetime and is assigned a reliability number of 1.

In the design, fabrication, and testing of the structure, a
number of requirements have been established to assure structural
reliability.   Structural design of the SSF will be performed using
an ultimate load design approach.   Structure must be sized to withstand an ultimate or design load equal to or greater than a factor
of safety times the limit load.   Limit loads are the maximum loads
expected on the structure during its service life. The SSF Program
requires a factor of safety of at least 1.5 on all structure, and as
high as 3.0 in certain critical areas.

Fracture control requirements will also be implemented to
assure the reliability of the SSF structure.   The design process
will include fracture control by combining fracture mechanics technology with non-destructive inspections.   Fracture control requirements may also dictate the performance of certain inspections during
the operational life of the SSF.

Several ground tests are required after the design is complete to further ensure the reliability of the structure.   Static
tests to ultimate load are to be performed to ensure that the factor
of safety requirements are met.   These tests may be on parts, components, or entire assemblies.   Fatigue life of the structure will
also be verified by test.   Pressure vessels will also be tested to
proof pressures to ensure meeting the required factor of safety.

The final requirement for structural reliability is the
inspection of structural parts during manufacturing. All structures
are to be screened for flaws which may be introduced during the
manufacturing process and defective parts are to be discarded.

4. STRUCTURAL VERIFICATION

Structural verification is the verification of all structural
models, procedures and forcing functions which are used to define
the design structural loading of the SSF.

Two structural models must be verified during the development
process.   One is the analytical stress model and the other is the
structural dynamic model.   The verification process is shown in
Figure 2.

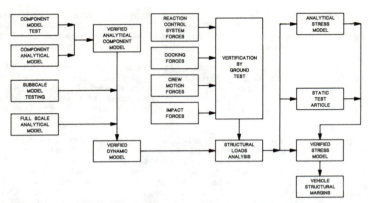

Figure 2.   Structural Model and Loads Verification Process

        A stress model is a highly detailed representation of the
vehicle structure which is typically a finite element model of the
structure from which stresses can be predicted at any desired point
when the structure is under load.   A finite element representation
of the Space Shuttle is shown in Figure 3.   A stress model is verified by comparing stress predictions from the model to stresses
measured on a static test article (STA).   The comparison is valid
when the loadings applied to the analytical model are the same as
those applied to the STA.   The model is verified when the analytical
model predictions match the STA results within an established,
acceptable error criterion.   For example, if the allowable difference between predictions and measurements is set at 10 percent, and
if all predictions are from 90 percent to 110 percent of measured
values, then the stress model is verified.   If not, then the analytical model must be refined and updated until predictions meet the
required criterion.

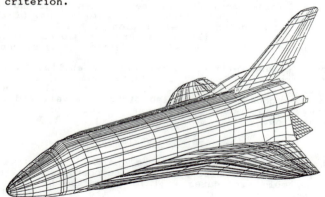

Figure 3.   Space Shuttle Finite Element Model

                                              Mackey

A static test article of the complete SSF is not planned.
Instead, components of the SSF will be tested, including: each of
the pressurized modules, the resource nodes, subassemblies of the
5-meter truss, rotary joints, and the photovoltaic array structure.

A structural dynamic model is a simplified structural model
which retains sufficient definition of the structure to allow predictions of the elastic body mode shapes and frequencies.   The
structural dynamic model is less detailed than the analytical stress
model and in many instances is a reduction of that finite element
model.

The structural dynamic model verification process begins with
component models and test articles.   Each of the test articles are
vibration tested to determine mode shapes and frequencies.   The test
article and test fixtures are modeled and the mode shapes and frequencies predicted.   If the predictions are within an allowable
error band (as with the stresses above), then the model of that component is verified.   All components or subassemblies of SSF structure will be tested in this manner.   These verified models will be
coupled to form an overall structural dynamic model of the SSF
orbiting structure.   Although no full-scale structural dynamic testing of the SSF is planned as a part of the verification, a subscale
model of the SSF orbiting structure is being developed by Langley
Research Center and may become a part of the SSF structural dynamic
verification.

All forcing functions which make up the structural design
loading conditions must be verified.   Typical of these are crew
motion forces, reaction control system (RCS) forces, docking forces,
micrometeroid or debris impact forces.   The magnitude and time
history of each of the forces will be verified by gound testing,
although additional verification of crew motion force time histories
may come from Space Shuttle orbital experiments and from zero-
gravity testing in the NASA 707 aircraft.

The RCS force magnitude will be verified by ground testing of
the motors in a vacuum environment.   The time history of the pulses
will be verified by ground representation of the SSF flight control
system.   Docking forces will be measured in ground docking tests.
Micrometeroid or debris forces will be determined by ground testing.

These ground-verified forces will be applied to the structural dynamic model to survey the large number of load conditions
possible during the SSF lifetime.   The most critical of these conditions are then chosen for detailed stress analysis and applied to
the finite element model for stress calculations and margin of
safety definition.

5. ASSEMBLY VERIFICATION

The SSF program will require the most extensive on-orbit
assembly operation ever attempted in space operations.   Verification
of these assembly procedures will be performed by ground testing
using the remote manipulator systems of the Space Shuttle Orbiter
and of the SSF.   This testing will be on air-bearing floors and in
large water tanks to simulate the zero-gravity environment.

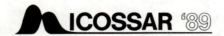 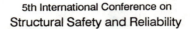
INCREMENTAL ASSEMBLY STAGE CONCEPTS
FOR SPACE STATION FREEDOM
STRUCTURAL DESIGN AND VERIFICATION

Alan J. Lindenmoyer

NASA Space Station Freedom Program Office
10701 Parkridge Blvd., Code SSE
Reston, VA 22091

ABSTRACT

The design and development of the U.S./International Space Station
Freedom poses extreme technical and management challenges to the NASA/
industry team. A comprehensive system engineering approach throughout all
phases of the program from design, ground test and verification, on–orbit integration and verification and on–orbit operations must be accomplished to assure adequate integration, safety and reliability of all S.S. Freedom program
elements. This paper describes the overall configuration and sequential on–orbit assembly concepts planned to complete Freedom's initial operating phase.
Also presented are analytical approaches utilized to evaluate the spacecraft's
structural adequacy and concepts for structural characterization and verification.

KEYWORDS

Space Station; Assembly Sequence; System Engineering; Loads Analysis;
Forcing Functions; Structural Identification; Verification.

1.0 INTRODUCTION

The design and development of Space Station Freedom poses significant engineering and management challenges due in part to the large number of organizational and physical interfaces. Four
NASA field centers are contributing to the development of the station's primary systems and structural
elements. The space agencies of Canada, Europe and Japan are providing additional laboratory modules and robotic equipment and the Kennedy Space Center will be supporting the launch activities.
With the support of the Grumman Corporation as the Space Station Engineering and Integration Contractor (SSEIC), the NASA Space Station Freedom Program Office (SSFPO) in Reston, Virginia is
responsible for the overall system engineering and integration tasks. Sophisticated and comprehensive
system engineering approaches are utilized at the SSEIC/SSFPO to assure adequate safety and reliability of all systems and elements.

2.0 CONFIGURATION ARCHITECTURE

S.S. Freedom's manned base consists of at least 21 major flight elements and 11 distributed systems providing required resources such as power, heat rejection and communications (see Table 1). Figure 1 is an illustration of the manned base configuration at the completion of assembly [1]. At assembly complete, Freedom is expected to have an equivalent mass of approximately 227,000 kg and external dimensions of nearly 152 meters in length and a solar array span of nearly 67 meters.

The Marshall Space Flight Center is responsible for the development of most of the manned pressurized modules including the U.S. habitation, laboratory and logistics modules. The Huntsville, Alabama based center will also provide the structure for the resource nodes which connect the modules and will develop the Environmental Control and Life Support System.

Table 1. Space Station Distributed Systems and Functions

DISTRIBUTED SYSTEM	PRIMARY FUNCTION
DATA MANAGEMENT SYTEM (DMS)	• DATA HANDLING, STORAGE, & TRANSMISSION
ELECTRICAL POWER SYSTEM (EPS)	• ELECTRICAL POWER GENERATION, STORAGE, CONDITIONING, AND DISTRIBUTION
COMMUNICATIONS AND TRACKING (C&T)	• SPACECRAFT-TO-EARTH COMMUNICATIONS AND TRACKING
ENVIRONMENTAL CONTROL & LIFE SUPPORT SYSTEM (ECLSS)	• CREW LIFE SUPPORT INCLUDING AIR REVITALIZATION & HANDLING, WATER RECLAMATION, WASTE MANAGEMENT, AND EVA SUPPORT
GUIDANCE, NAVIGATION, & CONTROL (GN&C)	• VEHICLE ATTITUDE AND POSITION CONTROL
FLUID MANAGEMENT SYSTEM (FMS)	• INTEGRATED NITROGEN, WATER, AND WASTE WATER DISTRIBUTION AND CONTROL
PROPULSION	• VEHICLE REBOOST AND MANEUVER PROPULSION
THERMAL CONTROL SYSTEM (TCS)	• WASTE HEAT COLLECTION, TRANSPORT, AND REJECTION, TEMPERATURE MAINTENANCE
MECHANISMS	• CONTROL OF HATCHES, DOCKING PORTS, AIRLOCKS, ETC.
STRUCTURES	• STATION STRUCTURAL INTEGRITY, LOAD DISTRIBUTION, ETC.
EXTRAVEHICULAR ACTIVITY SYSTEMS	• EVA SUPPORT, CONSUMABLES RESUPPLY, ETC.

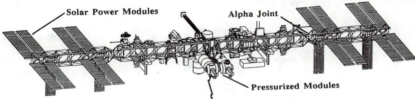

Figure 1. Assembly Complete Configuration

The Johnson Space Center in Houston, Texas is developing the integrated truss assembly which consists primarily of 21 5-meter cubed truss bays and two alpha joints which permit power module articulation for solar tracking. This transverse boom provides utility distribution and is Freedom's backbone upon which most all other elements are integrated. This center is also responsible for most of the station's distributed systems development.

The Goddard Space Flight Center, located in Greenbelt, Maryland, is developing the attached payload accommodation equipment. This equipment, including a payload pointing system, will allow the integration of large stellar, solar, and earth pointed scientific payloads to be integrated onto the truss assembly. Goddard is also leading the program's advanced automation and robotics effort with the development of a Flight Telerobotic Servicer, a device that will be used to assist with the assembly and maintenance of the vehicle.

Located in Cleveland, Ohio, the Lewis Research Center is developing the electrical power system. The current approach utilizes eight photovoltaic solar array panels to generate 75 kW of power. These large panels, each measuring approximately 9.75 by 30.5 meters, are clustered in power modules on truss bays outside the alpha joints. In addition to articulation about the transverse boom's longitudinal axis, each panel also provides an additional axis of rotation which allows adjustments in array pointing to compensate for seasonal variations in sun angles.

2.1 INCREMENTAL ASSEMBLY STAGES

Space Station Freedom is planned to be launched into orbit by the United States Space Transportation System. The Space Shuttle's payload capacity, approximately 18,144 kg to an assembly altitude of 354 km, requires that the spacecraft be launched in stages. Currently, 20 flights are needed to launch and assemble the station to complete the initial operating configuration. Space station assembly will be one of the most intensive space operations of the coming decade. Previous manned and unmanned systems relied on ground integration and testing without extensive on–orbit assembly [2].

Table 2 lists the major flight elements associated with the assembly sequence. (This particular sequence is known as "20/13" which signifies 20 total flights to assembly complete and a permanently manned capability after flight 13.) A man tended capability will offer early science efforts after the delivery of the U. S. Laboratory on the fourth flight. A permanently manned capability will be available after the thirteenth flight at which time a crew of four will remain on–orbit and operate the station. The complete assembly will accommodate a crew of eight. Figures 2 though 4 illustrate the interim stage configurations after the first, fourth and thirteenth flights.

Table 2. 20/13 Assembly Sequence

1	FEL	MB-1	18.75 STBD PV MODULE, ALPHA JOINT, STBD TRUSS (2 BAYS), AVIONICS PALLET, ANTENNA PALLET, TANK FARM # 3, RCS MODULES (2), AWP W/MOBILE TRANSPORTER
2		MB-2	STBD TRUSS (6 BAYS), STBD TCS SYSTEM , FTS/SHELTER, CMG PALLET, TANK FARM # 1 (W/WEU), PMAD PALLET, TDRSS ANTENNA, RCS MODULE
3		MB-3	AFT STBD NODE, MODULE SUPPORT STRUCT, MSC PHASE 1, PRESS. DOCKING MODULE, FMAD PALLET, STINGER/RESISTOJET
4	EMTC	MB-4	U.S. LAB MODULE
5		MB-5	18.75 PORT PV MODULE, ALPHA JOINT, PORT TRUSS, TANK FARM # 2 (W/WEU), RCS MODULE, PORT ANTENNA PALLET, UNPRESS. LOG. BERTHING MECH.
6		OF-1	PRESS. LOG. MOD., MODULE OUTFITTING
7		UOF-1	EXTENDED DURATION ORBITER (EDO), ATTACHED PAYLOADS & EQUIP.
8		MB-6	AFT PORT NODE, AIRLOCK (SSEMU VERF. UNIT/SSEMU), PRESS. DOCKING MODULE, ATTACHED PAYLOADS & EQUIP
9		MB-7	U.S. HAB MODULE
10		MB-8	PORT & STBD FORWARD NODES, CUPOLA, MODULE OUTFITING
11		MB-9	AIRLOCK (3 SSEMU), PORT TCS SYSTEM, TANK FARM # 4, CUPOLA, ATTACHED PAYLOADS & EQUIP
12		CF-2	PRESS. LOG. MOD., MODULE OUTFITTING, SPDM, 2nd TRUSS UTILITIES,
13	PMC	MB-10	CREW (4), PRESS. LOG. MOD., UNPRESS. LOG. CARRIER, LOGISTICS
14		MB-11	PORT & STBD OUTBOARD PV MODULES (37.5 KW)
15		MB-12	SSRMS-2, MMD PHASE 1, ATTACH PAYLOADS & EQUIP., LOGISTIC SPARES
16		L-1	PRESS. LOG. MOD., UNPRESS. LOG. CARRIER, LOGISTICS
17		MB-13	JEM MODULE, JEM EXPOSED FACILITY #1, LOGISTIC SPARES
18		MB-14	ESA MODULE, LOGISTIC SPARES
19		L-2	PRESS. LOG. MOD., UNPRESS. LOG. CARRIER, LOGISTICS
20	AC	MB-15	JEM EXPOSED FACILITY #2, JEM ELM, INT'L EQUIPMENT & PAYLOADS, LOGISTIC SPARES

2.2 ASSEMBLY STAGE SYSTEM ENGINEERING CHALLENGES

The requirement to launch S.S. Freedom in incremental assembly stages translates into a complex set of derived design requirements. Each stage of the assembly sequence must be considered a fully functional, distinct spacecraft. A program–level system engineering process has been established in recognition of this fact and features the following: identification and development of stage–specific design requirements; development of integrated spacecraft functional requirements; allocation of resources for each stage; establishment of incremental verifiability both on ground and on orbit [3].

One of the most significant engineering challenges involves the decomposition S.S. Freedom into launch packages. All station hardware must be packaged as space shuttle cargo elements and are

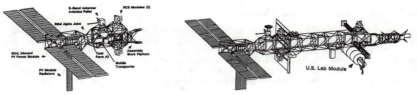

Figure 2. First Element Launch   Figure 3. Man Tended Capability

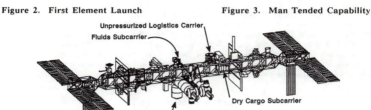

Figure 4. Permanently Manned Stage

subjected to the many constraints imposed on shuttle payloads. Each cargo element must be certified for flight, including structural testing, analysis and verification in its launch configuration. This process is in addition to the structural certification required for on orbit configurations. The combination of mass property and cargo bay volume constraints coupled with on orbit requirements to develop many distinct, fully functional spacecraft compose the complex system engineering challenge.

3.0 INTEGRATED DESIGN AND ANALYSIS

The construction of integrated math models is critical to the performance of spacecraft stage design and analysis. Integrated models will be constructed for each major discipline, fluid/thermal, structures/control, avionics, etc.

3.1 PROGRAM OFFICE INTEGRATED ANALYSES

Since the ultimate design integration of each manned base spacecraft stage resides with the Station Freedom Program Office, a highly integrated approach to the development of analytical models and the evaluation of design concepts is required. The development of design/analysis tools is driven by the following special needs: 1) A large number of design concepts must be evaluated with very rapid model/analyses turnaround. 2) Model input and data is submitted from a variety of sources, a standard communication format is called for. 3) The characterization of space station design concepts spans several engineering disciplines. It is mandatory that these evaluations use consistent data. 4) Full traceability is required for analytical models and their results. An automated, integrated system is required to facilitate the generation, storage and retrieval of this information in an efficient and organized manner.

3.2 IDEAS**2: INTEGRATED TOOLS FOR MODELING AND ANALYSIS

In response to these needs stated above, the IDEAS**2 multidisciplinary analysis system was created by integrating the commercial I-DEAS mechanical computer aided engineering package from

the Structural Dynamics Research Corporation (SDRC) with a set of NASA codes similarly known as
IDEAS. The NASA codes supply spacecraft–specific capabilities such as orbit lifetime predictions,
rigid–body control dynamics and plume impingement analysis. The SDRC *I–DEAS* software provides
the solids modeler, relational database, finite element pre/post processor, structural analysis and plotting capabilities as well as the open architecture and utilities which allowed the codes to be integrated
into a single package with a common user interface [4]. The Grumman SSEIC is fostering the continuing development and enhancement of *IDEAS**2*. A functional flowchart of the resulting capability now in use is shown in Figure 5.

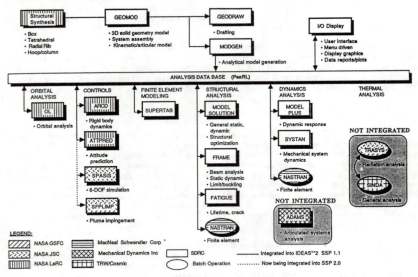

Figure 5. IDEAS2 Architecture**

A key customized feature of this tool in the structural dynamics area is the automatic finite element beam model generation capabilities. This feature allows finite element models representing each
stage as defined by the solids modeler to be automatically generated while preserving the solid's mass
properties and connectivity. The utility of this capability is appreciated when one considers the manual effort otherwise required to represent all 20 assembly stages and several configurations are being
evaluated simultaneously.

4.0 STRUCTURAL MODELING AND PRELIMINARY LOADS ANALYSES

The technique described above was the basic approach utilized during the preliminary loads
analysis. That is, simple beam finite element models were generated from solid models representing
several configurations of the assembly sequence. At this point in the program, detailed designs were
not available and the models generated by this process were deemed appropriate. The ultimate approach to S.S. Freedom structural modeling will involve the acquisition of reduced modal models of
each major component to be provided by the appropriate design center or international partner.
These models will be integrated under the direction of the program office and available for program-

wide use. During early studies, loads and dynamic responses must be evaluated for each stage of the assembly sequence until critical design cases are established.

4.1 PRELIMINARY ON ORBIT LOADS ANALYSIS

Interface loads and dynamic responses are required to be calculated to support several preliminary design activities. In addition to the forces and moments required to guide the preliminary design of the hardware, the dynamic response is equally important to evaluate other areas. Specifically, the station's many control systems require response time histories to evaluate the robustness of their design. Also the transient vibration environment at internal experiment racks is of particular interest to the users requiring microgravity conditions.

The preliminary on orbit loads analysis focused on two configurations, the permanently manned and assembly complete baselines. The sensitivity of the orientation of the large solar arrays was investigated by analyzing models with the arrays in the vertical and horizontal positions. It is anticipated that during the next loads cycle most if not all of the assembly stage configurations will be evaluated to determine critical cases.

4.1.1 FORCING FUNCTIONS

The applied loads to be considered during the preliminary design phase of development represents a significant effort since most of these disturbances may only be estimated by analysis at this time. The forcing functions evaluated to date for loads include those resulting from orbiter docking, logistics module berthing, internal and external crew motion and reaction control system reboost jet firings. Other forcing functions such as fluid and mechanism motion are required when evaluating the vehicle's microgravity environment, but are not significant for design load cases. Another class of forcing functions not yet investigated involve those loads resulting from system failures. At this point it was assumed that Freedom's systems will be sufficiently redundant such that system failures will not design primary structure. These failure modes and effects scenarios will be resolved during future load cycles as the program develops.

The docking, berthing and external crew motion forces have peak magnitudes of 1,112 N or greater. Reboost jet firings were assumed to be on the order of 445 N. Nominal internal crew motion such as hard mounted treadmill jogging and translational push-offs are typically less than 445 N. As the program matures, ground tests will be developed to better define these and other applied forces.

4.1.2 ON ORBIT LOADS RESULTS

Analysis of the assembly complete configuration models with the solar arrays in both the horizontal and vertical orientations resulted in 160 modes below five Hz. (the cutoff frequency of the analysis). Typically, natural frequencies of the two solar array configurations are only slightly affected by their orientation. Natural frequencies were as low as 0.07 Hz for the arrays and 0.144 Hz for the truss in bending.

Load indicators were provided by the various NASA organizations and international partners. A couple of loads results are presented here to provide an understanding of their magnitude. It was determined that the axial, shear and torsional responses (635 lb, 1520 lb and 45161 in–lb) between the pressurized modules were largest for the berthing forcing function and the largest compression load in the truss struts of 865 lb was experienced from the orbiter docking function.

Of the forcing functions studied, the berthing forcing function induced the highest force responses at all module/resource node interfaces. It also generated the highest solar array base and at-

tached payload accelerations. The orbiter docking forcing function created the largest loads in the truss members. The external crew motion activity caused the largest responses at the alpha gimbals. A complete description of the preliminary loads analysis may be found in Reference [5].

4.2 COUPLED ASSEMBLY LOADS

In addition to on orbit loads analyses, both launch and landing as well as on orbit assembly loads with the orbiter attached must be considered during preliminary design. The on orbit assembly loads with the coupled orbiter dynamics included are particularly important since it is the orbiter flight control system that is providing attitude control during the early flights. Preliminary results indicate that severe control/structure interaction may occur on these early flights should the orbiter's primary reaction control system be required in a contingency mode [6].

5.0 CONCEPTS FOR ON ORBIT STRUCTURAL VERIFICATION

Due to the large size of the space station structure and the number of flight elements, it is not practical to require integrated ground structural verification testing. Each flight element and distributed system will, however, through some form or combination of analysis and test be certified for on orbit operations. With this constraint, it is most important that component math models accurately reflect the dynamic characteristics of their representative structures. Component modal testing is expected to occur for each flight element with particular emphasis placed on boundary conditions. For most cases, flight hardware assume different boundary conditions depending on their cargo and on orbit flight configurations.

Since the assembly stages will be integrated only on orbit, all systems design requiring structural dynamic characteristics must rely on dynamic math models. The Johnson Space Center and their contractors are planning to instrument the station to measure mode shapes and natural frequencies. This data acquisition may be used to adjust the control systems parameters.

5.1 SPACE STATION STRUCTURAL CHARACTERIZATION EXPERIMENT

Although not formally part of the S.S. Freedom program or its verification requirements, the Space Station Structural Characterization Experiment (SSSCE) is being considered as an option for supplementing vibration data currently planned to be measured. The SSSCE is a space flight experiment being developed by the Langley Research Center in Hampton, Virginia sponsored by NASA's Office of Aeronautics and Space Technology. The experiment concept is to use Space Station Freedom as a test bed to study and verify techniques for identifying the structural dynamic characteristics of large space structures. Data from the experiment will be used to evaluate analytical models and assess adequacy of modeling technology of large space structures.

Although numerous excitation sources were considered, it was determined that the reboost maneuvers would provide the most favorable ambient disturbance for modal identification. Early studies indicate that a modulating reboost jet firing may pulse the jets near the frequency of several low target modes in order to increase the response of these modes and condition the free decay responses. Baseline instrumentation consists of servo accelerometers connected to the S.S. Freedom instrumentation system [7].

5.2 DYNAMIC SCALE MODEL TECHNOLOGY PROGRAM

Also under development at NASA Langley is the Dynamic Scale Model Technology Program (DSMT). The objective of this project is to evaluate test and analysis data for large space structures based on a scale models of potential space station configurations. The approach is to utilized the DSMT as a space station dynamics test bed to: evaluate the effectiveness of using component verified

dynamics to predict mated dynamics, evaluate modal prediction accuracy of mated finite element modeling, examine optimum placement and number of sensors, evaluate planned excitation procedures and validate planned on orbit signal handling procedures [8].

At this time, the Space Station Freedom Program Office is studying the potential uses of the data that may be provided by the above mentioned experiments. It is believed that dynamic modeling technology will be advanced with the integrated efforts of the Space Station Freedom Program and these efforts.

6.0 SUMMARY

The Space Station Freedom program poses significant system engineering challenges due to its many hardware and management interfaces. The assembly of the station is currently planned to be accomplished by 20 Space Shuttle flights. Each assembly stage may be considered a distinct, fully functional spacecraft requiring sophisticated and comprehensive system engineering approaches. The Space Station Freedom Program Office is responsible for the integration of the vehicle and has implemented the use of IDEAS**2 to assist with multidisciplinary analyses. A preliminary loads analysis has been completed providing interface loads to be used to guide the initiation of preliminary structural design. Since the assembly stages are not able to be integrated until actual assembly on orbit, verified math models will be used to design those systems requiring system dynamic characteristics. The Space Station Structural Characterization Experiment and the Dynamic Scale Model Technology Program are two projects that may supplement model verification efforts.

7.0 REFERENCES

[1]   Snyder, R. *Space Station Freedom Design and Development Concepts: A Presentation to the NASA/DoD Space Station Freedom Technical Interchange Meeting.* Space Station Freedom Program Office, Reston, Virginia, December 6, 1988.

[2]   Crabb, T., J. Kaidy and W. Bastedo, Jr. "Designing for Operations and Support", *Aerospace America*, November 1988, pp. 18–20.

[3]   Snyder, R. *Program Systems and Elements Interactions and Technical Challenges: A Presentation to NRC Workshop on Space Station Freedom Engineering Design Issues.* Space Station Freedom Program Office, Reston, Virginia, November 1988.

[4]   Lindenmoyer, A. and J. Habermeyer. *An Automated, Integrated Approach to Space Station Structural Modeling.* 30th AIAA Structures, Structural Dynamics and Materials Conference, Mobile, Alabama, April 3–5, 1989.

[5]   _*Preliminary Loads Analysis.* McDonnell Douglas Space System Company–Space Station Division, February 1989.

[6]   Singh, S. and A. Lindenmoyer. *Preliminary Control/Structure Interaction of Coupled Space Station Freedom/Assembly Work Platform/Orbiter.* 27th AIAA Aerospace Sciences Meeting, Reno, Nevada, January 9–12, 1989.

[7]   *Space Station Structural Characterization Experiment Concept Definition Study.* McDonnell Douglas Space Systems Company–Space Station Division, November 4, 1989.

[8]   McGowan, P. *Dynamic Scale Model Technology Program Technical Status & Relation to Space Station Dynamics* (A presentation to the SSFPO). NASA Langley Research Center, Hampton, Virginia, January 10, 1989.

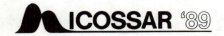
SPACE STATION FREEDOM PRESSURIZED MODULE
METEOROID/DEBRIS PROTECTION

Sherman L. Avans* and Jennifer R. Horn*

* National Aeronautics and Space Administration
  Marshall Space Flight Center, Huntsville,
  Alabama

ABSTRACT

Protection of spacecraft from the effects of a
meteoroid/debris particle impact can add
significant weight to the required structure. This
paper discusses the meteoroid and debris
environments and their implication on the design of
the SSF habitable modules. Protection
requirements, testing, analyses, and primary and
secondary effects of hypervelocity impacts are also
addressed. Finally, recommendations are made for
future activities related to the development of a
hypervelocity impact protection system design.

KEYWORDS

Space Debris; Debris Protection; Pressurized
Module; Hypervelocity Impact; Space Station Freedom

1. Introduction

The Space Station Freedom (SSF) pressurized module structural
design must address a new and major challenge: protecting the
crew and the SSF vital systems from the effects of a meteoroid or
space debris impact. The debris environment is now the critical
design environment for most spacecraft. This debris environment
cannot be ignored by any future NASA program; particularly those
exposed to it for extended periods and/or that have large exposed
areas. The program level protection requirements for SSF were
developed to satisfy the above safety concerns. These

requirements have been defined so as to adequately protect the SSF
but not adversely affect the SSF design.

The environment definition, protection requirements, testing,
analysis, and effects of an impact all interact and affect the
final protection system design solution. These parameters and
their relation to the SSF meteoroid/space debris protection system
design will be discussed in the following sections.

2. The Meteoroid and Debris Environment

The meteoroid environment has been considered a threat to
space vehicles since the beginning of man's activity in space.
Composed of icy dust, the majority of meteoroids are cometary
particles with densities approximately 1/5 that of aluminum. The
average meteoroid particle velocity is 20 km/s.

In spite of the immense energy of meteoroid particles, the
threat of these particles to spacecraft integrity is giving way to
the threat of man-made orbital debris particles. Debris
particles, which have the average material density of aluminum,
have an average particle velocity of 10 km/s. Although the energy
of debris particles is less than that of meteoroids, debris
particles possess a greater damage potential to a spacecraft in
orbit. This potential is a result of the existence of more debris
particles in orbit than meteoroid particles which are greater than
0.3 cm in diameter (figure 1). For most space vehicles, the
particles greater than 0.3 cm in diameter are severely damaging.

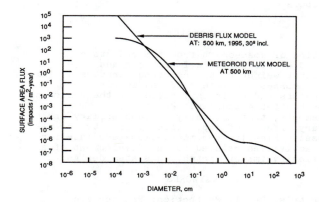

Figure 1. Comparison of Meteoroid Flux with Debris Flux
           JSC 20001, 1985

The number of spacecraft launches worldwide increased
significantly in the past decade. Accompanying this increased

activity was the inevitable growth in the number of orbiting,
man-made objects, both trackable (figure 2) and non-trackable.
Sources of space debris are fuel particulates from rocket firings,
spacecraft coating materials, satellite breakups, spent stages and
military experimentation. The number of particles can continue to
grow even without further launches. Collisions between the
existing debris is probable and could cause the number of debris
particles to exponentially increase.

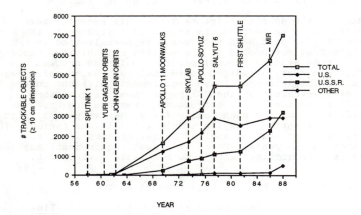

Figure 2.   Space Objects Cataloged (In Orbit),
U.S. Space Command Data

    Currently, efforts are underway to revise the debris
environment definition used to design protective shielding
systems. The revision will be based on measurements of impacts on
returned satellites and on radar tracking data, and will include
other factors which affect the debris growth rate. These factors
include the solar cycle and the expected growth based on an
average number of launches per year, as well as the orbital
altitude, orientation, and inclination of the spacecraft. A NASA
debris environment defintiion publication is imminent and will
soon be available for use by the aerospace technical community.

3.   Requirements Development

    The pressurized module meteoroid/debris protection system
design requirements were developed considering several factors.
Because the SSF will be manned continuously, a primary
consideration is the consequence of debris or meteoroid
penetration into a module. The current requirement is to protect
the module from a penetration with a reliability of 99.55 percent
for 10 years. This probability number directly determines the
protection level that the protection system must provide.
Variations in this number can directly translate into weight

changes in the protection system. Determination of an acceptable
risk must be carefully considered to avoid unnecessary weight
penalty for any spacecraft.

Other factors to be considered in developing requirements are
cost, schedule, and effects of a penetration. How a "typical"
penetration affects crew safety, mission success, and functional
capability must be addressed in SSF design. The testing and
analyses program, discussed later, must include the necessary
elements that allow these effects to be quantified.

Because the debris environment has only recently been
determined to be worse than the meteoroid environment, the
requirements for past NASA programs have been limited to
protection from meteoroids to a specified reliability (table 1).
The protection systems for meteoroids usually do not add
significant structural weight above that for the normally required
structure. Comparison of these requirements with the current SSF
requirement reveals that the SSF requirement is much more
difficult to achieve. The current requirements are based on
maintaining a risk for each module for a 10-year period which is
roughly equivalent to a Skylab module for a 1-year period. The
requirements also state a goal of achieving this level of
protection for 30 years. These stringent requirements may not be
achievable when combined with a much harsher environment.

Table 1 - Requirements History

Program	Requirements/Probability	Time
Apollo (Command/ Service Module)	Meteoroid/0.996	8.3 days
Skylab (each Module)	Meteoroid/0.995	1 year
Shuttle	Meteoroid/0.95	100 missions
Hubble Space Telescope	Meteoroid/0.95	2 years
SSF (1984)	Meteoroid/0.95	10 years
SSF (1984)	Meteoroid & Debris/0.95	10 years
SSF (each module)	Meteoroid & Debris/0.9955	10 years

4. Effects of Meteoroid and Orbital Debris Impacts on Module Design

It is common practice to design a protection system to assure
there is an extremely low probability of encountering particles
which damage the vehicle. Although the occurrence of a

penetration is not probable, it is important to understand the
resulting effects on the SSF manned elements. Functional failure,
loss of a structural element, or danger to crew member life are
all possible consequences.

Depending on the parameters of the impacting particle, a
complete penetration of the pressure wall could lead to a slow
pressure leak or to a rapid decompression of the module. It may
be possible to characterize the rate of depressurization of a
module as a function of the impacting particle parameters. This
part of the study is very important to help develop recommended
crew action for evacuation of a penetrated module and for repair
procedures.

Spall results from the peeling away of a portion of the
pressure wall, caused by the reflection of shock waves through the
wall. Like penetration, spall released internally from the
pressure wall must be treated as a threat to crew safety. These
jagged metallic pieces could cause serious personal injury since
their speeds approach or enter the hypervelocity regime. It may
be possible to limit the danger from spall by careful location of
internal equipment racks and life critical equipment. Design
studies have begun to investigate relocating the crew eating and
sleeping areas away from the external module walls, reducing the
direct spall threat to the astronauts.

Light flashes and internal pressure pulses are impact effects
that can cause injury as well. These two threats are caused by
the tremendous amount of energy released when a hypervelocity
particle impacts a pressure vessel. The light flash could
temporarily or permanently blind a person who is looking directly
at the impact point. Internal pressure pulses could cause
temporary or permanent deafness. Since no conclusive data has
been seen as of this writing, further work is planned to better
characterize these effects.

For a structure under stress, an impact can act as a stress
concentrator leading to crack growth. This is a concern with the
SSF pressurized modules for partial as well as complete
penetrations. The stresses caused by the internal pressure can
cause rapid tearing of the wall after impact of a meteoroid or
debris particle. This tearing may be avoided by using materials
that can resist crack growth while still providing adequate
protection from particle penetrations.

MLI (multi-layer insulation) damage occurs with any complete
penetration of the bumper. After this penetration, the particle
and a portion of the bumper break up to form a cone-shaped spray
of particulates that projects onto the MLI. This phenomenon is
helpful in preventing penetration of the pressure wall but is
obviously harmful to the insulation's efficiency. As the spray
reaches the MLI, it can penetrate and tear away large areas of
MLI, causing "hot spots" on the pressure wall. These could
indirectly cause personal injury to an astronaut through physical
contact or, if a large area of MLI is damaged, may cause internal
temperature changes for which the life support system will not be

able to compensate. NASA has just begun to study MLI damage caused by bumper penetrations, as well as replaceability of the MLI blankets on orbit.

The damaging effects of particulates which ricochet off impacted surfaces should be evaluated for any spacecraft design. For the SSF, these ricochet particles can cause contamination of windows, solar arrays or external experiments. They can also cause functional failure of equipment located near the impact site by penetrating that equipment. In the most optimistic view, if the ricochet particles completely miss any other SSF structure, the particles become additions to the existing debris environment. Bumper designs have been proposed which could possibly inhibit generation of ricochet particles, but further study is required before acceptance of these designs for SSF.

The overall SSF protection system may consist of more than the baseline bumper/wall configuration. An on-board system could detect and locate a penetration or MLI damage and perhaps estimate the amount of damage done. This system may also be used to continually measure the orbital debris environment and revise the definition. It could monitor the expected growth of the debris environment around SSF due to SSF-related activity, and potentially extend the life of the Station. A protection system which includes a multi-plate variation of the baseline configuration concept may be more effective and/or weigh less than the current baseline.

5. Effects of Orbital Debris Flux Growth on the SSF Module Design

As stated earlier, the orbital debris environment is becoming worse. Some experts predict at least a five percent per year growth of the number of debris particles orbiting the Earth, based on the growth rate over past years and on the expected launch rate for this coming decade. This growth can drastically affect the SSF design, thus driving the overall design weight up approaching the allowable Shuttle System payload launch weight.

Preliminary analyses using the environment predicted for the next 20 years show the protection system weight per module could easily exceed 15,000 pounds. However, new technologies and the careful focusing of requirements may bring the weight down to a reasonable amount. Very few new materials and configurations have been investigated for improved SSF protection from the growing debris environment. Work is now proceeding in this area to select a final "best" design.

6. SSF Testing and Analysis

Over 700 hypervelocity impact tests have been completed at NASA/Marshall Space Flight Center for various designs of the SSF pressurized module protection system (figure 3). These tests simulate the effects of impacts of meteoroid and debris particles.

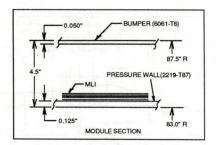

Figure 3.   Baseline Pressurized Module Protection System Concept

    Besides penetration of the pressure wall, there are several
"secondary" issues that must be considered in the design.   As
discussed earlier, spall, light flash, pressure pulse, insulation
degradation, fracture mechanics and ricochet must all be carefully
considered in the design of the pressure modules and to plan for
maintenance and refurbishment activities (figure 4).

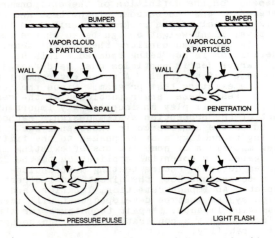

Figure 4.   Examples of Effects of Hypervelocity Impact

    Thus far, most SSF protection system configuration tests and
analyses have focused on the prevention of pressure wall
penetration alone. The combination of particle mass and velocity
which is stopped by a given structure just before complete
penetration is referred to as the "ballistic limit" of that
structure. Figure 5 shows the ballistic limit curve developed for
an example SSF module protection system.

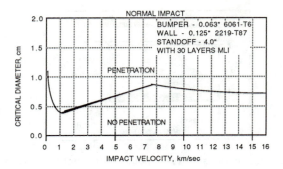

Figure 5. Ballistic Limit Curve for an Example Module
           Protection System

7. Conclusion

     The final design of the SSF meteoroid/space debris protection
system will depend on the definition of the environments, the SSF
program requirements, and the outcome of the test and analysis
program. It is the goal of the program described in this paper to
yield an effective, light-weight design based on the current set
of parameters. As described earlier, the initial debris
environment definition is now being updated. This update
indicates that larger debris particles must be defeated to protect
the SSF to the same safety level. A worse environment will cause
a revamping of the existing test program. Since the technology
does not exist to test these larger particles at the required
velocities, analyses will play an increasingly important roll in
the development of the protection system design. Methods must be
developed to add to the confidence in these analyses. In
addition, testing technology must be pushed to the limit to obtain
more data. New materials or combinations of existing materials
must be investigated for potential application to the SSF
protection system design. Future testing must include
instrumentation to measure both the direct and indirect results of
a particle impact. An impact detection, location, and
characterization system must be developed that will identify
potentially damaging impacts. Repair techniques for the pressure
wall are required so that a penetrated module can be brought back
to operational capability.

     The SSF meteoroid/space debris protection system design is a
difficult technical and programmatic challenge. It is obvious
that the design of any future spacecraft will have to account for
the hazard introduced by the orbital debris environment. The
technological advances made through the efforts of the SSF
meteoroid/debris research activities will prove beneficial to
these future projects.

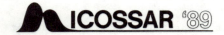
Pressure Vessels for Space Station Freedom

Glenn M. Ecord* and Dr. Henry W. Babel**

*   Technical Integration Manager, National Aeronautics
    and Space Administration, Lyndon B. Johnson Space
    Center, Houston, Texas

** Manager, M&P Space Station, McDonnell Douglas Space
    Systems Company, 5301 Bolsa Avenue, Huntington Beach,
    California

ABSTRACT

Space Station Freedom (SSF) will utilize more than 10
different types of pressure vessels for storage of
oxygen, hydrogen, nitrogen and mixed gases at operating
pressures ranging from 1.03 to 41.4 MPa (150 to 6000
psi). The number and size of the vessels makes maximum
weight efficiency in their design a necessity. Life
expectation of 10 to 30 years requires rigorous analysis
and design verification to assure reliability and safety
during use. Graphite/epoxy overwrapped composite
vessels with metal liners have been baselined for many
of the high pressure applications. The approach for
design, certification and use of space station pressure
vessels is presented.

KEYWORDS

Pressure vessels, graphite/epoxy, overwrap, metal liner,
composite tank, Space Station, high pressure gas
storage, long life, space environment, stress rupture.

1. INTRODUCTION

There are a variety of pressure vessel applications on SSF.
The single largest application is the storage of high pressure
oxygen and hydrogen gases for periodically reboosting the space
station. The current design utilizes four pallets of vessels,
providing an efficient means of packaging and manifesting for launch

and for convenience in space station installation. Each pallet
currently contains six hydrogen and three oxygen tanks. An early
configuration with 11 tanks is shown in Figure 1. These tanks are
currently baselined at 2.6 to 3.0 m

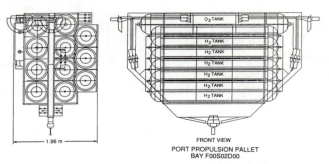

Figure 1.   Typical Propulsion Pallet (Early Design)

(8.6 to 9.8 ft.) length and 56 cm (22 in.) diameter. Although there
are many other pressure vessels on SSF, the hydrogen and oxygen
tanks will be used to exemplify the structural safety and
reliability issues generally associated with high pressure vessels.
The cost of delivering a pound of payload to orbit and the
manifesting requirements for transporting space station components
to orbit places a premium on developing the most weight efficient
structures possible, but without compromising safety and
reliability. Graphite/epoxy metal lined pressure vessels have
emerged as the most weight efficient designs for high pressure gases
when compared to other designs as shown in Table 1.

Table 1. Weight Comparison for Different Materials of Construction

Materials of Construction	Liner Weight kg (lb)	Yarn Weight kg (lb)	Total Weight kg (lb)
Graphite T-40/ 6061-T6 Al liner	61.2 (135)*	55.3 (122)*	117 (257)*
Kevlar 49/ 6061-T6 Al liner	57.6 (127)*	98.0 (216)*	156 (343)*
All metal 2219-T87	* Courtesy of SCI		385 (848)

There is an extensive successful history for such vessels in various
applications, but data are lacking for long life applications and
for the environment that exists in low earth orbit (LEO).

The major requirement that must be met for any pressure
vessel is that catastrophic failure, must never occur. At 20.7 MPa
(3000 psi) explosive failure would be equivalent to several pounds
of TNT and could seriously jeopardize the space station. The
certification process requires the examination of every credible
failure mode, the cause of that type of failure and development of a
design which prevents that failure from happening or, at worst,
would result in depressurization of the vessel without catastrophic
rupture. This process, including tests to verify the design,
applies to all pressure vessels independent of the materials of
construction or location.

2. LIFE CERTIFICATION

The credible failure modes for the composite pressure
vessels have been defined. Those that are the same for terrestrial
and space applications are sustained load failure (stress rupture)
of the overwrap and fatigue failure of the metal liner. Sustained
load failures familiar to the authors are a function of materials,
time, temperature and level of stress in the overwrap and have
always been catastrophic in nature. Therefore, through design,
materials selection, process control during manufacture and
selection of the appropriate operating stress, this type of failure
can be precluded. Exposure of the vessels to the vacuum environment
of space is not believed to have a significant effect on their
performance. Fracture control technology will be used to insure
that the failure mode of liners will be a stable through crack
allowing gas to escape without fragmentation.

The major difference between terrestrial and space
applications is the presence of atomic oxygen and the possibility of
a hypervelocity impact (HVI) in space. Atomic oxygen will react
with the overwrap materials but will be prevented from contact by
thin metal skins. The greater concern is HVI. Shields designed to
stop the largest HVI particle predicted for 10 years duration will
be employed but there are many uncertainties in the predictive
models which are used to establish this particle size.[1,2]
Additional shields are also a possibility and could be retrofitted
to existing shields. The latter approach is specified in the
requirements for the space station after 10 years of service.[3]
This could be done earlier if so decided.

Following is a brief description of the certification
approach to ensure a sound design and prevent a catastrophic
failure.

3. DESIGN BURST

The burst pressure is required to be twice the maximum
operating pressure. The maximum operating pressure for the hydrogen
and oxygen tanks in the propulsion system is 20.7 MPa (3000 psi) and
their minimum burst pressure is therefore 41.4 MPA (6000 psi). The
Space Station Program will use the statistical procedures described

in MIL-HDBK-5E for calculating design allowables. True "A" values will not be achieved because the requirements for the number of test samples and different lots of material cannot realistically be satisfied. Industry practice has been the use of subscale pressure vessels to determine the design allowables for various yarn/resin combinations. It is believed that properties determined on subscale vessels can be used for larger size vessels without adjustment because of size effects. One vessel manufacturer does reduce allowables by approximately 10% to account for differences in larger vessels. Results obtained on subscale pressure vessels for various graphite yarns are given in Table 2. These values will be used to design pressure vessels having a minimum burst of 41.4 MPa (6000 psi). Typical burst pressures on the order of 48.3 MPa (7000 psi) would be expected depending on the yarn selected. To further increase confidence in design burst capability, semi-scaled pressure vessels will be tested with the specific wrap pattern and yarn selected for the space station applications. In addition, full scale vessels will be tested as a part of the certification test program to verify design performance of space station pressure vessels.

		Yarn						
		A	B	C	D	E	F	G
Calculated from Static Burst Tests	No. of tests	5	5	5	5	5	5	5
	"A" basis* Fiber Stress GPa (ksi)	4.15 (602)	4.90 (710)	4.05 (587)	2.95 (428)	4.17 (605)	3.16 (459)	3.23 (469)
Calculated from Static Burst Tests + Sustained Load Tests	No. of tests	16	21	17	15	17	15	13
	"A" basis* Fiber Stress GPa (ksi)	4.32 (626)	5.47 (794)	4.16 (603)	3.39 (491)	4.17 (605)	3.54 (513)	361 (523)

*Per the procedures of Mil HDBK 5E.

Table 2. Ultimate Fiber Stress/"A" Basis

4. SUSTAINED LOAD

Graphite yarns do not appear to be as susceptible to sustained load failures as are Kevlar or glass yarns as shown in Figure 2. When these yarns are used in pressure vessel applications, the performance of Kevlar is better than in single yarn tests.[4] Unreported McDonnell Douglas and NASA data also show that vessel life is superior to yarn life for the stress levels tested. No comparable data were found for glass. A conservative analysis, which uses Shaffer's strand (yarn) data [5], McDonnell Douglas pressure vessel data and Weibull statistics predicts that for a survival probability of 0.9999 over a 30 year life no sustained load failure should occur for operating stresses less than 63% of ultimate. The value of 63% compares with the space station maximum stress which will be less than 50% of the determined ultimate. The analysis indicates that sustained load performance is not an issue even though data are limited and assumptions must be

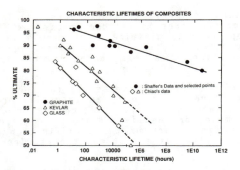

Figure 2.   Impregnated Strand Data

made relative to the applicability of the Weibull statistics.

        To obtain further verification on the sustained load
behavior of the graphite yarn and resin selected for production, a
separate test program is under consideration.   The program would
involve "fleet leader" tests.   Subscale vessels would be placed in
test at two pressures, i.e., maximum operating and a somewhat higher
pressure.   Each year these vessels would be pressure cycled an
appropriate number of times at the coldest anticipated service
temperature.   A sampling of vessels may be burst at periodic
intervals and compared with virgin burst values.   Data are available
for 10 year old glass wrapped vessels which show that at 33% of the
design burst pressure there was no statistically significant change
in burst strength.[6]   Tests of Kevlar/epoxy vessels at the Lyndon
B. Johnson Space Center (JSC) have shown no reduction in burst
strength at 50% of design burst pressure after 9 years.   These
vessels have also been periodically pressure cycled.   The same
behavior is expected for graphite overwrapped vessels although resin
cracking because of thermal cycles may result in some reduction in
burst strength after long periods of time.   These tests would be
started as far in advance of space station application as possible
so that a problem, if detected, would precede the actual use by a
substantial amount of time.   This approach has been successfully
applied for Kevlar/epoxy wrapped pressure vessels used on the Space
Shuttle Orbiter.[7]

5. LINER INTEGRITY

        The liner candidates are aluminum alloy for the hydrogen
tanks and 301 cryoformed stainless steel, Inconel 718 and Monel K500
for the oxygen tanks.   A major consideration in the selection of
materials has been compatibility with the fluids to be contained.
Embrittlement by hydrogen and flammability characteristics with

oxygen have been primary considerations. The other major consideration for the liner is fatigue crack growth due to pressure cycling. One approach to minimize crack growth in metal liners is to minimize the stress in the liner at operating pressure.[8] In addition, minimization of stress concentrations such as occur at welds is essential.

No specific development tests are foreseen for ensuring the integrity of the liner. There is an extensive body of knowledge using fracture mechanics principles which can be used to ensure a leakage failure mode for the liner design. Pressure cycle testing will be conducted as a part of the qualification test program to verify that a capability of four times the required cyclic life is achieved.

6. HYPERVELOCITY IMPACT

Debris and micrometeoroids are a new dimension that has been added to the design of pressure vessels for long term service in LEO. The debris problem is considerably the more severe problem. When attention was first directed toward this subject in the 1950's only a micrometeoroid environment existed. Since that time several spacecraft and satellites have come apart in orbit causing an accumulation of debris in LEO. Efforts are underway to obtain international agreements that will halt the proliferation of space debris. Currently it is difficult to predict what the environment might be 10, 20 or 30 years in the future.

The probability of being struck by space debris depends on the duration of the mission, the exposed area and the particle size of interest. For most non-redundant structures shield designs are being developed for the largest single particle that can be expected to strike the shield during the specified design lifetime. Unlike the typically small (1 mm) particle size for the micrometeoroids there are many large debris particles in space of sufficient size to cause serious damage. Because of the projected increases in the debris environment and weight considerations, the shields to protect structure from impact will be designed for a less than 30 year service life with the ability to retrofit additional shielding later. Initial analysis by McDonnell Douglas has shown that approximately 2320 kg (5100 lb.) of shielding weight would be required to protect the propulsion tank pallets from impact for 10 years with the specified probability of 0.9955. This analysis used the new environmental debris model developed at JSC which contains a modest 5% debris growth.[9] Analysis using the JSC model indicates that the initial shielding required for the propulsion tank pallets should be designed to prevent penetration of particles up to a size of 2 cm (0.8 in.). Particles larger than this size have a lower probability of impact than is required for space station design. It may also be possible to avoid particles larger than 10 cm (4 in.).

Should debris larger than 2 cm strike and penetrate the shield a number of events can occur. One is that a vessel is

impacted by debris from the shield or colliding particle but
penetration of the vessel itself does not take place. A means for
detecting such an event must be identified. An impacted vessel
could be isolated, depressurized and taken out of service. Another
possibility is that the debris from the shield or particle
penetrates the vessel. In this event the failure must not result in
an explosion or other catastrophic occurrence. Gas escaping from a
vessel penetration would cause a thrust requiring that the vessel be
structurally constrained and that the space station attitude
control be capable of handling the resultant thrust force. Since
the basic shield design will stop the largest particle size to be
encountered in 10 years to a probability of 0.9955, vessel
penetration represents an unlikely occurrence. Nevertheless
attention must be given to the failure mode of the vessels and the
effects of a failure should penetration occur.

There are a number of HVI certification concepts that are
being considered. One is to take a representative vessel and
subject it to various impacts and investigate the HVI tolerance of
the design. There is no analytical methodology for predicting
failure mode of a pressure vessel under HVI, therefore testing is
the only sure method to establish damage tolerance. Another
approach would be to design the vessels with lower operating stress
(higher safety factor) to increase the tolerance to HVI shield
penetrations. Damage tolerance of reduced stress designs must also
be determined by testing. Development of test programs is
complicated at this time by the many different shield designs being
considered. Another option might be an increase in shield
capability to minimize the size of particle that could penetrate the
shield and hence the damage or hole size that would be created in
the tank. These concepts will receive further review along with
others before the final certification approach is selected.

CLOSING REMARKS

Many pressure vessels will be used in various applications
on SSF. Not all will be overwrapped designs but the certification
approach for safety and reliability will be similar for all. For
the overwrapped designs, attention must also be given to development
of design allowables and the sustained load (stress rupture) life of
the fiber/resin composite. Damage tolerance may be a major issue
for overwrapped or all-metal vessels exposed to the LEO environment.

This paper has shown that for the design burst, sustained
load behavior (stress rupture) and liner fatigue resistance a very
high confidence can be developed for the pressure vessels on SSF.
In regard to the debris problem in LEO it is clear that shielding to
protect against impact for 30 years is not possible at the outset of
the program. Shielding can be developed to stop the largest
particle expected to impact in 10 years within defined probability,
providing a high degree of confidence that a catastrophic event will
not occur due to impact. The only area where a consensus does not
exist concerns vessel survivability if impacted by particles that

may penetrate the shields. Studies to establish specific requirements pertaining to damage tolerance will continue.

REFERENCES

[1] Johnson, N.L. and McNight, D.S.: "Artificial Space Debris," Orbit Book Co III, Malabar, Florida, 1987.

[2] Babel, H.W.: "Design Considerations for Space Debris, An Industry Viewpoint," In Space Technology Experiments Workshop, December 6-9, 1988, (To be published).

[3] Anon.,: "Space Station Requirements Document", JSC 31000, Rev. C, Paragraph 2.1.2.1.1.2.2, March 6, 1987.

[4] Barlow, R.E., Toland, R.H. and Freeman, T.: "Stress Rupture Life of Kevlar/Epoxy Spherical Pressure Vessels", Lawrence Livermore Laboratory Report UCID-17755, Part 3, February 23, 1979.

[5] Shaffer, J.T.: "Stress Rupture of Carbon Fiber Composite Materials," 18th International SAMPE Conference, pp 613-622, October 7-9, 1986.

[6] Personal Communication With Structural Composites Industries, Azusa, California, 1988.

[7] Schmidt, W.W. and Ecord, G.M.: "Static Fatigue Life of Kevlar Arimid/Epoxy Pressure Vessels at Room and Elevated Temperatures," AIAA Paper 83-1328, Propulsion Conference, June 27-29, 1983.

[8] McClymonds, K.A., Babel, H.W. and Ryan, D.P.: "Design of Light-Weight Impact Resistant Pressure Vessels for Space Station Fluid and Propulsion Systems," AIAA Paper 88-2466, (Also MDC H2647), AIAA SDM Issues of the International Space Station Symposium, Williamsburg, Virginia, April 21-22, 1988.

[9] Anon.,: "Space Station Program Natural Environment Definition for Design", NASA SSP 30425, (Revision Pending), January 15, 1987.

SPACE STATION TRUSS LONGEVITY

D. W. O'Neal and K. B. Kempster

McDonnell Douglas Space Systems Company
Space Station Division
M/S A95-J849-11-3
5301 Bolsa Avenue
Huntington Beach, California 92647

ABSTRACT

This paper describes the approaches taken to ensure Space Station truss longevity. The expected environmental and operational threats to structural integrity and the specific solutions to meeting the 30-year service lifetime of the truss structure are discussed.

KEYWORDS

Truss Structure, Reliability, Maintainability, Low Earth Orbit (LEO), Space Station, Fail-Safe

1. Introduction

When the Space Station Freedom (Figure 1) is launched and assembled in the mid 1990s, the main truss structure will have a certified service lifetime of 30 years. This requirement is met by a judicious combination of built-in reliability and maintainability features that are designed into the truss and demonstrated through a rigorous verification program. Reliability is built-in by selecting a redundant truss configuration that can carry limit loads with one strut out, by selecting materials that are rugged enough to withstand the expected manned activities, and by accounting for the environmental factors that could reduce the capabilities of the structure such as thermal cycling induced microcracking, micrometeoroids/debris impact, and atomic oxygen. Maintainability is enhanced by designing the truss to allow replacement of individual struts and by providing inspection methods for detecting damage. Finally, the truss verification program demonstrates that with the built-in features the lifetime will be 30 years.

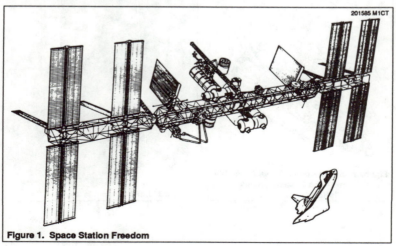

201585 M1CT

Figure 1. Space Station Freedom

The Space Station is assembled out of the NSTS Orbiter payload bay using an Assembly Work Platform (AWP) as shown in Figure 2. Truss struts, with node assemblies already attached to some of them, are packaged into the strut boxes for launch, are removed from the boxes by astronauts on orbit, and are attached to each other using an astronaut-actuated joint. As each cubic bay of the truss is assembled, the equipment located in that truss bay is installed. When the truss bay assembly operations are completed, the Mobile Transporter (MT) base, located at the top of the AWP, moves the assembly up one truss bay length. The next bay is then assembled by repetition. In this way, the suited astronauts are always near the Orbiter for personal safety.

2. Reliability

Reliability is built-in by selecting a redundant truss configuration that can carry maximum expected (limit) loads with one strut out, by selecting materials that are rugged enough to withstand the expected extravehicular activity (EVA) of crew, and by accounting for the environmental factors that could reduce the capabilities of the structure such as microcracking from thermal cycles, atomic oxygen erosion, and micrometeoroids/debris impacts.

2.1 Erectable Joint Reliability

Joint reliability is related to four different factors: crew interface, functioning, stiffness, and fatigue. The crew interface must be designed so that the astronauts can correctly position and actuate the joint with minimum hand fatigue. Once latched, the joint must then be capable of being disassembled at any time over the 30-year life of the structure. The stiffness of the joint must be linear to insure stable and predictable Space Station dynamic responses. Over its 30-year lifetime, the truss structure is subjected to 175,000 thermal cycles as well as a cyclic load spectrum with most loadings less than 200 pounds (890 Newtons). To meet safe life requirements, the erectable joint must be designed and certified to withstand four times the expected environment. Our current baseline erectable joint is a design developed at NASA Langley Research Center (NASA LaRC), shown in Figure 3. Preliminary tests have shown this design performs the best in the areas of stiffness, crew interface, and functioning.

2.2 One-Strut-Out Capability

Structural safety is provided by designing to a minimum factor of safety of 1.5 when subjected to the maximum expected loadings on orbit. In addition, the truss design must withstand all expected loadings when one strut member is missing or severely damaged, providing a fail-safe design. Because the truss can be assembled and disassembled by an astronaut, it can always be restored to its original structural capability by replacing damaged members.

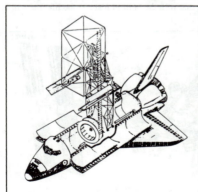

Figure 2. Space Station Truss Assembly with the Assembly Work Platform

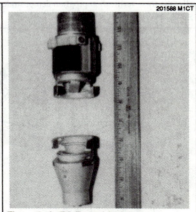

Figure 3. LaRC Erectable Joint

Degraded performance is acceptable with a member missing as long as applied loads can be resisted without failure. The worst case effect of losing a strut in a truss bay is to double the load in a strut of the adjacent bays. This situation is local in nature and the remainder of the truss maintains full structural capability. Degraded performance is then relegated to a minor change in structural dynamic characteristics of the truss from reduction in either local bending or torsional stiffness. Safety from a load resistance standpoint is always maintained.

During early Space Station studies, two cubic trusses, the Warren truss and orthogonal tetrahedral truss (OTT) shown in Figure 4, were extensively analyzed to determine structural performance with one strut missing. NASTRAN finite element models were created, and single struts were systematically removed. Load redistribution in the truss and local, one-bay, stiffness reductions were calculated when bending moments and torsion were applied to the truss models. At the same time, missing struts that caused the maximum load redistribution and stiffness change were noted. Both the Warren and OTT trusses doubled the load in a remaining strut when the worst location strut was removed.

The Warren truss was selected over the OTT as the baseline truss design because it has a bending stiffness equal to the OTT, it has a superior torsional stiffness, and it has higher residual bending and torsional stiffnesses with one strut missing. These performance characteristics are compared in Table 1.

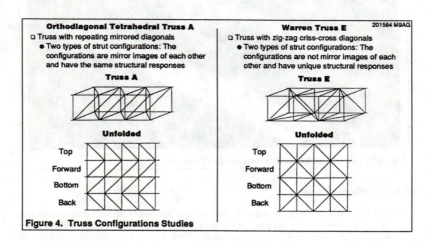

Figure 4. Truss Configurations Studies

Table 1. Truss Stiffness Study Results

Truss	Normalized Axial Stiffness	Normalized Torsional Stiffness	Normalized Bending Stiffness
Fully effective truss			
OTT	1.00	1.00	1.00
Warren	1.14	1.50	1.10
Residual stiffness with worst case missing strut in each bay			
OTT	1.00	1.00	1.00
Warren*	1.27	1.05	1.14

*Warren truss does not experience stiffness reduction for any or all batten strut removals

H5184: T1 08/04/89

2.3 Damage Resistance

The graphite epoxy strut tubes are the most susceptible component of the truss assembly. These components must withstand ground environments, such as handling and transportation, as well as launch loadings and a 30-year orbital lifetime. Struts will be individually packaged in a strut box designed to protect them during Orbiter launch loadings and to allow the astronauts to remove them one at a time for orbital assembly. Ruggedness for ground loadings and for orbital lifetime from equipment impacts/abrasions and crew EVA will be provided by designing the graphite epoxy strut to withstand local concentrated loads of up to 187 pounds (832 Newtons) without failure. Testing has shown that a thickness of 0.066 inch (0.13 mm) with our current composite laminate layup of $[\pm 10°, \pm 30°, \pm 10°]_s$ is adequate. In addition, as discussed in paragraph 2.4.2, a 0.005 inch (0.13 mm) aluminum foil is bonded over the exterior of the tube for environmental protection as shown in Figure 5.

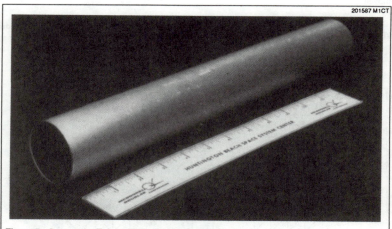

Figure 5. Composite Tube with Protective Foil Overwrap

2.4 Environmental Factors

The low earth orbit (LEO) environment exposes the truss structure to repeated temperature cycles, space radiation, free atomic oxygen, and hypervelocity particle (20 km/sec) impacts from micrometeoroids of cometary origin and man-made orbital debris. Space Station operations also expose the truss to a variety of external loadings for a 30-year period. Structural reliability of the truss to withstand these exposures must be verified on the ground before the Space Station is placed into orbit.

2.4.1 Thermal Cycling

Because of the low Space Station orbit (150 to 220 nmi, 270 to 408 km), the truss structure is exposed to 175,000 thermal cycles over its 30-year life. When subjected to this environment, the graphite epoxy laminates used in the truss struts are susceptible to microcracking. Microcracking of the tubes can change the coefficient of thermal expansion and reduce tube stiffness. These changes reduce the performance of the truss structure by making thermal and dynamic responses less stable and predictable.

The truss struts will be designed to preclude microcracking. Testing and analyses have been performed by McDonnell Douglas Corporation (MDC) and others to identify the parameters and threshold values that lead to the onset of microcracking[1]. For the fiber/resin systems being considered for Space Station, results show that this phenomenon is caused by laminate designs that result in internal thermal stresses exceeding the shear capability of the resin. Using ply angles ≤30° reduces the chances of microcracking to near zero. During the development of the truss struts, sensitivity of the struts to microcracking will be verified by accelerated thermal cycle testing. However, full life testing is not practical. Even with accelerated testing, testing for 175,000 cycles would take almost 4 years non-stop and testing for the 700,000 cycle safe life certification requirement would take over 15 years. Because full life thermal cycling is not pratical, the truss tube design will preclude microcracking by utilitizing ply angles that are ≤30° and fiber/resin systems that are less prone to microcracking, as indicated by rigorous thermal cycle testing.

2.4.2 Atomic Oxygen

Atomic oxygen is free oxygen in LEO. It forms a chemical reaction with organic materials and has the effect of eroding the material thickness. The sensitivity of specific organic materials is still under investigation by on-orbit tests and simulated ground tests. Baseline truss struts are made with an organic graphite epoxy material. A thin aluminum foil, 0.005 inch, is bonded on the tube exterior to prevent oxygen erosion. The maximum degradation of the foil will be an oxidation coating which normally forms on aluminum. However, the durability of the foil as a protective coating must be verified to assure an orbital lifetime of 30 years.

2.4.3 Micrometeoroid/Debris Impact

Two types of particles exist in a LEO environment. The first type is referred to as micrometeoroids, is of natural consequence, and is cometary in origin. These particles have been cataloged as a potentially harmful environment since 1969 in NASA SP 8013 and were a design consideration for long-term spacecraft such as Skylab in the early 1970s.

More recently, man-made orbital debris has been identified in LEO. This debris is the result of loose launch vehicle parts as a result of spacecraft separation, accidental pressure vessel explosions, and deliberate particle generations from such events as anti-satellite weapon experiments. As shown in Figure 6, orbital debris particles provide a very severe design environment compared to the micrometeoroids. The debris mostly lies in earth orbits that correspond to the most popular satellite orbit inclinations.

Both types of particles move at very high velocities, approximately 10 to 20 km/sec. Impacts with the truss will cause damage (Figure 7) that can penetrate the protective aluminum foil and even locally damage the truss struts. Table 2 shows the predicted number of damaging impacts over 30-year period, utilitizing both the current orbital debris design model and the recently proposed model shown in Figure 6. The effect of damaging impacts on structural integrity must be verified in a ground test program.

3. Maintainability

Maintainability is enhanced by allowing the replacement of individual truss members and by providing inspection methods for detecting damage.

3.1 Member Replacement/Repair

Though the truss is designed for 30 year exposure to the LEO environment, unforeseen circumstances may result in unacceptable damage to truss members. As stated previously, the erectable joints connecting the struts to the truss nodes are designed to be assembled and disassembled over the 30 year life of the truss structure. This allows struts to be replaced by a suited astronaut to quickly restore the truss to full capability.

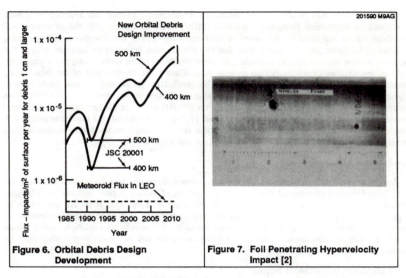

Figure 6. Orbital Debris Design
Development

Figure 7. Foil Penetrating Hypervelocity
Impact [2]

Table 2. Number of Damaging Impacts Over 30-Year Period for Two Debris Models

Period	Foil Perforations per Longeron	Foil Perforations per Diagonal	Foil Perforations per sq cm
30 years (current model) 0.005 in. (0.13 mm) thick 5052-H34 foil	141	204	0.0191
30 years (proposed model) 0.005 in. (0.13 mm) thick 5052-H34 foil	1633	2366	0.2215

H5184: T2 08/04/89

3.2 Damage Detection

Although the truss is the backbone of the Space Station, it also serves as a translation path and tether point for EVA astronauts. Under the loadings associated with EVA, a damaged truss strut that is undetected could fail and could cause injury to the EVA astronaut or separation from the Station. In addition, if damage went undetected for a long period of time, a second strut in the same truss bay could become damaged and exceed the damage tolerance of the truss structure. Damage detection is necessary so that the truss can be restored to full capability.

The current means for damage detection is remote visual inspection by video camera and on-the-spot inspection by EVA astronauts. Damaged areas should be readily visible as the dark composite will contrast sharply with the aluminum protective coating as shown in Figure 7. Analysis and testing has indicated that the truss struts can sustain a loss of 5.4 in^2 (35cm^2) of material (Figures 8 and 9) without any reduction in axial load-carrying capability[3]. Active methods for damage detection are being investigated with cost, monitoring simplicity, and ease of assembly and checkout being the main trade study drivers.

4. Verification

Reliablity is assured before launch by performing ground tests that confirm analytically predicted performance and assure structural integrity. Current areas of concern for truss longevity are fatigue of erectable joints and 30-year life of composite materials in LEO.

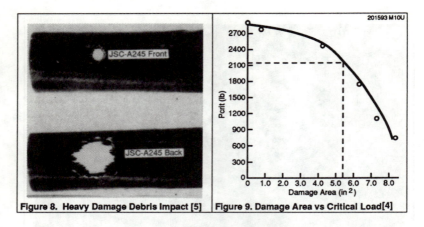

| Figure 8. Heavy Damage Debris Impact [5] | Figure 9. Damage Area vs Critical Load[4] |

It is highly desirable for the erectable joint to exhibit a linear response over the 30-year lifetime. On-orbit operational scenarios for the Space Station will be used to derive thermal and mechanical loading cycles the joint must withstand. A ground test will be devised to conservatively cyclic test the joint with static performance measured at selected intervals to identify any degradation in stiffness or linearity.

The 30-year life in LEO is a more difficult design parameter to verify because combined effects of the environment must be considered. Micrometeoroid/orbital debris impacts will be verified in a sequential fashion. Hypervelocity impact, up to 7 km/sec, will be performed using the light gas gun at NASA Johnson Space Center. Three particle sizes will be selected on the basis of probability of penetration of the aluminum foil. The intent of the test program is to identify a range of foil-coating penetrations from just penetrating the coating, highest probability of occurrence, to severe damage to the tube structure, least probability. Results of all three damage tests will be used as specimens in simulated ground tests for the on-orbit atomic oxygen environment. A more severe atomic oxygen flux will be used in the ground tests to accelerate the effects of 30 years of orbital exposure. The most severely damaged tube segment will then be subjected to structural testing to ascertain worst case effects on structural integrity of the truss tube.

Maintainability by replacement of a strut will be verified by having astronauts perform the removal/replacement operation in simulated zero gravity. As shown in Figure 10, underwater test facilities at McDonnell Douglas and Johnson Space Center can simulate zero gravity by providing neutral buoyancy of the astronauts and the equipment . Since maintainability is a design issue, tube replacement will be verified early in the program at the same time the truss assembly process is verified.

5. Summary

Ensuring that the Space Station truss structure will have a certified lifetime of 30 years requires that a combination of built-in reliability and maintainability features be designed into the truss and demonstrated through a rigorous verification program. The truss design incorporates a fail-safe truss diagonal configuration by selecting materials that are resistant to impact damage from crew activities and to thermally induced microcracking, and by providing a coating to protect against atomic oxygen and ultraviolet radiation. To resist the micrometeoroid/debris environment, the truss design can sustain a large amount of damage and still function and, when the damage becomes too great, on-orbit replacement of truss members restores the truss to its original capability.

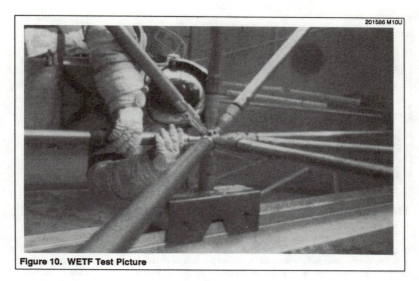

Figure 10. WETF Test Picture

References

[1] Babel, H.W., Shumate, T.P. and Thompson, D.F.: "Microcrack Resistant Structural Composite Tubes for Space Applications", SAMPE Journal, pp. 43-48, May/June 1987.

[2] Eagle Engineering: "Investigation of Hypervelocity Impact Damage to Space Station Truss Tubes", Eagle Report No. 88-176, pg. 101, 15 Feb 1988.

[3] pg. 148 of reference 2.

[4] pg. 228 of reference 2.

[5] pg. 136 of reference 2.

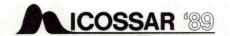
SYSTEM AND UTILIZATION STUDY OF JAPANESE EXPERIMENT MODULE
FOR INTERNATIONAL SPACE STATION

by

Yoshinori Fujimori,Kuniaki Shiraki
and
Tatsuo Matsueda
National Space Development Agency of Japan, Tokyo Japan

ABSTRACT

The paper briefly reviews the overall system study
of Japanese Experiment Module (JEM) which consists of
Pressurized Module, Exposed Facility and Small Remote
Manipulator System (RMS). JEM design is conducted based
on Japanese technology and effort is primarily directed
toward the mechanical and electrical performances.
Capabilities regarding the crew habitation and life
support subsystem rely upon NASAs Module.

Another aspect of the Space Station is its
scientific utilization. Potential uses are explored in
the areas such as astronomy, astrophysics, earth
observation,life science, material & fluid sciences and
technology development. Also experimental equipments
and facilities are under study.

Emphasis would be placed upon the safety and
reliability considerations to the hardware and software
designs.
KEYWORDS: Space Station,Japanese Experiment Module,
System Engineering, Space Utilization.

1.INTRODUCTION

Upon the call of United States President Reagan to
western nations for participation in the Space Station
Program January 1984, the Japanese Space Advisory
Committee (JSAC) recommended Space Technology Agency
(STA)to collaborate with National Aeronautics and Space
Administration (NASA) on the ground to enhance space
development, utilization promotion and international
cooperations. After heavy negotiations among the partners,
Japan, European Space Agency (ESA) and Canada reached the
agreement to start phase B study in 1985.

In Japan, National Space Development Agency of Japan (NASDA) takes responsibility to develop Japanese Experiment Module(JEM) and to integrate overall JEM utilization.The current progress covers the preliminary design of JEM system and requirement analysis of design reference missions. The report summarizes JEM system outline and its utilization.

2.JEM SYSTEM DESIGN

JEM system configuration and characteristics is briefly described. JEM is attached to the space station (SS) core and composed of the Pressurized Module (PM), Exposed Facility(EF), and Experiment Logistics Module(ELM). The small remote manipulator system (JEM RMS) is provided on the PM with the airlock. The EF consists of two identical units, EF-1 & EF-2, connected in tandem and one end being attached to PM. The ELM has two compartments, one the Pressurized Section (ELM-PS), another the Exposed Section (ELM-ES).

The JEM system consists of the following subsystems; (1)Structure,(2)Mechanism,(3)Electrical Power,(4)Communication and Control,(5)Thermal control,(6)Environmental Control and Life Support,(7)Experiment Support,(8)Remote manipulator.On-orbit configuration is shown in Figure 1.

Pressurized Module (PM)

PM is a pressurized laboratory in which crew can conduct experiments in a shirt-sleeve environment. In this laboratory, primarily material science (processing) and life science experiments are to be carried out.The control console for JEM RMS and working area for the airlock are provided in this laboratory.

Exposed Facility (EF)

EF Provides the installation bench for astronomical observation, earth observation, communication and technology development experimentations. It is suited for the missions that require wide field of view and evacuation from PM for the safety reasons. Mission payloads con be mounted on the upper and sidelong surfaces of the 2 identical units, EF-1 & EF-2, the attachment is made by Equipment Exchange Unit (EPU) of the payload on the mounting port. Some of the EF mounting ports may be occupied by the system operation equipments such as the gas supply unit or JEM RMS tool box.

Experiment Logistics Module(ELM)

ELM consisting of the two compartments; the pressurized section(ELM-PS) and the exposed section (ELM-ES), provides the storage area on orbit and the transport container to and from the orbit for the experiment payloads, specimens, consumables and spare parts of JEM.The unique feature of ELM is the emergency evacuation capability of ELM-PS to accommodate 2 crew survival condition for 3 hours.

Remote Manipulator System (JEM RMS)

JEM RMS is mounted on the aft end corner of PM and has the capability to attach/detach payloads on/off the EF and to assist EVA crew. As is shown in Fig.2, the entire system primarily consists of the main arm,the small fine arm, vision device, control device and so forth. The main and small arms,10m and 2m in length respectively have the similar characteristics in general performance but the small one is superior in positioning and inferior in handling mass. In addition to the system requirements the stowed configuration when JEM, RMS is packaged for launch gives the impact to the structural design. The grapple fixture and the end effector are two basic pieces to connect the two parts,the connection can be done in both electrical and mechanical ways.

The RMS operation can be executed by hand controllers for the main arm and a master arm for the small fine arm. All the necessary control and interface electronics are installed in the PM. Because the manipulation control essentially relies on the visual information to the operator, besides the window to get the direct view, many TV cameras as possible are installed all over the structure; the two on the elbow and wrist of the main arm,stereoscopic one on the small fine arm, the five on PM and upper edge lines of EF and so forth.

3 JEM UTILIZATION

In the late 1990s, it is expected that the space experiments aboard JEM will be being performed using the unique space environment such as micrograrity, high vacuum and wide field of view. As the space station program has passed the preliminary design phase and is currently proceeding to the basic design phase,the users requirements need to be properly implemented in the development of JEM and its payloads.

Mission requirements for JEM design have been studied by the Space Activities Commissions Ad Hoc committee through the channel of the Space Station Utilization Workshops composed and managed by the Science and Technology Agency (STA), the National Space Development Agency of Japan (NASDA) and the Japan Space utilization Promotion Center (JSUP).

Japans Space Station utilization planning was initiated as early as in June 1984 when the Ad Hoc Committee on Space Station program executed overall surveys. Mission themes were proposed by the universities, national research institutes and private companies totaling approximately 150 in that year and the double in the successive year.

Those proposed themes from users were categorized into the following six disciplines:

1 Scientific Observation
2 Earth Observation
3 Communication
4 Material Processing
5 Life Science
6 Technology Development

Grouping similar amd analogous ones, the proposed experiments were temporarily classified into 39 missions.The results were reported to NASA and included in the Mission Requirement Data Base serving a strawman in the international study among the partners.

The further study has been continued to work out mission scenarios out of the proposed mission themes. Elaboration of mission scenarios has been evaluated and polished through the closed sessions or open workshops yielding the tentative plan in terms of design reference missions at initial operational capability(IOC). Regarding the locations, material pressing and life sciences experiments are primarily carried out in PM, observation mission and technology development experiments are obviously conducted on EF. The workshops discussion mostly emphasized the step-by-step implementation of the missions, i,e, from fundamental /simple to advanced / complicated themes.
The most probable model missions considered suitable for IOC are being worked out.
The proposed missions necessitate the development of experimental payloads, i,e, flight hardwares, and payload accommodation.Procedure to implement the mission requirements in the hardware is as follows:

-Characterize the missions by the category of the
 similarity
-Investigate the functions of the hardware required
 by the mission
-Break down functional requirements and translate them
 into the specification stipulation.
-Clarify the parameters to specify the requirements
 for the development of the flight hardware.

    Generally, the experiment equipments is divided
roughly into two categories, user unique and generic
ones. The former may be found in the scientific / earth
observation, communication and technology development.
The latter mostly in material processing and life science
area.The generic equipment has the nature of multipurpose,
thus many users can use them for different missions.
Some of the generic equipments deemed necessary for
advancement of basic science and technology in Japan
will be considered for development by NASDA.

4.TECHNICAL ISSUES

    4.1 Issues on JEM system Integration & Interface
            with Space Station Core
    Since the design and construction of JEM itself
levies the heavy burden technically on the current
engineering capability of Japan, there seem a number
of concerns to be clarified, not to speak of to be
resolved. Some of them are too difficult to be handled
routinely on the ground sometimes, nevertheless it is
pertinent to describe the effort for the better system
design.

Factors of Safety (FS) and Damage Tolerance
    Strength aspect of the structure renders the basis
of the system safety, thus FS can be determined rather
independently by the designers depending on the
experiences or level of confidence in engineering
judgement.The number of FS itself could be ones deemed
appropriate in aerospace community and the guideline of
design requirements provided by NASA would be paid a full
regard in the process of the JEM design.
    The damage tolerance design usually applied for
the aircraft can be adopted by the spacecraft in a
similar but little bit different manner, as is obvious
when looked at by the timeline stress trend.

Orbital Debris

Man-made debris or natural meteorite are potential cause of hazard on orbit, even if the problem associated with those has not been so explicit in actual space activities.Especially the due consideration should be directed to the reduction of artificial debris before it is too late.

NASDA has taken a number of actions in designing the spacecrafts so that the number of space debris will not increase.Explosion-preventive measures to the uppers-tage of the launcher, the remaining gas evacuation valve to the plenum chamber or the fuel tank are installed.And all the satellites have the device to retain the bolts & nuts at the separation or paddle deployment.

Since JEM, as one of the space station elements, should have the equivalent safety level like other modules, the protection design requirement of NASA agaignst the space debris is also applied. The currently recommended requirement states;

The probability that the damage by space debris will not jeopardize the life of the crew or continuation of the space station mission shoulds be at least 0.9955 within 30 years operation.

However, due to the gradual degradation of orbital debris environment and uncertainty of material properies at the moment of collision, the target at the initial degign phase is required to be 0.9955 for 10 years.

Design approach of JEM follows a number of steps:
(1)Presise Modeling of the orbital debris based on NASA supplied data.
(2)Fundamental experiment execution to check the material & structure capability.
(3)Structure & bumper design.

The current bumper structure concept has the double wall configuration outside the pressure vessel. The thermal isolation materials are placed between the skin wall of the PM and the inner bumper.In case of visible penetration, the safety mesaures are conducted in the operational procedures such as damage location detection, crew evacuation, isolati-on of decompressed module and damage repair.

Atomic Oxygen
     Some of the STS flights have revealed the damaging
effect of atomic exygen to the number of materials.
Degradation in optical or electrical characteristics,
reduction in strength and rigidity due to the recesion
of the surface of the structural members are reported
in the open publication. Although they gave warning to the
designers of the space station, there would not be any
immediate remedy if the material is inherently susceptible
to the atomic oxygen. Design approach of JEM takes the
following;

*Use the oxygen-resistant materials or develop such
 materials,
*Design the layout so that the surface of susceptible
 materials may not be exposed directly to space or
 provide protective measures through outfitting process,
*Furnish the coating of the surface by the oxygenresistant
 materials.

Product Assurance
     One of the major significances of International
Space Station is its joint activity of all partners
toward the safety, quality control and reliability of
all softwares and hardwares. This activity is called
product assurance.It entails utmost requirements on safety,
reliability, maitanability and quality assuranre as well
as the evaluation by review & audit over the entire life
cycle of the space station. Presumably, the motivation
of its activity stems form the complexity of the Space
Station system and STS accident in 1986.
The current JEM activity covers the basic planning of
works to be done such as;
(1)Basic Principle to implement the philosophy to the
    JEM program.
(2)Documentation tree
(3)Technical evaluation of safety requirements
(4)Coordination, domestic & international
(5)Safety review & audit
(6)Evaluation of development activites
(7)Overall planning & execution

4.2 Issues on Mission Execution and Mission Payload Development

Other than technical issues for the system, there are more issues for the payloads and mission execution. Because utilization of space station is diverse, the number of concerns obviously gets very large. Here only prominent ones are to be listed.

*Late/ Early access
*Crew utilization
*Toxic and Reactive substance handling
*Experiment rack commonality
*Rapid sample return
*Specimen characterization
*High data rate requirements
*u-G Environment level.

5. CONCLUDING REMARKS

Current status of JEM development and its utilization is briefly summarized. The real outcome of what is described in this report will be subject to change depending on the progress of the program.

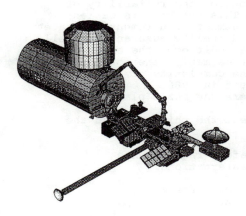

Figure 1

JEM On-Orbit Configuration

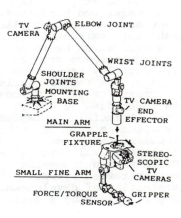

Figure 2

JEM RMS Arm Configuration

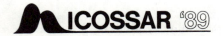
MODE TRUNCATION EFFECTS ON FLEXIBLE SPACE STRUCTURES

H. Miura*, M. Kobayakawa*, K. Ogasawara**, and H. Imai***

* Dept. of Aeronautical Engineering, Kyoto Univ., Kyoto, Japan
** Mitsubishi Heavy Industries Ltd., Nagoya, Japan
*** Dept. of Mechanical Engineering, Setsunan Univ., Osaka, Japan

ABSTRACT

This paper deals with the problem of active vibration control of flexible space structures. The main difficulty associated with this problem is that the structure has infinite vibration modes so that higher modes have to be truncated in designing controllers. It is possible that the modal truncation causes so-called spillover instability. This paper provides a new method for designing a stabilizing controller for flexible space structures. The method is based on a pole assignment algorithm which offers some design freedom beyond assignment of closed loop poles. The freedom is utilized to avoid spillover instability. A simulation study is presented for demonstrating the usefulness of the proposed algorithm.

KEYWORDS

Vibration control; flexible space structure; large space structure; modal analysis; mode truncation; spillover instability.

1. INTRODUCTION

During these two decades, many studies on the problem of vibration control of flexible space structures have been published; see for example [1] and its references. Major difficulties associated with this problem are originated from the absence of aerodynamic damping which is common to airplanes and from the complicated treatments of infinite oscillatory modes.

The governing equation of flexible space structures is a combination of ordinary and partial differential equations. Since such an equation cannot be solved analytically except for extremely simple models, a discretization through modal analysis is common to the most of investigations. In this situation, modeling errors due to truncation of higher modes might cause the spillover phenomenon which affects the stability of structures [1][2].

The purpose of this paper is to provide a new technique for the design of vibration controllers for flexible space structures without spillover instability. In this study, we assume an

observer based controller; such a controller is characterized by feedback and observer gain matrices. Those matrices are designed by means of a pole assignment algorithm. It is well known that, in multi input-multi output systems, the feedback and observer gain matrices to achieve a specified pole assignment are not unique and they retain some degrees of freedom [3]. Those degrees of freedom can be shown to be utilized to minimize norms of the gain matrices so as to improve the stability of the closed loop system.

2. MATHEMATICAL MODEL

Consider a flexible space structure illustrated in Fig. 1. The equation of motion of this structure is given by

$$\frac{Eh^2}{1-\nu^2}\nabla^4 z(x,y,t) = -\rho h\frac{\partial^2 z(x,y,t)}{\partial t^2} + g(x,y,t) \tag{1}$$

where $z(x,y,t)$ is the deflection and h the thickness of the panel. The last term $g(x,y,t)$ is the control force distribution. The solution to (1) can be written as

$$z(x,y,t) = \sum_{k=1}^{N} \tau_k(t) z_k(x,y) \tag{2}$$

where $z_k(x,y)$'s are mode functions and $\tau_k(t)$'s are mode amplitudes. The mode functions and natural frequencies can be obtained by Rayleigh-Ritz method. The number N of modes is, in theory, equal to infinity. However, a finite but large N is sufficient for $z(x,y,t)$ to approximate the true solution to (1). The control force is provided by two actuators,

$$g(x,y,t) = \sum_{k=1}^{2} u_k(t) b_k(x,y) \tag{3}$$

where $u_k(t)$'s are the actuator force amplitudes and $b_k(x,y)$'s the influence functions.

Then, Eq. (1) can be written in the following state space form

$$\dot{\mathbf{x}} = A\mathbf{x} + B\mathbf{u} \tag{4}$$

where $\mathbf{x} = \left[\tau_1,\dot{\tau}_1,\tau_2,\dot{\tau}_2,...,\tau_N,\dot{\tau}_N\right]^T$ and $\mathbf{u} = \left[u_1\ u_2\right]^T$, and where A and B are matrices of appropriate sizes whose components include structural damping coefficients, natural frequencies and the positions of actuators. Also, we assume two position sensors. Then, the output equation is

$$\mathbf{y} = C\mathbf{x} \tag{5}$$

where C is a $2\times2N$ output matrix.

3. SPILLOVER INSTABILITY

We note that implementable controllers are difficult to design for the above system when N is large. To avoid this difficulty, we divide the state vector as $\mathbf{x} = \left[\mathbf{x}_1^T,\mathbf{x}_2^T\right]^T$, where $\mathbf{x}_1$ is the control mode state vector and $\mathbf{x}_2$ the truncated mode state vector. Then, Eqs. (4) and (5) are written as

$$\begin{bmatrix}\dot{\mathbf{x}}_1\\\dot{\mathbf{x}}_2\end{bmatrix} = \begin{bmatrix}A_1 & 0\\0 & A_2\end{bmatrix}\begin{bmatrix}\mathbf{x}_1\\\mathbf{x}_2\end{bmatrix} + \begin{bmatrix}B_1\\B_2\end{bmatrix}\mathbf{u} \tag{6a}$$

$$\mathbf{y} = C_1\mathbf{x}_1 + C_2\mathbf{x}_2 \tag{6b}$$

The following observer-based controller is assumed in this study

$$\mathbf{u} = G\overline{\mathbf{x}}_1 \tag{7a}$$

$$\dot{\overline{\mathbf{x}}}_1 = A_1\overline{\mathbf{x}}_1 + B_1\mathbf{u} + K(\mathbf{y} - C_1\overline{\mathbf{x}}_1) \tag{7b}$$

where G is the feedback gain matrix and K the observer gain matrix. The controller is to be designed so as to stabilize the following reduced order system, which models the controlled mode,

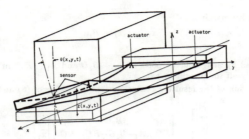

Fig. 1. Flexible space structure.

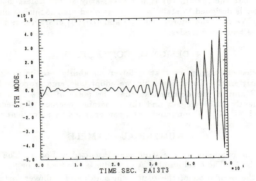

Fig. 2. Oscillation of the panel exhibiting spillover instability.

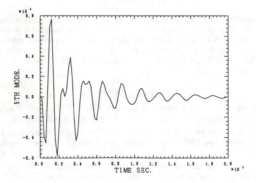

Fig. 3. Oscillation of the panel stabilized by the proposed scheme.

with sufficient stability margin.

$$\dot{\mathbf{x}}_1 = A_1\mathbf{x}_1 + B_1\mathbf{u}$$
$$\mathbf{y} = C_1\mathbf{x}_1 \tag{8}$$

This can be achieved by assigning the poles of $A_1 + B_1 G$ and $A_1 - KC_1$ in the left half complex plane and sufficiently off the imaginary axis.

The state equation of the resulting control system consisting of (6) and (7) is written as follows:

$$\begin{bmatrix} \dot{\mathbf{x}}_1 \\ \dot{\overline{\mathbf{x}}}_1 \\ \dot{\mathbf{x}}_2 \end{bmatrix} = \begin{bmatrix} A_1 & B_1 G & 0 \\ KC_1 & A_1+B_1G-KC_1 & KC_2 \\ 0 & B_2 G & A_2 \end{bmatrix} \begin{bmatrix} \mathbf{x}_1 \\ \overline{\mathbf{x}}_1 \\ \mathbf{x}_2 \end{bmatrix} \tag{9}$$

It is easily verified that the system (9) is not necessarily stable unless $B_2 G = 0$ or $KC_2 = 0$, whereas the controller is designed to stabilize the controlled mode. Such a phenomenon is called spillover instability. We note that norms $||KC_2||$ and $||B_1 G||$ characterize interactions between the controlled and the truncated modes.

4. DESIGN OF CONTROLLER

The above discussion suggest us that spillover instability can be avoided if $||KC_2||$ and $||B_2 G||$ are sufficiently small. To achieve this, we adopt a scheme proposed by Masui et al. [3]: the gain matrices G and K are designed so that the poles of $A_1 + B_1 G$ and $A_1 - KC_1$ are assigned in a sufficiently stable region and the remaining degrees of freedom are utilized to minimize $||KC_2||$ and $||B_2 G||$. The minimization can be made by a gradient method.

5. NUMERICAL EXAMPLE

The above scheme is applied to the structure illustrated in Fig. 1. The aspect ratio of the structure is taken to be 5. The total number N of modes is assumed to be 6.

Fig. 2 depicts the oscillation of the panel which exhibits spillover instability while Fig. 3 illustrates the oscillation which is controlled by the proposed scheme. It is observed in Fig. 3 that spillover instability does not take place and the vibration converges to the equilibrium state.

6. CONCLUSION

This paper proposes a new method for reduction of spillover instability in active vibration control systems for flexible space structures. The method is based on a pole assignment algorithm. The degrees of freedom offered by the algorithm is utilized to minimize the interaction between the controlled and the truncated modes. The proposed method is applied to a flexible panel which is oscillating in bending and torsional modes. The result of a simulation study is satisfactory.

REFERENCES

[1] M. J. Balas, "Trends in Large Space Structure Control Theory: Fondest Hopes, Wildest Dreams," IEEE Trans. on Automat. Control, Vol. AC-27, No. 3, pp. 522-535, 1982.

[2] K. Ogasawara, M. Kobayakawa, and H. Imai, "On the Mode Truncation Effects on the Flexible Space Structure," Proc. the 16th International Symposium on Space Technology and Science, pp. 1355-1361, 1988.

[3] K. Masui, M. Kobayakawa, and H. Imai, "Aircraft Longitudinal Motion Control by Utilizing Degrees of Freedom in ι ٮle Assignment," Trans. Japan Soc. Aero. Space Sci., Vol. 26, No. 74, pp. 198-215, 1984.

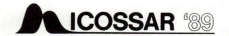
Reliability Consistent Load and Resistance Factors

J. H. Lee* and S. Y. Son**

Abstract

A practical algorithm to find the partial safety factors consistent with target reliability in LRFD format is proposed. The method is based on Second Order Second Moment method and Reliability Condition is also included.

Introduction

Conventional working stress design is, by and large, replaced by the Load and Resistance Factor Design(LRFD) to account for the uncertainties in the load and resistance variables. These factors are mostly based on the Advanced First Order Second Moment(AFOSM). However, the use of AFOSM may result in probability of failure which can be different from the target probability of failure for non-normally distributed basic variables. This shortcomings are primarily due to the fact that those factors are determined at the design point which is introduced for the reliability assessment. By employing the reliability conditioned(RC) method[1], the design factors may be calculated directly from the failure point by preassigning the probability weight at failure to basic variables as follows: firstly, the Second Order Second Moment(SOSM) method[2] is used to determine $\bar{R}$. Secondly, an efficient algorithm for determining the most likely failure points is set up.

Algorithm for reliability consistent design factors

In the process of calculating $\bar{R}$, fitting points are derived directly from the inverse of cumulative distribution of each variable. The design point is used as the initial value for the most likely failure point. In adapting Newton's iteration scheme, the increment of the basic variable is calculated using percentile requirement on the loads. Figure 1 shows the flow of computation.

---

* Dept. of Civil Eng., Sungkyunkwan Univ., Suweon, KOREA
** Dept. Mech. Eng., Ajou Univ., Suweon, KOREA

Fig.1 Flow of Computation

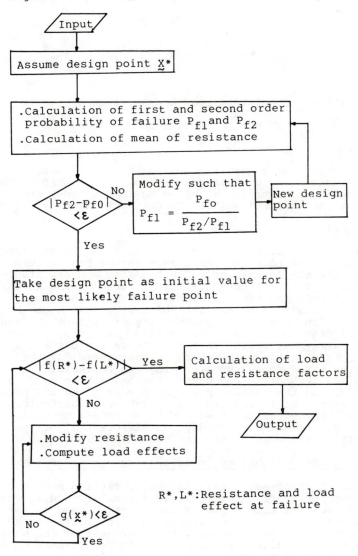

R*,L*:Resistance and load
       effect at failure

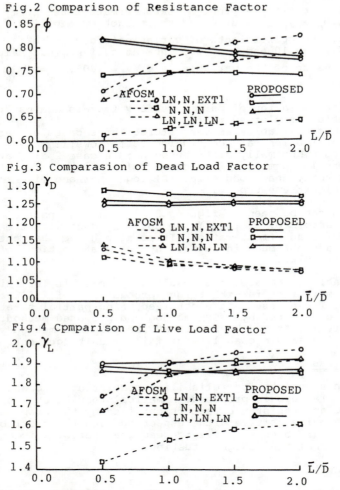

Fig.2 Comparison of Resistance Factor

Fig.3 Comparasion of Dead Load Factor

Fig.4 Cpmparison of Live Load Factor

Table 1. Probability of Failure of a Beam in Flexure
(Target probability Pf_o= 0.00135)

$\bar{L}$ / $\bar{D}$	AFOSM	PROPOSED
0.5	0.00155	0.00193
1.0	0.00189	0.00146
1.5	0.00243	0.00146
2.0	0.00281	0.00144

Numerical examples

In the example problem, LRFD format of the form $\phi R_n \geq \gamma_D D_n + \gamma_L L_n$ is used and the target probability P_{fo} is 0.00135. The coefficients of variation(c.o.v.) are : $\Omega_R=0.16$, $\Omega_D=0.10$ and $\Omega_L=0.26$. The ratio of nominal value to mean value of the variables are: $v_R=0.893$, $v_D=0.952$ and $v_L=0.833$.

The variation of partial safety factors against the load ratio is plotted in figures 2, 3 and 4 with the distribution of variables as parameters. Three cases of variable distributions are considered as follows; in case 1 all variables are normally distributed(N); in case 2 all variables are lognormally distributed(LN); in case 3 resistance (R)is normal, dead load effect(D) lognormal and live load effect(L) extreme type I.

As can be seen in figures 2,3 and 4, the partial safety factors ϕ, γ_D, and γ_L do not vary much with the increase of load effect ratio. It is also noted that these factors are not much influenced by the distributions of the basic variables.

Table 1 compares the resulting probability of failure of a beam in flexure with target probability of failure 0.00135. These results are obtained using partial safety factors calculated by AFOSM method and proposed method, respectively. It may be seen that the proposed method yields practically same probability of failure as target one.

Conclusions

An algorithm for reliability consistent load and resistance factors is proposed. Numerical results show that the load and resistance factors obtained by the proposed method are insensitive with respect to both the distribution of basic variables and live-to-dead load ratio, which is most desirable in design situation.

References

[1] Ayyub, B.M. and White, G.J.: "Reliability-Conditioned partial Safety Factors", J. of Struct. Eng., ASCE, Vol. 113, No. 2, pp.279-294, 1987
[2] Kiureghian, A.K., Lin, H-z.,and Hwang, S-J.: "Second-Order Reliability Approximations", J. of Eng. Mech., ASCE, Vol. 113, No. 8., pp. 1208-1225, 1987

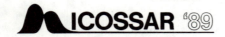
Reliability Assessment of Pressurized Fuselages
with Multiple-Site Fatigue Cracks

Hiroo Asada[1]

[1]Head, Full-Scale Test Section, Airframe Division,
National Aerospace Laboratory, Mitaka, Tokyo 181, Japan.

Abstract

Multiple-site damage could cause a catastrophic
structural failure of aged aircraft. A reliability
analysis is proposed for an assessment of the
capability of visual inspection on multiple-site
damage. This paper describes the procedure of the
reliability analysis to estimate an optimal
inspection schedule. Several numerical examples
are being computed to evaluate the proposed
analysis by using a Monte Carlo technique.

Introduction

As mentioned in the FAA Advisory Circular No.25.571-1A and
the Advisory Circular No.91-56, the evaluation of multiple-site
damage(MSD) due to fatigue cracks along a longitudinal rivet
splice in a pressurized fuselage is required to maintain the
structural integrity of aged aircraft. According to the MSD
concept, a number of small cracks might suddenly coalesce to form
a single critical crack leading to a catastrophic failure of
fuselage structure (Swift 1987) even if these crack lengths exceed
a rivet head but less than the length easily detected during
visual in-service inspections. In addition, it is pointed out that
the vigilance of inspectors tends to diminish as they inspect
thousands of fasteners and that, even if probability of detecting
a single crack increases, the efficiency of inspectors decreases.
Typical accidents due to MSD have recently occurred in the aft
pressure bulkhead of Japan Airlines B747 in 1985 and the fuselage
of Aloha Airlines B737 in 1988.

In order to prevent a catastrophic failure due to MSD, it is
argued that a full-scale fatigue test of at least two lifetimes
should be conducted to make sure that the probability of MSD
occurring is very low and that teardown inspections of the tested
aircraft and high-time aircraft should be performed to find out
critical rivet splices with MSD. The reliability analysis (Asada
et al. 1985; Shinozuka et al. 1981) is also one of effective

methods to evaluate capability of visual in-service inspection for detecting MSD.

This study emphasizes to propose a methodology of reliability analysis for estimating an optimal inspection schedule for aged aircraft and subsequent aircraft with a fleet control which can adequately detect MSD. This paper presents the procedure of this analysis with various factors to simulate failure processes of MSD.

Structural Model

A three-row rivet splice shown in Fig.1 is a typical example of rivet splice for a single lap joint of skin along a longeron of a pressurized fuselage. Multiple-site damage emanates along the first rivet row for the upper skin and the third row for the lower skin because of their highest stresses. It is recommended to take a two-bay width $2W$ (W:one frame spacing) as a critical crack length for the fail-safe design. In order to analyze the rivet row subjected to the highest stress, the row is replaced by a single row of rivet holes depicted in Fig.2. This row is subjected to cyclic loading of constant stress amplitude ΔS which is derived from cyclic hoop stress taking into account bearing and by-passing loads and is equal to the maximum stress S_{max}. As ΔS is assumed to be constant without depending on crack length, the ligament stress is increasing while cracks are propagating.

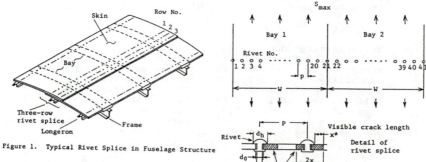

Figure 1. Typical Rivet Splice in Fuselage Structure

Figure 2. Model of Longitudinal Rivet Splice

The following assumptions are made for the present model:

1. Coalescence of cracks : When a ligament between two crack tips becomes smaller than the half length of the smaller of the two cracks, they coalesce to form a single large crack. A rivet hole and a crack coalesce when the ligament between both becomes smaller than the hole radius,

2. Residual strength : Residual strength of a rivet splice with MSD is limited by net stress yielding. Therefore, it is defined that a fracture of that splice occurs when the ligament stress of two bayes under a limit load S_{LL} reaches a material yield stress S_{ty}. A critical crack length x_c under this condition is given by

$$x_c = 2W(1 - S_{LL}/S_{ty}) - (n-1)d_0 \qquad (1)$$

in which d_0 and n are the rivet diameter and the number of rivet holes within two bayes,

3. Service and limit loads : In accordance with FAR 25.571, the relation between ΔS and S_{LL} is expressed by

$$S_{LL} = \{(1.1\Delta P + P_A)/(\Delta P + P_A)\}\Delta S \qquad (2)$$

where ΔP and P_A denote the normal operating differential pressure and the external aerodynamic pressure,

4. Fatigue crack initiation and propagation : Crack initiation times at both sides of a rivet hole are correlated each other. A joint probability density of the first crack initiation time at one side of a rivet hole and the second crack initiation time at the far side of the hole is assumed to be a two-dimentional Weibull distribution $f_{12}(N_1,N_2 : \alpha_1,\beta_1; \alpha_2,\beta_2,\rho)$ in this analysis. Two Weibull random variables correlated each other are approximately generated as follows (Grigoriu 1984). First, two random variables are generated from the standardized joint normal distribution with a correlation coefficient ρ, which is equal to the correlation coefficient of the Weibull distribution. Then, two random variables of the Weibull distribution are derived from transforming directly these two normal random variables.

The following inverse hyperbolic tangent equation is used to calculate fatigue crack propagation with the stress ratio R=0,

$$dx/dN = 10^{C_1}[\log\{(K_0/\Delta K)^2\}/\log\{(\Delta K/K_c)^2\}]^{C_2/(2\log e)} \qquad (3)$$

$$\Delta K = \Delta S \sqrt{\pi x \cdot g(x)}$$

in which a crack length 2x is defined as shown in Fig.3, C_1 is normally distributed with a mean and a standard deviation of μ_{c1} and σ_{c1}, and $g(x)$ is a modification coefficient which is a function of x, and , C_2, K_0 and K_c are parameters.

5. Detection of MSD and repair of detected cracks : Periodic visual inspections are carried out at an inspection interval ΔN to detect cracks whose lengths exceed a rivet head. In this analysis, it is assumed that MSD is detected if at least one crack out of multiple-site fatigue cracks is detected. A two-bay rivet splice with detected MSD is repaired and MSD will not be initiated again within a design life. Cracks propagating along rivet holes are interrupted by rivet heads, so that small cracks should be detected under low probability of detection during visual inspection. However, the probability of detecting MSD increases because of the condition that MSD is detected if at least one crack is detected out of many small cracks as mentioned above.

A three-parameter Weibull distribution function is applied to the probability of detecting a visible crack of length x* in Fig.2

$$D(x^*) = 1 - \exp[-\{(x^* - x_{min})/(\xi - x_{min})\}^\zeta] \qquad (4)$$

where x_{min} is the minimum detectable crack length, and, ξ and ζ are parameters.

6. Yield stress and critical crack length : Yield stress S_{ty} is normally distributed. The probability density function of a critical crack length x_c can be derived from eq.(1) and the above

normal distribution.

Probability of Detecting MSD under Periodic Inspection

A Monte Carlo technique is applied to compute the probability of failure P[Failure:Ni] in a two-bay splice and the probability of detecting MSD P[Detection:Ni] within [0, Ni]. The i-th inspection is performed at Ni cycles and the periodic inspection interval is ΔN. When m is a number of two-bay critical splices due to MSD in an aircraft, the probability of first failure in m splices within [0, Ni] is expressed by

$$P_m(N_i) = 1 - \{1 - P[Failure : N_i]\}^m \qquad (5)$$

Figure 3 shows numerical examples of probabilities of failure without inspection and of detecting MSD under the periodic inspection of the interval ΔN=2000 flts. The parameter values in the two-dimentional Weibull distribution of crack initiation are $\alpha_1 = \alpha_2 = 6$, $\beta_1 = \beta_2 = 78000$, 90000 and 102000 flts, and ρ=0.85. According to this model, the failure due to MSD can be prevented by periodic inspection effectively.

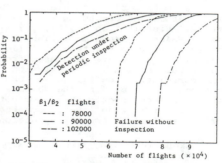

Figure 3. Effect of Periodic Inspection on Probability of Failure

Conclusions

The reliability analysis for detecting multiple-site fatigue cracks under periodic visual inspection is discussed with the structural model in this paper. The computation is now in progress by using the Monte Carlo technique.

Acknowledgments

The author is grateful to Mr. T. Swift, the Federal Aviation Administration, for his suggestion on MSD, and he also would like to thank Prof. M. Shinozuka, Princeton University, and Prof. H. Itagaki, Yokohama National University, for their valuable comments for this analysis.

References

Asada, H., Itagaki, H., and Itoh, S. (1985). "Effect of sampling inspection on aircraft structural reliability.", Proc. ICOSSAR'85, 1, 87-96.
Grigoriu, M. (1984). "Crossings of non-Gaussian translation processes." J. Engrg. Mech., ASCE, 110(4), 610-620.
Shinozuka, M., Itagaki, H., and Asada, H. (1981). "Reliability assessment of structures with latent cracks." Proc. US-Japan Cooperative Seminar on Fracture Tolerance Evaluation, 237-247.
Swift, T. (1987). "Damage tolerance in pressurized fuselages." Proc. 14th Sympo. of the International Committee on Aeronautical Fatigue, 1-77.

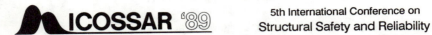
A Reliability Analysis of Aerospace Structures
Subject to Thermal Mechanical and Flight Loads

M.E. Artley* and K.C. Chou**

*Civil Engineering Department, Villanova University,
Villanova, PA 19085

**Department of Civil Engineering, Syracuse University
Syracuse, NY 13244

ABSTRACT

Aerospace structures are subjected to various
flight loads throughout their service life.   In
certain anticipated missions, significant thermal
stresses will be generated through atmospheric exit
and re-entry. It is the purpose of this study to
develop a method to assess the reliability of the
structural components of these aircraft. Due to the
loss of strength of the aerospace materials under
high temperature, the structural response of some
flight maneuvers may be nonlinear. Thus it becomes
necessary to include the the analysis of a
non-linear structural response process.
In this study, a preliminary selection of
stochastic models for the thermal and flight loads
is noted. Nonlinear load exceedance processes will
then be modified to analyze the load model developed
and a reliability analysis will be conducted.

KEYWORDS

Random Processes, load exceedance, nonlinear
response, high temperature.

INTRODUCTION

Designers of airframes currently under development must
ensure that their structures can withstand the stress of
variable amplitude flight loadings in combination with varying
thermal loads. In particular, selective structural components
of the National Aerospace Plane (NASP) are expected to
experience maximum temperatures ranging from 200°C to 650°C
during a typical mission [1]. These temperatures limit the
types of material which can be selected for structural
components.   The design criteria includes the requirement to

withstand the design stresses with strains within tolerance at
these high temperatures. This research focuses on calculating
the probability of exceeding a given strain knowing the entire
stress-strain law of the material and assuming stochastic
modeling of the loads.

    In this preliminary investigation, the temperature is
considered to be elevated and constant. The flight maneuver
loads are described by a generic load process. The condition
of constant temperature alleviates the difficulty in modeling a
yield point which varies with temperature. An assumption of
constant temperature is justified because this condition lends
a degree of conservatism to the results. The strength values
will be underestimated and the strain values will be
overestimated. A steady-state stress is introduced to the
component by the thermal load. The flight maneuver loads,
described by a generic load process, are superimposed on top.
The effect of the load level is dependent on the condition of
the material. A given maneuver load may illicit an elastic
response at 22° C, while the same load may illicit an inelastic
response at 540°C. Such a response is load history dependent,
therefore; a load sequence accountable method is necessary.

                            APPROACH

    A design spectra for the NASP airplane was selected as a
basis for making the assumptions in this study. A
load-temperature-time profile is shown in Figure 1 for a
typical mission [2]. Notice that the maximum temperature on
selected components is over 200°C 100% of the time. It is over
540°C 45% of the time. The 60 minutes of flight shown here
represent two missions per flight hour. The first 30 minutes
of this flight represents take-off, exit from the atmosphere, a
brief orbital cruise followed by re-entry into the atmosphere.
The second 30 minutes represents a cruise mission with high
altitude maneuvers. The plane does not leave the atmosphere in
this mission. The flight maneuver loads are predominantly
positive, and only occasionally negative. Compression loads
occur at take-off and landing.

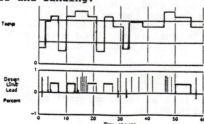

        Figure 1.   Load-temperature-time profiles.

Material selections can be made knowing the design requirements of the vehicle. Titanium 6%AL - 4%V was selected as a representative material with which to illustrate the model. It is a material whose properties are well documented and for which data is available in the open literature (e.g. [3,4]. It has an essentially bilinear stress-strain relationship, with a yield point which is dependent on temperature. For example, the yield point at 22°F occurs at approximately 965 MPa (true stress) and 0.9 percent (true strain). The yield point at 540°C occurs at approximately 414 MPa (true stress) and 0.8 percent (true strain), after 12 hour exposure and strain rate of 0.002 per minute (Price 1959). The compressive properties are similar to the tensile properties at 22°C, but tend to diverge beyond 120°C. The yield point at 400°C in compression is 482 MPa with a true strain of 0.6 percent.

In this preliminary study, the material properties in compression are assumed to be the same as they are in tension. The unloading is assumed to be parallel to the loading in the initial elastic range. In order to describe the nonlinear structural response due to the compression cycles, a typical reversible cyclic load process along with its corresponding response is used as developed by Chou [5].

ILLUSTRATION

Consider the example load sequence shown on the left in Figure 2. Due to the load reversal, not all of the hierarchical peaks become the exceedances. For the typical load process shown in Figure 2, the response for the first five (tension) loads is the same as that described in Chou, Corotis and Karr [6]. Due to the load-response relationship assumed, load 6 does not fall onto the unloading path, path b, defined by load 5 but falls onto the nonlinear part of the compression response portion. Thus load 6, though numerically smaller than load 5, will define a new loading path, path c, for subsequent loads. From this illustration, one can perceive that if one wishes to analyze the number of load exceedances, it may require the knowledge of the entire reponse process. The mathematics involved are undoubtedly prohibitively complex. An approach utilizing the Markov chain appears to have merit, and was pursued.

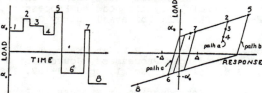

Figure 2. Sustained Load Process and Nonlinear Response.

For this example, we are interested in the number of
flight loads exceeding a predetermined threshold of
deformation, either on the compressive zone (say $-\Delta$) or on the
tensile zone (say $+\Delta$) or both. If the probability density
function (pdf) of the input is known, the exceedance output can
be calculated. In this case, the stress in a component is
directly related to the applied load, and can be determined.
Since we can deteministically calculate stress from the load,
we can also determine the pdf of the stress in the component,
given the pdf of the load, and from there determine the
exceedance of a threshold of strain. It is not critical to
know the precise response of any load, only the probability of
exceeding the threshold. One needs only to know the status of
the current load (causing an exceedance or not) in order to
predict the status of the future one. Given that a load is not
an exceedance, what is the probability that the subsequent load
would become one. Hence, states can be set up which expedite
the calculation of the probability of having K exceedances in
the lifetime of a structure. A one-step Markov chain
process[7] is used. The details of the analysis are presented
in [5].

REFERENCES

1.   Saff, C.R. and Talya, T.,   "Damage Tolerance Analysis
for Manned Hypervelocity Vehicles", Third Interim Technical
Report, USAF, AFWAL Contract NO. F33615-86-C-3208, December,
1987.

2.   Saff, C.R. and Talya, T., "Damage Tolerance Analysis
for Manned Hypervelocity Vehicles", Fifth Interim Technical
Report, USAF, AFWAL Contract No. F33615-86-C-3208, December,
1988.

3.   Price, H.L. "Tensile Properties of 6A1-4V Titanium
Alloy Sheet under Rapid-Heating and Constant-Temperature
Conditions", Technical Note, NASA-TN-D-121, November, 1959.

4.   Lindholm, U.S., Yeakley, L.M., Bessey, R.L. "An
Investigation of the Behavior of Materials under High Rates of
Deformation", Technical Report, AFML-TR-68-194, July, 1968.

5.   Chou, K.C. "Reliability Study of Nonlinear Structural
Response under Reversible Cyclic Loading Processes", Final
Report, USAF-UES AFPR/GSRP, Universal Energy System, Dayton,
Ohio, 1988.

6.   Chou, K.C., Corotis, R.B. and Karr, A.F. "Nonlinear
Response to Sustained Load Processes", Journal of Structural
Engineering, ASCE, Vol. 111, No. 1, pp. 142-157, 1985.

7.   Cinlar, E. Introduction of Stochastic Processes,
Prentice-Hall, New Jersey, 1975.

SUBJECT INDEX

Page number refers to first page of paper.

Passive control, 471
Pedestrians, 1799
Penstocks, 1855
Performance rating, 863, 911
Perturbation, 1097, 1113
Peru, 589
Physical properties, 839
Piers, 231
Pile driving, 231, 287
Pile foundations, 143
Pile load tests, 291
Pile stresses, 291
Pipe joints, 493
Pipe networks, 919
Pipelines, 533, 549, 677
Piping systems, 1507
Plastic deformation, 1217
Plastic hinges, 999, 1341
Plasticity, 223, 831, 1011, 1691
Plates, 1303, 1467, 2079
Platforms, 887
Poisson ratio, 439
Polynomials, 1255, 1627
Post tensioning, 2239
Power spectral density, 343, 557, 573, 1411
Power spectrum analysis, 167
Pressure pipes, 1747, 1779
Pressure vessels, 1747, 1779, 2187, 2195, 2337
Probabilistic methods, 21, 55, 175, 183, 247, 279, 431, 605, 613, 621, 629, 637, 717, 747, 805, 951, 967, 1041, 1113, 1121, 1311, 1475, 1515, 1639, 1667, 1847, 1863, 1911, 1919, 1965, 2211, 2231, 2267, 2275, 2283, 2291, 2299, 2307
Probabilistic models, 1499, 1523, 1799, 1827, 1973, 2115
Probability, 855
Probability density functions, 1137, 1209, 1349, 2255
Probability distribution, 111, 167, 959, 1011, 1507, 1667, 1675, 1783, 1973
Probability distribution functions, 1791, 2291
Probability theory, 239, 399, 501, 565, 1241, 1357, 1659, 2035
Production engineering, 1763
Progressive failure, 243, 255, 779, 2195
Protective structures, 2123

Quadratic forms, 1383
Quadratic formulas, 1435
Quality assurance, 1855, 1927

Rail transportation, 1295
Rails, 1643

Rainfall frequency, 1739
Random processes, 183, 1129, 1271, 1311, 1357, 1531, 1631, 1871, 2373
Random variables, 287, 747, 805, 855, 959, 991, 1007, 1065, 1089, 1643, 1667, 1675, 2275
Random vibration, 63, 71, 127, 271, 1209, 1233, 1241, 1287, 1303, 1341, 1367, 1371, 1995, 2003, 2247, 2251, 2299
Random waves, 127, 143, 159
Ratings, 2135, 2235
Reconstruction, 541
Redundancy, 763, 863, 935, 967, 975, 2171, 2239
Regional analysis, 295, 303, 311
Regression models, 255
Rehabilitation, 2235
Reinforcement, 1707
Reliability, 541, 1491, 1643, 1699, 1855, 2055, 2365
Reliability analysis, 21, 55, 79, 111, 135, 199, 223, 243, 247, 431, 439, 509, 677, 755, 779, 831, 871, 887, 895, 935, 959, 999, 1003, 1007, 1015, 1041, 1081, 1089, 1177, 1185, 1193, 1247, 1363, 1483, 1571, 1587, 1603, 1635, 1675, 1683, 1715, 1775, 1779, 1863, 2043, 2051, 2063, 2067, 2071, 2107, 2143, 2195, 2231, 2235, 2239, 2251, 2369, 2373
Relief valves, 2203
Repairing, 207, 1619, 2099, 2107, 2123
Residual strength, 967, 1049, 1723
Residual stress, 1579
Resource allocation, 645
Response spectra, 463, 573, 605, 613, 621, 629, 637
Response time, 455
Restoration, 685
Retrofitting, 645
Rigid frames, 1987
Rigid piers, 1987
Risk acceptance, 1911, 1919
Risk analysis, 21, 151, 175, 391, 399, 431, 447, 605, 613, 621, 629, 637, 645, 717, 747, 903, 919, 1659, 1739, 1903, 1911, 1919, 1927, 2211
Road surface roughness, 2247, 2251
Rock joints, 247
Rock properties, 247
Rotation, 1279
Rubble-mound breakwaters, 207

Safety analysis, 21, 123, 279, 415, 621, 629, 637, 771, 903, 1739, 1927, 2163, 2231
Safety factors, 45, 1927, 1949, 2063, 2083, 2135, 2235, 2365

AUTHOR INDEX

Page number refers to first page of paper.